Wireless Communications

Wireless Communications

Andreas F. Molisch
Fellow, IEEE

IEEE PRESS
IEEE Communications Society, Sponsor

JOHN WILEY & SONS, LTD

Copyright © 2005 John Wiley & Sons Ltd, The Atrium, Southern Gate, Chichester,
West Sussex PO19 8SQ, England

Telephone (+44) 1243 779777

Email (for orders and customer service enquiries): cs-books@wiley.co.uk
Visit our Home Page on www.wiley.com

Other Wiley Editorial Offices

John Wiley & Sons, Inc., 111 River Street, Hoboken, NJ 07030, USA

Jossey-Bass, 989 Market Street, San Francisco, CA 94103-1741, USA

Wiley-VCH Verlag GmbH, Boschstr. 12, D-69469 Weinheim, Germany

John Wiley & Sons Australia Ltd, 42 McDougall Street, Milton, Queensland 4064, Australia

John Wiley & Sons (Asia) Pte Ltd, 2 Clementi Loop #02-01, Jin Xing Distripark, Singapore 129809

John Wiley & Sons Canada Ltd, 22 Worcester Road, Etobicoke, Ontario, Canada M9W 1L1

Wiley also publishes its books in a variety of electronic formats. Some content that appears
in print may not be available in electronic books.

IEEE Communications Society
Sponsor COMMS-S Liaison to IEEE Press, Mostafa Hashem Sherif

British Library Cataloguing in Publication Data

A catalogue record for this book is available from the British Library

ISBN-13 978-0-470-84887-6 (HB)   ISBN-13 978-0-470-84888-3 (PB)
ISBN-10 0-470-84887-1 (HB)   ISBN-10 0-470-84888-X (PB)

Project management by Originator, Gt Yarmouth, Norfolk (typeset in 9/11pt Times).
Printed and bound in Great Britain by Antony Rowe Ltd, Chippenham, Wiltshire.
This book is printed on acid-free paper responsibly manufactured from sustainable forestry
in which at least two trees are planted for each one used for paper production.

Contents

Preface

When, in 1994, I wrote the very first draft of this book in the form of lecture notes for a wireless course, the preface started by justifying the need for giving such a course at all. I explained at length why it is important that communications engineers understand wireless systems, especially digital cellular systems. Now, more than 10 years later, such a justification seems slightly quaint and outdated. Wireless industry has become the fastest growing sector of the telecommunications industry, and there is hardly anybody in the world who is not a user of some form of wireless technology. From the ubiquituous cellphones, to wireless LANs, to wireless sensors that are proliferating – we are surrounded by wireless communications devices.

One of the key challenges in studying wireless communications is the amazing breadth of topics that impacts this field. Traditionally, communications engineers have concentrated on, for example, digital modulation and coding theory, while the world of antennas and propagation studies was completely separate – "and never the twain shall meet". However, such an approach does not work for wireless communications. We need an understanding of *all* aspects that impact the performance of systems, and make the whole system work. This book is an attempt to provide such an overview, concentrating as it does on the physical layer of wireless communications.

Another challenge is that not only practical wireless systems, but also the science on which they are based is constantly changing. It is often claimed that while wireless systems rapidly change, the scientific basis of wireless communications stays the same, and thus engineers can rely on knowledge acquired at a given time to get them through many cycles of system changes, with just minor adjustments to their skillsets. This thought is comforting – and unfortunately false. For example, 10 years ago, topics like multiple-antenna systems, OFDM, turbocodes and LDPC codes, and multiuser detection, were mostly academic curiosities, and would at best be treated in PhD-level courses; today, they dominate not only mainstream research and system development, but represent vital, basic knowledge for students and practicing engineers. I hope that, by treating both new aspects as well as more "classical" topics, my book will give today's students and researchers knowledge and tools that will prove useful for them in the future.

The book is written for advanced undergraduate and graduate students, as well as for practicing engineers and researchers. Readers are assumed to have an understanding of elementary communication theory, like modulation/demodulation as well as of basic aspects of electromagnetic theory, though a brief review of these fields is given at the beginning of the corresponding chapters of the book. The core material of this book tries to get students to a stage where they can read more advanced monographs, and even research papers; for all those readers who want to dig deeper, the majority of chapters include a "further reading" section that cites the most important references. The text includes both mathematical formulations, and intuitive explanations. I firmly believe that such a dual approach leads to the deepest understanding of the material. In addition to being a textbook, the text is also intended to serve as a reference tool for researchers and practitioners. For this reason, I have tried to make it easier to read isolated chapters. All acronyms are explained the first time they occur in each chapter (not just at their first occurrence in the book); a list of symbols (see p. xxxix) explains the meaning of symbols used in the equations. Also, frequent cross-references should help for this purpose.

Synopsis

The book is divided into five parts. The first part, the introduction, gives a high-level overview of wireless communications. Chapter 1 first gives a taxonomy of different wireless services, and then describes the requirements for data rate, range, energy consumption, etc., that the various applications impose. This chapter also contains a brief history, and a discussion of the economic and social aspects of wireless communications. Chapter 2 describes the basic challenges of wireless communications, like multipath propagation and limited spectrum resources. Chapter 3 then discusses how noise and interference limit the capabilities of wireless systems, and how link budgets can serve as simple system-planning tools that give a first idea about the achievable range and performance.

The second part describes the various aspects of wireless propagation channels and antennas. As the propagation channel is the medium over which communication happens, understanding it is vital to understanding the remainder of the book. Chapter 4 describes the basic propagation processes: free space propagation, reflection, transmission, diffraction, diffuse scattering, and waveguiding. We find that the signal can get from the transmitter to the receiver via many different propagation paths that involve one or more of these processes, giving rise to many multipath components. It is often convenient to give a statistical description of the effects of multipath propagation. Chapter 5 gives a statistical formulation for narrowband systems, explaining both small-scale (Rayleigh) and large-scale fading. Chapter 6 then discusses formulations for wideband systems, and systems that can distinguish the directions of multipath components at the transmitter and receiver. Chapter 7 then gives specific models for propagation channels in different environments, covering pathloss as well as wideband and directional models. Since all realistic channel models have to be based on (or confirmed by) measurements, Chapter 8 summarizes techniques that are used for measuring channel impulse responses. Finally, Chapter 9 briefly discusses antennas for wireless applications, especially with respect to different restrictions at base stations and mobile stations.

The third part of the book deals with the structure and theory of wireless transceivers. After a short summary of the components of a RF transceiver in Chapter 10, Chapter 11 then describes the different modulation formats that are used for wireless applications. The discussion not only includes mathematical formulations and signal space representations, but also an assessment of their advantages and disadvantages for various purposes. The performance of all these modems in flat-fading as well as frequency-selective channels is then the topic of Chapter 12. One critical observation we make here is the fact that fading leads to a drastic increase in error probability, and that increasing the transmit power is not a suitable way of improving performance. This motivates the next two Chapters, which deal with diversity and channel coding, respectively. We find that both these measures are very effective in reducing error probabilities in a fading channel. The coding chapter also includes a discussion of near-Shannon-limit-achieving codes (turbocodes and low-density parity check codes), which have gained great popularity in recent years. Since voice communication is still the most important application for cellphones and similar devices, Chapter 15 discusses the various ways of digitizing speech, and compressing information so that it can be transmitted over wireless channels in an efficient way. Chapter 16 finally discusses equalizers, which can be used to reduce the detrimental effect of frequency selectivity of wideband wireless channels. All the chapters in this part deal with a single link – i.e., the link between one transmitter and one receiver.

The fourth part then takes into account our desire to operate a number of wireless links simultaneously in a given area. This so-called *multiple-access* problem has a number of different solutions. Chapter 17 discusses frequency domain multiple access (FDMA) and time domain multiple access (TDMA), as well as packet radio, which has gained increasing importance for data transmission. This chapter also discusses the cellular principle, and the concept of frequency reuse that forms the basis not only for cellular, but also many other high-capacity wireless systems. Chapter 18 then describes spread spectrum techniques, in particular CDMA, where different users can be distinguished by different spreading sequences. This chapter also discusses multiuser detection, a very advanced receiver scheme that can greatly decrease the impact of multiple-access interference. Another topic of Part IV is "advanced transceiver techniques". Chapter 19 describes OFDM (orthogonal frequency domain multiplexing), which is a modulation method that can sustain very high data rates in channels with large delay spread. Chapter 20 finally discusses multiple-antenna techniques: "smart antennas", typically placed at the base station, are multiple-antenna elements with sophisticated signal processing that can (among other benefits) reduce interference and thus increase the capacity of cellular systems. MIMO (multiple-input–multiple-output) systems go one step further, allowing the transmission of parallel datastreams from multiple-antenna elements at the transmitter, which are then received and demodulated by multiple-antenna elements at the receiver. These systems achieve a dramatic capacity increase even for a single link.

The last part of the book describes standardized wireless systems. Standardization is critical so that devices from different manufacturers can work together, and also systems can work seamlessly across national borders. The book describes the most successful cellular wireless standards – namely, GSM (Global System for Mobile communications), IS-95 and its advanced form CDMA 2000, as well as Wideband CDMA (also known as UMTS) in Chapters 21, 22, and 23, respectively. Furthermore, Chapter 24 describes the most important standard for wireless LANs – namely, IEEE 802.11.

A companion website (*www.wiley.com/go/molisch*) contains some material that I deemed as useful, but which would have made the printed version of the book overly bulky. In particular, the appendices to the various chapters, as well as supplementary material on the DECT (Digital Enhanced Cordless Telecommunications) system, the most important cordless phone standard, can be found there.

Suggestions for courses

The book contains more material than can be presented in a single-semester course, and spans the gamut from very elementary to quite advanced topics. This gives the instructor the freedom to tailor teaching to the level and the interests of students. The book contains worked examples in the main text, and a large number of homework exercises at the end of the book. Solutions to these exercises, as well as presentation slides, are available to instructors on the companion website of this book.

A few examples for possible courses include:

- Introductory course:

    —introduction (Chapters 1–3);
    —basic channel aspects (Sections 4.1–4.3, 5.1–5.4, 6.1, 6.2, 7.1–7.3);

—elementary signal processing (Chapters 10, 11, and Sections 12.1, 12.2.1, 12.3.1, 13.1, 13.2, 13.4, 14.1–14.3, 16.1-16.2);

—multiple access and system design (Chapters 17, 22 and Sections 18.2, 18.3, 21.1–21.7).

- Wireless propagation:

  —introduction (Chapter 2);

  —basic propagation effects (Chapter 4);

  —statistical channel description (Chapters 5 and 6);

  —channel modeling and measurement (Chapters 7 and 8);

  —antennas (Chapter 9).

  This course can also be combined with more basic material on electromagnetic theory and antennas.

- Advanced topics in wireless communications:

  —introduction and refresher: should be chosen by the instructor according to audience;

  —CDMA and multiuser detection (Sections 18.2, 18.3, 18.4);

  —OFDM (Chapter 19);

  —ultrawideband communications (Sections 18.5, 6.6);

  —multiantenna systems (Sections 6.7, 7.4, 8.5, 13.5, 13.6, and Chapter 20);

  —advanced coding (Sections 14.5, 14.6).

- Current wireless systems:

  —TDMA-based cellular systems (Chapter 21);

  —CDMA-based cellular systems (Chapters 22 and 23);

  —cordless systems (supplementary material on companion website);

  —wireless LANs (Chapter 24); and

  —selected material from previous chapters for the underlying theory, according to the knowledge of the audience.

Acknowledgements

This book is the outgrowth of many years of teaching and research in the field of wireless communications. During that time, I worked at two universities (Technical University Vienna, Austria and Lund University, Sweden) and three industrial research labs (FTW Research Center for Telecommunications Vienna, Austria; AT&T (Bell) Laboratories–Research, Middletown, NJ, U.S.A.; and Mitsubishi Electric Research Labs., Cambridge, MA, U.S.A.), and cooperated with my colleagues there, as well as with numerous researchers at other institutions in Europe, the U.S.A., and Japan. All of them had an influence on how I see wireless communications, and thus, by extension, on this book. To all of them I owe a debt of gratitude. First and foremost, I want to thank Ernst Bonek, the pioneer and doyen of wireless communications in Austria, who initiated this project, and with whom I had countless discussions on technical as well as didactic aspects of this book and the lecture notes that preceded it (these lecture notes served for a course that we gave jointly at TU Vienna). Without his advice and encouragement, this book would never have seen the light of day. I also want to thank my colleagues and students at TU Vienna, particularly Paulina Erätuuli, Josef Fuhl, Alexander Kuchar, Juha Laurila, Gottfried Magerl, Markus Mayer, Thomas Neubauer, Heinz Novak, Berhard P. Oehry, Mario Paier, Helmut Rauscha, Alexander Schneider, Gerhard Schultes, and Martin Steinbauer, for their help. At Lund University, my colleagues and students also greatly contributed to this book: Peter Almers, Ove Edfors, Fredrik Floren, Anders Johanson, Johan Karedal, Vincent Lau, Andre Stranne, Fredrik Tufvesson, and Shurjeel Wyne. They contributed not only stimulating suggestions on how to present the material, but also figures and examples; in particular, most of the exercises and solutions were created by them and Section 19.5 is based on the ideas of Ove Edfors. A special thanks to Gernot Kubin from Graz University of Technology, who contributed Chapter 15 on speech coding. My colleagues and managers at FTW, AT&T, and MERL – namely, Markus Kommenda, Christoph Mecklenbraueker, Helmut Hofstetter, Jack Winters, Len Cimini, Moe Win, Martin Clark, Yang-Seok Choi, Justin Chuang, Jin Zhang, Kent Wittenburg, Richard Waters, Neelesh Mehta, Phil Orlik, Zafer Sahinoglu, Daqin Gu (who greatly contributed to Chapter 24), Giovanni Vanucci, Jonathan Yedidia, Yves-Paul, and Hongyuan Zhang, also greatly influenced this book. Special thanks and appreciation to Larry Greenstein, who (in addition to the many instances of help and advice) took an active interest in this book and provided invaluable suggestions.

A special thanks also to the reviewers of this book. The manuscript was critically read by anonymous experts selected by the publisher, as well as several of my friends and colleagues at various research institutions: John B. Anderson (Chapters 11–13), Anders Derneryd (Chapter 9), Larry Greenstein (Chapters 1–3, 7, 17–19); Steve Howard (Chapter 22), Thomas Kaiser (Chapter 20), Achilles Kogantis (Chapter 23), Gerhard Kristensson (Chapter 4), Thomas Kuerner (Chapter 21), Gerald Matz (Chapter 5–6), Neelesh B. Mehta (Chapter 20), Bob O'Hara (Chapter 24), Phil Orlik (Section 17.4), John Proakis (Chapter 16), Said Tatesh (Chapter 23), Reiner Thomae (Chapter 8), Chintha Tellambura (Chapters 11–13), Giorgia Vitetta (Chapter 16), Jonathan Yedidia (Chapter 14). To all of them goes my deepest appreciation. Of course, the responsibility for any possible remaining errors rests with me.

Mark Hammond as publisher, Sarah Hinton as project editor, and Olivia Underhill as assistant editor, all from John Wiley & Sons, Ltd, guided the writing of the book with expert advice and considerable patience. Manuela Heigl and Katalin Stibli performed many typing and drawing tasks with great care and cheerful enthusiasm. Originator expertly typeset the manuscript.

Abbreviations

2G	Second Generation
3G	Third Generation
3GPP	Third Generation Partnership Project
3GPP2	Third Generation Partnership Project 2
3SQM	Single Sided Speech Quality Measure
A/D	Analog to Digital
AB	Access Burst
AC	Access Category; Administration Centre
ACCH	Associated Control CHannel
ACELP	Algebraic Code Excited Linear Prediction
ACF	Auto Correlation Function
ACK	ACKnowledgement
ACLR	Adjacent Channel Leakage Ratio
ACM	Address Complete Message
AD	Access Domain
ADC	Analog to Digital Converter
ADDTS	ADD Traffic Stream
ADF	Average Duration of Fades
ADPCM	Adaptive Differential Pulse Code Modulation
ADPM	Adaptive Differential Pulse Modulation
ADPS	Angular Delay Power Spectrum
AGC	Automatic Gain Control
AGCH	Access Grant CHannel
AICH	Acquisition Indication CHannel
AIFS	Arbitration Inter Frame Spacing
ALOHA	random access packet radio system
AMPS	Advanced Mobile Phone System
AMR	Adaptive Multi Rate
AN	Access Network
ANSI	American National Standards Institute
AP	Access Point
APS	Angular Power Spectrum
ARFCN	Absolute Radio Frequency Channel Number
ARIB	Association of Radio Industries and Businesses (Japan)
ARQ	Automatic Repeat reQuest

ASIC	Application Specific Integrated Circuit
ATDPICH	Auxiliary forward Transmit Diversity PIlot CHannel
ATM	Asynchronous Transfer Mode
AUC	AUthentication Center
AWGN	Additive White Gaussian Noise
BAM	Binary Amplitude Modulation
BAN	Body Area Network
BCC	Base station Color Code
BCCH	Broadcast Control CHannel
BCF	Base Control Function
BCH	Bose–Chaudhuri–Hocquenghem (code)
BCH	Broadcast CHannel
BCJR	Initials of the authors of Bahl et al. [1974]
BEC	Backward Error Correction
BER	Bit Error Rate
BFI	Bad Frame Indicator
BFSK	Binary Frequency Shift Keying
BLAST	Bell labs LAyered Space Time
Bm	Traffic channel for full-rate voice coder
BN	Bit Number
BNHO	Barring all outgoing calls except those to Home PLMN
BPPM	Binary Pulse Position Modulation
BPSK	Binary Phase Shift Keying
BS	Base Station
BSC	Base Station Controller
BSI	Base Station Interface
BSIC	Base Station Identity Code
BSS	Base Station Subsystem
BSS	Basic Service Set
BSSAP	Base Station Application Part
BTS	Base Transceiver Station
BU	Bad Urban
CA	Cell Allocation
CAP	Controlled Access Period
CB	Citizens' Band
CBCH	Cell Broadcast CHannel
CC	Country Code
CCBS	Completion of Calls to Busy Subscribers
CCCH	Common Control CHannel
CCF	Cross Correlation Function
CCI	Co Channel Interference
CCK	Complementary Code Keying
CCPCH	Common Control Physical Channel
CCPE	Control Channel Protocol Entity
CCTrCH	Coded Composite Traffic CHannel
cdf	cumulative distribution function
CDMA	Code Division Multiple Access
CELP	Code Excited Linear Prediction
CEPT	European Conference of Postal and Telecommunications Administrations
CF-Poll	Contention-free Poll
CFB	Contention Free Burst

CFP	Contention Free Period
CI	Cell Identify
CM	Connection Management
CMA	Constant Modulus Algorithm
CMOS	Complementary Metal Oxide Semiconductor
CN	Core Network
CND	Core Network Domain
CNG	Comfort Noise Generation
CONP	Connect Number Identification Presentation
COST	European COoperation in the field of Scientific and Technical research
CP	Contention Period; Cyclic Prefix
CPCH	Common Packet CHannel
CPFSK	Continuous Phase Frequency Shift Keying
CPICH	Common PIlot CHannel
CRC	Cyclic Redundancy Check; Cyclic Redundancy Code
CS-ACELP	Conjugate Structure–Algebraic Code Excited Linear Prediction
CSI	Channel State Information
CSMA	Carrier Sense Multiple Access
CSMA/CA	Carrier Sense Multiple Access with Collision Avoidance
CTT	Cellular Text Telephone
CUG	Closed User Group
CW	Contention Window
D-BLAST	Diagonal BLAST
DAB	Digital Audio Broadcasting
DAC	Digital-to-Analog Converter
DAM	Diagnostic Acceptability Measure
DB	Dummy Burst
DBPSK	Differential Binary Phase Shift Keying
DC	Direct Current
DCCH	Dedicated Control CHannel
DCF	Distributed Coordination Function
DCH	Dedicated (transport) CHannel
DCM	Directional Channel Model
DCS1800	Digital Cellular System at the 1800-MHz band
DCT	Discrete Cosine Transform
DDDPS	Double Directional Delay Power Spectrum
DECT	Digital Enhanced Cordless Telecommunications (ETSI)
DFE	Decision Feedback Equalizer
DFT	Discrete Fourier Transform
DIFS	Distributed Inter Frame Space
DLL	Data Link Layer
DLP	Direct Link Protocol
DM	Delta Modulation
DMT	Discrete Multi Tone
DNS	Domain Name Server
DOA	Direction Of Arrival
DOD	Direction Of Departure
DPCCH	Dedicated Physical Control CHannel
DPDCH	Dedicated Physical Data CHannel
DPSK	Differential Phase Shift Keying
DQPSK	Differential Quadrature Phase Shift Keying

DRM	Discontinuous Reception Mechanisms
DRT	Diagnostic Rhyme Test
DRX	Discontonuous Reception
DS	Direct Sequence
DS-CDMA	Direct Sequence–Code Division Multiple Access
DS-SS	Direct Sequence–Spread Spectrum
DSCH	Downlink Shared Channel
DSI	Digital Speech Interpolation
DSL	Digital Subscriber Line
DSMA	Data Sense Multiple Access
DSR	Distributed Speech Recognition
DTAP	Direct Transfer Application Part
DTE	Data Terminal Equipment
DTMF	Dual Tone Multi Frequency (signalling)
DTX	Discontinuous Voice Transmission
DUT	Device Under Test
DVB	Digital Video Broadcasting
ECL	Emitter Coupled Logic
EDCA	Enhanced Distributed Channel Access
EDGE	Enhanced Data rates for GSM Evolution
EDPRS	Enhanced Data rate GPRS
EDSCD	Enhanced Data rate Circuit Switched Data
EFR	Enhanced Full Rate
EGC	Equal Gain Combining
EIA	Electronic Industries Alliance (U.S.A.)
EIFS	Extended Inter Frame Space
EIR	Equipment Identify Register
EIRP	Equivalent Isotropically Radiated Power
ELP	Equivalent Low Pass
EMS	Enhanced Messaging Service
EN	European Norm
ERLE	Echo Return Loss Enhancement
ESN	Electronic Serial Number
ESPRIT	Estimation of Signal Parameters by Rotational Invariance Techniques
ETS	European Telecommunication Standard
ETSI	European Telecommunications Standards Institute
EVD	Eigen Value Decomposition
EVM	Error Vector Measurement
EVRC	Enhanced Variable Rate Coder
F-APICH	Forward dedicated Auxiliary PIlot CHannel
F-BCCH	Forward Broadcast Control CHannel
F-CACH	Forward Common Assignment CHannel
F-CCCH	Forward Common Control CHannel
F-CPCCH	Forward Common Power Control CHannel
F-DCCH	Forward Dedicated Control CHannel
F-PDCCH	Forward packet Data Control CHannel
F-PDCH	Forward Packet Data CHannel
F-QPCH	Forward Quick Paging CHannel
F-SCH	Forward Supplemental CHannel
F-SYNC	Forward SYNChronization channel
F-TDPICH	Forward Transmit Diversity PIlot CHannel

F0	Fundamental frequency
FAC	Final Assembly Code
FACCH	Fast Associated Control CHannel
FACCH/F	Full-rate FACCH
FACCH/H	Half-rate FACCH
FACH	Forward Access CHannel
FB	Frequency correction Burst
FBI	Feed Back Information
FCC	Federal Communications Commission
FCCH	Frequency Correction CHannel
FCH	Fundamental CHannel
FCS	Frame Check Sequence
FDD	Frequency Domain Duplexing
FDMA	Frequency Division Multiple Access
FDTD	Finite Difference Time Domain
FEC	Forward Error Correction
FEM	Finite Element Method
FFT	Fast Fourier Transform
FH	Frequency Hopping
FHMA	Frequency Hopping Multiple Access
FIR	Finite Impulse Response
FM	Frequency Modulation
FN	Frame Number
FOMA	Japanese version of the UMTS standard
FQI	Frame Quality Indicator
FR	Full Rate
FS	Federal Standard
FSK	Frequency Shift Keying
GGSN	Gateway GPRS Support Node
GMSC	Gateway Mobile Services Switching Centre
GMSK	Gaussian Minimum Shift Keying
GPRS	General Packet Radio Service
GPS	Global Positioning System
GSC	Generalized Selection Combining
GSCM	Geometry-based Stochastic Channel Model
GSM	Global System for Mobile communications
GSM PLMN	GSM Public Land Mobile Network
GSM1800	Global System for Mobile communications at the 1800-MHz band
GTP	GPRS Tunneling Protocol
H-BLAST	Horizontal BLAST
H-S/MRC	Hybrid Selection/Maximum Ratio Combining
HC	Hybrid Coordinator
HCCA	HCF (Hybrid Coordination Function) Controlled Channel Access
HCF	Hybrid Coordination Function
HDLC	High Level Data Link Control
HF	High Frequency
HIPERLAN	HIgh PERformance Local Area Network
HLR	Home Location Register
HMSC	Home Mobile-services Switching Centre
HNM	Harmonic + Noise Modeling
hostid	host address

HR	Half Rate
HR/DS or HR/DSSS	High Rate Direct Sequence PHY
HRTF	Head Related Transfer Function
HSCSD	High Speed Circuit Switched Data
HSDPA	High Speed Downlink Packet Access
HSN	Hop Sequence Number
HT	Hilly Terrain
HTTP	Hyper Text Transfer Protocol
I/O	Input/Output
IAM	Initial Address Message
ICB	Incoming Calls Barred
ID	Identification
IDFT	Inverse Discrete Fourier Transform
I_e	Equipment impairment factor
IEEE	Institute of Electrical and Electronics Engineers
IETF	Internet Engineering Task Force
IF	Intermediate Frequency
IFFT	Inverse Fast Fourier Transformation
IFS	Inter Frame Space
iid	independent identically distributed
IIR	Infinite Impulse Response
ILBC	Internet Low Bit-rate Codec
IMBE	Improved Multi Band Excitation
IMEI	International Mobile station Equipment Identity
IMSI	International Mobile Subscriber Identity
IMT	International Mobile Telecommunications
IMT-2000	International Mobile Telecommunications 2000
INMARSAT	INternational MARitime SATellite System
IO	Interacting Object
IP	Internet Protocol
IPO	Initial Public Offering
IR	Impulse Radio
IRIDIUM	Project
IRS	Intermediate Reference System
IS-95	Interim Standard 95 (the first CDMA system adopted by the American TIA)
ISDN	Integrated Services Digital Network
ISI	Inter Symbol Interference
ISM	Industrial, Scientific, and Medical
ISPP	Interleaved Single Pulse Permutation
ITU	International Telecommunications Union
IWF	Inter Working Function
IWU	Inter Working Unit
JD-TCDMA	Joint Detection–Time/Code Division Multiple Access
Kc	Cipher Key
Ki	Key used to calculate SRES
Kl	Location Key
Ks	Session Key
LA	Location Area
LAC	Location Area Code
LAI	Location Area Identity
LAN	Local Area Network

LAP-Dm	Link Access Protocol on Dm Channel
LAR	Logarithmic Area Ratio
LBG	Linde–Buzo–Gray algorithm
LCR	Level Crossing Rate
LD-CELP	Low Delay–Code Excited Linear Prediction
LDPC	Low Density Parity Check
LEO	Low Earth Orbit
LLC	Logical Link Control
LLR	Log Likelihood Ratio
LMMSE	Linear Minimum Mean Square Error
LMS	Least Mean Square
LNA	Low Noise Amplifier
LO	Local Oscillator
LOS	Line Of Sight
LP	Linear Prediction; Linear Predictor
LPC	Linear Predictive Coding; Linear Predictive voCoder
LR	Location Register
LS	Least Squares
LSF	Line Spectral Frequency
LSP	Line Spectrum Pair
LTI	Linear Time Invariant
LTP	Long Term Prediction; Long Term Predictor
LTV	Linear Time Variant
M-QAM	M-ary Quadrature Amplitude Modulation
MA	Mobile Allocation; Multiple Access
MAC	Medium Access Control
MACN	Mobile Allocation Channel Number
MAF	Mobile Additional Function
MAHO	Mobile Assisted Hand Over
MAI	Multiple Access Interference
MAIO	Mobile Allocation Index Offset
MAP	Maximum A Posteriori
MAP	Mobile Application Part
MBE	Multi Band Excitation
MBOA	Multi Band OFDM Alliance
MC-CDMA	Multi Carrier Code Division Multiple Access
MCC	Mobile Country Code
ME	Maintenance Entity; Mobile Equipment
MEA	Multiple Element Antenna
MEF	Maintenance Entity Function
MEG	Mean Effective Gain
MELP	Mixed Excitation Linear Prediction
MFEP	Matched Front End Processor
MIC	Mobile Interface Controller
MIME	Multipurpose Internet Mail Extensions
MIMO	Multiple Input Multiple Output system
MIPS	Million Instructions Per Second
ML	Maximum Likelihood
MLSE	Maximum Likelihood Sequence Estimators (or Estimation)
MMS	Multimedia Messaging Service
MMSE	Minimum Mean Square Error

MNC	Mobile Network Code
MNRU	Modulated Noise Reference Unit
MOS	Mean Opinion Score
MoU	Memorandum of Understanding
MP3	Motion Picture Experts Group-1 layer 3
MPC	Multi Path Component
MPDU	MAC Protocol Data Unit
MPEG	Motion Picture Experts Group
MPSK	M-ary Phase Shift Keying
MRC	Maximum Ratio Combining
MS	Mobile Station
MS ISDN	Mobile Station ISDN Number
MSC	Mobile-services Switching Centre
MSCU	Mobile Station Control Unit
MSDU	MAC Service Data Unit
MSE	Mean Square Error
MSIN	Mobile Subscriber Identification Number
MSISDN	Mobile Station ISDN Number
MSK	Minimum Shift Keying
MSL	Main Signaling Link
MSRN	Mobile Station Roaming Number
MSS	Mobile Satellite Service
MT	Mobile Terminal; Mobile Termination
MTP	Message Transfer Part
MUMS	Multi User Mobile Station
MVM	Minimum Variance Method
NAV	Network Allocation Vector
NB	Narrow Ban; Normal Burst
NBIN	A parameter in the hopping sequence
NCELL	Neighbouring (adjacent) Cell
NDC	National Destination Code
netid	network address
NF	Network Function
NLOS	Non Line Of Sight
NLP	Non Linear Processor
NM	Network Management
NMC	Network Management Centre
NMSI	National Mobile Station Identification number
NMT	Nordic Mobile Telephone
Node-B	Base station
NRZ	Non Return to Zero
NSAP	Network Service Access Point
NSS	Network and Switching Subsystem
NT	Network Termination
NTT	Nippon Telephone and Telegraph
O&M	Operations & Maintenance
OACSU	Off Air Call Set Up
OCB	Outgoing Calls Barred
ODC	Ornithine DeCarboxylase
OEM	Original Equipment Manufacturer
OFDM	Orthogonal Frequency Division Multiplexing

OMC	Operations & Maintenance Center
OPT	Operator Perturbation Technique
OQAM	Offset Quadrature Amplitude Modulation
OQPSK	Offset Quadrature Phase Shift Keying
OS	Operating Systems
OSS	Operation Support System
OTD	Orthogonal Transmit Diversity
OVSF	Orthogonal Variable Spreading Factor
P-CCPCH	Primary Common Control Physical CHannel
P-CCPCH	Primary Common Control Physical CHannel
P/S	Parallel/Serial (conversion)
PABX	Private Automatic Branch eXchange
PACCH	Packet Associated Control CHannel
PACS	Personal Access Communications System
PAD	Packet Assembly/Disassambly facility
PAGCH	Packet Access Grant CHannel
PAM	Pulse Amplitude Modulation
PAN	Personal Area Network
PAR	Peak-to-Average Ratio
PARCOR	PARtial CORrelation
PBCCH	Packet Broadcast Control CHannel
PC	Point Coordinator
PCCCH	Packet Common Control CHannel
PCF	Point Coordination Function
PCG	Power Control Group
PCH	Paging CHannel
PCM	Pulse Code Modulated
PCPCH	Physical Common Packet Channel
PCS	Personal Communication System
PDA	Personal Digital Assistant
PDC	Pacific Digital Cellular (Japanese system)
PDCH	Packet Data CHannel
pdf	probability density function
PDN	Public Data Network
PDP	Power Delay Profile
PDSCH	Physical Downlink Shared CHannel
PDTCH	Packet Data Traffic CHannel
PESQ	Perceptual Evaluation of Speech Quality
PHS	Personal Handyphone System
PHY	PHYsical layer
PIC	Parallel Interface Cancellation
PICH	Page Indication Channel
PIFA	Planar Inverted F Antenna
PIFS	Priority Inter Frame Space
PIN	Personal Identification Number
PLCP	Physical Layer Convergence Procedure
PLL	Physical Link Layer
PLMN	Public Land Mobile Network
PN	Pseudo Noise
PNCH	Packet Notification CHannel
POP	Peak to Off Peak ratio

POTS	Plain Old Telephone Service
PPCH	Packet Paging CHannel
PPDU	Physical Layer Protocol Data Unit
PPM	Pulse Position Modulation
PRACH	Packet Random Access CHannel; Physical Random Access CHannel
PRake	Partial Rake
PRMA	Packet Reservation Multiple Access
PSDU	Physical Layer Service Data Unit
PSK	Phase Shift Keying
PSMM	Pilot Strength Measurement Message
PSPDN	Public Switched Public Data Network
PSQM	Perceptual Speech Quality Measurement
PSTN	Public Switched Telephone Network
PTCCH-D	Packet Timing advance Control CHannel-Downlink
PTCCH-U	Packet Timing advance Control CHannel-Uplink
PTM	Point To Multipoint
PTM-M	Point To Multipoint Multicast
PTM-SC	Point To Multipoint Service Center
PTO	Public Telecommunications Operators
PUK	Personal Unblocking Key
PWI	Prototype Waveform Interpolation
PWT	Personal Wireless Telephony
QAM	Quadrature Amplitude Modulation
QAP	QoS Access Point
QCELP	Qualcomm Code Excited Linear Prediction
QFGV	Quadratic Form Gaussian Variable
QOF	Quasi Orthogonal Function
QoS	Quality of Service
QPSK	Quadrature Phase Shift Keying
QSTA	QoS STAtion
QSTA	Quality-of-service STAtion
R-ACH	Reverse Access CHannel
R-ACKCH	Reverse ACKnowledgement CHannel
R-CCCH	Reverse Common Control CHannel
R-CQICH	Reverse Channel Quality Indicator CHannel
R-DCCH	Reverse Dedicated Control CHannel
R-EACH	Reverse Enhanced Access CHannel
R-FCH	Reverse Fundamental CHannel
R-PICH	Reverse PIlot CHannel
R-SCH	Reverse Supplemental CHannel
RA	Random Mode Request information field
RA	Routing Area
RA	Rural Area
RAB	Random Access Burst
RACH	Random Access CHannel
RAN	Radio Access Network
RC	Raised Cosine
RCDLA	Radiation Coupled Dual L Antenna
RE	Radio Environment
RF	Radio Frequency
RFC	Radio Frequency Channel; Request For Comments

RFL	Radio Frequency subLayer
RFN	Reduced TDMA Frame Number
RLC	Radio Link Control
RLP	Radio Link Protocol
RLS	Recursive Least Squares
RNC	Radio Network Controller
RNS	Radio Network Subsystem
RNTABLE	Table of 128 integers in the hopping sequence
RPE	Regular Pulse Excitation (Voice Codec)
RPE-LTP	Regular Pulse Excited with Long Term Prediction
RS	Reed–Solomon (code)
RSC	Recursive Systematic Convolutional
RSSI	Received Signal Strength Indication
RTP	Real-time Transport Protocol
rv	random variable
RX	Receiver
RXLEV	Received Signal Level
RXQUAL	Received Signal Quality
S-CCPCH	Secondary Common Control Physical CHannel
S/P	Serial/Parallel (conversion)
SABM	Set Asynchronous Balanced Mode
SACCH	Slow Associated Control CHannel
SAGE	Space Alternating Generalized Expectation – maximization
SAP	Service Access Point
SAPI	Service Access Point Identifier; Service Access Points Indicator
SAR	Specific Absorption Rate
SB	Synchronization Burst
SC-CDMA	single-carrier CDMA
SCCP	Signalling Connection Control Part
SCH	Synchronisation CHannel
SCN	Sub Channel Number
SDCCH	Standalone Dedicated Control CHannel
SDCCH/4	Standalone Dedicated Control CHannel/4
SDCCH/8	Standalone Dedicated Control CHannel/8
SDMA	Space Division Multiple Access
SEGSNR	SEGmental Signal-to-Noise Ratio
SEP	Symbol Error Probability
SER	Symbol Error Rate
SFIR	Spatial Filtering for Interference Reduction
SFN	System Frame Number
SGSN	Serving GPRS Support Node
SIC	Successive Interference Cancellation
SID	SIlence Descriptor
SIFS	Short Infer Frame Space
SIM	Subscriber Identity Module
SINR	Signal-to-Interference-and-Noise Ratio
SIR	Signal-to-Interference Ratio
SISO	Soft Input Soft Output; Single Input Single Output
SMS	Short Message Service
SMTP	Short Message Transfer Protocol
SN	Serial Number

SNDCP	Subnetwork Dependent Convergence Protocol 21
SNR	Signal-to-Noise Ratio
SOLT	Short Open Loss Termination
SQNR	Signal-to-Quantization Noise Ratio
SR	Spatial Reference; Selective Rake
SRMA	Split-channel Reservation Multiple Access
SSA	Small Scale Averaged
STA	STAtion
STBC	Space Time Block Code
STC	Sinusoidal Transform Coder
STDCC	Swept Time Delay Cross Correlator
STP	Short Term Prediction; Short Term Predictor
STS	Space Time Spreading
STTC	Space Time Trellis Code
SV	Saleh–Valenzuela model
TA	Terminal Adapter
TAC	Type Approval Code
TAF	Terminal Adapter Function1
TBF	Temporary Block Flow
TC	Traffic Category
TCH	Traffic CHannel
TCM	Trellis Coded Modulation
TDD	Time Domain Duplex
TDMA	Time Division Multiple Access
TE	Terminal Equipment; Transversal Electric
TETRA	TErrestrial Trunked RAdio
TFCI	Transmit Format Combination Indicator
TFI	Transport Format Indicator
TH-IR	Time Hopping Impulse Radio
TIA	Telecommunications Industry Association (U.S.)
TM	Transversal Magnetic
TMSI	Temporary Mobile Subscriber Identity
TPC	Transmit Power Control
TR	Technical Report (ETSI); Temporal Reference; Transmitted Reference
TS	Technical Specification; Traffic Stream
TSPEC	Traffic SPECifications
TTS	Text To Speech synthesis
TU	Typical Urban
TX	Transmitter
TXOP	Transmission OPportunity
U-NII	Unlicensed National Information Infrastructure
UARFCN	UTRA Absolute Radio Frequency Channel Number
UCPCH	Uplink Common Packet CHannel
UE	User Equipment
UE-ID	User Equipment in-band IDentification
UED	User Equipment Domain
ULA	Uniform Linear Array
UMTS	Universal Mobile Telecommunications System
UP	User Priority
US	Uncorrelated Scatterer
USB	Universal Serial Bus

USF	Uplink Status Flag
USIM	User Service Identity Module
UTRA	UMTS Terrestrial Radio Access
UTRAN	UMTS Terrestrial Radio Access Network
UWB	Ultra Wide Bandwidth system
UWC	Universal Wireless Communications
VAD	Voice Activity Detection; Voice Activity Detector
VCDA	Virtual Cell Deployment Area
VLR	Visitor Location Register
VoIP	Voice over Internet Protocol
VQ	Vector Quantization, Vector Quantizer
VSELP	Vector Sum Excited Linear Prediction
WAP	Wireless Application Protocol
WB	Wide Band
WCDMA	Wideband Code Division Multiple Access
WF	Whitening Factor
WG	Working Group
WI	Waveform Interpolation
WiFi	Wireless Fidelity
WLAN	Wireless Local Area Network
WLL	Wireless Local Loop
WM	Wireless Medium
WSS	Wide Sense Stationary
WSSUS	Wide Sense Stationary Uncorrelated Scatterer

Symbols

This list gives a brief overview of the use of variables in the text. Due to the large number of quantities occurring, the same letter might be used in different chapters for different quantities. For this reason, this list of symbols also gives the chapter number (at line ends) in which the variable is primarily used, though they can occur in other chapters as well. Those variables that are used only locally, and explained directly at their occurrence, are not mentioned here.

Lowercase symbols:

a_p, a_m	auxiliary variables	4
a_1	amplitudes of the MPCs	5, 6, 7, 8
$a(h_m)$	auxiliary function	7
$\mathbf{a}(\phi)$	steering vector	8
$a_{n,m}$	amplitudes of components from direction n, delay m	13
b_m	mth bit	11, 12, 13, 14, 16, 17
cdf	cumulative distribution function	5
$c_{i,k}$	amplitudes of resolvable MPCs; tap weights for tapped delay lines	7
c_0	speed of light	5, 13, 19
c_m	complex transmit symbols	11, 12, 13, 16
d	distance BS–MS	4
d	distance in signal space diagram	12, 13
d_R	Rayleigh distance	4
d_{break}	distance BS–breakpoint	4
d_{layer}	thickness of layer	4
d_{direct}	direct pathlength	4
d_{refl}	length of reflected path	4
d_p	distance to previous screen	4
d_n	distance to subsequent screen	4
d_0	distance to reference point	5
d_a	distance between antenna elements	8, 13
d_w	distance between turns of helix antenna	8
d_{km}	euclidean distance between signal points with index k and m	11
$d(\vec{x}, \vec{y})$	distance of codewords	14
d_H	Hamming distance	14
d_{cov}	coverage distance	3
d_{div}	diversity order	20

q_m	impulse response of channel + equalizer	16
$\mathbf{r}$	position vector	4
r	absolute value of fieldstrength	5
$r_{LP}(t)$	low-pass representation of received signal	12
$\mathbf{r}$	received signal vector	12
$r(t)$	received signal	14, 15, 16
$s(t)$	sounding signal	8
$s_1(t)$	auxiliary signal	8
$s_{LP,BP}(t)$	low-pass (bandpass) signal	11
$\mathbf{s}_{LP,BP}$	signal vector in low-pass (bandpass)	11
$\mathbf{s}_{synd}$	syndrome vector	14
$\mathbf{s}$	vector of signals at antenna array	8
t	absolute time	2, 11, 12, 13, 16, 17
t_0	start time	6
t_s	sampling time	12
u	auxiliary variable	11
u_m	sequence of sample values at equalizer input	16
$\mathbf{u}$	vector of information symbols	14
v	velocity	5
w_l	antenna weights	8, 13, 20
x	x-coordinate	4
x	general variable	5
$x(t)$	input signal	6
$\mathbf{x}$	code vector	14
y	y-coordinate	4
$y(t)$	output signal	6
z	z-coordinate	

Uppercase symbols:

$\mathbf{A}$	steering matrix	8
A_{RX}	antenna area of receiver	4
$A(d_{TX}, d_{RX})$	amplitude factors for diffraction	4
ADF	average duration of fades	
A	amplitude of dominant component	7
A	state in the trellis diagram	14
$B(\nu, f)$	Doppler-variant transfer function	6
B_{coh}	coherence bandwidth	6
B	bandwidth	11
BER	bit error probability	12
B_n	noise bandwidth	12
B_r	receiver bandwidth	12
B	state in the trellis diagram	14
B_G	bandwidth of Gaussian filter	11
C	capacity	20
$\mathbf{C}$	covariance matrix	
C_{crest}	crest factor	8
C	proportionality constant	
C	state in the trellis diagram	14
D	diffraction coefficient	5
D_W	diameter of helix antenna	8

D	quadratic form	12
D	maximum distortion	16
D	state in the trellis diagram	14
D_{leav}	interleaver separation	14
D	antenna directivity	
E	electric fieldstrength	4
E_{diff}	fieldstrength of diffuse field	4
E_{inc}	fieldstrength of incident field	4
$E\{\ \}$	expectation	4, 13, 18
$E1, E2$	fieldstrength of multipath components	5
E_0	normalization fieldstrength	13
E_{S}	Symbol energy	11, 12, 13
E_{B}	bit energy	11, 12, 13
E_{C}	chip energy	18
$E_{s,k}$	energy of kth signal	11, 18
$E(f)$	transfer function of equalizer	16
$F(\nu_{\text{F}})$	Fresnel integral	4
$\tilde{F}$	modified Fresnel integral	4
F	local mean of field strength	5
$F(z)$	factorization of the transfer function of the equivalent time discrete channel	16
F	noise figure	3
G_{RX}	antenna gain of receive antenna	3, 4
G_{TX}	antenna gain transmit antenna	4
$G(\gamma), G(\varphi, \theta)$	antenna pattern	5
$G(\nu, \nu_1, \nu_2)$	Gaussian function	7
G_{max}	maximum gain	
G_{R}	spectrum of rectangular pulse	11
G_{N}	spectrum of Nyquist pulse	11
G_{NR}	spectrum of root Nyquist pulse	11
$\mathbf{G}$	generator matrix	14
G_{code}	code gain	14
$\mathbf{G}_{\text{G}}$	matrix with iid Gaussian entries	20
G	gain of an amplifier stage	3
H	transfer function of the channel	5, 6, 19, 20
$H_R(f)$	transfer function of receive filter	12, 18
$\mathbf{H}$	parity check matrix	14
$\mathbf{H}_{\text{had}}$	Hadamard matrix	18, 19
$I(t)$	in-phase component	5
I_0	modified Bessel function	5, 12
$|J|$	Jacobi determinant	5
J_0	Bessel function	7
K_r	Rice factor	5
$K(t, \tau)$	kernel function	6
K	number of resolvable directions	8
K_I	system margin	3
K	number of bits in a symbol	11
K	number of information symbols in a codeword	14
$2K + 1$	number of equalizer taps	16
K	scaling constant for STDCC	8
L	number of clusters	7
L_{msd}	multiscreenloss	

$\mathbf{R}_{xx}$	correlation matrix of x	8
R_{rad}	radiation resistance	8
R_{S}	symbol rate	11
R_{B}	bit rate	11
R_{c}	code rate	14
R_{e}	rank of error matrix	17
R_h	impulse response correlation function	6
$\widetilde{R}_{yy}(t,t')$	autocovariance signal of received signal	7
SIR	signal-to-interference ratio	5
$S(f)$	power spectrum	5, 6, 12
$S(\nu,\tau)$	spreading function	6
S_τ	delay spread	6
$S_{\mathrm{D}}(\nu,\tau)$	Doppler spectrum	7
$S_{\mathrm{LP,BP}}(f)$	power spectrum of LP (BP) signal	11
SER	symbol error probability	12
S_{N}	noise power-spectral density	
S_ϕ	angular spread	6, 7, 13
T_{B}	bit duration	
T	transmission factor	4
T_{m}	mean delay	6
$T_{\mathrm{m}}(t)$	instantaneous mean delay	6
T_{rep}	repetition time of pulse signal	8
TB	time bandwidth product	8
T_{slip}	slip period	8
$\mathbf{T}$	auxiliary matrix	8
T	duration (general)	11
T_{per}	periodicity	11
T_{S}	symbol duration	11
T_{p}	packet duration	17
T_{cp}	duration of cyclic prefix	19
T_{C}	chip duration	18
T_{e}	temperature of environment	3
T_{d}	delay of pulse in PPM	11
T_{coh}	coherence time	6
T_{g}	group delay	12
$\mathbf{U}$	unitary matrix	8, 20
W	correlation spectrum	4
W_a	delay window	6
W	system bandwidth	
$\mathbf{W}$	unitary matrix	19
X	complex Gaussian random variable	12
$X(x)$	code polynomial	14
Y	complex Gaussian random variable	12
Z	complex Gaussian random variable	12

Lowercase Greek:

α	dielectric length	4
α	complex attenuation	12
α	rolloff factor	11
α	steering vector	8

β	decay time constant	7
γ	SNR	
$\bar{\gamma}$	mean SNR	
γ_{MRC}	SNR at output of maximum ratio combiner	13
γ_{EGC}	SNR at output of equal gain combiner	13
γ	angle for Doppler shift	5
γ_S	E_S/N_0	12
γ_B	E_B/N_0	12, 13, 16, 17
δ	complex dielectric constant	4
δ_{ik}	Kronecker delta	13, 16, 19
$\delta(\tau)$	Dirac function	12, 13, 16, 18
ϵ	dielectric constant	4
ϵ_{r}	relative dielectric constant	4
ϵ_{eff}	effective relative dielectric constant	4
ε_m	error signal	16
ε	error vector of a code symbol	14
φ	orientation of a street	7
φ	phase of an MPC	5
$\tilde{\varphi}$	deterministic phaseshift	7
$\varphi_m(t)$	base functions for expansion	11
ϕ	azimuth angle of arrival	6, 7
$\eta(t)$	$g(t) * h(t)$	16
κ	auxiliary variable	13
λ, λ_0	wavelength	
λ_{p}	packet transmission rate	17
λ_i	ith eigenvalue	
μ	metric	12
μ	stepwidth of LMS	16
ν	Doppler shift	6
ν_{max}	maximum Doppler shift	7
ν_{m}	mean Doppler shift	5
ν_{F}	Fresnel parameter	5
ω	angular frequency	4
ρ	position vector	
ρ_{km}	correlation coefficients between signals	11
σ_{c}	conductivity	4
σ_{h}	standard deviation of height	4
σ	standard deviation	5
σ_F	standard deviation of local mean	5
σ_{G}	standard deviation of Gaussian pulse	11
σ_{S}^2	power in symbol sequence	11
τ	delay	4, 5
τ_{Gr}	group delay	5, 12
τ_i	delay of the ith MPC	7
τ_{max}	maximum excess delay	
$\chi_i(t)$	distortion of the i-th pulse	13
ζ	ACF of $\eta(t)$	16
$\xi_n(t)$	noise correlation function	12
$\xi_s(t)$	FT of the normalized Doppler spectrum	12
$\xi_s(\nu, \tau)$	scattering function	12
$\xi_h(t, \tau)$	FT of the scattering function	12

Uppercase Greek:

Δh_b	$= h_b - h_{roof}$	7
Δx_s	distance between measurement points	8
$\Delta \tau_{min}$	minimum resolvable τ	8
Δf_{chip}	difference in chip frequency	8
$\Delta \varphi$	angle difference of paths	4
$\Delta \tau$	runtime difference	5
$\Delta \nu$	Doppler shift	5
Δ	phaseshift between two antenna elements	8
Δ_C	determinant of $\mathbf{C}$	13
$\Delta \phi$	angular range	13
Φ_H	phase of the channel transfer function	5, 12
$\Phi_{CPFSK}(t)$	phase of transmit signal for CPFSK signal	11
Λ	matrix of eigenvalues	
$\Phi_{TX}(t)$	phase of transmit signal	
Ω	mean quadratic power Nakagami	5
Ω	direction of departure	
Ω_n	nth moment of Doppler spectrum	5, 13
Θ_e	angle of incidence	4
Θ_r	angle of reflection	4
Θ_t	angle of transmission	4
Θ_n	transmission phase of nth bit	
Θ_d	diffraction angle	
ϕ_{TX}	angle TX wedge	4
ϕ_{RX}	angle RX wedge	4
ϕ	azimuth	7
ϕ_0	nominal DOA	7, 13
ϕ_i	DOA of ith wave	13
ψ	auxiliary angle	4, 13
ψ	angle of incidence $90 - \Theta_e$	4

$\bar{x}$	$= E\{x\}$	
$\dot{r}$	$= dr/dt$	
$\mathbf{U}^\dagger$	Hermitian transpose	
$\mathbf{U}^T$	transpose	
$\mathbf{x}^*$	complex conjugate	
Ξ	FT of ζ_m	

Part I

INTRODUCTION

In the first part of this book, we give an introduction to the basic applications of wireless communications, as well as to the technical problems inherent in this communication paradigm. After a brief history of wireless, Chapter 1 describes the different types of wireless services, and works out their fundamental differences. Subsequently, we look at the same problem from a different angle: What data rates, ranges, etc., occur in practical systems, and especially, what combination of performance measures are demanded (e.g., what data rates need to be transmitted over short distances; what data rates are required over long distances)? Chapter 2 then describes the fundamental technical challenges of communicating without wires, putting special emphasis on fading and co-channel interference. Chapter 3 describes the most elementary problem of designing a wireless system – namely, to set up a link budget in either a noise-limited or an interference-limited system.

After studying this part of the book, the reader should have an overview of different types of wireless services, and understand the technical challenges involved in each of them. Solutions to these challenges will be described in the later parts of this book.

1

Applications and requirements of wireless services

Wireless communications is one of the big engineering success stories of the last 20 years – not only from a scientific point of view, where the progress has been phenomenal, but also in terms of market size and impact on society. Companies that were completely unknown 20 years ago are now household names all over the world, due to their wireless products, and in several countries the wireless industry is dominating the whole economy. Working habits, and even more generally the ways we all communicate, have been changed by the possibility of talking "anywhere, anytime".

For a long time, wireless communications has been associated with cellular telephony, as this is the biggest market segment, and has had the highest impact on everyday lives. In recent times, wireless computer networks have also led to a significant change in working habits and mobility of workers – answering emails in a coffee shop has become an everyday occurrence. But besides these widely publicized cases, also a large number of less obvious applications have been developed, and are starting to change our lives. Wireless sensor networks monitor factories; wireless links replace the cables between computers and keyboards; and wireless positioning systems monitor the location of trucks that have goods identified by wireless RF (Radio Frequency) tags. This variety of new applications causes the technical challenges for the wireless engineers to become bigger with every day. This book aims to give an overview of the solution methods for current, as well as future challenges.

Quite generally, there are two paths to developing new technical solutions: engineering-driven, and market-driven. In the first case, the engineers come up with a brilliant scientific idea – without having an immediate application in mind. As time progresses, the market finds applications enabled by this idea.[1] In the other approach, the market demands a specific product, and the engineers try to develop a technical solution that fulfills this demand. In this first chapter, we will describe these market demands. We start out with a brief history of wireless communications, in order to convey a feeling of how the science, as well as the market, has developed in the past 100 years. Then follows a description of the types of services that constitute the majority of the wireless market today. Each of these services makes specific demands in terms of data rate, range, number of users, energy consumption, mobility, and so on. We will discuss all these aspects in Section 1.3. We wrap up this section with a description of

---

[1] The second chapter then gives a summary of the main technical challenges in wireless communications – i.e., the basis for the engineering-driven solutions. Chapters 3–20 discuss the technical details of these challenges and the scientific basis, while Chapters 21–24 expound specific systems that have been developed in recent years.

Wireless Communications   Andreas F. Molisch
© 2005 John Wiley & Sons, Ltd

the interaction between the engineering of wireless devices and the behavioral changes in society induced by them.

1.1 History

1.1.1 How it all started

When looking at the history of communications, we find that wireless communications is actually the oldest form – shouts and jungle drums did not require any wires or cables to function. Even the oldest "electromagnetic" (optical) communications are wireless: smoke signals are based on propagation of optical signals along a line-of-sight connection. However, wireless communications as we know it started only with the work of Maxwell and Hertz, who laid the basis for our understanding of the transmission of electromagnetic waves. It was not long after their groundbreaking work that Tesla demonstrated the transmission of information via these waves – in essence, the first wireless communications system. In 1898, Marconi made his well-publicized demonstration of wireless communications from a boat to the Isle of Wight in the English Channel. It is noteworthy that while Tesla was the first to succeed in this important endeavour, Marconi had the better public relations, and is widely cited as the inventor of wireless communications, receiving a Nobel prize in 1909.[2]

In the subsequent years, radio (and later television) became widespread throughout the world. While in the "normal" language we usually do not think of radio or TV as "wireless communications", they certainly are in a scientific sense – information transmission from one place to the other by means of electromagnetic waves. They can even constitute "mobile communications", as evidenced by car radios. A lot of basic research – especially concerning wireless propagation channels – was done for entertainment broadcasting. By the late 1930s, a wide network of wireless information transmission – though unidirectional – was in place.

1.1.2 The first systems

At the same time, the need for bi-directional mobile communications emerged. Police departments and the military had obvious applications for such two-way communications, and were the first to use wireless systems with closed user groups. Military applications drove a lot of the research during, and shortly after, the Second World War. This was also the time when much of the theoretical foundations for communications in general were laid. Claude Shannon's [1948] groundbreaking work "A mathematical theory of communications" appeared during that time, and established the possibility of error-free transmission under restrictions for the data rate and the signal-to-noise ratio. Some of the suggestions in that work, like the use of optimum power assignment in frequency-selective channels, are only now being introduced into wireless systems.

The 1940s and 1950s saw several important developments: CB (*Citizens' Band*) radios became widespread, establishing a new way of communicating between cars on the road. Communicating with these systems was useful for transferring vital traffic information and related aspects within the closed community of the drivers owning such devices, but it lacked an interface to the public telephone system, and the range was limited to some 100 kilometers, depending on the power of the (mobile) transmitters. In 1946, the first mobile telephone system was installed in the U.S.A. (St. Louis). This system did have an interface to the PSTN (*Public Switched Telephone Network*), the landline phone system, though this interface was not automated, but rather consisted of human telephone operators. However, with a total of six

---

[2] Marconi's patents were actually overturned in the 1940s.

speech channels for the whole city, the system soon met its limits. This motivated investigations of how the number of users could be increased, even though the allocated spectrum would remain limited. Researchers at AT&T's Bell Labs found the answer: the cellular principle, where the geographical area is divided into cells; different cells might use the same frequencies. To this day, this principle forms the basis for the majority of wireless communications.

Despite the theoretical breakthrough, cellular telephony did not experience significant growth during the 1960s. However, there were exciting developments on a different front: in 1957, the Soviet Union had launched the first satellite (*Sputnik*), and the U.S.A. soon followed. This development fostered research in the new area of satellite communications.[3] Many basic questions had to be solved, including the effects of propagation through the atmosphere, the impact of solar storms, the design of solar panels and other long-lasting energy sources for the satellites, and so on. To this day, satellite communications is an important area of wireless communications (though not one that we will address specifically in this book). The most widespread application lies in satellite TV transmission.

1.1.3 Analog cellular systems

The 1970s saw a revived interest in cellular communications. In scientific research, these years saw the formulation of models for pathloss, Doppler spectra, fading statistics, and other quantities that determine the performance of analog telephone systems. A highlight of that work was Jakes' book *Microwave Mobile Radio*, that summed up the state of the art in this area [Jakes 1974]. The 1960s and 1970s also saw a lot of basic research that was originally intended for landline communications, but later also proved to be instrumental for wireless communications. For example, the basics of adaptive equalizers, as well as multicarrier communications, were developed during that time.

For the practical use of wireless telephony, the progress in device miniaturization made the vision of "portable" devices more realistic. Companies like Motorola and AT&T vied for the leadership in this area, and made vital contributions. NTT (Nippon Telephone and Telegraph) established a commercial cellphone system in Tokyo in 1979. However, it was a Swedish company that built up the first system with large coverage and automated switching: up to that point, Ericsson AB had been mostly known for telephone switches, while radio communications was of limited interest to them. However, it was just that expertise in switching technology, and the (for that time daring) decision to use digital switching technology that allowed them to combine the different cells in a large area into a single network, and establish the *Nordic Mobile Telephone* (NMT) system [Meurling and Jeans 1994]. Note that while the switching technology was digital, the radio transmission technology was still analog, and the systems became therefore known as analog systems. Subsequently, other countries developed their own analog phone standards. The system in the U.S.A., for example, was called AMPS (*Advanced Mobile Phone System*).

An investigation of NMT also established an interesting method for estimating market size: business consultants equated the possible number of mobile phone users with the number of Mercedes 600 (the top-of-the-line luxury car at that time) in Sweden. Obviously, mobile telephony could never become a mass market, could it? Similar thoughts must have occurred to the management of the inventor of cellular telephony, AT&T. Upon advice from a consulting company, they decided that mobile telephony could never attract a significant number of participants, and stopped business activities in cellular communications.[4]

---

[3] Satellite communications – specifically by geostationary satellites – had already been suggested by science fiction writer Arthur C. Clark in the 1940s.

[4] These activities were restarted in the early 1990s, when the folly of the original decision became clear. AT&T then paid more than 10 billion dollars to acquire McCaw, which it renamed "AT&T Wireless".

The analog systems paved the way for the wireless revolution. During the 1980s, they grew at a frenetic pace, and reached market penetrations of up to 10% in Europe, though their impact was somewhat less in the U.S.A. In the beginning of the 1980s, the phones were "portable", but definitely not handheld. In most languages, they were just called "carphones", because the battery and transmitter were stored in the trunk of the car, and were too heavy to be carried around. But at the end of the 1980s, handheld phones with good speech quality and quite acceptable battery lifetime abounded. The quality had become so good that in some markets, digital phones had difficulties to establish themselves – there just did not seem to be a need for further improvements.

1.1.4 GSM, and the worldwide cellular revolution

Even though the public did not see a need for changing from analog to digital, the network operators knew better. Analog phones have a bad spectral efficiency (we will see why in Chapter 3), and due to the rapid growth of the cellular market, operators had a high interest in making room for more customers. Also, research in communications had started its inexorable turn to digital communications, and that included digital wireless communications as well. In the late 1970s and the 1980s, research into spectrally efficient modulation formats, the impact of channel distortions and temporal variations on digital signals, as well as multiple access schemes and much more was explored in research labs throughout the world. It thus became clear to the cognoscenti that the real-world systems would soon follow the research.

Again, it was Europe that led the way. The *European Telecommunications Standards Institute* (ETSI) group started the development of a digital cellular standard that would become mandatory throughout Europe, and was later adopted in most parts of the world: GSM (*Global System for Mobile communications*). The system was developed throughout the 1980s; deployment started in the early 1990s, and user acceptance was swift. Due to additional features, better speech quality, and the possibility for secure communications, GSM-based services overtook analog services typically within 2 years of their introduction. In the U.S.A., the change to digital systems was somewhat slower, but by the end of the 1990s, also this country was overwhelmingly digital.

Digital phones turned cellular communications, which was already on the road to success, into a blockbuster. In the year 2004, market penetration in Western Europe exceeded 80%, with some Scandinavian countries approaching the 100% mark (many people have two or three cellphones). Also the U.S.A. is exceeding the 50% mark, and Japan has about 70%. In absolute numbers, China has become the single biggest market, with some 300 million subscribers in 2004.

The development of wireless systems also made clear the necessity of standards. Devices can only communicate if they are compatible, and each receiver can "understand" each transmitter – i.e., if they follow the same standard. But how should these standards be set? Different countries developed different approaches. The approach in the U.S.A. is "hands-off": allow a wide variety of standards, and let the market establish the winner (or several winners). When frequencies for digital cellular communications were auctioned off in the 1990s, the buyers of the spectrum licences could choose the system standard they would use. For this reason, three different standards are now being used in the U.S.A. A similar approach was used by Japan, where two different systems fought for the market of Second Generation (2G) cellular systems. In both Japan and the U.S.A., the networks based on different standards work in the *same* geographical regions, allowing consumers to choose between different technical standards.

The situation was different in Europe. When digital communications were introduced, usually only one operator *per country* (typically, the incumbent public telephone operators) existed. If each of these operators would adopt a different standard, the result would be high market fragmentation (i.e., a small market for each standard), without the benefit of competition between the operators. Furthermore, roaming from country to country, which for obvious geographical regions is much more frequent in Europe than in the U.S.A. or Japan, would be impossible. It was thus logical to establish a single common standard for all of Europe. This decision proved to be beneficial for wireless communications in general, as it provided the economy of scales that decreased cost, and thus increased the popularity, of the new services.

1.1.5 New wireless systems and the burst of the bubble

Though cellular communications defined the picture of wireless in the general population, a whole range of new services was introduced in the 1990s. Cordless telephones started to replace the "normal" telephones in many homes. The first versions of these phones used analog technology; however, also for this application, digital technology proved to be superior. Among other aspects, the possibility of listening in to analog conversations, and the possibility for neighbors to "highjack" an analog cordless base station and make calls at other people's expense, led to a shift to digital communications. While cordless phones never achieved the spectacular market size of cellphones, they constitute a solid market.

Another market that seemed to have great promise in the 1990s was *fixed wireless access* and *Wireless Local Loop* (WLL) – in other words, replacing the copper lines to the homes of the users by wireless links, but without the specific benefit of mobility. A number of technical solutions were developed, but all of them ultimately failed. The reasons were as much economical and political as they were technical. The original motivation for WLL was to give access to customers for alternative providers of phone services, bypassing the copper lines that belonged to the incumbents. However, regulators throughout the world ruled in the mid-1990s that the incumbents *have* to lease their lines to the alternative providers, often at favorable prices. This eliminated much of the economic basis for WLL. Similarly, fixed wireless access was touted as the scheme to provide broadband data access at competitive prices. However, the price war between DSL (Digital Subscriber Line) technology and cable TV has greatly dimmed the economic attractiveness of this approach.

The biggest treasure thus seemed to lie in a further development of cellular systems, establishing the "Third Generation (3G)" (after the analog systems and 2G systems like GSM) [Bi et al. 2001]. 2G systems were essentially pure voice transmission systems (though some simple data services, like the *Short Message Service* – SMS – were included as well). The new systems were to provide data transmission at rates comparable with the ill-fated ISDN (*Integrated Services Digital Network*) (144 kbit/s), and even up to 2 Mbit/s, at speeds of up to 500 km/h. After long deliberations, two standards were established: 3GPP (*Third Generation Partnership Project*) (supported by Europe, Japan, and some American companies), and 3GPP2 (supported by another faction of American companies). The new standards also required a new spectrum allocation in most of the world, and the selling of this spectrum became one of the big bonanzas for the treasuries of several countries.

The development of 3GPP, and the earlier introduction of the IS-95 CDMA (*Code Division Multiple Access*) system in the U.S.A., sparked a lot of research into CDMA and other spread spectrum techniques (see Chapter 18) for wireless communications; by the end of the decade, multicarrier techniques (Chapter 19) had also gained a strong footing in the research community. Multiuser detection – i.e., the fact that the effect of interference can be greatly mitigated by exploiting its structure – was another area that many researchers concentrated on, particularly in the early 1990s. Finally, the field of multiantenna systems (Chapter 20) saw an enormous growth since 1995, and for some time accounted for almost half of all published research in the area of the physical layer design of wireless communications.

The spectrum sales for 3G cellular systems, and the IPOs (Initial Public Offerings) of some wireless startup companies represented the peak of the "telecom bubble" of the 1990s. In 2000/ 2001, the market collapsed with a spectacular crash. Developments on many of the new wireless systems (like fixed wireless) were stopped as their proponents went bankrupt, while the deployment of other systems, including 3G cellular systems, was slowed down considerably. Most worrisome, many companies slowed or completely stopped research, and the general economic malaise led to decreased funding of academic research as well.

1.1.6 Wireless revival

Since 2003, several developments have led to a renewed interest in wireless communications. The first one is a continued growth of cellular communications, stimulated by new markets (China, India), new applications (camera phones), and the slow but steady takeoff of 3G systems (where Japan is leading the way). Secondly, wireless computer networks (wireless Local Area Networks – LANs) have become an unexpected success. Devices following the IEEE (Institute of Electrical and Electronics Engineers) 802.11 standard (Chapter 24) have enabled computers to be used in a way that is almost as versatile and mobile as cellphones. The standardization process had already started in the mid-1990s, but it took several versions, and the impact of intense competition from manufacturers, to turn this into a mass product. Thirdly, wireless sensor networks offer new possibilities of monitoring and controlling factories and even homes from remote sites. Of course, the renewed interest in military and security applications is also creating interest in new wireless products.

In general, the "wireless revival" is based on three tendencies: (i) a much broader range of products, (ii) data transmission with a higher rate for already existing products, and (iii) higher user densities. These trends determine the directions of research in the field, and provide a motivation for many of the more recent scientific developments.

1.2 Types of services

1.2.1 Broadcast

The first wireless service was broadcast radio. In this application, information is transmitted to different, possibly mobile, users, see Fig. 1.1. Four properties differentiate broadcast radio from,

BROADCAST

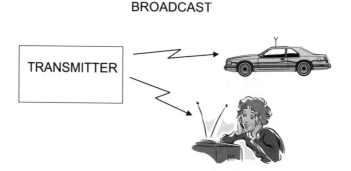

Figure 1.1   Principle of broadcast transmission.

for example, cellular telephony:

1. The information is only sent in one direction. It is only the broadcast station that sends information to the radio or TV receivers; the listeners (or viewers) do not transmit any information back to the broadcast station.
2. The transmitted information is the same for all users.
3. The information is transmitted continuously.
4. In many cases, multiple transmitters send the same information. This is especially true in Europe, where national broadcast networks cover a whole country, and broadcast the same program in every part of that country.[5]

The above properties lead to a great many simplifications in the design of broadcast radio networks. The transmitter does not need to have any knowledge or consideration about the receivers. There is no requirement to provide for duplex channels (i.e., for bringing information from the receiver to the transmitter). The number of possible users of the service does not influence the transmitter structure either – irrespective of whether there are millions of users, or just a single one, the transmitter sends out the same information.

The above description has been mainly true for traditional analog broadcast TV and radio. Satellite TV and radio differ by the fact that often the transmissions are intended only for a subset of all possible users (pay-TV or pay-per-view customers), and therefore, encryption of the content is required in order to prevent unauthorized viewing. Note, however, that this "privacy" problem is different from regular cellphones: for pay-TV, the content should be accessible to all members of the authorized user group, while for cellphones, each call should be accessible only for the single person it is intended for, and not to all customers of a network provider.

Despite their undisputed economic importance, broadcast networks will not be at the center of interest for this book – space restrictions prevent a more detailed discussion. Still, it is useful to keep in mind that they are a specific case of wireless information transmission, and recent developments, like simulcast digital TV and interactive TV, tend to obscure the distinction from cellular telephony even more.

1.2.2 Paging

Similar to broadcast, paging systems are unidirectional wireless communications systems. They are characterized by the following properties (see also Fig. 1.2):

1. the user can only receive information, but cannot transmit. Consequently, a "call" (message) can only be initiated by the call center, not by the user.
2. the information is intended for, and received by, only a single user.
3. the amount of transmitted information is very small. Originally, the received information consisted of a single bit of information, which indicated to the user that "somebody has sent you a message". The user then had to make a phone call (usually from a payphone) to the call center, where a human operator repeated the content of the waiting message. Later paging systems became more sophisticated, allowing the transmission of short messages (e.g., a different phone number that should be called, or the nature of an emergency). Still, the amount of information was rather limited.

Due to the unidirectional nature of the communications, and the small amount of

---

[5] The situation is slightly different in the U.S.A., where a "local station" usually covers only a single metropolitan area, often with a single transmitter.

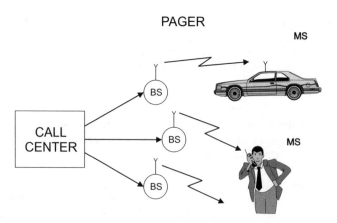

Figure 1.2 Principle of a pager.

information, the bandwidth required for this service is small. This in turn allows the service to operate at lower carrier frequencies – e.g., 150 MHz – where only small amounts of spectrum are available. As we will see later on, such lower carrier frequencies make it much easier to achieve good coverage of a large area with just a few transmitters.

Pagers were very popular during the 1980s and early 1990s. For some professional groups, like doctors, they were essential tools of the trade, allowing them to react to emergencies in shorter time. However, the success of cellular telephony has considerably reduced their appeal. Cellphones allow provision of all the services of a pager, plus many other features as well. The main appeal of paging systems after the year 2000 lies in the better area coverage that they can achieve.

1.2.3 Cellular telephony

Cellular telephony is the economically most important form of wireless communications. It is characterized by the following properties:

1. The information flow is bi-directional. A user can transmit and receive information at the same time.
2. The user can be anywhere within a (nationwide or international) network. Neither (s)he nor the calling party need to know the user's location; it is the network that has to take the mobility of the user into account.
3. A call can originate from either the network, or the user. In other words, a cellular customer can be called, or initiate a call.
4. A call is intended only for a single user; other users of the network should not be able to listen in.
5. High mobility of the users. The location of a user can change significantly during a call.

Figure 1.3 shows a block diagram of a cellular system. A mobile user is communicating with a *Base Station* (BS) that has a good radio connection with that user. The BSs are connected to a mobile switching center, which is connected to the public telephone system.

Since each user wants to transmit or receive different information, the number of active users in a network is limited. The available bandwidth must be shared between the different users; this can be done via different "multiple access" schemes (see also Chapters 17–20). This is an

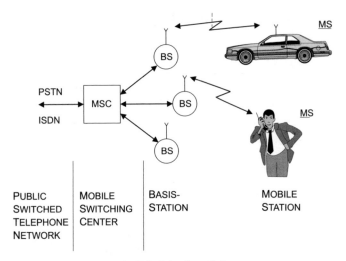

Figure 1.3 Principle of a cellular system.

important difference from broadcast systems, where the number of users (receivers) is unlimited, since they all receive the same information.

In order to increase the number of possible users, the *cellular principle* is used: the area served by a network provider is divided into a number of subareas, called *cells*. Within each cell, different users have to share the available bandwidth: let us consider in the following the case that each user occupies a different carrier frequency. Even users in neighboring cells have to use different frequencies, in order to keep co-channel interference low. However, for cells that are sufficiently far apart, the same frequencies can be used, because the signals get weaker with increasing distance from their transmitter. Thus, within one country, there can be hundreds or thousands of cells that are using the same frequencies.

Another important aspect of cellular telephony is the unlimited mobility. The user can be anywhere within the coverage area of the *network* (i.e., is not limited to a specific cell), in order to be able to communicate. Also, (s)he can move from one cell to the other during one call. The cellular network interfaces with the PSTN, as well as with other wireless systems.

As mentioned in our brief wireless history, cellphones started to become popular in the 1980s, and are now a dominant form of communications, with more than a billion users worldwide. Due to this reason, this book will often draw its examples from cellular telephony, even though the general principles are applicable to other wireless systems as well. Chapters 21–23 will give a detailed description of the most popular cellular systems.

1.2.4 Trunking radio

Trunking radio systems are an important variant of cellular phones, where there is no connection between the wireless system and the PSTN; therefore, it allows the communications of closed user groups. Obvious applications include police departments, fire departments, taxis, and similar services. The closed user group allows implementation of several technical innovations that are not possible (or more difficult) in normal cellular systems:

1. *Group calls*: a communication can be sent to several users simultaneously, or several users can set up a conference call between multiple users of the system.

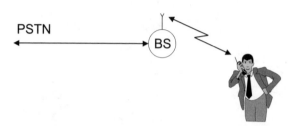

Figure 1.4   Principle of a simple cordless phone.

2. *Call priorities*: a normal cellular system operates on a "first-come, first-serve" basis. Once a call is established, it cannot be interrupted.[6] This is reasonable for cellphone systems, where the network operator cannot ascertain the importance or urgency of a call. However, for the trunk radio system of, e.g., a fire department, this is not an acceptable procedure. Notifications of emergencies have to go through to the affected parties, even if that means interrupting an existing, lower priority call. A trunking radio system thus has to enable a prioritization of calls, and has to allow dropping a low-priority call in favour of a high-priority one.
3. *Relay networks*: the range of the network can be extended by using each *Mobile Station* (MS) as a relay station for other MSs. Thus, an MS that is out of the coverage region of the BS might send its information to another MS that is within the coverage region, and that MS will forward the message to the BS; the system can even use multiple relays to finally reach the BS. Such an approach increases the effective coverage area and the reliability of the network. However, it can only be used in a trunking radio system and not in a cellular system – normal cellular users would not want to have to spend "their" battery power on relaying messages for other users.

1.2.5 Cordless telephony

Cordless telephony describes a wireless link between a handset and a BS that is directly connected to the public telephone system. The main difference from a cellphone is that the cordless telephone is associated with, and can communicate with, only a single BS (see Fig. 1.4). There is thus no *mobile switching center*; rather, the BS is directly connected to the PSTN. This has several important consequences:

1. The BS does not need to have any network functionality. When a call is coming in from the PSTN, there is no need to find out the location of the MS. Similarly, there is no need to provide for handover between different BSs.
2. There is no central system. A user typically has one BS for his/her apartment or business under control, but no influence on any other BSs. For that reason, there is no need for (and no possibility for) frequency planning.
3. The fact that the cordless phone is under the control of the user also implies a different pricing structure: there are no network operators that can charge fees for connections from the MS to the BS; rather, the only occurring fees are the fees from the BS into the PSTN.

---

[6] Except for interrupts due to technical problems, like the user moving outside the coverage region.

WIRELESS PABX

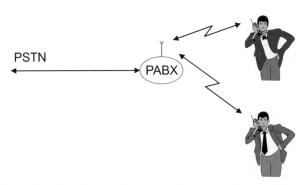

Figure 1.5 Principle of a wireless private automatic branch exchange.

In many other respects, the cordless phone is similar to the cellular phone: it allows mobility *within* the cell area; the information flow is bi-directional; calls can originate from either the PSTN or the mobile user, and there have to be provisions such that calls cannot be intercepted or listened to by unauthorized users and no unauthorized calls can be made.

Cordless systems have also evolved into wireless *PABXs* (*Private Automatic Branch eXchanges*) (see Fig. 1.5). In its most simple form, a PABX has a single BS that can serve several handsets simultaneously – either connecting them to the PSTN, or establishing a connection between them (for calls within the same company or house). In its more advanced form, the PABX contains several BSs that are connected to a central control station. Such a system has essentially the same functionality as a cellular system; it is only the size of the coverage area that distinguishes such a full functionality wireless PABX from a cellular network.

The first cordless phone systems were analog systems that just established a simple wireless link between a handset and a BS; often, they did not even provide rudimentary security (i.e., stopping unauthorized calls). Current systems are digital, and provide more sophisticated functionality. In Europe, the Digital Enhanced Cordless Telecommunications (DECT) system (see the companion website at *www.wiley.com/go/molisch*) is the dominant standard; Japan has a similar system called the Personal Handyphone System (PHS) that provides both the possibility for cordless telephony and an alternative cellular system (a full functionality PABX system that covers most of Japan, and that provides the possibility of public access). Both systems operate in the 1,800-MHz band, using a spectrum specifically dedicated to cordless applications. In the U.S.A., digital cordless phones mainly operate in the 2.45-GHz ISM (*Industrial, Scientific, and Medical*) band, which they share with many other wireless services.

1.2.6 Wireless local area networks and personal area networks

The functionality of WLANs is very similar to that of cordless phones – connecting a single mobile user device to a public landline system. The "mobile user device" in this case is usually a laptop computer, and the public landline system is the Internet. As in the cordless phone case, the main advantage is convenience for the user, allowing mobility. Wireless LANs can even be useful for connecting fixed-location computers (desktops) to the Internet, as they save the costs for laying cables to the desired location of the computer.

A major difference between wireless LANs and cordless phones is the required data rate. While cordless phones need to transmit (digitized) speech, which requires at most 64 kbit/s, wireless LANs should be at least as fast as the Internet that they are connected to. For consumer (home) applications, this means between 700 kbit/s (the speed of DSLs in the U.S.A.) to 3–5 Mbit/s (speed of cable providers in the U.S.A. and Europe) to ≥20 Mbit/s (speed of DSLs in Japan). For companies that have faster Internet connections, the requirements are proportionately higher. In order to satisfy the need for these high data rates, a number of standards have been developed, all of which carry the identifier IEEE 802.11. The original IEEE 802.11 standard enabled transmission with 1 Mbit/s; the very popular 802.11b standard (also known under the name WiFi) allows up to 11 Mbit/s, and the 802.11a standard extends that to 55 Mbit/s. Even higher rates will be realized by the 802.11n standard whose development started in 2004.

WLAN devices can, in principle, connect to any BS (access point) that uses the same standard. However, the owner of the access point can restrict the access – e.g., by appropriate security settings.

An alternative form of WLANs are *ad hoc networks* (see Fig. 1.6). In these networks, several computers set up a network in which all devices have the same functionality: communicating with each other. These networks therefore function without a BS and without any Internet connection at all. While the actual transmission of the data (i.e., physical-layer communication) is almost identical to that of normal WLANs, the medium access and the networking functionalities are very different. Ad hoc networks are usually restricted to a few devices and to a range of 10 m or less.

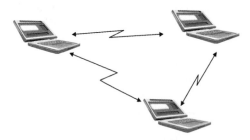

Figure 1.6   Principle of an ad hoc network.

1.2.7 Personal area networks

When the coverage area becomes even smaller than that of WLANs, we speak of *Personal Area Networks* (PANs). Such networks are mostly intended for simple "cable replacement" duties. For example, devices following the *Bluetooth* standard allow to connect a hands-free headset to a cellular phone without requiring a cable; in that case, the distance between the two devices is less than a meter. Similarly, wireless links between components in an entertainment system (DVD player to TV), between computer and peripheral devices (printer, mouse), and similar applications can be covered by PANs. A number of standards for PANs has been developed by the IEEE 802.15 group.

1.2.8 Fixed wireless access

Fixed wireless access systems can also be considered as a derivative of cordless phones or WLANs, essentially replacing a dedicated cable connection between the user and the public landline system. The main difference from a cordless system is that (i) there is no mobility of the user devices, and (ii) the BS almost always serves multiple users. Furthermore, the distances bridged by fixed wireless access devices are much larger (between 100 m and several tens of kilometers) than those bridged by cordless telephones.

The purpose of fixed wireless access lies in providing users with telephone and data connections without having to lay cables from a central switching office to the office or apartment the user is in. Considering the high cost of labour for the cable-laying operations, this can be an economical approach. However, it is worth keeping in mind that most buildings, especially in the urban areas of developed countries, are already supplied by some form of cable – regular telephone cable, cable TV, or even optical fiber. Rulings of the telecom regulators in various countries have stressed that incumbent operators (owners of these lines) have to allow competing companies to use these lines. As a consequence, fixed wireless access has its main market for covering rural areas, and for establishing connections in developing countries that do not have any wired infrastructure in place. In general, the business cases for fixed wireless has been disappointing (see "Burst of the bubble" in Section 1.1.5). The 802.16 (WiMax) standard tries to alleviate that problem by allowing some limited mobility in the system, and thus blur the distinction from cellular telephony.

1.2.9 Satellite cellular communications

Besides TV, which creates the biggest revenues in the satellite market, cellular communications are a second important application of satellites. Satellite cellular communications have the same operating principles as land-based cellular communications. However, there are some key differences.

The distance between the "BS" (i.e., the satellite), and the MS is *much* larger: for geostationary satellites, that distance is 36,000 km; for *Low Earth Orbit* (LEO) satellites, it is several hundred kilometers. Consequently, the transmit powers need to be larger, high-gain antennas need to be used on the satellite (and in many cases also on the MS), and communications from within buildings is almost impossible.

Another important difference from the land-based cellular system lies in the cell size: due to the large distance between the satellite and the Earth, it is impossible to have cells with diameters less than 100 km even with LEO satellites; for geostationary satellites, the cell areas are even larger. This large cellsize is the biggest advantage, as well as the biggest drawback, of the satellite systems. On the positive side, it makes it easy to have good coverage even of large, sparsely populated areas – a single cell might cover most of the Sahara region. On the other hand, the area spectral efficiency is very low, which means that (given the limited spectrum assigned to this service) only a few people can communicate at the same time.

The costs of setting up a "BS" – i.e., a satellite – are much higher than for a land-based system. Not only is the launching of a communications satellite very expensive, but it is also necessary to build up an appropriate infrastructure of ground stations for linking the satellites to the PSTN.

As a consequence of all these issues, the business case for satellite communications systems is quite different: it is based on supplying a small number of users with vital communications at a much higher price. Emergency workers and journalists in disaster and war areas, ship-based communications, and workers on offshore oil drilling platforms are typical users for such systems. The INMARSAT system is the leading provider for such communications. In the late 1990s, the IRIDIUM project attempted to provide lower priced satellite communications services by means of some 60 LEO satellites, but ended in bankruptcy.

1.3 Requirements for the services

A key to understanding wireless design is to realize that different applications have different requirements in terms of data rate, range, mobility, energy consumption, and so on. It is *not* necessary to design a system that can sustain Gigabit/s data rates over a 100-km range when the user is moving at 500 km/h. We stress this fact because there is a tendency among engineers to design a system that "does everything but wash the dishes"; while appealing from a scientific point of view, such systems tend to have a high price, and low spectral efficiency. In the following, we list the range of requirements encountered in system design, and we enumerate which requirements occur in which applications.

1.3.1 Data rate

Data rates for wireless services span the gamut from a few bits per second to several Gigabit/s, depending on the application:

- *Sensor networks* usually require data rates from a few bits per second to about 1 kbit/s. Typically, a sensor measures some critical parameter, like temperature, speed, etc., and transmits the current value (which corresponds to just a few bits) at intervals that can range from milliseconds to several hours. Higher data rates are often required for the central nodes of sensor networks that collect the information from a large number of sensors, and forward it for further processing. In that case, data rates of up to 10 Mbit/s can be required. These "central nodes" show more similarity to WLANs or fixed wireless access.
- *Speech communications* usually require between 5 kbit/s and 64 kbit/s, depending on the required quality and the amount of compression. For cellular systems, which require higher spectral efficiency, source data rates of about 10 kbit/s are standard. For cordless systems, less elaborate compression, and therefore higher data rates (32 kbit/s) are used.
- *Elementary data services* require between 10 and 100 kbit/s. One category of these services uses the display of the cellphone to provide Internet-like information. Since the displays are smaller, the required data rates are often smaller than for conventional Internet applications. Another type of data service provides a wireless mobile connection to laptop computers. In this case, speeds that are at least comparable with dial-up (around 50 kbit/s) are demanded by most users, though elementary services with 10 kbit/s (exploiting the same type of communications channels foreseen for speech) are sometimes used as well.
- *Communications between computer peripherals* and similar devices: for the replacement of cables that link computer peripherals, like mouse and keyboard, to the computer (or similarly for cellphones), wireless links with data rates around 1 Mbit/s are used. The functionality of these links is similar to the previously popular infrared links, but usually provide higher reliability.
- *High-speed data services*: WLANs and 3G cellular systems are used to provide fast Internet access, with speeds that range from 0.5 Mbit/s to 100 Mbit/s (currently under development).
- *Personal Area Networks* (PANs) is a newly coined term that refers mostly to the range of a wireless network (up to 10 m), but often also has the connotation of high data rates (over 100 Mbit/s), mostly for linking the components of consumer entertainment systems (streaming video from computer or DVD player to a TV) or high-speed computer connections (wireless USB).

1.3.2 Range and number of users

Another distinction among the different networks is the range and the number of users that they serve. By "range", we mean here the distance between one transmitter and receiver. The coverage area of a system can be made almost independent of the range, by just combining a larger number of BSs into one big network:

- *Body Area Networks* (BANs) cover the communication between different devices attached to one body – e.g., from a cellphone in a hip holster to a headset attached to the ear. The range is thus on the order of 1 m. BANs are often subsumed into PANs.
- *Personal Area Networks* include networks that achieve distances up to about 10 m, covering the "personal space" of one user. Examples are networks linking components of computers and home entertainment systems. Due to the small range, the number of devices within a PAN is small, and all are associated with a single "owner". Also, the number of overlapping PANs (i.e., sharing the same space or room) is small – usually less than five. That makes cell planning and multiple access much simpler.
- *Wireless Local Area Networks*, as well as cordless telephones, cover still larger ranges, up to 100 m. The number of users is usually limited to about 10. When much larger numbers occur (e.g., at conferences or meetings), the data rates for each user decrease. Similarly, cordless phones have a range of up to 300 m, and the number of users connected to one BS is on the same order as for WLANs. Note, however, that wireless PABXs can have much larger ranges and user numbers – as mentioned before, they can be seen essentially as small private cellular systems
- *Cellular systems* have a range that is larger than, e.g., the range of WLANs. Microcells typically cover cells with 500 m radius, while macrocells can have a radius of 10 or even 30 km. Depending on the available bandwidth and the multiple access scheme, the number of active users in a cell is usually between 5 and 50. If the system is providing high-speed data services to one user, the number of active users usually shrinks.
- *Fixed wireless access services* cover a range that is similar to that of cellphones – namely, between 100 m and several tens of kilometers. Also, the number of users is of a similar order as for cellular systems.
- *Satellite systems* provide even larger cell sizes, often covering whole countries and even continents. Cellsize depends critically on the orbit of the satellite: geostationary satellites provide larger cellsizes (1,000-km radius) than LEOs.

Figure 1.7 gives a graphical representation of the link between data rate and range. Obviously, higher data rates are easier to achieve if the required range is smaller. One exception is fixed wireless access, which demands a high data rate at rather large distances.

1.3.3 Mobility

Wireless systems also differ in the amount of mobility that they have to allow for the users. The ability to move around while communicating is one of the main charms of wireless communication for the user. Still, within that requirement of mobility, different grades exist:

- *Fixed devices* are placed only once, and after that time communicate with their BS, or each other, always from the same location. The main motivation for using wireless transmission techniques for such devices lies in avoiding the laying of cables. Even though the devices are not mobile, the propagation channel they transmit over can change with time: both due to people walking by, and due to changes in the environment (rearranging of machinery,

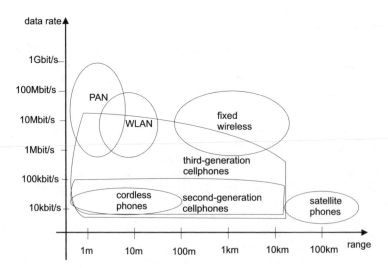

Figure 1.7   Data rate versus range for various applications.

furniture, etc.). Fixed wireless access is a typical case in point. Note also that all wired communications (e.g., the PSTN) fall into this category.

- *Nomadic devices*: nomadic devices are placed at a certain location for a limited duration of time (minutes to hours) and then moved to a different location. This means that during one "drop" (placing of the device), the device is similar to a fixed device. However, from one drop to the next, the environment can change radically. Laptops are typical examples: people do not operate their laptops while walking around, but place them on a desk to work with them. Minutes or hours later, they might bring them to a different location, and operate them there.
- *Low mobility*: many communications devices are operated at pedestrian speeds. Cordless phones, as well as cellphones operated by walking human users are typical examples. The effect of the low mobility is a channel that changes rather slowly, and – in a system with multiple BSs – handover from one cell to another is a rare event.
- *High mobility* usually describes speed ranges from about 30 km/h to 150 km/h. Cellphones operated by people in moving cars are one typical example.
- *Extremely high mobility* is represented by high-speed trains and planes, which covers speeds between 300 km/h and 1,000 km/h. These speeds pose unique challenges both for the design of the physical layer (Doppler shift, see Chapter 5), and for the handover between cells.

Figure 1.8 shows the relationship between mobility and data rate.

1.3.4 Energy consumption

Energy consumption is a critical aspect for wireless devices. Most wireless devices use (one-way or rechargeable) batteries, as they should be free of *any* wires – both the ones used for communication, and the ones providing the power supply.

- *Rechargeable batteries*: nomadic and mobile devices, like laptops, cellphones, and cordless phones, are usually operated with rechargeable batteries. Standby times as well as operating times are one of the determining factors for customer satisfaction. Energy consumption is

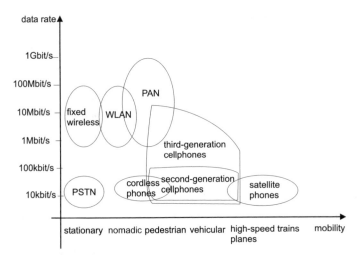

Figure 1.8 Data rate versus mobility for various applications.

determined on one hand by the distance over which the data have to be transmitted (remember that a minimum Signal-to-Noise Ratio – SNR – has to be maintained), and on the other hand by the amount of data that are to be transmitted (the SNR is proportional to the energy per bit). The energy density of batteries has increased slowly over the past 100 years, so that the main improvements in terms of operating and standby time stem from reduced energy consumption of the devices. For cellphones, talk times of more than 2 hours, and standby times of more than 48 hours are minimum requirements. For laptops, power consumption is not mainly determined by the wireless transmitter, but rather by other factors like harddrive usage and processor speed.

- *One-way batteries*: sensor network nodes often use one-way batteries, which offer higher energy density at lower prices. Furthermore, changing the battery is often not an option; rather the sensor including the battery and the wireless transceiver is often discarded after the battery has run out. It is obvious that in this case energy-efficient operation is even more important than for devices with rechargeable batteries.
- *Power mains*: BSs and other fixed devices can be connected to the power mains. Therefore, energy efficiency is not a major concern for them. It is thus desirable – if possible – to shift as much functionality (and thus energy consumption) from the MS to the BS.

User requirements concerning batteries are also important sales issues, especially in the market for cellular handsets:

- The weight of an MS is determined mostly (70–80%) by the battery. Weight and size of a handset are critical sales issues. It was in the mid-1980s that cellphones were commonly called "carphones", because the MS could only be transported in the trunk of a car, and was powered by the car battery. By the end of the 1980s, the weight and dimensions of the batteries had decreased to about 2 kg, so that it could be carried by the user in a backpack. By the year 2000, the battery weight had decreased to about 200 g. Part of this improvement stems from more efficient battery technology, but to a large part, it is caused by the decrease of the power consumption of the handsets.

- Also the costs of a cellphone (raw materials) are determined to a considerable degree by the battery.
- Users require standby times of several days, as well as talk times of at least 2 hours before recharging.

These "commercial" aspects determine the maximum size (and thus energy content) of the battery, and consequently, the admissible energy consumption of the phone during standby and talk operation.

1.3.5 Use of spectrum

Spectrum can be assigned on an exclusive basis, or on a shared basis. That determines to a large degree the multiple access scheme and the interference resistance that the system has to provide:

- *Spectrum dedicated to service and operator*: in this case, a certain part of the electromagnetic spectrum is assigned, on an exclusive basis, to a service provider. A prime point in case is cellular telephony, where the network operators buy or lease the spectrum on an exclusive basis (often for a very high price). Due to this arrangement, the operator has control over the spectrum and can plan the use of different parts of this spectrum in different geographical regions, in order to minimize interference.
- *Spectrum allowing multiple operators*:
  —*Spectrum dedicated to a service*: in this case, the spectrum can be used only for a certain service (e.g., cordless telephones in Europe and Japan), but is not assigned to a specific operator. Rather, users can set up qualified equipment without a license. Such an approach does not require (or allow) interference planning. Rather, the system must be designed in such a way that it avoids interfering with other users in the same region. Since the only interference *can* come from equipment of the same type, coordination between different devices is relatively simple. Limits on transmit power (identical for all users) are a key component of this approach – without them, each user would just increase the transmit power to drown out interferers, leading essentially to an "arms race" between users.
  —*Free spectrum* is assigned for different services as well as different operators. The ISM band at 2.45 GHz is the best known example – it is allowed to operate microwave ovens, WiFi LANs, and Bluetooth wireless links, among others, in this band. Also for this case, each user has to adhere to strict emission limits, in order not to interfere too much with other systems and users. However, coordination between users (in order to minimize interference) becomes almost impossible – the different systems cannot exchange coordination messages with each other, and often even have problems determining the exact characteristics (bandwidth, duty cycle) of the interferers.

After 2000, two new approaches have been promulgated, but are not yet in use:

- *Ultrawideband systems* spread their information over a very large bandwidth, while at the same time keeping a very low power spectral density. Therefore, the transmit band can include frequency bands that have already been assigned to other services, without creating significant interference.
- *Adaptive spectral usage*: another approach relies on first determining the current spectrum usage at a certain location, and then employing unused parts of the spectrum.

1.3.6 Direction of transmission

Not all wireless services need to convey information in both directions.

- *Simplex systems* send the information only on one direction – e.g., broadcast systems and pagers.
- *Semi-duplex systems* can transmit information in both directions. However, only one direction is allowed at any time. Walkie-talkies, which require the user to push a button in order to talk, are a typical example. Note that one user must signify (e.g., by using the word "over'") that (s)he has finished his/her transmission; then the other user knows that now (s)he can transmit.
- *Full duplex systems* allow simultaneous transmission in both directions – e.g., cellphones and cordless phones.
- *Asymmetric duplex systems*: for data transmission, we often find that the required data rate in one direction (usually the downlink) is higher than in the other direction. However, even in this case, full duplex capability is maintained.

1.3.7 Service quality

The requirements for service quality also differ vastly for different wireless services. A first main indicator for service quality is *speech quality* for speech services and *file transfer speed* for data services. Speech quality is usually measured by the *Mean Opinion Score* (MOS). It represents the average of a large number of (subjective) human judgements (on a scale from 1 to 5) about the quality of received speech (see also Chapter 15). The speed of data transmission is simply measured in bit/s – obviously, a higher speed is better.

An even more important factor is the availability of a service. For cellphones and other speech services, the *service quality* is often computed as the complement of "fraction of blocked calls[7] plus 10 times the fraction of dropped calls". This formula takes into account that the dropping of an active call is more annoying to the user than the inability to make a call at all. For cellular systems in Europe and Japan, this service quality measure usually exceeds 95%; in the U.S.A., the rate is closer to 50%.[8]

For emergency services and military applications, service quality is better measured as the complement of "fraction of blocked calls plus fraction of dropped calls". In emergency situations, the inability to make a call is as annoying as the situation of having a call interrupted. Also, the systems must be planned in a much more robust way, as service qualities better than 99% are required.

A related aspect is also the *admissible delay (latency)* of the communication. For voice communications, the delay between the time when one person speaks, and the other hears the message, must not be larger than about 100 ms. For streaming video and music, delays can be larger, as buffering of the streams (up to several tens of seconds) is deemed acceptable by most users. In both voice and streaming video communications, it is important that the data transmitted first are also the ones made available to the receiving user first. For data files, the acceptable delays can be usually larger, and the sequence with which the data arrive at the receiver is not critical (for example, when downloading email from a server, it is not important whether the first or the seventh of the emails is the first to arrive). However, there are some data applications where small latency is vital – e.g., for control applications, security and safety monitoring, etc.

---

[7] Here, "blocked calls" encompasses all failed call attempts, including those that are caused by insufficient signal strength, as well as insufficient network capacity.

[8] The reason for this discrepancy is partly historical and economical, and partly geographical.

1.4 Economic and social aspects

1.4.1 Economic requirements for building wireless communications systems

The design of wireless systems does not only aim to optimize performance for specific applications, but also to do that at reasonable cost. As economic factors impact the design, scientists and engineers have to have at least a basic understanding of the constraints imposed by marketing and sales divisions. Some of the guidelines for the design of wireless *devices* are:

- Move as much functionality as possible from the (more expensive) analog components to digital circuitry. The costs for digital circuits decrease much faster with time than those of analog components.
- For mass-market applications, try to integrate as many components onto one chip as possible. Most systems strive to use only two chips; one for analog RF circuitry, and one for digital (baseband) processing. Further integration into a single chip (system on a chip) is desirable. Exceptions are niche market products, which typically try to use general-purpose processors, Application Specific Integrated Circuits (ASICs), or off-the-shelf components, as the number of sold units does not justify the design of more highly integrated chips.
- As human labour is very expensive, any circuit that requires human intervention (e.g., tuning of RF elements), is to be avoided. Again, this aspect is more important for mass-market products.
- In order to increase the efficiency of the development process and production, the same chips should be used in as many systems as possible.

When it comes to the design of wireless *systems and services*, we have to distinguish between two different categories:

- Systems where the mobility is of value by itself – e.g., in cellular telephony. Such services can charge a premium to the customer – i.e., be more expensive than equivalent, wired systems. Cellular telephony is a case in point: the per-minute price has been higher than that of landline telephony in the past, and is expected to remain so (especially when compared, e.g., with Voice-over-Internet Protocol – VoIP – telephone services). Despite this fact, the services might compete (and ultimately edge out) traditional wired services if the price difference is not too large. The years since 1990 have certainly seen such a trend, with many consumers (and even companies) canceling wired services, and relying on cellular telephony alone.
- Services where wireless access is only intended as a cheap cable replacement, without enabling additional features – e.g., fixed wireless access. Such systems have to be especially cost-conscious, as the buildup of the infrastructure has to remain cheaper than the laying of new wired connections, or buying access to existing ones.

1.4.2 The market for wireless communications

Cellphones are a highly dynamic market that has grown tremendously. Still, different countries show different market penetrations. Some of the factors influencing this penetration are:

- *Price of the offered services*: the price of the services is in turn influenced by the amount of competition, the willingness of the operators to accept losses in order to gain greater market penetration, and the external costs of the operators (especially, the cost of spectrum licenses). However, the price of the services is not always the decisive factor for market penetration: Scandinavia, with its relatively high prices, has the highest market penetration in the world.

- *Price of the MSs*: the MSs are usually subsidized by the operators, and are either free, or sold at a nominal price, if the consumer agrees to a long-term contract. Exceptions are "prepaid" services, where a user buys a certain number of minutes of service usage (in that case, the handsets are sold to the consumer at full cost); at the other end of the market spectrum, high-end devices usually require a significant co-payment by the consumer.
- *Attractiveness of the offered services*: in many markets, the price of the services offered by different network operators is almost identical. Operators try to distinguish themselves by different features, like better coverage, text and picture message service, etc. The offering of these improved features also helps to increase the market size in general, as it allows customers to find services tailored to their needs.
- *General economic situation*: obviously, a good general economic situation allows the general population to spend more money on such "non-essential" things as mobile communications services. In countries where a very large percentage of the income goes to basic necessities like food and housing, the market for cellular telephony is obviously more limited.
- *Existing telecom infrastructure*: in countries or areas with a bad existing landline-based telecom infrastructure, cellular telephony and other wireless services can be the only way of communicating. This would enable high market penetration. Unfortunately, these areas are usually also the ones that have the bad economic situation mentioned above (large percentage of income goes to basic necessities). This fact has hindered especially the development of fixed wireless services.[9]
- *Predisposition of the population*: there are several social factors that can increase the cellular market: (i) people have a positive attitude to new technology (gadgets) – e.g., Japan and Scandinavia; (ii) people consider communication an essential part of their lives – e.g., China; and (iii) high mobility of the population, with people being absent from their offices or homes for a significant part of the day – e.g., U.S.A.

Wireless communications has become such a huge market that most of the companies in this business are not even known to most of the consumers. Consumers tend to know network operators and handset manufacturers. However, component suppliers and other auxiliary industries abound:

- *Infrastructure manufacturers for cellular telephony*: most of the major handset manufacturers also provide infrastructure (BSs, switches, etc.) to the network operators.
- *Component manufacturers*: most handset manufacturers buy chips, batteries, antennas, etc., from external suppliers. This trend was accelerated by the fact that many manufacturers and system integrators spun off their semiconductor divisions. There are even handset companies that do not manufacture anything, but are just design and marketing operations.
- *Software suppliers*: softward and applications are becoming an increasingly important part of the market. For example, ringtones for cellphones have become a multibillion dollar (euro) market. Similarly, operating systems and applications software for cellphones have become increasingly important as cellphones acquire more and more the functionalities of PDAs (*Personal Digital Assistants*).
- *Systems integrators*: WLANs and sensor networks need to be integrated either into larger networks, or combined with other hardware (e.g., sensor networks have to be integrated

---

[9] To put it succinctly: "the market for this product is the people who cannot afford to pay for it."

into a factory automation system). This offers new fields of business for OEMs and system integrators.

1.4.3 Behavioral impact

Engineering does not happen in a vacuum – the demands of people change what the engineers develop, and the products of their labor influences how people behave. Cellphones have enabled us to communicate anytime – something that most people think of as desirable. But we have to be aware that it changes our lifestyle. In former times, one did not call a person, but rather a location. That meant a rather clean separation between professional or personal life. Due to the cellphone, anybody can be reached at any time – somebody from work can call in the evening; a private acquaintance can call in the middle of a meeting; in other words, professional and private life get intermingled. On the positive side, this also allows new and more convenient forms of working and increased flexibility.

Another important behavioral impact is the development of (or lack of) *cellphone etiquette.* Most people tend to agree that hearing a cellphone ring during an opera performance is exasperating – and still there is a significant number of people who are not willing to turn their phones to the "silent" mode for such occasions. There also seems to be an innate reluctance in humans to just ignore a ringing cellphone. People are willing to interrupt whatever they are doing in order to answer a ringing phone. Caller identification, automatic callback features, etc., are solutions that engineers can provide to alleviate these problems.

On a more serious note, wireless devices, and especially cellphones, can be a matter of life and death. Being able to call for help in the middle of the wilderness after a mountaineering accident is definitely a lifesaving feature. Location devices for victims of avalanches have a similar beneficial role. On the downside, drivers that are distracted by their cellphone conversations constitute a serious hazard on the roads. The problem is, again, not purely a technical one, as a multitude of solutions (including the "off" button) have been developed to solve this issue. Rather, it is a matter of behavioral changes by the users, and the question of what the engineers can do to further these changes.

2

Technical challenges of wireless communications

In the previous chapter, we have described the requirements for wireless communications systems, stemming from the applications and user demands. In this chapter, we will give a high-level description of the physical challenges to wireless communications systems. Most notably, they are:

- multipath propagation: i.e., the fact that a transmit signal can reach the receiver via different paths (e.g., reflections from different houses or mountains);
- spectrum limitations;
- energy limitations;
- user mobility.

This will set the stage for the rest of the book, where these challenges, as well as remedies, will be discussed in more detail.

As a first step, it is useful to investigate the differences between wired and wireless communications. Let us first repeat some important properties of wired and wireless systems, as summarized in Table 2.1.

2.1 Multipath propagation

For wireless communications, the transmission medium is the radio channel between transmitter TX and receiver RX. The signal can get from the TX to the RX via a number of different propagation paths. In some cases, a Line Of Sight (LOS) connection might exist between TX and RX. Furthermore, the signal can get from the TX to the RX by being reflected at or diffracted by different *Interacting Objects* (IOs) in the environment: houses, mountains (for outdoor environments), windows, walls, etc. The number of these possible propagation paths is very large. As shown in Fig. 2.1, each of the paths has a distinct amplitude, delay (runtime of the signal), direction of departure from the TX, and direction of arrival; most importantly, the components have different *phase shifts* with respect to each other. In the following, we will discuss some implications of the multipath propagation for system design.

Wireless Communications   Andreas F. Molisch
© 2005 John Wiley & Sons, Ltd

Table 2.1 Wired and wireless communications.

Wired communications	Wireless communications
The communication takes place over a more or less stable medium like copper wires or optical fibers. The properties of the medium are well-defined, and time-invariant.	Due to user mobility as well as multipath propagation, the transmission medium varies strongly with time.
Increasing the transmission capacity can be achieved by using a different frequency on an existing cable, and/or by stringing new cables.	Increasing the transmit capacity must be achieved by more sophisticated transceiver concepts and smaller cell sizes (in cellular systems), as the amount of available spectrum is limited.
The range over which communications can be performed without repeater stations is mostly limited by attenuation by the medium (and thus noise); for optical fibers, the distortion of transmitted pulses can also limit the speed of data transmission.	The range that can be covered is limited both by the transmission medium (attenuation, fading, and signal distortion) and by the requirements of spectral efficiency (cell size).
Interference and crosstalk from other users either do not happen, or the properties of the interference are stationary.	Interference and crosstalk from other users is inherent in the principle of cellular communications. Due to the mobility of the users, they also are time-variant.
The delay in the transmission process is also constant, determined by the length of the cable, and the group delay of possible repeater amplifiers.	The delay of the transmission depends mostly on the distance between base station and mobile station, and is thus time-variant.
The *Bit Error Rate* (BER) decreases strongly (approximately exponentially) with increasing *Signal-to-Noise Ratio* (SNR). This means that a relatively small increase in transmit power can greatly decrease the error rate.	For simple systems, the average BER decreases only slowly (linearly) with increasing average SNR. Increasing the transmit power usually does not lead to a significant reduction in BER. However, more sophisticated signal processing helps.
Due to the well-behaved transmission medium, the quality of wired transmission is generally high.	Due to the difficult medium, transmission quality is generally low unless special measures are used.
Jamming and interception of wired transmission is almost impossible without consent by the network operator.	Jamming a wireless link is straightforward, unless special measures are taken. Interception of the on-air signal is possible. Encryption is therefore necessary to prevent unauthorized use of the information.
Establishing a link is *location*-based. In other words, a link is established from one outlet to another, independent of which *person* is connected to the outlet.	Establishing a connection is based on the (mobile) equipment, usually associated with a specific person. The connection is not associated with a fixed location.
Power is either provided through the communications network itself (e.g., for traditional landline telephones), or from traditional power mains (e.g., fax). In neither case is energy consumption a major concern for the designer of the device.	Mobile stations use rechargable or one-way batteries. Energy efficiency is thus a major concern.

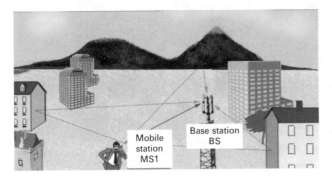

Figure 2.1 Multipath propagation.

2.1.1 Fading

A simple RX cannot distinguish between the different *Multi Path Components* (MPCs); it just adds them up, so that they interfere with each other. The interference between them can be constructive or destructive, depending on the phases of the MPCs, see Fig. 2.2. The phases, in turn, depend mostly on the runlength of the MPC, and thus on the position of the mobile station and the IOs. For this reason, the interference, and thus the amplitude of the total signal, changes with time if either TX, RX, or IOs are moving. This effect – namely, the changing of the total signal amplitude due to interference of the different MPCs – is called *small-scale fading*. At 2-GHz carrier frequency, a movement by less than 10 cm can already effect a change from constructive to destructive interference and vice versa. In other words, even a small movement can result in a large change in signal amplitude. A similar effect is known to all owners of car radios – moving the car by less than 1 meter (e.g., in stop-and-go traffic) can greatly affect the quality of the received signal. For cellphones, it can often be sufficient to move one step in order to improve signal quality.

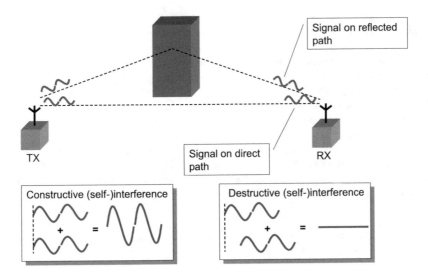

Figure 2.2 Principle of small-scale fading.

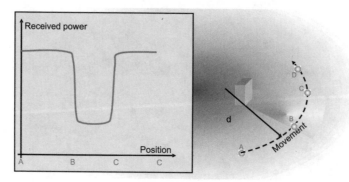

Figure 2.3 The principle of shadowing.

As an additional effect, the amplitudes of each separate MPC change with time (or with location). Obstacles can lead to a shadowing of one or several MPCs. Imagine, for example, the MS (Mobile Station) in Fig. 2.3 that at first (at position A) has LOS to the Base Station (BS). As the MS moves behind the high-rise building (at position B), the amplitude of the component that propagates along the direct connection (LOS) between BS and MS greatly decreases. This is due to the fact that the MS is now in the radio shadow of the high-rise building, and any wave going through or around that building is greatly attenuated – an effect called *shadowing*. Of course, shadowing can occur not only for a LOS component, but for *any* MPC. Note also that obstacles do not throw "sharp" shadows: the transition from the "light" (i.e., LOS) zone to the "dark" (shadowed) zone is gradual.[1] The MS has to move over large distances (from a few meters, up to several hundreds of meters) to move from the light to the dark zone. For this reason, shadowing gives rise to *large-scale fading*.

Large-scale and small-scale fading overlap, so that the received signal amplitude can look like the one depicted in Fig. 2.4. Obviously, the transmission quality is low at the times (or places) with low signal amplitude. This can lead to bad speech quality (for voice telephony), high Bit Error Rate (BER) and low data rate (for data transmission), and – if the quality is too low for an extended period of time – to termination of the connection.

It is well known from conventional digital communications that for non-fading communications links, the BER decreases approximately exponentially with increasing Signal-to-Noise Ratio (SNR) if no special measures are taken. However, in a fading channel, the SNR is not constant; rather, the probability that the link is in a *fading dip* (i.e., location with low SNR) dominates the behavior of the BER. For this reason, the average BER decreases only *linearly* with increasing average SNR. Consequently, improving the BER often cannot be achieved by simply increasing transmit power. Rather more sophisticated transmission and reception schemes have to be used. Most of the third and fourth part of this book (Chapters 13, 14, 16, 18, 19, 20) is devoted to such techniques.

Due to fading, it is almost impossible to exactly predict the received signal amplitude at arbitrary locations. For many aspects of system development and deployment, it is considered sufficient to predict the mean amplitude, and the *statistics* of fluctuations around that mean. Completely deterministic predictions of the signal amplitude – e.g., by solving approximations to

---

[1] This is due to (i) diffraction effects, as also explained in more detail in Chapter 4, and (ii) the fact that secondary radiation sources like houses are spatially extended (compare how a long fluorescent tube never throws a sharp shadow).

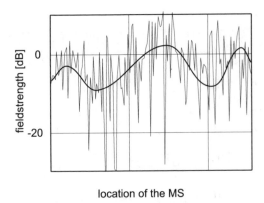

Figure 2.4   Typical example of fading. The thin line is the (normalized) instantaneous fieldstrength; the thick line is the average over a 1-m distance.

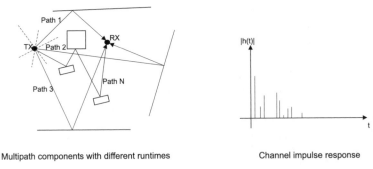

Figure 2.5   Multipath propagation and resulting impulse response.

Maxwell's equations[2] in a given environment – usually show errors of between 3 and 10 dB (for the total amplitude), and are even less reliable for the properties of individual MPCs. More details on fading can be found in Chapters 5 to 7.

2.1.2 Intersymbol interference

The runtimes for different MPCs are different. We have already mentioned above that this can lead to different phases of MPCs, which leads to interference in narrowband systems. In a system with large bandwidth, and thus good resolution in the time domain,[3] the major consequence is signal dispersion: in other words, the impulse response of the channel is not a single delta pulse, but rather a sequence of pulses (corresponding to different MPCs), each of which has a distinct arrival time in addition to having a different amplitude and phase (see Fig. 2.5). This signal dispersion leads to intersymbol interference at the RX. MPCs with long runtimes, carrying information from bit k, and MPCs with short runtimes, carrying contributions from bit $k + 1$ arrive at the RX at the same time, and interfere with each other

---

[2] The most popular of these deterministic prediction tools are "ray tracing" and "ray launching", which are discussed in Chapter 7.

[3] Strictly speaking, we refer here to resolution in the delay domain. An explanation for the difference between the time domain and delay domain will be given in Chapters 5 and 6.

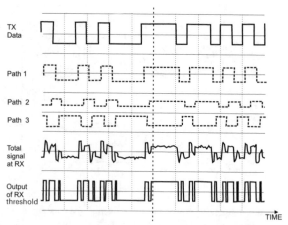

Figure 2.6 Intersymbol interference.

(see Fig. 2.6). Assuming that no special measures[4] are taken, this Inter Symbol Interference (ISI) leads to errors that cannot be eliminated by simply increasing the transmit power, and are therefore often called *irreducible errors*.

ISI is essentially determined by the ratio between symbol duration and the duration of the impulse response of the channel. This implies that ISI is not only more important for higher data rates, but also for multiple access methods that lead to an increase in transmitted *peak* data rate (e.g., time division multiple access, see Chapter 17). Finally, it is also noteworthy that ISI can even play a role when the duration of the impulse response is *shorter* (but not *much* shorter) than bit duration (see Chapters 12 and 16).

2.2 Spectrum limitations

The spectrum available for wireless communications services is limited, and regulated by international agreements. For this reason, the spectrum has to be used in a highly efficient manner. Two approaches are used: regulated spectrum usage, where a single network operator has control over the usage of the spectrum, and unregulated spectrum, where each user can transmit without additional control, as long as (s)he complies with certain restrictions on the emission power and bandwidth. In the following, we first review the frequency ranges assigned to different communications services. We then discuss the basic principle of frequency reuse for both regulated and unregulated access.

2.2.1 Assigned frequencies

The frequency assignment for different wireless services is regulated by the ITU (*International Telecommunications Union*), a suborganization of the United Nations. In its tri-annual conferences (*World Radio Conferences*), it establishes worldwide guidelines for the usage of spectrum in different regions and countries. Further regulations are issued by the frequency regulators of individual countries, including the FCC (*Federal Communications Commission*) in the U.S.A., the ARIB (*Association of Radio Industries and Businesses*) in Japan, and the CEPT (*European Conference of Postal and Telecommunications Administrations*) in Europe. While the

---

[4] Special measures include equalizers (Chapter 16), Rake receivers (Chapter 18), and OFDM (Chapter 19).

exact frequency assignments differ, similar services tend to use the same frequency ranges all over the world:

- Below 100 MHz: at these frequencies, we find CB (Citizens' Band) radio, pagers, and analog cordless phones.
- 100–800 MHz: these frequencies are mainly used for broadcast (radio and TV) applications.
- 400–500 MHz: a number of cellular and trunking radio systems make use of this band. It is mostly systems that need good coverage, but show low user density.
- 800–1,000 MHz: several cellular systems use this band (analog systems as well as second-generation cellular). Also some emergency communications systems (trunking radio) make use of this band.
- 1.8–2.0 GHz: this is the main frequency band for cellular communications. The current (second-generation) cellular systems operate in this band, as will most of the third-generation systems. Many cordless systems also operate in this band.
- 2.4–2.5 GHz: the ISM (Industrial, Scientific, and Medical) band. Cordless phones, Wireless Local Area Network (WLANs) and wireless PANs (Personal Area Networks) operate in this band; they share it with many other devices, including microwave ovens.
- 3.3–3.8 GHz is envisioned for fixed wireless access systems.
- 4.8–5.8 GHz: in this range, most WLANs can be found. Also, the frequency range between 5.7 and 5.8 GHz can be used for fixed wireless access, complementing the 3-GHz band.
- 11–15 GHz: in this range can we find the most popular satellite TV services, which use 14.0–14.5 GHz for the uplink, and 11.7–12.2 GHz for the downlink.

More details about the exact frequencies for specific services can be obtained from the national frequency regulators, as well as from the ITU.

Different frequency ranges are optimum for different applications. Low carrier frequencies usually propagate more easily (see also Chapter 4), so that a single BS can cover a large area. On the other hand, absolute bandwidths are smaller, and also the frequency reuse is not as efficient as it is at higher frequencies.[5] For this reason, low frequency bands are best for services that require good coverage, but have a small aggregate rate of information that has to be exchanged. Typical cases in point are paging services, and television; paging is suitable because the amount of information transmitted to each user is small, while in the latter case, only a single information stream is sent to *all* users. For cellular systems, low carrier frequencies are ideal for covering large regions with low user density (rural areas in the Midwest of the U.S.A. and in Russia, Northern Scandinavia, Alpine regions, etc.). For cellular systems with high user densities, as well as for WLANs, higher carrier frequencies are usually more desirable.[6]

The amount of spectrum assigned to the different services does not always follow technical necessities, but rather historical developments. For example, the amount of precious low-frequency spectrum assigned to TV stations is much higher than would be justified by technical requirements. Using appropriate frequency planning and different transmission techniques (including simulcast), a considerable part of the spectrum below 1 GHz could be freed up for alternative usage. However, broadcast stations fight such a development, as it would require modifications in their transmitters. As these stations have a considerable influence on public opinion as well as lobbying power, frequency regulators are hesitant to enforce appropriate rule changes.

---

[5] As we will see in the next subsection, frequency reuse requires that a signal is attenuated strongly outside the cell it is assigned to. However, low carrier frequency results in good propagation so that the signal can remain strong far outside its assigned cell.

[6] However, the carrier frequency should not become *too* high: at extremely high frequencies, it becomes difficult to cover even small areas.

It is also noticeable that the financial terms on which spectrum is assigned to different services differs vastly – from country to country, from service to service, and even depending on the time at which the spectrum is assigned. Obviously, spectrum is assigned to public safety services (police, fire department, military) without monetary compensation. Even television stations usually get the spectrum assigned for free. In the 1980s, spectrum for cellular telephony was often assigned for a rather small fee, in order to encourage the development of this then-new service. In the mid- and late-1990s, spectrum auctions were seen by some countries as a method to increase the country's revenues (consider the frequency auctions for the PCS band in the U.S. in 1995, and the auctions of the UMTS bands in the UK and Germany around 2000). Other countries chose to assign spectrum based on a "beauty contest", where the applicant had to guarantee a certain service quality, coverage, etc., in order to obtain a license. Unregulated services, like WLANs, are assigned spectrum without fees.

2.2.2 Frequency reuse in regulated spectrum

Since spectrum is limited, the *same* spectrum has to be used for *different* wireless connections in *different* locations. To simplify the discussion, let us consider in the following a cellular system where different connections (different users) are distinguished by the frequency channel (band around a certain carrier frequency) that they employ. If an area is served by a single BS, then the available spectrum can be divided into N frequency channels that can serve N users simultaneously. If more than N users are to be served, multiple BSs are required, and frequency channels have to be reused in different locations.

For this purpose, we divide the area (a region, a country, or a whole continent) into a number of *cells*; we also divide the available frequency channels into several groups. The channel groups are now distributed among the cells. The important thing is that channel groups can be used multiple times. The only requirement is that cells that use the same frequency group do not interfere with each other *significantly*.[7] It is fairly obvious that the same carrier frequency can be used for different connections in, say, Rome and Stockholm, at the same time. The large distance between the two cities makes sure that a signal from the MS in Stockholm does not reach the BS in Rome, and can therefore not cause any interference at all. But in order to achieve high efficiency, frequencies must actually be reused much more often – typically, several times within each city. Consequently, *intercell interference* (also known as *co-channel interference*) becomes a dominant factor that limits transmission quality. More details on co-channel interference can be found in Part IV.

Spectral efficiency describes the effectiveness of reuse – i.e., the traffic density that can be achieved per unit bandwidth and unit area. It is therefore given in units of $\mathrm{Erlang}/(\mathrm{Hz\,m^2})$ for voice traffic, and $\mathrm{bit}/(\mathrm{s\,Hz\,m^2})$ for data. Since the area covered by a network provider, as well as the bandwidth that it can use, are fixed, increasing the spectral efficiency is the only way to increase the number of customers that can be served, and thus revenue. Methods for increasing this spectral efficiency are thus at the center of wireless communications research.

Since a network operator buys a license for a spectrum, it can use that spectrum according to its own planning – i.e., network planning can make sure that the users in different cells do not interfere with each other significantly. The network operator is allowed to use as much transmit power as it desires; it can also dictate limits on the emission power of the MSs of different users.[8] The operator can also be sure that the only interference in the network is created by its own network and users.

---

[7] The threshold for *significant* interference (i.e., the admissible signal-to-interference ratio) is determined by modulation and reception schemes, as well as by propagation conditions.

[8] There are some exceptions to that rule – for example, emission limits dictated by health concerns, as well as limits imposed by the standard of the system used by the operator (e.g., GSM).

2.2.3 Frequency reuse in unregulated spectrum

In contrast to regulated spectrum, several services use frequency bands that are available to the general public. For example, some WLANs operate in the 2.45-GHz band, which has been assigned to "ISM" services. Anybody is allowed to transmit in these bands, as long as they (i) limit the emission power to a prescribed value, (ii) follow certain rules for the signal shape and bandwidth, and (iii) use the band according to the (rather broadly defined) purposes stipulated by the frequency regulators.

As a consequence, a WLAN receiver can be faced with a large amount of interference. This interference can either stem from other WLAN transmitters, or from microwave ovens, cordless phones, and other devices that operate in the ISM band. For this reason, a WLAN link must have the capability to deal with interference. That can be achieved by selecting a frequency band within the ISM band at which there is little interference, by using spread spectrum techniques (see Chapter 18), or some other appropriate technique.

There are also cases where the spectrum is assigned to a specific service (e.g., DECT), but not a specific operator. In that case, receivers might still have to deal with strong interference, but the structure of this interference is known. This allows the use of special interference mitigation techniques like dynamic frequency assignment, see the material on DECT on the companion website (*www.wiley.com/go/molisch*).

2.3 Limited energy

Truly wireless communications requires not only that the information is sent over the air (not via cables), but also that the MS is powered by one-way or rechargeable batteries. Otherwise, an MS would be tied to the "wire" of the power supply. Batteries in turn impose restrictions on the power consumption of the devices. The requirement for small energy consumption results in several technical imperatives:

- The power amplifiers in the transmitter have to have high efficiency. As power amplifiers account for a considerable fraction of the power consumption in an MS, mainly amplifiers with an efficiency above 50% should be used in MSs. Such amplifiers – specifically, class-C or class-F amplifiers – are highly nonlinear.[9] As a consequence, wireless communications tend to use modulation formats that are insensitive to nonlinear distortions. For example, constant envelope signals are preferred (see Chapter 11).
- Signal processing must be done in an energy-saving manner. This implies that the digital logic should be implemented using power-saving semiconductor technology like CMOS, while the faster but more power-hungry approaches like ECL-logic do not seem suitable for MSs. This restriction has important consequences for the algorithms that can be used for interference suppression, combatting of intersymbol interference, etc.
- The RX (especially at the BS) needs to have high sensitivity. For example, GSM (Global System for Mobile communications) is specified so that even a received signal power of −100 dBm leads to an acceptable transmission quality. Such an RX is several orders of magnitude more sensitive than a TV RX. If the GSM standard had defined −80 dBm instead, then the transmit power would have to be higher by a factor of 100 in order to achieve the same coverage. This in turn would mean that – for identical talktime – the battery would have to be 100 times as large – i.e., 20 kg instead of the current 200 g. But the high requirements on RX sensitivity have important consequences for the construction of the RX (low-noise amplifiers, sophisticated signal processing to fully exploit the received signal) as well as for network planning.

---

[9] Linear amplifiers, like class A, class B, or class AB, have efficiencies of less than 30%.

- Maximum transmit power should be used only when required. In other words, transmit power should be adapted to the channel state, which in turn depends on the distance between TX and RX (*power control*). If the MS is close to the BS, and thus the channel has only a small attenuation, transmit power should be kept low. Furthermore, for voice transmission, the MS should only transmit if the user at the MS actually talks, which is the case only about 50% of the time (*Discontinuous Voice Transmission* – DTX).
- For cellular phones, and even more so for sensor networks, an energy-efficient "standby" or "sleep" mode has to be defined.

Several of the mentioned requirements are contradictory. For example, the requirement to build an RX with high sensitivity (and thus, sophisticated signal processing), is in contrast to the requirement of having energy-saving (and thus slow) signal processing. Engineering tradeoffs are thus called for.

2.4 User mobility

Mobility is an inherent feature of most wireless systems, and has important consequences for system design. Fading was already discussed in Section 2.1.1. A second important effect is particular to mobile users in cellular systems: the system has to know at any time which cell a user is in:[10]

- If there is an incoming call for a certain MS (user), the network has to know in which cell the user is located. The first requirement is that an MS emits a signal at regular intervals, informing nearby BSs that it is "in the neighborhood". Two databanks then employ this information: the *Home Location Register* (HLR) and the *Visitor Location Register* (VLR). The HLR is a central database that keeps track of the location a user is currently at; the VLR is a database associated with a certain BS that notes all the users that are currently within the coverage area of this specific BS. Consider user *A*, who is registered in San Francisco, but is currently located in Los Angeles. It informs the nearest BS (in Los Angeles) that it is now within its coverage area; the BS enters that information into its VLR. At the same time, the information is forwarded to the central HLR (located, e.g., in New York). If now somebody calls user *A*, an enquiry is sent to the HLR to find out the current location of the user. After receiving the answer, the call is rerouted to Los Angeles. For the Los Angeles BS, user *A* is just a "regular" user, whose data are all stored in the VLR.
- If an MS moves across a cell boundary, a different BS becomes the *serving BS*; in other words, the MS is *handed over* from one BS to another. Such a handover has to be performed without interrupting the call; as a matter of fact, it should not be noticeable at all to the user. This requires complicated signalling. Different forms of handover are described in Chapter 18 for code-division-multiple-access-based systems, and in Chapter 21 for GSM.

---

[10] This effect is not relevant, e.g., for simple cordless systems: either a user is within the coverage region of the (one and only) BS, or (s)he is not.

3

Noise- and interference-limited systems

3.1 Introduction

This chapter explains the principles of link budgets, and the planning of wireless systems with one or multiple users. In Section 3.2, we set up link budgets for noise-limited systems, and compute the minimum transmit power (or maximum range) that can be achieved in the absence of interference. Such computations give a first insight into the basic capabilities of wireless systems, and also have practical applications. For example, Wireless Local Area Networks (WLANs) and cordless phones often operate in a noise-limited mode, if no other Base Station (BS) is in the vicinity. Even cellular systems sometimes operate in that mode if the user density is low (this happens, e.g., during the buildup phase of a network).

In Section 3.3, we discuss interference-limited systems. As we described in the last two chapters, the unregulated use of spectrum leads to interference that cannot be controlled by the user. When spectrum is regulated, the network operator can determine the location of BSs, and thus impact the signal-to-interference ratio. For either case, it is important to set up the link budgets that take the presence of interference into account; Section 3.3 will describe these link budgets. In Chapter 17, we will then see how these calculations are related to the cellular principle and the reuse of frequencies in different cells.

3.2 Noise-limited systems

Wireless systems are required to provide a certain minimum transmission quality (see Section 1.3). This transmission quality in turn requires a minimum *Signal-to-Noise Ratio* (SNR) at the receiver (RX). Consider now a situation where only a single BS transmits, and a Mobile Station (MS) receives; thus, the performance of the system is determined only by the strength of the (useful) signal and the noise. As the MS moves further away from the BS, the received signal power decreases, and at a certain distance, the SNR does not achieve the required threshold for reliable communications. Therefore, the range of the system is noise-limited; equivalently we can call it signal-power-limited. Depending on the interpretation, it is too much noise or too little signal power that leads to bad link quality.

Wireless Communications   Andreas F. Molisch
© 2005 John Wiley & Sons, Ltd

Let us assume for the moment that the received power decreases with d^2, the square of the distance between BS and MS. More precisely, let the received power P_{RX} be:

$$P_{RX} = P_{TX}G_{RX}G_{TX}\left(\frac{\lambda}{4\pi d}\right)^2 \tag{3.1}$$

where G_{RX} and G_{TX} are the gains of the receive and transmit antennas, respectively,[1] λ is the wavelength, and P_{TX} is the transmit power (see Chapter 4 for more details).

The noise that disturbs the signal can consist of several components, as follows

1. *Thermal noise*: the power spectral density of thermal noise depends on the environment temperature T_e that the antenna "sees". The temperature of the Earth is around 300 K, while the temperature of the (cold) sky is approximately $T_e \approx 4$ K (the temperature in the direction of the Sun is of course much higher). As a first approximation, it is usually assumed that the environmental temperature is isotropically 300 K. Noise power spectral density is then:

$$N_0 = k_B T_e \tag{3.2}$$

where k_B is Boltzmann's constant, $k_B = 1.38 \cdot 10^{-23}$ Joules/Kelvin, and the noise power is:

$$P_n = N_0 B \tag{3.3}$$

where B is the RX bandwidth (in units of Hz). It is common to write Eq. (3.2) using logarithmic units:

$$N_0 = -174 \, \text{dBm/Hz} \tag{3.4}$$

This means that the noise power contained in a 1-Hz bandwidth is -174 dBm. The noise power contained in bandwidth B is

$$-174 + 10 \log_{10}(B) \, \text{dBm} \tag{3.5}$$

The logarithm of bandwidth B, specifically $10 \log_{10}(B)$, has the units dBHz.

2. *Man-made noise*: we can distinguish two types of man-made noise:

   (a) *Spurious emissions*: many electric appliances as well as radio transmitters designed for other frequency bands have spurious emissions over a large bandwidth that includes the frequency range in which wireless communications systems operate. For urban outdoor environments, car ignitions and other impulse sources are an especially significant source of noise. In contrast to thermal noise, the noise created by impulse sources decreases with frequency (see Fig. 3.1). At 150 MHz, it can be 20 dB stronger than thermal noise; at 900 MHz, it is typically 10 dB stronger. At UMTS (Universal Mobile Telecommunications System) frequencies, Neubauer et al. [2001] measured 5-dB noise enhancement by man-made noise in urban environments, and about 1-dB in rural environments. Note that frequency regulators in most countries impose limits on "spurious" or "out-of-band" emissions for *all* electrical devices. Furthermore, for communications operating in licensed bands, such spurious emissions are the only source of man-made noise. It lies in the nature of the license (for which the license holder usually has paid) that no other intentional emitters are allowed to operate in this band. In contrast to thermal noise, man-made noise is not necessarily

---

[1] Roughly speaking, "receive antenna gain" is a measure for how much more power we can receive (from a certain direction) by using a specific antenna, compared with the use of an isotropic antenna; the definition for transmit antennas is similar. See Chapter 9 and/or [Stutzman and Thiele 1997] for details.

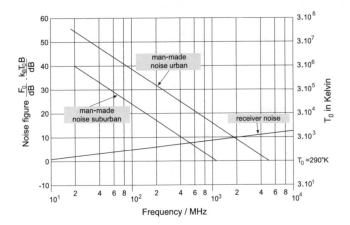

Figure 3.1 Noise as a function of frequency.

Reproduced with permission from Jakes [1974] © IEEE.

Gaussian-distributed. However, as a matter of convenience, most system-planning tools, as well as theoretical designs, assume *Gaussianity* anyway.

(b) *Other intentional emission sources*: several wireless communications systems operate in unlicensed bands. In these bands, everybody is allowed to operate (emit electromagnetic radiation) as long as certain restrictions with respect to transmit power, etc. are fulfilled. The most important of these bands is the 2.45-GHz ISM band. The amount of interference in these bands can be considerable.

3. *Receiver noise*: the amplifiers and mixers in the RX are noisy, and thus increase the total noise power. This effect is described by the noise figure F, which is defined as the SNR at the RX output (typically after downconversion to baseband) divided by the SNR at the RX input. As the amplifiers have gain, noise added in the later stages does not have as much of an impact as noise added in the first stage of the RX. Mathematically, the total noise figure F_{eq} of a cascade of components is:

$$F_{eq} = F_1 + \frac{F_2 - 1}{G_1} + \frac{F_3 - 1}{G_1 G_2} + \cdots \tag{3.6}$$

where the F_i and G_i are noise figures and gains of the individual stages in absolute units (not in dB).

For a digital system, the transmission quality is often described in terms of the *Bit Error Rate* (BER) probability. Depending on the modulation scheme, coding, and a range of other factors (discussed in Part III of this book), there is a relationship between SNR and BER for each digital communications systems. A minimum transmission quality can thus be linked to the minimum SNR, SNR_{min}, by this mapping (see Fig. 3.2). Thus, the planning methods of all analog and digital links in noise-limited environments are the same: the goal is to determine the minimum signal power P_S:

$$P_S = SNR_{min} + P_n \tag{3.7}$$

where all quantities are in dB. However, note that the actual *values* will be different for different systems.

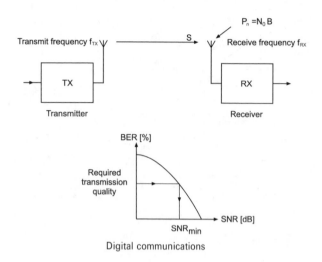

Figure 3.2   Noise-limited systems.
Reproduced with permission from Oehrvik [1994] © Ericsson AB.

3.2.1 Link budget

A link budget is the clearest and most intuitive way of computing the required transmitter (TX) power [Sklar 2001, ch. 5]. In this link budget, we write all equations that connect the TX power to the received SNR in tabular form. As most factors influencing the SNR enter in a multiplicative way, it is convenient to write all equations in logarithmic form – specifically, in dB (decibels). It has to be noted, however, that the link budget gives only an approximation (often a worst case estimate) for the total SNR, because some interactions between different effects are not taken into account.

Before showing some examples, the following points should be stressed:

- In Chapters 4 and 7, we will give extensive discussions of pathloss – i.e., the attenuation due to propagation effects, between TX and RX. For the purpose of this chapter, we will use a simple model, the so-called "breakpoint" model. For distances $d < d_{\text{break}}$, the power is proportional to d^{-2}, according to Eq. (3.1). Beyond that point, the power is proportional to d^{-n}, where n typically lies between 3.5 and 4.5. The received power is thus:

$$P_{\text{RX}}(d) = P_{\text{RX}}(d_{\text{break}}) \left(\frac{d}{d_{\text{break}}} \right)^{-n} \qquad \text{for } d > d_{\text{break}} \qquad (3.8)$$

- Wireless systems, especially mobile systems, suffer from temporal and spatial variations of the transmission channel (*fading*) (see Section 2.1). In other words, even if the distance is fixed, the received power can change significantly with time, or with the movement of the MS (keeping d effectively constant). The power computed from Eq. (3.8) is only a *mean* value. If it is used as the basis for the link budget, then the transmission quality will be above the threshold only in approximately 50% of the times and locations.[2] This is completely unacceptable coverage. Therefore, we have to add a *fading margin*, which makes sure that the minimum receive power is exceeded in at least, e.g., 90% of all cases (see Fig. 3.3). The value of the fading margin depends on the amplitude statistics of the fading and will be discussed in more detail in Chapter 5.

---

[2] It would lie above the threshold in exactly 50% of the cases if Eq. (3.8) represented the *median* power.

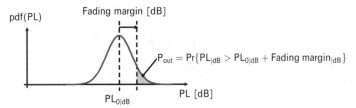

Figure 3.3 Fading margin to guarantee a certain outage probability.

- Uplink (MS to BS) and downlink (BS to MS) are reciprocal, in the sense that the voltage and currents at the antenna ports are reciprocal (as long as uplink and downlink use the same carrier frequency). However, the noise figures of BSs and MSs are typically quite different. As MSs have to be produced in quantity, it is desirable to use low-cost components, which typically have higher noise figures. Furthermore, battery lifetime considerations dictate that BSs can emit more power than MSs. Finally, BSs and MSs differ with respect to antenna diversity, how close they are to interferers, etc. Thus, the link budgets of uplinks and downlinks are different.

Example 3.1 *Link budget*

Consider the downlink of a GSM system (see also Chapter 21). The carrier frequency is 950 MHz, the RX sensitivity is (according to GSM specifications) -102 dBm (dBm means dB relative to 1 mW). The output power of the TX amplifier is 30 W. The antenna gain of the TX antenna is 10 dB, the losses in connectors, combiners, etc., are 5 dB. The fading margin is 12 dB, the breakpoint d_{break} is at a distance 100 m. What distance can be covered?

TX side:

TX power	P_{TX}	30 W	45 dBm
Antenna gain	G_{TX}	10	10 dB
Losses (combiner, connector, etc.)	L_f		-5 dB
EIRP (Equivalent Isotropically Radiated Power)			50 dBm

RX side:

RX sensitivity	P_{min}	-102 dBm
Fading margin		12 dB
Minimum RX power (median)		-90 dBm
Admissible pathloss (difference EIRP and min. RX power)		140 dB
Pathloss at $d_{break} = 100$ m	$[\lambda/(4\pi d)^2]$	72 dB
Pathloss beyond breakpoint	$\propto d^{-n}$	68 dB

Depending on the pathloss exponent:

$$n = 1.5...2.5 \text{ (line of sight)}^3$$

$$n = 3.5...4.5 \text{ (non-line-of-sight)}$$

---

[3] Note that LOS cannot exist beyond a certain distance even in environments that have no buildings or hills. The curvature of the earth cuts off the LOS at a distance that depends on the heights of the BS and the MS.

we obtain the coverage distance:

$$d_{\text{cov}} = 100 \cdot 10^{68/(10n)} \text{m} \tag{3.9}$$

If, for example, $n = 3.5$, then the coverage distance is 8.8 km.

This example was particularly easy, because RX sensitivity was prescribed by the system specifications. If it is not available, the computations at the RX become more complicated.

Example 3.2 *Link budget*

Consider a mobile radio system at 900-MHz carrier frequency, and with 25-kHz bandwidth, that is affected by thermal noise only (temperature of the environment $T_e = 300$ K). Antenna gains at TX and RX sides are 8 dB and -2 dB,[4] respectively. Losses in cables, combiners, etc. at the TX are 2 dB. The noise figure of the receiver is 7 dB, the 3-dB bandwidth of the signal is 25 kHz. The required operating SNR is 18 dB, the desired range of coverage is 2 km. The breakpoint is at 10-m distance; beyond that point, the pathloss exponent is 3.8, and the fading margin is 10 dB. What is the minimum TX power?

Due to the formulation of this problem, we can work our way backwards from the RX to the TX.

Noise spectral density	$k_B T_e$	-174 dBm/Hz
Bandwidth		44 dBHz
$\vdots$		
Thermal noise power at the RX		-130 dBm
RX excess noise		7 dB
Required SNR		18 dB
$\vdots$		
Required RX power		-105 dBm
Pathloss from 10 m to 2-km distance	$(200^{3.8})$	87 dB
Pathloss from TX to breakpoint at 10 m	$(\lambda/(4\pi d))^2$	52 dB
Antenna gain at the MS G_{RX}	(2-dB loss)	$-(-2)$ dB
Fading margin		10 dB
Required EIRP		46 dBm
TX antenna gain G_{TX}	(8-dB gain)	-8 dB
Losses in cables, combiners, etc. at TX	L_f	2 dB
Required TX power (amplifier output)		40 dBm

The required TX power is thus 40 dBm, or 10 W. The link budget is also represented in Fig. 3.4.

3.3 Interference-limited systems

Consider now the case that the interference is so strong that it completely dominates the performance, so that the noise can be neglected. Let a BS cover an area (cell) with radius R around the BS. Furthermore, there is an interfering TX at distance D from the "desired" BS,

---

[4] In most link budgets, the antenna gain for the MS is assumed to be 0 dB. However, recent measurements have shown that absorption and reflection by the head and body of the user reduce the antenna gain, leading to losses up to 10 dB. This will be discussed in more detail in Chapter 9.

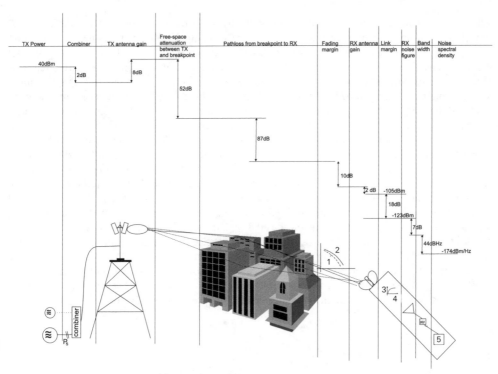

Figure 3.4 Link budget of Example 3.2.

1 = 10% decile; 2 = Median; 3 = MOS; 4 = SNR; 5 = Detector.

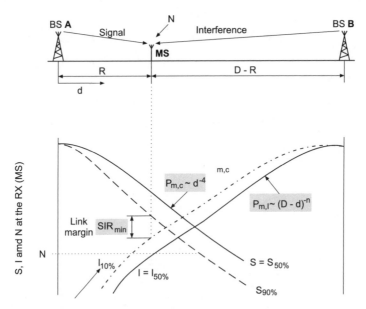

Figure 3.5 Relationship between cell radius and reuse distance. Solid lines: median values. Dashed lines: 90%-decile of the desired signal. Dash-dotted lines: 10%-decile of the interfering signal.

which operates at the same frequency, and with the same transmit power. How large does D have to be in order to guarantee satisfactory transmission quality 90% of the time, assuming that the MS is at the cell boundary (worst case)? The computations follow the link budget computations of the previous section. As a first approximation, we treat the interference as Gaussian. This allows us to treat the interference as equivalent noise, and the minimum signal to interference ratio, SIR_{min}, takes on the same value as SNR_{min} in the noise-limited case.

One difference between interference and noise lies in the fact that interference suffers from fading, while the noise power is typically constant (averaged over a short time interval). For determination of the fading margin, we thus have to account for the fact that (i) the desired signal is weaker than its median value during 50% of the time, and (ii) the interfering signal is stronger than its median value 50% of the time. Mathematically speaking, the cumulative distribution function of the signal-to-interference ratio is the probability that the ratio of two random variables is larger than a certain value in x% of all cases (where x is the percentage of locations in which transmission quality is satisfactory), see Chapter 5. As a first approximation, we can add the fading margin for the desired signal (i.e., the additional power we have to transmit to make sure that the desired signal level exceeds a certain value x% of the time, instead of 50%), and the fading margin of the interference – i.e., the power *reduction* to make sure that the interference exceeds a certain value only $(100 - x)$% of the time, instead of 50% of the time (see Fig. 3.5). This results in an overestimation of the true fading margin. Therefore, if we use that value in system planning, we are on the safe side.

Part II

WIRELESS PROPAGATION CHANNELS

A wireless propagation channel is the medium linking the transmitter and the receiver. Its properties determine the information-theoretic capacity – i.e., the ultimate performance limit of wireless communications, and also determine how specific wireless systems behave. It is thus essential that we know and understand the wireless propagation channel and apply this knowledge to system design. Part II of this book, consisting of Chapters 4–9, is intended to provide such an understanding.

Wireless channels differ from wired channels by *multipath propagation* (see also Chapter 2) – i.e., the existence of a multitude of propagation paths from transmitter to receiver, where the signal can be reflected, diffracted, or scattered along its way. One way to understand the channel is to consider all these different *propagation phenomena*, and how they impact each *Multi Path Component* (MPC). The propagation phenomena will be at the center of Chapter 4; Section 7.5 will explain how to apply this knowledge to deterministic channel models and prediction (ray tracing).

The alternative is a more phenomenological view. We consider important channel parameters, like received power, and analyze their *statistics*. In other words, we do not care how the channel looks in a specific location, or how it is influenced by specific MPCs. Rather, we are interested in describing the *probability* that a channel parameter attains a certain value. The parameter of highest interest is the received power or fieldstrength; it determines the behavior of narrowband systems in both noise- and interference-limited systems (see Chapter 3). We will find that received power *on average* decreases with distance. The physical reasons for this decrease are described in Sections 4.1 and 4.2; models are detailed in Section 7.1. However, there are variations around that mean that can be modeled stochastically. These stochastic variations are described in detail in Chapter 5.

The interference of the different MPCs creates not only fading (i.e., variations in received power with time and/or place), but also delay dispersion. Delay dispersion means that if we transmit a signal of duration T, the received signal has a *longer* duration T'. Naturally, this leads to *Inter Symbol Interference* (ISI). While this effect was not relevant for earlier, analog, systems, it is very important for digital systems like GSM (Global System for Mobile communications). Third-generation cellular systems and wireless local area networks are also influenced by the delay dispersion, and even angular characteristics, of the wireless channel. This requires the introduction of new parameters and description methods to quantify these characteristics, as described in Chapter 6. Sections 7.3 and 7.4 describe typical values for these parameters in outdoor and indoor environments.

For both the understanding of the propagation phenomena, and for the stochastic description of the channel, we need measurements. Chapter 8 describes the measurement equipment, and how the output from that equipment needs to be processed. While measuring the received power is fairly straightforward, finding the delay dispersion and angular characteristics of the channel is much more involved. Finally, Chapter 9 describes antennas for wireless channels. The antennas represent the interface between the transceivers and the propagation channel, and determine how the signal is sent out into the propagation channel, and collected from it.

4

Propagation mechanisms

In this chapter, we describe the basic mechanisms that govern the propagation of electro-magnetic waves, emphasizing those aspects that are relevant for wireless communications. The simplest case is free space propagation – in other words, one transmit and one receive antenna, in free space. In a more realistic scenario, there are dielectric and conducting obstacles (*Interacting Objects* – IOs). If these IOs have a *smooth* surface, waves are *reflected*, and a part of the energy penetrates the IO (*transmission*). If the surfaces are *rough*, the waves are diffusely *scattered*. Finally, waves can also be *diffracted* at the edges of the IOs. These effects will now be discussed one after the other.[1]

4.1 Free space loss

We start with the simplest possible scenario: a transmit and a receive antenna in free space.

Energy conservation dictates that the integral of the power density over any closed surface surrounding the transmit antenna must be equal to the transmitted power. If the closed surface is a sphere of radius d, centered at the transmitter (TX) antenna, and if the TX antenna radiates isotropically, then the power density on the surface is $P_{TX}/(4\pi d^2)$. The receiver (RX) antenna has an "effective area" A_{RX}. We can envision that all power impinging on that area is collected by the RX antenna [Stutzman and Thiele 1997]. Then the received power is:

$$P_{RX}(d) = P_{TX}\frac{1}{4\pi d^2}A_{RX} \tag{4.1}$$

If the transmit antenna is not isotropic, then the energy density has to be multiplied with the antenna gain G_{TX} in the direction of the receive antenna.[2] The product of transmit power and gain in the considered direction is also known as EIRP (*Equivalent Isotropically Radiated Power*). The effective antenna area is proportional to the power that can be extracted from the antenna connectors for a given energy density. For a parabolic antenna, for example, the effective antenna area is roughly the geometrical area of the surface. However, also antennas with very small geometrical area – e.g., dipole antennas – can have a considerable effective area.

---

[1] In the literature, the IOs are often called "scatterers", even if the interaction process is not scattering.
[2] If not stated otherwise, this book always defines the antenna gain as the gain over an isotropic radiator.

Wireless Communications Andreas F. Molisch
© 2005 John Wiley & Sons, Ltd

It can be shown that there is a simple relationship between effective area and antenna gain [Stutzmann and Thiele 1997]:

$$G_{RX} = \frac{4\pi}{\lambda^2} A_{RX} \tag{4.2}$$

Most noteworthy in this equation is the fact that – for a fixed antenna area – the antenna gain increases with frequency. This is also intuitive, as the directivity of an antenna is determined by its size in terms of wavelengths.

Substituting Eq. (4.2) into Eq. (4.1) gives the received power P_{RX} as a function of the distance d in free space, also known as *Friis' law*:

$$P_{RX}(d) = P_{TX} G_{TX} G_{RX} \left(\frac{\lambda}{4\pi d} \right)^2 \tag{4.3}$$

The factor $(4\pi d/\lambda)^2$ is also known as the *free space loss factor*.

Friis' law seems to indicate that the "attenuation" in free space increases with frequency. This is counterintuitive, as the energy is not lost, but rather redistributed over a sphere surface of area $4\pi d^2$. This mechanism has to be independent of the wavelength. This seeming contradiction is caused by the fact that we assume the *antenna gain* to be independent of the wavelength. If we assume, on the other hand, that the effective *antenna area* of the RX antenna is independent of frequency, then the received power becomes independent of the frequency (see Eq. (4.1)). For wireless systems, it is often useful to assume constant gain, as different systems (e.g., operating at 900 and 1,800 MHz) use the same antenna *type* (e.g., $\lambda/2$ dipole or monopole), and not the same antenna *size*.

The validity of Friis' law is restricted to the far field of the antenna – i.e., the TX and RX antennas have to be at least one *Rayleigh distance* apart. The Rayleigh distance (also known as the *Fraunhofer distance*) is defined as:

$$d_R = \frac{2L_a^2}{\lambda} \tag{4.4}$$

where L_a is the largest dimension of the antenna; furthermore the far field requires $d_d \gg \lambda$ and $d_d \gg L_a$.

Example 4.1 *Compute the Rayleigh distance of a square antenna with 20-dB gain.*

The gain is 100 on a linear scale. In that case, the effective area is approximately

$$A_{RX} = \frac{\lambda^2}{4\pi} G_{RX} = 8\lambda^2 \tag{4.5}$$

For a square-shaped antenna with $A_{RX} = L_a^2$, the Rayleigh distance is:

$$d_R = \frac{2 \cdot 8\lambda^2}{\lambda} = 16\lambda \tag{4.6}$$

For setting up link budgets, it is advantageous to write Friis' law on a logarithmic scale. Equation (4.3) then reads:

$$P_{RX}|_{dBm} = P_{TX}|_{dBm} + G_{TX}|_{dB} + G_{RX}|_{dB} + 20\log\left(\frac{\lambda}{4\pi d} \right) \tag{4.7}$$

where $|_{dB}$ means "in units of dB". In order to better point out the distance dependence, it is advantageous to first compute the received power at 1-m distance:

$$P_{RX}(1\,m) = P_{TX}|_{dBm} + G_{TX}|_{dB} + G_{RX}|_{dB} + 20\log\left(\frac{\lambda|_m}{4\pi \cdot 1} \right) \tag{4.8}$$

The last term on the r.h.s. of Eq. (4.8) is about $-32\,\mathrm{dB}$ at 900 MHz and $-38\,\mathrm{dB}$ at 1,800 MHz. The actual received power at a distance d (in meters) is then:

$$P_{\mathrm{RX}}(d) = P_{\mathrm{RX}}(1\,\mathrm{m}) - 20\log(d|_{\mathrm{m}}) \tag{4.9}$$

4.2 Reflection and Transmission

4.2.1 Snell's law

Electromagnetic waves are often reflected at one or more interacting objects before arriving at the RX. The reflection coefficient of the IO, as well as the direction into which this reflection occurs, determines the power that arrives at the RX position. In this section, we deal with *specular reflections*. This type of reflection occurs when waves are incident onto smooth, large (compared with the wavelength) objects. A related mechanism is the *transmission* of waves – i.e., the penetration of waves into and through an IO. Transmission is especially important for wave propagation inside buildings. If the Base Station (BS) is either outside the building, or in a different room, then the waves have to penetrate a wall (dielectric layer) in order to get to the RX.

To get a more precise mathematical formulation, consider the following setup. Let a homogeneous plane wave be incident onto a dielectric halfspace. The dielectric material is characterized by its dielectric constant $\varepsilon = \varepsilon_0 \varepsilon_r$ (where ε_0 is the vacuum dielectric constant $8.854 \cdot 10^{-12}$ Farad/m, and ε_r is the relative dielectric constant of the material) and conductivity σ_e. Furthermore, we assume that the material is isotropic, and has a relative permeability $\mu_r = 1$.[3] The dielectric constant and conductivity can be merged into a single parameter, the *complex* dielectric constant:

$$\delta = \varepsilon_0 \delta_r = \varepsilon - j\frac{\sigma_e}{2\pi f_c} \tag{4.10}$$

where f_c is the carrier frequency, and j is the imaginary unit. Though this definition is strictly valid only for a single frequency, it can actually be used for all *narrowband* systems, where the bandwidth is much smaller than the carrier frequency, as well as much smaller than the bandwidth over which the quantities σ_e and ε vary significantly.[4]

The plane wave is incident on the halfspace at an angle Θ_e, which is defined as the angle between the wave vector $\mathbf{k}$ and the unit vector that is orthogonal to the dielectric boundary. We have to distinguish between the TM (*Transversal Magnetic*) case, where the magnetic field component is parallel to the boundary between the two dielectrics, and the TE (*Transversal Electric*) case, where the electric field component is parallel (see Fig. 4.1).

The reflection and transmission coefficients can now be computed from postulating incident, reflected, and transmitted plane waves, and enforcing the continuity conditions at the boundary – see, e.g., Ramo et al. [1967]. From these considerations, we obtain Snell's law: the angle of incidence is the same as the reflected angle:

$$\Theta_r = \Theta_e \tag{4.11}$$

and the angle of the transmitted wave is given by:

$$\frac{\sin \Theta_t}{\sin \Theta_e} = \frac{\sqrt{\delta_1}}{\sqrt{\delta_2}} \tag{4.12}$$

where subscripts 1 and 2 index the considered medium.

---

[3] This is approximately true for most materials influencing mobile radio wave propagation.

[4] Note that this means "narrowband" in the RF (Radio Frequency) sense. We will later encounter a different definition for narrowband that is related to delay dispersion in wireless channels.

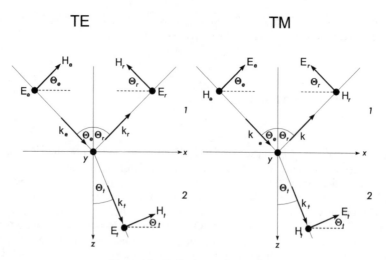

Figure 4.1   Reflection and transmission.

The reflection and transmission coefficients are different for TE and for TM waves. For TM polarization:

$$\rho_{\mathrm{TM}} = \frac{\sqrt{\delta_2}\cos\Theta_e - \sqrt{\delta_1}\cos(\Theta_t)}{\sqrt{\delta_2}\cos\Theta_e + \sqrt{\delta_1}\cos(\Theta_t)} \tag{4.13}$$

$$T_{\mathrm{TM}} = \frac{2\sqrt{\delta_1}\cos(\Theta_e)}{\sqrt{\delta_2}\cos\Theta_e + \sqrt{\delta_1}\cos(\Theta_t)} \tag{4.14}$$

and for TE polarization:

$$\rho_{\mathrm{TE}} = \frac{\sqrt{\delta_1}\cos(\Theta_e) - \sqrt{\delta_2}\cos(\Theta_t)}{\sqrt{\delta_1}\cos(\Theta_e) + \sqrt{\delta_2}\cos(\Theta_t)} \tag{4.15}$$

$$T_{\mathrm{TE}} = \frac{2\sqrt{\delta_1}\cos(\Theta_e)}{\sqrt{\delta_1}\cos(\Theta_e) + \sqrt{\delta_2}\cos(\Theta_t)} \tag{4.16}$$

Note that the reflection coefficient has both an amplitude and a phase. Figure 4.2 shows both for a dielectric material with complex dielectric constant $\delta = 4 - 0.25j$. It is noteworthy that for both TE and TM waves, the reflection coefficient becomes -1 (magnitude 1, phase shift of $180°$) at grazing incidence ($\Theta_e \to 90°$). This is the same reflection coefficient that would occur for reflection on an ideally conducting surface. We will see later on that this has important consequences for the impact of ground-reflected waves in wireless systems.

In highly lossy materials, the transmitted wave is not a homogeneous plane wave, so that Snell's law is not applicable anymore. Instead, there is a guided wave at the dielectric boundary. However, these considerations have more theoretical than practical use.

4.2.2 Reflection and transmission for layered dielectric structures

The previous section discussed reflection and transmission in a dielectric halfspace. This is of interest, e.g., for ground reflections, and reflections by terrain features, like mountains. A related problem is the problem of *transmission through* a dielectric layer. It occurs when a user inside a building is communicating with an outdoor BS, or in a picocell where the MS (Mobile Station)

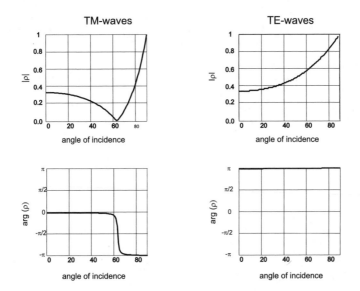

Figure 4.2   Reflection coefficient for a dielectric material with complex dielectricity constant $\delta = (4 - 0.25j)\varepsilon_0$.

and the BS are in different rooms. In that case, we are interested in the attenuation and phaseshift of a wave transmitted through a wall. Fortunately, the basic problem of dielectric layers is well-known from other areas of electrical engineering – e.g., optical thin film technology [Heavens 1965], and the results can be easily applied to wireless communications.

The most simple, and practically most important, case occurs when the dielectric layer is surrounded on both sides by air. The reflection and transmission coefficients can be determined by summation of the partial waves, resulting in a total transmission coefficient:

$$T = \frac{T_1 T_2 e^{-j\alpha}}{1 + \rho_1 \rho_2 e^{-2j\alpha}} \tag{4.17}$$

and a reflection coefficient:

$$\rho = \frac{\rho_1 + \rho_2 e^{-j2\alpha}}{1 + \rho_1 \rho_2 e^{-2j\alpha}} \tag{4.18}$$

where T_1 is the transmission coefficient of a wave from air into a dielectric halfspace (with the same dielectric properties as the considered layer) and T_2 is the transmission coefficient from dielectric into air; they can be computed from the results of Section 4.2.1. The quantity α is the electrical length of the dielectric as seen by waves that are at an angle Θ_t with the layer:

$$\alpha = \frac{2\pi}{\lambda} \sqrt{\varepsilon_{r,2}} d_{\text{layer}} \cos(\Theta_t) \tag{4.19}$$

where d_{layer} is the geometrical length of the layer. Note also that there is a waveguiding effect in lossy materials (see discussion in the previous section), so that the results of this section are not strictly applicable for dielectrics with losses.

In multilayer structures, the problem becomes considerably more complicated [Heavens 1965]. However, in practice, even multilayer structures are described by "effective" dielectric constants or reflection/transmission constants. These are measured directly for the composite structure. The alternative of measuring the dielectric properties for each layer separately and computing the resulting effective dielectric constant is prone to errors, as the measurement errors for the different layers add up.

Example 4.2   *Compute the effective ρ and T for a 50-cm-thick brick wall at 4-GHz carrier frequency for perpendicularly incident waves.*

Since $\Theta_e = 0$, Eqs. (4.11) and (4.12) imply that also $\Theta_r = \Theta_t = 0$. At $f = 4\,\text{GHz}$ ($\lambda = 7.5\,\text{cm}$), brick has a relative permittivity ε_r of 4.44 [Rappaport 1996]; we neglect the conductivity. With the air having $\varepsilon_{\text{air}} = \varepsilon_1 = \varepsilon_0$, Eqs. (4.13)–(4.16) give the reflection and transmission coefficients for the surface between air and brick as:

$$\left.\begin{aligned}
\rho_{1,\text{TM}} &= \frac{\sqrt{\varepsilon_2} - \sqrt{\varepsilon_1}}{\sqrt{\varepsilon_2} + \sqrt{\varepsilon_1}} = 0.36 \\[2mm]
\rho_{1,\text{TE}} &= \frac{\sqrt{\varepsilon_1} - \sqrt{\varepsilon_2}}{\sqrt{\varepsilon_2} + \sqrt{\varepsilon_1}} = -0.36 \\[2mm]
T_{1,\text{TM}} &= \frac{2\sqrt{\varepsilon_1}}{\sqrt{\varepsilon_2} + \sqrt{\varepsilon_1}} = 0.64 \\[2mm]
T_{1,\text{TE}} &= \frac{2\sqrt{\varepsilon_1}}{\sqrt{\varepsilon_2} + \sqrt{\varepsilon_1}} = 0.64
\end{aligned}\right\} \tag{4.20}$$

and between brick and air as:

$$\left.\begin{aligned}
\rho_{2,\text{TM}} &= \rho_{1,\text{TE}} \\[1mm]
\rho_{2,\text{TE}} &= \rho_{1,\text{TM}} \\[1mm]
T_{2,\text{TM}} &= T_{1,\text{TE}} \cdot \sqrt{\varepsilon_2/\varepsilon_1} = 1.36 \\[1mm]
T_{2,\text{TE}} &= T_{1,\text{TM}} \cdot \sqrt{\varepsilon_2/\varepsilon_1} = 1.36
\end{aligned}\right\} \tag{4.21}$$

Note that the transmission coefficient can become larger than unity (e.g., $T_{2,\text{TE}}$). This is not a violation of energy conservation: the transmission coefficient is defined as the ratio of the amplitudes of the incident and reflected field. Energy conservation only dictates that the flux (energy) density of a reflected and transmitted field equal that of the incident field.

The electrical length of the wall, α, is determined using Eq. (4.19) as:

$$\alpha = \frac{2\pi}{\lambda}\sqrt{\varepsilon_r}d = \frac{2\pi}{0.075}\sqrt{4.44} \cdot 0.5 = 88.26 \tag{4.22}$$

Finally the total reflection and transmission coefficients can be determined using Eqs. (4.17), (4.18), which gives:

$$\left.\begin{aligned}
T_{\text{TM}} &= \frac{T_{1,\text{TM}}T_{2,\text{TM}}e^{-j\alpha}}{1 + \rho_{1,\text{TM}}\rho_{2,\text{TM}}e^{-2j\alpha}} = \frac{0.64 \cdot 1.356 e^{-j88.26}}{1 - 0.36^2 e^{-j176.53}} = 0.90 - 0.36j \\[3mm]
\rho_{\text{TM}} &= \frac{\rho_{1,\text{TM}} + \rho_{2,\text{TM}}e^{-2j\alpha}}{1 + \rho_{1,\text{TM}}\rho_{2,\text{TM}}e^{-2j\alpha}} = \frac{0.36\left(1 - e^{-j176.53}\right)}{1 - 0.36^2 e^{-j176.53}} = 0.086 + 0.22j \\[3mm]
T_{\text{TE}} &= \frac{T_{1,\text{TE}}T_{2,\text{TE}}e^{-j\alpha}}{1 + \rho_{1,\text{TE}}\rho_{2,\text{TE}}e^{-2j\alpha}} = \frac{T_{2,\text{TM}}T_{1,\text{TM}}e^{-j\alpha}}{1 + \rho_{2,\text{TM}}\rho_{1,\text{TM}}e^{-2j\alpha}} = T_{\text{TM}} \\[3mm]
\rho_{\text{TE}} &= \frac{\rho_{1,\text{TE}} + \rho_{2,\text{TE}}e^{-2j\alpha}}{1 + \rho_{1,\text{TE}}\rho_{2,\text{TE}}e^{-2j\alpha}} = \frac{\rho_{2,\text{TM}} + \rho_{1,\text{TM}}e^{-2j\alpha}}{1 + \rho_{2,\text{TM}}\rho_{1,\text{TM}}e^{-2j\alpha}} = \frac{-\rho_{1,\text{TM}} - \rho_{2,\text{TM}}e^{-2j\alpha}}{1 + \rho_{2,\text{TM}}\rho_{1,\text{TM}}e^{-2j\alpha}} = -\rho_{\text{TM}}
\end{aligned}\right\} \tag{4.23}$$

It is easily verified that in both cases $|\rho|^2 + |T|^2 = 1$.

4.2.3 The d^{-4} power law

One of the "folk laws" of wireless communications says that the received signal power is inversely proportional to the *fourth* power of the distance between TX and RX. This law is often justified by computing the received power for the case that only a direct (line-of-sight, LOS) wave, plus a ground-reflected wave, exists. For this specific case, the following equation is derived in Appendix 4.A (see *www.wiley.com/go/molisch*):

$$P_{RX}(d) \approx P_{TX} G_{TX} G_{RX} \left(\frac{h_{TX} h_{RX}}{d^2} \right)^2 \tag{4.24}$$

where h_{TX} and h_{RX} are the height of the transmit and the receive antenna, respectively; it is valid for distances larger than:

$$d_{break} \gtrsim 4 h_{TX} h_{RX} / \lambda \tag{4.25}$$

This equation, which replaces the standard Friis' law, implies that the received power becomes independent of frequency. Furthermore, it follows from Eq. (4.24) that the received power increases with the square of the height of both BS and MS.

For the link budget, it is useful to rewrite the power law on a logarithmic scale. Assuming that the power decays as d^{-2} until a breakpoint d_{break}, and from there with d^{-n}, then the received power is (see also Chapter 3):

$$P_{RX}(d) = P_{RX}(1 \, m) - 20 \log(d_{break}|_m) - n10 \log(d/d_{break}) \tag{4.26}$$

Figure 4.3 shows the received power when there is a direct wave and a ground-reflected wave, both from the exact formulation (see Appendix 4.A) and Eq. (4.24). We find that the transition between the attenuation coefficient $n = 2$ and $n = 4$ is actually not a sharp breakpoint, but rather smooth. It is thus not possible to strictly test statements about the onset of the d^{-4} law. According to Eq. (4.25), the breakpoint is at $d = 90 \, m$; this seems to be approximated reasonably well in the plot.

The equations derived above (and in Appendix 4.A) are self-consistent, but it has to be emphasized that they are not a *universal* description of wireless channels. They do not agree with measurement results in realistic channels in several respects:

1. $n = 4$ is not a universally valid decay exponent. $n = 2$ is fulfilled close to the transmit antenna, but at larger distances, values between $1.5 < n < 5.5$ have been measured, and

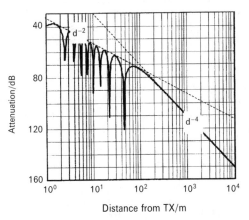

Figure 4.3 Propagation over ideally reflecting ground. Height of BS: 5 m. Height of MS: 1.5 m.

the actual value strongly depends on the surrounding environment. After all, propagation does not always happen over a flat Earth. Rather, multiple propagation paths are possible, LOS connections are often shadowed by IOs, etc. The value of $n = 4$ is, at best, a mean value of various environments.

2.   The transition between $n = 2$ and $n = 4$ almost never occurs at the theoretically predicted d_{break}.

3.   Measurements also show that there is a second breakpoint, beyond which an $n > 6$ exponent is valid. This effect is not predicted at all by the above model. For some situations, the effect can be explained by the radio horizon (i.e., the curvature of the Earth), which is not included in the model above [Parsons 1992].

Summarizing, the explanation for d^{-4} seems to be more a case of finding – after the fact – an explanation for a frequently measured value of n.

4.3 Diffraction

All the equations derived up to now deal with infinitely extended IOs. However, real IOs like buildings, cars, etc., have a finite extent. Now it is well known that a finite-sized object does not create sharp shadows (the way geometrical optics would have it), but rather there is diffraction due to the wave nature of electromagnetic radiation. In the limit of very small wavelength (large frequency), geometrical optics become exact.

In the following, we first treat two canonical diffraction problems: diffraction of a homogeneous plane wave (i) by a knife edge or screen, and (ii) by a wedge, and derive the diffraction coefficients that tell us how much power can be received in the shadow region behind an obstacle. Subsequently, we consider the effect of a concatenation of several screens.

4.3.1 Diffraction by a single screen or wedge

The diffraction coefficient

The simplest diffraction problem is the diffraction of a homogeneous plane wave by a semi-infinite screen, as sketched in Fig. 4.4. Diffraction can be understood from Huygen's principle that each point of a wavefront can be considered the source of a spherical wave. For a homogeneous plane wave, the superposition of these spherical waves results in another homogeneous plane wave, see transition from plane A' to B'. If, however, the screen eliminates parts of the point sources (and their associated spherical waves), the resulting wavefront is not plane anymore, see the transition from plane B' to C'. Constructive and destructive interferences occur in different directions.[5]

The electric field at any point to the right of the screen ($x \geq 0$) can be expressed in a form that involves only a standard integral, the *Fresnel integral*. With the incident field represented as $\exp(-jk_0 x)$, the total field becomes [Vaughan and Andersen 2003]:

$$E_{total} = \exp(-jk_0 x)\left(\frac{1}{2} - \frac{\exp(j\pi/4)}{\sqrt{2}} F(\nu_F)\right) = \exp(-jk_0 x)\widetilde{F}(\nu_F) \qquad (4.27)$$

---

[5] For more accurate considerations, it is noteworthy that Huygen's principle is not exact. A derivation from Maxwell's equations, which also includes a discussion of the necessary assumptions, is given, e.g., in Marcuse [1991].

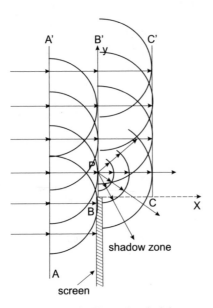

Figure 4.4 Huygen's principle.

where $\nu_F = -2y/\sqrt{\lambda x}$ and the Fresnel integral $F(\nu_F)$ is defined as:

$$F(\nu_F) = \int_0^{\nu_F} \exp\left(-j\pi\frac{t^2}{2}\right) dt \tag{4.28}$$

Figure 4.5 plots this function. It is interesting that $\widetilde{F}(\nu_F)$ can become larger than unity for some values of ν_F. This implies that the received power at a specific location can actually be *increased* by the presence of the screen. Huygen's principle again provides the explanation: some spherical waves that would normally interfere destructively in a specific location are blocked off. However, note that the *total* energy (integrated over the whole wavefront) can *not* be increased by the screen.

Consider now the more general geometry of Fig. 4.6. The TX is at height h_{TX}, the RX at h_{RX} and the screen extends from $-\infty$ to h_s. The diffraction angle θ_d is thus:

$$\theta_d = \arctan\left(\frac{h_s - h_{TX}}{d_{TX}}\right) + \arctan\left(\frac{h_s - h_{RX}}{d_{RX}}\right) \tag{4.29}$$

and the Fresnel parameter ν_F can be obtained from θ_d as:

$$\nu_F = \theta_d \sqrt{\frac{2d_{TX}d_{RX}}{\lambda(d_{TX} + d_{RX})}} \tag{4.30}$$

The fieldstrength can again be computed from Eq. (4.27), just using the Fresnel parameter from Eq. (4.30).

Note that the result given above is approximate in the sense that it neglects the polarization of the incident field. More accurate equations for both the TE and the TM case can be found in Bowman et al. [1987].

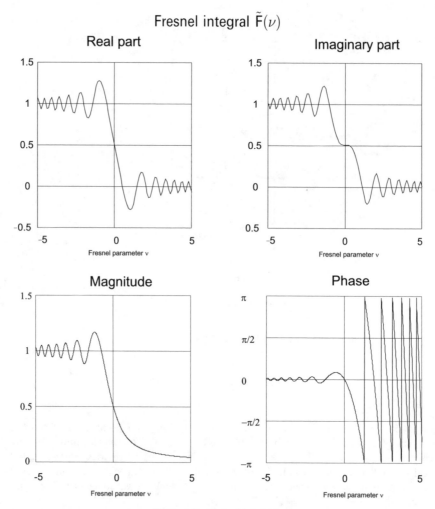

Figure 4.5   Fresnel integral.

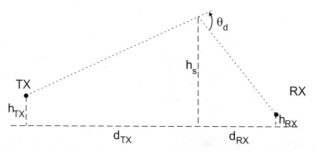

Figure 4.6   Geometry for the computation of the Fresnel parameters.

Example 4.3 *Consider diffraction by a screen with* $d_{TX} = 200\,m$, $d_{RX} = 50\,m$, $h_{TX} = 20\,m$, $h_{RX} = 1.5\,m$, $h_s = 40\,m$, *at a center frequency of 900 MHz. Compute the diffraction coefficient.*

A center frequency of 900 MHz implies a wavelength $\lambda = 1/3$ m. Computing the diffraction angle θ_d from Eq. (4.29) gives:

$$
\theta_d = \arctan\left(\frac{h_s - h_{TX}}{d_{TX}}\right) + \arctan\left(\frac{h_s - h_{RX}}{d_{RX}}\right)
$$

$$
= \arctan\left(\frac{40 - 20}{200}\right) + \arctan\left(\frac{40 - 1.5}{50}\right) = 0.756 \text{ rad} \tag{4.31}
$$

Then, the Fresnel parameter is given by Eq. (4.30) as:

$$
\nu_F = \theta_d \sqrt{\frac{2d_{TX}d_{RX}}{\lambda(d_{TX} + d_{RX})}} = 0.756\sqrt{\frac{2 \cdot 200 \cdot 50}{1/3 \cdot (200 + 50)}} = 11.71 \tag{4.32}
$$

Evaluation of Eq. (4.28), with **MATLAB** or Abramowitz and Stegun [1965], yields:

$$
F(11.71) = \int_0^{\nu_F} \exp\left(-j\pi\frac{t^2}{2}\right) dt \approx 0.527 - j0.505 \tag{4.33}
$$

Finally, Eq. (4.27) gives the total received field as:

$$
E_{\text{total}} = \exp(-jk_0 x)\left(\frac{1}{2} - \frac{\exp(j\pi/4)}{\sqrt{2}} F(11.71)\right)
$$

$$
= \exp(-jk_0 x)\left(\frac{1}{2} - \frac{\exp(j\pi/4)}{\sqrt{2}}(0.527 - j0.505)\right) = (-0.016 - j0.011)\exp(-jk_0 x) \tag{4.34}
$$

Fresnel zones

The impact of an obstacle can also be assessed qualitatively, and intuitively, by the concept of *Fresnel zones*. Figure 4.7 shows the basic principle. Draw an ellipsoid whose foci are the BS and the MS locations. According to the definition of an ellipsoid, all rays that are reflected at points on this ellipsoid have the same runlength (equivalent to runtime). The eccentricity of the ellipsoid determines the extra runlength compared with the LOS – i.e., the direct connection between the two foci. Ellipsoids where this extra distance is an integer multiple of $\lambda/2$ are called "Fresnel ellipsoids". Now extra runlength also leads to an additional phaseshift, so that the ellipsoids can be described by the phaseshift that they cause. More specifically, the *i*th Fresnel ellipsoid is the one that results in a phaseshift of $i \cdot \pi$.

How large does an obstacle have to be to shield off a Fresnel ellipsoid? The answer can be found from the following equation: let the *i*th Fresnel ellipsoid intersect with a plane that is orthogonal to the BS–MS connection line, and has a distance d_{TX} from the TX and d_{RX} from the RX. The intersection is then a circle with a radius:

$$
\sqrt{\frac{i\lambda d_{TX} d_{RX}}{d_{TX} + d_{RX}}} \tag{4.35}
$$

Fresnel zones can also be used for explanation of the d^{-4} law. The propagation follows a free space law up to the distance where the first Fresnel ellipsoid touches the ground. At this distance, which is the breakpoint distance, the phase difference between the direct and the reflected ray becomes π.

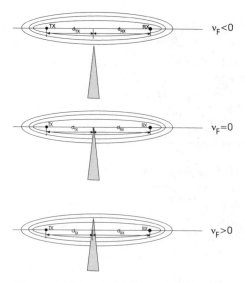

Figure 4.7 The principle of Fresnel ellipsoids.

Diffraction by a wedge

The semi-infinite absorbing screen is a useful tool for the explanation of diffraction, since it is the simplest possible configuration. However, many obstacles especially in urban environments are much better represented by a wedge structure, as sketched in Fig. 4.8. The problem of diffraction by a wedge has been treated for some 100 years, and is still an area of active research. Depending on the boundary conditions, solutions can be derived that are either valid at arbitrary observation points, or approximate solutions that are only valid in the far field (i.e., far away from the wedge). These latter solutions are usually much simpler, and will thus be the only ones considered here.

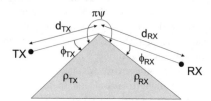

Figure 4.8 Geometry for wedge diffraction.

The part of the field that is created by diffraction can be written as the product of the incident field with a phase factor $\exp(-jk_0 d_{RX})$, a geometry factor $A(d_{TX}, d_{RX})$ that depends only on the distance of TX and RX from the wedge, and the diffraction coefficient $D(\phi_{TX}, \phi_{RX})$ that depends on the diffraction angles [Vaughan and Andersen 2003]:

$$E_{\text{diff}} = E_{\text{inc},0} D(\phi_{TX}, \phi_{RX}) A(d_{TX}, d_{RX}) \exp(-jk_0 d_{RX}) \tag{4.36}$$

The diffracted field has to be added to the field as computed by geometrical optics.[6]

---

[6] If the field incident on the wedge can be written as $E_{\text{inc},0} = E_0 \exp(-jk_0 d_{TX})/d_{TX}$ the above equations become completely symmetrical with respect to d_{TX} and d_{RX}.

The definition of the geometry parameters is shown in Fig. 4.8. The geometry factor is given by:

$$A(d_{\mathrm{TX}}, d_{\mathrm{RX}}) = \sqrt{\frac{d_{\mathrm{TX}}}{d_{\mathrm{RX}}(d_{\mathrm{TX}} + d_{\mathrm{RX}})}} \qquad (4.37)$$

The diffraction coefficient D depends on the boundary conditions – namely, the reflection coefficients ρ_{TX} and ρ_{RX}. Explicit equations are given in Appendix 4.B (see *www.wiley.com/go/ molisch*).

4.3.2 Diffraction by multiple screens

Diffraction by a single screen is a problem that has been widely studied, because it is amenable to closed-form mathematical treatment, and forms the basis for the treatment of more complex problems. However, in practice, we usually encounter situations where *multiple IOs* are located between TX and RX. Such a situation occurs, e.g., for propagation over the rooftops of an urban environment. As we see in Fig. 4.9, such a situation can be well approximated by diffraction by multiple screens. Unfortunately, diffraction by multiple screens is an extremely challenging mathematical problem, and – except for a few special cases – no exact solutions are available. Still, a wealth of approximate methods has been proposed in the literature, of which we give an overview in the remainder of this section.

Bullington's method

Bullington's method replaces the multiple screens by a single, "equivalent" screen. This equivalent screen is derived in the following way: put a tangential straight line from the TX to the real obstacles, and select the steepest one (i.e., the one with the largest elevation angle), so that all obstacles either touch this tangent, or lie below it. Similarly, take the tangents from the RX to the obstacles, and select the steepest one. The equivalent screen is then determined by the intersection of the steepest TX tangent and the steepest RX tangent (see Fig. 4.10). The field resulting from diffraction at this single screen can be computed according to Section 4.3.1.

The major attraction of Bullington's method is its simplicity. However, this simplicity also leads to considerable inaccuracies. Most of the physically existing screens do not impact the location of the equivalent screen. Even the highest obstacle might not have an impact. Consider Fig. 4.10: if the highest obstacle lies between screens 01 and 02, it could lie below the tangential lines, and thus not influence the "equivalent" screen, even though it is higher than either screen 01 or screen 02. In reality, these high obstacles *do* have an effect on propagation loss, and cause

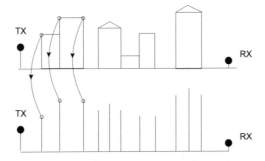

Figure 4.9 Approximation of multiple buildings by a series of screens.

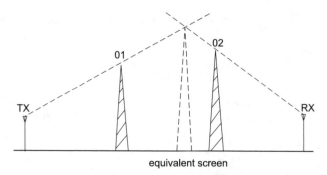

Figure 4.10   Equivalent screen after Bullington.

an additional attenuation. The Bullington method thus tends to give optimistic predictions of the received power.

The Epstein–Petersen method

The low accuracy of the Bullington method is due to the fact that only two obstacles determine the equivalent screen, and thus the total diffraction coefficient. This problem can be somewhat mitigated by the Epstein–Petersen method [Epstein and Petersen 1953]. This approach computes the diffraction losses for each screen separately. The attenuation of a specific screen is computed by putting a virtual "transmitter" and "receiver" on the tips of the screens to the left and right of this considered screen (see Fig. 4.11). The diffraction coefficient, and the attenuation, of this one screen can be easily computed from the principles of Section 4.3.1. Attenuations by the different screens are then added up (on a logarithmic scale). The method thus includes the effects of *all* screens.

Despite this more refined modeling, the method is still only approximate. It uses the diffraction attenuation (Eq. 4.27) that is based on the assumption that the RX is in the far field of the screen. If, however, two of the screens are close together, this assumption is violated, and significant errors can occur.

The inaccuracies caused by this "far-field assumption" can be reduced considerably by the *slope diffraction method*. In this approach, the field is expanded into a Taylor series. In addition to the zeroth-order term (far field), which enforces continuity of the electrical field at the screen, also the first-order term is taken into account, and used to enforce continuity of the first derivative of the field. This results in modified coefficients A and D, which are determined by recursion equations [Andersen 1997].

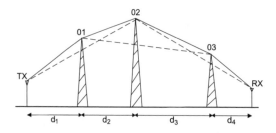

Figure 4.11   The Epstein–Petersen method.

Deygout's method

The philosophy of Deygout's method is similar to that of the Epstein–Petersen method, as it also adds up the attenuations caused by each screen [Deygout 1966]. However, the diffraction angles are defined in the Deygout method by a different algorithm:

- In a first step, determine the attenuation between TX and RX if only the ith screen is present (for all i).
- The screen that causes the largest attenuation is defined as the "main screen" – its index is defined as i_{ms}.
- Compute the attenuation between the TX and the tip of the main screen caused by the jth screen (with j running now from 1 to i_{ms}). The screen resulting in the largest attenuation is called the "subsidiary main screen". Similarly, compute the attenuation between the main screen and the RX, caused by the jth screen ($j > i_{ms} + 1$).
- Optionally, repeat that procedure to create "subsidiary subsidiary screens", and so on.
- Add up the losses (in dB) from all considered screens.

The Deygout method works well if there is actually one dominant screen that creates most of the losses. Otherwise, it can create considerable errors.

Example 4.4 *There are three screens, 20 m apart from each other and 30, 40, and 25 m high. The first screen is 30 m from the TX, the last screen is 100 m from the RX. The TX is 1.5 m high, the RX is 30 m high. Compute the attenuation due to diffraction at 900 MHz by the Deygout method.*

The attenuation L caused by a certain screen is given as:

$$L = -20 \log \widetilde{F}(\nu_F) \tag{4.38}$$

where $\widetilde{F}(\nu_F)$ as defined in Eq. (4.27) is given by:

$$\widetilde{F}(\nu_F) = \frac{1}{2} - \frac{\exp(j\pi/4)}{\sqrt{2}} F(\nu_F) \tag{4.39}$$

First, determine the attenuation caused by screen 1. The diffraction angle θ_d, Fresnel parameter ν_F, and attenuation L are given by:

$$\left. \begin{aligned} \theta_d &= \arctan\left(\frac{30 - 1.5}{30}\right) + \arctan\left(\frac{30 - 30}{140}\right) = 0.760 \, \text{rad} \\[2mm] \nu_F &= 0.760 \sqrt{\frac{2 \cdot 30 \cdot 140}{1/3 \cdot (30 + 140)}} = 9.25 \\[2mm] L_1 &= -20 \log\left(\left| \frac{1}{2} - \frac{\exp(j\pi/4)}{\sqrt{2}} F(9.25) \right|\right) \\[2mm] &\approx -20 \log\left(\left| \frac{1}{2} - \frac{\exp(j\pi/4)}{\sqrt{2}} \cdot (0.522 - j0.527) \right|\right) = 32.28 \, \text{dB} \end{aligned} \right\} \tag{4.40}$$

where the Fresnel integral $F(\nu_F)$ is numerically evaluated. Similarily, the attenuation caused by screens 2 and 3 is $L_2 = 33.59$ dB and $L_3 = 25.64$ dB, respectively. Hence, the main attenuation is caused by screen 2, which therefore becomes the "main screen".

Next, the attenuation from the TX to the tip of screen 2, as caused by screen 1, is determined. The diffraction angle θ_d, Fresnel parameter ν_F, and attenuation L becomes:

$$
\left.
\begin{aligned}
\theta_d &= \arctan\left(\frac{30 - 1.5}{30}\right) + \arctan\left(\frac{30 - 40}{20}\right) = 0.296\,\text{rad} \\[2mm]
\nu_F &= 0.296\sqrt{\frac{2 \cdot 30 \cdot 20}{1/3 \cdot (30 + 20)}} = 2.51 \\[2mm]
L_4 &= -20\log\left(\frac{1}{2} - \frac{\exp(-j\pi/4)}{\sqrt{2}}F(2.51)\right) \\[2mm]
&\approx -20\log\left(\frac{1}{2} - \frac{(\exp(-j\pi/4)}{\sqrt{2}} \cdot (0.446 - j0.614)\right) = 21.01\,\text{dB}
\end{aligned}
\right\}
\tag{4.41}
$$

Similarily, the attenuation from the tip of screen 2 to the RX, as caused by screen 3, is $L_5 = 0.17\,\text{dB}$. The total attenuation caused by diffraction is then determined as the sum of all attenuations – that is:

$$
\begin{aligned}
L_{\text{total}} &= L_2 + L_4 + L_5 \\
&= 33.59 + 21.01 + 0.17 = 54.77\,\text{dB}
\end{aligned}
$$

Empirical models

The ITU (*International Telecommunications Union*) proposed an extremely simple, semi-empirical model for diffraction losses (i.e., losses in addition to free space attenuation):

$$
L_{\text{total}} = \sum_{i=1}^{N} L_i + 20\log C_N
\tag{4.42}
$$

where L_i is the diffraction loss from each separate screen (in dB), and C_N is defined as:

$$
C_N = \sqrt{\frac{P_a}{P_b}}
\tag{4.43}
$$

where

$$
\left.
\begin{aligned}
P_a &= d_{p1}\prod_{i=1}^{N} d_{ni}\left(d_{p1} + \sum_{j=1}^{N} d_{nj}\right) \\[2mm]
P_b &= d_{p1}d_{nN}\prod_{i=1}^{N}(d_{pi} + d_{ni})
\end{aligned}
\right\}
\tag{4.44}
$$

Here, d_{pi} is the (geometrical) distance to the preceding screen tip, and d_{ni} is the distance to the following one. Li et al. [1997] showed that this equation leads to large errors and proposed a modified definition:

$$
C_N = \frac{P_a}{P_b}
\tag{4.45}
$$

Comparison of the different methods

The only special case where an exact solution can be easily computed is the case when all screens have the same height, and are at the same height as the TX and RX antennas. In that case,

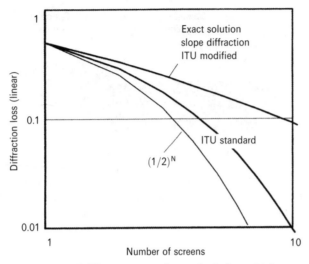

Figure 4.12   Comparison of different computation methods for multiple-screen diffraction.

diffraction loss (on a linear scale!) is proportional to the number of screens, $1/(N_{\text{screen}} + 1)$. Let us now check whether the above approximation methods give this result (see Fig. 4.12):

- The Bullington method is independent of the number of screens, and thus obviously gives a wrong functional dependence.
- The Epstein–Petersen method adds the attenuations *on a logarithmic scale* and thus leads to an exponential increase of the total attenuation on a linear scale.
- Similarly, the Deygout method and the ITU-R method predict an exponential increase of the total attenuation as the number of screens increases.
- The slope diffraction method (up to 15 screens), and the *modified* ITU method lead to a linear increase in total attenuation, and thus predict the trend correctly.

However, note that the above comparison considers a specific, limiting case. For a small number of screens of different height, both the Deygout and the Epstein–Petersen method can be used successfully.

4.4 Scattering by rough surfaces

Scattering on rough surfaces (Fig. 4.13) is a process that is very important for wireless communications. Scattering theory usually assumes roughness to be *random*. However, in wireless communications it is common to also describe *deterministic*, possibly periodic, structures (e.g., bookshelves or windowsills) as rough. For ray-tracing predictions (see Section 7.5), "roughness" thus describes all (physically present) objects that are not included in the used maps and building plans. The justifications for this approach are rather heuristic: (i) the errors made are smaller than some other error sources in ray-tracing predictions, and (ii) there is no better alternative.

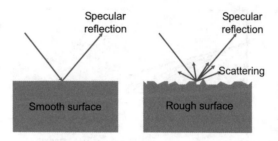

Figure 4.13 Scattering by a rough surface.

That being said, the remainder of this section will consider the mathematical treatment of genuinely rough surfaces. This area has been investigated extensively in the last 30 years, mostly due to its great importance in radar technology. Two main theories have evolved: the Kirchhoff theory, and the perturbation theory.

4.4.1 The Kirchhoff theory

The Kirchhoff theory is conceptually very simple, and requires only a small amount of information – namely, the probability density function of surface amplitude (height). The theory assumes that height variations are so small that different *scattering points* on the surface do not influence each other – in other words, that one point of the surface does not "cast a shadow" onto other points of the surface. This assumption is actually not fulfilled very well in wireless communications.

Assuming that the above condition is actually fulfilled, surface roughness leads to a reduction in power of the specularly reflected ray, as radiation is also scattered in other directions (see r.h.s. of Fig. 4.13). This power reduction can be described by an *effective* reflection coefficient ρ_{rough}. In the case of Gaussian height distribution, this reflection factor becomes:

$$\rho_{\text{rough}} = \rho_{\text{smooth}} \exp\left[-2(k_0 \sigma_h \sin \psi)^2\right] \tag{4.46}$$

where σ_h is the standard deviation of the height distribution, k_0 is the wavenumber $2\pi/\lambda$, and ψ is the angle of incidence (defined as the angle between the wavevector and the surface). The term $2k_0 \sigma_h \sin \psi$ is also known as Rayleigh roughness. Note that for grazing incidence ($\psi \approx 0$), the effect of the roughness vanishes, and the reflection becomes specular again.

4.4.2 Perturbation theory

The perturbation theory generalizes the Kirchhoff theory, using not only the probability density function of the surface height, but also its spatial correlation function. In other words, it takes into account the question "how fast does the height vary if we move a certain distance along the surface?" (see Fig. 4.14).

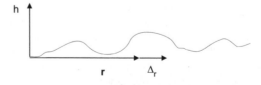

Figure 4.14 Geometry for perturbation theory of rough scattering.

Mathematically, the spatial correlation function is defined as:

$$\sigma_{\mathrm{h}}^2 W(\Delta_r) = E_{\mathbf{r}}\{h(\mathbf{r})h(\mathbf{r}+\Delta_r)\} \tag{4.47}$$

where $\mathbf{r}$ and Δ_r are (two-dimensional) location vectors, and $E_{\mathbf{r}}$ is expectation with respect to $\mathbf{r}$. We need this information to find whether one point on the surface can "cast a shadow" onto another point of the surface. If extremely fast amplitude variations are allowed, shadowing situations are much more common. The above definition enforces spatial statistical stationarity – i.e., the correlation is independent of the absolute location $\mathbf{r}$. The correlation length L_c is defined as the distance so that $W(L_c) = 0.5 \cdot W(0)$.[7]

The effect of surface roughness on the amplitude of a specularly reflected wave can be described by an "effective" (complex) dielectric constant δ_{eff}, which in turn gives rise to an "effective" reflection coefficient, as computed from Snell's law.[8] For vertical polarization, the δ_{eff} is given by [Vaughan and Andersen 2003]:

$$\frac{1}{\sqrt{\delta_{\mathrm{r,eff}}}} = \begin{cases} \dfrac{1}{\sqrt{\varepsilon_{\mathrm{r}}}} + j\dfrac{k_0\sigma_{\mathrm{h}}^2\sin(2\psi)}{2L_{\mathrm{c}}}\displaystyle\int_0^\infty \dfrac{1}{x}\dfrac{d\hat{W}(x)}{dx}dx & , \quad k_0L_{\mathrm{c}} \ll 1 \\[3mm] \dfrac{1}{\sqrt{\varepsilon_{\mathrm{r}}}} + (k_0\sigma_{\mathrm{h}})^2(\sin\psi)^3 & , \quad k_0L_{\mathrm{c}} \gg 1, \psi \gg \dfrac{1}{\sqrt{k_0L_{\mathrm{c}}}} \\[3mm] \dfrac{1}{\sqrt{\varepsilon_{\mathrm{r}}}} - \dfrac{\sigma_{\mathrm{h}}^2}{2L_{\mathrm{c}}}\dfrac{\sqrt{k_0L_{\mathrm{c}}}}{\sqrt{2\pi}}\exp(j\pi/4)\displaystyle\int_0^\infty \dfrac{1}{x\sqrt{x}}\dfrac{d\hat{W}(x)}{dx}dx & , \quad k_0L_{\mathrm{c}} \gg 1, \psi \ll \dfrac{1}{\sqrt{k_0L_{\mathrm{c}}}} \end{cases} \tag{4.48}$$

where $\hat{W}(x) = W(x/L_{\mathrm{c}})$, while for horizontal polarization, it is:

$$\frac{1}{\sqrt{\delta_{\mathrm{r,eff}}}} = \begin{cases} \dfrac{1}{\sqrt{\varepsilon_{\mathrm{r}}}} + j\dfrac{k_0\sigma_{\mathrm{h}}^2}{2L_{\mathrm{c}}}\displaystyle\int_0^\infty \dfrac{1}{x}\dfrac{d\hat{W}(x)}{dx}dx & , \quad k_0L_{\mathrm{c}} \ll 1 \\[3mm] \dfrac{1}{\sqrt{\varepsilon_{\mathrm{r}}}} + (k_0\sigma_{\mathrm{h}})^2\sin\psi & , \quad k_0L_{\mathrm{c}} \gg 1, \psi \gg \dfrac{1}{\sqrt{k_0L_{\mathrm{c}}}} \\[3mm] \dfrac{1}{\sqrt{\varepsilon_{\mathrm{r}}}} - \dfrac{(k_0\sigma_{\mathrm{h}})^2}{\sqrt{k_0L_{\mathrm{c}}}}\dfrac{2}{\sqrt{2\pi}}\exp(-j\pi/4)\displaystyle\int_0^\infty \dfrac{1}{\sqrt{x}}\dfrac{d\hat{W}(x)}{dx}dx & , \quad \sqrt{k_0L_{\mathrm{c}}} \gg 1, \psi \ll \dfrac{1}{\sqrt{k_0L_{\mathrm{c}}}} \end{cases} \tag{4.49}$$

Comparing these results to the Kirchhoff theory, we find that there is good agreement in the case $k_0L_{\mathrm{c}} \gg 1$, $\psi \gg 1/\sqrt{k_0L_{\mathrm{c}}}$. This agrees with our above discussion of the limits of the Kirchhoff theory: by assuming that the coherence length is long compared with the wavelength, there cannot be a diffraction by a sudden "spike" in the surface. And by fulfilling $\psi \gg 1/\sqrt{k_0L_{\mathrm{c}}}$, it is assured that a wave incident under angle ψ cannot cast a shadow onto other points of the surface.

---

[7] A more extensive discussion of expectation values and correlation functions will be given in Chapter 6.

[8] We assume here that the material is not conductive, $\sigma_{\mathrm{e}} = 0$. The imaginary contribution arises from surface roughness.

4.5 Waveguiding

Another important process is propagation in (dielectric) waveguides. This process models propagation in street canyons, corridors, and tunnels. The basic equations of dielectric waveguides are well-established [Collin 1991], [Marcuse 1991]. However, those waveguides occuring in wireless communications deviate from the idealized assumptions of theoretical work:

- The materials are lossy.
- Street canyons (and most corridors) do not have continuous walls, but are interrupted at more or less regular intervals by cross-streets. Furthermore, street canyons lack the "upper" wall of the waveguide.
- The surfaces are rough (window sills, etc.).
- The waveguides are not empty, but filled with metallic (cars) and dielectric (pedestrian) IOs.

Propagation prediction can be done either by computing the waveguide modes, or by a geometric optics approximation. If the waveguide cross section as well as the IOs in it are much larger than the wavelength, the latter method gives good results [Klemenschits and Bonek 1994].

Conventional waveguide theory predicts a propagation loss that increases exponentially with distance. Some measurements in corridors observed a similar behavior. The majority of measurements, however, fitted a d^{-n} law, where n varies between 1.5 and 5. Note that a loss exponent smaller than 2 does not contradict energy conservation or any other laws of physics. The d^{-2} law in free space propagation just stems from the fact that energy is spread out over a larger surface as distance is increased. If energy is guided, even d^0 becomes theoretically possible.

4.6 Appendices

Please go to *www.wiley.com/go/molisch*

Further reading

Several books on wireless propagation processes give an overview of all the phenomena discussed in this chapter [Bertoni 2000], [Blaunstein 1999], [Parsons 1992], and [Vaughan and Andersen 2003]. The basic propagation processes, especially reflection and transmission, are described in any number of classic textbooks on electromagnetics – e.g., Ramo et al. [1967], which is a personal favourite – but also many others. More involved results can be found in Bowman et al. [1987] and Felsen and Marcuvitz [1973]. The description of transmission through multilayer dielectric films can be read in Heavens [1965]. The different theories for multiple knife edge diffraction are nicely summarized in Parsons [1992]; however, continuous research is being done on that topic, and a number of new approximations is constantly being published. Scattering by rough surfaces, mostly inspired by radar problems, is described in Bass and Fuks [1979], de Santo and Brown [1986], Ogilvy [1991]; an excellent summary is given in Vaughan and Andersen [2003]. The theory of dielectric waveguides is described in Collin [1991], Marcuse [1991], and Ramo et al. [1967].

5

Statistical description of the wireless channel

5.1 Introduction

In many circumstances, it is too complicated to describe all reflection, diffraction, and scattering processes that determine the different Multipath Components (MPCs). Rather, it is often preferable to describe the *probability* that a channel parameter attains a certain value. The most important parameter is channel gain, as it determines received power or fieldstrength; it will be at the center of our interest in this chapter. Note that we talk of channel gain (and not attenuation) because it is directly proportional to receive power (receive power is transmit power times channel gain); of course, this gain is smaller than unity.[1]

In order to arrive at a better understanding of the channel, let us first look at a typical plot depicting received power as a function of distance (Fig. 5.1). The first thing we notice is that received power can vary strongly – by 100 dB or more. We also see that variations happen on different spatial scales:

- On a very-short-distance scale, power fluctuates around a (local) mean value, as can be seen from the insert in Fig. 5.1. These fluctuations happen on a scale that is comparable with one wavelength, and are therefore called *small-scale fading*. The reason for these fluctuations is interference between different MPCs, as mentioned in Chapter 2. Fluctuations in fieldstrength can be well described statistically – namely, by the (local) mean value of the power, and the statistics of the fluctuations around this mean.
- Mean power, averaged over about 10 wavelengths, itself shows fluctuations. These fluctuations occur on a larger scale – typically a few hundred wavelengths. These variations can be seen most clearly when moving on a circle around the transmitter (Fig. 5.2). The reason for these variations is shadowing by large objects (Chapter 2), and is thus fundamentally different from the interference that causes small-scale fading. However, this *large-scale fading* can also be described by a mean and the statistics of fluctuations around this mean.
- The large-scale mean itself depends monotonically on the distance between transmitter and receiver. This effect is related to free space pathloss or some variation thereof. This effect is usually described in a deterministic manner, and was already treated in Chapter 4 (further details and models can be found in Chapter 7).

---

[1] In the following, we will often assume unit transmit power, and use "receive power" and "channel gain" interchangeably.

Wireless Communications   Andreas F. Molisch
© 2005 John Wiley & Sons, Ltd

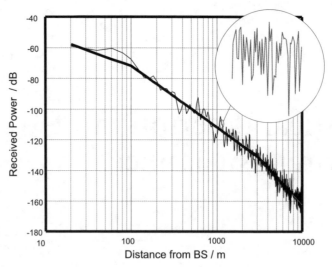

Figure 5.1 Received power as a function of distance from the transmitter.

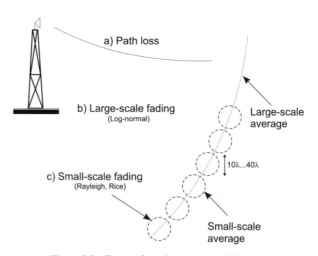

Figure 5.2 Types of receive power variations.

Chapter 5 only concentrates on the statistical variations of the channel gain; delay dispersion and other effects will be treated later. It is thus sufficient for this chapter to consider channel gain for an unmodulated (sinusoidal) carrier signal, though the considerations are also valid for *narrowband* systems (as defined in Section 6.1). Sections 5.2 and 5.3 explain the *two-path model* – it is the most simple model explaining fading effects. Sections 5.4 and 5.5 generalize these considerations to more realistic channel models, and give statistics of the received fieldstrength for various scenarios. The next two sections describe the statistics of fading dips (instances of very low received power), the frequency of their occurrence, and their average duration. Finally, the statistical variations due to shadowing effects are described in Section 5.8.

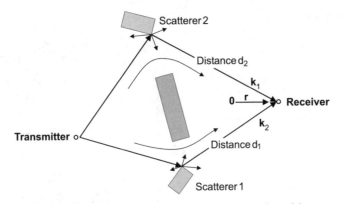

Figure 5.3 Geometry of the time-invariant two-path model.

5.2 The time-invariant two-path model

As an introduction to the rather involved subject of multipath propagation and fading, we consider the simplest possible case – time-invariant propagation along two paths. We transmit a sinusoidal waveform, and determine the (complex) transfer function at the location of the receiver.

First, consider a single wave. Let the transmit signal be a sinusoidal wave:

$$E_{\text{TX}}(t) \propto \cos(2\pi f_c t) \tag{5.1}$$

Let the received signal be approximated as a homogeneous plane wave. If the runlength between transmitter and receiver is d, the received signal can be described as:

$$E(t) = E_0 \cdot \cos(2\pi f_c t - k_0 d) \tag{5.2}$$

where k_0 is the wavenumber $2\pi/\lambda$. Using complex baseband notation,[2] this reads

$$E = E_0 \exp(-jk_0 d) \tag{5.3}$$

Note that the real part of the field in complex representation, $\text{Re}\{E\}$ is equal to the instantaneous value of the fieldstrength at time $t = 0$.

Now consider the case that the transmit signal gets to the receiver via two different propagation paths, created by two different Interacting Objects (IOs, see Fig. 5.3). These paths have different runtimes:

$$\tau_1 = d_1/c_0, \quad \text{and} \quad \tau_2 = d_2/c_0 \tag{5.4}$$

The receiver is in the far field of the IOs, so that the arriving waves are homogeneous planewaves. We assume furthermore that both waves are vertically polarized, and have amplitudes $E1$ and $E2$ at the reference position (origin of the coordinate system) $\mathbf{r} = 0$. We get the following expression for the superposition of two planewaves:

$$E(\mathbf{r}) = E1 \exp(-j\mathbf{k}_1\mathbf{r}) + E2 \exp(-j\mathbf{k}_2\mathbf{r}) \tag{5.5}$$

We assume here that the two waves arriving at the receiver position $\mathbf{r}$ are two plane waves whose amplitudes do not vary as a function of receiver position (we vary receiver positions only within an area that has less than about $10\,\lambda$ diameter).

---

[2] The bandpass signal – i.e., the physically existing signal – is related to the complex baseband (lowpass) representation as $s_{\text{BP}}(t) = \text{Re}\{s_{\text{LP}}(t)\exp[j2\pi f_c t]\}$ (see also Section 11.1).

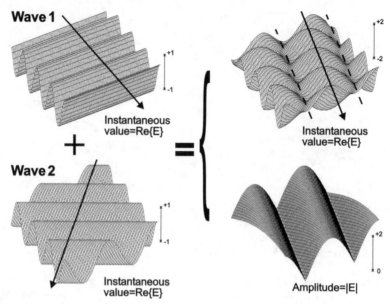

Figure 5.4   Interference of two planewaves with $E1 = E2 = 1$ and $\arg(\mathbf{k}_1, \mathbf{k}_2) = 30°$.

The upper part of Fig. 5.4 depicts the real part of $E(\mathbf{r})$, and the lower part shows the magnitude, which is proportional to the square root of received power. We see locations of both constructive and destructive interference – i.e., location-dependent fading. There are fading dips where the phase differences due to the different runtimes are exactly 180°. If the receiver is at a point of destructive interference, it sees a signal whose total amplitude is the difference of the amplitudes of the constituting waves. If the amplitudes of the constituting waves are equal, destructive interference can be complete. In the points of constructive interference, the total amplitude is the sum of the amplitudes of the constituting waves. This phenomenon can be seen in the lower right part of Fig. 5.4.

5.3 The time-variant two-path model

In general, the runtime (pathlength) difference between the different propagation paths changes with time. This change can be due to movements of the transmitter (TX), the receiver (RX), the IOs, or any combination thereof; to simplify discussion, we will henceforth assume only movement of the RX. The RX then "sees" a time-varying interference pattern; we can imagine that the receiver moves through the "mountains and valleys" of the fieldstrength plot. Spatially varying fading thus becomes time-varying fading. Since fading dips are approximately half a wavelength apart (corresponding to 16 cm at a carrier frequency of 900 MHz), this fading is called *small-scale fading*, also known as *short-term fading*, or *fast fading*.[3] The fading rate (number of fading dips per second) depends on the speed of the receiver.

---

[3] Unfortunately, the expression *fast* fading is often used for two completely different phenomena. On one hand, it is used as a synonym for small-scale fading, independent of the actual *temporal* scale of the fading (which depends on the movement speed of the TX, RX, and IOs). On the other hand it is used to denote channel variations within the duration of a symbol length (in contrast to the "quasi-static" channel). It is thus preferable to call the fading due to interference effects *small-scale fading*, as this is unambiguous.

The movement of the RX also leads to a shift of the received frequency, called the Doppler shift. In order to explain this phenomenon, let us first revert to the case of a *single* sinusoidal wave reaching the RX, and also revert to real passband notation. If the RX moves away from the TX with speed v the distance d between TX and RX increases with that speed. Thus:

$$E(t) = E_0 \cdot \cos(2\pi f_c t - k_0[d_0 + vt])$$

$$= E_0 \cdot \cos\left(2\pi t\left[f_c - \frac{v}{\lambda}\right] - k_0 d_0\right) \tag{5.6}$$

where d_0 is the distance at time $t = 0$. The frequency of the received oscillation is thus decreased by v/λ – in other words, the Doppler shift is:

$$\nu = -\frac{v}{\lambda} = -f_c \cdot \frac{v}{c_0} \tag{5.7}$$

Note that the Doppler shift is negative when the TX and RX move away from each other. Since the speed of the movement is always small compared with the speed of light, the Doppler shifts are relatively small.

In the above example, we had assumed that the direction of RX movement is aligned with the direction of wave propagation. If that is not the case, the Doppler shift is determined by the speed of movement *in the direction of wave propagation*, $v \cos(\gamma)$ (see Fig. 5.5). The Doppler shift is then:

$$\nu = -\frac{v}{\lambda} \cos(\gamma) = -f_c \cdot \frac{v}{c_0} \cos(\gamma) = -\nu_{\max} \cos(\gamma) \tag{5.8}$$

The maximum Doppler shift $\nu_{\max}$ typically lies between 1 Hz and 1 kHz. Note that in general the relationship $\nu_{\max} = f_c \cdot v/c_0$ is based on several assumptions – e.g., static IOs, no double reflections on moving objects, etc.

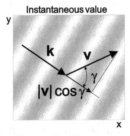

Figure 5.5 Projection of velocity vector **v** onto the direction of propagation **k**.

Since the Doppler shifts are so small, it seems natural to ask whether they have a significant influence on the radio link. If all constituent waves were Doppler-shifted by the same amount – e.g., 100 Hz – the effect on radio link performance would really be negligible – the local oscillator in the receiver could easily compensate for such a shift. The important point is, however, that the different MPCs have *different* Doppler shifts. The superposition of several Doppler-shifted waves creates the sequence of fading dips. Again, this can be demonstrated using the two-path model. As the receiver moves, it receives two waves that are each Doppler-shifted, but by different amounts. By Fourier transformation to the time domain, we get the well-known effect of *beating* of two oscillations with slightly different frequencies (see Fig. 5.6). The frequency of this beating envelope is equal to the frequency difference between the two carriers – i.e., the difference of the two Doppler shifts. The receiver thus sees a signal whose amplitude undergoes periodic variations – this is exactly the fast fading that is created by traversing the "mountains and valleys" of the fieldstrength plot of Fig. 5.4.

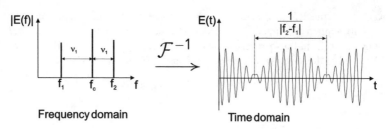

Figure 5.6   Superposition of two carriers with different frequencies (beating).

Summarizing, we can obtain the fading rate in the two-path model by two equivalent considerations:

1. We superimpose two incident waves, plot the resulting interference pattern (fieldstrength "mountains and valleys"), and count the number of fading dips per second that a RX sees when moving through that pattern.
2. Alternatively, we can think of superimposing two signals with different Doppler shifts at the receive antenna, and determine the fading rate from the beat frequency – i.e., the difference of the Doppler shifts of the two waves.

Doppler frequency is thus an important parameter of the channel, even though it is so small:

- Doppler frequency is a measure for the rate of change of the channel, as we have discussed above.
- Furthermore, the superposition of many slightly Doppler-shifted signals leads to phaseshifts of the total received signal that can impair the reception of angle-modulated signals (Chapters 11 and 12). These phaseshifts lead to a random frequency modulation (FM) of the received signal (see Section 5.7), and are especially important for signals with low bit rates.

5.4 Small-scale fading without a dominant component

Following these basic considerations utilizing the two-path model, we now investigate a more general case of multipath propagation. We consider a radio channel with several IOs and a moving RX. Due to the large number of IOs, a deterministic description of the radio channel is not efficient any more, which is why we take refuge in stochastic description methods. This stochastic description is essential for the whole field of wireless communications, and is thus explained in considerable detail. We start out with a computer experiment in Section 5.4.1, followed by a more exact mathematical derivation in Section 5.4.2.

5.4.1 A computer experiment

We start out with a simple computer experiment. The signals from several IOs are incident onto a RX that moves over a small area. The IOs are distributed approximately uniformly around the receiving area. They are also assumed to be sufficiently far away so that all received waves are homogeneous plane waves, and that movements of the RX within the considered area do not change the amplitudes of these waves. The different distances and strength of the interactions are taken into account by assigning a random phase and a random amplitude to each wave. We are then creating eight constituting waves E_i with absolute amplitudes $|a_i|$, angle of incidence (with respect to the x-axis) ϕ_i and phase φ_i:

	$|a_i|$	ϕ_i	φ_i
$E_1(x,y) = 1.0 \exp[-jk_0(x\cos(169°)+y\sin(169°))]\exp(j311°)$	1.0	169°	311°
$E_2(x,y) = 0.8 \exp[-jk_0(x\cos(213°)+y\sin(213°))]\exp(j\,32°)$	0.8	213°	32°
$E_3(x,y) = 1.1 \exp[-jk_0(x\cos(\,87°)+y\sin(\,87°))]\exp(j161°)$	1.1	87°	161°
$E_4(x,y) = 1.3 \exp[-jk_0(x\cos(256°)+y\sin(256°))]\exp(j356°)$	1.3	256°	356°
$E_5(x,y) = 0.9 \exp[-jk_0(x\cos(\,17°)+y\sin(\,17°))]\exp(j191°)$	0.9	17°	191°
$E_6(x,y) = 0.5 \exp[-jk_0(x\cos(126°)+y\sin(126°))]\exp(j\,56°)$	0.5	126°	56°
$E_7(x,y) = 0.7 \exp[-jk_0(x\cos(343°)+y\sin(343°))]\exp(j268°)$	0.7	343°	268°
$E_8(x,y) = 0.9 \exp[-jk_0(x\cos(297°)+y\sin(297°))]\exp(j131°)$	0.9	297°	131°

We now superimpose the constituting waves, using the complex baseband notation again. The total complex fieldstrength E thus results from the sum of the complex fieldstrengths of the constituting waves. We can also interpret this as adding up complex random phasors. Figure 5.7 shows the instantaneous value of the total fieldstrength at time $t = 0$ – i.e., Re$\{E\}$, in an area of size $5\lambda \cdot 5\lambda$.

Let us now consider the statistics of the fieldstrengths occurring in that area. As can be seen from the histogram (Fig. 5.8), the values of Re$\{E\}$ follow, to a good approximation, a zero-mean Gaussian distribution. This is a consequence of the central limit theorem: when superimposing N statistically independent random variables, none of which is dominant, the associated *probability density function* (pdf) approaches a normal distribution for $N \to \infty$ (see Appendix 5.A – at *www.wiley.com/go/molisch* – for a more exact formulation of this statement). The conditions for the validity of the central limit theorem are fulfilled approximately: the eight constituting waves have random angles of incidence and phases, and none of the amplitudes a_i is dominant. Figures 5.9 and 5.10 show that the imaginary part of the fieldstrength, Im$\{E\}$, is also normally distributed.[4]

The behavior of most receivers is determined by the received (absolute) *amplitude (magnitude)*. We thus need to investigate the distribution of the envelope of the received signal, corresponding to the magnitude of the (complex) fieldstrength phasor. Figure 5.12 shows the fieldstrength seen by a receiver that moves along the y-axis of Fig. 5.7 or Fig. 5.8. The left diagram of Fig. 5.11 shows the complex fieldstrength phasor, and the right side the absolute amplitude of the received signal.

Figures 5.12 and 5.13 show a three-dimensional representation of the amplitudes, and the statistics of the amplitude over that area, respectively. Figure 5.13 also exhibits a plot of a Rayleigh pdf. A Rayleigh distribution describes the magnitude of a complex stochastic variable whose real and imaginary parts are independent normally distributed; more details will be given below. Figure 5.14 shows that the *phase* is approximately uniformly distributed.

5.4.2 Mathematical derivation of the statistics of amplitude and phase

After these pseudo-experimental considerations, we now turn to a more detailed, and more mathematically sound, derivation of Rayleigh distribution. Consider a scenario where N homogeneous plane waves (MPCs) have been created by reflection/scattering from different IOs. The IOs and the TX do not move, and the RX moves with a velocity v. As above, we assume that the absolute amplitudes of the MPCs do not change overthe region of observation.

---

[4] The imaginary part represents the instantaneous value of the fieldstrength at a time that corresponds to a quarter-period of the radio frequency oscillation, $\omega t = \pi/2$.

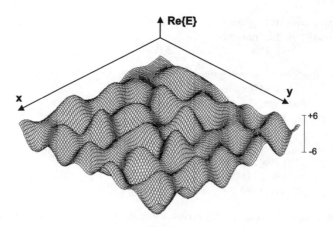

Figure 5.7 Instantaneous value of the fieldstrength at time $t = 0$—i.e. Re$\{E\}$. Superposition of the eight constituting waves in the area $0 < x < 5\lambda$, $0 < y < 5\lambda$.

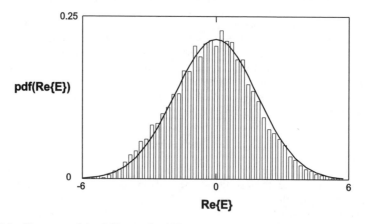

Figure 5.8 Histogram of the fieldstrength of Fig. 5.7. A Gaussian pdf is shown for comparison.

The sum of the squared amplitudes is thus:

$$\sum_{i=1}^{N} |a_i|^2 = C_\mathrm{P} \tag{5.9}$$

where C_P is a constant. However, the phases φ_i vary strongly, and are thus approximated as random variables that are uniformly distributed in the range $[0, 2\pi]$. The real part of the received fieldstrength due to the ith MPC is thus $a_i \cos(\varphi_i)$, the imaginary part is $|a_i| \sin(\varphi_i)$.

Furthermore, we need to consider the Doppler shift for computation of the total fieldstrength $E(t)$. If we look at an unmodulated carrier, we get (in real passband notation):

$$E(t) = \sum_{i=1}^{N} |a_i| \cos[2\pi f_\mathrm{c} t - 2\pi \nu_{\max} \cos(\gamma_i) t + \varphi_i] \tag{5.10}$$

Rewriting this in terms of in-phase and quadrature-phase components in real passband notation, we obtain:

$$E_{\mathrm{BP}}(t) = I(t) \cdot \cos(2\pi f_\mathrm{c} t) - Q(t) \cdot \sin(2\pi f_\mathrm{c} t) \tag{5.11}$$

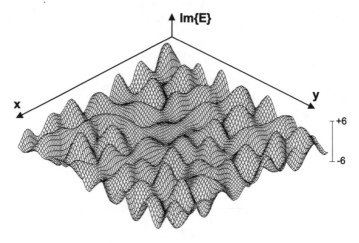

Figure 5.9 Imaginary part of E.

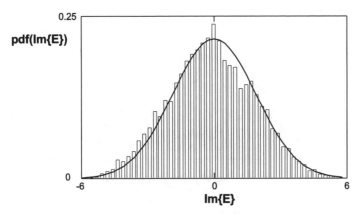

Figure 5.10 Histogram of the fieldstrength of Fig. 5.9. A Gaussian pdf is shown for comparison.

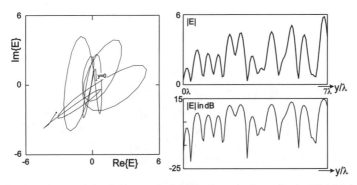

Figure 5.11 Complex phasor of the fieldstrength (left), and received amplitude $|E|$ (right) for a receiver moving along the y-axis.

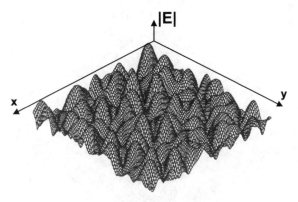

Figure 5.12   Amplitude of the fieldstrength.

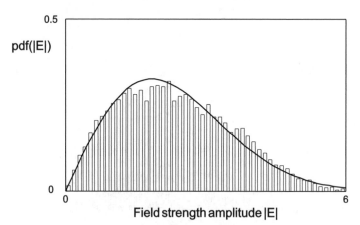

Figure 5.13   Pdf of the received amplitude.

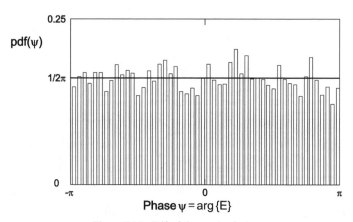

Figure 5.14   Pdf of the received phase.

with

$$I(t) = \sum_{i=1}^{N} |a_i| \cos[-2\pi\nu_{\mathrm{max}} \cos(\gamma_i)t + \varphi_i] \tag{5.12}$$

$$Q(t) = \sum_{i=1}^{N} |a_i| \sin[-2\pi\nu_{\mathrm{max}} \cos(\gamma_i)t + \varphi_i] \tag{5.13}$$

Both the in-phase and the quadrature-phase component are the sum of many random variables, none of which dominates (i.e., $|a_i| \ll C_P$). It follows from the central limit theorem that the pdf of such a sum is a normal (Gaussian) distribution, regardless of the exact pdf of the constituent amplitudes – i.e., we do not need knowledge of the a_i, or their distributions!!! (see Appendix 5.A – at *www.wiley.com/go/molisch* – for more details). A zero-mean Gaussian random variable has the pdf:

$$pdf_x(x) = \frac{1}{\sqrt{2\pi}\sigma} \exp\left(-\frac{x^2}{2\sigma^2}\right) \tag{5.14}$$

where σ^2 denotes the variance.

Starting with the statistics of the real and imaginary parts, Appendix 5.B derives the statistics of amplitude and phase of the received signal. The pdf is a product of a pdf for ψ – namely, a uniform distribution:

$$pdf_\psi(\psi) = \frac{1}{2\pi} \tag{5.15}$$

and a pdf for r – namely, a Rayleigh distribution:

$$pdf_r(r) = \frac{r}{\sigma^2} \cdot \exp\left[-\frac{r^2}{2\sigma^2}\right] \qquad 0 \le r < \infty \tag{5.16}$$

For $r < 0$ the pdf is zero, as amplitudes are by definition positive.

5.4.3 Properties of the Rayleigh distribution

A Rayleigh distribution has the following properties, which are also shown in Fig. 5.15:

$$\left. \begin{array}{rl} \text{Mean value} & \bar{r} = \sigma\sqrt{\dfrac{\pi}{2}} \\[2mm] \text{Mean square value} & \overline{r^2} = 2\sigma^2 \\[2mm] \text{Variance} & \overline{r^2} - (\bar{r})^2 = 2\sigma^2 - \sigma^2\dfrac{\pi}{2} = 0.429\sigma^2 \\[2mm] \text{Median value} & r_{50} = \sigma\sqrt{2 \cdot ln2} = 1.18\,\sigma \\[2mm] \text{Location of maximum} & \max\{pdf(r)\} \text{ occurs at } r = \sigma \end{array} \right\} \tag{5.17}$$

where the bar denotes expected value (we deviate here from our usual notation $E\{\}$ in order to avoid confusion with the fieldstrength E).

The *cumulative distribution function*, $cdf(x)$, is defined as the probability that the realization of the random variable has a value smaller than x. The cdf is thus the integral of the pdf:

$$cdf(r) = \int_{-\infty}^{r} pdf(u)du \tag{5.18}$$

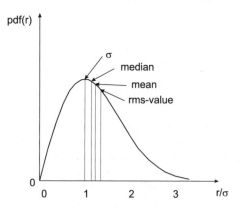

Figure 5.15   Pdf of a Rayleigh distribution.

Applying this equation to the Rayleigh pdf, we get:

$$cdf(r) = 1 - \exp\left(-\frac{r^2}{2\sigma^2}\right) \tag{5.19}$$

For small values of r this can be approximated as:

$$cdf(r) \approx \frac{r^2}{2\sigma^2} \tag{5.20}$$

It is straightforward to check whether a measured ensemble of fieldstrength values follows a Rayleigh distribution: its empirical cdf is plotted on so-called Weibull paper (see Fig. 5.16). The cdf of the Rayleigh distribution is a straight line on it. For small values of r, an increase of r by 10 dB has to increase the value of the cdf by 10 dB.

The Rayleigh distribution is widely used in wireless communications. This is due to several reasons:

- It is an *excellent approximation* in a large number of practical scenarios, as confirmed by a multitude of measurements. However, it is noteworthy that there *are* scenarios where it is not valid. These can occur, e.g., in *Line Of Sight* (LOS) scenarios, some indoor scenarios, and in (ultra) wideband scenarios (see Chapters 6 and 7).
- It describes a *worst case scenario* in the sense that there is no dominant signal component, and thus there is a large number of fading dips. Such a worst case assumption is useful for the design of robust systems.[5]
- It depends only on a *single parameter*, the mean received power – once this parameter is known, the complete signal statistics are known. It is easier, and less error-prone, to obtain this single parameter either from measurements or deterministic prediction methods than to obtain the multiple parameters of more involved channel models.
- *Mathematical convenience*: computations of error probabilities and other parameters can often be done in closed form when the fieldstrength distribution is Rayleigh.

---

[5] As we will see later on, there are some fading distributions that show a larger number of fading dips – e.g., Nakagami distributions with $m < 1$; furthermore, the large number of MPCs can also be an advantage for specific systems, see Section 20.2.

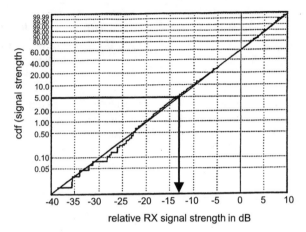

Figure 5.16 Measured cdf of the (normalized) receive power of an indoor non-line-of-sight scenario [Gahleitner 1993].

5.4.4 Fading margin for Rayleigh-distributed fieldstrength

Knowledge of the fading statistics is extremely important for the design of wireless systems. We saw in Chapter 3 that for noise-limited systems the received fieldstrength determines the performance. As fieldstrength is a random variable, even a large mean fieldstrength does not guarantee successful communications at *all* times. Rather, the fieldstrength exceeds a minimum value only in a certain *percentage* of situations. The task is therefore to answer the following question: "Given a minimum receive power or fieldstrength required for successful communications, how large does the mean power have to be in order to ensure that communication is successful in $x\%$ of all situations?" In other words, how large does the *fading margin* have to be?

The cdf gives by definition the probability that a certain fieldstrength level is not exceeded. In order to achieve an $x\%$ outage probability, it follows that:

$$x = cdf(r_{\min}) \approx \frac{r_{\min}^2}{2\sigma^2} \tag{5.21}$$

where the r.h.s. follows from Eq. (5.20). From this, we can immediately compute the mean square value of fieldstrength $2\sigma^2$ as $2\sigma^2 = r_{\min}^2/x$.

Example 5.1 *For a signal with Rayleigh-distributed amplitude, what is the probability that the received signal power is 20, 6, 3 dB or lower relative to the mean power. Compare the exact result and the result from the approximate formulation Eq. (5.20).*

From a Rayleigh-distributed signal envelope r:

$$\overline{r^2} = 2\sigma^2 \tag{5.22}$$

A power level 20 dB below the mean power corresponds to $\dfrac{r_{\min}^2}{2\sigma^2} = \dfrac{1}{100}$:

$$\Pr\{r < r_{\min}\} = 1 - \exp\left(-\frac{1}{100}\right) = 9.95 \cdot 10^{-3} \tag{5.23}$$

Similarly, the exact results for 6 and 3 dB are 0.221 and 0.393, respectively.

The approximate formulation Eq. (5.20) gives $\frac{r_{\min}^2}{2\sigma^2} = 0.01$, 0.25, and 0.5, respectively. It is thus reasonably accurate for power levels 6 dB below the mean power, but breaks down for higher values of $r_{\min}$.

For the interference-limited case, the situation is somewhat more complicated: not only does the desired signal fade, but so do the interferers. It is thus necessary to compute the statistics of the *Signal to Interference Ratio* (SIR). If both the desired signal and the interference are Rayleigh-fading, the pdf of the SIR (i.e., the pdf of the ratio of two random variables, each of which is Rayleigh-distributed) is [Molisch et al. 1996]:

$$pdf_{SIR}(r) = \frac{2\tilde{\sigma}^2 r}{(\tilde{\sigma}^2 + r^2)^2} \tag{5.24}$$

where $\tilde{\sigma}^2 = \sigma_1^2/\sigma_2^2$ is the ratio of mean signal power to mean interference power. The associated cdf is:

$$cdf_{SIR}(r) = 1 - \frac{\tilde{\sigma}^2}{(\tilde{\sigma}^2 + r^2)} \tag{5.25}$$

This formulation is essential for computation of the reuse distance (see Chapters 3 and 17).

5.5 Small-scale fading with a dominant component

5.5.1 A computer experiment

Fading statistics change when a dominant MPC – e.g., an LOS component or a dominant specular component – is present. We can gain some insights by repeating the computer experiment of Section 5.4.1, but now adding an additional wave with the (dominant) amplitude 5:

	$\lvert a_9 \rvert$	ϕ_9	φ_9
$E_9(x,y) = 5.0 \exp[-jk_0(x\cos(0°) + y\sin(0°))]\exp(j0°)$	5.0	0°	0°

Figure 5.17 shows the real part of E, and the contribution from the dominant component is visible; while Fig. 5.18 shows the absolute value. The histogram of the absolute value of the fieldstrength is shown in Fig. 5.19. It is clear that the probability of deep fades is much smaller than in the Rayleigh-fading case.

5.5.2 Derivation of the amplitude and phase distribution

The pdf of the amplitude can be computed in a way that is similar to our derivation of the Rayleigh distribution (Appendix 5.B – see *www.wiley.com/go/molisch*). Without restriction of generality, we assume that the LOS component has zero phase, so that it is purely real. The real part thus has a non-zero-mean Gaussian distribution, while the imaginary part has a zero-mean Gaussian distribution. Performing the variable transformation as in Appendix 5.B, we get the joint pdf of amplitude r and phase ψ [Rice 1947]:

$$pdf_{r,\psi}(r,\psi) = \frac{r}{2\pi\sigma^2}\exp\left(-\frac{r^2 + A^2 - 2rA\cos(\psi)}{2\sigma^2}\right) \tag{5.26}$$

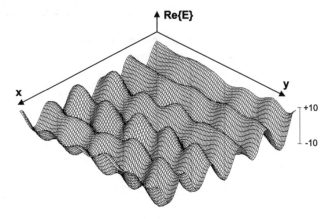

Figure 5.17   Re(E)—i.e., the instantaneous value at $t = 0$, in the presence of a dominant MPC.

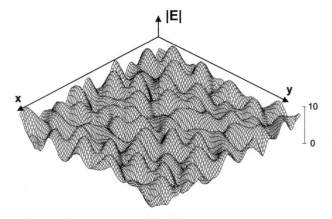

Figure 5.18   Magnitude of the electric fieldstrength, $|E|$, in an example area, in the presence of a dominant MPC.

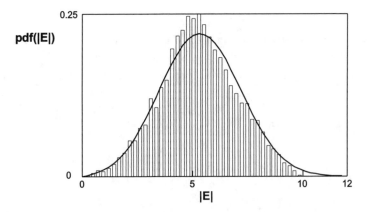

Figure 5.19   Histogram of the amplitudes in the presence of a dominant MPC.

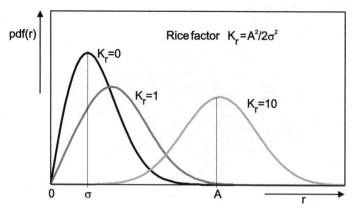

Figure 5.20 Rice distribution for three different values of K_r—i.e., ratio between the power of the LOS component and the diffuse components.

where A is the amplitude of the dominant component. In contrast to the Rayleigh case, this distribution is not separable. Rather, we have to integrate over the phases to get the amplitude pdf, and vice versa.

The pdf of the amplitude is given by the *Rice distribution* (solid line in Fig. 5.19):

$$pdf_r(r) = \frac{r}{\sigma^2} \cdot \exp\left[-\frac{r^2 + A^2}{2\sigma^2}\right] \cdot I_0\left(\frac{rA}{\sigma^2}\right) \qquad 0 \leq r < \infty \tag{5.27}$$

$I_0(x)$ is the modified Bessel function of the first kind, zero order [Abramowitz and Stegun 1965]. The mean square value of a Rice-distributed random variable r is:

$$\overline{r^2} = 2\sigma^2 + A^2 \tag{5.28}$$

The ratio of the power in the LOS component to the power in the diffuse component, $A^2/(2\sigma^2)$, is called the Rice factor K_r.

Figure 5.20 shows the Rice distribution for three different values of the Rice factor. The stronger the LOS component, the rarer the occurrence of deep fades. For $K_r \to 0$, the Rice distribution becomes a Rayleigh distribution, while for large K_r it approximates a Gaussian distribution with mean value A.

Example 5.2 *Compute the fading margin for a Rice distribution with $K_r = 0.3, 3,$ and 20 dB so that the outage probability is less than 5%.*

Recall that the outage probability can be expressed in terms of the *cdf* of the Ricean envelope:

$$P_{out} = cdf(r_{min}) \tag{5.29}$$

For the Ricean distribution the cdf is given as:

$$cdf(r_{min}) = \int_0^{r_{min}} \frac{r}{\sigma^2} \cdot \exp\left[-\frac{r^2 + A^2}{2\sigma^2}\right] \cdot I_0\left(\frac{rA}{\sigma^2}\right) dr \qquad 0 \leq r < \infty$$

$$= 1 - Q_M\left(\frac{A}{\sigma}, \frac{r_{min}}{\sigma}\right) \tag{5.30}$$

where $Q_M(a, b)$ is Marcum's Q-function (see also Chapter 12) given by:

$$Q_M(a, b) = e^{-\left(a^2 + b^2\right)/2} \sum_{n=0}^{\infty} \left(\frac{a}{b}\right)^n I_n(ab) \tag{5.31}$$

$I_n(\cdot)$ is the modified Bessel function of the first kind, order n. The fading margin is:

$$\frac{\overline{r^2}}{r_{\min}^2} = \frac{2\sigma^2(1 + K_r)}{r_{\min}^2} \tag{5.32}$$

The Rice power *cdf* is plotted in Fig. 5.21 as a function of the normalized envelope. The required fading margins at different K_r can be found from that figure: they are 11.5, 9.7, and 1.1 dB, for Rice factors of 0.3, 3, and 20 dB, respectively.

The presence of a dominant component also changes the *phase distribution*. This becomes intuitively clear by recalling that for a very strong dominant component, the phase of the total signal must be very close to the phase of the dominant component – in other words, the phase distribution converges to a delta function. For the general case (remember that we define the phase of the LOS component as $\psi = 0$), the pdf of the phase becomes [Rice 1947]:

$$pdf_\psi(\psi) = \frac{1}{2\pi} \exp\left(-\frac{A^2}{2\sigma^2}\right) \left[1 + \sqrt{\frac{\pi}{2}} \frac{A\cos(\psi)}{\sigma} \exp\left(\frac{A^2 \cos(\psi)^2}{2\sigma^2}\right)\right] \left[1 + \mathrm{erf}\left(\frac{A\cos(\psi)}{\sigma\sqrt{2}}\right)\right] \tag{5.33}$$

where $erf(x)$ is the error function [Abramowitz and Stegun 1965]:

$$erf(x) = (2/\sqrt{\pi}) \int_0^x \exp(-t^2)\, dt$$

Figure 5.22 shows the phase distribution for $\sigma = 1$ and different values of A.

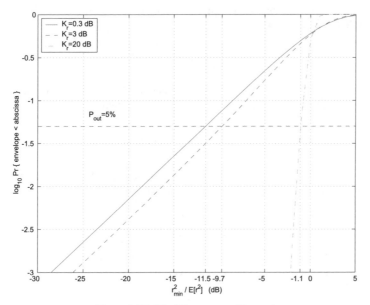

Figure 5.21 The Rice power *cdf*, $\sigma = 1$.

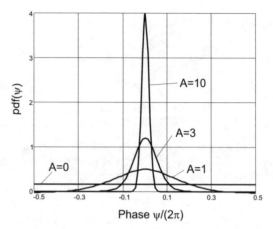

Figure 5.22 Pdf of the phase of a non-zero-mean complex Gaussian distribution, with $\sigma = 1$, $A = 0, 1, 3, 10$.

From a historical perspective, it is interesting to note that all the work about Rice distributions was performed without the slightest regard for wireless channels. The classical paper of Rice [Rice 1947] considered the problem of a sinusoidal wave in additive white Gaussian noise. However, from a mathematical point of view, this is just the problem of a deterministic phasor (giving rise to a non-zero-mean) added to a zero-mean complex Gaussian distribution – exactly the same problem as in the fieldstrength computation. Existing results thus just had to be reinterpreted by wireless engineers. This fact is so interesting because there are probably other wireless problems that can be solved by such "reinterpretation" methods.

5.5.3 Nakagami distribution

Another probability distribution for fieldstrength that is in widespread use is the Nakagami m-distribution. The pdf is given as:

$$pdf_r(r) = \frac{2}{\Gamma(m)} \left(\frac{m}{\Omega}\right)^m r^{2m-1} \exp\left(-\frac{m}{\Omega} r^2\right) \tag{5.34}$$

for $r \geq 0$ and $m \geq 1/2$; $\Gamma(m)$ is Euler's Gamma function [Abramowitz and Stegun 1965]. The parameter Ω is the mean square value $\Omega = \overline{r^2}$, and the parameter m is:

$$m = \frac{\Omega^2}{\overline{(r^2 - \Omega)^2}} \tag{5.35}$$

It is straightforward to extract these parameters from measured values. If the amplitude is Nakagami-fading, then the power follows a Gamma distribution:

$$pdf_P(P) = \frac{m}{\Omega \Gamma(m)} \left(\frac{mP}{\Omega}\right)^{m-1} \exp\left(-\frac{mP}{\Omega}\right) \tag{5.36}$$

Nakagami and Rice distribution have a quite similar shape, and one can be used to approximate the other. For $m > 1$ the m-factor can be computed from K_r by [Stueber 1996]:

$$m = \frac{(K_r + 1)^2}{(2K_r + 1)} \tag{5.37}$$

while

$$K_r = \frac{\sqrt{m^2 - m}}{m - \sqrt{m^2 - m}} \tag{5.38}$$

While Nakagami and Rice pdfs show good "general" agreement, they have different slopes close to $r = 0$. This in turn has an important impact on the achievable diversity order (see Chapter 13).

The main difference between the two pdfs is that the Rice distribution gives the *exact* distribution of the amplitude of a non-zero-mean complex Gaussian distribution – this implies the presence of one dominant component, and a large number of non-dominant components. The Nakagami distribution describes *in an approximate way* the amplitude distribution of a vector process where the central limit theorem is not necessarily valid (e.g., ultrawideband channels – see Chapter 7).

5.6 Doppler spectra

Section 5.3 showed us the physical interpretation of the frequency shift by movement – i.e., the Doppler effect. If the Mobile Station (MS) is moving, then different directions of the MPCs arriving at the MS give rise to different frequency shifts. This leads to a broadening of the received spectrum. The goal of this section is to derive this spectrum, assuming that the transmit signal is a sinusoidal signal (i.e., the narrowband case). The more general case of a wideband transmit signal will be treated in Chapter 6.

Let us first repeat the expressions for Doppler shift when a wave only comes from a single direction. Let γ denote the angle between the velocity vector $\mathbf{v}$ of the MS and the direction of the wave at the location of the MS. As shown in Eq. (5.8), the Doppler effect leads to a shift of the received frequency f by the amount ν, so that the received frequency is:

$$f = f_c \left[1 - \frac{v}{c_0} \cos(\gamma) \right] = f_c - \nu \tag{5.39}$$

where $v = |\mathbf{v}|$. Obviously, the frequency shift depends on the direction of the wave, and must lie in the range $f_c - \nu_{max} \cdots f_c + \nu_{max}$, where $\nu_{max} = f_c v / c_0$.

If there are multiple MPCs, we need to know the distribution of power of the incident waves as a function of γ. As we are interested in the *statistical* distribution of the received signal, we consider the pdf of the received power; in a slight abuse of notation, we call it the pdf of the incident waves $pdf_\gamma(\gamma)$. The MPCs arriving at the receiver are also weighted by the antenna pattern of the MS; therefore, an MPC arriving in the direction γ has to be multiplied by the pattern $G(\gamma)$. The received power spectrum as a function of direction is thus:

$$S(\gamma) = \overline{\Omega} [pdf_\gamma(\gamma) G(\gamma) + pdf_\gamma(-\gamma) G(-\gamma)] \tag{5.40}$$

where $\overline{\Omega}$ is the mean power of the arriving field. In Eq. (5.40), we have also exploited the fact that waves from the direction γ and $-\gamma$ lead to the same Doppler shift, and thus need not be distinguished for the purpose of deriving a Doppler spectrum.

In a final step, we have to perform the variable transformation $\gamma \longrightarrow \nu$. The Jacobian can be determined as:

$$\left| \frac{d\gamma}{d\nu} \right| = \left| \frac{1}{\frac{d\nu}{d\gamma}} \right| = \frac{1}{\left| \frac{v}{c_0} f_c \sin(\gamma) \right|} = \frac{1}{\sqrt{\left(f_c \frac{v}{c_0} \right)^2 - (f - f_c)^2}} = \frac{1}{\sqrt{\nu_{max}^2 - \nu^2}} \tag{5.41}$$

so that the Doppler spectrum becomes:

$$S_D(\nu) = \begin{cases} \overline{\Omega}\left[pdf_\gamma(\gamma)G(\gamma) + pdf_\gamma(-\gamma)G(-\gamma)\right]\dfrac{1}{\sqrt{\nu_{max}^2 - \nu^2}} & \text{for } -\nu_{max} \leq \nu \leq \nu_{max} \\ 0 & \text{otherwise} \end{cases} \quad (5.42)$$

For further use, we also define:

$$\Omega_n = (2\pi)^n \int_{-\nu_{max}}^{\nu_{max}} S_D(\nu)\nu^n d\nu \quad (5.43)$$

as the nth moment of the Doppler spectrum.

More specific equations can be obtained for specific angular distributions and antenna patterns. A very popular model for the angular spectrum at the MS is that the waves are incident uniformly from all azimuthal directions, and all arrive in the horizontal plane, so that:

$$pdf_\gamma(\gamma) = \frac{1}{2\pi} \quad (5.44)$$

This situation corresponds to the case when there is no LOS connection, and a large number of IOs are distributed uniformly around the MS (see also Chapter 7). Assuming furthermore that the antenna is a vertical dipole, with an antenna pattern $G(\gamma) = 1.5$ (see Chapter 9), the Doppler spectrum becomes:

$$S_D(\nu) = \frac{1.5\overline{\Omega}}{\pi\sqrt{\nu_{max}^2 - \nu^2}} \quad (5.45)$$

This spectrum, known as the *classical* or *Jakes* spectrum, is depicted in Fig. 5.23. It has the characteristic "bathtub" shape – i.e., (integrable) singularities at the minimum and maximum Doppler frequencies $\nu = \pm\nu_{max} = \pm f_c v/c_0$. These singularities thus correspond to the direction of the movement of the MS, or its opposite. It is remarkable that a uniform directional distribution can lead to a highly *non*uniform Doppler spectrum.

Naturally, *measured* Doppler spectra do not show singularities; even if the model and underlying assumptions would be strictly valid, it would require an infinite number of measurement samples to arrive at a singularity. Despite this fact, the classical Doppler

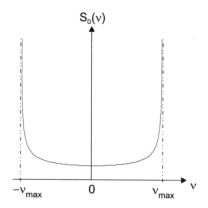

Figure 5.23 Classical Doppler spectrum.

spectrum is the most widely used model. Alternative models include:

- Aulin's model (see Parsons [1992]), which limits the amplitude of the Jakes spectrum at and near its singularities);
- Gaussian spectrum;
- uniform spectrum (this corresponds to the case when all waves are incident uniformly in all three dimensions, and the antenna has an omnidirectional pattern).

As already mentioned in Section 5.2, the Doppler spectrum has two important interpretations:

1. It describes *frequency dispersion*. For narrowband systems, as well as OFDM, such frequency dispersion can lead to transmission errors. This will be discussed in more detail in Chapters 12 and 19. It has, however, no *direct* impact on most other wideband systems (like single-carrier TDMA or CDMA systems).
2. It is a measure for the *temporal variability* of the channel. As such, it is important for *all* systems.

The temporal dependence of fading is best described by the autocorrelation function of fading. The normalized correlation between the in-phase component at time t, and the in-phase component at time $t + \Delta t$ can be shown to be:

$$\frac{\overline{I(t)I(t+\Delta t)}}{\overline{I(t)}^2} = J_0(2\pi\nu_{\max}\Delta t) \tag{5.46}$$

which is proportional to the Fourier transform of the Doppler spectrum $S_D(\nu)$, while the correlation $\overline{I(t)Q(t+\Delta t)} = 0$ for all values of Δt. The normalized correlation function of the envelope is:

$$\frac{\overline{r(t)r(t+\Delta t)}}{\overline{r(t)}^2} = J_0^2(2\pi\nu_{\max}\Delta t) \tag{5.47}$$

More details about the autocorrelation function can be found in Chapters 6 and 13.

Example 5.3   *Assume that an MS is located in a fading dip. On average, what minimum distance should the MS move so that it is no longer influenced by this fading dip?*

As a first step, we plot the envelope correlation function (Eq. 5.47) in Fig. 5.24. If we now define "no longer influenced" as "envelope correlation coefficient equals 0.5," then we see that (on average) the receiver has to move through 0.18λ. If we want complete decorrelation from the fading dip, then moving through 0.38λ is required.

5.7 Temporal dependence of fading

5.7.1 Level crossing rate

The Doppler spectrum is a complete characterization of the temporal statistics of fading. However, it is often desirable to have a different formulation that allows more direct insights into system behavior. A quantity that allows immediate interpretation is the occurrence rate of fading dips – this occurrence rate is known as the *Level Crossing Rate* (LCR). Obviously, it depends on which level we are considering (i.e., how a fading dip is defined): falling below a level

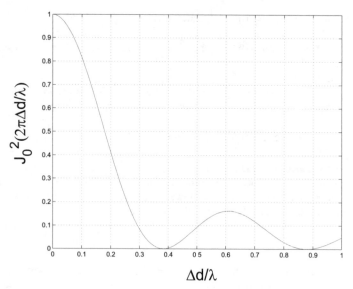

Figure 5.24 Amplitude correlation as a function of displacement of the receiver.

that is 30 dB below the mean happens more rarely than falling 3 dB below this mean. As the admissible depth of fading dips depends on the mean fieldstrength, as well as on the considered system, we want to derive the LCR for arbitrary levels (i.e., depth of fading dips).

Providing a mathematical formulation, the LCR is defined as the expected value of the rate at which the received fieldstrength crosses a certain level r in the positive direction. This can also be written as:

$$N_R(r) = \int_0^\infty \dot{r} \cdot pdf_{r,\dot{r}}(r, \dot{r}) d\dot{r} \tag{5.48}$$

where $\dot{r} = dr/dt$ is the temporal derivative, and $pdf_{r,\dot{r}}$ is the joint pdf of r and $\dot{r}$.

In Appendix 5.C (see *www.wiley.com/go/molisch*), we derive the LCR as:

$$N_R(r) = \sqrt{\frac{\Omega_2}{\pi\Omega_0}} \frac{r}{\sqrt{2\Omega_0}} \exp\left(-\frac{r^2}{2\Omega_0}\right) \tag{5.49}$$

Note that $\sqrt{2\Omega_0}$ is the root-mean-square value of amplitude. Figure 5.25 shows the LCR for a Rayleigh-fading amplitude and Jakes Doppler spectrum.

5.7.2 Average duration of fades

Another parameter of interest is the *Average Duration of Fades* (ADF). In previous sections, we have already derived the rate at which the fieldstrength goes below the considered threshold (i.e., the LCR), and the total percentage of time the fieldstrength is lower than this threshold (i.e., the cdf of the fieldstrength). The average duration of fades can be simply computed as the quotient of these two quantities:

$$ADF(r) = \frac{cdf_r(r)}{N_R(r)} \tag{5.50}$$

A plot of the ADF is shown in Fig. 5.26.

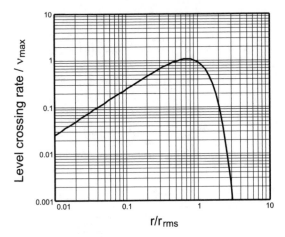

Figure 5.25 Level crossing rate normalized to the maximum Doppler frequency as a function of the normalized level r/r_{rms}, for a Rayleigh-fading amplitude and Jakes spectrum.

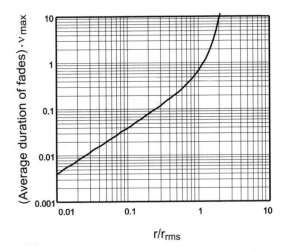

Figure 5.26 Average duration of fades normalized to the maximum Doppler frequency as a function of the normalized level r/r_{rms}, for a Rayliegh-fading amplitude and Jakes spectrum.

Example 5.4 *Assume a multipath environment where the received signal has a Rayleigh distribution and the Doppler spectrum has the classical bathtub (Jakes) shape. Compute the LCR and the ADF for a maximum Doppler frequency* $v_{max} = 50\,Hz$, *and amplitude thresholds:*

$$r_{min} = \frac{\sqrt{2\Omega_0}}{10}, \frac{\sqrt{2\Omega_0}}{2}, \sqrt{2\Omega_0}$$

The LCR is computed from Eq. (5.49). For a Jakes scenario the second moment of the Doppler spectum is given as:

$$\Omega_2 = \tfrac{1}{2}\Omega_0 (2\pi v_{max})^2 \qquad (5.51)$$

therefore

$$N_R(r_{min}) = \sqrt{2\pi} \cdot v_{max} \cdot \frac{r_{min}}{\sqrt{2\Omega_0}} \exp\left(-\frac{r_{min}^2}{2\Omega_0}\right) \tag{5.52}$$

The average duration of fade is computed from Eq. (5.50) where:

$$cdf(r_{min}) = 1 - \exp\left(-\left(\frac{r_{min}}{\sqrt{2\Omega_0}}\right)^2\right) \tag{5.53}$$

These expressions are evaluated for different values of the threshold, r_{min}. The results are tabulated in Table 5.1.

Table 5.1   Effect of threshold on ADF and LCR.

r_{min}	$N_R(r_{min})$	$cdf(r_{min})$	ADF (msec)
$\dfrac{\sqrt{2\Omega_0}}{10}$	12.4	0.01	0.8
$\dfrac{\sqrt{2\Omega_0}}{2}$	48.8	0.22	4.5
$\sqrt{2\Omega_0}$	46.1	0.63	13.7

5.7.3 Random frequency modulation

A random channel leads to a random phaseshift of the received signal; in a time-variant channel, these phaseshifts are time-variant as well. By definition, a temporally varying phaseshift is a FM. In this section, we will compute this *random FM*.

The pdf of the instantaneous frequency $\dot{\psi}$ can be computed from the joint pdf $pdf_{r,\dot{r},\psi,\dot{\psi}}$ of Appendix 5.C (see *www.wiley.com/go/molisch*). We now have to integrate over the variables $r, \dot{r},$ and ψ. This results in:

$$pdf_{\dot{\psi}}(\dot{\psi}) = \frac{1}{2}\sqrt{\frac{\Omega_0}{\Omega_2}}\left(1 + \frac{\Omega_0}{\Omega_2}\dot{\psi}^2\right)^{-3/2} \tag{5.54}$$

This is a "student's t-distribution" with two degrees of freedom [Mardia et al. 1979]. The cumulative distribution is given as:

$$cdf_{\dot{\psi}}(\dot{\psi}) = \frac{1}{2}\left[1 + \sqrt{\frac{\Omega_0}{\Omega_2}}\dot{\psi}\left(1 + \frac{\Omega_0}{\Omega_2}\dot{\psi}^2\right)^{-1/2}\right] \tag{5.55}$$

It is remarkable that all instantaneous frequencies in a range $-\infty$ to ∞ can occur – they are not restricted to the range of possible Doppler frequencies!

Strong FM most likely occurs in fading dips: if the signal level is low, very strong relative changes can occur. This intuitively pleasing result can be shown mathematically by considering the pdf of the instantaneous frequency conditioned on the amplitude level r_0 [Jakes 1974]:

$$pdf(\dot{\psi}|r_0) = \frac{r_0}{\sqrt{2\pi\Omega_2}} \exp\left(-r_0^2 \frac{\dot{\psi}^2}{2\Omega_2}\right) \tag{5.56}$$

This is a zero-mean Gaussian distribution with variance Ω_2/r_0^2. The variance is the smaller the larger the signal level.

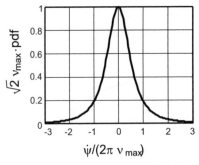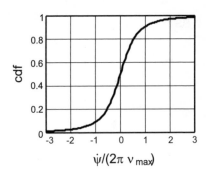

Figure 5.27   Normalized pdf (left-hand side) and cdf (right-hand side) of a random FM.

We thus see that fading dips can create errors in two ways: on one hand, the lower signal level leads to higher susceptibility to noise. On the other hand, they increase the probability of strong random FMs, which introduces errors in any system that conveys information by means of the *phase* of the transmitted signal. Fading dips are also related to intersymbol interference (see Chapter 12).

5.8 Large-scale fading

Small-scale fading, created by the superposition of different multipath components, changes rapidly over the spatial scale of a few wavelengths. If the fieldstrength is averaged over a small area (e.g., ten by ten wavelengths), we obtain the *Small Scale Averaged* (SSA) fieldstrength.[6] In previous sections, we treated the SSA fieldstrength as a constant. However, as explained in the introduction, it varies when considered on a larger spatial scale.

Many experimental investigations have shown that the SSA fieldstrength F, *plotted on a logarithmic scale*, shows a Gaussian distribution around a mean μ. Such a distribution is known as lognormal, and its pdf is:

$$pdf_F(F) = \frac{20/\ln(10)}{F\sigma_F\sqrt{2\pi}} \cdot \exp\left[-\frac{(20\log_{10}(F) - \mu_{dB})^2}{2 \cdot \sigma_F^2}\right] \tag{5.57}$$

where σ_F is the standard deviation of F, and μ_{dB} is the mean of the values of F expressed in dB.[7] Typical values of σ_F are 4 to 10 dB. Another interesting property of the lognormal distribution is that the sum of lognormally distributed values is also approximately lognormally distributed. More details on how to compute the mean and the variance of the sum of such variables can be found in Stueber [1996].

The lognormal variations of F have commonly been attributed to shadowing effects. Consider the situation in Fig. 5.28. If the MS moves, it changes the angle γ, and thus the diffraction parameter ν_F. If the distance between the edge and the MS is now large, the MS must move over a large distance (e.g., several tens of wavelengths) for the fieldstrength to change noticeably. Note that this effect changes the absolute amplitude of the MPC, $|a|$, and has nothing to do with

---

[6] Strictly speaking, this is only an approximation to the SSA fieldstrength, as there can only be a finite number of statistically independent sample values within the finite area over which we average.

[7] Also, power is distributed lognormally. However, there is an important fine point: when fitting the logarithm of small-scale averaged *power* to a normal distribution, we find that the median value of this distribution, $\mu_{P,dB}$, is related to the median of the fieldstrength distribution μ_{dB} as $\mu_{P,dB} = \mu_{dB} + 10\log(4/\pi)$ dB.

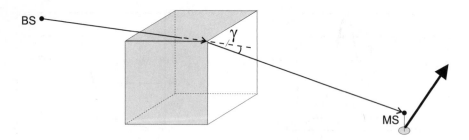

Figure 5.28   Shadowing by a building.

an interference effect. Consider now the situation where an MPC undergoes several of these or similar processes on its way from the transmitter and receiver, each of which contributes a certain attenuation. The received fieldstrength then depends on the *product* of the attenuations; on a logarithmic scale, these attenuations add up. We can thus model the effect as the sum of random variables *on a dB scale* – i.e., a lognormal distribution. However, the mechanism just described is not necessarily valid in all physical situations. An alternative explanation, based on double-scattering processes, has recently been proposed [Andersen 2002]. But, independently of the mechanism that leads to its creation, many measurements have confirmed that the pdf of F is well-approximated by a lognormal function.

Finally, consider the statistics of the fieldstrength based on samples taken from a large area (i.e., including both lognormal fading and interference effects). The probability density function of these samples is given by the so-called Suzuki distribution, which follows in a straightforward manner from the laws of conditional statistics. The expected value of the fieldstrength, taken over a small area, is $\sigma\sqrt{\pi/2}$, and the distribution of the local value of the fieldstrength conditioned on σ is:

$$pdf(r|\sigma) = \frac{\pi r}{2\sigma^2}\exp\left(-\frac{\pi r^2}{4\sigma^2}\right) \tag{5.58}$$

The local mean (i.e., the expected value used above) is distributed according to lognormal statistics. Unconditioning results in the pdf of the fieldstrength:

$$pdf_r(r) = \int_0^\infty \frac{\pi r}{2\sigma^2}\exp\left(-\frac{\pi r^2}{4\sigma^2}\right)\frac{20/\ln(10)}{\sigma\,\sigma_F\sqrt{2\pi}}\exp\left(-\frac{(20\log_{10}(\sigma)-\mu_{dB})^2}{2\sigma_F^2}\right)d\sigma \tag{5.59}$$

This pdf is also known as the Suzuki distribution, an example is shown in Fig. 5.29. A similar function can be defined if the small-scale statistics are Rician or Nakagami.

Since both large-scale and small-scale fading occurs in practical situations, the fading margin must account for the combination of the two effects (see also Chapter 3). One possibility is to just add up the fading margin for the Rayleigh distribution and the fading margin for a lognormal distribution. This method is commonly used because of its simplicity, but overestimates the required fading margin. The more accurate method is based on the cdf of the Suzuki distribution, which can be obtained by integrating Eq. (5.59) from $-\infty$ to x. This then allows computation of the necessary mean fieldstrength r_0 for a given admissible outage probability (e.g., 0.05), as shown in the following example.

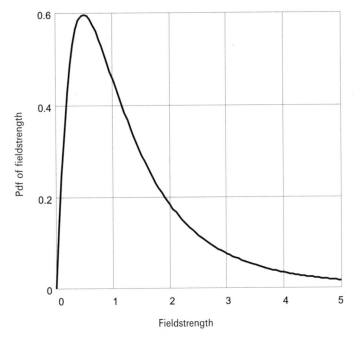

Figure 5.29 Suzuki distribution with $\sigma_F = 5.6\,\mathrm{dB}$, $\mu = 1.4\,\mathrm{dB}$.

Example 5.5 *Consider a channel with $\sigma_F = 6\,dB$ and $\mu = 0\,dB$. Compute the fading margin for a Suzuki distribution so that the outage probability is smaller than 5%. Also compute the fading margin for Rayleigh fading and shadow fading separately.*

For the Suzuki distribution:

$$P_{\mathrm{out}} = cdf(r_{\mathrm{min}}) = \int\limits_{0}^{r_{\mathrm{min}}} pdf_r(r)dr$$

$$= \int\limits_{0}^{\infty} \left(1 - \exp\left(-\frac{\pi r_{\mathrm{min}}^2}{4\sigma^2}\right)\right) \frac{20/\ln(10)}{\sigma\,\sigma_F\sqrt{2\pi}} \exp\left(-\frac{(20\log_{10}\sigma - \mu_{dB})^2}{2\sigma_F^2}\right) d\sigma \qquad (5.60)$$

Inserting $\sigma_F = 6\,\mathrm{dB}$ and $\mu_{dB} = 0$, the *cdf* is expressed as:

$$cdf(r_{\mathrm{min}}) = \frac{20/\ln 10}{\sqrt{2\pi}\cdot 6} \int_0^\infty \frac{1}{\sigma} \exp\left(-\frac{(20\log\sigma - 0)^2}{2\cdot 36}\right)\left(1 - \exp\left(-\frac{\pi}{4}\cdot\frac{r_{\mathrm{min}}^2}{\sigma^2}\right)\right) d\sigma \qquad (5.61)$$

The fading margin is defined as:

$$M = \frac{\mu^2}{r_{\mathrm{min}}^2} \qquad (5.62)$$

The *cdf* plot is obtained by evaluating Eq. (5.61) for different values of r_{min}, the result is shown in Fig. 5.30). A fading margin of 15.5 dB is required to get an outage probability of 5%.

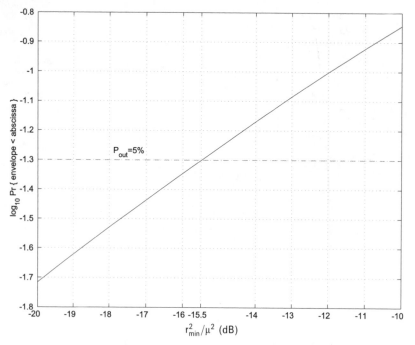

Figure 5.30   The Suzuki cdf, $\sigma_F = 6\,\mathrm{dB}$ and $\mu_{dB} = 0$.

The outage probability for Rayleigh fading only is evaluated as:

$$P_{\mathrm{out}} = cdf(r_{\min}) = 1 - \exp\left(-\frac{r_{\min}^2}{2\sigma^2}\right)$$

$$= 1 - \exp(-1/M) \tag{5.63}$$

After some manipuation, we get:

$$M_{\mathrm{Rayleigh,dB}} = -10\log_{10}(-\ln(1 - P_{\mathrm{out}}))$$

$$= 12.9\,\mathrm{dB} \tag{5.63}$$

For shadow fading, the pdf of the fieldstrength values in dB is a standard Gaussian distribution. The cdf is thus given by a Q-function defined as:

$$Q(a) = \frac{1}{\sqrt{2\pi}}\int_a^\infty \exp\left(-\frac{x^2}{2}\right)dx$$

The outage probability is given as:

$$P_{\mathrm{out}} = Q\left(\frac{M_{\mathrm{large-scale_{dB}}}}{\sigma_F}\right) \tag{5.65}$$

Inserting $\sigma_F = 6$ dB and $P_{\mathrm{out}} = 0.05$ into Eq. (5.65) we get:

$$M_{\mathrm{large-scale_{dB}}} = 6 \cdot Q^{-1}(0.05)$$

$$= 9.9\,\mathrm{dB} \tag{5.66}$$

When computing the fading margin as the sum of the margin for Rayleigh fading and shadowing, we obtain:

$$M_{dB} = 12.9 + 9.9 \, dB = 22.8 dB \qquad (5.67)$$

Compared with the fading margin obtained from the Suzuki distribution, this is a more conservative estimate.

5.9 Appendices

Please see companion website *www.wiley.com/go/molisch*

Further reading

The statistical basis for the derivations in this chapter is described in a number of standard textbooks on statistics and random processes, especially the classical book of Papoulis [1991]. The statistical model for the amplitude distribution and the Doppler spectrum was first described in Clarke [1968]. A comprehensive exposition of the statistics of the channel, derivation of the Doppler spectrum, LCR, and ADFs, for the Rayleigh case, can be found in Jakes [1974]. Since Rayleigh fading is based on Gaussian fading of the I- and Q-component, the rich literature on Gaussian multivariate analysis is applicable [Muirhead 1982]. Derivation of the Nakagami distribution, and many of its statistical properties, can be found in Nakagami [1960]; a physical interpretation is given in [Braum and Dersch 1991]; the Rice distribution is derived in Rice [1947]. The Suzuki distribution is derived in Suzuki [1977]. More details about the lognormal distribution, especially the summing of several lognormally distributed variables, can be found in Stueber [1996] and Cardieri and Rappaport [2001]. The combination of Nakagami small-scale fading and lognormal shadowing is treated in Thjung and Chai [1999]. The fading statistics, including ADF and LCR, of the Nakagami and Rice distributions are summarized in Abdi et al. [2000]. Another important aspect of statistical channel descriptions is the generation of random variables according to a prescribed Doppler spectrum. An extensive description of this area can be found in Paetzold [2002].

6

Wideband and directional channel characterization

6.1 Introduction

In the previous chapter, we considered the effect of multipath propagation on the received fieldstrength and the temporal variations of a sinusoidal transmit signal. These considerations are also valid for all systems where the bandwidth of the transmitted signal is "very small" (see below for a more precise definition). However, most current and future wireless systems use a large bandwidth, either because they are intended for high data rates, or because of their multiple access scheme (see Chapters 17–19). We therefore have to describe variations of the channel over a large bandwidth as well. These description methods are the topic of this chapter.

The impact of multipath propagation in wideband systems can be interpreted in two different ways: (i) the transfer function of the channel varies over the bandwidth of interest (this is called the *frequency selectivity* of the channel), or (ii) the impulse response of the channel is not a delta function; in other words, the arriving signal has a longer duration than the transmitted signal (this is called *delay dispersion*). These two interpretations are equivalent, as can be shown by performing Fourier transformations to move from the delay (time) domain to the frequency domain.

In this chapter, we first explain the basic concepts of wideband channels using again the most simple channel – namely, the two-path channel. We then formulate the most general statistical description methods for wideband, time-variant channels (Section 6.3), and discuss its most common special form, the WSSUS – Wide Sense Stationary Uncorrelated Scatterer – model (Section 6.4). Since these description methods are rather complicated, condensed parameters are often used for a more compact description (Section 6.5). Section 6.6 considers the case when the channel is not only wide enough to show appreciable variations of the transfer function, but even so wide that the bandwidth becomes comparable with the carrier frequency – this case is called "ultrawideband".

Systems operating in wideband channels have some important properties:

- They suffer from intersymbol interference. This can be most easily understood from the interpretation of delay dispersion. If we transmit a symbol of length T_S, the arriving signal corresponding to that symbol has a longer duration, and therefore interferes with the subsequent symbol (Chapter 2). Section 12.3 describes the effect of this Inter Symbol Interference (ISI) on the bit error rate if no further measures are taken; Chapter 16 describes equalizer structures that can actively combat the detrimental effect of the ISI.
- They can reduce the detrimental effect of fading. This effect can be most easily understood in the frequency domain: even if some part of the transmit spectrum is strongly attenuated,

there are other frequencies that do not suffer from attenuation. Appropriate coding and signal processing can exploit this fact, as we will explain in Chapters 16, 18, and 19.

The properties of the channel can vary not only depending on the frequency at which we consider it, but also depending on the location. This latter effect is related to the directional properties of the channel – i.e., the directions from which the Multi Path Components (MPCs) are incident. Section 6.7 discusses the stochastic description methods for these directional properties; they are especially important for antenna diversity (Chapter 13) and multielement antennas (Chapter 20).

6.2 The causes of delay dispersion

6.2.1 The two-path model

Why are there variations of the channel over a given frequency range? The most simple picture arises again from the two-path model, as introduced at the beginning of Chapter 5. The transmit signal gets to the receiver via two different propagation paths with different runtimes:

$$\tau_1 = d_1/c_0 \quad \text{and} \quad \tau_2 = d_2/c_0 \tag{6.1}$$

We assume now that runtimes do not change with time (this occurs when neither transmitter (TX), receiver (RX), nor IOs, move). Consequently, the channel is linear and time-invariant, and has an impulse response:

$$h(\tau) = a_1 \, \delta(\tau - \tau_1) + a_2 \, \delta(\tau - \tau_2) \tag{6.2}$$

where again the complex amplitude $a = |a| \exp(j\varphi)$. A Fourier transformation of the impulse response gives the transfer function $H(j\omega)$:

$$H(f) = \int_{-\infty}^{\infty} h(\tau) \, \exp[-j2\pi f\tau] \, d\tau = a_1 \, \exp[-j2\pi f\tau_1] + a_2 \, \exp[-j2\pi f\tau_2] \tag{6.3}$$

The magnitude of the transfer function is:

$$|H(f)| = \sqrt{|a_1|^2 + |a_2|^2 + 2|a_1||a_2|\cos(2\pi f \cdot \Delta\tau - \Delta\varphi)}$$

$$\text{with } \Delta\tau = \tau_2 - \tau_1 \text{ and } \Delta\varphi = \varphi_2 - \varphi_1 \tag{6.4}$$

Figure 6.1 shows the transfer function for a typical case. We observe first that the transfer function depends on the frequency, so that we have *frequency selective fading*. We also see that there are dips (*notches*) in the transfer function at the so-called *notch frequencies*. In the two-path model, the notch frequencies are those frequencies where the phase difference of the two arriving waves becomes 180°. The frequency difference between two adjacent notch frequencies is:

$$\Delta f_{\text{Notch}} = \frac{1}{\Delta\tau} \tag{6.5}$$

The destructive interference between the two waves is stronger the more similar the amplitudes of the two waves are.

Channels with fading dips distort not only the amplitude, but also the phase of the signal. This can be best seen by considering the group delay, which is defined as the derivative of the phase of the channel transfer function $\phi_H = \arg(H(f))$:

$$\tau_{\text{Gr}} = -\frac{1}{2\pi} \frac{d\phi_H}{df} \tag{6.6}$$

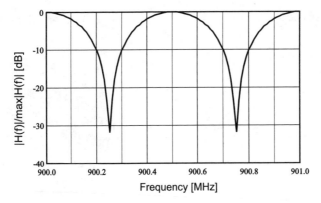

Figure 6.1 Normalized transfer function for $|a_1| = 1.0$, $|a_2| = 0.95$, $\Delta\varphi = 0$, $\tau_1 = 4\,\mu s$, $\tau_2 = 6\,\mu s$ at the 900-MHz carrier frequency.

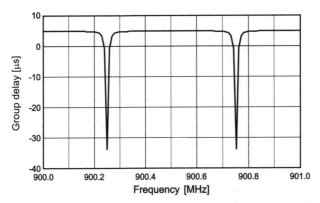

Figure 6.2 Group delay as a function of frequency (same parameters as in Fig. 6.1).

As can be seen in Fig. 6.2 group delay can become very large in fading dips. As we will see later, this group delay can be related to intersymbol interference.

6.2.2 The general case

After the simple two-path model, we now progress to the more general case where Interacting Objects (IOs) can be at any place in the plane. Again, the scenario is static, so that neither TX, RX, nor IOs move. We now draw the ellipses that are defined by their focal points – TX and RX – and the eccentricity determining the runtime.[1] All rays that undergo a single interaction with an object on a specific ellipse arrive at the receiver at the same time. Signals that interact with objects on different ellipses arrive at different times. Thus, the channel is *delay-dispersive* if the IOs in the environment are not all located on a single ellipse.

It is immediately obvious that in a realistic environment, IOs never lie exactly on a single ellipse. The next question is thus: How strict must this "single ellipse" condition be fulfilled so that the channel is still "effectively" non-dispersive? The answer depends on the system

---

[1] These ellipses are thus quite similar to the Fresnel ellipses described in Chapter 4. The difference is that in Chapter 4 we were interested in excess runtimes that introduce a phaseshift of $i \cdot \pi$, while here we are interested in delays that are typically much larger.

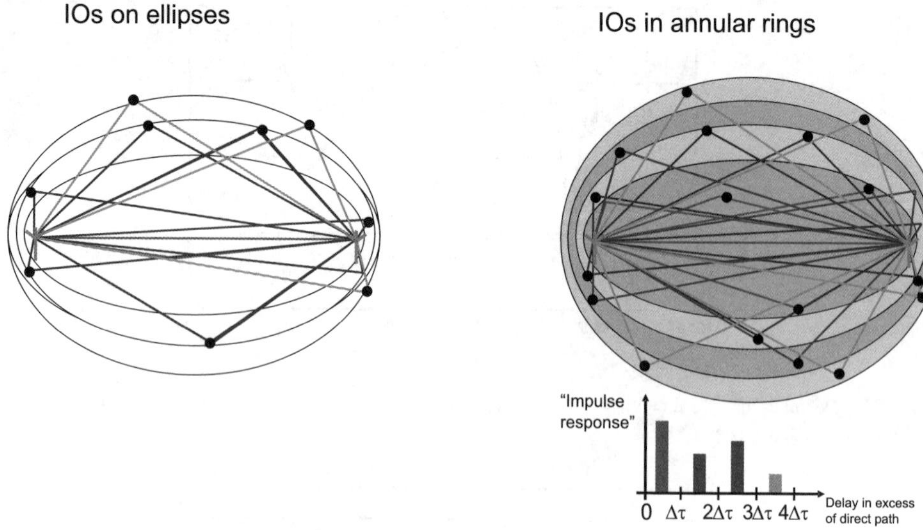

IOs on ellipses   IOs in annular rings

Figure 6.3   Scatterers located on the same ellipses lead to the same delays.

bandwidth. A receiver with bandwidth W cannot distinguish between echoes arriving at τ and $\tau + \Delta\tau$, if $\Delta\tau \ll 1/W$ (for many qualitative considerations, it is sufficient to consider the above condition with $\Delta\tau = 1/W$). Thus echoes that are reflected in the donut-shaped region corresponding to runtimes between τ and $(\tau + \Delta\tau)$ arrive at "effectively" the same time (see Fig. 6.3).

A time-discrete approximation to the impulse response of a wideband channel can thus be obtained by dividing the impulse response into bins of width $\Delta\tau$ and then computing the sum of echoes within each bin. If enough non-dominant IOs are in each donut-shaped region, then the MPCs falling into each delay bin fulfill the central limit theorem. In that case, the amplitude of each bin can be described statistically, and the probability density function (pdf) of this amplitude is Rayleigh or Rician. Thus, all the equations of Chapter 5 are still valid; but now they apply for the fieldstrength *within* one delay bin. We furthermore define the minimum delay as the runtime of the direct path between the Base Station (BS) and the MS $c_0 \cdot d$, and we define the maximum delay as the runtime from the BS to the MS via the farthest "significant" IO – i.e., the farthest IO that gives a measurable contribution to the impulse response.[2] The *maximum excess delay* $\tau_{\max}$ is then defined as the difference between minimum and maximum delay.

The above considerations also lead us to a mathematical formulation for *narrowband* and *wideband* from a time domain point of view: a system is narrowband if the inverse of the system bandwidth $1/W$ is much larger than the maximum excess delay $\tau_{\max}$. In that case, all echoes fall into a single delay bin, and the amplitude of this delay bin is $\alpha(t)$. A system is wideband in all other cases. In a wideband system, the *shape* and duration of the arriving signal is different from the shape of the transmitted signal; in a narrowband system, they stay the same.

If the impulse response has a finite extent in the delay domain, it follows from the theory of Fourier transforms that the transfer function $\mathcal{F}\{h(\tau)\} = H(f)$ is frequency-dependent. Delay dispersion is thus equivalent to *frequency selectivity*. A frequency-selective channel cannot be described by a simple attenuation coefficient, but rather the details of the transfer function must be modeled. Note that any real channel is frequency-selective if analyzed over a large enough

---

[2] We see from this definition that the maximum delay is a quantity that is extremely difficult to measure, and depends on the measurement system.

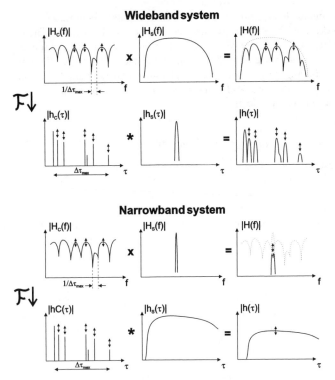

Figure 6.4 Narrow- and wideband systems. $H_C(f)$ .... channel transfer function; $h_C(\tau)$ .... channel impulse response.

bandwidth; in practice, the question is whether this is true over the bandwidth of the considered system. This is equivalent to comparing the maximum excess delay of the channel impulse response with the inverse system bandwidth. Figure 6.4 sketches these relationships, demonstrating the variations of wideband systems in the delay and frequency domain.

We stress that the definition of a wideband wireless system is fundamentally different from the definition of "wideband" in the usage of Radio Frequency (RF) engineers. The RF definition of wideband implies that the system bandwidth becomes comparable with carrier frequency.[3] In wireless communications, on the other hand, we compare the properties of the channel with the properties of the system. It is thus possible that the same system is wideband in one channel, but narrowband for another.

6.3 System-theoretic description of wireless channels

As we have seen in the previous section, a wireless channel can be described by an impulse response; it thus can be interpreted as a linear filter. If the BS, MS, and IOs are all static, then the channel is time-invariant, with an impulse response $h(\tau)$. In that case, the well-known theory of LTI (*Linear Time Invariant*) systems [Oppenheim and Schaefer 1985] is applicable. In general,

---

[3] In wireless communications, it has become common to denote systems whose bandwidth is larger than 20% of the carrier frequency as *Ultra Wide Bandwidth* (UWB) systems (see Section 6.6). This definition is similar to the RF definition of wideband.

however, wireless channels are time-variant, with an impulse response $h(t, \tau)$ that changes with time; we have to distinguish between the absolute time t and the delay τ. Thus, the theory of the *Linear Time Variant* (LTV) system must be used. This is not just a trivial extension of the LTI theory, but gives rise to considerable theoretical challenges, and causes the breakdown of many intuitive concepts. Fortunately, most wireless channels can be classified as *slowly* time-variant systems, also known as *quasi-static*. In that case, many of the concepts of LTI systems can be retained with only minor modifications.

6.3.1 Characterization of deterministic linear time-variant systems

As the impulse response of a time-variant system, $h(t, \tau)$, depends on two variables, τ and t, we can perform Fourier transformations with respect to either (or both) of them. This results in four different, but equivalent, representations. In this section, we will investigate these representations, their advantages, and drawbacks.

From a system-theoretic point of view, it is most straightforward to write the relationship between the system input (transmit signal) $x(t)$ and the system output (received signal) $y(t)$ as:

$$y(t) = \int_{-\infty}^{\infty} x(\tau) K(t, \tau) d\tau \tag{6.7}$$

where $K(t, \tau)$ is the *kernel* of the integral equation, which can be related to the impulse response. For LTI systems, the well-known relationship $K(t, \tau) = h(t - \tau)$ holds. Generally, we define the time-variant impulse response:

$$h(t, \tau) = K(t, t - \tau) \tag{6.8}$$

so that

$$y(t) = \int_{-\infty}^{\infty} x(t - \tau) h(t, \tau) d\tau \tag{6.9}$$

We can now perform a Fourier transformation with respect to the variable t, τ, or both.

An intuitive interpretation is possible if the impulse response changes only slowly with time – more exactly, the duration of the impulse response (and the signal) should be much shorter than the time over which the channel changes significantly. Then we can consider the behavior of the system at one time t like that of an LTI system. The variable t can thus be viewed as "absolute" time that tells us which impulse response $h(\tau)$ is currently valid. Such a system is also called *quasi-static*.

Fourier transforming the impulse response with respect to the variable τ results in the *time-variant transfer function* $H(t, f)$:

$$H(t, f) = \int_{-\infty}^{\infty} h(t, \tau) \exp(-j2\pi f \tau) d\tau \tag{6.10}$$

The input–output relationship is given by:

$$y(t) = \int_{-\infty}^{\infty} X(f) H(t, f) \exp(j2\pi f t) df \tag{6.11}$$

The interpretation is straightforward for the case of the quas-istatic system – the spectrum of the input signal is multiplied by the spectrum of the "currently valid" transfer function, to give the spectrum of the output signal. If, however, the channel is quickly time-varying, then Eq. (6.11) is a purely mathematical relationship. The spectrum of the output signal is given by a double integral

$$Y(\tilde{f}) = \int_{-\infty}^{\infty} \int_{-\infty}^{\infty} X(f) H(t, f) \exp(j2\pi f t) \exp(-j2\pi \tilde{f} t) df \, dt \tag{6.12}$$

which does *not* reduce to $Y(f) = H(f) X(f)$ [Matz and Hlawatsch 1998].

A Fourier transformation with respect to t results in a different representation – namely, the *delay Doppler function*, better known as *spreading function* $S(\nu, \tau)$:

$$S(\nu, \tau) = \int_{-\infty}^{\infty} h(t, \tau) \exp(-j2\pi\nu t)\, dt \tag{6.13}$$

This function describes the spreading of the input signal in the delay and Doppler domains.

Finally, the function $S(\nu, \tau)$ can be transformed with respect to the variable τ, resulting in the *Doppler-variant transfer function* $B(\nu, f)$:

$$B(\nu, f) = \int_{-\infty}^{\infty} S(\nu, \tau) \exp(-j2\pi f\tau)\, d\tau \tag{6.14}$$

A summary of the interrelations between the system functions is given in Fig. 6.5. Figure 6.6 shows an example of a measured impulse response; Figure 6.7 shows the spreading function computed from it.

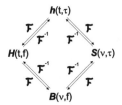

Figure 6.5   Interrelation between deterministic system functions.

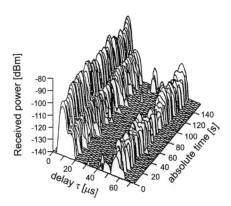

Figure 6.6   Squared magnitude of the impulse response $|h(t, \tau)|^2$ measured in hilly terrain near Darmstadt, Germany. Measurement duration 140 s; center frequency 900 MHz. τ denotes the excess delay.

6.3.2 Stochastic system functions

We now return to the stochastic description of wireless channels. Interpreting them as time-variant stochastic systems, a complete description requires the multidimensional pdf of the impulse response – i.e., the joint pdf of the complex amplitudes at all possible values of delay and time. However, this is usually much too complicated in practice. Instead, we restrict our attention to a second-order description – namely, the *Auto Correlation Function* (ACF).

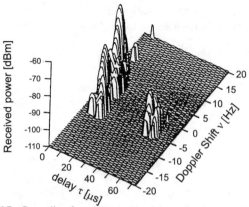

Figure 6.7 Spreading function computed from the data of Fig. 6.6.

Reproduced with permission from U. Liebenow.

Let us first repeat some facts about the ACFs of one-dimensional stochastic processes (i.e., processes that depend on a single parameter t). The ACF of a stochastic process y is defined as:

$$R_{yy}(t, t') = E\{y^*(t)y(t')\} \tag{6.15}$$

where the expectation is taken over the *ensemble of possible realizations* (for the definition of this ensemble see Appendix 6.A at *www.wiley.com/go/molisch*). The ACF describes the relationship between the second-order moments of the amplitude pdf of the signal y at different times. If the pdf is zero-mean Gaussian, then the second-order description contains all the required information. If the pdf is non-zero-mean Gaussian, the mean

$$\bar{y}(t) = E\{y(t)\} \tag{6.16}$$

together with the autocovariance function

$$\tilde{R}_{yy}(t, t') = E\{[y(t) - \bar{y}(t)]^*[y(t') - \bar{y}(t')]\} \tag{6.17}$$

constitutes a complete description. If the channel is non-Gaussian, then the first- and second-order statistics are not a complete description of the channel. In the following, we will mainly concentrate on zero-mean Gaussian channels.

Let us now revert to the problem of giving a stochastic description of the channel. Inserting the input–output relationship into Eq. (6.15), we obtain the following expression for the ACF of the received signal:

$$R_{yy}(t, t') = E\left\{ \int_{-\infty}^{\infty} x^*(t - \tau)h^*(t, \tau)\, d\tau \int_{-\infty}^{\infty} x(t' - \tau')h(t', \tau')\, d\tau' \right\} \tag{6.18}$$

The system is linear, so that expectation can be interchanged with integration. Furthermore, the transmit signal can be interpreted as a stochastic process that is independent of the channel, so that expectations over the transmit signal and over the channel can be performed independently. Thus, the ACF of the received signal is:

$$\begin{aligned} R_{yy}(t, t') &= \int_{-\infty}^{\infty} \int_{-\infty}^{\infty} E\{x^*(t - \tau)x(t' - \tau')\}E\{h^*(t, \tau)h(t', \tau')\}\, d\tau\, d\tau' \\ &= \int_{-\infty}^{\infty} \int_{-\infty}^{\infty} R_{xx}(t - \tau, t' - \tau')R_h(t, t', \tau, \tau')\, d\tau\, d\tau' \end{aligned} \tag{6.19}$$

i.e., a combination of the ACF of the transmit signal and the ACF of the channel:

$$R_h(t, t', \tau, \tau') = E\{h^*(t, \tau)h(t', \tau')\} \tag{6.20}$$

Note that the ACF of the channel depends on four variables since the underlying stochastic process is two-dimensional.

We observe a formal similarity of the channel ACF to the impulse response of a deterministic channel: we can form stochastic system functions by Fourier transformations. In contrast to the deterministic case, we now have to perform a *double* FT, with respect to the pair of variables t, t' and/or τ, τ'. From that, we obtain in an elementary way the relationships between the different formulations of the ACFs of input and output – e.g., $R_s(\nu, \nu', \tau, \tau') = E\{s^*(\nu, \tau)s(\nu', \tau')\} = \int\int R_h(t, t', \tau, \tau') \cdot \exp(-j2\pi\nu t) \exp(+j2\pi\nu' t') \, dt \, dt'$.

6.4 The WSSUS model

The correlation functions depend on four variables, and are thus a rather complicated form for the characterization of the channel. Further assumptions about the physics of the channel can lead to a simplification of the correlation function. The most frequently used assumptions are the so-called WSS (*Wide Sense Stationary*) assumption and the U.S. (*Uncorrelated Scatterers*) assumption. A model using both assumptions simultaneously is called a *WSSUS* model.

6.4.1 Wide-sense stationarity

The mathematical definition of *wide-sense stationarity* is that the ACF depends not on the two variables t, t' separately, but only on their difference $t - t'$. Consequently, *second-order amplitude statistics* do not change with time.[4] We can thus write:

$$R_h(t, t', \tau, \tau') = R_h(t, t + \Delta t, \tau, \tau') = R_h(\Delta t, \tau, \tau') \tag{6.21}$$

Physically speaking, wide-sense stationarity means that the *statistical properties* of the channel do not change with time. This must not be confused with a static channel, where fading *realizations* do not change with time. For the simple case of a flat Rayleigh-fading channel, WSS means that the mean power and the Doppler spectrum do not change with time, while the instantaneous amplitude can change.

According to the mathematical definition, WSS has to be fulfilled for any arbitrary time t. In practice, this is not possible: as the MS moves over larger distances, the mean received power changes because of shadowing and variations in pathloss. Rather, WSS is typically fulfilled over an area of about 10λ diameter (compare also Section 5.1). We can thus define quasi-stationarity over a finite time interval (associated with a movement distance of the MS), over which statistics do not change noticeably.

The Doppler-variant impulse response $s(\nu, \tau)$ provides a further interpretation of the WSS assumption. Inserting Eq. (6.21) into the definition of R_s, we get:

$$R_s(\nu, \nu', \tau, \tau') = \int_{-\infty}^{\infty} \int_{-\infty}^{\infty} R_h(t, t + \Delta t, \tau, \tau') \exp[2\pi j(\nu t - \nu'(t + \Delta t))] \, dt \, dt' \tag{6.22}$$

which can be rewritten as:

$$R_s(\nu, \nu', \tau, \tau') = \int_{-\infty}^{\infty} \exp[2\pi j t(\nu - \nu')] \, dt \int_{-\infty}^{\infty} R_h(\Delta t, \tau, \tau') \exp[-2\pi j \nu' \Delta t] \, d\Delta t \tag{6.23}$$

---

[4] Strict sense stationarity means that fading statistics of arbitrary order do not change with time. For Gaussian channels, wide-sense stationarity implies strict sense stationarity.

The first integral is an integral representation of the delta function $\delta(\nu - \nu')$. Thus, R_s can be factored as:

$$R_s(\nu, \nu', \tau, \tau') = \widetilde{\widetilde{P}}_s(\nu, \tau, \tau')\delta(\nu - \nu') \tag{6.24}$$

This implies that contributions undergo uncorrelated fading if they have different Doppler shifts. The function $\widetilde{\widetilde{P}}_s()$, which is implicitly defined by Eq. (6.24), will be discussed below in more detail.

Analogously, we can write R_B as:

$$R_B(\nu, \nu', f, f') = P_B(\nu, f, f')\delta(\nu - \nu') \tag{6.25}$$

6.4.2 Uncorrelated scatterers

The U.S. assumption is defined as "contributions with different delays are uncorrelated", which is written mathematically as:

$$R_h(t, t', \tau, \tau') = P_h(t, t', \tau)\delta(\tau - \tau') \tag{6.26}$$

or for R_s as:

$$R_s(\nu, \nu', \tau, \tau') = \widetilde{P}_s(\nu, \nu', \tau)\delta(\tau - \tau') \tag{6.27}$$

The U.S. condition is fulfilled if the phase of an MPC does not contain any information about the phase of another MPC with a different delay. If scatterers are distributed randomly in space, phases change in an uncorrelated way even when the MS moves only a small distance.

For the transfer function, the U.S. condition means that R_H does not depend on the absolute frequency, but only on the frequency difference:

$$R_H(t, t', f, f + \Delta f) = R_H(t, t', \Delta f) \tag{6.28}$$

6.4.3 WSSUS assumption

The U.S. assumption is dual to the WSS assumption: U.S. defines contributions with different delays as uncorrelated, while WSS defines contributions with different Doppler shifts as uncorrelated. Alternatively, we can state that U.S. means that R_H depends only on the frequency difference, while WSS means that R_H depends only on the time difference. It is thus obvious to combine these two definitions in the WSSUS condition. This imposes the following restrictions on the ACF:

$$R_h(t, t + \Delta t, \tau, \tau') = \delta(\tau - \tau')P_h(\Delta t, \tau) \tag{6.29}$$

$$R_H(t, t + \Delta t, f, f + \Delta f) = R_H(\Delta t, \Delta f) \tag{6.30}$$

$$R_s(\nu, \nu', \tau, \tau') = \delta(\nu - \nu')\delta(\tau - \tau')P_s(\nu, \tau) \tag{6.31}$$

$$R_B(\nu, \nu', f, f + \Delta f) = \delta(\nu - \nu')P_B(\nu, \Delta f) \tag{6.32}$$

In contrast to the ACFs, which depend on four variables, the P-functions on the r.h.s. depend only on *two* variables. This greatly simplifies their formal description, parameterization, and application in further derivations. Because of their importance, they have been given distinct names. Following Kattenbach [1997], we define:

- $P_h(\Delta t, \tau)$ as *delay cross power spectral density*;
- $R_H(\Delta t, \Delta f)$ as *time frequency correlation function*;
- $P_s(\nu, \tau)$ as *scattering function*;
- $P_B(\nu, \Delta f)$ as *Doppler cross power spectral density*.

The scattering function has special importance because it can be easily interpreted physically. If only single interactions occur, then each differential element of the scattering function corresponds to a physically existing IO. From the Doppler shift we can determine the direction of arrival; the delay determines the radii of the ellipse on which the scatterer lies.

The WSSUS assumption is very popular, but not always fulfilled in practice. Appendix 6.A (see *www.wiley.com/go/molisch*) gives a more detailed discussion of the assumptions and their validity.

6.4.4 Tapped delay line models

A WSSUS channel can be represented as a tapped delay line, where the coefficients multiplying the output from each tap vary with time. The impulse response is then written as:

$$h(t, \tau) = \sum_{i=1}^{N} c_i(t) \delta(\tau - \tau_i) \tag{6.33}$$

where N is the number of taps, $c_i(t)$ are the time-dependent complex coefficients for the taps, and τ_i is the delay of the ith tap. For each tap, a Doppler spectrum determines the changes of the coefficients with time. This spectrum can be different for each tap, though many models assume the same spectrum for each tap, see also Chapter 7.

We can look at this representation in two ways: one is as a physical representation of the multipath propagation in the channel. Each of the N components corresponds to one group of closely spaced MPCs: the model would be purely deterministic only if the arriving signals consisted of *completely resolvable* echoes from discrete IOs. However, in most practical cases, the resolution of the receiver is not sufficient to resolve all MPCs. We thus write the impulse response as:

$$h(t, \tau) = \sum_{i=1}^{N} \sum_{k} a_{i,k}(t) \delta(\tau - \tau_i) = \sum_{i=1}^{N} c_i(t) \delta(\tau - \tau_i) \tag{6.34}$$

Note that the second part of this equation makes sense only in a band-limited system. In that case, each complex amplitude $c_i(t)$ represents the sum of several MPCs, which fades. WSSUS implies that all the taps are fading independently, and that their average power does not depend on time.

Another interpretation of the tapped delay line is the following: any wireless system, and thus the channel we are interested in, is band-limited. Therefore, the impulse response can be represented by a sampled version of the continuous impulse response $\tilde{h}_{\text{bl}}(t, \tau) = \sum A_\ell(t) \delta(\tau - \tau_\ell)$; similarly, the scattering function, correlation functions, etc., can be represented by their sampled versions. Commonly, the samples are spaced equidistantly, $\tau_\ell = \ell \cdot \Delta\tilde{\tau}$, where the distance between the taps $\Delta\tilde{\tau}$ is determined by the Nyquist theorem. The continuous version of the impulse response can be recovered by interpolation:

$$h_{\text{bl}}(t, \tau) = \sum_{\ell} A_\ell(t) \operatorname{sinc}(W(\tau - \tau_\ell)) \tag{6.35}$$

where W is the bandwidth. If the *physical* IOs fulfill the WSSUS condition, but are not equidistantly spaced, then the tap weights $A_\ell(t)$ are *not* necessarily WSSUS.

Many of the standard models for wireless channels (see Chapter 7) were developed with a specific system and thus a specific system bandwidth in mind. It is often necessary to adjust the tap locations to a different sampling grid for a discrete simulation: in other words, a discrete simulation requires a channel representation $h(t, \tau) = \sum A_\ell(t) \delta(\tau - \ell T_s)$, but τ_ℓ / T_s is a non-integer. The following methods are in widespread use:

1. *Rounding to the nearest integer.* This method leads to errors, which are smaller the higher the new sampling rate is.
2. *Splitting the tap energy.* The energy is divided between the two adjacent taps $kT_s < \tau_\ell < (k+1)T_s$, possibly weighted by the distance to the original tap.
3. *Resampling.* This can be done by using the interpolation formula – i.e., resampling $h_{bl}(t, \tau)$ in Eq. (6.35) at the desired rate. Alternatively, we can describe the channel in the frequency domain and transform it back (with a discrete Fourier transform) with the desired tap spacing.

6.5 Condensed parameters

The correlation functions are a somewhat cumbersome way of describing wireless channels. Even when the WSSUS assumption is used, they are still *functions* of two variables. A preferable representation would be a function of one variable, or even better, just a single parameter. Obviously, such a representation implies a serious loss of information, but this is a sometimes acceptable price for a compact representation.

6.5.1 Integrals of the correlation functions

A straightforward way of getting from two variables to one is to integrate over one of them. Integrating the scattering function over the Doppler shift ν gives the *delay power spectral density* $P_h(\tau)$, more popularly known as the *Power Delay Profile* (PDP). The PDP contains information about how much power (from a transmitted delta pulse with unit energy) arrives at the receiver with a delay between $[\tau, \tau + d\tau]$, irrespective of a possible Doppler shift. The PDP can be obtained from the complex impulse responses $h(t, \tau)$ as:

$$P_h(\tau) = \int_{-\infty}^{\infty} |h(t, \tau)|^2 dt \tag{6.36}$$

if ergodicity holds. Note that in practice the integral will not extend over infinite time, but rather over the timespan during which quasi-stationarity is valid (see above).

Analogously, integrating the scattering function over τ results in the *Doppler power spectral density* $P_B(\nu)$.

The *frequency correlation function* can be obtained from the time frequency correlation function by setting $\Delta t = 0$ – i.e., $R_H(\Delta f) = R_H(0, \Delta f)$. It is noteworthy that this frequency correlation function is the Fourier transform of the PDP. The *temporal correlation function* $R_H(\Delta t) = R_H(\Delta t, 0)$ is the Fourier transform of the Doppler power spectral density.

6.5.2 Moments of the power delay profile

The PDP is a quantity that can be plotted as the final product of measurement in one stationarity region. However, it is still a *function*. For obtaining a quick overview of measurement results, it is preferable to have each measurement campaign described by a single *parameter*. While there is a large number of possible parameters, normalized moments of the PDP are the most popular.

We start out by computing the zeroth-order moment – i.e., time-integrated power:

$$P_m = \int_{-\infty}^{\infty} P_h(\tau) \, d\tau \tag{6.37}$$

The normalized first-order moment, the *mean delay*, is:

$$T_{\mathrm{m}} = \frac{\int_{-\infty}^{\infty} P_h(\tau)\tau\, d\tau}{P_{\mathrm{m}}} \qquad (6.38)$$

The normalized second-order central moment is known as *rms delay spread* and defined as:

$$S_\tau = \sqrt{\frac{\int_{-\infty}^{\infty} P_h(\tau)\tau^2\, d\tau}{P_{\mathrm{m}}} - T_{\mathrm{m}}^2} \qquad (6.39)$$

The rms delay spread has obtained a special stature among all parameters. It has been shown that under some specific circumstances, the error probability due to delay dispersion is proportional to the rms delay spread only (see Chapter 12), while the actual *shape* of the PDP does not have a significant influence. In that case, S_τ is all we need to know about the channel. It cannot be stressed enough, however, that this is true only under specific circumstances, and that the rms delay spread is not a "solve-it-all". It is also noteworthy that S_τ does not attain finite values for all physically reasonable signals. A channel with $P_h(\tau) \propto 1/(1+\tau^2)$ is physically possible, and does not contradict energy conservation, but $\int_{-\infty}^{\infty} P_h(\tau)\tau^2\, d\tau$ does not converge.

Example 6.1 *Compute the rms delay spread of a two-spike profile*
$$P_h(\tau) = \delta(\tau - 10\ \mu s) + 0.3\delta(\tau - 17\ \mu s).$$

The time-integrated power is given by Eq. (6.37):

$$P_{\mathrm{m}} = \int_{-\infty}^{\infty} \left(\delta(\tau - 10^{-5}) + 0.3\delta(\tau - 1.7 \cdot 10^{-5})\right) d\tau$$

$$= 1.30 \qquad (6.40)$$

and the mean delay is given by Eq. (6.38):

$$T_{\mathrm{m}} = \int_{-\infty}^{\infty} \left(\delta(\tau - 10^{-5}) + 0.3\delta(\tau - 1.7 \cdot 10^{-5})\right)\tau\, d\tau / P_{\mathrm{m}}$$

$$= \left(10^{-5} + 0.3 \cdot 1.7 \cdot 10^{-5}\right)/1.3 = 1.16 \cdot 10^{-5} s \qquad (6.41)$$

Finally, the rms delay spread is computed according to Eq. (6.39):

$$S_\tau = \sqrt{\frac{\int_{-\infty}^{\infty}(\delta(\tau - 10^{-5}) + 0.3\delta(\tau - 1.7 \cdot 10^{-5}))\tau^2 d\tau}{P_{\mathrm{m}}} - T_{\mathrm{m}}^2}$$

$$= \sqrt{\frac{(10^{-5})^2 + 0.3(1.7 \cdot 10^{-5})^2}{1.3} - (1.16 \cdot 10^{-5})^2} = 3\ \mu s. \qquad (6.42)$$

6.5.3 Moments of the Doppler spectra

Moments of the Doppler spectra can be computed in complete analogy to the moments of the PDP. The integrated power is:

$$P_{\mathrm{B,m}} = \int_{-\infty}^{\infty} P_{\mathrm{B}}(\nu)\, d\nu \qquad (6.43)$$

where obviously $P_{B,m} = P_m$. The mean Doppler shift is:

$$\nu_m = \frac{\int_{-\infty}^{\infty} P_B(\nu)\nu\,d\nu}{P_{B,m}} \tag{6.44}$$

The rms Doppler spread is:

$$S_\nu = \sqrt{\frac{\int_{-\infty}^{\infty} P_B(\nu)\nu^2\,d\nu}{P_{B,m}} - \nu_m^2} \tag{6.45}$$

6.5.4 Coherence bandwidth and coherence time

In a frequency-selective channel, different frequency components fade differently. Obviously, the correlation between fading at two different frequencies is the smaller the more these two frequencies are apart. The *coherence bandwidth* B_{coh} defines the frequency difference that is required so that the correlation coefficient is smaller than a given threshold.

A mathematically exact definition starts out with the frequency correlation function $R_H(0, \Delta f)$, assuming WSSUS (see Fig. 6.8a for an example). The coherence bandwidth can then be defined as:

$$B_{coh} = \frac{1}{2}\left[\arg\max_{\Delta f > 0}\left(\frac{R_H(0, \Delta f)}{R_H(0,0)} = 0.5\right) - \arg\min_{\Delta f < 0}\left(\frac{R_H(0, \Delta f)}{R_H(0,0)} = 0.5\right)\right] \tag{6.46}$$

This is essentially the 3-dB bandwidth of the correlation function. The somewhat complicated formulation stems from the fact that the correlation function need not decay *monotonically*. Rather, there can be local maxima that exceed the threshold. A precise definition thus uses the bandwidth that encompasses all parts of the correlation function exceeding the threshold.[5]

The rms delay spread S_τ and the coherence bandwidth B_{coh} are obviously related: S_τ is derived from the PDP $P_h(\tau)$ while B_{coh} is obtained from the frequency correlation function, which is the Fourier transform of the PDP. Based on this insight, Fleury [1996] derived an "uncertainty relationship":

$$B_{coh} \gtrsim \frac{1}{2\pi S_\tau} \tag{6.47}$$

Equation (6.47) is an inequality and therefore does *not* offer the possibility to obtain one parameter from the other. The question thus arises whether B_{coh} or S_τ better reflects the channel properties. An answer to that question can only be given for a specific system. For a Frequency Division Multiple Access (FDMA) or Time Division Multiple Access (TDMA) system without an equalizer, the rms delay spread is the quantity of interest, as it is related to the Bit Error Rate (BER) (see Chapter 12), though generally it overemphasizes long-delayed echoes. For OFDM systems (Chapter 19), where the information is transmitted on many parallel carriers, the coherence bandwidth is obviously a better measure.

The temporal correlation function is a measure of how fast a channel changes. The definition of the coherence time T_{coh} is thus analogous to the coherence bandwidth; it also has an uncertainty relationship with the rms Doppler spread.

Figure 6.9 summarizes the relationships between system functions, correlation functions, and special parameters. Ergodicity is assumed throughout this figure.

---

[5] An alternative definition would define the coherence bandwidth as the second central moment of the correlation function; this would circumvent all problems with local maxima. Unfortunately, this second moment becomes infinite in the practically important case that the correlation function is Lorentzian $1/(1 + \Delta f^2)$, which corresponds to an exponential PDP.

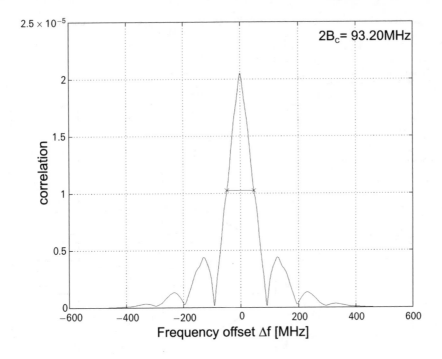

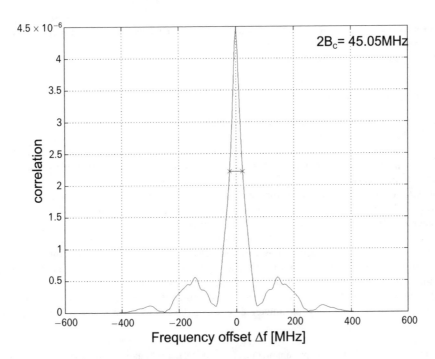

Figure 6.8 Typical frequency correlation function.

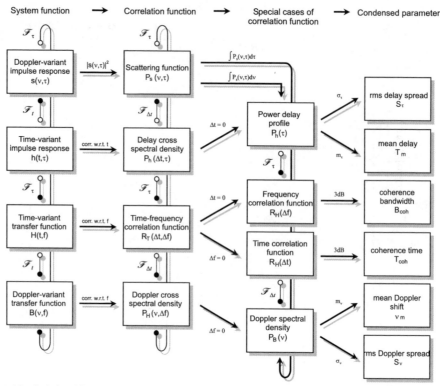

Figure 6.9 Relationships between system functions, correlation functions, and condensed parameters for ergodic channel impulse responses.

Reproduced with permission from Kattenbach [1997] © Shaker Verlag.

6.5.5 Window parameters

Another useful set of parameters are the so-called *window parameters* [de Weck 1992], more precisely the *interference quotient* Q_T and the *delay window* W_Q. They are a measure for the percentage of energy of the average PDP arriving within a certain delay interval. In contrast to the delay spread and coherence bandwidth, the window parameters are mostly useful in the context of specific systems.

The *interference quotient* Q_T is the ratio between the signal power arriving within a time window of duration T, relative to the power arriving outside that window. The delay window characterizes the self-interference due to delay dispersion. If, for example, a system has an equalizer that can process MPCs with a delay up to T, then every MCP within the window is "useful", while energy outside the window creates interference – these components carry information about bits that cannot be processed, and thus act as independent interferers.[6] For the Global System for Mobile communications (GSM) (see Chapter 21), an equalizer length of four-symbol duration, corresponding to a 16-μs delay, is required by the specifications. It is thus common to define a parameter Q_{16} – i.e., the interference quotient for $T = 16\,\mu s$.

---

[6] The interpretation is only an approximate one. There is no sharp "jump" from "useful" to "interference" when the delay of a multipath component exceeds the equalizer length. Rather, there is a smooth transition, similar to the effect of delay dispersion in unequalized systems, see Chapter 12.

A mathematical definition of the interference quotient is given as:

$$Q_T = \frac{\int_{t_0}^{t_0+T} P_h(\tau)\,d\tau}{P_m - \int_{t_0}^{t_0+T} P_h(\tau)\,d\tau} \tag{6.48}$$

This quotient depends not only on the PDP and the duration T, but also on the starting delay of the window t_0. This latter dependence is often eliminated by either setting the starting delay to the minimum excess delay (i.e., the first MPC determines the start of the window) $t_0 = \tau_{\min}$. Alternatively, the t_0 can be chosen to maximize Q_T:

$$Q_T = \max_{t_0} \left\{ \frac{\int_{t_0}^{t_0+T} P_h(\tau)\,d\tau}{P_m - \int_{t_0}^{t_0+T} P_h(\tau)\,d\tau} \right\} \tag{6.49}$$

This definition makes sense because a receiver can often adapt equalizer timing to optimize performance.

A related parameter is the delay window W_Q (see Fig. 6.10). This defines how long a window has to be so that the power within that window is a factor of Q larger than the power outside the window. The defining equations are the same as for the interference quotient. The difference is just that now T is considered as variable, and Q as fixed.

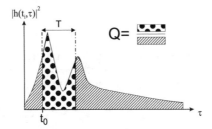

Figure 6.10   Definition of window parameters.
Reproduced with permission from Molisch [2000] © Prentice Hall.

Example 6.2   *For an exponential PDP, $P_h(\tau) = \exp(-\tau/2\,\mu s)$, compute the delay window so that the interference quotient becomes 10 and 20 dB, respectively – i.e., so that 91% and 99% of the energy are contained in the window. Do the same for the two-spike profile $\delta(\tau - 10\mu s) + 0.3\delta(\tau - 17\mu s)$.*

Let the starting delay be equal to the minimum excess delay. For an exponential PDP, the interference quotient, given by Eq. (6.48), is:

$$Q_T = \frac{\int_0^T e^{-\tau/2\cdot 10^{-6}}\,d\tau}{\int_0^\infty e^{-\tau/2\cdot 10^{-6}}\,d\tau - \int_0^T e^{-\tau/2\cdot 10^{-6}}\,d\tau}$$

Solving for T yields:

$$T = 2 \cdot 10^{-6}\ln(Q_T + 1)$$

and the 91% and 99% windows are thus $T_{91\%} = 4.8\,\mu s$ and $T_{99\%} = 9.2\,\mu s$, respectively. For the two-spike profile, the starting delay is $t_0 = 10^{-5}$ and the energy within the window is:

$$\int_{10^{-5}}^{T+10^{-5}} \left(\delta(\tau - 10^{-5}) + 0.3\delta(\tau - 1.7\cdot 10^{-5})\right)d\tau = \begin{cases} 1, & 0 < T < 7\,\mu s \\ 1.3, & T > 7\,\mu s \\ 0, & \text{otherwise} \end{cases}$$

Hence, the interference quotient is greater than 10 dB and/or 20 dB for $T > 7\,\mu s$.

6.6 Ultrawideband channels

The above models are wideband in the sense that they model the delay dispersion caused by multipath propagation. However, they are still based on the following two assumptions:

1. The reflection, transmission, and diffraction coefficients of the interacting objects are constant over the considered bandwidth.
2. The relative bandwidth of the system (bandwidth divided by carrier frequency) is *much* smaller than unity.

Note that these conditions are met for the bandwidth of most currently used wireless systems. However, in recent years, a technique called *Ultra Wide Band (UWB) transmission* (see also Chapter 25) has gained increased interest. UWB systems have a relative bandwidth of more than 20%.[7] In that case, the different frequency components contained in the transmitted signal "see" different propagation environments. For example, the diffraction coefficient of a building corner is different at 100 MHz compared with 1 GHz; similarly, the reflection coefficients of walls and furniture can vary over the bandwidth of interest. Channel impulse realization is then given by:

$$h(\tau) = \sum_{i=1}^{N} a_i \chi_i(\tau) \otimes \delta(\tau - \tau_i) \qquad (6.50)$$

where $\chi_i(\tau)$ denotes the distortion of the ith MPC by the frequency selectivity of interacting objects. Expressions for these distortions are given, e.g., in Molisch [2005] and Qiu [2002]; one example for a distortion of a short pulse by diffraction by a screen is shown in Fig. 6.11.

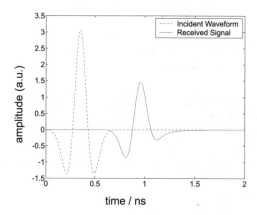

Figure 6.11   UWB pulse diffracted by a semi-infinite screen.
Reproduced with permission from Qiu [2002] © IEEE.

For UWB systems, free space pathloss can also show frequency dependence. As explained in Chapter 4, pathloss is a function of frequency if the antennas have constant gain. Thus, the higher frequency components of the transmitted signal are attenuated more strongly by the combination of antenna and channel. Also this effect leads to a distortion of individual MPCs since *any* frequency dependence of the transfer function leads to delay dispersion, and thus distortion of an MPC.

---

[7] An alternative definition describes UWB systems as having an absolute bandwidth of more than 500 MHz.

Distortion of the MPCs can also be interpreted in an alternative way: when representing the impulse response as a tapped delay line, the fading at the different taps becomes correlated. The MPC distortion makes each multipath contribution influence several subsequent taps. From that, it follows that the fading of one component influences the amplitudes of several taps, and therefore causes correlation.

6.7 Directional description

The wideband description we have given in the previous sections is not a completely general description of the channel:

- It models only the temporal properties of the channel. The direction of the MPCs does not enter the description. These directional properties are important for spatial diversity (Chapter 13) and multielement antennas (Chapter 20).
- The impulse response does not actually describe just the propagation channel, but also includes the effect of the antennas. It adds up the weighted MPCs, where the weighting depends on the specific antenna used. Thus, changing the antenna changes the impulse response, even though the true propagation channel remains unchanged.

These problems are eliminated by using the most fundamental deterministic description of the propagation channel, the *double-directional impulse response*,[8] which consists of the sum of contributions from MPCs:

$$h(t, \tau, \Omega, \Psi) = \sum_{\ell=1}^{N(t)} h_\ell(t, \mathbf{r}_\mathrm{X}, \mathbf{r}_\mathrm{X}, \tau, \Omega, \Psi) \tag{6.51}$$

The impulse response depends on the delay τ, the *Direction Of Departure* (DOD) Ω, the *Direction Of Arrival* (DOA) Ψ, and the number of MPCs, $N(t)$, for the specific time. The $h_\ell(t, \tau, \Omega, \Psi)$ is the contribution of the ℓth MPC, modeled as:

$$h_\ell(t, \tau, \Omega, \Psi) = |a_\ell| e^{j\varphi_\ell} \delta(\tau - \tau_\ell) \delta(\Omega - \Omega_\ell) \delta(\Psi - \Psi_\ell) \tag{6.52}$$

which essentially just adds the directional properties to the tapped delay line model. Besides the absolute amplitude $|a|$ and the delay, also the DOA and DOD vary slowly (over many wavelengths), while again phase φ varies quickly.

The single-directional impulse response can be obtained by integrating the double-directional impulse response (weighted by the transmit antenna pattern) over the DODs. Integrating the single-directional impulse response (weighted by the receiver antenna pattern) over the DOAs results in the conventional impulse response.

The stochastic description of directional channels is analogous to the non-directional case. The autocorrelation function of the impulse response can be generalized to include directional dependence so that it depends on six or eight variables. We can also introduce a "generalized WSSUS condition" so that contributions coming from different directions fade independently. Note that the directions of the MPCs at the MS, on one hand, and Doppler spreading, on the other hand, are linked, and thus ν and Ψ are not independent variables anymore (we assume in the following that Ψ are the directions at the MS).

Analogously to the non-directional case, we can then define condensed descriptions of the wireless channel. We first define:

---

[8] To be completely general, we would have to include a description of polarization as well. To avoid the cumbersome matrix notation, we omit this case here and only briefly discuss it in Section 7.4.4.

$$E\{s^*(\Omega, \Psi, \tau, \nu)s(\Omega', \Psi', \tau', \nu')\} = P_s(\Omega, \Psi, \tau, \nu)\delta(\Omega - \Omega')\delta(\Psi - \Psi')\delta(\tau - \tau')\delta(\nu - \nu') \quad (6.53)$$

from which the *Double Directional Delay Power Spectrum* (DDDPS) is derived as:

$$DDDPS(\Omega, \Psi, \tau) = \int P_s(\Psi, \Omega, \tau, \nu)\, d\nu \tag{6.54}$$

From this, we can establish the *Angular Delay Power Spectrum* (ADPS) as seen from the BS antenna:

$$ADPS(\Omega, \tau) = \int DDDPS(\Psi, \Omega, \tau)G_{MS}(\Psi)d\Psi \tag{6.55}$$

where G_{MS} is the antenna power pattern of the MS. The ADPS is usually normalized as:

$$\int \int ADPS(\tau, \Omega)\, d\tau\, d\Omega = 1 \tag{6.56}$$

The *Angular Power Spectrum* (APS) is given by:

$$APS(\Omega) = \int APDS(\Omega, \tau)\, d\tau \tag{6.57}$$

Note also that an integration of the ADPS over Ω recovers the PDP.

The *azimuthal spread* is defined as the second central moment of the APS if all MPCs are incident in the horizontal plane, so that $\Omega = \phi$. In many papers it is defined in a form analogous to Eq. (6.39) – namely:

$$S_\phi = \sqrt{\frac{\int APS(\phi)\phi^2\, d\phi}{\int APS(\phi)\, d\phi} - \left(\frac{\int APS(\phi)\phi\, d\phi}{\int APS(\phi)\, d\phi}\right)^2} \tag{6.58}$$

However, this definition is ambiguous because of the periodicity of the azimuthal angle: by this definition, $APS = \delta(\phi - \pi/10) + \delta(\phi - 19\pi/10)$ would have a different angular spread from $APS = \delta(\phi - 3\pi/10) + \delta(\phi - \pi/10)$, even though the two APSs differ just by a constant offset. A better definition is given in Fleury [2000]:

$$S_\phi = \sqrt{\frac{\int |\exp(j\phi) - \mu_\phi|^2 APS(\phi)d\phi}{\int APS(\phi)d\phi}} \tag{6.59}$$

with

$$\mu_\phi = \frac{\int \exp(j\phi)APS(\phi)d\phi}{\int APS(\phi)d\phi} \tag{6.60}$$

Example 6.3 *Consider the APS defined as $APS = 1$ for $0° < \varphi < 90°$ and $340° < \varphi < 360°$, compute the angular spread according to the definitions of Eqs. (6.58) and (6.59), respectively.*

According to Eq. (6.58) we have:

$$S_\phi = \sqrt{\frac{\int_0^{\pi/2} \phi^2 d\phi + \int_{17\pi/9}^{2\pi} \phi^2 d\phi}{\int_0^{\pi/2} d\phi + \int_{17\pi/9}^{2\pi} d\phi} - \left(\frac{\int_0^{\pi/2} \phi d\phi + \int_{17\pi/9}^{2\pi} \phi d\phi}{\int_0^{\pi/2} d\phi + \int_{17\pi/9}^{2\pi} d\phi}\right)^2}$$

$$= 2.09 \text{ rad} = 119.7° \tag{6.61}$$

In contrast, Eqs. (6.59) and (6.60) yield:

$$\left.\begin{aligned}
\mu_\phi &= \frac{18}{11\pi} \int_{-\pi/9}^{\pi/2} \exp(j\phi)d\phi = 0.7 + 0.49j \\
S_\phi &= \sqrt{\frac{18}{11\pi} \int_{-\pi/9}^{\pi/2} \left((\cos(\phi) - Re(\mu_\phi))^2 + (\sin(\phi) - Im(\mu_\phi))^2 \right) d\phi} \\
&= 0.521 \text{ rad} = 29.9°
\end{aligned}\right\} \quad (6.62)$$

The values obtained from the two methods differ radically. We can easily see that the second value – namely, 29.9°, makes more sense: the APS extends continuously from −20° to 90°. The angular spread should thus be the same as for an APS that extends from 0° to 111° degrees. Inserting this modified APS in Eq. (6.58), we obtain an angular spread of 32°, while the value from Eq. (6.59) remains at 29.9°. It is also interesting that this value is close to $(\phi_{\max} - \phi_{\min})/(2\sqrt{3})$ – and remember that for a rectangular power delay profile, the relationship between rms delay spread and maximum excess delay is also $S_\tau = (\tau_{\max} - \tau_{\min})/(2\sqrt{3})$.

Similarly to delay spread, angular spread also is only a partial description of angular dispersion. It has been shown that the correlation of signals at the elements of a uniform linear array depends only on the rms angular spread and not on the shape of the APS; however, this is valid only under some very specific assumptions.

Directional channel descriptions are especially valuable in the context of multiantenna systems (Chapter 20). In that case, we are often interested in obtaining joint impulse responses at the different antenna elements. The impulse response thus becomes a matrix if we have antenna arrays at both link ends, and a vector if there is an array at one link end. We denote the transmit and receive element coordinates as $\mathbf{r}_{TX}^{(1)}, \mathbf{r}_{TX}^{(2)}, \ldots, \mathbf{r}_{TX}^{(N_t)}$, and $\mathbf{r}_{RX}^{(1)}, \mathbf{r}_{RX}^{(2)}, \ldots, \mathbf{r}_{RX}^{(N_r)}$, respectively, so that the impulse response from the ith transmit to the jth receive element becomes:

$$\begin{aligned}
h_{i,j} &= h\left(\mathbf{r}_{TX}^{(i)}, \mathbf{r}_{RX}^{(j)} \right) \\
&= \sum_\ell h_\ell(\mathbf{r}_{TX}^{(1)}, \mathbf{r}_{RX}^{(1)}, \tau, \Omega_\ell, \Psi_\ell) \widetilde{G}_{TX}(\Omega_\ell) \widetilde{G}_{RX}(\Psi_\ell) \exp\left(j\langle \mathbf{k}(\Omega_\ell), (\mathbf{r}_{TX}^{(i)} - \mathbf{r}_{TX}^{(1)}) \rangle \right) \\
&\quad \times \exp\left(-j\langle \mathbf{k}(\Psi_\ell), (\mathbf{r}_{RX}^{(i)} - \mathbf{r}_{RX}^{(1)}) \rangle \right)
\end{aligned} \quad (6.63)$$

Letting h_ℓ depend on the location of the reference antenna elements $\mathbf{r}_{TX}^{(1)}$ and $\mathbf{r}_{RX}^{(1)}$ instead of on absolute time t. Here $\widetilde{G}_{TX}$ and $\widetilde{G}_{RX}$ are the complex (amplitude) patterns of the transmit and receive antenna elements, respectively, $\{\mathbf{k}\}$ is the unit wave vector in the direction of the ℓth DOD or DOA, and $\langle \cdot, \cdot \rangle$ denotes the dot product. We thus see that it is always possible to obtain the impulse response matrix from a double-directional impulse response.

If the receive arrays are uniform linear arrays, we can write Eq. (6.63) as:

$$\mathbf{H} = \int \int h(\tau, \Omega, \Psi) \widetilde{G}_{TX}(\Omega) \widetilde{G}_{RX}(\Psi) \mathbf{a}_{RX}(\Psi) \mathbf{a}_{TX}^\dagger(\Omega) \, d\Psi \, d\Omega \quad (6.64)$$

where we used the *steering vectors*:

$$\mathbf{a}_{TX}(\Omega) = \frac{1}{\sqrt{N_t}} \left[1, \exp(-j2\pi \frac{d_a}{\lambda} \sin(\Omega)), \ldots, \exp(-j2\pi(N_t - 1)\frac{d_a}{\lambda} \sin(\Omega)) \right]^T$$

and analogously defined $\alpha_{RX}(\Psi)$. Ω and Ψ are measured from the antenna broadside.

6.8 Appendices

Please see companion website *www.wiley.com/go/molisch*

Further reading

The theory of linear time-variant systems is described in the classical paper of Bello [1963]; further details are discussed in Kozek [1997], and Matz and Hlawatsch [1998]. The theory of WSSUS systems was established in Bello [1963], and further investigated in Hoeher [1992], Kattenbach [1997], Molnar et al. [1996], Paetzold [2002]; more considerations about the validity of WSSUS in wireless communications can be found in Fleury [1990], Kattenbach [1997], Kozek [1997], Molisch and Steinbauer [1999]. A method for characterizing non-WSSUS channels is described in Matz [2003]. An overview of condensed parameters, including delay spread, is given in Molisch and Steinbauer [1999]. Generic descriptions of UWB channels can be found in Molisch [2005], Molisch et al. [2004a], Qiu [2004]. Description methods for spatial channels are discussed in Durgin [2003], Ertel et al. [1998], Molisch [2002], Yu and Ottersten [2002]. Generalizations of the WSSUS approach to directional models are discussed in Fleury [2000], Kattenbach [2002].

7

Channel models

7.1 Introduction

For the design, simulation, and planning of wireless systems we need *models* for the propagation channels. In the previous chapters, we have discussed some basic properties of wireless channels, and how they can be described mathematically – amplitude-fading statistics, scattering function, delay spread, etc. In this chapter, we will discuss in a more concrete way how these mathematical description methods can be converted into generic simulation models, and how to parameterize these models.

There are two main applications for channel models:

- For the design, testing, and type approval of *wireless systems*, we need simple channel models that reflect the important properties of propagation channels – i.e., properties that have an impact on system performance. This is usually achieved by simplified channel models that describe the statistics of the impulse response in parametric form. The number of parameters is small, and *independent of specific locations*. Such models sometimes lead to insights due to closed-form relationships between channel parameters and system performance. Furthermore, they can easily be implemented by system designers for testing purposes.
- The designers of *wireless networks* are interested in optimizing a given system in a certain geographical region. Locations of base stations (BSs) and other network design parameters should be optimized on the computer, and not by field tests, and trial and error. For such applications, *location-specific channel models* that make good use of available geographical and morphological information are desirable. However, the models should be robust with respect to small errors in geographical databases.

The following three modeling methods are in use for these applications:

- *Stored channel impulse responses*: a channel sounder (see Chapter 8) measures, digitizes, and stores impulse responses $h(t, \tau)$. The main advantage of this approach is that the resulting impulse responses are realistic. Furthermore, system simulations using the stored data are reproducible, as the data remain available and can be reused indefinitely, even for simulations of different systems. This is an important distinction from field trials of whole systems, where there can be no guarantee that the impulse response remains constant over time. The disadvantages of using stored impulse responses are (i) the large effort in acquiring and storing the data, and (ii) the fact that the data characterize only a certain area, and need not be typical for a propagation environment.

- *Deterministic channel models*: these models use the geographical and morphological information from a database for a deterministic solution of Maxwell's equation or some approximation thereof. The basic philosophy is the same as for stored impulse responses: determining the impulse response in a certain geographic location. Both of these methods are therefore often subsumed as *site-specific models*. The drawbacks of deterministic (computed) channel models compared with stored (measured) impulse responses are (i) the large computational effort, and (ii) the fact that the results are inherently less accurate, due to inaccuracies in the underlying databases and the approximate nature of numerical computation methods. The main advantage is that computer simulations are easier to perform than measurement campaigns. Furthermore, certain types of computation methods (e.g., ray tracing, see Section 7.5) allow the effects of different propagation mechanisms to be isolated.
- *Stochastic channel models* model the probability density function of the channel impulse response (or equivalent functions). These methods do not attempt to correctly predict the impulse response in one specific location, but rather to predict the probability density function (pdf) over a large area. The simplest example of this approach is the Rayleigh-fading model: it does not attempt to correctly predict the fieldstrength at each location, but rather attempts to correctly describe the pdf of the fieldstrength over a large area. Stochastic wideband models can be created in the same spirit.

Generally speaking, stochastic models are used more for the design and comparison of systems, while site-specific models are preferable for network planning and system deployment. Furthermore, deterministic and stochastic approaches can be combined to enhance the efficiency of a model: for example, large-scale averaged power can be obtained from deterministic models, while the variations within an averaging area are modeled stochastically.

It is obvious that none of the above models can achieve perfect accuracy. Establishing a criterion for "satisfactory accuracy" is thus important:

- From a purely scientific point of view, any inaccuracy is unsatisfactory. From an engineering point of view, however, there is no point in increasing modeling accuracy (and thus effort) beyond a certain point.[1]
- For deterministic modeling methods, inaccuracies in the underlying databases lead to unavoidable errors. For stochastic models that are derived from measurements, the finite number of underlying measurement points, as well as measurement errors, limit the possible accuracy. Ideally, errors due to a specific modeling method should be smaller than errors due to these unavoidable inaccuracies.
- Requirements on modeling accuracy can be relaxed even more by the following pragmatic criterion: the inaccuracies in the model should not "significantly" alter a system design or deployment plan. For this definition, the system designer has to determine what "significant" is.

---

[1] Many research papers use the words "satisfactory accuracy" as a rhetorical tool to emphasize the value of a new modeling method. If a method decreases the deviation between theory and measurement from 12 to 9 dB, it will consider 9 dB as "satisfactory". A subsequent paper that decreases the error to 6 dB will consider the same 9 dB as "unsatisfactory".

7.2 Narrowband models

7.2.1 Modeling of small-scale and large-scale fading

For a narrowband channel, the impulse response is a delta function with a time-varying attenuation, so that for slowly time-varying channels:

$$h(t, \tau) = \alpha(t)\delta(\tau) \tag{7.1}$$

As mentioned in Chapter 5, the variations in amplitude over a small area are typically modeled as a random process, with an autocorrelation function that is determined by the Doppler spectrum. The complex amplitude is modeled as a zero-mean, circularly symmetric complex Gaussian random variable. As this gives rise to a Rayleigh distribution of the absolute amplitude, we will henceforth refer to this case simply as "Rayleigh-fading".

When considering variations in a somewhat larger area, the small-scale averaged amplitude F obeys a lognormal distribution, with standard deviation σ_F; typically values of σ_F are 4 to 10 dB. The spatial autocorrelation function of lognormal shadowing is usually assumed to be a double-sided exponential, with correlation distances between 5 and 100 m, depending on the environment.

7.2.2 Pathloss models

Next, we consider models for the received fieldstrength, averaged over both small-scale and the large-scale fading. This quantity is modeled completely deterministically. The most simple models of that kind are the free space pathloss model, and the "breakpoint" model (with $n = 2$ valid for distances up to $d < d_{\text{break}}$, and $n = 4$ beyond that, as described in Chapter 4). In more sophisticated models, described below, pathloss depends not only on distance, but also on some additional external parameters like building height, measurement environment (e.g., suburban environment), etc.

The Okumura–Hata model

The Okumura–Hata model is by far the most popular model in that category. Pathloss (in dB) is written as:

$$PL = A + B\log(d) + C \tag{7.2}$$

where A, B, and C are factors that depend on frequency and antenna height. Factor A increases with carrier frequency, and decreases with increasing height of the BS and Mobile Station (MS). Also, the pathloss exponent (proportional to B) decreases with increasing height of the BS. Appendix 7.A – see *www.wiley.com/go/molisch* – gives details of these correction factors. The model is only intended for large cells, with the BS being placed higher than the surrounding rooftops.

The COST[2] 231–Walfish–Ikegami model

The COST 231–Walfish–Ikegami model is also suitable for microcells and small macrocells, as it has restrictions on the distance between the BS and MS and the antenna height.

---

[2] European COoperation in the field of Scientific and Technical research.

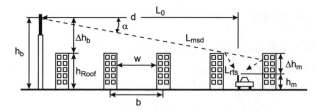

Figure 7.1 Parameters in the COST 231–Walfish–Ikegami model.
Reproduced with permission from Damosso and Correia [1999] © European Union.

In this model, total pathloss consists of the free space pathloss PL_0, *multiscreen loss* L_{msd} along the propagation path, and attenuation from the last roof-edge to the MS, L_{rts} (*rooftop-to-street diffraction and scatter loss*). Free space loss depends on carrier frequency and distance, while the rooftop-to-street diffraction loss depends on frequency, the width of the street, and the height of the MS, as well as on the orientation of the street with respect to the connection line BS–MS. Multiscreen loss depends on the distance between buildings and the distance between the BS and MS, as well as on carrier frequency, BS height, and rooftop height. The model assumes a Manhatten street grid (streets intersecting at right angles), constant building height, and flat terrain. Furthermore, the model does not include the effect of waveguiding through street canyons, which can lead to an underestimation of the received fieldstrength. Details of the model can be found in Appendix 7.B (see companion website *www.wiley.com/go/molisch*).

The Motley–Keenan model

For indoor environments, wall attenuation plays an important role. Based on this consideration, the Motley–Keenan model suggests that pathloss (expressed in dB) can be written as [Motley and Keenan 1988]:

$$PL = PL_0 + 10n\log(d/d_0) + F_{wall} + F_{floor}$$

where F_{wall} is the sum of attenuations by the walls that a Multi Path Component (MPC) has to penetrate on its way from the transmitter (TX) to the receiver (RX); similarly, F_{floor} describes the summed-up attenuation of the floors that are located between the BS and MS. Depending on the building material, attenuation by one wall can lie between 1 and 20 dB in the 300-MHz–5-GHz range, and can be much higher at higher frequencies.

The Motley–Keenan model is a site-specific model, in the sense that it requires knowledge of the location of the BS and MS, and the building plan. It is, however, not very accurate, as it neglects propagation paths that "go around" the walls. For example, propagation between two widely separated offices can occur either through many walls (quasi-Line of Sight – LOS), or through a corridor (signal leaves the office, propagates down a corridor, and enters from there into the office of the RX). The latter type of propagation path can often be more efficient, but is not taken into account by the Motley–Keenan model.

7.3 Wideband models

7.3.1 Tapped delay line models

The most commonly used wideband model is an N-tap Rayleigh-fading model. This is a fairly generic structure, and is basically just the tapped delay line structure of Chapter 6, with the added restriction that the amplitudes of all taps are subject to Rayleigh fading. Adding an LOS component does not pose any difficulties; the impulse response then just becomes:

$$h(t, \tau) = a_0 \delta(\tau - \tau_0) + \sum_{i=1}^{N} c_i(t) \delta(\tau - \tau_i) \tag{7.3}$$

where the LOS component a_0 does *not* vary with time, while the $c_i(t)$ are zero-mean complex Gaussian random processes, whose autocorrelation function is determined by their associated Doppler spectra (e.g., Jakes spectra). In most cases, $\tau_0 = \tau_1$, so the amplitude distribution of the first tap is Rician.

The model is further simplified when the number of taps is limited to $N = 2$, and no LOS component is allowed. This is the simplest stochastic fading channel exhibiting delay dispersion,[3] and thus very popular for theoretical analysis. It is alternatively called the *two-path channel, two-delay channel,* or *two-spike channel.*

Another popular channel model consists of a purely deterministic LOS component, plus *one* fading tap ($N = 1$) whose delay τ_0 can differ from τ_1. This model is widely used for satellite channels – in these channels, there is almost always an LOS connection, and the reflections from buildings near the RX give rise to a delayed fading component. The channel reduces to a flat-fading Rician channel when $\tau_0 = \tau_1$.

7.3.2 Models for the power delay profile

It has been observed in many measurements that the Power Delay Profile (PDP) can be approximated by a one-sided exponential function:

$$P_h(\tau) = P_{\mathrm{sc}}(\tau) = \begin{cases} \exp(-\tau/S_\tau) & \tau \geq 0 \\ 0 & \text{otherwise} \end{cases} \tag{7.4}$$

In a more general model (see also Section 7.3.3), the PDP is the sum of several delayed exponential functions, corresponding to multiple *clusters* of Interacting Objects (IOs):

$$P_h(\tau) = \sum_l \frac{P_l^{\mathrm{c}}}{S_{\tau,l}^{\mathrm{c}}} P_{\mathrm{sc}}(\tau - \tau_{0,l}^{\mathrm{c}}) \tag{7.5}$$

where P_l^{c}, $\tau_{0,l}^{\mathrm{c}}$, S_{τ}^{c} are the power, delay, and delay spread of the lth cluster, respectively. The sum of all cluster powers has to add up to the narrowband power described in Section 7.1.

For a PDP of form Eq. (7.4), the rms delay spread characterizes delay dispersion. In the case of multiple clusters Eq. (7.5), the rms delay spread is defined mathematically, but often has a limited physical meaning. Still, the vast majority of measurement campaigns available in the literature use just this parameter for characterization of delay dispersion.

Typical values of the delay spread for different environments are (see Molisch and Tufvesson [2004] for more details and extensive references):

- *Indoor residential buildings*: 5–10 ns are typical; but up to 30 ns have been measured.
- *Indoor office environments* show typical delay spreads of between 10 and 100 ns, but even 300 ns have been measured. Room size has a clear influence on delay spread. Building size and shape have an impact as well.
- *Factories and airport halls* have delay spreads that range from 50 to 200 ns.
- In *microcells*, delay spreads range from around 5–100 ns (for LOS situations) to 100–500 ns (for non-LOS).

---

[3] Note that each of the taps of this channel exhibits Rayleigh fading; the channel is thus different from the two-path model used in Chapters 5 and 6 for purely didactic reasons.

- *Tunnels and mines*: empty tunnels typically show a very small delay spread (on the order of 20 ns), while car-filled tunnels exhibit larger values (up to 100 ns).
- *Typical urban and suburban environments* show delay spreads between 100 ns and 800 ns, although values up to 3 μs have also been observed.
- *Bad urban and hilly terrain environments* show clear examples of multiple clusters that lead to much larger delay spreads. Delay spreads up to 18 μs, with cluster delays of up to 50 μs have been measured in various European cities, while American cities show somewhat smaller values. Cluster delays of up to 100 μs occur in mountainous terrain.

The delay spread is a function of the distance BS–MS, increasing with distance approximately as d^ε, where $\varepsilon = 0.5$ in urban and suburban environments, and $\varepsilon = 1$ in mountainous regions. The delay spread also shows considerable large-scale variations. Several papers find that the delay spread has a lognormal distribution with a variance of typically 2–3 dB in suburban and urban environments. A comprehensive model containing all these effects was first proposed by Greenstein et al. [1997].

7.3.3 Models for the arrival times of rays and clusters

In the previous section, we described models for the PDP. The modeled PDPs were continuous functions of the delay; this implies that the RX bandwidth was so small that different discrete MPCs could not be resolved, and were "smeared" into a continuous PDP. For systems with higher bandwidth, MPCs can be resolved. In that case, it is advantageous to describe the PDP by the arrival times of the MPCs, plus an "envelope" function that describes the power of the MPCs as a function of delay.

In order to statistically model the arrival times of MPCs, a first-order approximation assumes that objects that cause reflections in an urban area are located randomly in space, giving rise to a *Poisson distribution* for excess delays. However, measurements have shown that MPCs tend to arrive in groups ("clusters"). Two models have been developed to reflect this fact: the $\Delta - K$ model, and the Saleh–Valenzuela (SV) model.

The $\Delta - K$ *model* has two states: S_1, where the mean arrival rate is $\lambda_0(t)$, and S_2, where the mean arrival rate is $K\lambda_0(t)$. The process starts in S_1. If a MPC arrives at time t, a transition is made to S_2 for the interval $[t, t + \Delta]$. If no further paths arrive in this interval, a transition is made back to S_1 at the end of the interval. Note that for $K = 1$ or $\Delta = 0$, the above-mentioned process reverts to a standard Poisson process.

The *Saleh–Valenzuela model* takes a slightly different approach. It assumes a priori the existence of cluster. *Within* each cluster, the MPCs are arriving according to a Poisson distribution, and the arrival times of the *clusters themselves* are Poisson-distributed (but with a different interarrival time constant). Furthermore, the powers of the MPCs within a cluster decrease exponentially with delay, and the power of the clusters follows a (different) exponential distribution (see Fig. 7.2).

Mathematically, the following discrete time impulse response is used:

$$h(\tau) = \sum_{l=0}^{L} \sum_{k=0}^{K} c_{k,l}(\tau)\delta(\tau - T_l - \tau_{k,l})$$

where the distribution of cluster arrival time and the ray arrival time are described (with a slight abuse of notation) as:

$$pdf(T_l|T_{l-1}) = \Lambda \exp[-\Lambda(T_l - T_{l-1})], \qquad l > 0$$

$$pdf(\tau_{k,l}|\tau_{(k-1),l}) = \lambda \exp[-\lambda(\tau_{k,l} - \tau_{(k-1),l})], \qquad k > 0$$

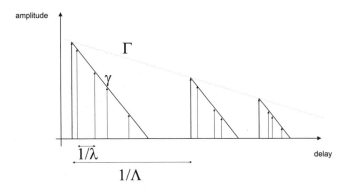

Figure 7.2 The Saleh–Valenzuela model.

where T_l is the arrival time of the first path of the lth cluster, $\tau_{k,l}$ is the delay of the kth path within the lth cluster relative to the first path arrival time of this (by definition, $\tau_{0,l} = 0$), Λ is the cluster arrival rate, and λ is the ray arrival rate – i.e., the arrival rate of paths within each cluster. All dependences of the parameters on absolute time have been suppressed in the above equations.

The PDP within each cluster is:

$$E\{|c_{k,l}|^2\} = \frac{P_l^c}{\gamma}\exp(-\tau_{k,l}/\gamma) \tag{7.6}$$

where P_l^c is the energy of the lth cluster, and γ is the intracluster decay time constant. Cluster power decreases exponentially as well:

$$P_l^c \exp(-T_l/\Gamma) \tag{7.7}$$

7.3.4 Standardized channel model

A special case of the tapped delay line model is the COST 207 model, which specifies the PDPs or tap weights and Doppler spectra for four typical environments. These PDPs were derived from numerous measurement campaigns in Europe. The model distinguishes between four different types of macrocellular environments – namely, *Typical Urban* (TU), *Bad Urban* (BU), *Rural Area* (RA), and *Hilly Terrain* (HT). Depending on the environment, the PDP has a single-exponential decay, or it consists of two single-exponential functions (clusters) that are delayed with respect to each other (see Fig. 7.3). The second cluster corresponds to groups of far-away high-rise buildings or mountains that act as efficient IOs, and thus give rise to a group of delayed MPCs with considerable power. More details can be found in Appendix 7.C (see companion website *www.wiley.com/go/molisch*).

The COST 207 models are based on measurements with a rather low bandwidth, and are applicable only for systems with 200-kHz bandwidth or less. For simulation of third-generation cellular systems, which have a bandwidth of 5 MHz, the *International Telecommunications Union* (ITU) specified another set of models that accounts for the larger bandwidth. This model distinguishes between pedestrian, vehicular, and indoor environments. Details can be found in Appendix 7.D (see *www.wiley.com/go/molisch*). Additional tapped delay line models were also derived for indoor wireless Local Area Network (LAN) systems and Personal Area Networks (PANs); for an overview, see Molisch and Tufvesson [2004].

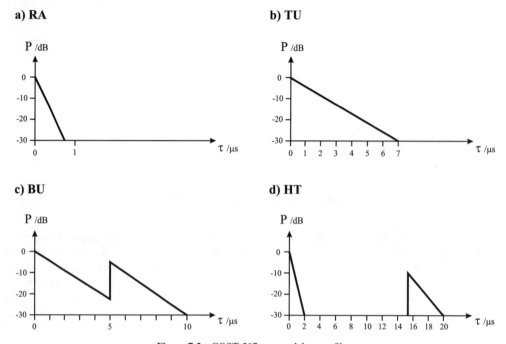

Figure 7.3 COST 207 power delay profiles.
Reproduced with permission from Molisch [2000] © Prentice Hall.

7.4 Directional models

7.4.1 General model structure and factorization

As discussed in Chapter 6, a fairly general model is based on the Double Directional Delay Power Spectrum (DDDPS), which depends on the three variables Direction Of Departure (DOD), Direction Of Arrival (DOA), and delay. An important simplification is obtained if the DDDPS can be factored into three functions, each of which depends on just a single parameter:

$$DDDPS(\Omega, \Psi, \tau) = APS^{\mathrm{BS}}(\Omega)APS^{\mathrm{MS}}(\Psi)P_h(\tau) \tag{7.8}$$

This implies that the Angular Power Spectrum (APS) at the BS is independent of delay, as is the angular power spectrum at the MS. Furthermore, the angular power spectrum at the MS is independent of the direction in which the BS transmits, and vice versa.

Such a factorization greatly simplifies theoretical computations, and also the parameterization of channel models. However, it does not always correspond to physical reality. A more general model assumes that the DDDPS consists of several *clusters*, each of which has a separable DDDPS:

$$DDDPS(\Omega, \Psi, \tau) = \sum_l P_l^c APS_l^{c,\mathrm{BS}}(\Omega)APS_l^{c,\mathrm{MS}}(\Psi)P_{h,l}^c(\tau) \tag{7.9}$$

where superscript ^{c} stands for "cluster". and l indexes the clusters. Obviously, this model reduces to Eq. (7.8) only if a single cluster exists.

In the remainder of this section, we assume that factorization is possible, and describe only models for the angular spectra $APS_l^{c,\mathrm{BS}}$ and $APS_l^{c,\mathrm{MS}}(\Psi)$ – i.e., the components in this factorization (the PDP has already been discussed in Section 7.3).

7.4.2 Angular dispersion at the base station

The most common model for the APS at the BS is a Laplacian distribution in azimuth [Pedersen et al. 1997]:

$$APS(\phi) \propto \exp\left[-\sqrt{2}\frac{|\phi - \phi_0|}{S_\phi}\right] \tag{7.10}$$

where ϕ_0 is the mean azimuthal angle. The elevation spectrum is usually modeled as a delta function (i.e., all radiation is incident in the horizontal plane) so that $\Omega = \phi$; alternatively, it has been modeled as a Laplacian function.

The following range of rms angular spreads (see Section 6.7) and cluster angular spreads can be considered typical [Molisch and Tufvesson 2004]:

- *Indoor office environments*: rms cluster angular spreads between $10°$ and $20°$ for non-LOS situations, and typically around $5°$ for LOS.
- *Industrial environments*: rms angular spreads between $20°$ and $30°$ for non-LOS situations.
- *Microcells*: rms angular spreads between $5°$ and $20°$ for LOS, and $10°$–$40°$ for non-LOS.
- *Typical urban and suburban environments*: measured rms angular spreads on the order of $3°$–$20°$ in dense urban environments. In suburban environments, the angular spread is smaller, due to a frequent occurrence of LOS.
- *Bad urban and hilly terrain environments*: rms angular spreads of $20°$ or larger, due to the existence of multiple clusters.
- *Rural environments*: rms angular spreads between $1°$ and $5°$ have been observed.

In outdoor environments, the distribution of the angular spread over large areas has also been found to be lognormal and correlated with the delay spread. This permits the logarithms of the spreads to be treated as correlated Gaussian random variables. The dependence of angular spread on distance is still a matter of discussion.

7.4.3 Angular dispersion at the mobile station

For outdoor environments, it is commonly assumed that radiation is incident from all azimuthal directions onto the MS, because the MS is surrounded by "local IOs" (cars, people, houses, etc.). This model dates back to the 1970s. However, recent studies indicate that the azimuthal spread can be considerably smaller, especially in street canyons. The APS is then again approximated as Laplacian; cluster angular spreads on the order of $20°$ have been suggested. Furthermore, the angular distribution is a function of delay. For MSs located in street canyons without LOS, small delays are related to over-the-rooftop propagation, which results in large angular spreads, while later components are waveguided through the streets and thus confined to a smaller angular range. In indoor environments with (quasi-) LOS, early components have a very small angular spread, while components with larger delay have an almost uniform APS.

For the outdoor elevation spectrum, MPCs that propagate over the rooftops have an elevation distribution that is uniform between 0 and the angle under which the rooftops are seen; later-arriving components, which have propagated through the street canyons, show a Laplacian elevation distribution.

7.4.4 Polarization

Most channel models analyze only the propagation of vertical polarization, corresponding to transmission and reception using vertically polarized antennas. However, there is increased interest in polarization diversity – i.e., antennas that are co-located, but receive waves with different polarizations. In order to simulate such systems, models for the propagation of dual-polarized radiation are required.

Transmission from a vertically polarized antenna will undergo interactions that result in energy being leaked into the horizontal polarization component before reaching the RX antenna (and vice versa). The fading coefficients for MPCs thus have to be written as a polarimetric 2×2 matrix, so that the complex amplitude $\mathbf{a}_\ell$ becomes:

$$\mathbf{a}_\ell = \begin{pmatrix} a_\ell^{VV} & a_\ell^{VH} \\ a_\ell^{HV} & a_\ell^{HH} \end{pmatrix} \tag{7.11}$$

where V and H denote vertical and horizontal polarization, respectively.

The most common polarimetric channel model assumes that the entries in the matrix are statistically independent, complex Gaussian fading variables. The *mean* powers of the VV and the HH components are assumed to be identical; similarly, the *mean* powers of the VH and HV components are the same. The cross-polarization ratio, XPD, which is the ratio (expressed in dB) of the mean powers in VV and VH, is modeled as a Gaussian random variable. The mean and variance of the XPD can depend on the propagation environment, and even on the delay of the considered components. Typical values for the mean of the XPD lie between 0 and 12 dB; for the variance, around 3–6 dB [Shafi et al. 2005].

7.4.5 Model implementations

The above sections have discussed a continuous model for the angular spectra. For a computer implementation, we usually need a discretized version. One way of implementing a directional channel model is a generalized tapped delay line. In this approach, the DDDPS is discretized, according to the same principles as described in Section 6.4.

An alternative is the so-called *Geometry-based Stochastic Channel Model* (GSCM). In this approach, it is not the strength and direction of the MPCs that is modeled stochastically, but rather the location of IOs and the strength of the interaction processes (see Fig. 7.4). Additionally, it is assumed that only single interaction processes can occur. The directionally resolved impulse response is then obtained in two steps:

1. Assign locations to the IOs, according to the pdf of their position.
2. Based on the assumption of single interaction only, determine the contributions of the IOs to the double-directional impulse response. Each MPC (corresponding to one IO) has a unique DOA, DOD, amplitude delay, and phaseshift.

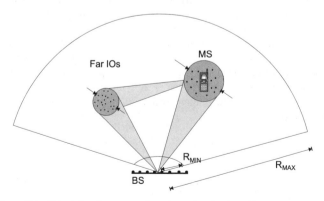

Figure 7.4 Principle of geometry-based stochastic channel model modeling.

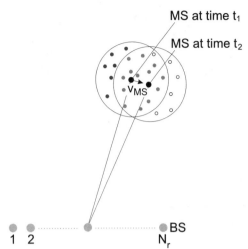

Figure 7.5 "Vanishing" and "appearing" of IOs when the MS moves. It is assumed that all the IOs are in a disk around the MS. Scatterers that are active only at time t_1 (t_2) are shown as black (empty) circles; scatterers that are active at both time instants are shown in grey.

Reproduced with permission from Fuhl et al. [1998[© IEE.

The simplest model is based on the assumption that all relevant IOs are close to the MS. This case occurs, e.g., in macrocells with regular building structures like suburban environments. In this case, radiation from the MS interacts with IOs around the MS, but can proceed without further interaction from those objects to the BS. Different models exist for the distribution of the IOs around the MS:

- Some papers place all IOs on a circle around the MS.
- Other papers suggest a uniform distribution within a disk. When the MS moves, the disk around the MS also moves. Some IOs thus "fall out" of the IO disk, while new IOs enter (see Fig. 7.5). This corresponds to the physical reality that IOs that are far away from the MS do not make significant contributions (although naturally they still "exist" physically).
- A one-sided Gaussian distribution $pdf(r) = \exp(-r^2/2\sigma^2)$, $r \geq 0$, has also been suggested. Computing the PDP and the APS from this distribution gives the results shown in Figs. 7.6 and 7.7. We see that these results are fairly similar to an exponential PDP and Laplacian APS.

The case when all IOs are close to the MS is the "single-cluster" case, with an Angular Delay Power Spectrum (ADPS) that is approximately given as:

$$ADPS(\tau, \phi) \propto \exp(-\tau/S_\tau) \exp(-\sqrt{2}|\phi - \phi_0|/S_\phi)$$

The generalization includes so-called *far IOs* (also known as far scatterers), which correspond to high-rise buildings or mountains. Such a far IO can be modeled either as a single specular reflector (corresponding, e.g., to a high-rise building with a smooth glass front) or a cluster of IOs. In contrast to the IOs around the MS, the location of such an IO cluster stays constant during a whole simulation process.

Geometric channel models have advantages especially when movement is to be simulated. Whenever the MS moves, adjustments to the parameters of the MPCs are automatically made. Thus, the correct fading correlation results automatically from movement; also the correlation between the DOAs at the MS and the Doppler shift is taken into account. Any

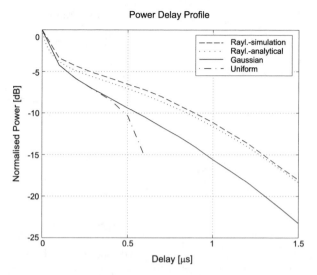

Figure 7.6 PDP for different distributions of scatterers.

Reproduced with permission from Laurila et al. [1998] © IEEE.

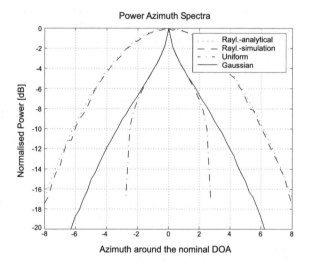

Figure 7.7 Angular power spectrum for different distributions of scatterers.

Reproduced with permission from Laurila et al. [1998] © IEEE.

changes in the mean DOAs, DODs, and delays due to large-scale movement of the MS are automatically included, while they would be difficult to model in tapped delay line models.

7.4.6 Standardized directional models

The European research initiative COST 259 developed a Directional Channel Model (DCM) that has gained widespread acceptance. It is very realistic, incorporating a wealth of effects and their interplay, for a number of different environments. As the model is rather involved, this section only points out some basic features. A more detailed description of the first version of the model is described in Steinbauer and Molisch [2001], and a full account is found in Asplund et al. [2004] and Molisch et al. [2004b].

The COST 259 DCM includes small-scale as well as continuous large-scale changes of the channel. This is achieved efficiently by distinguishing between three different layers:

1. At the top layer, there is a distinction between different *Radio Environments* (REs) – i.e., environmental classes with similar propagation characteristics (e.g., "Typical Urban"). All in all, there are 13 REs: 4 macrocellular REs (i.e., BS height above rooftop), 4 microcellular (outdoor, BS height below rooftop) REs, and 5 picocellular (indoor) REs.
2. *Large-scale effects* are described by their pdfs, whose parameters differ for different REs. For example, delay spread, angular spread, shadowing, and the Rice factor change as the MS moves over large distances. Each realization of large-scale fading parameters determines a DDDPS.
3. On a third layer, double-directional impulse responses are realizations of the DDDPS, created by the *small-scale fading*.

Large-scale effects are described in a mixed geometrical–stochastic fashion, applying the concept of IO clusters as described above. At the beginning of a simulation, IO clusters (one local cluster around the MS, and several far IO clusters) are distributed at random in the coverage area; this is the stochastic component. During the simulation, the delays and angles between the clusters are obtained deterministically from their position and the positions of the BS and MS; this is the geometrical component. Each of the clusters has a small-scale averaged DDDPS that is exponential in delay, Laplacian in azimuth and elevation at the BS, and uniform or Laplacian in azimuth and elevation at the MS. Double-directional complex impulse responses are then obtained from the average ADPS either directly, or by mapping it onto an IO distribution and obtaining impulse responses in a geometrical way.

In macrocells the positions of clusters are random. In micro- and picocells the positions are deterministic, using the concept of *Virtual Cell Deployment Areas* (VCDAs). A VCDA is essentially a map of a virtual town or office building, with the route of the MS prescribed in it. This approach is similar to the ray-tracing approach but differs in two important respects: (i) the "city maps" need not reflect an actual city and can thus be made to reflect a cross-section of different cities; (ii) only the cluster *positions* are determined by ray tracing, while the behavior *within* one cluster is still treated stochastically.

Another double-directional channel model was standardized by the 3GPP (Third Generation Partnership Project) and 3GPP2, the standardization organizations for third-generation cellular systems (see Chapter 23). This model, which is similar to the COST 259 model, is described in detail in Calcev et al. [2004].

7.4.7 Multiple-input multiple-output matrix models

The previous sections have described models that include the directional information of MPCs. An alternative concept that is popular in the context of multiantenna systems is to stochastically model the impulse response matrix (see Section 6.7) of a Multiple Input Multiple Output (MIMO) channel. In this case, the channel is characterized not only by the amplitude statistics of each matrix entry (which is usually Rayleigh or Rician), but also by the *correlation* between these entries. The correlation matrix (for each delay tap) is defined by first "stacking" all the entries of the channel matrix in one vector

$\mathbf{h}_{\text{stack}} = [h_{1,1}, h_{2,1}, \ldots, h_{N_r,1}, h_{1,2}, \ldots, h_{N_r N_t}]^T$ and then computing the correlation matrix as $\mathbf{R} = E\{\mathbf{h}_{\text{stack}} \mathbf{h}_{\text{stack}}^{\dagger}\}$ where superscript † denotes the Hermitian transpose. One popular simplified model assumes that the correlation matrix can be written as a Kronecker product $\mathbf{R} = \mathbf{R}_{\text{TX}}^T \otimes \mathbf{R}_{\text{RX}}$, where $\mathbf{R}_{\text{TX}} = E\{H^T H^*\}$ and $\mathbf{R}_{\text{RX}} = E\{HH^{\dagger}\}$. This model implies that the correlation matrix at the RX is independent of the direction of transmission; this is equivalent to assuming that the DDDPS can be factored into independent APSs at the BS and at the MS. In that case, the channel transfer function matrix can be generated as:

$$\mathbf{H} = \mathbf{R}_{\text{RX}}^{1/2} \mathbf{G}_{\text{G}} \mathbf{R}_{\text{TX}}^{1/2} \tag{7.12}$$

where $\mathbf{G}_{\text{G}}$ is a matrix with independent identically distributed (iid) complex Gaussian entries.

7.5 Deterministic channel-modeling methods

In principle, a wireless propagation channel can be viewed as a deterministic channel. Maxwell's equations, together with electromagnetic boundary conditions (location, shape, and dielectric and conductive properties of all objects in the environment), allow determination of the fieldstrength at all points and times. For outdoor environments, such purely deterministic channel models have to take into account all the geographical and morphological features of a propagation environment; for indoor environments, the building structure, wall properties, and even furniture should be taken into consideration. In this section, we will outline the basic principles of channel models based on such a deterministic point of view.

Deterministic models suffer from *two main problems*: they require (i) large amounts of computer time, and (ii) exact knowledge of boundary conditions:

- Up to about 1990, such deterministic models could not be used because the requirements in terms of *computer time and storage* were prohibitive. However, this has changed since then. On one hand, computers have become so much faster that tasks that seemed unfeasible even with supercomputers in 1990 are realistic options on a personal computer nowadays. On the other hand, the development of more efficient deterministic algorithms has improved the situation as well.
- Another fundamental problem of deterministic models is the requirement of *exact knowledge of boundary conditions*. This implies that the position and the electromagnetic properties of the whole "relevant" environment have to be known (we will discuss later what "relevant" means in this context). The creation of digital terrain maps and city plans, based on satellite images or building plans, has also made considerable progress in the last few years.

The most accurate solution (given an environment database) is a "brute force" solution of Maxwell's equations, employing either integral or differential equation formulations. Integral equations are most often variations of the well-known *Method of Moments*, where the unknown currents induced in the IOs are represented by a set of basis functions. In their most simple form, basis functions are rectangular functions, extending over a fraction of a wavelength. Differential equation formulations include the *Finite Element Method* (FEM), or the increasingly popular *Finite Difference Time Domain* (FDTD) method.

All these methods are highly accurate, but the computational requirements are prohibitive in most environments. It is thus much more common to use approximations to Maxwell's equations as a basis for solution. The most widespread approximation is the *high-frequency* approximation (also known as *ray approximation*).[4] In this approximation, electromagnetic waves are modeled as rays that follow the laws of geometrical optics (Snell's laws for reflection and transmission). Further refinements allow to include diffraction and diffuse scattering in an approximate way.

7.5.1 Ray launching

In the *ray-launching* approach, the transmit antenna sends (launches) rays in different directions. Typically, the total spatial angle 4π is divided into N units of equal magnitude, and each ray is sent in the direction of the center of one such unit (i.e., uniform sampling of the spatial angle) (see Fig. 7.8). The number of launched rays is a tradeoff between accuracy of the method and computation time.

The algorithm follows the propagation of each ray until it either hits the RX, or becomes too weak to be significant (e.g., drops below the noise level). When following a ray, a number of effects have to be taken into account:

- *Free space attenuation*: as each ray represents a certain spatial angle, the energy *per unit area* decreases as d^{-2} along the path of the ray.
- *Reflections* change the direction of a ray, and cause an additional attenuation. Reflection coefficients can be computed from Snell's laws (see Chapter 4) depending on the angle of incidence and possibly the polarization of the incident ray.
- *Diffraction and diffuse scattering* are included in more advanced models. In those cases, a ray that is incident on an IO gives rise to several new rays. The amplitudes of diffracted rays are usually computed from the geometrical or uniform theory of diffraction, as discussed in Chapter 4.

The *ray-splitting* algorithm is an important improvement in the accuracy of the method. The algorithm is based on the premise that the effective cross-section of the ray should never exceed a certain size (e.g., the size of a typical IO). Thus, if a ray has propagated too far from the TX, it is subdivided into two rays. Let us explain that principle in more detail using Fig. 7.9. To simplify the discussion, we consider only the two-dimensional case. Each ray represents not only a certain angle, but rather an angular *range* of width ϕ – corresponding to the angle between two launched rays. The intersection of such an angular range with a circle of radius d has a length of approximately ϕd. Thus, the farther we get away from the TX, the larger the length that is covered by the ray. It is this length that should not become too large. As soon as it reaches a length $\overline{L}$, the ray is split (thus reducing the length to $\overline{L}/2$). The resulting subrays (which again represent a whole angular range of width ϕ) then propagate until they reach a length $\overline{L}$, and so on.

Ray launching gives the channel characteristics in the whole environment – i.e., for many different RX positions and a given TX position. In other words, once we have decided on a BS location, we can compute coverage, delay spread, and other channel characteristics in the whole envisioned cell area. Furthermore, a preprocessing scheme allows the inclusion of multiple TX locations. The environment (the IOs) is subdivided into "tiles" (areas of finite size, typically the same size as the maximum effective area of a ray), and the interaction between all tiles is computed. Then, for each TX position, only the interaction between the TX and the tiles that can act as first IOs has to be computed [Hoppe et a. 2003].

---

[4] In the literature, such methods are often generally known as *ray tracing*. However, the expression "ray tracing" is also used for one specific implementation method (described below). We will thus stick with the name "high frequency approximation" for the general class of algorithms.

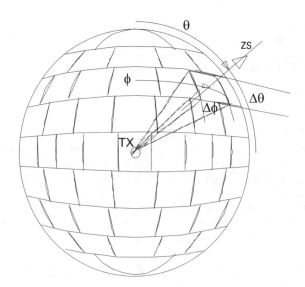

Figure 7.8 Principle of ray launching.

Reproduced with permission from Damosso and Correia [1999] © European Union.

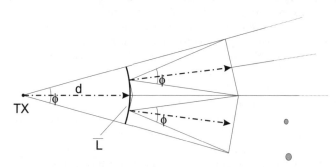

Figure 7.9 Principle of ray splitting.

7.5.2 Ray tracing

Classical *ray tracing* determines all rays that can go from *one* TX location to *one* RX location. The method operates in two steps:

1. First, all rays that can transfer energy from the TX location to the RX location are determined. This is usually done by means of the image principle. Rays that can get to the RX via a reflection show the same behavior as rays from a virtual source that is located where an image of the original source (with respect to the reflecting surface) would be located (see Fig. 7.10).
2. In a second step, attenuations (due to free space propagation and finite reflection coefficients) are computed, thus providing the parameters of all MPCs.

Ray tracing allows fast computation of single- and double-reflection processes, and also does not require ray splitting. On the downside, effort increases exponentially with the order of reflections that are included in the simulation. Also, the inclusion of diffuse scattering and

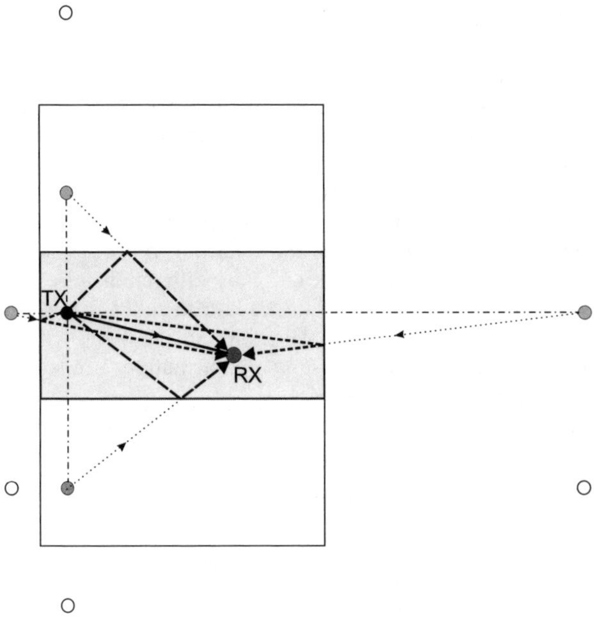

Figure 7.10 The image principle. Grey circles: virtual sources corresponding to a single reflection. White circles: virtual sources corresponding to double reflections. Dotted lines: rays from the virtual sources to the RX. Dashed lines: actual reflections. Solid lines: line of sight.

diffraction is non-trivial. Finally, the method is less efficient than ray launching for the computation of channel characteristics over a wide area.

7.5.3 Efficiency considerations

Both for ray launching and ray tracing, it is almost impossible to correctly predict the *phases* of arriving rays. Such a prediction would require a geographical and building database that is accurate to within a fraction of a wavelength. It is thus preferable to assume that all rays have uniformly distributed random phases. In this case, it is only possible to deterministically predict the small-scale *statistics* of channel characteristics; realizations of the impulse responses are obtained by ascribing random phases to MPCs. This is another form of the mixed deterministic–stochastic approach mentioned at the beginning of this chapter.

A further method to reduce the computational effort is to perform ray tracing not in all three dimensions, but rather only in two dimensions. It depends on the propagation environment whether this simplification is admissible:

- *Indoor*: indoor environments practically always require three-dimensional considerations. Even when the BS and the MS are on the same floor, reflections at floors and ceilings represent important propagation paths.
- *Macrocells*: by definition, the BS antenna is considerably above the rooftops. Propagation thus occurs mostly over the rooftops to points that are close to the MS. From these points, they then reach the MS, possibly via a diffraction or a reflection from the wall of the house opposite. Ray tracing in the vertical plane alone can thus be sufficient for *some* cases. This is especially true when ray tracing should only predict the received power and delay spread.

On the other hand, such a purely vertical ray tracing will not correctly predict the directions of the rays at the MS.

- *Microcells, small distance BS–MS*: as both BS and MS antennas are below the rooftop, the diffraction loss of over-the-rooftop propagation is large. Propagation in the horizontal plane – i.e., through street canyons – can be a much more efficient process. Under these conditions, ray tracing in just the horizonthal plane can be sufficient.

- *Microcells, large distance BS–MS*: in this case, the relative power of rays propagating in the horizontal plane (compared with over-the-rooftop components) is smaller. Horizontal components undergo multiple diffraction and reflection processes, while losses from over-the-rooftop components are mostly determined by diffraction losses near the BS and the MS, and thus depend less on distance. In this case, a so-called *2.5-dimensional model* can be used: only propagation in the horizontal plane, on one hand, and only in the vertical plane, on the other hand, is simulated, and these two contributions are added together.

2.5-dimensional modeling can also be used for macrocells. However, both in macrocells, and in microcells with the BS antenna close to the rooftop height, there are propagation processes that cannot be correctly modeled by 2.5-dimensional ray tracing. For example, reflections at a far IO cluster (high-rise building) are not accounted for in this approach (see Fig. 7.11).

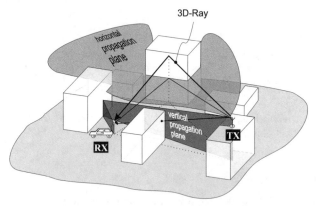

Figure 7.11   Two- and three-dimensional modeling.

7.5.4 Geographical databases

The foundation of all deterministic methods is the information about the geography and morphology of the environment. The accuracy of that information determines the achievable accuracy of any deterministic channel model.

For *indoor environments*, that information can usually be obtained from building plans, which nowadays are often available in digital form.

In *rural areas*, geographical databases are available with a resolution of 10–100 m. These databases are often created by means of satellite observations. In many countries, morpholgical information (land usage) is also available; however, obtaining this information in an automated and consistent way can be quite challenging.

In *urban areas*, digital databases use two different types of data: vector data and pixel data. For vector data, the actual location of building endpoints are stored. For pixel data, a regular grid of points is superimposed on the area, and for each pixel it is stated whether it falls on "free space" (streets, parks, etc.) or is covered by a building. In both cases, building heights and materials might be included in the database.

7.6 Appendices

Please see companion website *www.wiley.com/go/molisch*

Further reading

There is a rich literature on channel models. Besides the original papers already mentioned in the main text, Andersen et al. [1995] and Molisch and Tufvesson [2004] give overviews of different models. For implementation of the tapped delay line model, we recommend Paetzold [2002]. Generalizations of the tapped delay line approach to the MIMO case were suggested by Xu et al. [2002]. Poisson approximations for arrival times were developed by Turin et al. [1972] and later improved and extended by Suzuki [1977] and Hashemi [1979]. The Saleh–Valenzuela model was first proposed in Saleh and Valenzuela [1987].

The Okumura–Hata model is based on the extensive measurements of Okumura et al. [1968] in Japan, and was brought into a form suitable for computer simulations by Hata [1980].

The COST 231–Walfish–Ikegami model was developed by the research and standardization group COST 231 [Damosso and Correia 1999] based on the work of Walfish and Bertoni [1988] and Ikegami et al. [1984].

The GSCM was proposed in one form or the other in Blanz and Jung [1998], Fuhl et al. [1998], Norklit and Andersen [1998], and Petrus et al. [2002], see also Liberti and Rappaport [1996]. More details about efficient implementations of a GSCM can be found in Molisch et al. [2003], while a generalization to multiple-interaction processes is described in Molisch [2004]. The Laplacian structure of the power azimuthal spectrum was first suggested in Pedersen et al. [1997], and though there has been some discussion about its validity, it is now in widespread use.

The description of the channels for MIMO systems by transfer function matrices stems from the classical work of Foschini and Gans [1998], and Winters [1987]. The Kronecker assumption was proposed in Kermoal et al. [2002]; a more general model encompassing correlations between DOAs and DODs was introduced by Weichselberger et al. [2003]; other generalized models were proposed by Gesbert et al. [2002] and Sayeed [2002].

The Method of Moments is described in the classical book by Harrington [1993]. Special methods that increase the efficiency of the method include *natural basis sets* [Moroney and Cullen 1995], the *fast multipole method* [Rokhlin 1990], and the *tabulated interaction method* [Brennan and Cullen 1998]. The FEM is described in Zienkiewicz and Taylor [2000], while the FDTD method is described in Kunz and Luebbers [1993].

Ray tracing originally comes from the field of computer graphics, and a number of books (e.g., Glassner [1989]) are available from that perspective; its application to wireless is described, e.g., in Valenzuela [1993]. Descriptions of the ray-launching algorithm can be found in Lawton and McGeehan [1994]. Ray splitting was introduced in Kreuzgruber et al. [1993]. There is also a considerable number of commercial software programs for radio channel prediction that use ray tracing.

As far as standardized models are concerned, a number of models that include directional information or have a larger bandwidth have been developed recently. The 3GPP model describes the spatial characteristics of the channel in outdoor cellular environments [Calcev et al. 2004], while the IEEE 802.11n model covers spatial models for indoor environments [Erceg et al. 2004]. The IEEE 802.15.3a and 4a channel models [Molisch et al. 2003a, 2004a] describe channel models for the ultrawideband case.

8

Channel sounding

8.1 Introduction

8.1.1 Requirements for channel sounding

Any channel model is based on measurement data. For stochastic channel models, parameter values have to be obtained from extensive measurement campaigns, while for deterministic models, the quality of the prediction has to be checked by comparisons with measured data. Measurement of the properties (impulse responses) of wireless channels, better known as *channel sounding*, is thus a fundamental task for wireless communications engineering.

As the systems, and the required channel models, become more complex, so do the tasks of channel sounding. Measurement devices in the 1960s only had to measure the received fieldstrength. The transition to wideband systems necessitated the development of a new class of channel sounders that could measure impulse responses – i.e., delay dispersion. The focus on directional propagation properties that arose in the 1990s, caused by the interest in multiantenna systems, also affects channel sounders. These sounders now have to be able to measure double-directional impulse responses.

In addition to the changes in measured quantities, the *environments* in which the measurements are done are changing. Up to 1990, measurement campaigns were usually performed in macrocells. Since then, microcells and especially indoor propagation has become the focus of interest. For channel sounders, this means that a finer delay resolution is required.

In the following, we discuss the most important channel-sounding approaches. After a discussion of the basic requirements of wideband measurements, different types of sounders are described. An outline of spatially resolved channel sounding concludes the chapter.

8.1.2 Generic sounder structure

The word *channel sounder* gives a graphic description of the functionality of such a measurement device. A transmitter (TX) sends out a signal that excites – i.e., "sounds" – the channel. The output of the channel is observed by the receiver (RX), and stored. From the knowledge of the transmit and the receive signal, the time-variant impulse response or one of the other (deterministic) system functions is obtained.

Figure 8.1 shows a block diagram of the channel sounder that is conceptually most simple.

Wireless Communications   Andreas F. Molisch
© 2005 John Wiley & Sons, Ltd

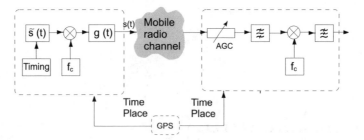

Figure 8.1 Principle of a channel sounder. Correct synchronization between transmitter and receiver is especially important.

The TX sends out a signal $s(t)$ that consists of modulated, periodically repeated pulses $p(t)$:

$$s(t) = \sum_{i=0}^{N-1} p(t - iT_{\text{rep}}) \tag{8.1}$$

where T_{rep} is the repetition interval of the transmitted pulses. One *measurement run* consists of N pulses that are transmitted at fixed intervals. The pulses are the convolution of a basis pulse $\widetilde{s}(t)$ created by a pulse generator and a transmit filter:

$$p(t) = \widetilde{s}(t) * g(t) \tag{8.2}$$

where $g(t)$ is the impulse response of the transmit filter. The waveform used for $p(t)$ depends on the type of sounder, and can greatly differ for sounders working in the time domain, compared with sounders working in the frequency domain, as will be discussed in Sections 8.2 and 8.3.

This block diagram is generic; the properties of the sounder are mostly determined by the choice of the sounding signal. In order to perform efficient measurements, the following requirements should be fulfilled by the sounding signal:

- *Large bandwidth*: the bandwidth is inversely proportional to the shortest temporal changes in the sounding signal and thus determines the achievable *delay resolution*.
- *Large time bandwidth product*: it is often advantagous if the sounding signal has a duration that is longer than the inverse of the bandwidth – i.e., a time bandwidth product TW larger than unity. For many systems, the transmit *power* is limited. In this case, a large TW allows the transmission of high *energy* in the sounding signal, and thus obtain a higher Signal to Noise Ratio (SNR) at the RX. Sounding schemes with large TW are related to spread spectrum systems (Chapter 18). At the RX, special signal processing (despreading) is required in order to exploit the benefits of large TW.
- *Signal duration*: the effective signal duration must be adapted to channel properties. On one hand, a long sounding signal can give a large time bandwidth product, which is beneficial (see above). On the other hand, the sounding signal should not be longer than the coherence time of the channel – i.e., the time during which the channel can be considered to be approximately constant. For practical reasons, the pulse repetition time T_{rep} should be larger than the duration of the constituent pulse $p(t)$ and the maximum excess delay of the channel.
- *Power-spectral density*: the power-spectral density of the sounding signal, $|P_{\text{TX}}(j\omega)|^2$, should be uniform across the bandwidth of interest. This allows us to have the same quality of the channel estimate at all frequencies. Due to efficiency considerations, little energy should be transmitted outside the bandwidth of interest.

- *Low crest factor*: signals with a low crest factor

$$C_{\text{crest}} = \frac{\text{Peak amplitude}}{\text{rms amplitude}} = \frac{\max\{s(t)\}}{\sqrt{\overline{s^2(t)}}} \qquad (8.3)$$

allow efficient use of the transmit power amplifier. A first estimate for the crest factor of *any* signal can be obtained from Felhauer et al. [1993]:

$$1 < C_{\text{crest}} \leq \sqrt{TW} \qquad (8.4)$$

- *Good correlation properties*: correlation-based channel sounders require signals whose auto-correlation function has a high *Peak to Off Peak ratio* (POP), and a zero mean. The latter property allows *unbiased estimates* (see Section 8.4.2). Correlation properties are critical for channel estimates that directly use the correlation function, while it is less important for parameter-based estimation techniques (see Section 8.5).

The design of optimum, digitally synthesized sounding signals thus proceeds in the following steps:

1. Choose the duration of the sounding signal according to the channel coherence time and the required time bandwidth product.
2. For a constant power-spectral density, all frequency components need to have the same absolute value.
3. Now the only remaining free parameters are the phases of the frequency components. These can be adjusted to yield a low crest factor.

8.1.3 Identifiability of wireless channels

The temporal variability of wireless channels has an impact on whether the channel can be identified (measured) in a unique way. A band-limited *time-invariant* channel can always be identified by appropriate measurement methods, the only requirement being that the RX fulfills the Nyquist theorem [Proakis 1995] in the delay domain – i.e., samples the received signal sufficiently fast. In a *time-variant* system, the repetition period T_{rep} of the sounding pulses $p(t)$ is of fundamental importance. The channel response to any excitation pulse $p(t)$ can be seen as one "snapshot" (sample) of the channel (see Fig. 8.2). In order to track changes in the channel, these snapshots need to be taken sufficiently often. Intuitively, T_{rep} must be smaller than the time over which the channel changes. This notion can be formalized by establishing a sampling theorem in the time domain. Just as there is a minimum sampling rate to identify a signal with a band-limited spectrum, so is there a minimum temporal sampling rate to identify a time-variant process with a band-limited Doppler spectrum. Thus, the temporal sampling frequency must be twice the maximum Doppler frequency ν_{max}:

$$f_{\text{rep}} \geq 2\nu_{\text{max}} \qquad (8.5)$$

Rewriting Eq. (8.5), and using the relationship between the movement speed of the MS and the Doppler frequency $\nu_{\text{max}} = f_0 v_{\text{max}}/c_0$ (see Eq. 5.8), the repetition frequency for the pulses can be written as:

$$T_{\text{rep}} \leq \frac{c_0}{2f_c v_{\text{max}}} \qquad (8.6)$$

In that case:

$$\frac{v}{\Delta x_s} \geq 2 \frac{v_{\text{max}}}{\lambda_c} \qquad (8.7)$$

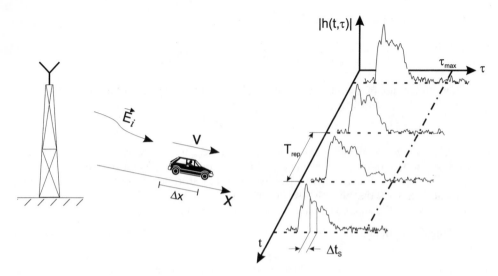

Figure 8.2 Time-variant impulse response of the channel and channel identifiability. A new snapshot can only be taken after the impulse response of the previous excitation has died down.

holds, so that the distance Δx_s between the locations at which the sounding has to take place is upper-bounded as:

$$\Delta x_s \leq \frac{v}{v_{max}} \frac{\lambda}{2} \leq \frac{\lambda}{2} \tag{8.8}$$

Equation (8.8) thus tells us that for an aliasing-free measurement at least two snapshots per wavelength are required.

For strongly time-varying channels, requirements for the design of sounding signals can become contradictory. On one hand, the repetition frequency T_{rep} has to be larger than the maximum excess delay of the channel τ_{max}; otherwise the impulse responses from the different excitation pulses start to overlap. On the other hand, we have just shown that the repetition frequency has to fulfill $T_{rep} \leq 1/2\nu_{max}$. Thus channels can be identified in an unambiguous way only if:

$$2\tau_{max}\nu_{max} \leq 1 \tag{8.9}$$

This equation is also known as the *two-dimensional Nyquist criterion*. A channel that fulfills these requirements is known as *underspread*. If Eq. (8.9) is not fulfilled, then the channel can only be identified by making specific assumptions – e.g., a certain parametric model. Fortunately, the overwhelming majority of wireless channels are underspread; in many cases, even $2\tau_{max}\nu_{max} \ll 1$ is fulfilled. We will assume that this is fulfilled in the following. This also implies that the channel is *slowly time-variant* (see Chapter 6), so that $h(t, \tau)$ can be interpreted as the impulse response $h(\tau)$ that is valid at a certain (fixed) time instant t.[1]

---

[1] Strictly speaking, use of the "slow time variance" concept also requires that the sounding signal has a duration that is much smaller than the coherence time of a channel. We will assume this in the following.

Example 8.1 *A channel sounder is in a car that moves along a street at 36 km/h. It measures the channel impulse response at a carrier frequency of 2 GHz. At what intervals does it have to measure? What is the maximum excess delay the channel can have so as to still remain underspread?*

$$\left.\begin{aligned} v &= 36 \text{ km/h} = 10 \text{ m/s} \\ \lambda_c &= \frac{c_0}{f_c} = \frac{3 \cdot 10^8}{2 \cdot 10^9} = 0.15 \text{ m} \end{aligned}\right\} \tag{8.10}$$

The channel must be sampled in the time domain at a rate that is, at minimum, twice the maximum Doppler shift. Using Eq. (8.5):

$$f_{\text{rep}} = 2 \cdot v_{\max} = 2 \cdot \frac{v}{\lambda_c} \tag{8.11}$$

The sampling interval T_{rep} is given as:

$$T_{\text{rep}} = \frac{1}{f_{\text{rep}}} = \frac{\lambda_c}{2 \cdot v} = 7.5 \text{ ms} \tag{8.12}$$

At a mobile speed of 36 km/h, this corresponds to a channel snapshot taken every 75 mm. To calculate the maximum excess delay $\tau_{\max}$, we make use of the fact that the channel must be underspread in order to be identifiable. Hence from Eq. (8.9):

$$\left.\begin{aligned} 2 \cdot \tau_{\max} \cdot v_{\max} &= 1 \\ \tau_{\max} &= \frac{1}{2 \cdot v_{\max}} = T_{\text{rep}} = 7.5 \text{ ms} \end{aligned}\right\} \tag{8.13}$$

This is orders of magnitude larger than the maximum excess delays that occur in typical wireless channels (see Chapter 7). However, note that this is only the theoretical maximum delay spread that still guarantees identifiability. A correlative channel sounder would show a significant degradation in estimation quality for this high $\tau_{\max}$.

If $\tau_{\max}$ is smaller, then the repetition period T_{rep} of the sounding pulse should be increased; this would allow averaging of the snapshots, and thus improvement of the SNR.

8.1.4 Influence on measurement data

When performing the measurements, we have to be aware of the fact that measured impulse responses carry undesired contributions as well. These are mainly:

- Interference from other (independent) signal sources that also use the channel.
- Additive white Gaussian noise.

Interference is created especially when measurements are done in an environment where other wireless systems are already active in the same frequency range. Wideband measurements in the 2-GHz range, for example, become quite difficult, as these bands are heavily used by various systems. If the number of interferers is large, the resulting interference can usually be approximated as equivalent Gaussian noise. This equivalent noise raises the noise floor, and thus decreases the dynamic range.

8.2 Time domain measurements

A time-domain measurement directly measures the (time-variant) impulse response. Assuming that the channel is slowly time-variant, the measured impulse response is the convolution of the *true channel impulse response* with the *impulse response of the sounder*:

$$h_{\text{meas}}(t_i, \tau) = \widetilde{p}(\tau) * h(t_i, \tau) \qquad (8.14)$$

where the effective *sounder impulse response* $\widetilde{p}(\tau)$ is the convolution of the transmitted pulseshape and the RX filter impulse response:

$$\widetilde{p}(\tau) = p_{\text{TX}}(\tau) * p_{\text{RX}}(\tau) \qquad (8.15)$$

if the channel and the transceiver are linear.[2] The sounder impulse response should be as close to an ideal delta (Dirac) function as possible.[3] This minimizes the impact of the measurement system on the results. If the impulse response of the sounder is not a delta function, it has to be eliminated from the measured impulse response by a deconvolution procedure, which leads to noise enhancement and other additional errors.

8.2.1 Impulse sounder

This type of channel sounder, which is comparable with an impulse radar, sends out a sequence of short pulses $p_{\text{TX}}(\tau)$. These pulses should be as short as possible, in order to achieve good spatial resolution, but also contain as much energy as possible, in order to obtain a good SNR. Figure 8.3 shows a rough sketch of a transmit pulse, and the received signal after this pulse has propagated through the channel.

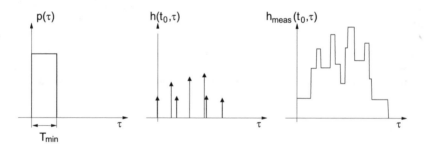

Figure 8.3   Principle of pulse-based measurements. The figure on the left shows one sample of a (periodically repeated) transmit pulse. The figure in the middle shows the impulse response of the channel. The figure on the right shows the output from the channel, as measured by the RX.

The receive filter is a bandpass filter – i.e., has a constant-magnitude spectrum in the frequency range of interest. Ideally, $p_{\text{RX}}(\tau)$ should not have an impact, so that:

$$p(\tau) = p_{\text{TX}}(\tau) \qquad (8.16)$$

When comparing the sounding signal with the requirements of Section 8.1.2, we find that the signal has a small time bandwidth product, a short signal duration, and a high crest factor.

---

[2] Strictly speaking, a sounder impulse response is also time-variant, due to second-order effects like temperature drift. However, it is more common to recalibrate the sounder, and consider it as time-invariant until the next calibration.

[3] This corresponds to a spectrum that is flat over all frequencies. For practical purposes, it is sufficient that the spectrum is flat over the bandwidth of interest.

The requirement of high pulse energy and short duration for transmitted pulses implies that the pulses have a very high peak power. Amplifiers and other Radio Frequency (RF) components that are designed for such high peak powers are expensive, or show other grave disadvantages (e.g., nonlinearities). A further disadvantage of an impulse sounder is its low resistance to interference. As the sounder interprets the received signal directly as the impulse response of the channel, any interfering signal – e.g., from a cellphone active in the band of interest – is interpreted as part of the channel impulse response.

8.2.2 Correlative sounders

The time bandwidth product can be increased by using correlative channel sounders. Eqs. (8.14) and (8.15) show that it is not the transmit pulseshape alone that determines the impact of the measurement system on the observed impulse response. Rather, it is the convolution of $p_{\text{TX}}(\tau)$ and $p_{\text{RX}}(\tau)$. This offers additional degrees of freedom for designing transmit signals that result in high delay resolution but low crest factors.

The first step is to establish a general relationship between the desired $p_{\text{TX}}(\tau)$ and $p_{\text{RX}}(\tau)$. As is well known from digital communications theory, the SNR of the RX filter output is maximized if the receive filter is the *matched filter* with respect to the transmit waveform [Barry et al. 2003, Proakis 1995].[4] Concatenation of the transmit and receive filters thus has an impulse response that is identical to the Auto Correlation Function (ACF) of the transmit filter:

$$\tilde{p}(\tau) = p_{\text{TX}}(\tau) * p_{\text{RX}}(\tau) = R_{p_{\text{TX}}}(\tau) \tag{8.17}$$

The sounding pulses thus should have an ACF that is a good approximation of a delta function; in other words, a high autocorrelation peak $R_{p_{\text{TX}}}(0)$, as well as low ACF sidelobes. The ratio between the height of the autocorrelation peak and the largest sidelobe is called the *Peak to Off Peak ratio* (POP) and is an important quantity for characterization of correlative sounding signals. Figure 8.4 shows an example of an ACF, and the delay resolution that can be achieved with such a signal [de Weck 1992].

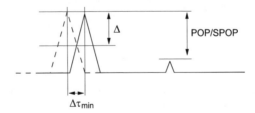

Figure 8.4   Definition of the *peak-to-offpeak ratio* and the delay resolution $\Delta\tau_{\min} \cdot (\Delta = 6\,\text{dB})$.

In practice, *Pseudo Noise* (PN) *sequences* or linearly frequency modulated signals (chirp signals) have become the prevalent sounding sequences. Maximum-length PN-sequences (*m-sequences*), which can be created by means of a shift register with feedback, are especially popular. Such sequences are well-known from Code Division Multiple Access (CDMA) systems, and have been extensively studied both in the mathematical and the communications engineering literature (see Chapter 18 for more details). The ACF of an m-sequence with periodicity M_c has

---

[4] Strictly speaking, the SNR at the output of the RX-matched filter is maximized if the receive filter is matched to the signal that is actually received at the RX antenna connector. However, that would require knowledge of the channel impulse response – the very quantity we are trying to measure. Thus, matching the filter to the transmit signal is the best we can do.

only a single peak of height M_c, and a POP of M_c. Following the CDMA literature, each of the M_c elements of such a sequence is called a *chip*.

For constant chip duration and increasing length of the m-sequence, the POP, as well as the time bandwidth product, increases: signal duration increases linearly with M_c, while bandwidth, which is approximately the inverse of chip duration, stays constant. The increased time bandwidth product improves immunity to noise and interference. More exactly, noise and interference are suppressed by a factor M_c. The reason for this will be discussed in more detail in Chapter 18, as the principle is identical to that of direct sequence CDMA.

Most channel sounders use the principle of correlative channel measurements. However, the interpretation of measurements in time-varying channels requires some extra care. The basic principle of correlative channel sounders is that $p_{TX}(\tau) * h(t, \tau) * p_{RX}(\tau)$ is identical to $[p_{TX}(\tau) * p_{RX}(\tau)] * h(t, \tau)$. In other words, we require that the channel at the beginning of the PN sequence is the same as the one at the end of the PN sequence. This is a good approximation for slowly time-variant channels. If this condition is not fulfilled, correction procedures need to be used [Matz et al. 2002].

8.3 Frequency domain analysis

The techniques described in the previous section directly estimate the impulse response of the channel in the time domain. Alternatively, we can try to directly estimate the transfer function – i.e., measure in the frequency domain. The fundamental relationship (Eq. 8.1) still holds. However, the shape of the waveform $p(t)$ is now different. The main criterion for its design is that it has a power spectrum $|P(j\omega)|^2$ that is approximately constant in the bandwidth of interest, and that it allows interpretation of the measurement result directly in the frequency domain.

One method of frequency domain analysis is based on chirping. The transmit waveform is given as:

$$p(t) = \exp\left[2\pi j\left(f_0 t + \Delta f \frac{t^2}{2T_{chirp}}\right)\right] \quad \text{for } 0 \le t \le T_{chirp} \tag{8.18}$$

Consequently, the instantaneous frequency is:

$$f_0 + \Delta f \frac{t}{T_{chirp}} \tag{8.19}$$

and thus changes linearly with time, covering the whole range Δf of interest. The receive filter is again a matched filter. Intuitively, the chirp filter "sweeps" through the different frequencies, measuring different frequencies at different times.

Alternatively, we can sound the channel on different frequencies at the same time. The conceptually most simple way is to generate different, sinusoidal sounding signals with different weights, phases and frequencies, and transmit them all from the TX antenna simultaneously:

$$p(t) = \sum_{i=1}^{N_{tones}} a_i \cdot \exp[2\pi j t (f_0 + i\Delta f / N_{tones}) + j\varphi_i] \quad \text{for } 0 \le t \le T_{ss} \tag{8.20}$$

Due to hardware costs, calibration issues, etc., analog generation of $p(t)$ using multiple oscillators to generate multiple frequencies is not practical. However, it is possible to generate $p(t)$ digitally, similar to the principles of Orthogonal Frequency Division Multiplexing (OFDM) described in Chapter 19, and then use just a single oscillator to upconvert the signal to the desired passband (and similarly at the RX).

8.4 Modified measurement methods

8.4.1 Swept Time Delay Cross Correlator (STDCC)

The STDCC is a modification of correlative channel sounders that aims to reduce the sampling rate at the RX. Normal correlative channel sounders require sampling at the Nyquist rate. In contrast, the STDCC uses just a single sample value for each m-sequence – namely, at the maximum of the autocorrelation function. The position of this maximum is changed for each repetition of the m-sequence, by shifting the time base of the RX with respect to the TX. Thus, K_{scal} transmissions of the m-sequence give the sampled values of a single impulse response $h(\tau_i)$, $i = 1, \ldots, K_{scal}$. The delay resolution is thus better, by a factor K_{scal}, than the inverse sampling rate. This drastically reduces the sampling rate and the requirements for subsequent processing and storing of the impulse response. On the downside, the duration of each measurement is increased by the same factor K_{scal}.

In an STDCC, shifting of the maximum of the autocorrelation function is achieved by using different time bases in the TX and RX. The delayed time base of the RX correlator (compared with the TX sequence) is achieved by using a frequency that is smaller by Δf. This results in a slow relative shift of the TX and RX sequences. During each repetition of the sequence, the correlation maximum corresponds to a different delay. After a duration:

$$T_{slip} = \frac{1}{f_{TX} - f_{RX}} = \frac{1}{\Delta f} \qquad (8.21)$$

the TX and RX signals are fully aligned again – i.e., the ACF maximum again occurs at delay $\tau = 0$. This means that (for a static channel) the output from the sampler is periodic with the so-called *slip rate*:

$$f_{slip} = f_{TX} - f_{RX} \qquad (8.22)$$

The ratio:

$$K_{scal} = \frac{f_{TX}}{f_{slip}} \gg 1 \qquad (8.23)$$

is the *scaling factor* K_{scal} of the impulse response. The actual impulse response can be obtained from the measured sample values as:

$$\hat{h}(t_i, k\Delta\tau) = C \cdot h_{STDCC}(t_i, k\Delta\tau K_{scal}) \qquad (8.24)$$

where C is a proportionality constant.

The drawback of the measurement method is increased measurement duration. Remember that a channel is identifiable only if it is underspread – i.e., $2\nu_{max}\tau_{max} < 1$, which is usually fulfilled in wireless channels. For an STDCC, this requirement changes to $2K_{scal}\nu_{max}\tau_{max} < 1$, which is *not* fulfilled for many outdoor channels and typical values of K_{scal}.

Example 8.2 *Consider an STDCC that performs measurements in an environment with 500-Hz maximum Doppler frequency and maximum excess delay of 1 μs. The sounder can sample at most with 1 Msample/s. What is the maximum delay resolution (inverse bandwidth) that the sounder can achieve?*

In order for the channel to remain identifiable (underspread) when measured with an STDCC, the following condition has to hold:

$$2 \cdot K_{scal} \cdot \tau_{max} \cdot \nu_{max} = 1$$

$$K_{scal} = 1/(2 \cdot \tau_{max} \cdot \nu_{max}) = 1,000$$

The sounder can take one sample for each repetition of the sounding pulse – i.e., one sample per μs. Hence, the STDCC sounder can resolve multipath components separated by a delay of $1\,\mu s/1{,}000 = 1\,ns$.

8.4.2 Inverse filtering

At first glance, it sounds paradoxical to use a filter that is *not* ideally matched to the transmit signal, and thus has a worse SNR. However, there can be good practical reasons for this approach. Small variations of the SNR are usually less important than the sidelobes of the ACF: while sidelobes can be eliminated by appropriate deconvolution procedures, they can give rise to additional errors. It is thus meaningful to optimize the receive filters with respect to the POP ratio, not with respect to the SNR.

A special case of mismatched filtering is inverse filtering. For the matched filter, the receive filter transfer function is chosen as $P_{TX}^*(f)$, so that the total filter transfer function $P_{MF}(f)$ (concatentation of transmit and receive filter) is given as:

$$P_{MF}(f) = P_{TX}(f) \cdot P_{TX}^*(f) \tag{8.25}$$

For inverse filtering, the receive filter transfer function is chosen as $1/P_{TX}(f)$ in the bandwidth of interest, so that the total transfer function is made as close to unity as possible:

$$P_{IF}(f) = P_{TX}(f) \cdot \frac{1}{P_{TX}(f)} \approx 1 \tag{8.26}$$

The inverse filter is thus essentially a *zero-forcing equalizer* (see Chapter 16) for compensation of distortions by the transmit filter. It is important that the transmit spectrum P_{TX} does not have any nulls in the bandwidth of interest. The inverse filter leads to noise enhancement, and thus to a worse SNR than a matched filter. On the positive side, the inverse filter is *unbiased*, so that the estimation error is zero-mean.

8.4.3 Averaging

It is common to average over several, subsequently recorded, impulse responses of transfer functions. Assuming that the channel does not change during the whole measurement time, and that the noise is statistically independent for the different measurements, then the averaging of M profiles results in an enhancement of the SNR by $10 \cdot \log_{10} M$ dB. However, note that the maximum measurable Doppler frequency decreases by a factor of M.

Averaging over different realizations of the channel is used to obtain, e.g., the Small Scale Averaged (SSA) power.

Example 8.3   *A Mobile Station (MS) moves along a straight line, and can take a statistically independent sample of the impulse response every 15 cm. Measurements are taken over a distance of 1.5 m, so that shadowing can be considered constant over this distance. The goal is estimation of channel gain (attenuation), averaged over small-scale fading. Assume first that the measurements are only taken at a single frequency. What is the standard deviation of the estimate of channel gain? What is the probability that the estimator is more than 20% off?*

Since measurement is only taken at a single frequency, we assume a flat Rayleigh-fading channel with mean power $\bar{P}$. Each sample of the power of the impulse response is then an exponentially distributed random variable with mean power $\bar{P}$. A reasonable estimate of mean power is:

$$\hat{P} = \frac{1}{N} \sum_{k=1}^{N} P_k$$

where P_k is the kth sample of the power of the impulse response and N is the number of samples. As a measure of the error we choose the normalized standard deviation, $\sigma_{\hat{P}}/\bar{P}$. Since the P_k are identically distributed and independent, we have:

$$\frac{\sigma_{\hat{P}}}{\bar{P}} = \frac{1}{\bar{P}} \sqrt{\mathrm{Var}\left(\frac{1}{N}\sum_{k=1}^{N} P_k\right)} = \frac{1}{\bar{P}}\sqrt{\frac{1}{N^2} N \bar{P}^2} = \frac{1}{\sqrt{N}}$$

Using 11 samples, the relative standard deviation is 0.3. Approximating the probability density function (pdf) of the estimator to be Gaussian with mean $\bar{P}$ and variance $\bar{P}^2/N$, the probability that the estimate is more than 20% off is then:

$$1 - \Pr\left(0.8\bar{P} < \hat{P} < 1.2\bar{P}\right) = \mathrm{erfc}\left(\frac{0.2\sqrt{N}}{\sqrt{2}}\right) \tag{8.27}$$

For $N = 11$ this probability becomes 0.5.

Example 8.4 *Consider now the case of wideband measurements, where measurements are done at ten independently fading frequencies. How do the results change?*

Again aiming to estimate the narrowband channel attenuation averaged over small-scale fading, usage of wideband measurements just implies that $N = 110$ measurements are now available. Modifying Eq. (8.27), we find that the probability for more than 20% error has decreased to 0.036.

8.4.4 Synchronization

The synchronization of TX and RX is a key problem for wireless channel sounding. It is required to establish synchronization in frequency and time at a TX and RX that can be separated by distances up to several kilometers. This task is made more difficult by the presence of multipath propagation and time variations of the channel. Several different approaches are in use:

1. In indoor environments, *synchronization by cables* is possible. For distances up to about 10 m, coaxial cables are useful; for larger distances, fiber-optic cables are preferable. In either case, the synchronization signal is transmitted on a known and well-defined medium from the TX to the RX.
2. For many outdoor environments, the *Global Positioning System* (GPS) offers a way of establishing common time and frequency references. The reference signals required by channel sounders are an integral part of the signals that GPS satellites transmit. An additional benefit lies in the fact that the measurement location is automatically recorded as well. The disadvantage is that this method requires that both TX and RX have line-of-sight connection to GPS satellites; the latter condition is rarely fulfilled in microcellular and indoor scenarios.

3. *Rubidium clocks* at the TX and RX are an alternative to GPS signals. They can be synchronized at the beginning of a measurement campaign; as they are extremely stable (relative drifts of 10^{-11} are typical), they retain synchronization for several hours.
4. *Measurements without synchronization*: it is possible to synchronize via the wireless link itself – i.e., the received signal self-triggers the recording at the RX by exceeding a certain threshold. The advantage of this technique is its simplicity. However, the drawbacks are twofold: (i) noise or interference can erroneously trigger the RX, and (ii) it is not possible to determine absolute delays.

8.4.5 Vector network analyzer measurements

The measurement techniques described in Sections 8.2 and 8.3, including the frequency domain techniques, require dedicated equipment that can be quite expensive, due especially to the high-speed components required for the generation of short pulses or sequences with short chip duration. An alternative measurement technique is based on a slow sweep in the frequency domain. In the following, we discuss measurements by means of a vector network analyzer, as these devices are present in many RF labs.

A vector network analyzer measures the S-parameters of a *Device Under Test* (DUT). The DUT can be a wireless channel, in which case the parameter S_{21} is the channel transfer function at the frequency that is used to excite the channel. By having the excitation signal sweep or step through the frequency band of interest, we obtain a continuous or sampled version of the transfer function $H(t,f)$.

In order to reduce the impact of the network analyzer itself, back-to-back calibration has to be performed. The necessity to do a calibration is common to all sorts of channel sounders, and is indeed a basic principle of a good measurement procedure. What is specific to network analyzers is the type of calibration, which is commonly *SOLT calibration* (Short Open Loss Termination). This calibration establishes the reference planes and measures the frequency response of the network analyzer. During subsequent measurement, the network analyzer compensates for this frequency response, so that it measures only the frequency response of the DUT. Note that calibration does not include the antennas. This is not a problem if the antennas are to be considered a part of the channel. If, however, antenna effects are to be eliminated, a separate calibration of the antennas has to be performed, and taken into account during evaluation of the measurements.[5]

Measurements using a vector network analyzer are usually accurate, and can be performed in a straightforward way. However, there are also important disadvantages:

- Such measurements are slow, so that repetition rates typically cannot exceed a few Hz (!). Since we require that the channel does not change significantly during one measurement, network analyzer measurements are limited to static environments.
- The TX and RX are often placed in the same casing. This puts an upper limit on the distance that TX and RX antennas can be spaced apart.

From these restrictions it follows that network analyzers are mainly suitable for indoor measurements.

---

[5] Note that corrections for antenna pattern are possible only if the directions of the Multi Path Component (MPCs) are known (see Section 8.5).

8.5 Directionally resolved measurements

We have seen in Chapters 6 and 7 that directionally resolved channel models are important for the design and simulation of multiantenna systems. To set up and verify such models, we first need directionally resolved measurements. In the first three subsections of this section, we will discuss how to make measurements that are directionally resolved at just the RX. These concepts are then generalized to Multiple Input Multiple Output (MIMO) measurements (directionally resolved at both link ends) at the end of this section.

Fortunately, it is not necessary to devise directional channel sounders from scratch. Rather, a clever combination of existing devices can be used to allow directional measurements. We can distinguish two basic approaches: *measurements with directional antennas*, and *array measurements*.

- *Measurements with directional antennas*: a highly directive antenna is installed at the RX. This antenna is then connected to the RX of a "regular" channel sounder. The output of the RX is thus the impulse response of the combination of the channel and the antenna pointing in a specific direction; that is:

$$h(t, \tau, \phi_i) = \int h(t, \tau, \phi) \widetilde{G}_{RX}(\phi - \phi_i) \, d\phi \qquad (8.28)$$

  where ϕ_i is the direction in which the maximum of the receive antenna pattern is pointing. By stepping through different values of ϕ_i, we can obtain an approximation of the directionally resolved impulse response. One requirement for this measurement is that the channel stays constant during the *total* measurement duration, which encompasses the measurements of all the different ϕ_i. As the antenna has to be rotated mechanically in order to point to a new direction, the total measurement duration can be several seconds or even minutes. The better the directional resolution, the longer the measurement duration.
- *Measurement with an antenna array*: an antenna array consists of a number of antenna elements, each of which has low (or no) directivity, which are spaced apart a distance d_a that is on the order of one wavelength. The impulse response is measured at all these antenna elements (quasi-) simultaneously. The resulting vector of impulse responses is either useful by itself (e.g., for the prediction of diversity performance), or the directional impulse response can be extracted from it by appropriate signal-processing techniques (*array processing*). Measurement of the impulse response at the different antenna elements can be done by means of three different approaches (see Fig. 8.5):

  —*real arrays*: in this case one demodulator chain exists for each receive antenna element. Measurement of the impulse response thus truly occurs at all antenna elements simultaneously. The drawbacks include high costs, as well as the necessity to calibrate multiple demodulator chains.
  —*multiplexed arrays*: in this technique, multiple antenna elements, but only one demodulator chain, exist. The different antenna elements are connected to a demodulator chain (conventional channel sounder) via a fast RF switch [Thomae et al. 2000]. The receiver thus first measures the impulse response at the first antenna element $h_{1,1}$, then it connects the switch to the second element, measures $h_{1,2}$, and so on.
  —*virtual array*: in this technique, there is only a single antenna element, which is moved mechanically from one position to the next, measuring the $h_{1,j}$.

A basic assumption for evaluation is again that the environment does not change during the measurement procedure. "Virtual arrays" (which – due to the need of mechanically moving an antenna – require a few seconds or even minutes for one measurement run) can thus

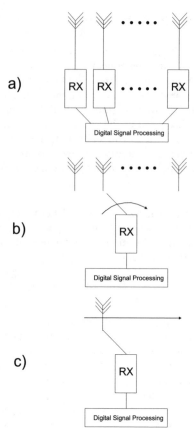

Figure 8.5 Types of arrays for channel sounding: real array (a), switched array (b), virtual array (c).
Reproduced with permission from Molisch and Tufvesson [2005] © Hindawi.

only be used in static environments. This precludes scenarios where cars or moving persons are significant Interacting Objects (IOs). In non-static environments, multiplexed arrays are usually the best compromise between measurement speed and hardware effort. A related aspect is the impact of frequency drift and loss of synchronization of the TX/RX. The longer a measurement lasts, the higher the impact of these impairments.

8.5.1 Data model for receive arrays

For the extraction of directional information from the impulse responses at the array elements, we first need a mathematical model for the array and incoming signals. We analyze the case of a *Uniform Linear Array* (ULA) consisting of N_r elements, where the signal $r_n(t)$ is detected at the nth element. To simplify the discussion, we will assume that all waves are propagating in the horizontal plane.

Consider now the case when plane waves from N different directions are incident on the array, where each wave is described by its direction of arrival ϕ_i. The relationship between incident signals s_i and the signal it creates at the first antenna element is simply:

$$r_1(t) = \sum_i a_{i,1} s_i(t - \tau_{i,1}) + n_1(t) \tag{8.29}$$

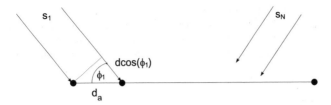

Figure 8.6 Plane waves incident at angle $\phi_1, \ldots, \phi_N$ on a uniform linear array.

where $\tau_{i,1}$ is the runtime between the source of the ith signal and the first antenna element, $a_{i,1}$ is the (complex) amplitude of the signal, and $n_1(t)$ is the noise at the first antenna element. Consider now the second antenna element. Here the received signal is:

$$r_2(t) = \sum_i a_{i,2} s_i(t - \tau_{i,2}) + n_2(t) \qquad (8.30)$$

If the ith source is in the far field, then $|a_{i,2}| = |a_{i,1}|$, and:

$$s_i(t - \tau_{i,2}) = s_i(t - \tau_{i,1}) \exp(-j(\tau_{i,2} - \tau_{i,1}) 2\pi f_c) \qquad (8.31)$$

This last equation assumes that the signal is narrowband in the RF sense – i.e., the bandwidth of the signal is much smaller than the carrier frequency. The physical interpretation of this fact is that the only influence of the antenna position is a phaseshift due to the additional runtime. This runtime difference is:

$$\tau_{i,2} - \tau_{i,1} = (d_a/c_0) \cos(\phi_i) \qquad (8.32)$$

The relationship between r_2 and s is thus:

$$r_1(t) = \sum_i \widetilde{s}_i(t) + n_1(t) \qquad (8.33)$$

$$r_2(t) = \sum_i \widetilde{s}_i(t) \exp(-j2\pi d_a \cos(\phi_i)/\lambda_0) + n_2(t) \qquad (8.34)$$

where $\widetilde{s}_i(t) = a_{i,1} s_i(t - \tau_{i,1})$. For the next antenna element, we get:

$$r_3(t) = \sum_i \widetilde{s}_i(t) \exp(-j2\pi 2 d_a \cos(\phi_i)/\lambda_0) + n_3(t) \qquad (8.35)$$

From this, we can conclude the general relationship between r and s:

$$\mathbf{r}(t) = \mathbf{A}\mathbf{s}(t) + \mathbf{n}(t) \qquad (8.36)$$

where $\mathbf{r}(t) = [r_1(t), r_2(t), \ldots, r_{N_r}(t)]^T$, $\mathbf{s}(t) = [s_1(t), s_2(t), \ldots, s_{N_r}(t)]^T$, $\mathbf{n}(t) = [n_1(t), n_2(t), \ldots, n_{N_r}(t)]^T$, and:

$$\mathbf{A} = \begin{pmatrix} 1 & 1 & \cdots & 1 \\ \exp(-jk_0 d_a \cos(\phi_1)) & \exp(-jk_0 d_a \cos(\phi_2)) & \cdots & \exp(-jk_0 d_a \cos(\phi_N)) \\ \exp(-j2k_0 d_a \cos(\phi_1)) & \exp(-j2k_0 d_a \cos(\phi_2)) & \cdots & \exp(-j2k_0 d_a \cos(\phi_N)) \\ \vdots & \vdots & \vdots & \vdots \\ \exp(-j(N_r-1)k_0 d_a \cos(\phi_1)) & \exp(-j(N_r-1)k_0 d_a \cos(\phi_2)) & \cdots & \exp(-j(N_r-1)k_0 d_a \cos(\phi_N)) \end{pmatrix}$$
$$(8.37)$$

is the *steering matrix*.

Additionally, we assume that the noise at the different antenna elements is independent (spatially white), so that the correlation matrix of the noise is a diagonal matrix with entries σ_n^2 on the main diagonal. If the signals from the antenna elements are linearly combined with time-variant, complex antenna weights w_n, then the combiner output can be written as:

$$y(t) = \mathbf{w}^T(t)\mathbf{r}(t) \tag{8.38}$$

where $\mathbf{w} = [w_1, w_2, ... w_{N_r}]^T$.

8.5.2 Beamforming

The most simple determination of the angle of incidence can be obtained by a Fourier transform of the signal vector $\mathbf{r}$. This gives the directions of arrival ϕ_i with a resolution that is determined by the size of the array, approximately $2\pi/N_r$. The advantage of this method is its simple implementability (requiring only an Fast Fourier Transform (FFT)); the drawback is its small resolution.

More exactly, the angular spectrum $P_{\mathrm{BF}}(\phi)$ is given as:

$$P_{\mathrm{BF}}(\phi) = \frac{\alpha^\dagger(\phi)\mathbf{R}_{rr}\alpha(\phi)}{\alpha^\dagger(\phi)\alpha(\phi)} \tag{8.39}$$

where $\mathbf{R}_{rr}$ is the correlation matrix of the incident signal and:

$$\alpha_{\mathrm{RX}}(\phi) = \begin{pmatrix} 1 \\ \exp(-jk_0 d_a \cos(\phi)) \\ \exp(-j2k_0 d_a \cos(\phi)) \\ \vdots \\ \exp(-j(N_r - 1)k_0 d_a \cos(\phi)) \end{pmatrix} \tag{8.40}$$

is the steering vector into direction ϕ (compare also Eq. 8.37).

8.5.3 High-resolution algorithms

The problem of low resolution can be eliminated by means of so-called *high-resolution methods*. The resolution of these methods is not limited by the size of the antenna array, but only by modeling errors and noise. This advantage is paid for by high computational complexity. Furthermore, there is now often a limit on the *number* of MPCs whose directions can be estimated.

High-resolution methods include:

- ESPRIT: it determines the signal subspace, and extracts the directions of arrival in closed form. A description of this algorithm, which is mainly suitable for uniform linear arrays, is given in Appendix 8.A (see *www.wiley.com/go/molisch*).
- MUSIC: this algorithm also requires determination of the signal and noise subspaces, but then uses a spectral search to find the directions of arrival.
- Minimum Variance Method (MVM: Capon's beamformer): this method is a pure spectral search method, determining an angular spectrum such that for each considered direction the sum of the noise and the interference from other directions is minimized. The modified spectrum is easy to compute, namely:

$$P_{\mathrm{MVM}}(\phi) = \frac{1}{\alpha^\dagger(\phi)\mathbf{R}_{rr}^{-1}\alpha(\phi)} \tag{8.41}$$

- Maximum-likelihood estimation of the parameters of incident waves. The problem of maximum-likelihood parameter extraction is its high computational complexity. An efficient, iterative implementation is the SAGE (Space Alternating Generalized Expectation) maximization algorithm. Since 2000, it has become the most popular method for channel-sounding evaluations. The drawback is that the iterations might converge to a local optimum, not the global one.

One problem that is common to all the algorithms is array calibration. Most of the algorithms use certain assumptions about the array: the antenna patterns of all elements are identical, no *mutual coupling* between antenna elements, and the distance between all antenna elements is identical. If the actual array does not fulfill the assumptions, calibration is required, so that appropriate corrections can be applied. Such calibrations have to be done repeatedly, as temperature drift, aging of components, etc., tend to destroy the calibration.

For many high-resolution algorithms (including subspace-based algorithms), it is required that the correlation matrix does not become singular. Such singular $\mathbf{R}_{rr}$ typically occur if the sources of the waves from the different directions are correlated. For channel sounding, all signals typically come from the same source, so that they are completely correlated. In that case, subarray averaging ("spatial smoothing" or "forward–backward" averaging) has to be used to obtain the correct correlation matrix [Haardt and Nossek 1995]. The drawback with subarray averaging is that it decreases the effective size of the array.

Example 8.5   *Three independent signals with amplitudes 1, 0.8, and 0.2 are incident from directions 10, 45, and 72 degrees, respectively. The noise level is such that the SNR of the first signal is 15 dB. Compute first the correlation matrix, and steering vectors for a five-element linear array with $\lambda/2$ spacing of the antenna elements. Then plot $P_{BF}(\phi)$ and $P_{MVM}(\phi)$.*

Let us first assume that the three signals arrive at the linear array with zero initial phase. Thus we have the following parameters for the three multipath components:

$$a_1 = 1 \cdot e^{j \cdot 0} \quad , \quad \phi_1 = 10\pi/180 \text{ rad}$$
$$a_2 = 0.8 \cdot e^{j \cdot 0}, \quad \phi_2 = 45\pi/180 \text{ rad} \qquad (8.42)$$
$$a_3 = 0.2 \cdot e^{j \cdot 0}, \quad \phi_3 = 72\pi/180 \text{ rad}$$

We assume that measurement noise has a complex Gaussian distribution, and is spatially white. The common variance σ_n^2 of the noise samples is given as:

$$\sigma_n^2 = \frac{1}{10^{\frac{15}{10}}} = 0.032 \qquad (8.43)$$

According to Eq. (8.37), the steering matrix for the 5-element linear array with element spacing $d = \dfrac{\lambda}{2}$ is given as:

$$\mathbf{A} = \begin{bmatrix} 1 & 1 & 1 \\ \exp(-j \cdot \pi \cdot \cos(\phi_1)) & \exp(-j \cdot \pi \cdot \cos(\phi_2)) & \exp(-j \cdot \pi \cdot \cos(\phi_3)) \\ \exp(-j \cdot 2 \cdot \pi \cdot \cos(\phi_1)) & \exp(-j \cdot 2 \cdot \pi \cdot \cos(\phi_2)) & \exp(-j \cdot 2 \cdot \pi \cdot \cos(\phi_3)) \\ \exp(-j \cdot 3 \cdot \pi \cdot \cos(\phi_1)) & \exp(-j \cdot 3 \cdot \pi \cdot \cos(\phi_2)) & \exp(-j \cdot 3 \cdot \pi \cdot \cos(\phi_3)) \\ \exp(-j \cdot 4 \cdot \pi \cdot \cos(\phi_1)) & \exp(-j \cdot 4 \cdot \pi \cdot \cos(\phi_2)) & \exp(-j \cdot 4 \cdot \pi \cdot \cos(\phi_3)) \end{bmatrix} \qquad (8.44)$$

We note that each column of this matrix is a steering vector corresponding to one multipath component. Inserting values for the directions of arrival gives:

$$\mathbf{A} = \begin{bmatrix} 1 & 1 & 1 \\ -0.9989 - 0.0477i & -0.6057 - 0.7957i & 0.5646 - 0.8253i \\ 0.9954 + 0.0953i & -0.2663 + 0.9639i & -0.3624 - 0.9320i \\ -0.9898 - 0.1427i & 0.9282 - 0.3720i & -0.9739 - 0.2272i \\ 0.9818 + 0.1898i & -0.8582 - 0.5133i & -0.7374 + 0.6755i \end{bmatrix} \qquad (8.45)$$

The three incident signals result in an observed array response given by:

$$\mathbf{r}(t) = \mathbf{A} \cdot \mathbf{s}(t) + \mathbf{n}(t) \qquad (8.46)$$

where $\mathbf{s}(t) = \begin{bmatrix} 1 & 0.8 & 0.2 \end{bmatrix}^T$ and $\mathbf{n}(t)$ is the vector of additive noise samples. We evaluate the correlation matrix $\mathbf{R}_{rr} = E\left[\mathbf{r}(t)\mathbf{r}^\dagger(t)\right]$ for 10,000 realizations (where the columns of the steering vectors have different phases φ_i, corresponding to independent realizations[6]) resulting in:

$$\mathbf{R}_{rr} = \begin{bmatrix} 1.7295 & -1.3776 + 0.5941i & 0.8098 - 0.6763i & -0.4378 + 0.3755i & 0.4118 + 0.1286i \\ -1.3776 - 0.5941i & 1.7244 & -1.3621 + 0.5957i & 0.7993 - 0.6642i & -0.4275 + 0.3666i \\ 0.8098 + 0.6763i & -1.3621 - 0.5957i & 1.7080 & -1.3493 + 0.5850i & 0.7932 - 0.6599i \\ -0.4378 - 0.3755i & 0.7993 + 0.6642i & -1.3493 - 0.5850i & 1.6903 & -1.3439 + 0.5780i \\ 0.4118 - 0.1286i & -0.4275 - 0.3666i & 0.7932 + 0.6599i & -1.3439 - 0.5780i & 1.6875 \end{bmatrix}$$
$$(8.47)$$

The angular spectrum for the conventional beamformer is given by Eq. (8.39), and it is plotted as the dashed line in Fig. 8.7. We see that the conventional beamformer fails to identify the three incident signals. The angular spectrum for the MVM (Capon's beamformer) is given by Eq. (8.41); its result is plotted as a solid line in Fig. 8.7. Three peaks can be identified in the vicinity of the true angles of arrival 10°, 45°, and 72°.

8.5.4 Multiple-input multiple-output measurements

The methods described above are intended for getting the directions of arrival at one link end. They can be easily generalized to double-directional or MIMO measurements. Antenna arrays can be used at *both* link ends. In this case, it is necessary to use transmit signals in such a way that the RX can determine which antenna they were transmitted from. This can be done, e.g., by sending signals at different times, on different frequencies, or modulated with different codes (see Fig. 8.8). Of these methods, using different times requires the least hardware effort, and is thus in widespread use in commercial MIMO channel sounders. The signal-processing techniques for determination of the directions at the two link ends are also fairly similar to the processing techniques for the one-dimensional case.

---

[6] We *assume* here that we have different realizations available. As discussed above, many channel-sounding applications require additional measures like subarray averaging to obtain those realizations.

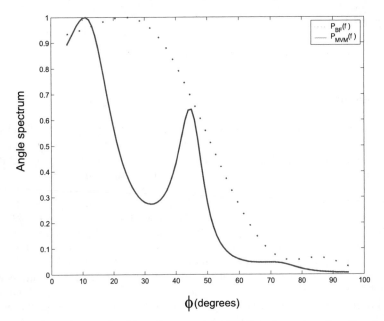

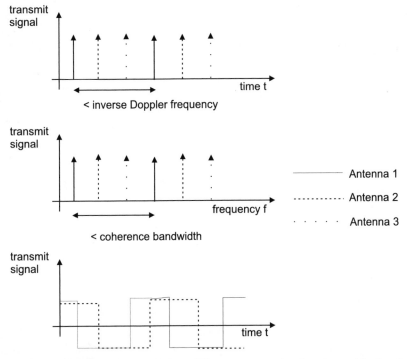

Figure 8.7 Comparison of conventional beamformer and the MVM angle spectrum for a five-element linear array. The true angles are 10°, 45°, and 72°.

Figure 8.8 Transmission of sounding signals from different antennas: signals orthogonal in time (top), frequency (middle), or code (bottom).

8.6 Appendix

Please see companion website *www.wiley.com/go/molisch*

Further reading

Overviews of different channel-sounding techniques are given in Parsons [1992] and Parsons et al. [1991]. Cullen et al. [1993] concentrates on correlative channel sounders; the STDCC is also described in detail in Cox [1972], where it was first introduced. Matz et al. [2002] discuss the impact of time variations of the channel on measurement results. Measurement procedures are also discussed by most papers presenting measurement results (especially back-to-back calibration and deconvolution). For the measurement of directional properties, the alternative method of using a rotating directional antenna is discussed, e.g., in Pajusco [1998].

The ESPRIT algorithm is described in Haardt and Nossek [1995] and Roy et al. [1986]; MUSIC in Schmidt [1986]; for the MVM (Capon's beamformer), see Krim and Viberg [1996]; SAGE is described in Fleury et al. [1999]. An elegant combination of the SAGE algorithm with the gradient method is described in Thomae et al. [2005]. Other high-resolution algorithms include the CLEAN algorithm that is especially suitable for the analysis of ultrawideband signals [Cramer et al. 2002], the JADE algorithm [van der Veen et al. 1997], and many others.

9

Antennas

9.1 Introduction

9.1.1 Integration of antennas into systems

Antennas are the interface between the wireless propagation channel and the transmitter (TX) and receiver (RX), and thus have a major impact on the performance of wireless systems. This chapter therefore discusses some important aspects of antennas for wireless systems, both at the Base Station (BS) and at the Mobile Station (MS). We concentrate mainly on the antenna aspects that are specific to practical wireless systems.

Antenna design for wireless communications is influenced by two factors: (i) performance considerations, and (ii) size and cost considerations. The latter aspect is especially important for antennas on MSs. These antennas must show not only good electromagnetic performance, but must be small, mechanically robust, and easy to produce. Furthermore, the performance of these antennas is influenced by the casing on which they are mounted, and by the person operating the handset. Antennas for BSs, on the other hand, are more similar to "conventional" antennas. Both size and cost are less restricted, and – at least for outdoor applications – the immediate surroundings of the antennas are clear of obstacles (some exceptions to this rule will also be discussed in this chapter).

The remainder of this chapter thus distinguishes between MS antennas and BS antennas. Different types of antennas will be discussed, as will the impact of the environment on antenna performance.

9.1.2 Characteristic antenna quantities

Directivity

The *directivity* D of an antenna is a measure for how much a transmit antenna concentrates the emitted radiation to a certain direction, or how much a receive antenna emphasizes radiation from a certain direction. More precisely, it is defined as [Vaughan and Andersen 2003]:

$$D(\Omega) = \frac{\text{Total power radiated per unit solid angle in a direction } \Omega}{\text{Average power radiated per unit solid angle}} \tag{9.1}$$

Wireless Communications   Andreas F. Molisch
© 2005 John Wiley & Sons, Ltd

Due to the principle of reciprocity, directivity is the same in the transmit and in the receive case. It is related to the far-field antenna power pattern $G(\Omega)$.[1] It is worth keeping in mind that the antenna power pattern is normalized so that:

$$\frac{1}{4\pi} \int G(\Omega)\, d\Omega = 1 \tag{9.2}$$

In many cases, antennas have different patterns for different polarizations. In such cases:

$$D(\phi_0, \theta_0) = \frac{G_\phi(\phi_0, \theta_0) + G_\theta(\phi_0, \theta_0)}{\dfrac{1}{4\pi} \int \int (G_\phi(\phi, \theta) + G_\theta(\phi, \theta)) \sin(\theta)\, d\theta\, d\phi} \tag{9.3}$$

where θ and ϕ are elevation and azimuth, respectively, and G_ϕ and G_θ are the power gains for radiation polarized in ϕ and θ, respectively.

The *gain* of an antenna in a certain direction is related to directivity. However, the gain also has to account for losses – e.g., ohmic losses. Those losses are described by the antenna efficiency, which is discussed in the next subsection.

Efficiency

Losses in the antenna can be caused by several different phenomena. First, ohmic losses (i.e., due to the finite conductivity of the antenna material) can occur. Second, polarizations between the receive antenna and the incident radiated field can be misaligned (see below). Finally, losses can occur because of imperfect matching. The efficiency can thus be written as:

$$\eta = \frac{R_{\mathrm{rad}}}{R_{\mathrm{rad}} + R_{\mathrm{ohmic}} + R_{\mathrm{match}}} \tag{9.4}$$

where R_{rad} is the radiation resistance; it is defined as the resistance of an equivalent network element so that the radiated power P_{rad} can be written as $0.5|I_0|^2 R_{\mathrm{rad}}$ where I_0 is the excitation current magnitude. For example, the input impedance of a half-wavelength dipole is $Z_0 = 73 + j42\ \Omega$, so that the radiation resistance is $R_{\mathrm{rad}} = 73\ \Omega$.

For antennas that are to operate on just a single frequency, perfect matching can be achieved. However, there are limits to how well an antenna can be matched over a larger band. The so-called Fano bound for the case that the antenna impedance can be modeled as a resistor and a capacitor in series reads:

$$\int \frac{1}{(2\pi f_c)^2} \ln\left(\frac{1}{|\rho(f)|}\right) df < \pi R C \tag{9.5}$$

where ρ is the reflection coefficient, and R and C are resistance and capacity, respectively.

High efficiency is a key criterion for any wireless antenna. From the point of view of the transmit antenna, high efficiency reduces the required power of the amplifier to achieve a given fieldstrength. From the point of view of the receive antenna, the achievable Signal-to-Noise Ratio (SNR) is directly proportional to antenna efficiency. Antenna efficiency thus enters the battery lifetime of the MS, as well as the transmission quality of a link. It is difficult, however, to define an absolute goal for efficiency. Ideally, $\eta = 1$ should be obtained over the whole bandwidth of interest. However, it must be noted that in recent years the antenna efficiency of MS antennas has *decreased*. It has been sacrificed mainly for cosmetic reasons – namely, to decrease the size of the antennas.

---

[1] Note that $G(\Omega)$ refers to antenna power, while we define $\widetilde{G}(\Omega)$ as the complex amplitude gain, so that $G(\Omega) = |\widetilde{G}(\Omega)|^2$. Both quantities are defined for the far field.

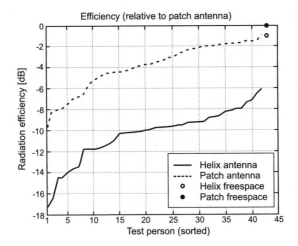

Figure 9.1 Efficiency of mobile station antennas close to a human body, relative to the efficiency of the antennas in free space. Different users and positions of the MS were used in the ensemble.

Reproduced with permission from Pedersen et al. [1998] © IEEE.

Another factor influencing radiation efficiency is the presence of dielectric and/or conducting material – namely, the user – in the vicinity of an MS. When computing antenna gain that can be used in a link budget, this material has to be taken into account, as it leads to strong distortions of the antenna pattern, and absorption of energy. Therefore, radiation efficiency is decreased when these effects are taken into account. The effective gain should be analyzed for a large number of users in order to eliminate the specific properties of any one user (at which angle with respect to the head is the MS being held?, is the user right-handed or left-handed?, what are the dielectric properties of the hand holding the device?, etc.). Figure 9.1 shows an exemplary cumulative distribution function for a patch antenna and a helical antenna. Note that the difference between effective antenna gain and theoretical values can exceed 10 dB. This has a large impact on network planning.

Q-factor

A fundamental quantity of antennas is their Q-factor, defined as [Vaughan and Andersen 2003]:

$$Q = 2\pi \frac{\text{Energy stored}}{\text{Energy dissipated per cycle}} \qquad (9.6)$$

The Q-factor can be related to input impedance as:

$$Q = \frac{f_c}{2R} \frac{\partial X}{\partial f} \qquad (9.7)$$

where input impedance $Z = R + jX$. For an antenna contained in a sphere with diameter L_a, the Q-factor is given as:

$$Q \geq \frac{1}{(k_0 L_a/2)^3} + \frac{1}{k_0 L_a/2} \qquad (9.8)$$

where k_0 is again $2\pi/\lambda$.

Mean effective gain

The antenna pattern, and thus antenna directivity, is defined for the far field – i.e., assumes the existence of homogeneous plane waves – and is a function of the considered direction. It is measured in an anechoic chamber, where all obstacles and distorting objects are removed from the vicinity of the antenna. The patterns and gains measured this way are very close to the numbers found in any book on antenna theory – e.g., a Hertzian dipole has a gain of 1.5. When the antenna is operated in a random scattering environment, it becomes meaningful to investigate the *Mean Effective Gain* (MEG), which is an average of the gain over different directions when the directions of incident radiation are determined by a random environment [Andersen and Hansen 1977, Taga 1990]. The MEG is defined as the ratio of the average power received at the mobile antenna and the sum of the average power of the vertically and horizontally polarized waves received by isotropic antennas.

Polarization

In some wireless systems (e.g., satellite TV), the polarizations of the radiation and the receive antenna should be carefully aligned. This alignment can be measured by the Poincaré sphere. Each polarization state is associated with a point on the sphere: right-hand circular and left-hand circular polarizations are the north and south poles, respectively, while the different linear polarization states are on the equator; all other points correspond to elliptical polarizations. The angle between two points on the sphere is a measure for their mismatch.

For non-line-of-sight scenarios, requirements for a specific polarization of the antennas are typically not very stringent. Even if the transmit antenna sends mainly with a single polarization, propagation through the wireless channel leads to depolarization (see Chapter 7), so that cross-polarization at the *receiver* is rarely higher than 10 dB. This fact is advantageous in many practical situations: as the orientation of MS antennas cannot be predicted (different users hold handsets in different ways), low sensitivity of the antenna to polarization of the incident radiation and/or uniform polarization of the incident radiation is advantageous.

One situation where good cross-polarization discrimination of antennas is required is the case of polarization diversity (see Chapter 13). In that case, the absolute polarization of the antennas is not critical; however, cross-polarization discrimination of two antenna elements plays an important role.

Bandwidth

Antenna bandwidth is defined as the bandwidth over which antenna characteristics (reflection coefficient, antenna gain, etc.) fulfill the specifications. Most wireless systems have a relative bandwidth of approximately 10%. This number already accounts for the fact that most systems use Frequency Domain Duplexing (FDD, see Chapter 17) – i.e., transmit and receive on different frequencies. For example, in a GSM1800[2] system (Chapter 21), bandwidth is about 200 MHz, while the carrier frequency is around 1.8 GHz. As it is desirable to have only a single antenna for TX and RX cases, the antenna bandwidth must be large enough to include both the TX and RX bands. As we have mentioned above, a large bandwidth implies that the matching circuits can no longer be perfect. This effect is the stronger, the smaller the physical dimensions of the antenna are.

Another interesting special case involves dual- or multimode devices. For example, most GSM phones have to be able to operate at both the 900- and the 1,800-MHz carrier frequencies. Antennas for such devices can be constructed by requiring good performance across the whole

---

[2] Global System for Mobile communications in the 1,800-MHz band.

0.9–2-GHz band. However, using such an antenna with 50% relative bandwidth would be "overkill", as the frequency range from 1 GHz to 1.7 GHz need not be covered. It is therefore better to just design antennas that have good performance in the required bands; this goal is easier to fulfill if the different carrier frequencies are integer multiples of each other. Still, the design of antennas that cover two or more bands with high efficiency is a very challenging task. For BS antennas, it is thus often preferable from a technical point of view to use different antennas for different frequency bands. However, multiband antennas are to be preferred for esthetical reasons and reduced visual impact.

9.2 Antennas for mobile stations

9.2.1 Monopole and dipole antennas

Linear antennas are the "classical" antennas for MSs, and have for long determined the typical "look" of these devices. The most common ones are electric monopoles, located above a conducting plane (the casing), and dipoles. The antenna pattern of a short (Hertzian) dipole oriented along the z-axis is uniform in azimuth, and sine-shaped in polar angle θ (measured from the z-axis):

$$\widetilde{G}(\varphi, \theta) \propto \sin(\theta) \tag{9.9}$$

with a maximum gain:

$$G_{\text{max}} = 1.5 \tag{9.10}$$

A $\lambda/2$ dipole has the following properties (see Fig. 9.2):

$$\widetilde{G}(\varphi, \theta) \propto \frac{\cos\left(\frac{\pi}{2}\cos(\theta)\right)}{\sin(\theta)} \tag{9.11}$$

and a maximum gain:

$$G_{\text{max}} = 1.64 \tag{9.12}$$

We can thus see that the patterns of the $\lambda/2$ dipole and the Hertzian dipole do not differ dramatically. However, the radiation resistance can be quite different. For a dipole with *uniform current distribution*, the radiation resistance is:

$$R_{\text{rad}}^{\text{uniform}} = 80\pi^2 (L_a/\lambda)^2 \tag{9.13}$$

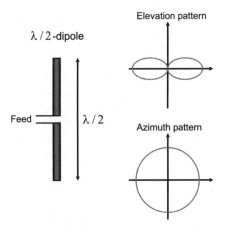

Figure 9.2   Shape and radiation pattern of a $\lambda/2$ dipole antenna.

For dipoles with a tapered current distribution (maximum at the feed, and linear decrease towards the end), the radiation resistance is $0.25 R_{\mathrm{rad}}^{\mathrm{uniform}}$.

From the image principle it follows that the radiation pattern of a monopole located above a conducting plane is identical to that of a dipole antenna in the upper half-plane. Since the energy transmitted into the upper half-space $0 \leq \theta \leq \pi/2$ is twice that of the dipole, the maximum gain is twice and the radiation resistance half that of the dipole. Note, however, that the image principle is valid only if the conducting plane extends infinitely – this is not the case for MS casings. We can thus also anticipate considerable radiation in the lower half-space even for monopole antennas.

The reduction in radiation resistance is often undesirable, since it makes matching more difficult, and leads to a reduction in efficiency due to ohmic losses. One way of increasing efficiency without increasing the physical size of the antenna is the use of *folded* dipoles. A folded dipole consists of a pair of half-wavelength wires that are joined at the non-feed end [Vaughan and Andersen 2003]; these increase the input impedance.

The biggest plus of monopole and dipole antennas is that they can be produced easily and cheaply. The relative bandwidth is sufficient for most applications in single-antenna systems. The disadvantage is the fact that a relatively long metal stick must be attached to the MS casing. In the 900-MHz band, a $\lambda/4$ monopole is 8 cm long – often longer than the MS itself. Even when realized as a retractable element, it is easily damaged. For this reason, shorter and/or integrated antennas, which are less efficient, are becoming increasingly widespread. This does not pose a significant problem in Europe and Japan, where coverage is usually good, and most systems are interference-limited anyway. However, in the U.S.A. and other countries where coverage is somewhat haphazard in many regions, this can have considerable influence on performance.

9.2.2 Helical antennas

The geometry of helical antennas is outlined in Fig. 9.3. A helical antenna can be seen as a combination of a loop antenna and a linear antenna, and thus has two modes of operation. The dimensions of the antenna determine which mode it is operating in. If the dimensions of the helix are much smaller than a wavelength, then the antenna is operating in normal mode. It behaves similar to a linear antenna, and has a pattern that is shaped mainly in the radial direction. This is the operating condition used in MS antennas. In general, the polarization is elliptical, though it can become circular if the ratio $2\lambda d/(\pi D)^2$ becomes unity [Balanis 1997]. The polarization becomes vertical if the helical antenna is arranged over a conducting plane, as the horizontal components of the actual antenna and its image cancel out. When the circumference of the helix is on the order of one wavelength, the antenna pattern has its maximum along the axis of the helix, and polarization is almost circular.

Since the helical antenna in normal mode is similar to a linear antenna, the number of turns the antenna makes does not influence the antenna pattern. However, the bandwidth, efficiency, and radiation resistance increase with increasing h. In general, a helical antenna has lower bandwidth and a smaller input impedance than a monopole antenna; however, a relative bandwidth of 10% can be achieved by appropriate matching circuits. The main advantage of the helical antenna is its smaller size; this has made it (together with linear antennas) the most widely used external antenna for MSs.

9.2.3 Microstrip antennas

A microstrip antenna (patch antenna) consists of a thin dielectric substrate, which is covered on one side by a thin layer of conducting material (ground plane), while on the other side there is a

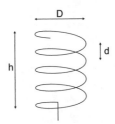

Figure 9.3   Geometry of a helical antenna.

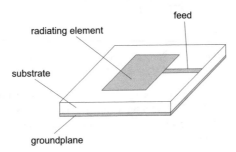

Figure 9.4   Geometry of a microstrip antenna.

patch of conducting material. The configuration is outlined in Fig. 9.3 (see also Fujimoto and James [2001] and Fujimoto et al. [1987]).

The properties of a microstrip antenna are determined by the shape and dimension of the metallic patch, as well as by the dielectric properties of the used substrate. Essentially, the patch is a resonator whose dimensions have to be multiples of the effective dielectric wavelength. Thus, a high dielectric constant of the substrate allows the construction of small antennas. The most commonly used patch shapes are rectangular, circular, and triangular.

The patch is usually fed either by a coaxial cable, or a microstrip line. It is also possible to feed the patch via electromagnetic coupling. This latter case uses a substrate where the ground plane is sandwiched between two layers of dielectric material. On the top of one material is the patch, while the feedline is at the bottom of the other dielectric layer. Coupling is effected through a slot (*aperture*) in the ground plane. These antennas are thus called aperture-coupled patch antennas. This design has the advantage that the dielectric properties of the two layers can be chosen differently, depending on the requirements for the patch and the feedline. Furthermore, this design shows a larger bandwidth than conventional patch antennas.

As mentioned above, the size and efficiency of the microstrip antenna are determined by the parameters of the dielectric substrate. A large ε_r reduces the size. This follows immediately from the fact that in a resonator the length of one side must be:

$$L = 0.5\lambda_{\text{substrate}} \tag{9.14}$$

where

$$\lambda_{\text{substrate}} = \lambda_0/\sqrt{\varepsilon_r} \tag{9.15}$$

Unfortunately, a reduction in physical size also leads to a smaller bandwidth, which is usually undesirable. For this reason, substrates used in practice usually have a very low ε_r – even air is used quite frequently. A further possibility for reducing the size of patch antennas is the use of short-circuited resonators, which reduces the required size of a resonator from $\lambda/2$ to $\lambda/4$ (see Fig. 9.5).

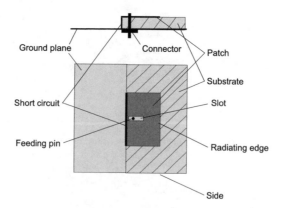

Figure 9.5 Short-circuited $\lambda/4$ patch antenna.

The bandwidth of microstrip antennas can be increased by various measures. The most straightforward one is an increase in antenna volume – i.e., the use of thicker substrates with a lower ε_r. Alternatives are the use of matching circuits and the use of parasitic elements.

Microstrip antennas have several important advantages for wireless applications:

- they are small and can be manufactured cheaply;
- the feedlines can be manufactured on the same substrate as the antenna;
- they can be integrated into the MS, without sticking out from the casing.

However, they also have serious weaknesses:

- they have a low bandwidth (usually just a few percent of the carrier frequency);
- they have low efficiency.

9.2.4 Planar inverted-F antenna

Some of the problems of microstrip antennas can be alleviated by a PIFA (*Planar Inverted F Antenna*). The shape of the PIFA is similar to that of a $\lambda/4$ short-circuited microstrip antenna (see Fig. 9.6). A planar, radiating element is located parallel to a ground plane. This element is

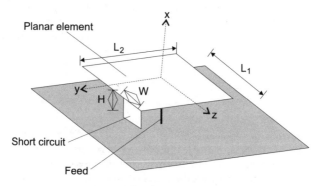

Figure 9.6 Planar inverted-F antenna.

short-circuited over a distance W. If W is chosen equal to the length of the edge L, then we obtain a short-circuited $\lambda/4$ microstrip antenna. If W is chosen smaller, then the resonance length increases, and the current distribution on the radiating element changes.

9.2.5 Radiation-coupled dual-L antenna

A further improvement is achieved by the so-called RCDLA (*Radiation Coupled Dual L Antenna*) (see Fig. 9.7 and Rasinger et al. [1990]). It consists of two L-shaped angular structures, only one of which is fed directly (conductively). The other L-shaped structure is fed by the first L by means of radiation coupling. This increases the bandwidth of the total arrangement. By optimally placing the antenna on the casing, relative bandwidths of up to 10% can be achieved, which is about twice the bandwidth of a PIFA.

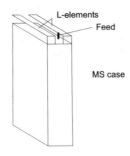

Figure 9.7   Radiation-coupled dual-L antenna.

9.2.6 Multiband antennas

Modern cellular handsets are anticipated to be able to handle different frequencies for communications. As discussed in Section 9.2.2 a GSM handset, for example, needs to be able to deal at least with 900 and 1,800 MHz foreseen in the specifications for most countries. As an added difficulty, many handsets should also be able to cope with the 1,900-MHz mode used in the U.S.A., as well as 2.4 GHz if Bluetooth connections (e.g., to a wireless headset) are required. The situation becomes even more complicated for dual-mode devices that can handle both GSM and Wideband Code Division Multiple Access (WCDMA) (see Chapter 23). The design of internal multiband antennas is very complicated, and few rules for a closed-form design are available. Figure 9.8 shows an example of a microstrip multiband antenna.

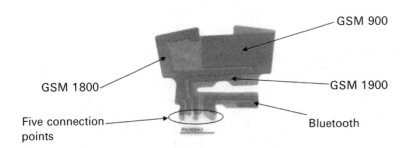

Figure 9.8   Integrated multiband antenna.

9.2.7 Antenna mounting on the mobile station

Antennas do not operate in empty space, but are placed on top of the casing, which can be considered to be part of the radiator. Furthermore, antenna characteristics are influenced by the hand and the head of the user; this influence also depends on the mounting of the antenna on the MS. It is therefore important to investigate different options for placing the antenna, and to see how this placement influences performance.

Linear and helical antennas are usually placed on the upper, narrow side of the casing – i.e., they stick out from that part of the casing. This has mainly ergonomic reasons – if they were sticking out from the lower side, they would feel uncomfortable to the user, and the hand of the user would often cover the antenna, leading to additional attenuation.

For microstrip antennas, PIFAs, and RCDLAs, there are more options for placements. These antennas are usually used as internal antennas – i.e., integrated into the casing, or enclosed within the casing. This greatly reduces the danger of mechanical damage. However, there is an increased probability that users will place their hands over the antenna, which increases absorption of the electromagnetic energy, and thus worsens link performance [Erätuuli and Bonek 1997]. Examples for the positioning of RCDLAs can be found in Fig. 9.9; other types of microstrip antennas can be placed in a similar way.

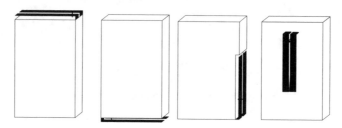

Figure 9.9   Placement of radiation-coupled dual-L antennas on the casing of a mobile station.

9.3 Antennas for base stations

9.3.1 Types of antennas

The design requirements for BS antennas are different from those of MS antennas. Cost has a smaller impact, as BSs are much more expensive in the first place. Also, size restrictions are less stringent: for macrocells, it is only required that (i) the mechanical stress on the antenna mast, especially due to wind forces, must remain reasonable, and (ii) the "cosmetic" impact on the surroundings must be small. For micro- and picocells, antennas need to be considerably smaller, as they are mounted on building surfaces, street lanterns, or on office walls. The desired antenna pattern is quite different for BS antennas compared with MS antennas. As the physical placement and orientation of the BS antenna is known, patterns should be shaped in such a way that no energy is wasted (see also Section 9.3.3). Such pattern shaping can be most easily implemented by multielement antennas.

For these reasons, macrocellular antennas typically are antenna arrays or Yagi antennas whose elements are linear antennas. For micro- and picocells, antenna arrays consisting of patch antennas are common. All considerations from Section 9.2 about bandwidth, efficiency, etc., remain valid, except for the size requirements, which are relaxed quite a bit.

9.3.2 Array antennas

Array antennas are often used for BS antennas. They result in an antenna pattern that can be more easily shaped. This shaping can either be done in a predetermined way (see Section 9.3.3), or adaptively (see Chapters 13 and 20). The pattern of an antenna array can be written as the product of the pattern of a single element and the array factor $M(\phi, \theta)$. In the plane of the array antenna, the array factor of a uniform linear array is (compare Section 8.5.1) [Stutzman and Thiele 1997]:

$$M(\phi, \theta) = \sum_{n=0}^{N_r-1} w_n \exp[-j\cos(\phi)2\pi d_a n/\lambda] \qquad (9.16)$$

where d_a is the distance between the antenna elements,[3] and w_n are the complex weights of element excitations. For the case that all $|w_n| = 1$, and $\arg(w_n) = n\Delta$, the array factor becomes:

$$|M(\phi, \theta)| = \left| \frac{\sin\left[\frac{N_r}{2}\left(\frac{2\pi}{\lambda}d_a\cos\phi - \Delta\right)\right]}{\sin\left[\frac{1}{2}\left(\frac{2\pi}{\lambda}d_a\cos\phi - \Delta\right)\right]} \right| \qquad (9.17)$$

The phaseshift Δ of the feed currents determines the direction of the antenna main lobe. This principle is well known from the theory of phased array antennas ([Hansen 1998], see also Chapter 8). By imposing $\varphi_n = n\Delta$ the degrees of freedom have reduced to one; while the main lobe can be put into an arbitrary direction, the placement of the sidelobes and the nulls follows uniquely from that direction.

Example 9.1   *Consider a BS antenna mounted at a height of 51 m, providing coverage for a cell with a radius of 1 km. The antenna consists of eight vertically stacked short dipoles, separated by $\lambda/2$. How much should the phaseshift Δ be so that the maximum points at the cell edge at an MS height of 1 m?*

In order to get constructive interference of the contributions from the different antennas in the direction θ_0, the angle Δ should fulfill:

$$\Delta = \frac{2\pi}{\lambda}d_a\cos\theta_0 \qquad (9.18)$$

Obviously, if $\theta_0 = \pi/2$, then $\Delta = 0$ – i.e., no phaseshift is necessary. In our example, we are looking for a tilt angle:

$$\theta_0 = \frac{\pi}{2} + \arctan\left(\frac{51-1}{1,000}\right) = \frac{\pi}{2} + 0.05 \qquad (9.19)$$

The phaseshift thus has to be:

$$\Delta = \frac{2\pi}{\lambda}d_a\cos\left(\frac{\pi}{2} + 0.05\right) = -0.05\pi \qquad (9.20)$$

The impact of the element pattern can be neglected.

---

[3] Distance is usually chosen as $d_a = \lambda/2$ ($d_a \leq \lambda/2$ is necessary to avoid spatial aliasing – i.e., periodicities in the antenna pattern). In the following, we assume that the elements of the linear array are on the x-axis. Obviously, the same principles are valid when antenna elements are stacked vertically, and the elevation pattern of the array should be shaped.

9.3.3 Modifying the antenna pattern

It is undesirable that a BS antenna radiates isotropically. Radiation emitted in a direction that has a large elevation angle (i.e., into the sky) is not only wasted, but actually increases interference with other systems. The optimum antenna pattern is achieved when the received power is constant within the whole cell area, and vanishes outside. The problem is then to synthesize such an elevation pattern. This can be achieved approximately by means of array antennas, where the complex weights are chosen in such a way as to minimize the mean-square error of the pattern. An even simpler approach is to use an antenna array with $w_n = 1$, but tilt the main lobe down by about 5°. This downtilt can be achieved either mechanically (by tilting the whole antenna array) or electronically.

The desired azimuthal antenna pattern is either (i) omnidirectional – i.e., uniform in $[0, 2\pi)$ – or (ii) uniform within a sector, and zero outside. The angular range of a sector is usually 60° or 120°. For omnidirectional antennas, which are mostly used in rural macrocells, linear antennas are used. The arrays consist of an array that extends only along the vertical axis. For sector antennas, either linear antennas with appropriately shaped reflectors, or microstrip antennas (which have inherently nonuniform antenna patterns) can be used.

9.3.4 Impact of the environment on antenna pattern

The antenna pattern of BS antennas is usually defined for the case when an antenna is in free space, or above an ideally conducting plane. This is what is measured in an anechoic chamber, and is also the easiest to compute. However, at its location of operation, a BS antenna is surrounded by different objects of finite extent and conductivity. The antenna patterns in such surroundings can deviate significantly from theoretical patterns. The antenna mast, which is usually metallic, and thus highly conductive, can lead to distortions. Similarly, roofs made out of certain building materials, like reinforced concrete, can distort antenna patterns (see Fig. 9.10).

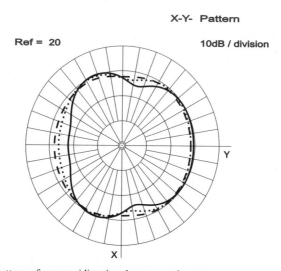

Figure 9.10 Antenna pattern of an omnidirectional antenna close to an antenna mast. Distance from the antenna mast 30 cm. Diameter of the mast: very small (solid), 5 cm (dashed), 10 cm (dotted).

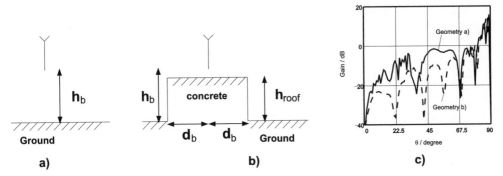

Figure 9.11   Distortion in the vertical antenna pattern. (a) Idealized geometry. (b) Real geometry, including the roof. (c) Distortions in antenna patterns. Properties of all materials: relative dielectric constant $\varepsilon_r = 2$; conductivity 0.01 S/m; $d_b = 20$ m.

Reproduced with permission from Molisch et al. [1995] © European Microwave Association.

Distortions in the vertical pattern are plotted in Fig. 9.11. They arise from two effects: the fact that the (finite extent) roof is much closer to the antenna than the ground, and that the dielectric properties of the roof are different from those of the ground.

Further reading

For a general introduction to antenna theory, we just refer to the many excellent books on antenna theory: Balanis [1997], Kraus and Marhefka [2002], Ramo et al. [1967], and Stutzmann and Thiele [1997], as well as the somewhat more advanced text of Collin [1985]; the latter also devotes a separate chapter to the properties of receiving antennas. For more details on antennas specifically for wireless communications, see Godara [2001] and Vaughan and Andersen [2003]. Antennas for the MS are discussed in Fujimoto and James [2001] and Hirasawa and Haneishi [1992]. Finally, phased array antennas are treated in detail in Hansen [1998] and Mailloux [1994]. Discussions on beamtilting can be found in Manholm et al. [2003].

Part III

TRANSCEIVERS AND SIGNAL PROCESSING

The ultimate performance limits of wireless systems are determined by the wireless propagation channels that we have investigated in the previous parts. The task of practical transceiver design now involves finding suitable modulation schemes, codes, and signal-processing algorithms so that these performance limits can be approximated "as closely as possible". This task always involves a tradeoff between performance and the hardware and software effort. As technology progresses, more and more complicated schemes can be implemented. For example, a third-generation cellphone has computation power that is comparable with a (year 2000) personal computer, and can thus implement signal-processing algorithms whose practical use was unthinkable in the mid-1990s. For this reason, this part of the book will not pay too much attention to algorithm complexity – what is too complex at the current time might well be a standard solution a few years down the road.

The part starts with a description of the general structure of a transceiver in Chapter 10. It describes the various blocks in a transmitter and receiver, as well as simplified models that can be used for system simulations and design. Next, we discuss the various *modulation formats*, and the specific advantages and disadvantages of their use in a wireless context. For example, we find that constant modulus modulation methods are especially useful for battery-powered transmitters, since they allow the use of high-efficiency amplifiers. Building on the formal mathematical description of these modulation formats, Chapter 12 then describes how to evaluate their performance in terms of *bit error probability* in different types of fading channels. We find that the performance of such modulations is mostly limited by two effects: fading, and delay dispersion. The effect of fading can be greatly mitigated by *diversity* – i.e., by transmitting the same signal via different paths. Chapter 13 describes the different methods of obtaining such different paths – e.g., by implementing multiple antennas, by repeating the signal at different frequencies, or at different times. The chapter also discusses the effect that diversity has on the performance of different modulation schemes. The negative effects of delay dispersion can also be combatted by diversity; however, it is more effective to use equalization. *Equalizers* not only combat intersymbol interference created by delayed echoes of the original signal, but they make use of them, exploiting the energy contained in such echoes. They can thus lead to a considerable improvement in performance, especially in systems with high data rates and/or systems operating in channels with large delay spreads. Chapter 16 describes different equalizer structures, from simple linear equalizers to the optimum (but highly complex) maximum-likelihood sequence detectors.

Diversity and equalizers are not always a sufficient or effective way of improving error probability. In many cases, *coding* can greatly enhance performance, and provide the transmission quality required by a specific application. Chapter 14 thus gives an overview of the different coding schemes that are most popular for wireless communications, including the near-optimum turbocodes and Low Density Parity Check (LDPC) codes that have drawn great attention since the early 1990s. These coding schemes are intended to correct errors introduced on the propagation channel. A different type of coding is source coding, which translates the information from the source into a bitstream that can be transmitted most efficiently over the wireless channel. Chapter 15 gives an overview of *speech coding*, which is the most important type of source coding for wireless applications.

Modulation, coding, and equalization for wireless communications is, of course, strongly related to digital communications in general. This part of the book is *not* intended as a textbook of digital communications, but rather assumes that the reader is already familiar with the topic from either previous courses, or one of the many excellent textbooks (e.g., Anderson [2005], Barry et al. [2003], Proakis [1995], Sklar [2001]). While the text gives summaries of the most salient facts, they are rather terse, and only intended as a reminder to the reader.

10

Structure of a wireless communication link

10.1 Transceiver block structure

In this section, we describe a block diagram of a wireless communication link, and give a brief description of the different blocks. More detailed considerations are left for the later chapters of Parts III and IV. We start out with a rough overview that concentrates on the *functionality* of the different blocks. Subsequently, we describe a block diagram that concentrates more on the different *hardware* elements.

Figure 10.1 shows the rough structure of a communications link. In most cases, the goal of a wireless link is the transmission of information from an analog information source (microphone, videocamera) via an analog wireless propagation channel to an analog information sink (loudspeaker, TV screen); the digitizing of information is done only in order to increase the reliability of the link. Chapter 15 will describe speech coding, which represents the most common form of digitizing analog information. For other transmissions – e.g., file transfer – information is already digital.

The transmitter (TX) can then add redundancy in the form of a forward error correction code, in order to make it more resistant to errors introduced by the channel (note that such encoding is done for most, but not all, wireless systems). The encoded data are then used as input to a modulator, which maps the data to output waveforms that can be transmitted. By transmitting these symbols on specific frequencies or at specific times, different users can be distinguished.[1] The signal is then sent through the propagation channel, which attenuates and distorts it, and adds noise, as discussed in Part II.

At the receiver (RX), the signal is received by one or more antennas (see Chapter 13 for a discussion on how to combine the signals from multiple antennas). The different users are separated (e.g., by receiving signals only at a single frequency). If the channel is delay-dispersive, then an equalizer can be used to reverse that dispersion, and eliminate intersymbol interference. Afterwards, the signal is demodulated, and a channel decoder eliminates (most of) the errors that are present in the resulting bitstream. A source decoder finally maps this bitstream to an analog information stream that goes to the information sink (loudspeaker, TV monitor, etc.); in the case when the information was originally digital, this last stage is omitted.

The above description of the blocks is of course oversimplified, and – especially in the RX – the separation of blocks need not be that clearcut. An optimum RX would use as its input the

---

[1] Alternative multiple-access methods are described in Chapters 18–20.

Wireless Communications   Andreas F. Molisch
© 2005 John Wiley & Sons, Ltd

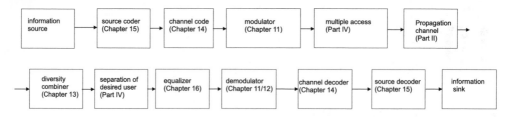

Figure 10.1 Block diagram of a transmitter and receiver, denoting which blocks are discussed in which chapter.

sampled received signal, and compute from it the signal that has been transmitted with the largest likelihood. While this is still too computationally intensive for most applications, joint decoding and demodulation and similar schemes have already been investigated in the literature.

Figures 10.2 and 10.3 show a more detailed block diagram of a digital TX and RX that concentrate on the hardware aspects and the interfaces between analog and digital components:

- As mentioned above, the *information source* provides an analog source signal and feeds it into the *source ADC* (Analog to Digital Converter). This ADC first band-limits the signal from the analog information source (if necessary), and then converts the signal into a stream of digital data at a certain sampling rate and resolution (number of bits per sample). For example, speech would typically be sampled at 8 ksamples/s, with 8-bit resolution, resulting in a datastream at 64 kbit/s. For the transmission of digital data, these steps can be omitted, and the digital source directly provides the input to interface "G" in Fig. 10.2.
- The *source coder* uses a priori information on the properties of the source data in order to reduce redundancy in the source signal. This reduces the amount of source data to be transmitted, and thus the required transmission time and/or bandwidth. For example, the Global System for Mobile communications (GSM) speech coder reduces the source data rate from 64 kbit/s mentioned above to 13 kbit/s. Similar reductions are possible for music and video (MPEG standards). Also, fax information can be compressed significantly. One thousand subsequent symbols "00" (representing "white" color), which have to be represented by 2,000 bit, can be replaced by the statement: "what follows now are 1,000 symbols 00", which requires only 12 bit. For a typical fax, compression by a factor of 10 can be achieved. The source coder increases the entropy (information per bit) of the data at interface F; as a consequence, bit errors have greater impact. For some applications, source data are *encrypted* in order to prevent unauthorized listening in.
- The *channel coder* adds redundancy in order to protect data against transmission errors. This increases the data rate that has to be transmitted at interface E – e.g., GSM channel coding increases the data rate from 13 kbit/s to 22.8 kbit/s. Channel coders often use information about the statistics of error sources in the channel (noise power, interference statistics) to design codes that are especially well-suited for certain types of channels (e.g., Reed–Solomon codes protect especially well against burst errors). Data can be sorted according to importance; more important bits then get stronger protection. Besides adding redundancy, interleaving helps to break up error bursts. Note that interleaving is mainly effective if it is combined with channel coding.
- *Signaling* adds control information for the establishing and ending of connections, for associating information with the correct users, synchronization, etc. Signaling information is usually strongly protected by error correction codes.

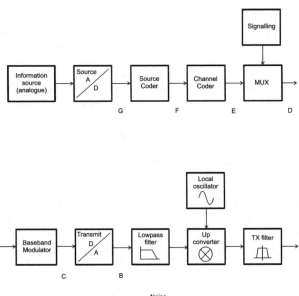

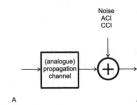

Figure 10.2   Block diagram of a radio link with digital transmitter and receiver, and analog propagation channel.

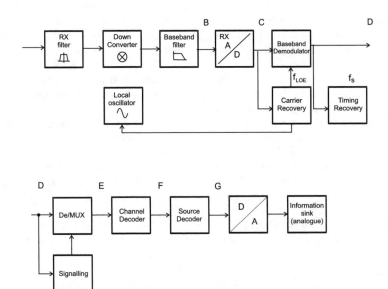

Figure 10.3   Block diagram of a digital receiver chain for mobile communications.

- The *multiplexer* combines user data and signaling information, and combines the data from multiple users.[2] If this is done by time-multiplexing, the multiplexing requires some time compression. In GSM, multiaccess multiplexing increases the data rate from 22.8 to 182.4 kbit/s (8 · 22.8) for the standard case of eight participants. The addition of signaling information increases the data rate to 271 kbit/s.

- The *baseband modulator* assigns the gross data bits (user data and signaling at interface D) to complex transmit symbols in the baseband. Spectral properties, intersymbol interference, peak-to-average ratio, and other properties of the transmit signal are determined by this step. The output from the baseband modulator (interface C) provides the transmit symbols in oversampled form, discrete in time and amplitude.

  Oversampling and quantization determine the aliasing and quantization noise. Therefore, high resolution is desirable, and the data rate at the output of the baseband modulator should be much higher than at the input. For a GSM system, an oversampling factor of 16, and 8-bit amplitude resolution result in a data rate of about 70 Mbit/s.

- The *TX Digital-to-Analog Converter* (DAC) generates a pair of analog, discrete amplitude voltages corresponding to the real and imaginary part of the transmit symbols, respectively.

- The *analog low-pass filter* in the TX eliminates the (inevitable) spectral components outside the desired transmission bandwidth. These components are created by the out-of-band emission of an (ideal) baseband modulator, which stem from the properties of the chosen modulation format. Furthermore, imperfections of the baseband modulator and imperfections of the D/A converter lead to additional spurious emissions that have to be suppressed by the TX filter.

- The *TX local oscillator* provides an unmodulated sinusoidal signal, corresponding to one of the admissible center frequencies of the considered system. The requirements for frequency stability, phase noise, and switching speed between different frequencies depend on the modulation and multiaccess method.

- The *upconverter* converts the analog, filtered baseband signal to a passband signal by mixing it with the local oscillator signal. Upconversion can occur in a single step, or in several steps. Finally, amplification in the Radio Frequency (RF) domain is required.

- The *RF TX filter* eliminates out-of-band emissions in the RF domain. Even if the low-pass filter succeeded in eliminating all out-of-band emissions, upconversion can lead to the creation of additional out-of-band components. Especially, nonlinearities of mixers and amplifiers lead to intermodulation products and "spectral regrowth" – i.e., creation of additional out-of-band emissions.

- The *(analog) propagation channel* attenuates the signal, and leads to delay and frequency dispersion. Furthermore, the environment adds noise (Additive White Gaussian Noise – AWGN) and co-channel interference.

- The *RX filter* performs a rough selection of the received band. The bandwidth of the filter corresponds to the total bandwidth assigned to a specific service, and can thus cover multiple communications channels belonging to the same service.

- The *low-noise amplifier* amplifies the signal, so that the noise added by later components of the RX chain has less effect on the Signal-to-Noise Ratio (SNR). Further amplification occurs in the subsequent steps of downconversion. *Automatic Gain Control* (AGC) reduces the dynamic fluctuations of signals.

- The *RX Local Oscillator* (LO) provides sinusoidal signals corresponding to possible signals at the TX local oscillator. The frequency of the LO can be fine-tuned by a carrier recovery

---

[2] Actually, only the multiplexer at a Base Station (BS) really combines the data from multiple users for transmission. At a Mobile Station (MS), the multiplexer only makes sure that the RX at the BS can distinguish between the datastreams from different users.

algorithm, to make sure that the LOs at the TX and the RX produce oscillations with the same frequency and phase.

- The *RX downconverter* converts the received signal (in one or several steps) into baseband. In baseband, the signal is thus available as a complex analog signal.
- The *RX lowpass filter* provides a selection of desired frequency bands for one specific user (in contrast to the RX bandpass filter that selects the frequency range in which the service operates). It eliminates adjacent channel interference as well as noise. The filter should influence the desired signal as little as possible.
- The RX *Analog-to-Digital Converter* converts the analog signal into values that are discrete in time and amplitude. The required resolution of the ADC is determined essentially by the dynamics of the subsequent signal processing. The sampling rate is of limited importance as long as the conditions of the sampling theorem are fulfilled. Oversampling increases the requirements for the ADC, but simplifies subsequent signal processing.
- *Carrier recovery* determines the frequency and phase of the carrier of the received signal, and uses it to adjust the RX LO.
- The *baseband demodulator* obtains *soft-decision* data from digitized baseband data, and hands them over to the decoder. The baseband demodulator can be an optimum, coherent demodulator, or a simpler differential or incoherent demodulator. This stage can also include further signal processing like equalization.
- If there are *multiple antennas*, then the RX either selects the signal from one of them for further processing, or the signals from all of the antennas have to be processed (filtering, amplification, downconversion). In the latter case, those baseband signals are then either combined before being fed into a conventional baseband demodulator, or they are fed directly into a "joint" demodulator that can make use of information from the different antenna elements.
- *Symbol-timing recovery* uses demodulated data to determine an estimate of the duration of symbols, and uses it to fine-tune sampling intervals.
- The *decoder* uses soft estimates from the demodulator to find the original (digital) source data. In the most simple case of an uncoded system, the decoder is just a hard-decision (threshold) device. For convolutional codes, *Maximum Likelihood Sequence Estimators* (MLSEs, such as the Viterbi Decoder) are used. Recently, iterative RXs that perform joint demodulation and decoding have been proposed. Remaining errors are either taken care of by repetition of a data packet (*Automatic Repeat reQuest* – ARQ), or are ignored. The latter solution is usually resorted to for speech communications, where the delay entailed by retransmission is unacceptable.
- *Signaling recovery* identifies the parts of the data that represent signaling information, and controls the subsequent demultiplexer.
- The *demultiplexer* separates the user data and signaling information, and reverses possible time compression of the TX multiplexer. Note that the demultiplexer can also be placed earlier in the transmission scheme; its optimum placement depends on the specific multiplexing and multiaccess scheme.
- The *source decoder* reconstructs the source signal from the rules of source coding. If the source data are digital, the output signal is transferred to the data sink. Otherwise, the data are transferred to the DAC, which converts the transmitted information into an analog signal, and hands it over to the information sink.

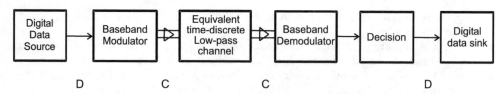

Figure 10.4   Mathematical link model for the analysis of modulation formats.

10.2 Simplified models

It is often preferable to have simplified models for the link. Figure 10.4 shows a model that is suitable for the analysis of modulation methods. The parts of the TX between the information source and the output of the TX multiplexer are subsumed into a "black box" digital data source. The analog radio channel, together with the upconverters, downconverters, RF elements (filters, amplifiers), and all noise and interference signals, is subsumed into an equivalent time-discrete lowpass channel, characterized by a time-variant impulse response and the statistics of additive disturbances. The criterion for judging the quality of the modulation format is the bit error probability at the interfaces D–D.

Other simplified models use a digital representation of the channel (e.g., binary symmetric channel), and are mainly suitable for the analysis of coding schemes.

Further reading

This chapter gave in a very brief form an overview of the structure of TXs and RXs. Many of the aspects of digital signal processing are discussed in subsequent chapters, so we refer the reader to them for details and further references.

An important aspect for power consumption as well as the manufacturing cost of high-speed wireless devices are the ADCs and DACs. van der Plasche [2003] gives many details for implementation in CMOS, which is the technology preferred by the majority of manufacturers. The RF hardware, including amplifiers, mixers, and synthesizers, is discussed in Pozar [2000], Razavi [1997], and Sayre [2001]. Another very important topic – synchronization – is discussed in Mengali and D'Andrea [1997], Meyr and Ascheid [1990], and Meyr et al. [1997].

For source coding, we only describe in this book (Chapter 15) the source coding of speech. The case of audio- and videodigitizing, including the popular MPEG standards, is not treated here – for more detailed information, we recommend Symes [2003] and Watkinson [2001].

11

Modulation formats

11.1 Introduction

As we saw in the previous chapter, the data that we want to transmit over the wireless propagation channel are digital – either because we really want to transmit data files, or because the source coder has rendered the source information into digital form. *Digital modulation* is the mapping of data bits to signal waveforms that can be transmitted over an (analog) channel. At the receiver (RX), the demodulator tries to recover the bits from the received waveform. This chapter gives a brief review of digital *modulation formats*, concentrating mostly on the *results*; for references with more details see the "further reading" at the end of this chapter. Chapter 12 will then describe optimum and suboptimum demodulators, and their performance in wireless channels.

The most simple modulation is binary modulation, where a $+1$-bit value is mapped to one specific waveform, while a -1-bit value is mapped to a different waveform. More generally, a group of K bits can be subsumed into a symbol, which in turn is mapped to one out of a set of $M = 2^K$ waveforms; in that case we speak of *M-ary modulation, higher order modulation, multilevel modulation*, or modulation with *alphabet size M*. In any case, different modulation formats differ in the waveforms that are transmitted, and in the way the mapping from bit groups to waveforms is achieved. Typically, the waveform corresponding to one symbol is time-limited to a time T_S, and waveforms corresponding to different symbols are transmitted one after the other. Obviously, the data (bit) rate is K times the transmitted symbol rate (signaling rate).

When choosing a modulation format in a wireless system, the ultimate goal is to transmit with a certain energy as much information a possible over a channel with a certain bandwidth, while assuring a certain transmission quality (Bit Error Rate – BER). From this basic requirement, some additional criteria follow logically:

- The *spectral efficiency* of the modulation format should be as high as possible. This can best be achieved by a higher order modulation format. This allows the transmission of many data bits with each symbol; however, the required power is higher.
- *Adjacent channel interference* must be small. This entails that the power spectrum of the signal should show a strong rolloff outside the desired band. Furthermore, the signal must be filtered before transmission.
- The *sensitivity with respect to noise* should be very small. This can be best achieved with a low-order modulation format, where (assuming equal average power) the difference between the waveforms of the alphabet is largest.

- *Robustness with respect to delay and Doppler dispersion* should be as large as possible. Thus, the transmit signal should be filtered as little as possible, as filtering creates delay dispersion that makes the system more sensitive to channel-induced delay dispersion.
- Waveforms should be *easy to generate* with hardware that is easy to produce and highly energy-efficient. This requirement stems from the practical requirements of wireless transmitters (TXs). In order to be able to use efficient class-C (or class-E and -F) amplifiers, modulation formats with constant envelopes are preferable. If, on the other hand, the modulation format is sensitive to distortions of the envelope, then the TX has to use linear (class-A or -B) amplifiers. In the former case, power efficiency can be up to 80%, while in the latter case, it is below 40%. This has important consequences for battery lifetime.

The above outline shows that some of these requirements are contradictory. There is thus no "ideal" modulation format for wireless communications. Rather, the modulation format has to be selected according to the requirements of a specific system and application.

11.2 Basics

The remainder of this chapter uses an equivalent baseband representation for the description of modulation formats in equivalent baseband. The bandpass signal – i.e., the physically existing signal – is related to the complex baseband (lowpass) representation as:[1]

$$s_{BP}(t) = \text{Re}\{s_{LP}(t) \exp[j2\pi f_c t]\} \tag{11.1}$$

11.2.1 Pulse amplitude modulation

Many modulation formats can be interpreted as Pulse Amplitude Modulation (PAM) formats, where a *basis pulse* $g(t)$ is multiplied with a modulation coefficient c_i:

$$s_{LP}(t) = \sum_{i=-\infty}^{\infty} c_i g(t - iT_S) \tag{11.2}$$

where T_S is symbol duration. Different PAM formats differ in how the data bits b_i are mapped to the modulation coefficients c_i. This will be analyzed in more detail in subsequent sections. For now, let use turn to the possible shapes of the basis pulse $g(t)$ and the impact on the spectrum. We will assume that basis pulses are normalized to unit average power, so that:

$$\int_{-\infty}^{\infty} (g(t))^2 \, dt = T \tag{11.3}$$

---

[1] Note that definition (11.1) results in an energy of the bandpass signal that is half the energy of the baseband signal. In order to achieve equal energy, the energy must be defined as $E = \|s_{BP}\|^2 = 0.5\|s_{LP}\|^2$. Furthermore, it is required that the impulse response of a filter $h_{\text{filter,BP}}(t)$ in the passband has the following equivalent baseband representation $h_{\text{filter,BP}}(t) = 2\text{Re}\{h_{\text{filter,LP}}(t) \exp[j2\pi f_c t]\}$, in order to assure that the output of the filter has the same normalization as the input signal. The normalization used is thus not very logical, but is used here because it is the one that is most commonly used in the literature.

Alternatively, some authors (e.g., Barry et al. [2003]) define $s_{BP}(t) = \sqrt{2}\text{Re}\{s_{LP}(t) \exp[j2\pi f_c t]\}$.

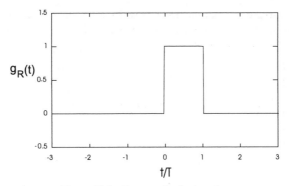

Figure 11.1   Rectangular basis pulse.

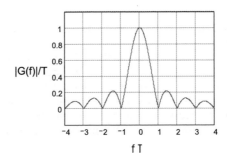

Figure 11.2   Spectrum of a rectangular basis pulse.

Rectangular basis pulses

The most simple basis pulse is a rectangular pulse with duration T. Figure 11.1 shows the pulse as a function of time, Fig. 11.2 the corresponding spectrum:

$$\left.\begin{array}{l} g_{\mathrm{R}}(t, T) = \begin{cases} 1 \ldots & \text{for } 0 \leq t \leq T \\ 0 \ldots & \text{otherwise} \end{cases} \\ G_{\mathrm{R}}(f, T) = \mathcal{F}\{g_{\mathrm{R}}(t, T)\} = T \operatorname{sinc}(\pi f T) \exp(-j\pi f T) \end{array}\right\} \tag{11.4}$$

where $\operatorname{sinc}(x) = \sin(x)/x$.

Nyquist pulse

The rectangular pulse has a spectrum that extends over a large bandwidth; the first sidelobes are only 13 dB weaker than the maximum. This leads to large adjacent channel interference, which in turn decreases the spectral efficiency of a cellular system. It is thus often required to obtain a spectrum that shows a stronger rolloff in the frequency domain. The most common class of pulses with strong spectral rolloffs are Nyquist pulses – i.e., pulses that fulfill the Nyquist criterion, and thus do not lead to intersymbol interference.[2]

---

[2] More precisely, Nyquist pulses do not create Inter Symbol Interference (ISI) assuming that the channel does not create ISI and sampling is done at ideal sampling times.

As an example, we present equations for a raised cosine pulse. This filter has a rolloff that follows a sinusoidal shape; the parameter determining the steepness of spectral decay is the rolloff factor α. Defining the functions:

$$
G_{N0}(f, \alpha, T) = \begin{cases} 1 & 0 \leq |2\pi f| \leq (1-\alpha)\dfrac{\pi}{T} \\[2mm] \dfrac{1}{2} \cdot \left(1 - \sin\left(\dfrac{T}{2\alpha}\left(|2\pi f| - \dfrac{\pi}{T}\right)\right)\right) & (1-\alpha)\dfrac{\pi}{T} \leq |2\pi f| \leq (1+\alpha)\dfrac{\pi}{T} \\[2mm] 0 & (1+\alpha)\dfrac{\pi}{T} \leq |2\pi f| \end{cases} \tag{11.5}
$$

the spectrum of the raised cosine pulse is:

$$
G_N(f, \alpha, T) = \frac{1}{\sqrt{1 - \dfrac{\alpha}{4}}} \cdot T \cdot G_{N0}(f, \alpha, T) \exp(-j\pi f T_S) \tag{11.6}
$$

and the temporal waveform follows from the inverse Fourier transformation (see Figs. 11.3 and 11.4).

For many applications, the *concatenation* of the TX filter and RX filter should result in a raised cosine shape. Due to the requirements of matching the receive filter to the received waveform, the spectrum of both the TX and the RX filter should be the square root of a raised cosine filter. Such a filter is known as a root-raised cosine filter, and henceforth denoted by subscript NR. Closed-form equations for the time domain functions of raised cosine and root-raised-cosine pulses can be found in Chennakeshu and Saulnier [1993].

Example 11.1 *Compute the ratio of signal power to adjacent channel interference when using (i) raised cosine pulses and (ii) root-raised cosine pulses with $\alpha = 0.5$ when the two considered signals have center frequencies 0 and 1.25/T.*

The two signals overlap slightly in the rolloff region and with $\alpha = 0.5$ the desired signal has spectral components in the band $-0.75/T \leq f \leq 0.75/T$. There is interfering energy between $0.5/T \leq f \leq 0.75/T$. Without loss of generality we assume $T = 1$. In the case of raised cosine pulses and assuming a matched filter at the RX, signal energy is proportional to:

$$
S = 2 \int_{f=0}^{0.25} |1|^4 df + 2 \int_{f=0.25}^{0.75} \left| \frac{1}{2} \cdot \left(1 - \sin\left(\frac{2\pi}{2\alpha}\left(|f| - \frac{1}{2}\right)\right)\right) \right|^4 df
$$

$$
= 0.77 \tag{11.7}
$$

while the interfering signal energy (assuming that there is an interferer at both the lower and the upper adjacent channel) is proportional to:

$$
I = 2 \int_{f=0.5}^{f=0.75} |G_N(f, \alpha, T) G_N(f - 1.25, \alpha, T)|^2 df
$$

$$
= 2 \int_{f=0.5}^{f=0.75} \left| \frac{1}{2} \cdot \left(1 - \sin\left[2\pi\left(|f| - \frac{1}{2}\right)\right]\right) \frac{1}{2} \cdot \left(1 - \sin\left[2\pi\left(|f - 1.25| - \frac{1}{2}\right)\right]\right) \right|^2 df
$$

$$
= 0.9 \cdot 10^{-4} \tag{11.8}
$$

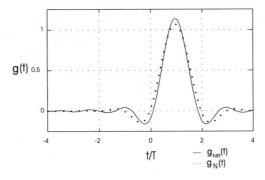

Figure 11.3   Raised cosine pulse and root-raised cosine pulse.

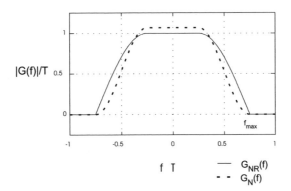

Figure 11.4   Spectrum of raised cosine pulse G_N and root-raised cosine pulse G_{NR}.

so the signal-to-interference ratio is $10 \log_{10}\left(\dfrac{S}{I}\right) \approx 39$ dB. Using root-raised cosine pulses, the signal energy is given by:

$$S = 2 \int\limits_{f=0}^{f=0.75/T} |G_N(f, \alpha, T)|^2 df = 0.875 \qquad (11.9)$$

and the interfering energy is given by:

$$I = 2 \int\limits_{f=0.5}^{f=-0.75/T} |G_N(f, \alpha, T) G_N(f - 1.25, \alpha, T)| \, df$$

$$= 2 \int\limits_{f=0.5}^{f=0.75} \left| \left| \frac{1}{2} \cdot \left(1 - \sin\left[2\pi\left(|f| - \frac{1}{2} \right) \right] \right) \right| \left| \frac{1}{2} \cdot \left(1 - \sin\left[2\pi\left(|f - 1.25| - \frac{1}{2} \right) \right] \right) \right| \right| df$$

$$= 5.6 \cdot 10^{-3} \qquad (11.10)$$

so the signal-to-interference ratio is $10 \log_{10}(SI) \approx 22$ dB. Obviously, the root-raised cosine filter, which has a flatter decay in the frequency domain, leads to a worse signal-to-interference ratio.

11.2.2 Multipulse modulation and continuous phase modulation

PAM (Eq. 11.2) can be generalized to multipulse modulation, where the signal is composed of a set of basis pulses; the pulse to be transmitted depends on the modulation coefficient c_i:

$$s_{LP}(t) = \sum_{i=-\infty}^{\infty} g_{c_i}(t - iT) \tag{11.11}$$

A typical example is M-ary Frequency Shift Keying (FSK): here the basis pulses have an offset from the carrier frequency $f_{mod} = i\Delta f/2$, where $i = \pm 1, \pm 3, \ldots, \pm(M-1)$. It is common to choose a set of orthogonal or bi-orthogonal pulses, as this simplifies the detector.

The situation is slightly different for *Continuous Phase Frequency Shift Keying* (CPFSK). In this case, the basis pulses follow each other in such a way that the phase of the total signal is continuous. The amplitude of the total signal is chosen as constant; the phase $\Phi(t)$ can be written as:

$$\Phi_{CPFSK}(t) = 2\pi h_{mod} \sum_{i=-\infty}^{\infty} c_i \int_{-\infty}^{t} \tilde{g}(u - iT)\, du \tag{11.12}$$

where u is the integration variable, h_{mod} is the modulation index, $\tilde{g}(t)$ is the *basis phase pulse*, which is normalized to:

$$\int_{-\infty}^{\infty} \tilde{g}(t)\, dt = 1/2 \tag{11.13}$$

Gaussian basis pulses

A Gaussian basis pulse is the convolution of a rectangular and a Gaussian function – in other words, the output of a filter with Gaussian impulse response that is excited by a rectangular waveform. Speaking mathematically, the impulse response of the filter is:

$$\frac{1}{\sqrt{2\pi}\sigma_G T} \exp\left(-\frac{t^2}{2\sigma_G^2 T^2}\right) \tag{11.14}$$

where

$$\sigma_G = \frac{\sqrt{\ln(2)}}{2\pi B_G T} \tag{11.15}$$

and B_G is the 3-dB bandwidth of the Gaussian filter. The transfer function of the Gaussian filter is:

$$\exp\left(-\frac{(2\pi f)^2 \sigma_G^2 T^2}{2}\right) \tag{11.16}$$

Using the normalization for phase basis pulse (11.13):

$$\tilde{g}_G(t) = \frac{1}{4T}\left[\text{erfc}\left(\frac{2\pi}{\sqrt{2\ln(2)}}B_G T\left(-\frac{t}{T}\right)\right) - \text{erfc}\left(\frac{2\pi}{\sqrt{2\ln(2)}}B_G T\left(1 - \frac{t}{T}\right)\right)\right] \tag{11.17}$$

where $\text{erfc}(x)$ is the complementary error function:

$$\text{erfc}(x) = \frac{2}{\sqrt{\pi}} \int_{x}^{\infty} \exp(-t^2)\, dt \tag{11.18}$$

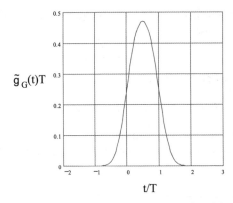

$\tilde{g}_G(t)T$

t/T

Figure 11.5 Shape of a Gaussian-phase basis pulse with $B_G T = 0.5$.

11.2.3 Power spectrum

Occupied bandwidth

The bandwidth of the transmitted signal is a very important characteristic of a modulation format, especially in the context of a wireless signal, where spectrum is at a premium. Before going into a deeper discussion, however, we have to clarify what we mean by "bandwidth", since various definitions are possible:

- *The noise bandwidth* is defined as the bandwidth of a system with a rectangular transfer function $|H_{rect}(f)|$ (and identical peak amplitude $\max(|H(f)|)$) that receives as much noise as the system under consideration.
- The *3 dB bandwidth* is the bandwidth at which $|H(f)|^2$ has decreased to a value that is 3 dB below its maximum value.
- The 90% *energy bandwidth* is the bandwidth that contains 90% of total emitted energy; analogous definitions are possible for the 99% energy bandwidth or other percentages of the contained energy.

Bandwidth efficiency is defined as the ratio of the data (bit) rate to the occupied bandwidth.

Power-spectral density of cyclostationary processes

A cyclostationary process $x(t)$ is defined as a stochastic process whose mean and autocorrelation functions are periodic with period T_{per}:

$$\left. \begin{aligned} E\{x(t + T_{per})\} &= E\{x(t)\} \\ R_{xx}(t + T_{per} + \tau, t + T_{per}) &= R_{xx}(t + \tau, t) \end{aligned} \right\} \tag{11.19}$$

These properties are fulfilled by most modulation formats; periodicity T_{per} is the symbol duration T_S. We furthermore assume that the data symbols are zero-mean and uncorrelated, so that the spectral density of the data symbols is white. Those properties are fulfilled by most PAM signals; either they are uncoded, or coded and scrambled. The power-spectral density of a modulated signal can then be computed as the product of the power spectrum of a basis pulse $|S_G(f)|^2$ and the spectral density of the data σ_S^2:

$$S_{LP}(f) = \frac{1}{T_{per}} \cdot |S_G(f)|^2 \cdot \sigma_S^2 \tag{11.20}$$

The power-spectral density of the basis pulse is simply the squared magnitude of the Fourier transform $G(f)$ of the basis pulse $g(t)$:

$$|S_G(f)|^2 = |G(f)|^2 \tag{11.21}$$

The power-spectral density in passband is:

$$S_{BP}(f) = \frac{1}{2}[S_{LP}(f - f_c) + S_{LP}(-f - f_c)] \tag{11.22}$$

Note that CPFSK signals have memory and thus correlation between the symbols, so that the computation of power-spectral densities is much more complicated. Details can be found in Proakis [1995].

11.2.4 Signal space diagram

One of the most important tools in the analysis of modulation formats is the signal space diagram. It leads to a graphical representation of signals that allows an intuitive and uniform treatment of different modulation methods. In Chapter 12, we will explain in more detail how the signal space diagram can be used to compute bit error probabilities. For now, we mainly use it as a convenient shorthand, and a method for representing different modulation formats.

To simplify explanations, we assume in the following rectangular basis pulses so that $g(t) = g_R(t, T)$. The signals $s(t)$ that can be present during the ith interval $iT_S < t < (i+1)T_S$ form a finite set of functions (in the following, we assume $i = 0$ without loss of generality). The size of this set is M. This representation covers both pulse amplitude modulation and multipulse modulation.

We now choose an orthogonal set (of size N) of expansion functions $\varphi_n(t)$, which are defined in the range $[0, T_S]$.[3] The set of expansion functions has to be complete, in the sense that all transmit signals can be represented as linear combinations of expansion functions. Such a complete set of expansion functions can be obtained by a Gram–Schmidt orthogonalization procedure (see, e.g., Wozencraft and Jacobs [1965]).

Given the set of expansion functions $\{\varphi_n(t)\}$, any transmit signal $s_m(t)$ can be represented as a vector $\mathbf{s}_m = (s_{m,1}, s_{m,2}, \ldots, s_{m,N})$ where:

$$s_{m,n} = \int_0^{T_S} s_m(t)\varphi_n^*(t)\, dt \tag{11.23}$$

where $*$ denotes complex conjugation. In particular, the vector components of a *bandpass signal* are computed as:

$$s_{BP,m,n} = \int_0^{T_S} s_{BP,m}(t)\varphi_{BP,n}(t)\, dt \tag{11.24}$$

Conversely, the actual transmit signal can be obtained from the vector components (both for passband and baseband) as:

$$s_m(t) = \sum_{n=1}^{N} s_{m,n}\varphi_n(t) \tag{11.25}$$

As each signal is represented by a vector $\mathbf{s}_m$, we can plot these vectors in the so-called *signal space diagram*.[4] A graphical representation of these points is especially simple for $N = 2$, which

---

[3] Expansion functions are often called "basis functions". However, we avoid this name in order to avoid confusion with "basis pulses". It is noteworthy that expansion functions are by *definition* orthogonal, while basis pulses can be orthogonal, but are not necessarily so.

[4] More precisely, the endpoints of the vectors starting in the origin of the coordinate system.

fortunately covers most important modulation formats. For the passband representation, the expansion functions for Quadrature Amplitude Modulation (QAM) are:

$$\left.\begin{array}{l} \varphi_{\text{BP},1}(t) = \sqrt{\dfrac{2}{T_{\text{S}}}} \cos(2\pi f_{\text{c}}t) \\[3mm] \varphi_{\text{BP},2}(t) = \sqrt{\dfrac{2}{T_{\text{S}}}} \sin(2\pi f_{\text{c}}t) \end{array}\right\} \tag{11.26}$$

Here it is assumed that $f_{\text{c}} \gg 1/T_{\text{S}}$; this implies that all products containing $\cos(2 \cdot 2\pi f_{\text{c}}t)$ are negligible and/or eliminated by a filter.

If starting from the *complex baseband representation*, the signal space diagram can be obtained with the expansion vectors:

$$\left.\begin{array}{l} \varphi_1(t) = \sqrt{\dfrac{1}{T_{\text{S}}}} \cdot 1 \\[3mm] \varphi_2(t) = \sqrt{\dfrac{1}{T_{\text{S}}}} \cdot j \end{array}\right\} \tag{11.27}$$

It is important that the signal space diagrams that result from bandpass and lowpass representation differ by a factor of $\sqrt{2}$. In the following, we will employ the signal constellations resulting from the bandpass signals, so that for a binary antipodal signal (e.g., Binary Phase Shift Keying – BPSK), where the signal is:

$$s_{\text{BP},\frac{1}{2}} = \pm\sqrt{\dfrac{2E_{\text{S}}}{T_{\text{S}}}} \cos(2\pi f_{\text{c}}t) \tag{11.28}$$

the points in the signal space diagram are located at $\pm\sqrt{E_{\text{S}}}$, with E_{S} being the symbol energy.

The *energy* contained in a symbol can be computed from the bandpass signal as:

$$E_{\text{S},m} = \int_0^{T_{\text{S}}} s_{\text{BP},m}^2(t)\, dt = \|s_{\text{BP},m}\|^2 \tag{11.29}$$

where $\|s\|$ denotes the L_2-norm (Euclidian norm) of s. When considering equivalent baseband, the signal energy is given as:

$$E_{\text{S},m} = \frac{1}{2}\int_0^{T_{\text{S}}} \|s_{\text{LP},m}(t)\|^2 dt \simeq \frac{1}{2}\|s_{\text{LP},m}\|^2 \tag{11.30}$$

and $s_{\text{LP},m}$ are the signal vectors obtained from the equivalent baseband representation. Note that for many modulation formats (e.g., BPSK), E_{S} is independent of m.

The *correlation coefficient* between $s_m(t)$ and $s_k(t)$ can be computed as:

$$\text{Re}\{\rho_{k,m}\} = \frac{s_{\text{BP},m} s_{\text{BP},k}}{\|s_{\text{BP},m}\|\, \|s_{\text{BP},k}\|} \tag{11.31}$$

for bandpass representation or

$$\rho_{k,m} = \frac{s_{\text{LP},m}\left(s_{\text{LP},k}\right)^*}{\|s_{\text{LP},m}\|\, \|s_{\text{LP},k}\|} \tag{11.32}$$

The *Euclidean distance* between the two signals is:

$$d_{k,m}^2 = \left[E_{\text{S},m} + E_{\text{S},k} - 2\sqrt{E_{\text{S},m}E_{\text{S},k}}\ \text{Re}\{\rho_{k,m}\}\right] \tag{11.33}$$

We will see later on that this distance is important for computation of the bit error probability.

11.3 Important modulation formats

In this section, we summarize in a very concise form the most important properties of different digital modulation formats. We will give the following information for each modulation format:

- bandpass and baseband signal as a function of time;
- representation in terms of PAM or multipulse modulation;
- signal space diagram;
- spectral efficiency.

11.3.1 Binary phase shift keying

BPSK modulation is the simplest modulation method: the carrier phase is shifted by $\pm\pi/2$, depending on whether a $+1$ or -1 is sent.[5] Despite this simplicity, two different interpretations of BPSK are possible. The first one is to see BPSK as a *phase modulation*, in which the datastream influences the phase of the transmit signal:

$$s_{BP}(t) = \sqrt{2E_B/T_B} \cos\left(2\pi f_c t + \frac{\pi}{2}\widetilde{p}_D(t)\right) \tag{11.34}$$

where the phase modulation sequence $\widetilde{p}_{D,BPSK}(t)$ is:

$$\widetilde{p}_D(t) = \sum_{i=-\infty}^{\infty} b_i g(t - iT) = b_i * g(t) \tag{11.35}$$

and

$$g(t) = g_R(t, T_B) \tag{11.36}$$

The above interprets BPSK as phase modulation, so that the phase basis pulses are rectangular pulses with amplitude $\pi/2$.

The second, and more popular interpretation, is to view BPSK as a *pulse amplitude modulation* where the basis pulses are rectangular pulses with amplitude 1, so that:

$$s_{BP}(t) = \sqrt{2E_B/T_B}\, p_D(t) \cos\left(2\pi f_c t + \frac{\pi}{2}\right)$$

where $p_D(t) = \widetilde{p}_D(t)$. Figure 11.6 shows the signal waveform and Fig. 11.7 the signal space diagram. In equivalent baseband, the complex modulation symbols are $\pm j$:

$$c_i = j \cdot b_i \tag{11.38}$$

so that the real part of the signal is:

$$\text{Re}\{s_{LP}(t)\} = 0 \tag{11.39}$$

and the imaginary part is:

$$\text{Im}\{s_{LP}(t)\} = \sqrt{\frac{2E_B}{T_B}} p_D(t) \tag{11.40}$$

The envelope has constant amplitude, except at times $t = iT_B$. The spectrum shows a very slow rolloff, due to the use of unfiltered rectangular pulses as basis pulses. The bandwidth efficiency is 0.59 bit/s/Hz when we consider the bandwidth that contains 90% of the energy, but only 0.05 bit/s/Hz when considering the 99% energy bandwidth (see also Fig. 11.8).

---

[5] Strictly speaking, the reference phase φ also has to be given, which determines the phase at $t = 0$. Without restriction of generality, we assume in the following $\varphi_{S,0} = 0$.

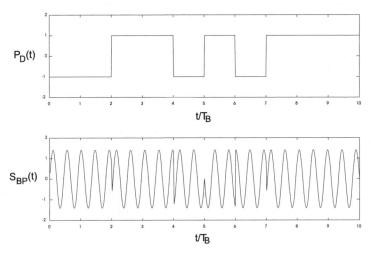

Figure 11.6 Binary-phase-shift-keying signal as function of time.

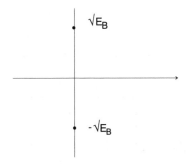

Figure 11.7 Signal space diagram for binary-phase shift keying.

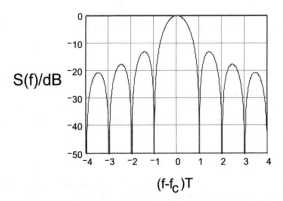

Figure 11.8 Normalized power-spectral density for binary-phase shift keying.

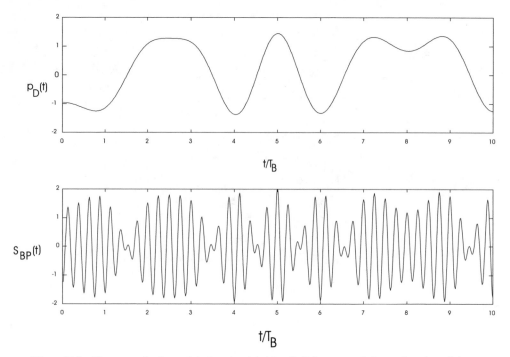

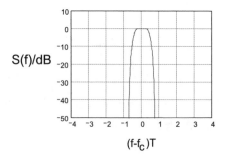

Figure 11.9 Binary amplitude modulation signal (with rolloff factor $\alpha = 0.5$) as a function of time.

Figure 11.10 Normalized power-spectral density of a binary amplitude modulation signal (with rolloff factor $\alpha = 0.5$).

Because of the low bandwidth efficiency of rectangular pulses, practical TXs often use Nyquist-shaped pulses as basis pulses.[6] Even the relatively mild filtering of $\alpha = 0.5$ leads to a dramatic increase in spectral efficiency: for 90% and 99% energy bandwidth, spectral efficiency becomes 1.02 and 0.79 bit/s/Hz, respectively. On the other hand, we see that the signal no longer has a constant envelope (see Figs. 11.9 and 11.10).

---

[6] Note that there is a difference between a PAM with Nyquist-shaped basis pulses, and a phase modulation with Nyquist-shaped phase pulses. In the following, we will consider only PAM with Nyquist basis pulses, and call it BAM (*Binary Amplitude Modulation*) for clarity.

An important variant is *Differentially encoded PSK* (DPSK). The basic idea is that the transmitted phase is not solely determined by the current symbol; rather, we transmit the phase of the previous symbol plus the phase corresponding to the current symbol. For BPSK, this reduces to a particularly simple form; we first encode the data bits according to:

$$\tilde{b}_i = b_i b_{i-1} \tag{11.41}$$

and then use $\tilde{b}_i$ instead of b_i in Eq. (11.35).

The advantage of differential encoding is that it enables a differential decoder, which only needs to compare the phases of two subsequent symbols in order to demodulate received signals. This obviates the need to recover the absolute phase of the received signal, and thus allows simpler and cheaper RXs to be built.

11.3.2 Quadrature-phase shift keying

A QPSK-modulated signal carries 1 bit on both the in-phase and quadrature-phase component. The original datastream is split into two streams, $b1_i$ and $b2_i$:

$$\left.\begin{array}{l} b1_i = b_{2i} \\ b2_i = b_{2i+1} \end{array}\right\} \tag{11.42}$$

each of which has a data rate that is half that of the original datastream:

$$R_S = 1/T_S = R_B/2 = 1/(2T_B) \tag{11.43}$$

Let us first consider the situation where basis pulses are rectangular pulses, $g(t) = g_R(t, T_S)$. Then we can give an interpretation of QPSK as either a phase modulation or as a pulse amplitude modulation. We first define two sequences of pulses (see Fig. 11.11):

$$\left.\begin{array}{l} p1_D(t) = \displaystyle\sum_{i=-\infty}^{\infty} b1_i g(t - iT_S) = b1_i * g(t) \\ p2_D(t) = \displaystyle\sum_{i=-\infty}^{\infty} b2_i g(t - iT_S) = b2_i * g(t) \end{array}\right\} \tag{11.44}$$

When interpreting QPSK as a *pulse amplitude modulation*, the bandpass signal reads:

$$s_{BP}(t) = \sqrt{E_B/T_B}[p1_D(t)\cos(2\pi f_c t) - p2_D(t)\sin(2\pi f_c t)] \tag{11.45}$$

Normalization is done in such a way that the energy within one symbol interval is $\int_0^{T_S} s_{BP}(t)^2 dt = 2E_B$, where E_B is the energy expended on transmission of a bit. Figure 11.13 shows the signal space diagram. The baseband signal is:

$$s_{LP}(t) = [p1_D(t) + jp2_D(t)]\sqrt{E_B/T_B} \tag{11.46}$$

When interpreting QPSK as a *phase modulation*, the low-pass signal can be written as $\sqrt{E_B/T_B}\exp(j\Phi_S(t))$ with:

$$\Phi_S(t) = \pi \cdot \left[\frac{1}{2} \cdot p2_D(t) - \frac{1}{4} \cdot p1_D(t) \cdot p2_D(t)\right] \tag{11.47}$$

It is obvious from this representation that the signal is constant envelope, except for the transitions at $t = iT_S$ (see Fig. 11.12).

The spectral efficiency of QPSK is twice the efficiency of BPSK, since both the in-phase and the quadrature-phase components are exploited for the transmission of information. This means that when considering the 90% energy bandwidth, the efficiency is 1.1 bit/s/Hz, while for the 99% energy bandwidth, it is 0.1 bit/s/Hz (see Fig. 11.14). The slow spectral rolloff motivates

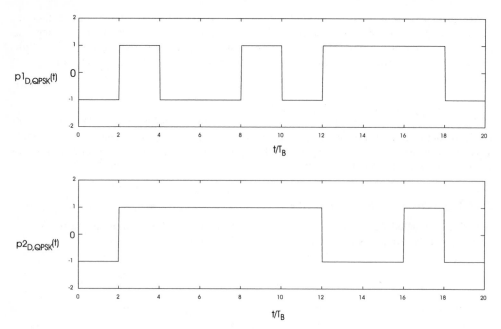

Figure 11.11   Datastreams of in-phase and quadrature-phase components in quadrature-phase shift keying.

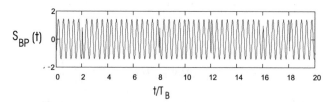

Figure 11.12   Quadrature-phase-shift-keying signal as a function of time.

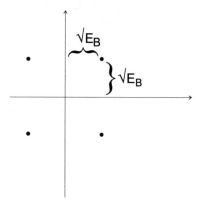

Figure 11.13   Signal space diagram of quadrature-phase shift keying.

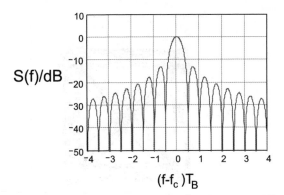

Figure 11.14 Normalized power-spectral density of quadrature-phase shift keying.

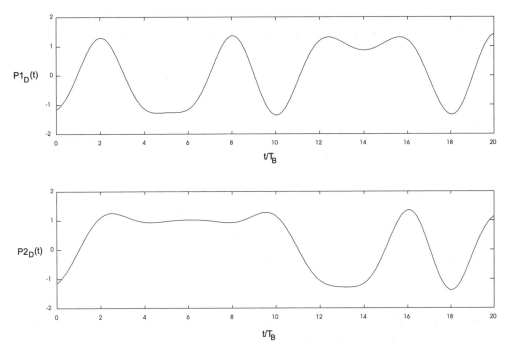

Figure 11.15 Quadrature amplitude modulation pulse sequence.

(similarly to the BPSK) the use of raised cosine basis pulses (see Fig. 11.15); we will refer to the resulting modulation format as QAM in the following. The spectral efficiency increases to 2.04 and 1.58 bit/s/Hz, respectively (see Fig. 11.17). On the other hand, the signal shows strong envelope fluctuations (Fig. 11.16).

11.3.3 $\pi/4$-differential quadrature-phase shift keying

Even though QPSK is nominally a constant envelope format, it has amplitude dips at bit transitions; this can also be seen by the fact that the trajectories in the I–Q diagram pass

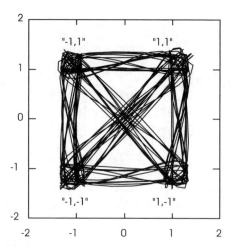

Figure 11.16 I–Q diagram of quadrature amplitude modulation with raised cosine basis pulses. Also shown are the four normalized points of the normalized signal space diagram, $(1, 1)$, $(1, -1)$, $(-1, -1)$, $(-1, 1)$.

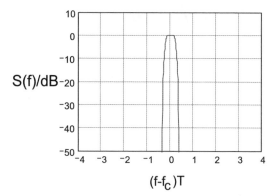

Figure 11.17 Normalized power-spectral density of quadrature amplitude modulation with raised cosine filters with $\alpha = 0.5$.

through the origin for some of the bit transitions. The duration of the dips is longer when non-rectangular basis pulses are used. Such variations of the signal envelope are undesirable, because they make the design of suitable amplifiers more difficult. One possibility for reducing these problems lies in the use of $\pi/4$-*DQPSK* ($\pi/4$ differential quadrature-phase shift keying). This modulation format is one of the most important for wireless communications – it is used in several American standards (IS-54, IS-136, PWT), as well as the Japanese cellphone (JDC) and cordless (PHS) standards, and the European trunk radio standard (TETRA).

The principle of $\pi/4$-DQPSK can be understood from the signal space diagram of DQPSK (see Fig. 11.18). There exist *two* sets of signal constellations: $(0, 90, 180, 270°)$ and $(45, 135, 225, 315°)$. All symbols with an even temporal index i are chosen from the first set, while all symbols with odd index are chosen from the second set. In other words: whenever t is a multiple integer of the symbol duration, the transmit phase is increased by $\pi/4$, in addition to the change of phase due to the transmit symbol. Therefore, transitions between subsequent signal constellations can never pass through the origin (see Fig. 11.21); in physical terms, this means much smaller fluctuations of the envelope.

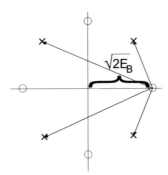

Figure 11.18 Allowed transitions in the signal space diagram of $\pi/4$ differential quadrature-phase shift keying.

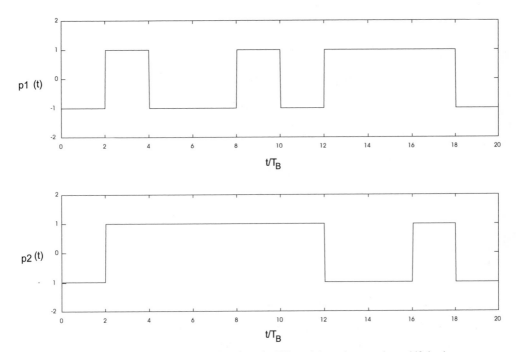

Figure 11.19 Sequence of basis pulses for $\pi/4$ differential quadrature-phase shift keying.

The signal phase is given by:

$$\Phi_s(t) = \pi\left[\frac{1}{2}p2_{\mathrm{D}}(t) - \frac{1}{4}p1_{\mathrm{D}}(t)p2_{\mathrm{D}}(t) + \frac{1}{4}\left\lfloor\frac{t}{T_{\mathrm{S}}}\right\rfloor\right] \tag{11.48}$$

where $\lfloor x \rfloor$ denotes the largest integer smaller or equal to x. Comparing this with Eq. (11.47), we can clearly see the change in phase at each integer multiple of T_{S}. Figure 11.19 shows the underlying data sequences, and Fig. 11.20 depicts the resulting bandpass signals when using rectangular or raised cosine basis pulses.

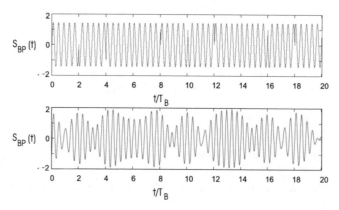

Figure 11.20 $\pi/4$ differential quadrature-phase shift keying signals as function of time for rectangular basis functions (top) and raised cosine basis pulses (bottom).

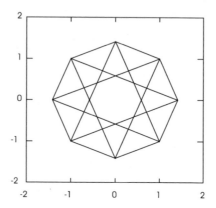

Figure 11.21 I–Q diagram of a $\pi/4$-differential-quadrature-phase-shift-keying signal with rectangular basis functions.

11.3.4 Offset quadrature-phase shift keying

Another way of improving the peak-to-average ratio in QPSK is to make sure that bit transitions for the in-phase and the quadrature-phase components occur at different time instants. This method is called OQPSK (offset QPSK). The bitstreams modulating the in-phase and quadrature-phase components are offset half a symbol duration with respect to each other (see Fig. 11.22), so that transitions for the in-phase component occur at integer multiples of the symbol duration (even integer multiples of the bit duration), while quadrature component transitions occur half a symbol duration (1-bit duration) later. Thus, the transmit pulsestreams are:

$$
\left.
\begin{aligned}
p1_D(t) &= \sum_{i=-\infty}^{\infty} b1_i g(t - iT_S) = b1_i * g(t) \\
p2_D(t) &= \sum_{i=-\infty}^{\infty} b2_i g\left(t - \left(i + \frac{1}{2}\right)T_S\right) = b2_i * g\left(t - \frac{T_S}{2}\right)
\end{aligned}
\right\}
\qquad (11.49)
$$

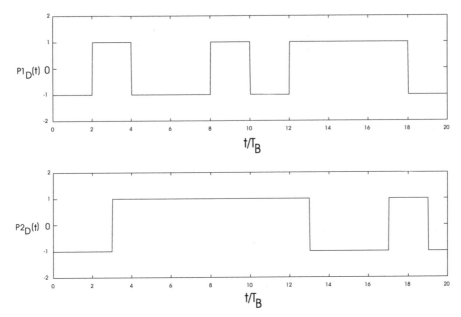

Figure 11.22   Sequence of basis pulses for offset quadrature-phase shift keying.

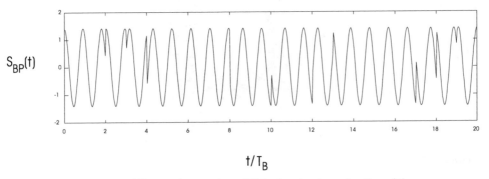

Figure 11.23   Offset quadrature-phase shift keying signal as a function of time.

These datastreams can again be used for interpretation as pulse amplitude modulation (Eq. 11.46) or as phase modulation, according to Eq. (11.47). The resulting bandpass signal is shown in Fig. 11.23.

The representation in the envelope diagram (Fig. 11.24) makes clear that there are no transitions passing through the origin of the coordinate system; thus this modulation format takes care of envelope fluctuations as well.

As for regular QPSK, we can use smoother basis pulses, like raised cosine pulses, to improve spectral efficiency. Figure 11.25 shows the resulting I–Q diagram.

11.3.5 Higher order modulation

Up to now, we have treated modulation formats that transmit at most 2 bits per symbol. In this section, we mention how these schemes can be generalized to transmit more information in each

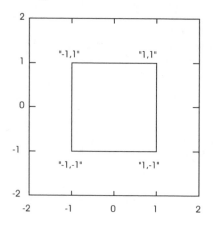

Figure 11.24 I–Q diagram for offset quadrature-phase shift keying with rectangular basis functions. Also shown are the four points of the normalized signal space diagram, $(1, 1)$, $(1, -1)$, $(-1, -1)$, $(-1, 1)$.

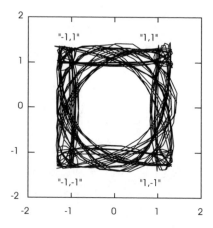

Figure 11.25 I–Q diagram for offset quadrature amplitude modulation with raised cosine basis pulses.

symbol interval. Such schemes result in higher spectral efficiency, but consequently also in higher sensitivity to noise and interference. They are therefore not used as often in wireless transmission schemes. Only recently has multilevel QAM found adaptation in the standards for wireless Local Area Networks (LANs) (see Chapter 25) and cellular packet radio (Chapter 22).

Higher order QAM

4-QAM (also known simply as QAM, or QPSK) transmits 1 bit on both the in-phase and the quadrature-phase component. It does so by sending a signal with positive or negative polarity, but fixed amplitude, on each component. This scheme can be generalized by allowing multiple amplitude levels. The resulting scheme is called *higher order QAM*. The mathematical representation is the same as for 4-QAM, just that in the pulse sequences of Eq. (11.44) we not only allow the levels ± 1, but $2m - 1 - \sqrt{M}$, with $m = 1, \ldots, \sqrt{M}$. The signal space diagram for 16-QAM is shown in Fig. 11.26.

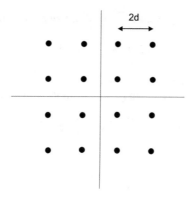

Figure 11.26 Signal space diagram for 16-QAM.

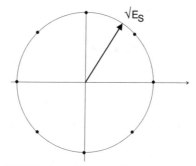

Figure 11.27 Envelope diagram of 8-PSK. Also shown are the eight points of the signal space diagram.

Larger constellations, including 64-QAM and 256-QAM, can be constructed according to similar principles. Naturally, the larger the peak-to-average ratio of the output signal, the larger the constellation is.

Example 11.2 *Relate the average energy of a 16-QAM signal to the distance between two adjacent points in the signal space diagram.*

Let us for simplicity consider the first quadrant of the signal space diagram. Assume that the signal points are located at $d + jd$, $d + j3d$, $3d + j$, and $3d + j3d$. The average energy is given by $E_s = \dfrac{d^2}{4}(2 + 10 + 10 + 18) = 10d^2$. The squared distance between two points is $4d^2$, so the ratio then becomes 2.5.

Higher order phase modulation

The drawback of higher order QAM is the fact that the resulting signals show strong variations of output amplitude; this implies that linear amplifiers need to be used. An alternative is the use of higher order phase shift keying, where the transmit signal can be written as:

$$s_{BP}(t) = \sqrt{2E_S/T_S} \cos\left(2\pi f_c t + \frac{2\pi}{M}(m-1)\right), \qquad m = 1, 2, \ldots, M \tag{11.50}$$

that is, the TX picks one of the M transmit phases (see Fig. 11.27); note that we normalize energy here in terms of *symbol* energy and *symbol* duration. The equivalent low-pass signal is:

$$s_{LP}(t) = \sqrt{2E_S/T_S} \exp\left(j\frac{2\pi}{M}(m-1)\right) \tag{11.51}$$

The correlation coefficient between two signals is:

$$\rho_{km} = \exp\left(j\frac{2\pi}{M}(m-k)\right) \tag{11.52}$$

11.3.6 Binary frequency shift keying

In *Frequency Shift Keying* (FSK), each symbol is represented by transmitting (for a time T_S) a sinusoidal signal whose frequency depends on the symbol to be transmitted. FSK cannot be represented as PAM. Rather, it is a form of multipulse modulation: depending on the bit to be transmitted, basis pulses with different center frequencies are transmitted:

$$g_m(t) = \cos[(2\pi f_c + b_m 2\pi f_{mod})t/T + \psi] \tag{11.53}$$

For a Continuous Phase FSK (CPFSK) signal, the phase is given as:

$$\Phi_S(t) = h_{mod} \int_{-\infty}^{t} \tilde{p}_{D,FSK}(\tau) \, d\tau \tag{11.54}$$

and the signal has a constant envelope, where h_{mod} is the modulation index. Using the normalization for phase pulses (Eq. 11.13) and assuming rectangular phase basis pulses:

$$\tilde{g}_{FSK}(t) = \frac{1}{2T_B} g_R(t, T_B) \tag{11.55}$$

the phase pulse sequence is:

$$\tilde{p}_{D,FSK}(t) = \sum_{i=-\infty}^{\infty} b_i \tilde{g}_{FSK}(t - iT_B) = b_i * \tilde{g}_{FSK}(t) \tag{11.56}$$

The instantaneous frequency is given as:

$$f(t) = f_c + b_i f_{mod}(t) = f_c + f_D(t) = f_c + \frac{1}{2\pi}\frac{d\Phi_S(t)}{dt} \tag{11.57}$$

Figure 11.28 shows a bandpass signal as a function of time. The real and imaginary parts of the equivalent baseband signal are then:

$$\text{Re}(s_{LP}(t)) = \sqrt{2E_B/T_B} \cos\left[2\pi h_{mod} \int_{-\infty}^{t} \tilde{p}_{D,FSK}(\tau) \, d\tau\right] \tag{11.58}$$

$$\text{Im}(s_{LP}(t)) = \sqrt{2E_B/T_B} \sin\left[2\pi h_{mod} \int_{-\infty}^{t} \tilde{p}_{D,FSK}(\tau) \, d\tau\right] \tag{11.59}$$

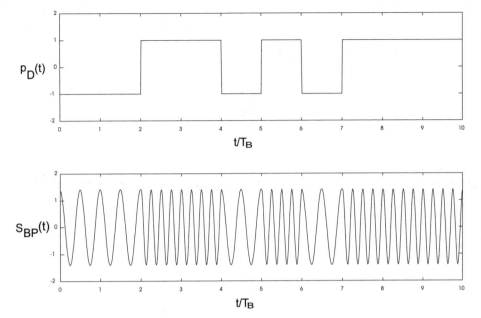

Figure 11.28   Frequency-shift-keying signal as a function of time.

For the signal space diagram, we can find two different representations:

1. Use sinusoidal oscillations at the two possible signal frequencies $f_c \pm f_{mod}$. Expressing this in terms of phase pulses, the expansion functions (in the passband) read:

$$\left. \begin{array}{l} \varphi_{BP,1}(t) = \sqrt{2/T_B} \cos(2\pi f_c t + 2\pi f_{mod} t) \\[2mm] \varphi_{BP,2}(t) = \sqrt{2/T_B} \cos(2\pi f_c t - 2\pi f_{mod} t) \end{array} \right\} \tag{11.60}$$

    In this case, the signal space diagram consists of two points on the two orthogonal axes (see Fig. 11.29). Note that we have made the implicit assumption that the two signals $\varphi_{BP,1}(t)$ and $\varphi_{BP,2}(t)$ are orthogonal to each other.
2. Use the in-phase and quadrature-phase component of the center frequency f_c. In this case, the signal shows up in the envelope diagram not as a discrete point, but as a continuous trajectory – namely, a circle (see Fig. 11.30). At any time instant, the transmit signal is represented by a different point in the diagram.

The power spectrum of FSK can be shown to consist of a continuous and a discrete (spectral lines) part:

$$S(f) = S_{cont}(f) + S_{disc}(f) \tag{11.61}$$

where (see Benedetto and Biglieri [1999]):

$$S_{cont}(f) = \frac{1}{2T} \left\{ \sum_{m=1}^{2} |G_m(f)|^2 - \frac{1}{2} \left| \sum_{m=1}^{2} G_m(f) \right|^2 \right\} \tag{11.62}$$

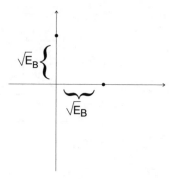

Figure 11.29 Signal space diagram of frequency shift keying when using $\cos[2\pi(f_c \pm f_D)t]$ as basis functions.

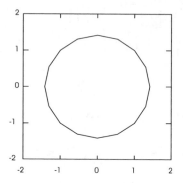

Figure 11.30 Envelope diagram of frequency shift keying. This is equivalent to a signal space diagram for FSK when using $\cos(2\pi f_c t)$ and $\sin(2\pi f_c t)$ as basis functions.

and

$$S_{\text{disc}}(f) = \frac{1}{(2T)^2} \left| \sum_{m=1}^{2} G_m(f) \right|^2 \sum_n \delta\left(f - \frac{n}{T}\right) \tag{11.63}$$

where $G_m(f)$ is the Fourier transform of $g_m(t)$. An example is shown in Fig. 11.31.

11.3.7 Minimum shift keying

Minimum Shift Keying (MSK) is one of the most important modulation formats for wireless communications. However, it can be interpreted in different ways, which leads to considerable confusion:

1. The first interpretation is as CPFSK with a modulation index:

    $$h_{\text{mod}} = 0.5, \qquad f_{\text{mod}} = 1/4T \tag{11.64}$$

    This implies that the phase changes by $\pm\pi/2$ during a 1-bit duration (see Fig. 11.32). The bandpass signal is shown in Fig. 11.33.

2. Alternatively, we can interpret MSK as *Offset QAM* (OQAM). As is shown in Appendix 11.A (see *www.wiley.com/go/molisch*), basis pulses are sinusoidal half-waves extending over a duration of $2T_B$ (see also Fig. 11.34):

    $$g(t) = \sin(2\pi f_{\text{mod}}(t + T_B))g_R(t, 2T_B) \tag{11.65}$$

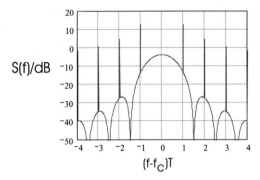

Figure 11.31 Power spectral density of (non-continuous phase) frequency shift keying with $h_{mod} = 1$.

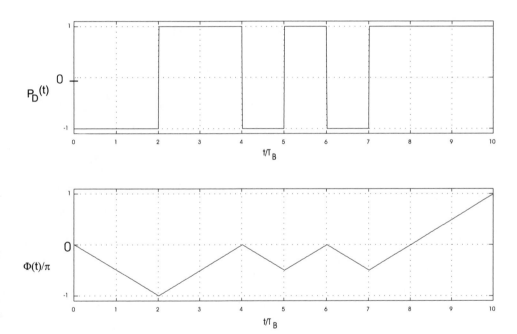

Figure 11.32 Phase pulse and phase as function of time for minimum-shift-keying signal.

Due to the use of smoother basis functions, the spectrum decreases faster than that of "regular" OQPSK:

$$S(f) = \frac{16T_B}{\pi^2} \left(\frac{\cos(2\pi f T_B)}{1 - 16f^2 T_B^2} \right)^2 \tag{11.66}$$

see also Fig. 11.35. On the other hand, MSK is only a binary modulation format, while OQPSK transmits 2 bits per symbol duration. As a consequence, MSK has lower spectral efficiency when considering the 90% energy bandwidth (1.29 bit/s/Hz), but still performs reasonably well when considering the 99% energy bandwidth (0.85 bit/s/Hz).

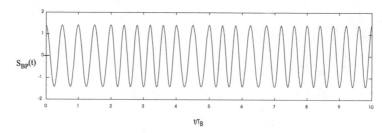

Figure 11.33   Minimum-shift-keying-modulated signal.

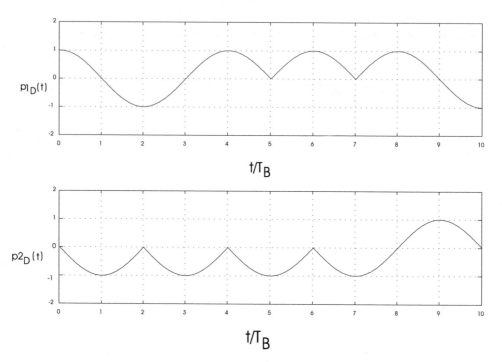

Figure 11.34   Composition of minimum shift keying from sinusoidal half-waves.

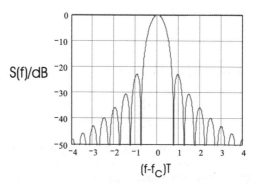

Figure 11.35   Power-spectral density of minimum shift keying.

Example 11.3 *Compare the spectral efficiency of MSK and QPSK with rectangular constitutent pulses. Consider systems with equal bit duration. Compute the out-of-band energy at $1/T_B$, $2/T_B$, and $3/T_B$.*

The power spectral density of MSK is given by:

$$S_{\text{MSK}}(f) = \frac{16T_B}{\pi^2} \left(\frac{\cos(2\pi f T_B)}{1 - 16 f^2 T_B^2} \right)^2 \tag{11.67}$$

whereas the power spectral density for QPSK with rectangular pulses is the same as for ordinary QAM given by:

$$S_{\text{OQPSK}}(f) = (2T_B \, \text{sinc}(2\pi f T_B))^2 \tag{11.68}$$

where it must be noted that $T_S = 2T_B$ for QPSK. The out-of-band power is, for MSK and $T_B = 1$, given by:

$$
\begin{aligned}
P_{\text{out}}(f_0) &= 2 \int\limits_{f=f_0}^{\infty} S(f) df = 2 \int\limits_{1}^{\infty} \frac{16}{\pi^2} \left(\frac{\cos(2\pi f)}{1 - 16 f^2} \right)^2 df \\
&= \frac{32}{\pi^2} \int_{1}^{\infty} \frac{\cos^2 2\pi f}{256 f^4 - 32 f^2 + 1} df = \frac{32}{\pi^2} \int_{1}^{\infty} \frac{\frac{1}{2}\cos 4\pi f + \frac{1}{2}}{256 f^4 - 32 f^2 + 1} df
\end{aligned}
\tag{11.69}
$$

and for QPSK with $T_B = 1$, given by:

$$P_{\text{out}}(f_0) = \int\limits_{1}^{\infty} \left(2 \frac{\sin(2\pi f)}{(2\pi f)} \right)^2 df \tag{11.70}$$

Solving these integrals numerically gives the following table:

	$1/T_B$	$2/T_B$	$3/T_B$
QPSK	0.050	0.025	0.017
MSK	0.0024	$2.8 * 10^{-4}$	$7.7 * 10^{-5}$

11.3.8 Demodulation of minimum shift keying

The different interpretations of MSK are not just useful for gaining insights into the modulation scheme, but also for building demodulators. Different demodulator structures correspond to different interpretations:

- *Frequency discriminator detection*: since MSK is a type of frequency shift keying, it is straightforward to check whether the instantaneous frequency is larger or smaller than the carrier frequency (larger or smaller than 0 when considering equivalent baseband). The instantaneous frequency can be sampled in the middle of the bit, or it can be integrated over (part of the) bit duration in order to reduce the effect of noise. This RX structure is simple, but suboptimum, since it does not exploit the continuity of the phase at bit transitions.
- *Differential detection*: the phase of the signal changes by $+\pi/2$ or $-\pi/2$ over a 1-bit duration, depending on the bit that was transmitted. An RX thus just needs to determine the phases at times iT and $(i+1)T$, in order to make a decision. It is remarkable that no differential encoding of the transmit signal is required; an erroneous estimate of the phase at one sampling time leads to two (but not more) bit errors.

- *Matched filter reception*: it is well-known that matched filter reception is optimum (see also Chapter 12). This is true both when considering MSK as OQPSK, and when considering it as multipulse modulation. However, it has to be noted that MSK is a modulation format with memory. Thus, bit-by-bit detection is suboptimum: consider the signal space diagram: four constellation points (at $0°$, $90°$, $180°$, and $270°$) are possible. For a bit-by-bit decision, the decision boundaries are thus the first and second main diagonals. The distance between the signal constellations and the decision boundary are $\sqrt{E}/\sqrt{2}$, and thus worse by $3\,\mathrm{dB}$ compared with BPSK. However, such a decision method has thrown away the information arising from the memory of the system: if the previous constellation point had been at $0°$, the subsequent signal constellations can only be at either $90°$ or $270°$. The decision boundary should thus be the x-axis, and the distance from the signal constellations to the decision boundary is $\sqrt{E}$ – i.e., equal to BPSK. Memory can be exploited either by a maximum-likelihood sequence estimation,[7] or, much simpler, by a derotation of signal constellations. Rotating the phase at each observation instant by $-90°$ essentially transforms the signal space diagram into that of BPSK; thus performance has to be identical.

11.3.9 Gaussian minimum shift keying

While MSK has many advantages, its spectral efficiency is rather low. This drawback is eliminated by *GMSK* (Gaussian MSK). GMSK is (like MSK) CPFSK with modulation index $h_{\mathrm{mod}} = 0.5$; the difference lies in the fact that phase basis pulses are Gaussian pulses:

$$\widetilde{g}(t) = g_{\mathrm{G}}(t, T_{\mathrm{B}}, B_{\mathrm{G}}T) \tag{11.71}$$

so that the sequence of transmit phase pulses is:

$$p_{\mathrm{D}}(t) = \sum_{i=-\infty}^{\infty} b_i \widetilde{g}(t - iT_{\mathrm{B}}) = b_i * \widetilde{g}(t) \tag{11.72}$$

(see Fig. 11.36). The spectrum is shown in Fig. 11.37.

GMSK is the modulation format most widely used in Europe. It is applied in the cellular Global System for Mobile communications (GSM) standard (with $B_{\mathrm{G}}T = 0.3$) and the cordless standard DECT (with $B_{\mathrm{G}}T = 0.5$) (see Chapters 21 and 24, respectively). It is also used in the Bluetooth (IEEE 802.15.1) standard for wireless personal area networks.[8]

It is noteworthy that GMSK *cannot* be interpreted as PAM. However, Laurent [1986] derived equations that allow the interpretation of GMSK as PAM with memory.

11.3.10 Pulse position modulation

Pulse Position Modulation (PPM) is another form of multipulse modulation. Remember that for FSK we used pulses that had different center frequencies as basis pulses. For PPM, we use pulses that have different delays. In the following we will consider M-ary PPM:

$$s_{\mathrm{LP}}(t) = \sqrt{\frac{2E_{\mathrm{B}}}{T_{\mathrm{B}}}} \sum_{i=-\infty}^{\infty} g_{c_i}(t - iT) \tag{11.73}$$

$$= \sqrt{\frac{2E_{\mathrm{B}}}{T_{\mathrm{B}}}} \sum_i g_{\mathrm{PPM}}(t - iT - c_i T_{\mathrm{d}}) \tag{11.74}$$

---

[7] It can be shown that for a maximum-likelihood sequence estimation, only three sampling values can influence the decision for a specific bit (see Benedetto and Biglieri [1999]).

[8] Strictly speaking, DECT and Bluetooth use GFSK with modulation index 0.5, which is equivalent to GMSK. However, deviations from the nominal modulation index are tolerated.

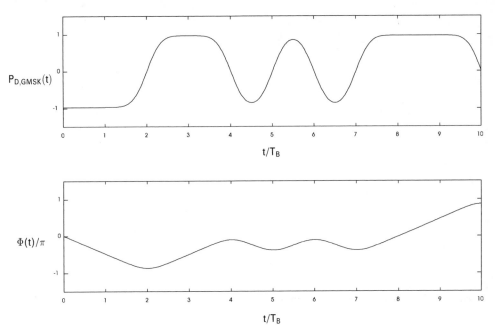

Figure 11.36   Pulse sequence and phase of Gaussian-minimum-shift-keying signal.

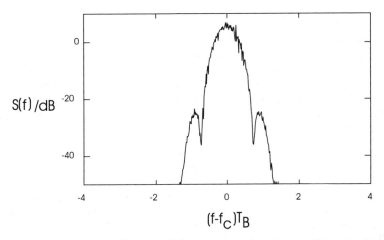

Figure 11.37   Gaussian-minimum-shift-keying power-spectral density (from simulations).

where T_d is the modulation delay. Note that the modulation symbols are directly mapped to the delay of the pulses, and are thus real.

The envelope shows strong fluctuations; however, it is still possible to use nonlinear amplifiers, since only two output amplitude levels are allowed.

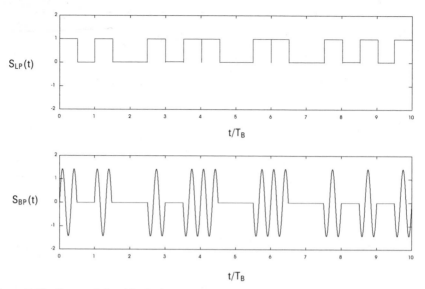

Figure 11.38   Temporal signal in the low-pass and bandpass domain for pulse position modulation.

Because PPM is a nonlinear modulation, the spectrum is not given by Eq. (11.20), but rather by the more complicated expression for multipulse modulation (see also Eq. 11.61):

$$S(f) = \frac{1}{M^2 T_S^2} \sum_{i=-\infty}^{+\infty} \left(\left| \sum_{m=0}^{M-1} G_m(f) \right|^2 \delta\left(f - \frac{i}{T_S}\right) \right) + \frac{1}{T_S} \left(\sum_{m=0}^{M-1} \frac{1}{M} |G_m(f)|^2 - \left| \sum_{m=0}^{M-1} \frac{1}{M} G_m(f) \right|^2 \right)$$

(11.75)

We note that the spectrum has a number of lines.

Up to now, we have assumed that the transmitted pulses are rectangular pulses, and offset with respect to each other by one pulsewidth. However, this is not spectrally efficient, as the pulses have a very broad spectrum. Other pulseshapes can be used as well. The performance of these different pulses depends on the detection method. When coherent detection is employed, then the key quantity determining performance is the correlation between the pulses representing $+1$ and -1 (assuming binary PPM):

$$\rho = \frac{\int g(t)g^*(t - T_d)\, dt}{\int |g(t)|^2 dt}$$

(11.76)

If $\rho = 0$, modulation is orthogonal as is the case for rectangular pulses if $T_d > T_B$. It is actually possible to choose pulseshapes so that $\rho < 0$. In that case, we have not only better spectral efficiency, but also better performance. However, when the RX uses *incoherent* (energy) detection, then it is best if the pulses do not have any overlap. The relevant correlation coefficient in this case is:

$$\rho_{env} = \frac{\int |g(t)||g(t - T_d)|\, dt}{\int |g(t)|^2 dt}$$

(11.77)

PPM can be combined with other modulation formats. For example, binary PPM can be combined with BPSK, so that each symbol represents 2 bits – 1 bit determined by the phase of the transmitted pulse, and 1 bit by its position.

Example 11.4 *Consider a PPM system with $g(t) = \sin(t/T)/t$. What is the correlation coefficient ρ and the correlation coefficient of the envelopes when T_d is (i) T, (ii) $5T$?*

The correlation coefficient is given by:

$$\rho = \frac{\int g(t)g(t - T_d)\, dt}{\int |g(t)|^2 dt} \tag{11.78}$$

Assuming $T = 1$, the denominator is:

$$\int |g(t)|^2 dt = \int_{-\infty}^{\infty} \frac{\sin^2(t)}{t^2}\, dt = \pi \tag{11.79}$$

For $T_d = 1$, using numerical integration, the numerator becomes:

$$\int g(t)g(t - T_d)\, dt = \int_{-\infty}^{\infty} \frac{\sin(t)}{t}\frac{\sin(t-1)}{t-1}\, dt \approx 2.63 \tag{11.80}$$

so that the correlation coefficient becomes $\rho \approx 0.84$. For $T_d = 5$, using numerical integration, the numerator becomes:

$$\int g(t)g(t - T_d)\, dt = \int_{-\infty}^{\infty} \frac{\sin(t)}{t}\frac{\sin(t-5)}{t-5}\, dt \approx -0.61 \tag{11.81}$$

so that the correlation coefficient becomes $\rho \approx -0.2$. For $T_d = 1$, the numerator for the envelope correlation is given by:

$$\rho \int_{-\infty}^{\infty} \left|\frac{\sin(t)}{t}\right| \left|\frac{\sin(t-1)}{t-1}\right| dt \approx 2.73 \tag{11.82}$$

so that the envelope correlation becomes $\rho \approx 0.87$. For $T_d = 5$, the numerator for the envelope correlation is given by:

$$\rho \int_{-\infty}^{\infty} \left|\frac{\sin(t)}{t}\right| \left|\frac{\sin(t-5)}{t-5}\right| dt \approx 1.04 \tag{11.83}$$

so that the envelope correlation becomes $\rho \approx 0.33$.

PPM is not used very often for wireless systems. This is due to its relatively low spectral efficiency, as well as to the effect of delay dispersion on a PPM system. Still, there are some emerging applications where PPM is used, especially impulse radio (see Section 18.5).

11.3.11 Summary of spectral efficiencies

Table 11.1 Spectral efficiency and envelope variations for different modulation schemes.

Modulation method	Spectral efficiency for 90% of total energy (bit/s/Hz)	Spectral efficiency for 99% of total energy (bit/s/Hz)
BPSK	0.59	0.05
BAM ($\alpha = 0.5$)	1.02	0.79
QPSK, OQPSK	1.18	0.10
MSK	1.29	0.85
GMSK ($B_G = 0.5$)	1.45	0.97
QAM ($\alpha = 0.5$)	2.04	1.58

11.4 Appendix

Please go to *www.wiley.com/go/molisch*

Further reading

The description of different modulation formats, and their representation in a signal space diagram, can be found in any of the numerous textbooks on digital communications – e.g., Anderson [2005], Barry et al. [2003], Proakis [1995], Sklar [2001], Wilson [1996], and Xiong [2000]. A book specifically dedicated to modulation for wireless communications is Burr [2001]. Quadrature amplitude modulation is described in Hanzo et al. [2000]. A description of the multiple interpretations of MSK, and the resulting demodulation structures, can be found in Benedetto and Biglieri [1999]. GMSK was invented by Murota and Hirade [1981]. An authoritative description of different forms of continuous-phase modulation is Anderson et al. [1986].

12

Demodulation

In this chapter, we describe how to demodulate the received signal, and discuss the performance of different demodulation schemes in *Additive White Gaussian Noise* (AWGN) channels as well as flat-fading channels and dispersive channels.

12.1 Demodulator structure and error probability in additive-white-Gaussian-noise channels

This section deals with basic demodulator structures for the modulation formats described in Chapter 11, and the computation of the Bit Error Rate (BER) and Symbol Error Rate (SER) in an AWGN channel. As in the previous chapter, we concentrate on the most important results; for more detailed derivations, we refer the reader to the monographs of Anderson [2005], Benedetto and Biglieri [1999], Proakis [1995], and Sklar [2001].

12.1.1 Model for channel and noise

The AWGN channel attenuates the transmit signal, causes phase rotation, and adds Gaussian-distributed noise. Attenuation and phase rotation are temporally constant, and are thus easily taken into account. Thus, the received signal (in complex baseband notation) is given by:

$$r_{\mathrm{LP}}(t) = \alpha s_{\mathrm{LP}}(t) + n_{\mathrm{LP}}(t) \tag{12.1}$$

where α is the (complex) attenuation $|\alpha| \exp(j\varphi)$, and $n(t)$ is a (complex) Gaussian noise process.

In order to derive the properties of noise, let us first consider noise in the bandpass system. We assume that over the bandwidth of interest (see Fig. 12.1a) the noise power-spectral density is constant. The value of the *two-sided* noise-spectral density is $N_0/2$. The (complex) equivalent low-pass noise has a power-spectral density (see Fig. 12.1a):

$$S_{\mathrm{n,LP}}(f) = \begin{cases} N_0 & |f| \leq B/2 \\ 0 & \text{otherwise} \end{cases} \tag{12.2}$$

Note that $S_{\mathrm{n,LP}}(f)$ is symmetric with respect to f – i.e., $S_{\mathrm{n,LP}}(f) = S_{\mathrm{n,LP}}(-f)$.

Wireless Communications   Andreas F. Molisch
© 2005 John Wiley & Sons, Ltd

Bandpass noise

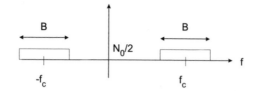

Equivalent lowpass noise

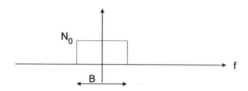

Figure 12.1 Bandpass noise, and equivalent low-pass noise power spectrum.

When considering noise in the time domain, we find that it is described by its autocorrelation function $R_{LP,nn}(\tau) = (1/2)E\{n_{LP}^*(t)n_{LP}(t+\tau)\}$.[1] For a system limited to the band $[-B/2, B/2]$, the ACF is:

$$R_{LP,nn}(\tau) = N_0 \frac{\sin(\pi B\tau)}{\pi\tau} \tag{12.3}$$

which becomes:

$$R_{LP,nn}(\tau) = N_0\delta(\tau) \tag{12.4}$$

as the bandwidth B tends to infinity. We also note that the correlations of the in-phase and the quadrature-phase components of noise, respectively, are both $R_{LP,nn}(\tau)$, while the cross-correlation between in-phase and quadrature-phase noise is zero.

The autocorrelation function of the bandpass signal is:

$$R_{BP,nn}(\tau) = \text{Re}\{R_{LP,nn}(\tau)\exp(j2\pi f_c\tau)\} \tag{12.5}$$

12.1.2 Signal space diagram and optimum receivers

We now derive the structure of the optimum receivers for digital modulation. In the process, we will also find an additional motivation for using signal space diagrams. For these derivations we make the following assumptions:

1. all transmit symbols are equally likely;
2. the modulation format does not have memory;
3. the channel is an AWGN channel, and both attenuation and phase rotation are completely known. Without loss of generality, we assume henceforth that phase rotation has been compensated completely, so that the channel attenuation is real.

---

[1] Note that this definition of the Auto Correlation Function (ACF) differs by a factor of 2 from the one used in Chapter 6. The reason for using this modified definition here is that it is commonly used in communications theory for BER computations.

The ideal detector, called the *Maximum A Posteriori* (MAP) detector, aims to answer the following question: "If a signal $r(t)$ was received then which symbol $s_m(t)$ was most likely transmitted?" In other words, which symbol maximizes m?:

$$\Pr[s_m(t)|r(t)] \tag{12.6}$$

Bayes' rules can be used to write this as (in other words, find the symbol m that achieves):

$$\max_m \Pr[n(t) = r(t) - \alpha s_m(t)] \Pr[s_m(t)] \tag{12.7}$$

where α is the attenuation of the channel. Since we assume that all symbols are equiprobable, the MAP detector becomes identical to the *Maximum Likelihood* (ML) detector:

$$\max_m \Pr[n(t) = r(t) - \alpha s_m(t)] \tag{12.8}$$

In Chapter 11, we introduced the signal space diagram for the representation of modulated transmit signals. There, we treated the signal space diagram just as a convenient shorthand. Now we will show how a transmit and received signal can be related, and how the signal space diagram can be used to compute error probabilities.

Remember that the transmit signal can be represented in the form:[2]

$$s_m(t) = \sum_{n=1}^{N} s_{m,n} \varphi_n(t) \tag{12.9}$$

where

$$s_{m,n} = \int_0^{T_s} s_m(t) \varphi_n^*(t) \, dt \tag{12.10}$$

Now we find that the received signal can be represented by a similar expansion. Using the same expansion functions $\varphi_n(t)$ we obtain:

$$r(t) = \sum_{n=1}^{\infty} r_n \varphi_n(t) \tag{12.11}$$

where

$$r_n = \int_0^{T_s} r(t) \varphi_n^*(t) \, dt \tag{12.12}$$

Since the received signal contains noise, it seems at first glance we need more terms in the expansion – infinitely many, to be exact. However, we find it useful to split the series into two parts:

$$r(t) = \sum_{n=1}^{N} r_n \varphi_n(t) + \sum_{n=N+1}^{\infty} r_n \varphi_n(t) \tag{12.13}$$

and similarly:

$$n(t) = \sum_{n=1}^{N} n_n \varphi_n(t) + \sum_{n=N+1}^{\infty} n_n \varphi_n(t) \tag{12.14}$$

Using these expansions, the expression that the ML receiver aims to maximize can be written as:

$$\max_m \Pr[\mathbf{n} = \mathbf{r} - \alpha \mathbf{s}_m] \tag{12.15}$$

where the received signal vector $\mathbf{r}$ is simply $\mathbf{r} = (r_1, r_2, \ldots)^T$, and similarly for $\mathbf{n}$. Since the noise components are independent, the probability density function of the received vector $\mathbf{r}$,

---

[2] Note that these equations are valid for both the baseband and the bandpass representation – it is just a matter of inserting the correct expansion functions.

assuming that $\mathbf{s}_m$ was transmitted, is given as:

$$p(\mathbf{r}_{\text{LP}}|\alpha\mathbf{s}_{\text{LP},m}) \propto \exp\left\{-\frac{1}{2N_0}||\mathbf{r}-\alpha\mathbf{s}_m||^2\right\} \tag{12.16}$$

$$= \prod_{n=1}^{\infty}\exp\left\{-\frac{1}{2N_0}(r_n-\alpha s_{m,n})^2\right\} \tag{12.17}$$

Since $s_{m,n}$ is nonzero only for $n \le N$, the ML detector aims to find:

$$\max_m \prod_{n=1}^{N}\exp\left\{-\frac{1}{2N_0}(r_n-\alpha s_{m,n})^2\right\} \prod_{n=N+1}^{\infty}\exp\left\{-\frac{1}{2N_0}(r_n)^2\right\} \tag{12.18}$$

A key realization is now that the second product in Eq. (12.18) is independent of s_m, and thus does not influence the decision. This is another way of saying that components of the noise (and thus of the received signal) that do not lie in the signal space of the transmit signal are irrelevant for the decision of the detector (Wozencraft's irrelevance theorem). Finally, as $\exp(.)$ is a monotonic function, we find that it is sufficient to minimize the metric:

$$\mu(\mathbf{s}_{\text{LP},m}) = ||\mathbf{r}_{\text{LP}}-\alpha\mathbf{s}_{\text{LP},m}||^2 \tag{12.19}$$

Geometrically, this means that the ML receiver decides for symbol m which transmit vector $\mathbf{s}_{\text{LP},m}$ has the smallest Euclidean distance to the received vector $\mathbf{r}_{\text{LP}}$. We need to keep in mind that this is an optimum detection method only in memoryless, uncoded systems. Vector $\mathbf{r}$ contains "soft" information – i.e., how sure the receiver is about its decision. This soft information, which is lost in the decision process of finding the nearest neighbor, is irrelevant when one bit does not tell us anything about any other bit. However, for coded systems and systems with memory (Chapters 14 and 16), this information can be very helpful.

The metric can be rewritten as:

$$\mu(\mathbf{s}_{\text{LP},m}) = ||\mathbf{r}_{\text{LP}}||^2 + ||\alpha\mathbf{s}_{\text{LP},m}||^2 - 2\alpha\text{Re}\left\{\mathbf{r}_{\text{LP}}\mathbf{s}_{\text{LP},m}^*\right\} \tag{12.20}$$

Since the term $||\mathbf{r}_{\text{LP}}||^2$ is independent of the considered $\mathbf{s}_{\text{LP},m}$, minimizing the metric is equivalent to maximizing:

$$\text{Re}\left\{\mathbf{r}_{\text{LP}}\mathbf{s}_{\text{LP},m}^*\right\} - \alpha E_m \tag{12.21}$$

(remember that $E_m = ||\mathbf{s}_{\text{LP},m}||^2/2$, see Chapter 11).

One important consequence of this decision rule is that the receiver has to know the value of channel attenuation α. This can be difficult in wireless systems, especially if channel properties quickly change (see Part II). Thus, modulation and detection methods that do not require this information are preferred. Specifically, the magnitude of channel attenuation does not need to be known if all transmit signals have equal energy, $E_m = E$. The phase rotation of the channel (argument of alpha) can be ignored if either incoherent detection or differential detection are used.

The beauty of the above derivation is that it is independent of the actual modulation scheme. The transmit signal is represented in the signal space diagram, which gives all the *relevant* information. The receiver structure of Fig. 12.2 is then valid for optimum reception for any modulation alphabet represented in a signal space diagram. The only prerequisite is that the conditions mentioned at the beginning of this chapter are fulfilled. This receiver can be interpreted as a correlator, or as a matched filter, matched to the different possible transmit waveforms.

In the following, we will find that the Euclidean distance of two points in the signal space diagram is a vital parameter for modulation formats. We find that in the bandpass

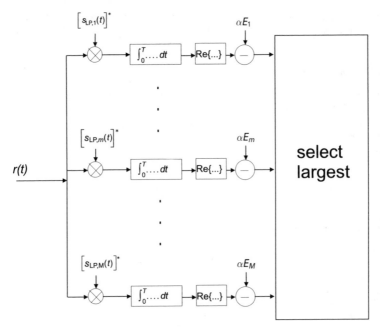

Figure 12.2   Structure of a generic optimum receiver.

representation for equal-energy signals, which we assumed in Chapter 11, the Euclidean distance is related to the energy of the signal component as:

$$d_{12}^2 = 2E(1 - \mathrm{Re}\{\rho_{12}\})$$ (12.22)

where ρ_{jk} is the correlation coefficient as defined in Chapter 11:

$$\mathrm{Re}\{\rho_{km}\} = \frac{s_k s_m}{|s_k||s_m|}$$ (12.23)

12.1.3 Methods for the computation of error probability

In this subsection, we discuss how to compute the performance that can be achieved with optimum receivers. Before going into details, let us define some important expressions: the *bit error rate* is, as its name says, a rate. Thus, it describes the number of bit errors per unit time, and has dimension s^{-1}. The *bit error ratio* is the number of errors divided by the number of transmitted bits; it thus is dimensionless. For the case when infinitely many bits are transmitted, this becomes the *bit error probability*. In the literature, there is a tendency to mix up these three expressions; specifically, the expression bit error rate is often used when the authors mean bit error probability. This misnomer has become so common that in the following we will use it as well.

Though we have treated a multitude of modulation formats in the previous chapter, they can all be classified easily:

1. Binary Phase Shift Keying (BPSK) and Minimum Shift Keying (MSK) with derotation, are antipodal signals.
2. Binary Frequency Shift Keying (BFSK), Binary Pulse Position Modulation (BPPM), and MSK with frequency discriminator detection are orthogonal signals.

3. Quadrature Phase Shift Keying (QPSK), $\pi/4$-DQPSK (Differential Quadrature Phase Shift Keying), and Offset Quadrature Phase Shift Keying (OQPSK) are bi-orthogonal signals.

Error probability for coherent receivers – general case

As mentioned above, coherent receivers compensate for phase rotation of the channel by means of *carrier recovery*. Furthermore, the magnitude of channel attenuation $|\alpha|$ is assumed to be known, and absorbed into the received signal, so that in the absence of noise, $\mathbf{r} = \mathbf{s}$ holds. The probability that symbol $\mathbf{s}_j$ is mistaken for symbol $\mathbf{s}_k$ that has Euclidean distance d_{jk} from $\mathbf{s}_j$ (*pairwise error probability*) is given as:

$$\Pr_{\text{pair}}(\mathbf{s}_j, \mathbf{s}_k) = Q\left(\sqrt{\frac{d_{jk}^2}{2N_0}}\right) = Q\left(\sqrt{\frac{E}{N_0}(1 - \text{Re}\{\rho_{jk}\})}\right) \qquad (12.24)$$

where the Q-function is defined as:

$$Q(x) = \frac{1}{\sqrt{2\pi}}\int_x^\infty \exp(-t^2/2)\,dt \qquad (12.25)$$

This is related to the complementary error function:

$$Q(x) = \frac{1}{2}\text{erfc}\left(\frac{x}{\sqrt{2}}\right) \qquad (12.26)$$

and

$$\text{erfc}(x) = 2Q(\sqrt{2}x) \qquad (12.27)$$

Equation (12.24) can be found by computing the probability that the noise is large enough to make the received signal geometrically closer to the point $\mathbf{s}_k$ in the signal space diagram, even though the signal $\mathbf{s}_j$ was transmitted.

Error probability for coherent receivers – binary orthogonal signals

As we saw in Chapter 11, a number of important modulation formats can be viewed as binary orthogonal signals – most prominently, binary FSK and binary PPM. Figure 12.3 shows the signal space diagram for this case. The figure also shows the decision boundary: if a received signal point falls into the shaded region, then it is decided that a $+1$ was transmitted, otherwise a -1 was transmitted.

Defining the Signal-to-Noise Ratio (SNR) for one symbol as $\gamma_S = E_S/N_0$, we get:

$$\Pr_{\text{pair}}(\mathbf{s}_j, \mathbf{s}_k) = Q\left(\sqrt{\gamma_S(1 - \text{Re}\{\rho_{jk}\})}\right) \qquad (12.28)$$

$$= Q(\sqrt{\gamma_S}) \qquad (12.29)$$

Note that since we are considering binary signaling $\gamma_S = \gamma_B$.

Error probability for coherent receivers – antipodal signaling

For antipodal signals, the pairwise error probability is:

$$\Pr_{\text{pair}}(\mathbf{s}_j, \mathbf{s}_k) = Q\left(\sqrt{\gamma_S(1 - \text{Re}\{\rho_{jk}\})}\right) \qquad (12.30)$$

$$= Q(\sqrt{2\gamma_S}) \qquad (12.31)$$

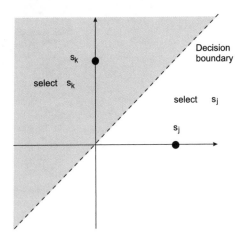

Figure 12.3 Decision boundary for the selection between s_k and s_j.

For binary signals with equal-probability transmit symbols, pairwise error probability is equal to symbol error probability, which in turn is equal to bit error probability. This is the case, e.g., for BPSK, as well as for MSK with ideal coherent detection (see Chapter 11), and the BER is given by Eq. (12.30). Note that MSK can be detected like FSK, but then does not exploit the continuity of the phase. In this case, it becomes an orthogonal modulation format, and the BER is given by Eq. (12.29). This means deterioration of the effective SNR by 3 dB.

Union bound and bi-orthogonal signaling

For M-ary modulation methods, exact computation of the BER is much more difficult. For this reason, the *union bound* is frequently used. The principle of this bounding technique is outlined in Fig. 12.4. The region of the signal space diagram that results in an erroneous decision consists of partial regions, each of which represents a pairwise error – i.e., confusing the correct symbol with another symbol. The symbol error probability is then written as the sum of these pairwise probabilities. Since the partial regions overlap, this represents an upper bound for true symbol error probabilities. The approximation improves as the SNR increases, because the SER is then mainly determined by regions close to the decision boundaries whereas overlap regions have little impact.

Great care must be taken when using equations for the BER of higher order modulation formats from the literature. There are several possible pitfalls:

- Does the equation give the symbol error probability or the bit error probability? Take, for example, the 4-QAM of Fig. 12.4, and assume furthermore that the signal constellations are Gray-coded, so that the four points represent the bit combinations 00, 01, 11, and 10 (when read clockwise). There is a high probability of errors occuring between two neighboring signal constellations. One symbol error (in one transmitted symbol) then corresponds to one bit error (one error in *two* transmitted bits). Thus, bit error probability is only about half symbol error probability.[3]

---

[3] This demonstrates the drawbacks of mixing up the expressions "bit error probability" and "bit error rate". In our example, the bit error rate is the same as the symbol error rate, while the bit error probability is only half the symbol error probability.

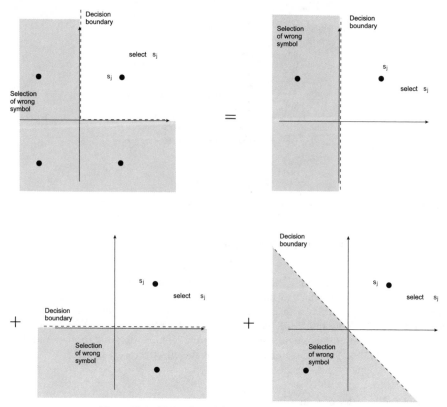

Figure 12.4 Union bound for symbol error probabilities.

- Does computation of the SNR use bit energy or symbol energy? A fair comparison between different modulation formats should be based on E_B/N_0.
- Some authors define the distance between the origin and (equal-energy) signal constellation points not as $\sqrt{E}$, but rather as $\sqrt{2E}$. This is related to a different normalization of the expansion functions, which is compensated by different values for n_n. The final results do not change, but this makes the combination of intermediate results from different sources much more difficult.

As an example, we compute in the following the BER of 4-QAM (Quadrature Amplitude Modulation). As it is a four-state modulation format (2 bit/symbol), we can invest twice the bit energy for each symbol. The signal points thus have Euclidean distance $d^2 = E_S = 2E_B$ from the origin; the points in the signal constellation diagram are thus at $\pm\sqrt{E_B}(\pm/\pm j)$, and the distance between two neighboring points is $d_{jk}^2 = 4E_B$.

We can now consider two types of union bounds:

1. For a "full" union bound, we compute the pairwise error probability with *all* possible signal constellation points, and add them up. This is shown in Fig. 12.4.
2. We compute pairwise error probability using nothing more than neighboring points. In this case, we omit the last decision region in Fig. 12.4 from our computations. As we can see, the union of the first two regions (for pairwise error with nearest neighbors) already covers the whole "erroneous decision region". In the following, we will use this type of union bound.

> **Example 12.1** *Compute the BER and SER of QPSK.*
>
> From Eq. (12.24) it follows that the pairwise error probability is:
>
> $$Q\left(\sqrt{2\frac{E_B}{N_0}}\right) = Q\left(\sqrt{2\gamma_B}\right) \tag{12.32}$$
>
> According to Fig. 12.4, the symbol error probability as computed from the union bound is twice as large:
>
> $$SER \approx 2Q(\sqrt{2\gamma_B}) \tag{12.33}$$
>
> Now, as discussed above, the BER is half the SER:
>
> $$BER = Q(\sqrt{2\gamma_B}) \tag{12.34}$$
>
> This is the same BER as for BPSK.
>
> For QPSK, it is also possible to compute the symbol error probability exactly:
>
> $$SER = 2Q(\sqrt{2\gamma_B})\left[1 - \frac{1}{2}Q(\sqrt{2\gamma_B})\right] \tag{12.35}$$
>
> which shows the magnitude of the error made by the union bound. The relative error is $0.5Q(\sqrt{2\gamma_B})$, which tends to 0 as γ_B tends to infinity.

The exact computation of the BER or SER of higher order modulation formats can be complicated. However, the union bound offers a simple and (especially for high SNRs) very accurate approximation.

Error probability for differential detection

Carrier recovery can be a challenging problem in many situations (see Meyr et al. [1997] and Proakis [1995]). Differential detection is an attractive alternative. In this approach, the receiver just needs to compare the phases (and possibly amplitudes) of two subsequent symbols. This phase difference is independent of the absolute phase. If the phase rotation introduced by the channel is slowly time-varying (and thus effectively the same for two subsequent symbols), it enters just the absolute phase, and thus need not be taken into account in the detection process.

For differential detection of Phase Shift Keying (PSK), the transmitter needs to provide differential encoding. For binary symmetric PSK, the transmit phase Φ_i of the ith bit is:

$$\Phi_i = \Phi_{i-1} + \begin{cases} +\dfrac{\pi}{2} b_i = +1 \\[2mm] -\dfrac{\pi}{2} b_i = -1 \end{cases} \tag{12.36}$$

Comparison of the difference between phases on two subsequent sampling instances determines whether the transmitted bit b_i was +1 or −1.[4]

For Continuous Phase Frequency Shift Keying (CPFSK), such differential encoding can be avoided. Remember that in the case of MSK (without differential encoding), the phase rotation over a 1-bit duration is $\pm\pi/2$. It is thus possible to determine the phases at two subsequent sampling points, take the difference, and conclude which bit has been transmitted. This can also be interpreted by the fact that the phase of the transmit signal is an integral over the uncoded bit

---

[4] Theoretically, differential detection could also be performed on a non-encoded signal. However, one bit error would then lead to a whole chain of errors.

sequence. Computing the phase difference is a first approximation to taking the derivative (it is exact if the phase change is linear), and thus reverses the integration.

For binary orthogonal signals, the BER for differential detection is [Proakis 1995]:

$$BER = \frac{1}{2}\exp(-\gamma_b) \tag{12.37}$$

For 4-PSK with Gray-coding, it is:

$$BER = Q_{\mathrm{M}}(a,b) - \frac{1}{2}I_0(ab)\exp\left(-\frac{1}{2}(a^2 + b^2)\right) \tag{12.38}$$

where

$$a = \sqrt{2\gamma_{\mathrm{B}}\left(1 - \frac{1}{\sqrt{2}}\right)}, \qquad b = \sqrt{2\gamma_{\mathrm{B}}\left(1 + \frac{1}{\sqrt{2}}\right)} \tag{12.39}$$

and $Q_{\mathrm{M}}(a,b)$ is Marcum's Q-function:

$$Q_{\mathrm{M}}(a,b) = \int_b^\infty x\exp\left[-\frac{a^2 + x^2}{2}\right]I_0(ax)\,dx \tag{12.40}$$

whose series representation is given in Eq. (5.31).

Error probability for noncoherent detection

When the carrier phase is completely unknown, and differential detection is not an option, then noncoherent detection can be used. For equal-energy signals, the detector tries to minimize the metric:

$$\left|\mathbf{r}_{\mathrm{LP}}\mathbf{s}_{\mathrm{LP},m}^*\right| \tag{12.41}$$

so that the optimum receiver has a structure according to Fig. 12.5. In this case, the BER can be computed from Eq. (12.38), but with a different definition of the parameters a and b:

$$a = \sqrt{\frac{\gamma_{\mathrm{B}}}{2}\left(1 - \sqrt{1 - |\rho|^2}\right)}, \qquad b = \sqrt{\frac{\gamma_{\mathrm{B}}}{2}\left(1 + \sqrt{1 - |\rho|^2}\right)} \tag{12.42}$$

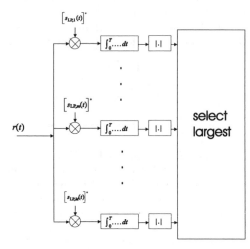

Figure 12.5 Optimum receiver structure for noncoherent detection.

The optimum performance is achieved in this case if $\rho = 0$ – i.e., the signals are orthogonal. For the case when $|\rho| = 1$, which occurs for PSK signals, including BPSK and 4-QAM, the BER becomes 0.5.

Example 12.2 *Compute the BER of binary FSK in an AWGN channel with $\gamma_B = 5\,dB$, and compare it with DBPSK and BPSK.*

For binary FSK in AWGN, the BER is given by Eq. (12.29) with $\gamma_S = \gamma_B$. Thus, with $\gamma_B = 5$ dB, we have:

$$\text{BER}_{\text{BFSK}} = Q(\sqrt{\gamma_B})$$

$$= 0.038 \tag{12.43}$$

For BPSK, the BER is instead given by Eq. (12.37), and thus:

$$\text{BER}_{\text{BPSK}} = Q\left(\sqrt{2\gamma_B}\right)$$

$$= 0.006$$

Finally, for differential BPSK, the BER is given by Eq. (12.37), which results in:

$$\text{BER}_{\text{DBPSK}} = (1/2)e^{-\gamma_B}$$

$$= 0.021 \tag{12.45}$$

12.2 Error probability in flat-fading channels

12.2.1 Average BER – classical computation method

In AWGN channels, the BER quickly decreases as the SNR increases. For binary modulation formats, a 10-dB SNR is sufficient to give a BER on the order of 10^{-4}, for 15 dB the BER is below 10^{-8}. The situation is completely different in fading channels. The received signal power is not constant, but changes as the fading of the channel changes. In order to determine the BER in such a channel, we have to proceed in three steps:

1. Determine the BER for any arbitrary SNR. We did this in the previous section.
2. Determine the probability that a certain SNR occurs in the channel – in other words, determine the probability density function (pdf) of the energy attenuation of the channel.
3. Average the BER over the distribution of SNRs.

We will see below that in a fading channel the BER decreases only linearly with the (average) SNR, while in AWGN channels it decreases exponentially. At first glance, this is astonishing: sometimes fading leads to high SNRs, sometimes it leads to low SNRs, but the mean power is the same, and it could be assumed that high and low values would compensate for each other. The important point here is that the relationship between (instantaneous) BER and (instantaneous) SNR is highly nonlinear.

Example 12.3 *BER in a two-state fading channel.*

Consider the following simple example: a fading channel has an average SNR of 10 dB, where fading causes the SNR to be $-\infty$ dB half of the time, while it is 13 dB the rest of the time. The BERs corresponding to the two channel states are 0.5 and 10^{-9} respectively (assuming antipodal modulation with differential detection). The mean BER is then $\overline{BER} = 0.5 \cdot 0.5 + 0.5 \cdot 10^{-9} = 0.25$. For an AWGN channel with a 10-dB SNR, the BER is $2 \cdot 10^{-5}$.

Following this intuitive explanation, we now turn to the mathematical details. Point 2 of the three-step procedure requires computation of the SNR distribution. In Chapter 5, we mostly concentrated on the distribution of *amplitude*. This has to be converted to distribution of received *power*. Such a transformation can be done by means of the Jacobian. We start out with a Rayleigh distribution for the amplitude of the received signal, $r = |r(t)|$:

$$pdf_r(r) = \frac{r}{\sigma^2} \exp\left(-\frac{r^2}{2\sigma^2}\right) \qquad \text{for } r \geq 0 \tag{12.46}$$

Define now the (instantaneous) received power as $P_{\text{inst}} = r^2$, and the mean power $P_{\text{m}} = 2\sigma^2$. Using the Jacobian $|dP_{\text{inst}}/dr| = 2r$, the pdf of the received power becomes:

$$pdf_{P_{\text{inst}}}(P_{\text{inst}}) = \frac{1}{P_{\text{m}}} \exp\left(-\frac{P_{\text{inst}}}{P_{\text{m}}}\right) \qquad \text{for } P_{\text{inst}} \geq 0 \tag{12.47}$$

Since the SNR is the received power scaled with $1/N_0$, the pdf of the SNR is:

$$pdf_{\gamma_{\text{B}}}(\gamma_{\text{B}}) = \frac{1}{\overline{\gamma_{\text{B}}}} \exp\left(-\frac{\gamma_{\text{B}}}{\overline{\gamma_{\text{B}}}}\right) \tag{12.48}$$

where $\overline{\gamma_{\text{B}}}$ is the mean SNR. For Rician-fading channels, a similar, but more tedious, computation gives [Rappaport 1996]:

$$pdf_{\gamma_{\text{B}}}(\gamma_{\text{B}}) = \frac{1+K_r}{\overline{\gamma_{\text{B}}}} \exp\left(-\frac{\gamma_{\text{B}}(1+K_r)+K_r\overline{\gamma_{\text{B}}}}{\overline{\gamma_{\text{B}}}}\right) I_0\left(\sqrt{\frac{4(1+K_r)K_r\gamma_{\text{B}}}{\overline{\gamma_{\text{B}}}}}\right) \tag{12.49}$$

where K_r is the Rice factor. Analoguous computations are possible for other amplitude distributions.

In the last step, the BER has to be averaged over the distribution of the SNR:

$$\overline{BER} = \int pdf_{\gamma_{\text{B}}}(\gamma_{\text{B}}) BER(\gamma_{\text{B}})\, d\gamma_{\text{B}} \tag{12.50}$$

For Rayleigh fading, a closed-form evaluation of Eq. (12.50) is possible. Using:

$$2 \int_0^\infty Q(\sqrt{2x})a \exp(-ax)dx = 1 - \sqrt{\frac{1}{1+a}} \tag{12.51}$$

the mean BER for coherent detection of binary antipodal signals is:

$$\overline{BER} = \frac{1}{2}\left[1 - \sqrt{\frac{\overline{\gamma_{\text{B}}}}{1+\overline{\gamma_{\text{B}}}}}\right] \approx \frac{1}{4\overline{\gamma_{\text{B}}}} \tag{12.52}$$

and the mean BER of coherent detection of binary orthogonal signals is:

$$\overline{BER} = \frac{1}{2}\left[1 - \sqrt{\frac{\overline{\gamma_{\text{B}}}}{2+\overline{\gamma_{\text{B}}}}}\right] \approx \frac{1}{2\overline{\gamma_{\text{B}}}} \tag{12.53}$$

These BERs are plotted in Fig. 12.6.

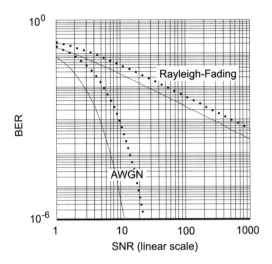

Figure 12.6 Bit error rate for binary signals with coherent detection: antipodal (solid) and orthogonal (dashed).

For differential detection, the averaging process is even simpler, as the BER curve is an exponential function. For binary antipodal signals:

$$\overline{BER} = \frac{1}{2(1 + \overline{\gamma}_{\mathrm{B}})} \approx \frac{1}{2\overline{\gamma}_{\mathrm{B}}} \tag{12.54}$$

and for the non-coherent detection of binary orthogonal signals:

$$\overline{BER} = \frac{1}{2 + \overline{\gamma}_{\mathrm{B}}} \approx \frac{1}{\overline{\gamma}_{\mathrm{B}}} \tag{12.55}$$

For differential detection, the BER in Rician channels can also be computed in closed form. For antipodal signals:

$$\overline{BER} = \frac{1 + K_{\mathrm{r}}}{2(1 + K_{\mathrm{r}} + \overline{\gamma}_{\mathrm{B}})} \exp\left(-\frac{K_{\mathrm{r}}\overline{\gamma}_{\mathrm{B}}}{1 + K_{\mathrm{r}} + \overline{\gamma}_{\mathrm{B}}}\right) \tag{12.56}$$

and for orthogonal signals:

$$\overline{BER} = \frac{1 + K_{\mathrm{r}}}{(2 + 2K_{\mathrm{r}} + \overline{\gamma}_{\mathrm{B}})} \exp\left(-\frac{K_{\mathrm{r}}\overline{\gamma}_{\mathrm{B}}}{2 + 2K_{\mathrm{r}} + \overline{\gamma}_{\mathrm{B}}}\right) \tag{12.57}$$

Example 12.4 *Compute the BER of DBPSK with $\overline{\gamma}_{\mathrm{B}} = 12\,dB$ and $K_{\mathrm{r}} = -3\,dB$, $0\,dB$, $10\,dB$.*

The BER is given by Eq. (12.56), and hence, with $K_{\mathrm{r}} = -3\,\mathrm{dB}$:

$$\overline{BER} = \frac{1 + K_{\mathrm{r}}}{2(1 + K_{\mathrm{r}} + \overline{\gamma}_{B})} \exp\left(-\frac{K_{\mathrm{r}}\overline{\gamma}_{B}}{1 + K_{\mathrm{r}} + \overline{\gamma}_{B}}\right)$$

$$= 0.027 \tag{12.58}$$

Then, with $K_{\mathrm{r}} = 0\,\mathrm{dB}$ we get $\overline{BER} = 0.023$, and with $K_{\mathrm{r}} = 10\,\mathrm{dB}$ we get $\overline{BER} = 0.000\,56$. Note the strong decrease in BER for high Rice factors, even though we kept the SNR constant.

12.2.2 Computation of average error probability – alternative method

In the late 1990s, a new method for computation of the average BER was proposed, and shown to be very efficient. It is based on an alternative representation of the Q-function, and allows easier averaging over different fading distributions ([Annamalai et al. 2000], [Simon and Alouini 2000]).

Alternative representation of the Q-function

When evaluating the classical definition of the Q-function

$$Q(x) = \frac{1}{\sqrt{2\pi}} \int_x^\infty \exp(-t^2/2) \, dt \tag{12.59}$$

there is the problem that the argument of the Q-function is in the integration limit, not in its integrand. This makes evaluation of the integrals of the Q-function much more difficult, especially because we cannot use the beloved trick of exchanging the sequence of integration in multiple integrals. This problem is particularly relevant in BER computations. The problem can be solved by using an alternative formulation of the Q-function:

$$Q(x) = \frac{1}{\pi} \int_0^{\pi/2} \exp\left(-\frac{x^2}{2\sin^2\theta}\right) d\theta \tag{12.60}$$

This representation now has the argument in the integrand (in a Gaussian form, $\exp(-x^2)$), and also has finite integration limits. We will see below that this greatly simplifies evaluation of the error probabilities.

It turns out that Marcum's Q-function, as defined in Eq. (12.40), also has an alternative representation:

$$Q_M(a,b) = \begin{cases} \dfrac{1}{2\pi} \displaystyle\int_{-\pi}^{\pi} \dfrac{b^2 + ab\sin\theta}{b^2 + 2ab\sin\theta + a^2} \exp\left(-\tfrac{1}{2}\left(b^2 + 2ab\sin\theta + a^2\right)\right) d\theta & \text{for } b > a \geq 0 \\[4mm] 1 + \dfrac{1}{2\pi} \displaystyle\int_{-\pi}^{\pi} \dfrac{b^2 + ab\sin\theta}{b^2 + 2ab\sin\theta + a^2} \exp\left(-\tfrac{1}{2}\left(b^2 + 2ab\sin\theta + a^2\right)\right) d\theta & \text{for } a > b \geq 0 \end{cases} \tag{12.61}$$

Error probability in additive-white-Gaussian-noise channels

For computation of the BER in AWGN channels, we find that the BER for BPSK can be written as (compare also Eq. 12.30):

$$BER = Q(\sqrt{2\gamma_S}) \tag{12.62}$$

$$= \frac{1}{\pi} \int_0^{\pi/2} \exp\left(-\frac{\gamma_S}{\sin^2\theta}\right) d\theta \tag{12.63}$$

For QPSK, the symbol error rate can be computed as (compare Eq. 12.35):

$$SER = 2Q(\sqrt{\gamma_S}) - Q^2(\sqrt{\gamma_S}) \tag{12.64}$$

$$= \frac{1}{\pi} \int_0^{3\pi/4} \exp\left(-\frac{\gamma_S}{\sin^2\theta}\sin^2(\pi/4)\right) d\theta \tag{12.65}$$

and quite generally for M-ary PSK:

$$SER = \frac{1}{\pi} \int_0^{(M-1)\pi/M} \exp\left(-\frac{\gamma_S}{\sin^2\theta}\sin^2(\pi/M)\right) d\theta \tag{12.66}$$

For binary orthogonal FSK, we finally find that:

$$BER = Q(\sqrt{\gamma_S}) \tag{12.67}$$

$$= \frac{1}{\pi} \int_0^{\pi/2} \exp\left(-\frac{\gamma_S}{2\sin^2\theta}\right) d\theta \tag{12.68}$$

Example 12.5 *Compare the BER for 8-PSK as computed from Eq. (12.66) with the value obtained from the union bound for $\gamma_S = 3$ and 10 dB.*

First, using Eq. (12.66) with $M = 8$ and $\gamma_S = 3$ dB we have:

$$SER = \frac{1}{\pi} \int_0^{(M-1)\pi/M} \exp\left(-\frac{\gamma_S}{\sin^2\theta}\sin^2(\pi/M)\right) d\theta$$

$$= 0.442 \tag{12.69}$$

and for $\gamma_S = 10$ dB we get $SER = 0.087$, where the integral has been evaluated numerically.

The 8-PSK constellation has a minimum distance of $d_{min} = 2\sqrt{E_S}\sin(\pi/8)$. The nearest neighbor union bound on the SER is then given by:

$$SER_{union\text{-}bound} = 2 \cdot Q\left(\frac{d_{min}}{\sqrt{2N_0}}\right)$$

$$= 2 \cdot Q\left(\frac{2\sqrt{E_S}\sin(\pi/8)}{\sqrt{2N_0}}\right)$$

$$= 2 \cdot Q\left(\sqrt{2\gamma_S}\sin(\pi/8)\right) \tag{12.70}$$

Thus, with $\gamma_S = 3$ dB we get $SER_{union\text{-}bound} = 0.445$, and with $\gamma_S = 10$ dB we get $SER_{union\text{-}bound} = 0.087$. In this example the union bound gives a very good approximation, even at the rather low SNR of 3 dB.

Error probability in fading channels

For AWGN channels, the advantages of the alternative representation of the Q-function are rather limited. They allow a simpler formulation for higher order modulation formats, but do not exhibit significant advantages for the modulation formats that are mostly used in practice. The real advantage emerges when we apply this description method as the basis for computations of the BER in fading channels. We find that we have to average over the probability density function of the SNR $pdf_\gamma(\gamma)$, as described in Eq. (12.50). We have now seen that the alternative representation of the Q-function allows us to write the SER (for a given SNR) in the generic form:

$$SER(\gamma) = \int_{\theta_1}^{\theta_2} f_1(\theta) \exp(-\gamma f_2(\theta)) \, d\theta \tag{12.71}$$

Thus, the average SER becomes:

$$\overline{SER} = \int_0^\infty pdf_\gamma(\gamma) SER(\gamma)\, d\gamma \tag{12.72}$$

$$= \int_0^\infty pdf_\gamma(\gamma) \int_{\theta_1}^{\theta_2} f_1(\theta) \exp(-\gamma f_2(\theta))\, d\theta\, d\gamma \tag{12.73}$$

$$= \int_{\theta_1}^{\theta_2} f_1(\theta) \int_0^\infty pdf_\gamma(\gamma) \exp(-\gamma f_2(\theta))\, d\gamma\, d\theta \tag{12.74}$$

Let us now have a closer look at the inner integral:

$$\int_0^\infty pdf_\gamma(\gamma) \exp(-\gamma f_2(\theta))\, d\gamma \tag{12.75}$$

We find that it is the moment-generating function of $pdf_\gamma(\gamma)$, evaluated at the point $-f_2(\theta)$. Remember (see also, e.g., Papoulis [1991]) that the moment-generating function is defined as the Laplace transform of the pdf of γ:

$$M_\gamma(s) = \int_0^\infty pdf_\gamma(\gamma) \exp(\gamma s)\, d\gamma \tag{12.76}$$

and the mean SNR is the first derivative, evaluated at $s = 0$:

$$\overline{\gamma} = \left. \frac{dM_\gamma(s)}{ds} \right|_{s=0} \tag{12.77}$$

Summarizing, the average SER can be computed as:

$$\overline{SER} = \int_{\theta_1}^{\theta_2} f_1(\theta) M_\gamma(-f_2(\theta))\, d\theta \tag{12.78}$$

The next step is then finding the moment-generating function of the distribution of the SNR. Without going into the details of the derivations, we find that for a Rayleigh distribution of the signal amplitude, the moment-generating function of the SNR distribution is:

$$M_\gamma(s) = \frac{1}{1 - s\overline{\gamma}} \tag{12.79}$$

for a Rice distribution it is:

$$M_\gamma(s) = \frac{1 + K_r}{1 + K_r - s\overline{\gamma}} \exp\left[\frac{K_r s\overline{\gamma}}{1 + K_r - s\overline{\gamma}} \right] \tag{12.80}$$

and for a Nakagami distribution with parameter m:

$$M_\gamma(s) = \left(1 - \frac{s\overline{\gamma}}{m} \right)^{-m} \tag{12.81}$$

Having now the general form of the SER (Eq. 12.78) and the form of the moment-generating function, the computations of the error probabilities become straightforward (if sometimes a bit tedious).

Example 12.6 *BER of BPSK in Rayleigh fading.*

As one example, let us go through the computation of the average BER of BPSK in a Rayleigh-fading channel – a problem for which we already know the result from Eq. (12.52). Looking at Eq. (12.63), we find that $\theta_1 = 0$, $\theta_2 = \pi/2$:

$$f_1(\theta) = \frac{1}{\pi} \tag{12.82}$$

$$f_2(\theta) = \frac{1}{\sin^2(\theta)} \tag{12.83}$$

Since we consider Rayleigh fading:

$$M_\gamma(-f_2(\theta)) = \frac{1}{1 + \dfrac{\overline{\gamma}}{\sin^2(\theta)}} \tag{12.84}$$

so that the total BER is, according to Eq. (12.78):

$$\overline{SER} = \frac{1}{\pi} \int_0^{\pi/2} \frac{\sin^2(\theta)}{\sin^2(\theta) + \overline{\gamma}} d\theta \tag{12.85}$$

which can be shown to be identical to the first (exact) expression in Eq. (12.52) – namely:

$$\overline{BER} = \frac{1}{2}\left[1 - \sqrt{\frac{\overline{\gamma}}{1 + \overline{\gamma}}}\right] \tag{12.86}$$

Many modulation formats and fading distributions can be treated in a similar way. An extensive list of solutions, dealing with coherent detection, partially coherent detection, and non-coherent detection in different types of channels, is given in the book by Simon and Alouini [2000].

12.2.3 Outage probability versus average error probability

In the previous section, we computed the average bit error probability, where averaging was done over small-scale fading. Similarly, we could also average this distribution over large-scale fading – the mathematics are quite similar. But what is the physical meaning of these averaged values? In order to understand *what* we are computing here, and what averaging makes sense, we first have to have a look at the different operating scenarios that can occur in a wireless system.

As a first step, we have to compare the length of the "memory" of the system with the coherence time of the channel. The "memory" of the system in this context could, e.g., be caused by:

1. perception of the human ear (error bursts on the order of a few microseconds or less tend to be "smoothed out" by the human ear); or
2. the size of typical data structures that are to be transmitted over the wireless connection;
3. further memory can also come from coding (in which case the block length of the block code or the constraint length of a convolutional code would determine memory duration), and interleaving, in which case the length of the interleaver would determine the memory (see also Chapter 14).

Let us first consider the case where system memory is much shorter than coherence time. To name but one example: if we wish to transmit a file from a (stationary) laptop, and the objects in the environment are not moving, the channel seen by the wireless connection is static for the duration of the transmission. The SNR seen by the demodulator is thus constant, but random, depending on the position. It is thus meaningful to investigate the pdf of the SNR, in order to see what SNR is available in what percentage of locations. For each of the locations – i.e., for each realization of the SNR – we compute the BER simply as the BER of an AWGN channel with that specific SNR. In such a case, we will also get a *distribution* of the BER; note that this distribution is for a fixed transmit power.

If the receiver (or transmitter) moves, then the channel changes within a finite time, and system memory can thus extend over many channel realizations. Such system memories might typically see many small-scale realizations of the channel, but still be short enough that no large-scale variations of the channel occur within memory duration. When we now want to see the average BER for a file transfer, we have to average the BER over the duration of that file transfer, and thus average it over the distribution of the SNRs seen during that time.[5] Finally, for faster movements of the receiver, there might also be situations where there are large-scale channel variations within the system memory. In that case, the BER has to be averaged over both small-scale and large-scale channel variations. It is then no longer a random variable, but a single, deterministic value.

From these considerations, we also obtain the concept of *outage probability*. For many applications, it is not important what the exact value of the BER is, as long as it stays below a certain threshold. For example, coded speech sounds acceptable to a user as long as a certain threshold BER (typically on the order of a few percent) is not exceeded. It is then meaningful to determine the percentage of time that acceptable speech quality will not be available. This percentage is known as outage probability.

Computation of outage probability becomes much simpler if we define not a maximum BER, but rather a minimum SNR, for the system to work properly. This mapping from BER to SNR can be done in a simple way: for example, for a system with short memory, Eqs. (12.30)–(12.40) describe this relationship between $BER_{\text{min-accept}}$ and γ_0. We can then find the outage probability as:

$$\text{Pr}_{\text{out}} = P(\gamma < \gamma_0) = \int_0^{\gamma_0} pdf_\gamma(\gamma)\, d\gamma \tag{12.87}$$

Outage probability can also be seen as another way of establishing a fading margin: we need to find the mean SNR that guarantees a certain outage.

12.3 Error probability in delay- and frequency-dispersive fading channels

12.3.1 Physical cause of error floors

In wireless propagation channels, transmission errors are caused not only by noise, but also by signal distortions. These distortions are created on one hand by delay dispersion (i.e., echoes of the transmit signal arriving with different delays), and on the other hand by frequency dispersion (Doppler effect – i.e., signal components arriving with different Doppler shifts). For high data rates, delay dispersion is dominant; at low data rates, the Doppler effect is the main reason for signal distortion errors. In either of these cases, an increase in transmit power does not lead to a reduction of the BER; for that reason, these errors are often called *error floor* or *irreducible errors*. Of course, these errors can also be reduced or eliminated, but this has to be done by

---

[5] For a coded system, where the codelength is larger than channel coherence time, computations are a bit trickier (as discussed in Chapter 14).

methods other than increasing power (e.g., equalization, diversity, etc.). In this section, we only treat the case when the receiver does not use any of these countermeasures, so that dispersion leads to increased error rates. Later chapters will discuss in detail the fact that dispersion can actually be a benign effect if specific receiver structures are used.

Frequency dispersion

We first consider errors due to frequency dispersion. For FSK, it is immediately obvious how frequency dispersion leads to errors: a Doppler shift can push a bit over the decision boundary. Assume that a +1 was sent (i.e., the frequency $f_c + f_{mod}$). Due to the Doppler effect, the frequency $f_c + f_{mod} - \nu$ is received. If this is smaller than f_c, the receiver opts for a -1. However, this interpretation is very dangerous. One could erroneously conclude that errors cannot occur when $|\Delta f| > \nu_{max}$. But *instantaneous* frequency shifts can be significantly larger than the maximum Doppler frequency. Consider the following equation for the instantaneous frequency:

$$f_{inst}(t) = \frac{\mathrm{Im}\left(r^*(t)\dfrac{dr(t)}{dt}\right)}{|r(t)|^2} \tag{12.88}$$

Obviously, this can become very large when the amplitude becomes very small. In other words, deep fading dips lead to large Doppler shifts. We can also interpret this effect in terms of random Frequency Modulation (FM) (see Section 5.7.3).

A somewhat different interpretation can be given for differential detection. As mentioned above, differential detection assumes that the channel does not change between two adjacent symbols. However, if there is a finite Doppler, then the channel *does* change – remember that the Doppler spectrum gives a statistical description of channel changes. Thus, a nonzero Doppler effect implies a wrong reference phase for differential detection. If this effect is strong, it can lead to erroneous decisions. In this case it is also true that channel changes are strongest near fading dips.[6]

For MSK with differential detection, Hirade et al. [1979] determined the BER due to the Doppler effect:

$$\overline{BER}_{Doppler} = \frac{1}{2}(1 - \xi_s(T_B)) \tag{12.89}$$

where $\xi_s(t)$ is the normalized autocorrelation function of the channel (so that $\xi_s(0) = 1$) – i.e., the Fourier transform of the normalized Doppler spectrum. For small $\nu_{max}T_B$ we then get a BER that is proportional to the squared magnitude of the product of Doppler shift and bit duration:

$$\overline{BER}_{Doppler} = \frac{1}{2}\pi^2(\nu_{max}T_B)^2 \tag{12.90}$$

This basic functional relationship also holds for other Doppler spectra and modulation formats; it is only the proportionality constant that changes.

From this relationship, we find that errors due to frequency dispersion are mainly important for systems with a low data rate. For example, paging systems and sensor networks exhibit data rates on the order of 1 kbit/s, while Doppler frequencies can be up to a few hundred Hz. Error floors of 10^{-2} are thus easily possible. This has to be taken into account when designing the coding for such systems. For high-data-rate systems (which include almost all current cellular,

---

[6] For general QAM, not only the reference phase, but also the reference amplitude is relevant. However, in the following we will restrict our considerations to PSK and FSK, and thus ignore amplitude distortions.

cordless, and WLAN systems), errors due to frequency dispersion do not play a noticeable role.[7] Even for the Japanese JDC cellular system, which has a symbol duration of 50 μs, the BER due to frequency dispersion is only on the order of 10^{-4}, which is negligible compared with errors due to noise.

Delay dispersion

In contrast to frequency dispersion, delay dispersion has great importance for high-data-rate systems. This becomes obvious when we remember that the errors in unequalized systems are determined by the ratio of symbol duration that is disturbed by Inter Symbol Interference (ISI) and the undisturbed part of the symbol. The maximum excess delay of a channel impulse response is determined by the environment, and independent of the system. Let us assume in the following a maximum excess delay of 1 μs. In a system with a symbol duration of 20 μs, the ISI can disturb 5% of each symbol, while it can disturb 20% if the symbol duration is 5 μs. Just as for frequency dispersion, errors mainly occur near fading dips. Section 12.3.2 will give an interpretation of this fact in terms of group delay, which reaches its largest values near fading dips (see also Chapter 5).

Many theoretical and experimental investigations have shown that the error floor due to delay dispersion is given by the following equation:

$$\overline{BER} = K \left(\frac{S_\tau}{T_B} \right)^2 \tag{12.91}$$

where S_τ is the rms delay spread of the channel (see Chapter 6). Note, however, that Eq. (12.91) is only valid if the maximum excess delay of the channel is much smaller than the symbol duration, and the channel is Rayleigh fading. The proportionality constant K depends on the modulation method, filtering at transmitter and receiver, the form of the average impulse response, and choice of the sampling instant, as we will discuss in the sections below.

Choice of the sampling instant In a flat-fading channel, the choice of sampling instant is obvious – sampling should always occur at the maximum opening of the eye pattern, $t_s = 0$, where the offset t_s is defined via the equation for general sampling times:

$$t'_{s,i} : t'_s = t_s + iT \tag{12.92}$$

For channels with delay dispersion, the choice is no longer that clear. Most theoretical derivations assume that either $t_s = 0$ (i.e., sampling occurs at the minimum excess delay),[8] or at the *average mean delay*. The latter actually is the optimum sampling time for some Power Delay Profiles (PDPs, see Chapter 6), as demonstrated in Fig. 12.7.

When the sampling instant is chosen adaptively, according to the instantaneous state of the channel, the error floor can be decreased considerably, and – in unfiltered systems – even completely eliminated.

Impact of the shape of the power delay profile To a first approximation, only the *rms delay spread* determines the BER due to delay dispersion. A closer look reveals, however, that the actual shape of the PDP also has an impact. The variation of the BER for different PDPs (for equal S_τ) is usually less than a factor of 2, and is thus often neglected. Figure 12.8 shows that a rectangular PDP leads to a slightly larger BER than a two-spike PDP; the difference is 75%.

---

[7] However, this does *not* mean that time variations in the channel are unimportant in such systems. Channel variations can also have an impact on coding, on the validity of channel estimation, etc.

[8] Without restriction of generality, we assume that $\tau_0 = 0$.

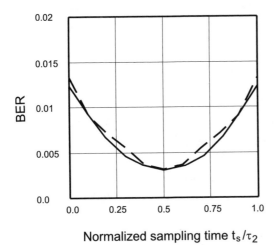

Figure 12.7   Dependence of delay-dispersion-induced error probability BER on choice of sampling instant in a two-spike channel.

Reproduced with permission from Molisch [2000] © Prentice Hall.

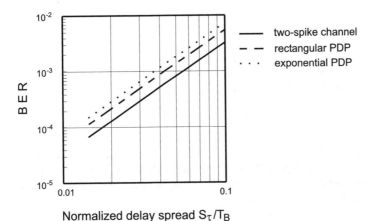

Figure 12.8   Impact of the shape of the PDP on the error floor due to delay dispersion. The modulation method is MSK with differential detection.

Reproduced with permission from Molisch [2000] © Prentice Hall.

An exponential PDP leads to an even larger error floor; however, for this shape, sampling at the average mean delay is not optimum.

Filtering   Filtering at the transmitter and/or receiver also leads to signal distortion, and thus makes the signal more susceptible to errors by the additional distortions caused by the channel. The narrower the filtering, the larger the error floor. Figure 12.9 shows the effect of filtering on QAM with raised cosine filters.

The question naturally arises as to the optimum filter bandwidth. Very narrow filters (bandwidth on the order of the inverse symbol duration or less) lead to strong ISIs by themselves. Even if all decisions can be made correctly in the absence of further disturbances,

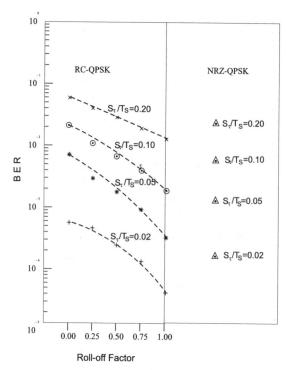

Figure 12.9 Error floor of quadrature-phase shift keying with coherent detection and raised cosine filters as a function of the rolloff factor. For comparison purposes, we also show the BER of conventional QPSK. Note that there is no rolloff factor where Raised Cosine (RC) QPSK reduces to conventional Non Return to Zero (NRZ) QPSK.

Reproduced with permission from Chuang [1987] © IEEE.

such a filter makes the system more susceptible to channel-induced ISI and noise. For wide filters, a lot of noise can pass through the filter, which also leads to high error rates. The optimum filter bandwidth depends on the ratio of delay dispersion to noise. Figure 12.9 shows an example of such a tradeoff.

Modulation method The modulation method also has an impact on the error floor: obviously, a modulation format is more sensitive to distortions by the channel the closer the signals are in the signal constellation diagram. QPSK shows a higher error floor than BPSK when the rms delay spread is normalized to the same *symbol* duration. Higher-order modulation formats fare better when equal *bit* durations are assumed.

12.3.2 Computation of the error floor using the group delay method

In this section, we present a very simple, approximate method for computing the BER due to delay dispersion. The method was introduced in Andersen [1991] for PSK and Crohn et al. [1993] for MSK. In the following, we present the method for differentially detected MSK.

Group delay T_g is defined as:

$$T_g = -\frac{\partial \Phi_c}{\partial \omega}\bigg|_{\omega=0} \tag{12.93}$$

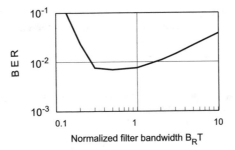

Figure 12.10 Bit error rate of filtered minimum shift keying with differential detection. The normalized rms delay spread is 0.1, and $SNR = 12\,\text{dB}$.

Reproduced with permission from Molisch [2000] © Prentice Hall.

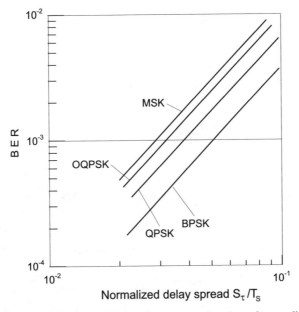

Figure 12.11 Error floor of different modulation formats as a function of normalized rms delay spread (normalized to symbol duration).

Reproduced with permission from Chuang [1987] © IEEE.

where ω is angular frequency. Here, $\Phi_{\text{c}}(\omega)$ is the phase of the channel transfer function, which can be expanded into a Taylor series:

$$\Phi_{\text{c}}(\omega) = \Phi_{\text{c}}(0) + \omega \frac{\partial \Phi_{\text{c}}}{\partial \omega}\bigg|_{\omega=0} + \frac{1}{2}\omega^2 \frac{\partial^2 \Phi_{\text{c}}}{\partial \omega^2}\bigg|_{\omega=0} + \cdots \tag{12.94}$$

The first term of this series corresponds to the *average mean delay* and can be omitted if the sampling is done at the average mean delay. Terminating the Taylor series after the linear term, we obtain:

$$\Phi_{\text{c}}(\omega) = -\omega T_{\text{g}} \tag{12.95}$$

Obviously, phase distortion at the sampling instant depends on the instantaneous frequency at these instants. If a $+1$ is transmitted, then the instantaneous frequency is $\pi/(2T_{\text{B}})$; otherwise, it is

$-\pi/(2T_B)$. When the same bits (i.e., a $+1$ followed by a $+1$, or a -1 followed by a -1), then the difference between the phase distortions is $\Delta\Phi = 0$; when different bits are transmitted, then:

$$\Delta\Phi = \pm\frac{\pi}{T_B}T_g \tag{12.96}$$

A decision error occurs when the size of channel-induced phase distortions (or rather, their difference at the sampling instants) is larger than $\pi/2$ – i.e., when:

$$|T_g| > T_B/2 \tag{12.97}$$

The statistics of group delay in a Rayleigh-fading channel is well-known [Bach Andersen et al. 1990] – namely, a *Student's t* distribution:

$$pdf_{T_g}(T_g) = \frac{1}{2S_\tau}\frac{1}{\left[1 + \left(T_g/S_\tau\right)^2\right]^{3/2}} \tag{12.98}$$

The probability for bit errors can thus be easily computed from Eq. (12.97). Furthermore, when averaging over the different possible bit combinations, we get:

$$BER = \frac{4}{9}\left(\frac{S_\tau}{T_B}\right)^2 \approx \frac{1}{2}\left(\frac{S_\tau}{T_B}\right)^2 \tag{12.99}$$

Example 12.7   *Consider a system using differentially detected MSK with $T_B = 35\,\mu s$, operating at 900 MHz, moving at 360 km/h (high-speed train), and an exponential power delay profile with $S_\tau = 10\,\mu s$. Compute the BER due to frequency dispersion and delay dispersion.*

The BER due to frequency dispersion can be computed by first calculating the maximum Doppler shift as:

$$\nu_{max} = f_c\frac{v}{c}$$

$$= 9\cdot 10^8\frac{100}{3\cdot 10^8}$$

$$= 300\,\text{Hz} \tag{12.100}$$

For MSK with differential detection, and assuming a classical Jake's Doppler spectrum, the BER due to frequency dispersion is given by Eq. (12.90), and hence:

$$\overline{BER}_{\text{Doppler}} = \frac{1}{2}\pi^2(\nu_{max}T_B)^2$$

$$= 5.4\cdot 10^{-4} \tag{12.101}$$

The BER due to time dispersion is given by Eq. (12.99), which means that:

$$\overline{BER}_{\text{Time}} = \frac{4}{9}\left(\frac{S_\tau}{T_B}\right)^2$$

$$= 3.6\cdot 10^{-2} \tag{12.102}$$

12.3.3 General fading channels: the quadratic-form-Gaussian-variable method

A general method for the computation of BERs in dispersive fading channels is the so-called QFGV method.[9] The channel is Rayleigh and Rice fading, suffering from delay dispersion and frequency dispersion, and adds AWGN.

Mathematically speaking, the QFGV method determines the probability that a variable D:

$$D = A|X|^2 + B|Y|^2 + CXY^* + C^* X^* Y \tag{12.103}$$

is smaller than 0. Here, X and Y are complex Gaussian variables, A and B are real constants, and C is a complex constant. Defining the auxiliary variables:

$$\left. \begin{aligned} w &= \frac{AR_{xx} + BR_{yy} + CR_{xy}^* + C^* R_{xy}}{4\left(R_{xx}R_{yy} - |R_{xy}|^2\right)\left(|C|^2 - AB\right)} \\[2ex] v_{1,2} &= \sqrt{w^2 + \frac{1}{4\left(R_{xx}R_{yy} - |R_{xy}|^2\right)\left(|C|^2 - AB\right)}} \mp w \\[2ex] \alpha_1 &= 2\left(|C|^2 - AB\right)\left(|\bar{X}|^2 R_{yy} + |\bar{Y}|^2 R_{xx} - \bar{X}^* \bar{Y} R_{xy} - \bar{X}\,\bar{Y}^* R_{xy}^*\right) \\[2ex] \alpha_2 &= A|\bar{X}|^2 + B|\bar{Y}|^2 + C\bar{X}^* \bar{Y} + C^* \bar{X}\,\bar{Y}^* \\[2ex] p_1 &= \frac{\sqrt{2v_1^2 v_2 (\alpha_1 v_2 - \alpha_2)}}{|v_1 + v_2|} \\[2ex] p_2 &= \frac{\sqrt{2v_1 v_2^2 (\alpha_1 v_1 + \alpha_2)}}{|v_1 + v_2|} \end{aligned} \right\} \tag{12.104}$$

where R_{xy} is the second central moment $R_{xy} = \frac{1}{2} E\{(X - \bar{X})(Y - \bar{Y})^*\}$, the error probability becomes:

$$P\{D < 0\} = Q_M(p_1, p_2) - \frac{v_2/v_1}{1 + v_2/v_1} I_0(p_1 p_2) \exp\left(-\frac{p_1^2 + p_2^2}{2}\right) \tag{12.105}$$

where Q_M is Marcum's Q-function (Eq. 12.40). If there is only Rayleigh fading, and no dispersion, then α_1 and α_2 are zero, so that $P(D < 0) = v_1/(v_1 + v_2)$. The biggest problem is often to formulate the error probability as $P\{D < 0\}$. This will be discussed in the following.

Canonical receiver

It is often helpful to reduce different receiver structures to a "canonical" receiver [Suzuki 1982], whose stucture is shown in Fig. 12.12. The received signal (after bandpass filtering) is multiplied by a reference signal: once by a "normal" reference signal, and once by the signal shifted by $\pi/2$ to get the quadrature component. The resulting signal is lowpass-filtered, and sent to the decision device. For coherent detection, the reference signal is obtained from a carrier recovery circuit; for differential detection, the reference signal is a delayed version of the incoming signal.

As an example, we consider the differential detection of MSK. It is fairly easy to describe a bit error in the form $D < 0$. An error is made if a $+1$ was transmitted, but the phase difference of the signals at the sampling instants, $X = r(t_s)$ and $Y = r(t_s - T)$, lies between π and 2π.

---

[9] This method is based on evaluation of certain quadratic forms of Gaussian variables. It is also often named after the groundbreaking papers of Bello and Nelin [1963] and Proakis [1968].

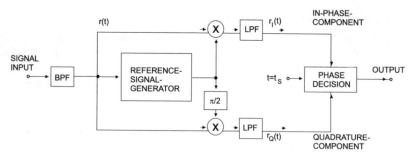

Figure 12.12 Canonical receiver structure.

BPF = bandpass filter; LPF = lowpass filter.
Reproduced with permission from Suzuki [1982] © IEEE.

The condition for an error is thus identical to:

$$\text{Re}\{b_0 X Y^* \exp(-j\pi/2)\} < 0 \tag{12.106}$$

where b_0 is the transmitted bit. Since $\text{Re}\{Z\} = (Z + Z^*)/2$, Eq. (12.106) is a quadratic form D with $A = B = 0$.

Bit error probability

A first step in the application of the QFGV method to differential detection is computation of the correlation between the quantities $X = r(t_s)$ and $Y = r(t_s - T)$. For detection using a frequency discriminator, we define X as the sample value $r(t_s)$ and define Y as the derivative $dr(t)/dt$ at time $t = t_s$. Explicit equations for the resulting correlation coefficient are given in Adachi and Parsons [1989] and Molisch [2000].

After the correlation coefficient has been found, the mean BER can be computed from Eq. (12.105). For the case of differential detection of binary FSK in Rayleigh fading, these equations can be simplified to:

$$\overline{BER} = \frac{1}{2} - \frac{1}{2} \frac{b_0 \, \text{Im}\{\rho_{XY}\}}{\sqrt{\text{Im}\{\rho_{XY}\}^2 + (1 - |\rho_{XY}|^2)}} \tag{12.107}$$

For $\pi/4$-DQPSK we obtain:

$$\overline{BER} = \frac{1}{2} - \frac{1}{4} \left\{ \frac{b_0 \, \text{Re}\{\rho_{XY}\}}{\sqrt{(\text{Re}\{\rho_{XY}\})^2 + (1 - |\rho_{XY}|^2)}} + \frac{b_0' \, \text{Im}\{\rho_{XY}\}}{\sqrt{(\text{Im}\{\rho_{XY}\})^2 + (1 - |\rho_{XY}|^2)}} \right\} \tag{12.108}$$

where b_0 and b_0' are the bits making up a symbol.

Summarizing, the BER can be computed in the following steps:

(i) reduce the actual receiver structure to canonical form;
(ii) formulate the condition for the occurence of errors as $D < 0$;
(iii) compute the mean and the correlation coefficients of X and Y;
(iv) compute the BER according to the general equations (12.105) and (12.104), or use the simplified equations (12.108) or (12.107) for MSK and $\pi/4$-DQPSK in Rayleigh-fading channels.

Further reading

The literature describing the bit error probability of digital modulation formats encompasses hundreds of papers. For the BER in AWGN, we again just refer to the textbooks on digital communications [Anderson 2005, Barry et al. 2003, Proakis 1995, and Sklar 2001]. Fundamental computation methods that are applicable both to non-fading and fading channels were proposed in Pawula et al. [1982], Proakis [1968], and Stein [1964]. The computation of the error probability in flat-fading channels is described in many papers: essentially, each modulation format, combined with the amplitude statistics of the channel, results in at least one paper. Some important examples in Rayleigh-fading channels include Chennakeshu and Saulnier [1993] for $\pi/4$-DQPSK, Varshney and Kumar [1991] and Yongacoglu et al. [1988] for GMSK, and Divsalar and Simon [1990] for differential detection of MPSK.

An important alternative to the classical computation of the BER is computation via the moment-generating function described in Simon and Alouini [2000], or via the characteristic function (see Annamalai et al. [2000]). Simon and Alouini's [2000] monograph gives a number of BER equations for different modulation formats, channel statistics, and receiver structures.

For delay-dispersive channels, a number of different computation methods have been proposed: we have already mentioned in the main part of the text the QFGV method ([Adachi and Parsons 1989], [Proakis 1968]), and the group delay method ([Andersen 1991], [Crohn et al. 1993]). Furthermore, a number of papers use the pdf of the angles between Gaussian vectors as derived by Pawula et al. [1982]. Chuang [1987] is a very readable paper comparing different modulation formats. A wealth of papers are also dedicated to modulation formats or detection methods that reduce the impact of delay dispersion in unequalized systems. A summary of these methods, and further literature, can be found in Molisch [2000].

13

Diversity

13.1 Introduction

13.1.1 Principle of diversity

In the previous chapter, we treated conventional transceivers that transmit an uncoded bitstream over fading channels. For Additive White Gaussian Noise (AWGN) channels, such an approach can be quite reasonable: the Bit Error Rate (BER) decreases exponentially as the Signal-to-Noise Ratio (SNR) increases, and a 10-dB SNR leads to BERs on the order of 10^{-4}. However, in Rayleigh fading the BER decreases only linearly with the SNR. We thus would need an SNR on the order of 40 dB in order to achieve a 10^{-4} BER, which is clearly unpractical. The reason for this different performance is the fading of the channel: the BER is mostly determined by the probability of channel attenuation being large, and thus of the instantaneous SNR being low. A way to improve the BER is thus to change the effective channel statistics – i.e., to make sure that the SNR has a smaller probability of being low. Diversity is a way to achieve this.

The principle of diversity is to ensure that the same information reaches the receiver (RX) on statistically independent channels. Consider the simple case of a receiver with two antennas. The antennas are assumed to be far enough from each other that small-scale fading is independent at the two antennas. The receiver always chooses the antenna that has instantaneously larger receive power.[1] As the signals are statistically independent, the probability that both antennas are in a fading dip *simultaneously* is low – certainly lower than the probability that one antenna is in a fading dip. The diversity thus changes the SNR statistics at the detector input.

Example 13.1 *Diversity reception in a two-state fading channel.*

To quantify this effect, let us consider a simple numerical example: the noise power within the RX filter bandwidth is 50 pW, the average received signal power is 1 nW, the SNR is thus 13 dB. In an AWGN channel, the resulting BER is 10^{-9}, assuming that the modulation is differentially detected Frequency Shift Keying (FSK). Now consider a fading channel where during 90% of the time the received power is 1.11 nW, and the SNR is thus 13.5 dB, while for the remainder, it is zero. This means that during 90% of the time, the BER is 10^{-10}; the remainder of the time, it is 0.5; the average BER is thus:

$$0.9 \cdot 10^{-10} + 0.1 \cdot 0.5 = 0.05 \tag{13.1}$$

---

[1] We will see later on that this scheme is only one of many different possible diversity schemes.

Wireless Communications   Andreas F. Molisch
© 2005 John Wiley & Sons, Ltd

For the case of diversity, the probability that the received signal power is 0 at both antennas simultaneously is $0.1 \cdot 0.1 = 0.01$. The probability that the received power is $1.11\,\text{nW}$ at both antennas simultaneously is $0.9 \cdot 0.9 = 0.81$; the probability that it is $1.11\,\text{nW}$ at one antenna and 0 at the other is 0.18. In both of the latter cases, the SNR at the detector is $13.5\,\text{dB}$. The total BER is thus:

$$0.01 \cdot 0.5 + 0.99 \cdot 10^{-10} = 0.005 \qquad (13.2)$$

This is approximately the square of the BER for a single-antenna system. If we have three antennas, then the probability that the signal power is 0 at all three antennas simultaneously is 0.1^3; the total BER is then $0.5 \cdot 0.001 + 0.999 \cdot 10^{-10} = 0.0005$; this is approximately the third power of the BER for a single-antenna system.

Later sections will give exact equations for the BER with diversity in Rayleigh-fading channels. The general concepts, however, are the same as in the simple example described above. With N_r diversity antennas, we obtain a bit error probability $\propto BER_{oc}^{N_r}$, where BER_{oc} is the BER with just one receive channel.[2]

In the following, we first describe the characterization of correlation coefficients between different transmission channels. We then give an overview about how transmission over independent channels can be realized – spatial antenna diversity described above is one, but certainly not the only approach. Next, we describe how signals from different channels can best be combined, and what performance can be achieved with the different combining schemes.

13.1.2 Definition of the correlation coefficient

Diversity is most efficient when the different transmission channels (also called diversity branches) carry independently fading copies of the same signal. This means that the joint probability density function of fieldstrength (or power) $pdf_{r_1, r_2, \dots}(r_1, r_2, \dots)$ is equal to the product of the marginal pdfs for the channels, $pdf_{r_1}(r_1), pdf_{r_2}(r_2), \dots$ Any correlation between the fading of the channels decreases the effectiveness of diversity.

The *correlation coefficient* characterizes the correlation between signals on different diversity branches. A number of different definitions is being used for this important quantity: complex correlation coefficients, correlation coefficient of the phase, etc. The most important one is the correlation coefficient of signal envelopes x and y:

$$\rho_{xy} = \frac{E\{x \cdot y\} - E\{x\} \cdot E\{y\}}{\sqrt{(E\{x^2\} - E\{x\}^2) \cdot (E\{y^2\} - E\{y\}^2)}} \qquad (13.3)$$

For two statistically independent signals, the relationship $E\{xy\} = E\{x\}E\{y\}$ holds; therefore, the correlation coefficient becomes zero. Signals are often said to be "effectively" decorrelated if ρ is below a certain threshold (typically 0.5 or 0.7).

13.2 Microdiversity

As mentioned in the introduction, the basic principle of diversity is that the RX has multiple copies of the transmit signal, where each of the copies goes through a statistically independent channel. This section describes different ways of obtaining these statistically independent copies.

---

[2] Since in a Rayleigh-fading channel, $BER_{oc} \propto SNR^{-1}$, we find that for a diversity system with N_r independently fading channels, $BER \propto SNR^{-N_r}$. Quite generally in fading channels with diversity, $BER \propto SNR^{-d_{\text{div}}}$, where d_{div} is known as *diversity order*.

We concentrate on methods that can be used to combat small-scale fading, which are therefore called "microdiversity". The five most common methods are:

1. Spatial diversity: several antenna elements separated in space.
2. Temporal diversity: repetition of the transmit signal at different times.
3. Frequency diversity: transmission of the signal on different frequencies.
4. Angular diversity: multiple antennas (with or without spatial separation) with different antenna patterns.
5. Polarization diversity: multiple antennas receiving different polarizations (e.g., vertical and horizontal).

When we speak of antenna diversity, we imply that there are multiple antennas at the *receiver*. Only in Section 13.6 (and later in Chapter 20) will we discuss how multiple *transmit* antennas can be exploited to improve performance.

The following important equation will come in handy: Consider the correlation coefficient of two signals that have a temporal separation τ and a frequency separation $f_1 - f_2$. As shown in Appendix 13.A (see *www.wiley.com/go/molisch*), the correlation coefficient is:

$$\rho_{xy} = \frac{J_0^2(k_0 v \tau)}{1 + (2\pi)^2 S_\tau^2 (f_2 - f_1)^2} \tag{13.4}$$

Note that for moving Mobile Stations (MSs), temporal separation can be easily converted into spatial separation, so that temporal and spatial diversity become mathematically equivalent. Eq. (13.4) is thus quite general in the sense that it can be applied to spatial, temporal, and frequency diversity. However, a number of assumptions were made in the derivation of this equation: (i) validity of the Wide Sense Stationary Uncorrelated Scatterer (WSSUS) model, (ii) no existence of Line Of Sight (LOS), (iii) exponential shape of the power delay profile, (iv) isotropic distribution of incident power, and (v) use of omnidirectional antennas.

Example 13.2 *Compute the correlation coefficient of two frequencies with separation (i) 30 kHz, (ii) 200 kHz, (iii) 5 MHz, in the "typical urban" environment, as defined in COST 207[3] channel models.*

For zero temporal separation, the Bessel function in Eq. (13.4) is unity, and the correlation coefficient is thus only dependent on rms delay spread and frequency separation. Rms delay spread is calculated according to Eq. (6.39), where the Power Delay Profile (PDP) of the COST 207 typical urban channel model is given in Section 7.6.3. We obtain:

$$S_\tau = \sqrt{\frac{\int_0^{7 \cdot 10^{-6}} e^{-\tau/10^{-6}} \tau^2 \, d\tau}{\int_0^{7 \cdot 10^{-6}} e^{-\tau/10^{-6}} \, d\tau} - \left(\frac{\int_0^{7 \cdot 10^{-6}} e^{-\tau/10^{-6}} \tau \, d\tau}{\int_0^{7 \cdot 10^{-6}} e^{-\tau/10^{-6}} \, d\tau}\right)^2}$$

$$= 0.977 \ \mu s \tag{13.5}$$

The correlation coefficient thus becomes:

$$\rho_{xy} = \frac{1}{1 + (2\pi)^2 (0.977 \cdot 10^{-6})^2 (f_1 - f_2)^2}$$

$$= \begin{cases} 0.97, & f_1 - f_2 = 30 \text{ kHz} \\ 0.4, & f_1 - f_2 = 200 \text{ kHz} \\ 1 \cdot 10^{-3}, & f_1 - f_2 = 5 \text{ MHz} \end{cases} \tag{13.6}$$

---

[3] European COoperation in the field of Scientific and Technical research.

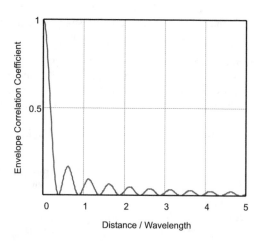

Figure 13.1 Envelope correlation coefficient as a function of antenna separation.

From this, we can see that the correlation between two neighboring 30-kHz bands – as used, e.g., in the IS-136[4] Time Division Multiple Access (TDMA) cellular system – is very high; correlation over 200-kHz bands (two neighboring channels in GSM are 200 kHz apart, see Chapter 21) is also appreciable, but two neighboring channels in Wideband Code Division Multiple Access (WCDMA) (5 MHz apart) are uncorrelated in this environment. Furthermore, carriers that are, e.g., 45 MHz apart – used for Base Station (BS)–MS and MS–BS communication in Global System for Mobile communications (GSM) by frequency duplex (see Chapter 21 and also Chapter 17) are completely uncorrelated.

13.2.1 Spatial diversity

Spatial diversity is the oldest and simplest form of diversity. Despite (or because) of this, it is also the most widely used. The transmit signal is received at several antenna elements, and the signals from these antennas are then further processed according to the principles that will be described in Section 13.4. But, irrespective of the processing method, performance is influenced by correlation of the signals between the antenna elements. A large correlation between signals at antenna elements is undesirable, as it decreases the effectiveness of diversity. A first important step in designing diversity antennas is thus to establish a relationship between antenna spacing and the correlation coefficient. This relationship is different for BS antennas and MS antennas, and thus will be treated separately.

1. *Mobile station in cellular and cordless systems*: it is a standard assumption that waves are incident from all directions at the MS (see also Chapter 5). Thus, points of positive and negative interference of Multi Path Component (MPCs) – i.e., points where we have high and low received power, respectively – are spaced approximately $\lambda/4$ apart. This is therefore the distance that is required for decorrelation of received signals. This intuitive insight agrees very well with the results from the exact mathematical derivation (Eq. 13.4, with $f_2 - f_1 = 0$), given in Fig. 13.1: decorrelation, defined as $\rho = 0.5$, occurs at an antenna separation of $\lambda/4$. Compare also Example 5.3.

---

[4] IS-136 is a cellular time in the time division multiplex access system.

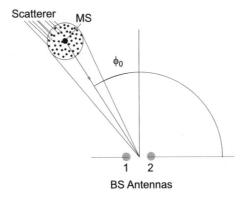

Figure 13.2   Scatterers concentrated around the mobile station.

The above considerations imply that the minimum distance for antenna elements in GSM (at 900 MHz) is about 8 cm, and for various cordless and cellular systems at the 1,800-MHz band it is about 4 cm. For Wireless Local Area Network (WLANs) (at 2.4 and 5 GHz), the distances are even smaller. It is thus clearly possible to place two antennas on an MS of a cellular system. This has also been demonstrated by the Japanese Pacific Digital Cellular (PDC) system, where antenna diversity at the MS is required in order to achieve acceptable quality.

2. *Base station in cordless systems and WLANs*: in a first approximation, the angular distribution of incident radiation at indoor BSs is also uniform – i.e., radiation is incident with equal strength from all directions. Therefore, the same rules apply as for MSs.

3. *Base stations in cellular systems*: for a cellular BS, the assumption of uniform directions of incidence is no longer valid. Interacting Object (IOs) are typically concentrated around the MS (Fig. 13.2, see also Chapter 7). Since all waves are incident essentially from one direction, the correlation coefficient (for a given distance between antenna elements d_a) is much higher. Expressed differently, the antenna spacing required to obtain sufficient decorrelation increases.

To get an intuitive insight, we start with the simple case when there are only two MPCs whose wavevectors are at an angle α with respect to each other (Fig. 13.3, see also

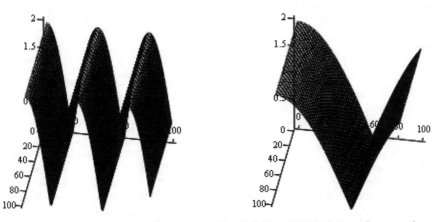

Figure 13.3   Interference pattern of two waves with 45° (left) and 15° (right) angular separation.

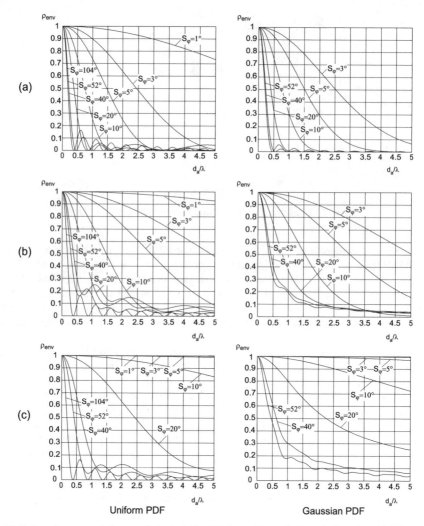

Figure 13.4 Envelope correlation coefficient at the BS for uniform (left) and Gaussian (right) probability density function (pdf) of the directions of arrival (a) $\phi_0 = 90°$, (b) $\phi_0 = 45°$, (c) $\phi_0 = 10°$, and different values of angular spread S_φ.

Reproduced with permission from Fuhl et al. [1998] © IEE.

Chapter 5). It is obvious that the distance between the maxima and minima of the interference pattern is larger the smaller α is. For very small α, the connection line between antenna elements lies on a "ridge" of the interference pattern and antenna elements are completely correlated.

Numerical evaluations of the correlation coefficient as a function of antenna spacing are shown in Fig. 13.4. The first column shows the results for rectangular angular power spectra; the results for Gaussian distributions are shown in the second column. We can see that antenna spacing has to be on the order of 2–20 wavelengths for angular spreads between $1°$ and $5°$ in order to achieve decorrelation. We also find that it is mostly rms angular spread that determines the required antenna spacing, while the shape of the angular power spectrum has only a minor influence.

13.2.2 Temporal diversity

As the wireless propagation channel is time-variant, signals that are received at different times are uncorrelated. For "sufficient" decorrelation, the temporal distance must be at least $1/(2\nu_{max})$, where ν_{max} is the maximum Doppler frequency. Temporal diversity can be realized in different ways:

1. *Repetition coding*: this is the simplest form. The signal is repeated several times, where the repetition intervals are long enough to achieve decorrelation. This obviously achieves diversity, but is also highly bandwidth-inefficient. Spectral efficiency decreases by a factor that is equal to the number of repetitions.
2. *Automatic repeat request (ARQ)*: here, the RX sends a message to the transmitter (TX) to indicate whether it received the data with sufficient quality (see Appendix 14.A at *www.wiley.com/go/molisch*). If this is not the case, then the transmission is repeated (after a wait period that achieves decorrelation). The spectral efficiency of ARQ is better than that of repetition coding, since it requires multiple transmissions only when the first transmission occurs in a bad fading state, while for repetition coding, retransmissions occur always. On the downside, ARQ requires a feedback channel.
3. *Combination of interleaving and coding*: a more advanced version of repetition coding is forward error correction coding with interleaving. The different symbols of a codeword are transmitted at different times, which increases the probability that at least some of them arrive with a good SNR. The transmitted codeword can then be reconstructed. For more details, see Chapter 14.

For the case when only the MS is moving, while the IOs and the BS are fixed, temporal correlation can be converted into spatial correlation.[5] From this it is clear that temporal diversity is useless in a static scenario where BSs, MSs, and IOs are immobile; such a situation can occur, e.g., for WLANs. In such a case, the correlation coefficient is $\rho = 1$ for all time intervals, and temporal diversity is useless.

13.2.3 Frequency diversity

In frequency diversity, the same signal is transmitted at different frequencies. If these frequencies are spaced apart by more than the coherence bandwidth of the channel, then their fading is approximately independent, and the probability is low that the signal is in a deep fade at both frequencies simultaneously. For an exponential PDP the correlation between two frequencies can be obtained from Eq. (13.4) by setting the numerator to unity as the signals at the two frequencies occur at the same time. Thus:

$$\rho = \frac{1}{1 + (2\pi)^2 S_\tau^2 (f_2 - f_1)^2} \qquad (13.7)$$

This again confirms that the two signals have to be at least one coherence bandwidth apart from each other. Figure 13.5 shows ρ as a function of the spacing between the two frequencies. For a more general discussion of frequency correlation see Chapter 6.

It is not common to actually repeat the same information at two different frequencies, as this would greatly decrease spectral efficiency. Rather, information is spread over a large bandwidth, so that small parts of the information are conveyed by different frequency components. The RX can then average over the different frequencies to recover the original information.

---

[5] A similar situation occurs when the BS and MS are fixed, and all the IOs move. The situation is more complicated if the MS and some of the IOs are moving.

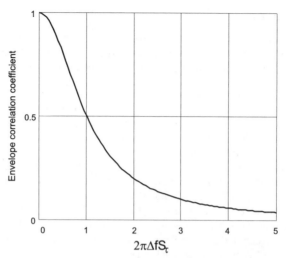

Figure 13.5 Correlation coefficient of the envelope as a function of normalized frequency spacing.

This spreading can be done by different methods:

- compressing the information in time – i.e., sending short bursts that each occupy a large bandwidth – TDMA (see Chapter 17);
- Code Division Multiple Access – CDMA (Section 18.2);
- multicarrier CDMA (Section 19.9) and coded orthogonal frequency division multiplexing (Section 19.4.3);
- frequency hopping in conjunction with coding: different parts of a codeword are transmitted on different carrier frequencies (Section 18.1).

These methods allow the transmission of information without wasting bandwidth and will be described in greater detail in Chapters 17–19. For the moment, we just stress that the use of frequency diversity requires the channel to be frequency-selective. In other words, frequency diversity (delay dispersion) can be exploited by the system to make it more robust, and decrease the effects of fading. This seems to be a contradiction to the results of Chapter 12, where we had shown that frequency selectivity leads to an *increase* of the BER, and even an error floor. The reason for this discrepancy is that Chapter 12 considered a very simple RX that takes no measures to combat (or exploit) the effects of frequency selectivity.

13.2.4 Angle diversity

A fading dip is created when MPCs, which usually come from different directions, interfere destructively. If some of these waves are attenuated or eliminated, then the location of fading dips changes. In other words, two co-located antennas with different patterns "see" differently weighted MPCs, so that the MPCs interfere differently for the two antennas. This is the principle of *angle diversity* (also known as *pattern diversity*).

Angular diversity is usually used in conjunction with spatial diversity; it enhances the decorrelation of signals at closely spaced antennas. Different antenna patterns can be achieved very easily. Of course, different types of antennas have different patterns. But even identical antennas can have different patterns when mounted close to each other (see Fig. 13.6). This effect is due to *mutual coupling*: antenna B acts as a reflector for antenna A, whose pattern is

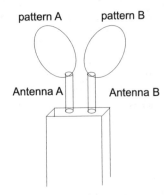

Figure 13.6   Angle diversity for closely spaced antennas.

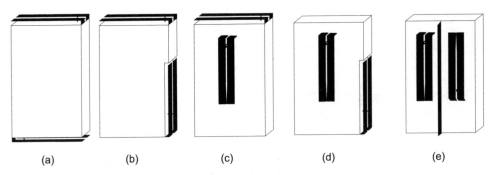

(a)   (b)   (c)   (d)   (e)

Figure 13.7   Configurations of diversity antennas at a mobile station.
Reproduced with permission from Erätuuli and Bonek [1997] © IEEE.

therefore skewed to the left.[6] Analogously, the pattern of antenna B is skewed to the right due to reflections from antenna A. Thus, the two patterns are different.

The different patterns are even more pronounced when the antennas are located on different parts of the casing. While dipole antennas are usually restricted to the top of the casing, patch antennas and inverted-F antennas (see Chapter 9) can be placed on all parts of the casing (see Fig. 13.7). In all of these cases, decorrelation is good even if the antennas are placed very closely to each other.

13.2.5 Polarization diversity

Horizontally and vertically polarized MPCs propagate differently in a wireless channel,[7] as the reflection and diffraction processes depend on polarization (see Chapters 4 and 7). Even if the transmit antenna only sends signals with a single polarization, the propagation effects in the channel lead to depolarization so that both polarizations arrive at the RX. The fading of signals with different polarizations is statistically independent. Thus, receiving both polarizations

---

[6] This arrangement can also be considered as a Yagi antenna. It depends on the spacing of the two elements whether antenna B acts as director or reflector, and thus which direction the pattern is skewed into.
[7] For simplicity, we henceforth speak of horizontal and vertical polarization. However, the considerations are valid for any two orthogonal polarizations.

using a dual-polarized antenna, and processing the signals separately, offers diversity. This diversity can be obtained without any requirement for a minimum distance between antenna elements.

Let us now consider more closely the situation where the transmit signal is vertically polarized, while the signal is received in both vertical and horizontal polarization. In that case, fading of the two received signals is independent, but the average received signal strength in the two diversity branches is *not* identical. Depending on the environment, the horizontal (i.e., cross-polarized) component is some 3–20 dB weaker than the vertical (co-polarized) component. As we will see later on, this has an important impact on the effectiveness of the diversity scheme. Various antenna arrangements have been proposed in order to mitigate this problem.

It has also been claimed that the diversity order that can be achieved with polarization diversity is up to 6: three possible components of the E-field and three components of the H-field can all be exploited [Andrews et al. 2001].[8] However, propagation characteristics as well as practical considerations prevent a full exploitation of that diversity order especially for outdoor situations. This is usually not a serious restriction for diversity systems, as we will see later on that going from diversity order 1 (i.e., no diversity) to diversity order 2 gives larger benefits than increasing the diversity order from 2 to higher values. However, it is an important issue for Multiple Input Multiple Output (MIMO) systems (see Section 20.2).

13.3 Macrodiversity and simulcast

The previous section described diversity methods that combat small-scale fading – i.e., the fading created by interference effects. As we have seen, spatial diversity for such cases requires antenna spacings on the order of only a few wavelengths. However, these diversity methods are not suitable for combating large-scale fading, which is created by shadowing effects. Shadowing affects different MPCs almost equally, so that frequency diversity or polarization diversity are not effective. Furthermore, the correlation distances for large-scale fading are on the order of tens or hundreds of meters, so that spatial diversity or temporal diversity (with reasonable antenna spacings and latency times, respectively), cannot be used either. In other words, if there is a hill between the TX and RX, adding antennas on either the BS or the MS does not help to eliminate the shadowing caused by this hill. The only way to circumvent the problem is to use a separate base station (BS2) that is placed in such a way that the hill is not in the connection line between the MS and BS2. This in turn implies a large distance between BS1 and BS2, which gives rise to the word *macrodiversity*.

The simplest method for macrodiversity is the use of *on-frequency repeaters* that receive the signal and retransmit an amplified version of it. *Simulcast* is very similar to this approach; the same signal is transmitted simultaneously from different BSs. In cellular applications the two BSs should be synchronized, and transmit the signals intended for a specific user in such a way that the two waves arrive at the RX almost simultaneously (timing advance).[9] Note that synchronization can only be obtained if the runtimes from the two BSs to the MS are known. Generally speaking, it is desirable that the synchronization error is no larger than the delay dispersion that the RX can handle. Especially critical are RXs in regions where the strengths of the signals from the two BSs are approximately equal.

---

[8] Note that this result is surrounded by some controversy, and (at the time of this writing) there is no general agreement in the scientific community whether diversity order 6 can really be achieved.

[9] If the RX cannot easily deal with delay dispersion, it is desirable that the two signals arrive exactly at the same time. For advanced RXs (Chapters 16–19) it can be desirable to have a small amount of delay dispersion; if the timing of the BSs and the runtimes of the signals to the RX are exactly known, this can also be achieved.

Simulcast is also widely used for broadcast applications, especially digital TV. In this case, the exact synchronization of all possible RXs is not possible – each RX would require a different timing advance from the TXs.

A disadvantage of simulcast is the large amount of signaling information that has to be carried on landlines. Synchronization information as well as transmit data have to be transported on landlines (or microwave links) to the BSs. This used to be a serious problem in the early days of digital mobile telephony, but the current wide availability of fiber-optic links has made this less of an issue.

The use of on-frequency repeaters is simpler than that of simulcast, as no synchronization is required. On the other hand, delay dispersion is larger, because (i) the runtime from BS to repeater, and repeater to MS is larger (compared with the runtime from a second BS), and (ii) the repeater itself introduces additional delays due to the group delays of electronic components, filters, etc.

13.4 Combination of signals

Now we turn our attention to the question of how to use diversity signals in a way that improves the total quality of the signal that is to be detected. To simplify the notation, we speak here only about the combination of signals from different antenna signals at the RX. However, the mathematical methods remain valid for other types of diversity signals as well. In general, we can distinguish two ways of exploiting signals from the multiple diversity branches:

1. *Selection diversity*, where the "best" signal copy is selected and processed (demodulated and decoded), while all other copies are discarded. There are different criteria for what constitutes the "best" signal.
2. *Combining diversity*, where all copies of the signal are combined (before or after the demodulator), and the combined signal is decoded. Again, there are different algorithms for combination of the signals.

Combining diversity leads to better performance, as all available information is exploited. On the downside, it requires a more complex RX than selection diversity. In most RXs, all processing is done in the baseband. Thus, an RX with combining diversity needs to downconvert all available signals, and combine them appropriately in the baseband. Thus, it requires N_r antenna elements as well as N_r complete Radio Frequency (RF) (downconversion) chains. An RX with selection diversity requires only *one* RF chain, as it processes only a single received signal at a time.

In the following, we give a more detailed description of selection (combination) criteria and algorithms. We assume that different signal copies undergo statistically independent fading – this greatly simplifies the discussion of both the intuitive explanations and the mathematics of the signal combination. Discussions on the impact of a finite correlation coefficient are relegated to Section 13.5.

In these considerations, we also have to keep in mind that the gain of multiple antennas is due to two effects: *diversity gain* and *beamforming gain*. Diversity gain reflects the fact that it is improbable that several antenna elements are in a fading dip simultaneously; the probability for very low signal levels is thus decreased by the use of multiple antenna elements. Beamforming gain reflects the fact that (for combining diversity) the combiner performs an averaging over the noise at different antennas. Thus, even if the signal levels at all antenna elements are identical, the combiner output SNR is larger than the SNR at a single antenna element.

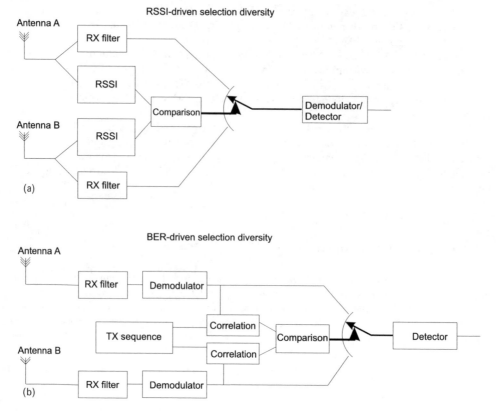

Figure 13.8 Selection diversity principle: (a) Received-signal-strength-indication-controlled diversity. (b) Bit-error-rate-controlled diversity.

13.4.1 Selection diversity

Received-signal-strength-indication-driven diversity

In this method, the RX selects the signal with the largest instantaneous power (or *Received Signal Strength Indication, RSSI*), and processes it further. This method requires N_r antenna elements, N_r RSSI sensors, and a N_r-to-1 multiplexer, but only one RF chain (see Fig. 13.8). The method allows simple tracking of the selection criterion even in fast-fading channels. Thus, we can switch to a better antenna as soon as the RSSI becomes higher there.

1. If the BER is determined by noise, then RSSI-driven diversity is the best of all the selection diversity methods, as maximization of the RSSI also maximizes the SNR.
2. If the BER is determined by co-channel interference, then RSSI is no longer a good selection criterion. High receive power can be caused by a high level of interference, such that the RSSI criterion makes the system select branches with a low signal-to-interference ratio. This is especially critical when interference is caused mainly by one dominant interferer – a situation that is typical for Frequency Division Multiple Access (FDMA) or TDMA systems.
3. Similarly, RSSI-driven diversity is suboptimum if the errors are caused by the frequency selectivity of the channel. RSSI-driven diversity can still be a reasonable approximation,

because we have shown in Chapter 12 that errors caused by signal distortion occur mainly in the fading dips of the channel. However, this is only an approximation, and it can be shown that (uncoded, unequalized) systems with RSSI-driven selection diversity have a BER that is higher by a constant factor compared with optimum (BER-driven) diversity.

For an exact performance assessment (Section 13.5), it is important to obtain the SNR distribution of the output of the selector. Assume that the instantaneous signal amplitude is Rayleigh-distributed, such that the SNR of the nth diversity branch, γ_n, is (see Eq. 12.48):

$$pdf_{\gamma_n}(\gamma_n) = \frac{1}{\bar{\gamma}}\exp\left(-\frac{\gamma_n}{\bar{\gamma}}\right) \tag{13.8}$$

where $\bar{\gamma}$ is the mean branch SNR (assumed to be identical for all diversity branches). The cumulative distribution function (cdf) is then:

$$cdf_{\gamma_n}(\gamma_n) = 1 - \exp\left(-\frac{\gamma_n}{\bar{\gamma}}\right) \tag{13.9}$$

The cdf is, by definition, the probability that the instantaneous SNR lies below a given level. As the RX selects the branch with the largest SNR, the probability that the chosen signal lies below the threshold is the product of the probabilities that the SNR at each branch is below the threshold. In other words, the cdf of the selected signal is the product of the cdfs of each branch:

$$cdf_{\gamma}(\gamma) = \left[1 - \exp\left(-\frac{\gamma}{\bar{\gamma}}\right)\right]^{N_r} \tag{13.10}$$

Example 13.3 *Compute the probability that the output power of a selection diversity system is 5 dB lower than the mean power of each branch, when using $N_r = 1, 2, 4$ antennas.*

The threshold is $\gamma_{|dB} = \bar{\gamma}_{|dB} - 5\,dB$, which in linear scale is $\gamma = \bar{\gamma} \cdot 10^{-0.5}$. Using Eq. (13.10), the probability that the output power is less than $\bar{\gamma} \cdot 10^{-0.5}$ becomes:

$$cdf_{\gamma}\left(\bar{\gamma} \cdot 10^{-0.5}\right) = \left[1 - \exp\left(-10^{-0.5}\right)\right]^{N_r}$$

$$= \begin{cases} 0.27, & N_r = 1 \\ 7.4 \cdot 10^{-2}, & N_r = 2 \\ 5.4 \cdot 10^{-3}, & N_r = 4 \end{cases}$$

Example 13.4 *Consider now the case that $N_r = 2$, and that the mean powers in the branches are $1.5\bar{\gamma}$ and $0.5\bar{\gamma}$, respectively. How does the result change?*

In this case, the probability is:

$$cdf_{\gamma}\left(\bar{\gamma} \cdot 10^{-0.5}\right) = \left[1 - \exp\left(-\frac{1}{1.5} \cdot 10^{-0.5}\right)\right]\left[1 - \exp\left(-\frac{1}{0.5} \cdot 10^{-0.5}\right)\right] \tag{13.12}$$

$$= 8.9 \cdot 10^{-2} \tag{13.13}$$

This demonstrates that diversity is less efficient when the average branch powers are different.

Bit-error-rate-driven diversity

For BER-driven diversity, we first transmit a *training sequence* – i.e., a bit sequence that is known at the RX. The RX then demodulates the signal from each receive antenna element and compares it with the transmit signal. The antenna whose associated signal results in the smallest BER is judged to be the "best", and used for the subsequent reception of data signals. A similar approach is the use of the mean square error of the "soft-decision" demodulated signal, or the correlation between transmit and receive signal.

If the channel is time-variant, the training sequence has to be repeated at regular intervals and selection of the best antenna has to be done anew. The necessary repetition rate depends on the coherence time of the channel.

BER-driven diversity has several drawbacks:

1. The RX needs either N_r RF chains and demodulators (which makes the RX more complex), or the training sequence has to be repeated N_r times (which decreases spectral efficiency), so that the signal at all antenna elements can be evaluated.
2. If the RX has only one demodulator, then it is not possible to continuously monitor the selection criterion (i.e., the BER) of all diversity branches. This is especially critical if the channel changes quickly.
3. Since the duration of the training sequence is finite, the selection criterion – i.e., bit error probability – cannot be determined exactly. The variance of the BER around its true mean decreases as the duration of the training sequence increases. There is thus a tradeoff between performance loss due to erroneous determination of the selection criterion, and spectral efficiency loss due to longer training sequences.

13.4.2 Switched diversity

The main drawback of selection diversity is that the selection criteria (power, BER, etc.) of *all* diversity branches have to be monitored in order to know when to select a different antenna. As we have shown above, this leads to either increased hardware effort or reduced spectral efficiency. An alternative solution, which avoids these drawbacks, is *switched diversity*. In this method, the selection criterion of just the active diversity branch is monitored. If it falls below a certain threshold, then the RX switches to a different antenna.[10] Switching only depends on the quality of the active diversity branch; it does not matter whether the other branch actually provides a better signal quality or not.

Switched diversity works well for those cases where there is sufficient signal quality on at least one diversity branch. If both branches have signal quality below the threshold, then the RX just switches back and forth between the branches. This situation can be avoided by introducing a hysteresis or hold time, so that the new diversity branch is used for a certain amount of time, independent of the actual signal quality. We thus have two free parameters: switching threshold, and hysteresis time. These parameters have to be selected very carefully: if the threshold is chosen too low, then a diversity branch is used even when the other antenna might offer better quality; if it is chosen too high, then it becomes probable that the branch the RX switches to actually offers lower signal quality than the currently active one. If hysteresis time is chosen too long, then a "bad" diversity branch can be used for a long time; if it is chosen too short, then the RX spends all the time switching between two antennas.

Summarizing, the performance of switched diversity is worse than that of selection diversity; we will therefore not consider it further.

---

[10] The method is mostly applied when two diversity branches are available.

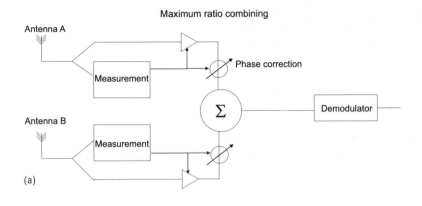

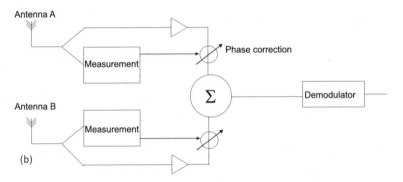

Figure 13.9 Combining diversity principle: (a) maximum ratio combining, (b) equal gain combining.

13.4.3 Combining diversity

Basic principle

Selection diversity wastes signal energy by discarding $(N_r - 1)$ copies of the received signal. This drawback is avoided by combining diversity, which exploits *all* available signal copies. Each signal copy is multiplied by a (complex) weight, and then added up. Each complex weight can be thought of as consisting of a phase correction, plus a (real) weight for the amplitude:

- Phase correction causes the *signal amplitudes* to add up, while, on the other hand, noise is added incoherently, so that *noise powers* add up.
- For amplitude weighting, two methods are widely used: *Maximum Ratio Combining* (MRC) weighs all signal copies by their amplitude. It can be shown that (using some assumptions) this is an optimum combination strategy. An alternative is *Equal Gain Combining* (EGC), where the signals are not weighted, but just undergo a phase correction. The two methods are outlined in Fig. 13.9.

Maximum ratio combining

MRC compensates for the phases, and weights the signals from the different antenna branches according to their SNR. This is the optimum way of combining different diversity branches –

if several assumptions are fulfilled. Let us assume a propagation channel that is slow-fading and flat-fading. The only disturbance is AWGN. Under these assumptions, each channel realization can be written as a time-invariant filter with impulse response:

$$h_n(\tau) = \alpha_n \delta(\tau) \tag{13.14}$$

where α_n is the (instantaneous) attenuation of diversity branch n. The antenna weights should then be chosen as:

$$w_{\mathrm{MRC}} = \alpha_n{}^* \tag{13.15}$$

i.e., the signals are phase-corrected and weighted by the amplitude. We can then easily see that the output SNR of the diversity combiner is the *sum* of the branch SNRs:

$$\gamma_{\mathrm{MRC}} = \sum_{n=1}^{N_r} \gamma_n \tag{13.16}$$

If the branches are statistically independent, then the moment-generating function of the total SNR can be computed as the product of the characteristic functions of the branch SNRs. If, furthermore, the SNR distribution in each branch is exponential (corresponding to Rayleigh fading), and all branches have the same mean SNR $\bar{\gamma}_i = \bar{\gamma}$, we find after some manipulations that:

$$pdf_\gamma(\gamma) = \frac{1}{(N_r - 1)!} \frac{\gamma^{N_r - 1}}{\bar{\gamma}^{N_r}} \exp\left(-\frac{\gamma}{\bar{\gamma}}\right) \tag{13.17}$$

and the mean SNR of the combiner output is just the mean branch SNR, multiplied by the number of diversity branches:

$$\bar{\gamma}_{\mathrm{MRC}} = N_r \bar{\gamma} \tag{13.18}$$

Figure 13.10 compares the statistics of the SNR for RSSI-driven selection diversity and MRC. Naturally, there is no difference between the diversity types for $N_r = 1$, since there is no diversity. We furthermore see that the slope of the distribution is the same for MRC and selection diversity, but that the difference in the mean values increases with increasing N_r. This is intuitively clear, as selection diversity discards $N_r - 1$ signal copies – something that increases with N_r. For $N_r = 3$, the difference between the two types of diversity is only about 2 dB.

Equal-gain combining

For EGC, we find that the SNR of the combiner output is:

$$\gamma_{\mathrm{EGC}} = \frac{\left(\sum_{n=1}^{N_r} \sqrt{\gamma_n}\right)^2}{N_r} \tag{13.19}$$

where we have assumed that noise levels are the same on all diversity branches. The mean SNR of the combiner output can be found to be:

$$\bar{\gamma}_{\mathrm{EGC}} = \bar{\gamma}\left(1 + (N_r - 1)\frac{\pi}{4}\right) \tag{13.20}$$

if all branches suffer from Rayleigh fading with the same mean SNR $\bar{\gamma}$. Remember that we only assume here that the *mean* SNR is the same in all branches, while instantaneous branch SNRs (representing different channel realizations) can be different. And in situations where branch SNRs are different, MRC should obviously perform better than EGC. It is thus quite remarkable that EGC performs worse than MRC by only a factor $\pi/4$ (in terms of mean

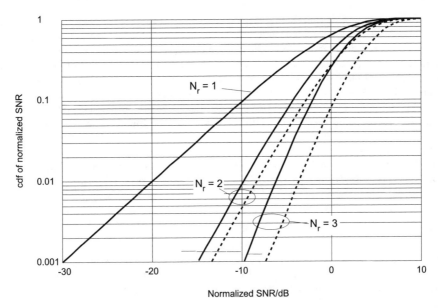

Figure 13.10 Cumulative distribution function of the normalized instantaneous signal-to-noise ratio $\gamma/\bar{\gamma}$ for received-signal-strength-indication-driven selection diversity (solid), and maximum ratio combining (dashed) for $N_r = 1, 2, 3$. Note that for $N_r = 1$, there is no difference between diversity types.

SNR). The performance difference between EGC and MRC becomes bigger when mean branch SNRs are also different.

Equations for the probability density function of the SNR can be computed from Eq. (13.19), but become unwieldy very quickly (results can be found, e.g., in Lee [1982]).

Optimum combining

One of the assumptions in the derivation of MRC was that only AWGN disturbs the signal. If it is interference that determines signal quality, then MRC is no longer the best solution. Weights should then be determined according to a strategy called *optimum combining*, first derived in the groundbreaking paper of Winters [1984]. The first step is the determination of the correlation matrix of noise and interference at the different antenna elements:

$$\mathbf{R} = \sigma_n^2 \mathbf{I} + \sum_{k=1}^{K} E\{\mathbf{r}_k^* \mathbf{r}_k^T\} \tag{13.21}$$

where expectation is over a time period when the channel remains constant, and $\mathbf{r}_k$ is the receive signal vector of the kth interferer. We furthermore need the complex transfer function (complex attenuation, since we assume flat-fading channels) of the N_r diversity branches; these are written into the vector $\mathbf{h}_d$. The vector containing the optimum receive weights is then:

$$\mathbf{w}_{\text{opt}} = \mathbf{R}^{-1} \mathbf{h}_d^* \tag{13.22}$$

These weights have to be adjusted as the channel changes. It is easy to see that for a noise-limited system the correlation matrix becomes a (scaled) identity matrix, and optimum combining reduces to MRC.

Optimum combining of signals from N_r diversity branches gives N_r degrees of freedom. This allows interference from $N_r - 1$ interferers to be eliminated. Alternatively, $N_s \leq N_r - 1$ interferers can be eliminated, while the remaining $N_r - N_s$ antennas behave like "normal" diversity antennas that can be used for noise reduction. This seemingly simple statement has great impact on wireless system design! As we discussed in Chapter 3, many wireless systems are interference-limited. The possibility of eliminating at least some of the interferers by appropriate diversity combining opens up the possibility of drastically improving the link quality of such systems, or of increasing their capacity (more details can be found in Section 20.1).

Example 13.5 *Consider a desired signal with Binary Phase Shift Keying (BPSK) propagating through a frequency-flat channel with $\mathbf{h}_d = [1, 0.5 + 0.7j]^T$ and a synchronuous, interfering BPSK signal with $\mathbf{h}_{int} = [0.3, -0.2 + 1.7j]^T$. The noise variance is $\sigma_n^2 = 0.01$. Show that weighting the received signal with the weights as computed from Eq. (13.22) leads to mitigation of interference.*

Let the desired transmitted signal be $s_d(t)$, the interfering transmitted signal be $s_{int}(t)$, and $n_1(t)$ and $n_2(t)$ be independent zero-mean noise processes. It is assumed that desired and interfering signals are uncorrelated, and that $E\{s_d s_d^*\} = 1$, and also $E\{s_{int} s_{int}^*\} = 1$.

The noise-plus-interference correlation matrix is computed according to Eq. (13.21):

$$\mathbf{R} = \sigma_n^2 \mathbf{I} + \sum_{k=1}^{K} E\{\mathbf{r}_k^* \mathbf{r}_k^T\} \tag{13.23}$$

$$= 0.01 \begin{pmatrix} 1 & 0 \\ 0 & 1 \end{pmatrix} + \begin{pmatrix} 0.3 \\ -0.2 - 1.7j \end{pmatrix} (0.3 \quad -0.2 + 1.7j) \tag{13.24}$$

$$= \begin{pmatrix} 0.1 & -0.06 + 0.51j \\ -0.06 - 0.51j & 2.94 \end{pmatrix} \tag{13.25}$$

Using Eq. (13.22) and normalizing the weights, we obtain:

$$\mathbf{w} = \frac{\mathbf{R}^{-1} \mathbf{h}_d^*}{|\mathbf{R}^{-1} \mathbf{h}_d^*|} \tag{13.26}$$

$$= \begin{pmatrix} 0.979 - 0.111j \\ 0.041 + 0.165j \end{pmatrix} \tag{13.27}$$

Determining now the inner product of these weights with the desired channel vector $\mathbf{h}_d$, we find $|\mathbf{w}\mathbf{h}_d|^2 = 0.782$, which indicates some attenuation of the desired signal, and thus worsening of the SNR. On the other hand, $|\mathbf{w}\mathbf{h}_{int}|^2 = 5 \cdot 10^{-5}$, which shows that interference has been completely suppressed.

Hybrid selection – maximum ratio combining

A compromise between selection diversity and full signal combining is the so-called hybrid selection scheme, where the best L out of N_r antenna signals are chosen, downconverted, and processed. This reduces the number of required RF chains from N_r to L, and thus leads to significant savings. The savings come at the price of a (usually small) performance loss compared with the full-complexity system. The approach is called *Hybrid Selection/Maximum Ratio Combining* (H-S/MRC), or sometimes also *Generalized Selection Combining* (GSC).

It is well known that the output SNR of MRC is just the sum of the SNRs at the different receive antenna elements. For H-S/MRC, the instantaneous output SNR of H-S/MRC looks deceptively similar to MRC – namely:

$$\gamma_{\text{H-S/MRC}} = \sum_{n=1}^{L} \gamma_{(n)} \tag{13.28}$$

The major difference from MRC is that the $\gamma_{(n)}$ are *ordered* SNRs – i.e., $\gamma_{(1)} > \gamma_{(2)} > \cdots > \gamma_{(N_r)}$. This leads to different performance, and poses new mathematical challenges for performance analysis. Specifically, we have to introduce the concept of *order statistics*. Note that selection diversity (where just one of N_r antennas is selected) and MRC are limiting cases of H-S/MRC with $L = 1$ and $L = N_r$, respectively.

H-S/MRC schemes provide good diversity gain, as they select the best antenna branches for combining. Actually, it can be shown that the diversity *order* obtained with such schemes is proportional to N_r, not to L. However, they do not provide full beamforming gain. If the signals at all antenna elements are completely correlated, then the SNR gain of H-S/MRC is only L, compared with N_r for an MRC scheme.

The analysis of H-S/MRC based on a chosen ordering of the branches at first appears to be complicated, since the SNR statistics of the ordered branches are *not* independent. However, we can alleviate this problem by transforming ordered branch variables into a new set of random variables. It is possible to find a transformation that leads to *independently distributed* random variables (termed *virtual branch variables*). The fact that the combiner output SNR can be expressed in terms of independent identically distributed (iid) virtual branch variables enormously simplifies performance analysis of the system. For example, the derivation of Symbol Error Probability (SEP) for uncoded H-S/MRC systems, which normally would require evaluation of nested N-fold integrals, essentially reduces to evaluation of a *single* integral with finite limits.

The mean and variance of the output SNR for H-S/MRC is thus:

$$\overline{\gamma}_{\text{H-S/MRC}} = L\left(1 + \sum_{n=L+1}^{N_r} \frac{1}{n}\right)\overline{\gamma} \tag{13.29}$$

and

$$\sigma_{\text{H-S/MRC}}^2 = L\left(1 + L\sum_{n=L+1}^{N_r} \frac{1}{n^2}\right)\overline{\gamma}^2 \tag{13.30}$$

13.5 Error probability in fading channels with diversity reception

In this section we determine the Symbol Error Rate (SER) in fading channels when diversity is used at the RX. We start with the case of flat-fading channels, computing the statistics of the received power and the BER. We then proceed to dispersive channels, where we analyze how diversity can mitigate the detrimental effects of dispersive channels on simple RXs.

13.5.1 Error probability in flat-fading channels

Classical computation method

Analogously to Chapter 12, we can compute the error probability of diversity systems by averaging the conditional error probability (conditioned on a certain SNR) over the distribution of the SNR:

$$\overline{SER} = \int_0^{\infty} pdf_\gamma(\gamma) SER(\gamma)\, d\gamma \tag{13.31}$$

As an example, let us compute the performance of BPSK with N_r diversity branches with MRC. The SER of BPSK in AWGN is (see Chapter 12):

$$SER(\gamma) = Q(\sqrt{2\gamma}) \tag{13.32}$$

Let us apply this principle to the case of MRC. When inserting Eqs. (13.17) and (13.32) into Eq. (13.31), we obtain an equation that can be evaluated analytically:

$$\overline{SER} = \left(\frac{1-b}{2}\right)^{N_r} \sum_{n=0}^{N_r-1} \binom{N_r - 1 + n}{n} \left(\frac{1+b}{2}\right)^n \tag{13.33}$$

where b is defined as:

$$b = \sqrt{\frac{\overline{\gamma}}{1 + \overline{\gamma}}} \tag{13.34}$$

For large values of $\overline{\gamma}$, this can be approximated as:

$$\overline{SER} = \left(\frac{1}{4\overline{\gamma}}\right)^{N_r} \binom{2N_r - 1}{N_r} \tag{13.35}$$

From this, we can see that (with N_r diversity antennas) the BER decreases with the N_r-th power of the SNR.

Computation via the moment-generating function

In the previous section, we averaged the BER over the distribution of SNRs, using the "classical" representation of the Q-function. As we have already seen (Chapter 12), there is an alternative definition of the Q-function, which can easily be combined with the moment-generating function $M_\gamma(s)$ of the SNR. Let us start by writing the SER conditioned on a given SNR in the form (see Chapter 12):

$$SER(\gamma) = \int_{\theta_1}^{\theta_2} f_1(\theta) \exp(-\gamma_{\mathrm{MRC}} f_2(\theta)) \, d\theta \tag{13.36}$$

Since

$$\gamma_{\mathrm{MRC}} = \sum_{n=1}^{N_r} \gamma_n \tag{13.37}$$

this can be rewritten as:

$$SER(\gamma) = \int_{\theta_1}^{\theta_2} f_1(\theta) \prod_{n=1}^{N_r} \exp(-\gamma_n f_2(\theta)) \, d\theta \tag{13.38}$$

Averaging over the SNRs in different branches then becomes:

$$\overline{SER} = \int d\gamma_1 pdf_{\gamma_1}(\gamma_1) \int d\gamma_2 pdf_{\gamma_2}(\gamma_2) \cdots \int d\gamma_{N_r} pdf_{\gamma_{N_r}}(\gamma_{N_r}) \int_{\theta_1}^{\theta_2} d\theta f_1(\theta) \prod_{n=1}^{N_r} \exp(-\gamma_n f_2(\theta)) \tag{13.39}$$

$$= \int_{\theta_1}^{\theta_2} d\theta f_1(\theta) \prod_{n=1}^{N_r} \int d\gamma_n pdf_{\gamma_n}(\gamma_n) \exp(-\gamma_n f_2(\theta)) \tag{13.40}$$

$$= \int_{\theta_1}^{\theta_2} d\theta f_1(\theta) \prod_{n=1}^{N_r} M_\gamma(-f_2(\theta)) \tag{13.41}$$

$$= \int_{\theta_1}^{\theta_2} d\theta f_1(\theta) \left[M_\gamma(-f_2(\theta))\right]^{N_r} \tag{13.42}$$

With that, we can write the error probability for BPSK in Rayleigh fading as:

$$\overline{SER} = \frac{1}{\pi} \int_0^{\pi/2} \left[\frac{\sin^2(\theta)}{\sin^2(\theta) + \overline{\gamma}} \right]^{N_r} d\theta \tag{13.43}$$

Example 13.6 *Compare the error probability of 8-PSK (Phase Shift Keying) and four available antennas with 10-dB SNRs when using $L = 1, 2, 4$ receive chains.*

The SER for M-ary Phase Shift Keying (MPSK) with H-S/MRC can be shown to be:

$$\overline{SER}_{e,\text{H-S/MRC}}^{\text{MPSK}} = \frac{1}{\pi} \int_0^{\pi(M-1)/M} \left[\frac{\sin^2 \theta}{\sin^2(\pi/M)\overline{\gamma} + \sin^2 \theta} \right]^L \prod_{n=L+1}^{N_r} \left[\frac{\sin^2 \theta}{\sin^2(\pi/M)\overline{\gamma}\frac{L}{n} + \sin^2 \theta} \right] d\theta \tag{13.44}$$

Evaluating Eq. (13.44) with $M = 8$, $\overline{\gamma} = 10\,\text{dB}$, $N_r = 4$, and $L = 1, 2, 4$ yields:

L	$\overline{SER}_{e,\text{H-S/MRC}}^{\text{MPSK}}$
1	0.0442
2	0.0168
4	0.0090

13.5.2 Symbol error rate in frequency-selective fading channels

We now determine the BER in channels that suffer from time dispersion and frequency dispersion. We assume here FSK with differential phase detection. The analysis uses the correlation coefficient ρ_{XY} between signals at two sampling times that was discussed in Chapter 12.

For binary FSK with selection diversity:

$$\overline{SER} = \frac{1}{2} - \frac{1}{2} \sum_{n=1}^{N_r} \binom{N_r}{n} (-1)^{n+1} \frac{b_0 \,\text{Im}\{\rho_{XY}\}}{\sqrt{(\text{Im}\{\rho_{XY}\})^2 + n\left(1 - |\rho_{XY}|^2\right)}} \tag{13.45}$$

where b_0 is the transmitted bit. This can be approximated as:

$$\overline{SER} = \frac{(2N_r - 1)!!}{2} \left(\frac{1 - |\rho_{XY}|^2}{2(\text{Im}\{\rho_{XY}\})^2} \right)^{N_r} \tag{13.46}$$

where $(2N_r - 1)!! = 1 \cdot 3 \cdot 5 \cdots (2N_r - 1)$.

For binary FSK with MRC:

$$\overline{SER} = \frac{1}{2} - \frac{1}{2} \frac{b_0 \,\text{Im}\{\rho_{XY}\}}{\sqrt{1 - (\text{Re}\{\rho_{XY}\})^2}} \sum_{n=0}^{N_r - 1} \frac{(2n - 1)!!}{(2n)!!} \left(1 - \frac{(\text{Im}\{\rho_{XY}\})^2}{1 - (\text{Re}\{\rho_{XY}\})^2} \right)^n \tag{13.47}$$

which can be approximated as:

$$\overline{SER} = \frac{(2N_r - 1)!!}{2(N_r!)} \left(\frac{1 - |\rho_{XY}|^2}{2(\text{Im}\{\rho_{XY}\})^2} \right)^{N_r} \tag{13.48}$$

This formulation shows that MRC improves the SER by a factor $N_r!$ compared with selection diversity. A further important consequence is that the errors due to delay dispersion and random Frequency Modulation (FM) are decreased in the same way as errors due to noise. This is

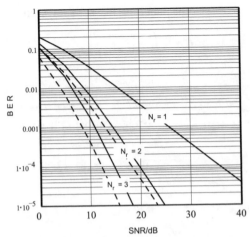

Figure 13.11 Bit error rate of MSK with received-signal-strength-indication-driven selection diversity (solid) and maximum ratio combining (dashed) as a function of the signal-to-noise ratio with M diversity antennas.
Reproduced with permission from Molisch [2000] © Prentice Hall.

shown by the expressions in parentheses that are taken to the N_r-th power. These terms subsume the errors due to all different effects. The SER with diversity is approximately the N_r-th power of the SER without diversity (see Figs. 13.11–13.13).

For Differential Quadrature Phase Shift Keying (DQPSK) with selection diversity, the average BER is:

$$\overline{BER} = \frac{1}{2} - \frac{1}{4} \sum_{n=1}^{N_r} \binom{N_r}{n} (-1)^{n+1} \left[\frac{b_0 \operatorname{Re}\{\rho_{XY}\}}{\sqrt{(\operatorname{Re}\{\rho_{XY}\})^2 + n\left(1 - |\rho_{XY}|^2\right)}} + \frac{b_0' \operatorname{Im}\{\rho_{XY}\}}{\sqrt{(\operatorname{Im}\{\rho_{XY}\})^2 + n\left(1 - |\rho_{XY}|^2\right)}} \right]$$

(13.49)

where (b_0, b_0') is the transmitted symbol.

For DQPSK with MRC:

$$\overline{BER} = \frac{1}{2} - \frac{1}{4} \sum_{n=0}^{N_r-1} \frac{(2n-1)!!}{(2n)!!} \left[\frac{b_0 \operatorname{Re}\{\rho_{XY}\}}{\sqrt{1 - (\operatorname{Im}\{\rho_{XY}\})^2}} \left(\frac{1 - |\rho_{XY}|^2}{1 - (\operatorname{Im}\{\rho_{XY}\})^2} \right)^n \right.$$

$$\left. + \frac{b_0' \operatorname{Im}\{\rho_{XY}\}}{\sqrt{1 - (\operatorname{Re}\{\rho_{XY}\})^2}} \left(\frac{1 - |\rho_{XY}|^2}{1 - (\operatorname{Re}\{\rho_{XY}\})^2} \right)^n \right]$$

(13.50)

More general cases can be treated with the QFGV method (see Section 12.3.3), where Eq. (12.105) is replaced by:

$$P(D < 0) = Q_M(p_1, p_2) - I_0(p_1 p_2) \exp[-\tfrac{1}{2}(p_1^2 + p_2^2)] + \frac{I_0(p_1 p_2) \exp[-\tfrac{1}{2}(p_1^2 + p_2^2)]}{(1 + v_2/v_1)^{2N_r - 1}} \sum_{n=0}^{N_r-1} \binom{2N_r - 1}{n} \left(\frac{v_2}{v_1}\right)^n$$

$$+ \frac{\exp[-\tfrac{1}{2}(p_1^2 + p_2^2)]}{(1 + v_2/v_1)^{2N_r - 1}} \cdot \sum_{n=1}^{N_r-1} I_n(p_1 p_2) \sum_{k=0}^{N_r-1-n} \binom{2N_r - 1}{k} \left[\left(\frac{p_2}{p_1}\right)^n \left(\frac{v_2}{v_1}\right)^k - \left(\frac{p_1}{p_2}\right)^n \left(\frac{v_2}{v_1}\right)^{2N_r-1-k} \right]$$

(13.51)

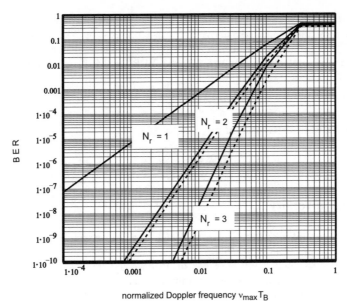

normalized Doppler frequency $v_{max}T_B$

Figure 13.12 Bit error rate of minimum shift keying with received-signal-strength-indication–driven selection diversity (solid) and maximum ratio combining (dashed) as a function of the normalized Doppler frequency with N_r diversity antennas.

Reproduced with permission from Molisch [2000] © Prentice Hall.

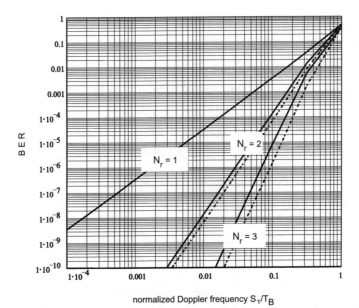

normalized Doppler frequency S_T/T_B

Figure 13.13 Bit error rate of minimum shift keying with received-signal-strength-indication-driven selection diversity (solid) and maximum ratio combining (dashed) as a function of the normalized rms delay spread with N_r diversity antennas.

Reproduced with permission from Molisch [2000] © Prentice Hall.

RSSI-driven diversity is not the best selection strategy when errors are mostly caused by frequency selectivity and time selectivity. It puts emphasis on signals that have a large amplitude, and not on those with the smallest distortion.[11] In these cases, BER-driven selection diversity is preferable. For $N_r = 2$, the BER of MSK with differential detection becomes:

$$\overline{BER} \approx \left(\frac{\pi}{4}\right)^4 \left(\frac{S_\tau}{T_B}\right)^4 \tag{13.52}$$

compared with the RSSI-driven result:

$$\overline{BER} \approx 3\left(\frac{\pi}{4}\right)^4 \left(\frac{S_\tau}{T_B}\right)^4 \tag{13.53}$$

13.6 Transmit diversity

For many situations, multiple antennas can be installed at just one link end (usually the BS). For the uplink transmission from the MS to BS, multiple antennas can act as receive diversity branches. For the downlink, any possible diversity originates at the *transmitter*. In this section, we will thus discuss ways of transmitting signals from several TX antennas and achieve a diversity effect with it. Note that this section specifically refers to antenna diversity (which encompasses spatial diversity, pattern diversity, and polarization diversity). Time diversity and frequency diversity inherently involve the TX, and thus need not be discussed again here.

The question of how transmit diversity can be combined with multiple antenna elements at the RX is discussed in Chapter 20.

13.6.1 Transmitter diversity with channel-state information

The first situation we analyze is the case where the TX knows the channel perfectly. This knowledge might be obtained from feedback from the RX, or from reciprocity principles; a more detailed discussion of this issue can be found in Section 20.1.6. In this case, we find that (at least for the noise-limited case) there is a complete equivalence between transmit diversity and receive diversity. In other words, the optimum transmission scheme linearly weights signals transmitted from different antenna elements with the complex conjugates of the channel transfer functions from the transmit antenna elements to the single receive antenna. This approach is known as *maximum ratio transmission*.

13.6.2 Transmitter diversity without channel-state information

In many cases, Channel State Information (CSI) is not available at the TX. We then cannot simply transmit weighted copies of the same signal from different transmit antennas, because we cannot know how they would add up at the RX. It is equally likely for the addition of different components to be constructive or destructive; in other words, we would just be adding up MPCs with random phases, which results in Rayleigh fading. We thus cannot gain any diversity (or beamforming).

In order to give benefits, transmission of the signals from different antenna elements has to be done in such a way that it allows the RX to distinguish different transmitted signal components. One way is *delay diversity*. In this scheme, signals transmitted from different antenna elements are delayed copies of the same signal. This makes sure that the effective impulse response is

---

[11] For a similar reason, MRC is not the best combining strategy.

delay-dispersive, even if the channel itself is flat-fading. So, in a flat-fading channel, we transmit datastreams with a delay of 1 symbol duration (relative to preceding antennas) from each of the transmit antennas. The effective impulse response of the channel then becomes:

$$h(\tau) = \frac{1}{\sqrt{N_r}} \sum_{n=1}^{N_t} h_n \delta(\tau - nT_S) \tag{13.54}$$

where the h_n are attenuations from the nth transmit antenna to the receive antenna, and the impulse response has been normalized so that total transmit power is independent of the number of antenna elements. The signals from different transmit antennas to the RX act effectively as delayed multipath components. If antenna elements are spaced sufficiently far apart, these coefficients fade independently. With an appropriate RX for delay-dispersive channels – e.g., an equalizer as described in Chapter 16, or a Rake receiver as described in Chapter 18 – we get a diversity order that is equal to the number of antenna elements.

If the channel from a single transmit antenna to the RX is already delay-dispersive, then the scheme still works, but care has to be taken in the choice of delays for different antenna elements. The delay between signals transmitted from different antenna elements should be at least as large as the maximum excess delay of the channel.

An alternative method is *phase-sweeping diversity*. In this method, which is especially useful if there are only two antenna elements, the same signal is transmitted from both antenna elements. However, one of the antenna signals undergoes a time-varying phaseshift. This means that at the RX the received signals add up in a time-varying way; in other words, we are artificially introducing temporal variations into the channel. The reason for this is that – even if the TX, RX, and the IOs are stationary – the signal does not remain stuck in a fading dip. If this scheme is combined with appropriate coding and/or interleaving, it improves performance.

Yet another possibility for achieving transmit diversity is *space–time coding*. This method will be discussed in Chapter 20.

13.7 Appendix

Please go to *www.wiley.com/go/molisch*

Further reading

Spatial (antenna) diversity is the oldest form of diversity, and is discussed in textbooks on wireless propagation channels (see, e.g., Vaughan and Anderson [2003]). Evaluations of the antenna correlation coefficient for different angular spectra can be found in Fuhl et al. [1998] and Roy and Fortier [2004]. For antennas on a handset, results are given in Ogawa et al. [2001]. Taga [1993] also shows the effect of mutual coupling on pattern diversity and thus the correlation coefficient, as do a number of recent papers written in the context of MIMO systems [Waldschmidt et al. 2004 and Wallace and Jensen 2004]. Polarization diversity is discussed in more detail, for example, in Narayanan et al. [2004] and Shafi et al. [2005]; joint spatial, polarization, and pattern (angle) diversity is discussed in Dietrich et al. [2001]. Different combination strategies for antenna signals, and the resulting channel statistics, are discussed in Proakis [1995], and in many papers in the primary literature. The situation is similar to that mentioned in Chapter 12: each type of fading statistics, and each combination strategy merits at least one paper. Application of these fading statistics to computation of BERs gives rise to an even greater variety of papers. A nice, unified treatment can be found in Simon and Alouini

[2000], who include different fading statistics and antenna combination strategies for a variety of modulation formats. For Rayleigh or Rice fading, the classical method of Proakis [1968] is applicable; Adachi and Ohno [1991] and Adachi and Parsons [1989] compute the performance for DQPSK and FSK, respectively. The virtual path method for HS-MRC was introduced in Win and Winters [1999]; Simon and Alouini [2000] also investigate this case. General mathematical aspects of order statistics can be found in David and Nagaraja [2003]. Discussions on the impact of diversity on systems in frequency-selective channels can be found in Molisch [2000, ch. 13]. Maximum ratio transmission was first suggested in Lo [1999]. Delay diversity was suggested by Winters [1994] and Wittneben [1993]. An overview of various transmit diversity techniques can be found in Hottinen et al. [2003].

14

Channel coding

14.1 Introduction

14.1.1 History and motivation

Chapter 12 demonstrated that Bit Error Rates (BERs) on the order of 10^{-2} can occur for Signal-to-Noise Ratios (SNRs) typically encountered in wireless systems. These high BERs are mostly due to the effect of multipath propagation. Advanced receiver structures can help to reduce these values: diversity combats fading dips, while equalizers and Rake receivers (see Chapters 16 and 18) improve performance in frequency-selective channels. However, even these advanced receivers might not sufficiently reduce the BER. Data communications often require BERs on the order of 10^{-6}–10^{-9}. Such low values can only be achieved by employing channel coding – i.e., introducing redundancy into the transmission. The use of error-correcting codes[1] leads to a reduction of the BER, or equivalently to a coding gain G_c – i.e., we have to use G_c dB less transmit power to achieve the target BER than in an uncoded system.[2]

The history of coding starts with the seminal work of Claude Shannon [Shannon 1948] on the mathematical theory of communication. He showed that it is possible to transmit data without errors as long as the bit rate is smaller than the *channel capacity*. The absence of errors is achieved by the use of "appropriate" codes. Shannon showed that (infinitely long) random codes achieve capacity. Unfortunately, such codes cannot be used in practice due to the enormous effort required for their decoding. For more than 50 years, the work of coding theorists mainly consisted in finding practical codes that come close to the Shannon limit – i.e., allow almost error-free communications with rates close to channel capacity. The last decade has finally seen these efforts come to fruition: *turbocodes* and *Low Density Parity Check* (LDPC) codes approach the Shannon limit within less than 1 dB.

One of the fundamental results of Shannon is that error-free communications is possible if the SNR for each data bit exceeds -1.6 dB. However, this also requires an extremely high bandwidth, and is thus often unsuitable for wireless applications. On a more general note, we find that not only is coding gain an important quantity in wireless systems, but so is the code rate, which in turn is related to spectral efficiency.

---

[1] In the remainder of this chapter, we will use *coding* when we mean *error-correcting coding*. Note that this is different from Chapter 15, where coding will refer to source coding.
[2] Note that coding gain can depend on the target error rate. Coding leads to a change in shape of the BER-over-SNR curve.

In this chapter, we will give a brief overview of error-correcting coding. The basic coding theory holds for temporally constant and temporally varying channels; Sections 14.2–14.6 thus do not distinguish between these cases. Rather, they lay out the theoretical background of the most important classes of codes and their decoding: block codes, convolutional codes, trellis-coded modulation, turbocodes, and LDPC codes. Section 14.7 then deals specifically with the idiosynchrasies of fading channels, and describes how coding structures need to be adapted for this case.

14.1.2 Classification of codes

One way of classifying codes is to distinguish between *block codes*, where redundancy is added to blocks of data, and *convolutional codes*, where redundancy is added continuously. Block codes are well-suited for correcting burst errors – something that frequently occurs in wireless communications; however, error bursts can also be converted to random errors by interleaving techniques. Convolutional codes have the advantage that they are easily decoded by means of a Viterbi decoder. They also offer the possibility of joint decoding and equalization by means of the same algorithm. Turbocodes and LDPC codes easily fit into this categorization. As we will find in Section 14.5, turbocodes use two parallel, interleaved, convolutional codes in order to encode information; however, the decoding structures employed are different because of the length of memory of the encoder. A similar thing can be said about LDPC codes: as their name implies, they are block codes, but their large size necessitates different decoding structures; for this reason, they are treated in a separate section.

Research in recent years has actually led to a "smearing" of the boundary between block codes and convolutional codes. It has been shown that Viterbi decoders (the classical solution for convolutional codes) can be used to detect block codes [Wolf 1978], while belief propagation algorithms [Loeliger 2004] (commonly used for decoding block codes) can be generalized to decode convolutional codes. However, these are very advanced research topics, and will not be treated further here. The interested reader is referred to the cited papers.

Another classification of codes can be based on whether coder input and/or output are *hard* or *soft* information. Hard information just tells us the (binary) value that we detect for a bit. Soft information also tells us the confidence we have in our decision for this bit.

Example 14.1   *Soft and hard information for repetition coding.*

Let us consider a simple example: a repetition code that repeats a bit three times. Let the transmitted bit sequence be 1 1 1. Let the demodulated signal at the receiver be −0.05, −0.1, 1.0. One strategy is now to make a hard decision on each bit before decoding. After the decision, the bits are then −1 −1 1. The decoder performs a majority decision on these hard decoder inputs, and decides that −1 was transmitted. A soft decoder, on the other hand, might add up the demodulated signals,[3] giving 0.85 as a soft decoder input, a hard decision based on this value concludes that +1 was transmitted.

For some applications (e.g., iterative and concatenated decoders), it is necessary that the decoder does not only use soft input information, but also that it *puts out* soft information, and not just hard bits. In any case, the use of soft input information always leads to a performance improvement compared with the use of hard input information.

---

[3] This is purely for demonstration purposes. We will discuss more intelligent soft combining strategies later on.

14.2 Block codes

14.2.1 Introduction

Block codes are codes that group the source data into blocks, and – from the values of the bits in that block – compute a longer codeword that is actually transmitted. The smaller the code rate – i.e., the ratio of the number of bits in the original datablock to that of the transmitted block – the higher the redundancy, and the higher the probability that errors can be corrected. The most simple codes are repetition codes with blocksize 1: for an input "block" x, the output block is xxx (for a repetition code with rate $1/3$).

After this intuitive introduction, let us now give a more precise description. First, we define some important terms and notation:

- For block coding, source data are parsed into blocks of K symbols. Each of these uncoded datablocks is then associated with a codeword of length N symbols.
- The ratio K/N is called the *code rate* R_c (assuming the symbol alphabet of coded and uncoded data is the same).
- *Binary codes* occur when the symbol alphabet is binary, using only "0" and "1". Almost all practical block codes are binary, with the exception of *Reed–Solomon (RS) codes* (see below). If not stated otherwise, the remainder of the chapter always talks about binary codes. Therefore, in the following, "sum" means "modulo-2 sum", and "+" denotes a modulo-2 addition.
- The *Hamming distance* $d_H(\mathbf{x}, \mathbf{y})$ between two codewords is the number of different bits:

$$d_H(\mathbf{x}, \mathbf{y}) = \sum_n |x_n - y_n| \qquad (14.1)$$

where $\mathbf{x}, \mathbf{y}$ are codevectors, $\mathbf{x} = [x_1 \ x_2 \ \cdots \ x_N]$, $x_n \in \{0, 1\}$. For example, the Hamming distance between the codewords 01001011 and 11101011 is 2. Note that it is common in coding theory to use row vectors instead of the column vectors commonly used in communication theory. In order to simplify cross-referencing to coding books, we follow this established notation in this chapter.

- The *Euclidean distance* between two codewords is the geometric distance between the codevectors $\mathbf{x}$ and $\mathbf{y}$:

$$d_E^2(\mathbf{x}, \mathbf{y}) = \sum_n |x_n - y_n|^2 \qquad (14.2)$$

- *Minimum distance*: the minimum distance $d_{\min}$ of a code is the minimum Hamming distance $\min(d_H)$, where the minimum is taken over all possible combinations of two codewords of the code. Note that this minimum distance is equal to the number of linearly independent columns in the parity check matrix (see below).
- *Weight*: the weight of a codeword is the distance from the origin – i.e., the number of 1's in the codeword. For example, the weight of codeword 01001011 is 4.
- *Systematic codes*: in a systematic code, the information-bearing bits and the parity check (redundant) bits are at fixed locations. For transmission over an ideal (noise-free, non-distorting channel), the codeword could be determined without any information from the parity check bits. A systematic $(7, 4)$ block code is created in the following way:

k	k	k	k	m	m	m	m

where k represents information symbols and m represents parity check symbols.

- *Linear codes (group codes)*: for these codes, the sum of any two codewords gives another valid codeword. Important properties can be derived from this basic fact:
  - (i) the all-zero word is a valid codeword;
  - (ii) all codewords (except the all-zero word) have a weight equal to or larger than d_{min};
  - (iii) the distribution of distances – i.e., the Hamming distances between valid codewords – is equal to the weight distribution of the code;
  - (iv) all codewords can be represented by a linear combination of basic codewords (*generator words*).
- *Cyclic codes* (Section 14.2.6): cyclic codes are a special case of linear codes, with the property that any cyclic shift of a codeword results in another valid codeword. Cyclic codes can be interpreted either by codevectors, or by *polynomials* of degree $\leq N - 1$. The nonzero coefficients correspond to the nonzero entries of the codevector; the variable x is a dummy variable.

  As an example, we show both representations of the codeword 011010:

$$\mathbf{x} = [0 \quad 1 \quad 1 \quad 0 \quad 1 \quad 0] \tag{14.3}$$

$$X(x) = 0 \cdot x^5 + 1 \cdot x^4 + 1 \cdot x^3 + 0 \cdot x^2 + 1 \cdot x^1 + 0 \cdot x^0 \tag{14.4}$$

- *Galois fields*: A Galois field $GF(p)$ is a finite field with p elements, where p is a prime integer. A field defines addition and multiplication for operating on elements, and it is closed under these operations (i.e., the sum of two elements is again a valid element, and similar for the product); it contains identity and inverse elements for the two operations; and the associative, commutative, and distributive laws apply. The most important example is $GF(2)$. It consists of the elements 0 and 1 and is the smallest finite field. Its addition and multiplication tables are as follows:

+	0	1		×	0	1
0	0	1		0	0	0
1	1	0		1	0	1

  Codes often use $GF(2)$ because it is easily represented on a computer by a single bit. It is also possible to define extension fields $GF(p^m)$, where again p is a prime integer, and m is an arbitrary integer.
- *Primitive polynomials*: we define as *irreducible* a polynomial of degree N that is not divisible by any polynomial of degree less than N and greater than 0. A primitive polynomial $g(x)$ of degree m is defined as primitive if it is an irreducible polynomial such that the smallest integer N for which $g(x)$ divides $(x^N + 1)$ is $N = 2^m - 1$.

14.2.2 Encoding

The most straightforward encoding is a mapping table: any K-valued information word is associated with an N-valued codeword; the table just checks the input, and reads out the associated codeword. However, this method is highly inefficient: it requires the storing of 2^K codewords. For linear codes, any codeword can be created by a linear combination of other codewords. For a K-valued information word, only K out of the 2^K codewords are linearly independent, and thus have to be stored. It is advantageous to select those codewords that have only a single 1 in the first K positions. This choice automatically leads to a systematic code. The encoding process can then best be described by:

$$\mathbf{x} = \mathbf{uG} \tag{14.5}$$

Here, $\mathbf{x}$ denotes the N-dimensional codevector, $\mathbf{u}$ the K-dimensional information vector, and $\mathbf{G}$ the $K \times N$-dimensional generator matrix. For a systematic code, the leftmost K columns of the generator matrix are a $K \times K$ identity matrix, while the right $N - K$ columns denote the parity check bits. The first K bits of **x** are identical to **u**. Note that – as discussed above – we use *row* vectors to represent codewords, and that a vector–matrix product is obtained by premultiplying the matrix with this row vector.

Example 14.2 *Encoding with a (7,4) Hamming code.*

To make things more concrete, let us now consider an example of a $(7,4)$ code. We encode the sourceword [1011] with a generator matrix:

$$\mathbf{G} = \begin{bmatrix} 1 & 0 & 0 & 0 & 1 & 1 & 0 \\ 0 & 1 & 0 & 0 & 1 & 0 & 1 \\ 0 & 0 & 1 & 0 & 0 & 1 & 1 \\ 0 & 0 & 0 & 1 & 1 & 1 & 1 \end{bmatrix} \tag{14.6}$$

Computing $\mathbf{x} = \mathbf{uG}$:

$$\mathbf{x} = \begin{bmatrix} 1 & 0 & 1 & 1 \end{bmatrix} \begin{bmatrix} 1 & 0 & 0 & 0 & 1 & 1 & 0 \\ 0 & 1 & 0 & 0 & 1 & 0 & 1 \\ 0 & 0 & 1 & 0 & 0 & 1 & 1 \\ 0 & 0 & 0 & 1 & 1 & 1 & 1 \end{bmatrix} \tag{14.7}$$

the codeword becomes:

$$\mathbf{x} = \begin{bmatrix} 1 & 0 & 1 & 1 & 0 & 1 & 0 \end{bmatrix} \tag{14.8}$$

14.2.3 Decoding

In order to decide whether the received codeword is a valid codeword, we multiply it by a *parity check matrix* **H**. This results in a $N - K$-dimensional *syndrome vector* $\mathbf{s}_{\text{synd}}$. If this vector has all-zero entries, then the received codeword is valid.

Example 14.3 *Syndrome for (7,4) Hamming code.*

Let us demonstrate the computation of the syndrome with our above example. Let the received bit sequence (after hard decision) be:

$$\hat{\mathbf{x}} = \begin{bmatrix} 1 & 0 & 0 & 0 & 1 & 0 & 1 \end{bmatrix} \tag{14.9}$$

A parity check matrix for the code of Example 14.2 is:

$$\mathbf{H} = \begin{bmatrix} 1 & 1 & 0 & 1 & 1 & 0 & 0 \\ 1 & 0 & 1 & 1 & 0 & 1 & 0 \\ 0 & 1 & 1 & 1 & 0 & 0 & 1 \end{bmatrix} \tag{14.10}$$

Below, we will describe a constructive method for obtaining **H** from **G**.

Let us now compute the expression $\mathbf{s}_{\text{synd}} = \hat{\mathbf{x}}\mathbf{H}^T$. Writing this expression component by component, we obtain three parity check equations, corresponding to the three parity check bits:

$$\mathbf{s}_{\text{synd}}^T = \hat{\mathbf{x}}\mathbf{H}^T = \begin{bmatrix} 0 & 1 & 1 \end{bmatrix} \tag{14.11}$$

The computation of the syndrome can thus be interpreted the following way:

- Received information: 1000.
- Computed parity check bits: 110 (parity of the transmitted codeword).
- Received parity check bits: 101 (parity of the received codeword).
- $\implies$ syndrome: 011.

Next, let us determine how to find a **H**-matrix. The relationship $\mathbf{H} \cdot \mathbf{G}^T = \mathbf{0}$ has to hold, as each row of the generator matrix is a valid codeword, whose product with the parity check matrix has to be 0.[4] Representing **G** as:

$$\mathbf{G} = (\mathbf{I} \quad \mathbf{P}) \tag{14.12}$$

the relationship $\mathbf{H} \cdot \mathbf{G}^T = \mathbf{0}$ reduces to:

$$\mathbf{H} \cdot \mathbf{G}^T = (\mathbf{H}_1 \quad \mathbf{H}_2) \begin{pmatrix} \mathbf{I} \\ \mathbf{P}^T \end{pmatrix} = (\mathbf{H}_1 + \mathbf{H}_2 \mathbf{P}^T) = \mathbf{0} \tag{14.13}$$

If we now select $\mathbf{H}_2 = \mathbf{I}$ and $\mathbf{H}_1 = -\mathbf{P}^T$ then the equation is certainly fulfilled: the subtraction of two identical matrices is the all-zero matrix. Note that different parity check matrices can exist for each generator matrix. We have described here just one possibility.

Example 14.4 *Computation of parity check matrix for Hamming code.*

The procedure for computing the parity check matrix for the $(7, 4)$ Hamming code (and actually for any systematic code) can also be described the following way: start with the $N - K$ rightmost columns of the generator matrix:

$$\begin{bmatrix} 1 & 1 & 0 \\ 1 & 0 & 1 \\ 0 & 1 & 1 \\ 1 & 1 & 1 \end{bmatrix} \tag{14.14}$$

and transpose them; note that in modulo-2 arithmetic, there is no difference between addition and subtraction. Appending now an $(N - K) \times (N - K)$ identity matrix gives **H**:

$$\mathbf{H} = \begin{bmatrix} 1 & 1 & 0 & 1 & 1 & 0 & 0 \\ 1 & 0 & 1 & 1 & 0 & 1 & 0 \\ 0 & 1 & 1 & 1 & 0 & 0 & 1 \end{bmatrix} \tag{14.15}$$

A well-known class of linear codes are the Hamming codes. A Hamming code can be defined most easily through its parity check matrix. The columns of **H** contain all possible 2^{N-K} bit combinations of length K, with the exception of the all-zero word. Consequently, all columns of **H** are distinct. Note that the parity check matrix of Eq. (14.15) fulfills the condition, as can be easily verified by the reader. The size of a Hamming code is $(2^m - 1, 2^m - 1 - m)$, where m is an integer.

14.2.4 Recognition and correction of errors

Due to the linear properties of the code, each received word can be interpreted as the sum of a codeword **x** and an error word ε. This implies that the syndrome depends only on the error

---

[4] In other words, **G** is the codespace, and **H** is the associated nullspace.

word, not on the transmitted codeword. If a word $\hat{\mathbf{x}}$ is received, then the receiver computes the syndrome vector:

$$\mathbf{s}_{\text{synd}} = \hat{\mathbf{x}}\mathbf{H}^T = (\mathbf{x} + \boldsymbol{\varepsilon})\mathbf{H}^T = \mathbf{x}\mathbf{H}^T + \boldsymbol{\varepsilon}\mathbf{H}^T = \mathbf{0} + \boldsymbol{\varepsilon}\mathbf{H}^T = \boldsymbol{\varepsilon}\mathbf{H}^T \qquad (14.16)$$

An error $(\boldsymbol{\varepsilon} \neq 0)$ is indicated by a nonzero syndrome. Decoding requires that we find the "correct" $\boldsymbol{\varepsilon}$ – by subtracting $\boldsymbol{\varepsilon}$ from $\hat{\mathbf{x}}$, we obtain the correct codeword. On the other hand, it is clear that a number of different error words $\boldsymbol{\varepsilon}$ can lead to the same syndrome; these error words are called a *coset*. The goal is thus to find the most probable $\boldsymbol{\varepsilon}$ corresponding to the observed syndrome. In a "reasonable" channel, the probability that a bit arrives correctly is higher than the probability that a bit arrives with an error. Therefore, the $\boldsymbol{\varepsilon}$ that has the minimum weight is the most probable $\boldsymbol{\varepsilon}$ in a given coset.

A different interpretation, which leads to an estimate of the number of detectable and correctable errors, starts with the representation in the *codespace* (see Fig. 14.1): Each of the 2^N possible bit combinations of a binary (N, K) block code corresponds to a point in the codespace. 2^K of these bit combinations correspond to a valid codeword, each of which is separated at least by a distance d_{min}. The other bit combinations can be interpreted as points located in "correction spheres" around the valid codewords. The minimum size of correction spheres t determines the number of corrigible errors.

For codes of small size, decoding via lookup tables is possible – in other words, all possible syndromes and their corresponding minimum-weight error words are stored in a lookup table. For each received codeword, this table is then used to determine the transmitted codeword. Unfortunately, codes with small size usually show bad performance. For this reason, alternative decoding schemes that can be used for larger codes as well have been investigated. The most important one, the *belief propagation* algorithm, is described in more detail in Section 14.6. Furthermore, special cases of linear codes that allow a simplified decoding, like cyclic codes, might be preferable.

From these considerations, the following conclusions can be drawn: a code with minimum distance d_{min} always allows the detection of $d_{\text{min}} - 1$ errors in a codeword. Only errors that influence d_{min} bits can lead to another valid codeword. Alternatively, such a code can be capable of correcting $\lfloor (d_{\text{min}} - 1)/2 \rfloor$ errors – in other words, the received bit sequence has to lie in the correction sphere of the true codeword ($\lfloor x \rfloor$ denotes the largest integer smaller than x). If all bit combinations can be uniquely assigned to a correction sphere, the code is called *perfect*.

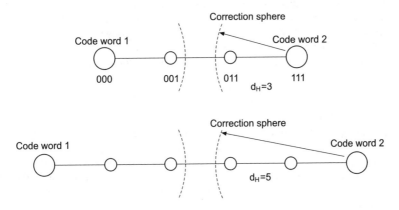

Figure 14.1 Cuts through a codespace (top figure) for a code with $d_H = 3$ and (bottom) a codespace for a code with $d_H = 5$ and a sketch of the correction spheres.

For a Hamming code the location of a single error can be uniquely determined. Since all the columns in the parity check matrix are different and linearly independent, and a Hamming code can correct exactly one error, the syndrome tells us the location of the error.

14.2.5 Concatenated codes

Error protection can be made stronger by using two codes: a so-called *inner code* protects the data in the usual way, as described above. Some errors can still remain (when the received word lies in the wrong correction sphere). We can thus interpret the combination of channel and inner code as a *superchannel* that exhibits a lower BER than the original channel. An *outer code* then provides protection against these errors. The errors of the superchannel usually occur in bursts; if the inner code is an (N, K) block code, then the length of the burst is K bits. Thus, the outer code is usually a code that is especially well-suited for combating burst errors. RS codes are efficient for these types of applications.[5]

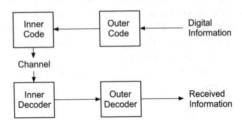

Figure 14.2   Concatenation of codes.

Appropriate concatenation of codes is quite a difficult task. In particular, the relative capabilities of the two codes need to be carefully balanced. If the inner code is not strong enough, then it is useless, and might even increase the BER that the outer code has to deal with.

14.2.6 Cyclic block codes

Basic properties

In a cyclic block code, the cyclic shift of a codeword creates another valid codeword. This does *not* mean that all codewords can be created by shifting a single basis word alone. However, it does imply that all codewords can be obtained from a single codeword by shifting (due to the cyclic property) and addition (due to linearity).

It is common to represent cyclic codes by their code polynomials (see Section 14.1.1); for example:

$$X(x) = 0 \cdot x^5 + 1 \cdot x^4 + 1 \cdot x^3 + 0 \cdot x^2 + 1 \cdot x^1 + 0 \cdot x^0 \tag{14.17}$$

for a bit sequence 011010.

For this polynomial representation, a cyclic shift corresponds to a multiplication of $X(x)$ by x, taken $\mathrm{mod}(x^N + 1)$. There is one codeword polynomial, called the *generator polynomial*, which has minimum degree – namely, $N - K$. It can be shown that this generator polynomial $G(x)$ must be a factor of $x^N + 1$. Multiplying $G(x)$ by x^i, $i = 1 \cdots K - 1$ (no modulo operation necessary here) gives a basis set of codewords from which all other codewords can be generated by linear combination.

---

[5] Code concatenation can also be applied when the inner code is a convolutional code, see below.

Systematic cyclic block codes

We can form systematic codes for cyclic block codes as well – i.e., codes that consist of the original codeword and parity check information. The codewords are determined the following way: let $U(x)$ be an information polynomial of degree $K - 1$ and $G(x)$ the generator polynomial of degree $N - K$. The code polynomial $X(x)$ is then computed in the following way. We form the expression:

$$\frac{(x^{N-K}U(x))}{G(x)} = Q(x) + \frac{P(x)}{G(x)} \tag{14.18}$$

where $Q(x)$ and $P(x)$ are quotient and remainder, respectively; the total codeword is $X(x) = P(x) + (x^{N-K}U(x))$. In other words, the data word is shifted to the left by $N - K$ bits, and the check bits (which correspond to division of the data word by the generator polynomial) are appended. The resulting code is thus obviously systematic.

For the decoding, we test whether the received bit sequence is a valid codeword. This is the case if the condition $[\widehat{X}(x)]_{\mathrm{mod}G(x)} = 0$ is fulfilled. Otherwise, the syndrome is nonzero:

$$S_{\mathrm{synd}}(x) = [\widehat{X}(x)]_{\mathrm{mod}G(x)} = [X(x) + \varepsilon(x)]_{\mathrm{mod}G(x)} = 0 + \varepsilon(x)_{\mathrm{mod}G(x)} = \varepsilon(x)_{\mathrm{mod}G(x)} \tag{14.19}$$

A major advantage of cyclic codes is that they allow simplified encoding structures: data encoding and error detection can be done by means of a shift register, and also finding the location of errors is greatly simplified (for details, see Lin and Costello [2004]).

14.2.7 Examples for cyclic block codes

Cyclic Hamming codes

For a Hamming code, it is possible to arrange the columns of the parity check matrix in such a way that the resulting code is a cyclic code (remember that the parity check matrix of a Hamming code contains all possible distinct columns of length $N - K$). The generator polynomial is an (arbitrary) primitive polynomial of degree $N - K$.

Bose–Chaudhuri–Hocquenghem code

BCH codes are cyclical codes that can be viewed as generalizations of Hamming codes. In contrast to Hamming codes, they can correct t errors in each block. They have the following properties: blocklength $N = 2^m - 1$, number of parity check bits $N - K \leq mt$, and minimum distance $d_{\mathrm{min}} \geq 2t + 1$. The generator polynomial is defined in the following way: let α be the primitive element in $\mathrm{GF}(2^m)$. The generator polynomial $G(x)$ is then the lowest degree polynomial over $\mathrm{GF}(2)$ that has $[\alpha, \alpha^1, \alpha^2, \ldots, \alpha^{2t}]$ as its roots. As a consequence, the syndrome can be computed by evaluating the received polynomial at these roots of the generator polynomial [Lin and Costello 2004].

Reed–Solomon codes

RS codes are a special form of non-binary BCH codes; they have the best error-correcting capability of any code of the same length and dimension. Defining them in a Galois field $\mathrm{GF}(q)$, where q typically is a power of 2, RS codes are BCH codes with blocklength $N = q - 1$. Having $N - K = 2t$ parity check bits, they can correct t errors. Due to their great practical importance, we will in the following briefly describe a decoding algorithm for RS codes [Steele and Hanzo 1999]:

1. As a first step, we determine the so-called partial syndromes:

$$S_i = \hat{X}(\alpha^i) \qquad \text{for } i = 1, \ldots, 2t \tag{14.20}$$

where again α is the primitive element of $GF(q)$. For later use, we state that we can formally write the first partial syndrome as:

$$S_1 = \sum_{j=1}^{N_e} M_j P_j \tag{14.21}$$

where M_j and P_j are the magnitude and location of the jth error, respectively, and N_e is the number of errors (at this point of the algorithm still unknown).

2. We next need to determine the number of errors N_e. We first assume that $N_e = t$ – i.e., the maximal corrigible number of errors. From this, we form a matrix $\mathbf{S}$ as:

$$\mathbf{S} = \begin{bmatrix} S_1 & S_2 & \cdots & S_{N_e} \\ S_2 & S_3 & \cdots & S_{N_e+1} \\ \vdots & \vdots & \cdots & \vdots \\ S_{N_e} & S_{N_e+1} & \cdots & S_{2N_e-1} \end{bmatrix} \tag{14.22}$$

we furthermore define the vectors $\mathbf{s}_s$ and $\mathbf{s}_L$:

$$\mathbf{s}_s = \begin{pmatrix} -S_{N_e+1} \\ -S_{N_e+2} \\ \vdots \\ -S_{2N_e} \end{pmatrix}, \qquad \mathbf{s}_L = \begin{pmatrix} L_{N_e} \\ L_{N_e-1} \\ \vdots \\ L_1 \end{pmatrix} \tag{14.23}$$

3. Compute $\det(\mathbf{S})$. If it is zero, we strip off the last row and last column of $\mathbf{S}$, and compute the determinant of this new $\mathbf{S}$. This procedure is repeated until $\det(\mathbf{S}) \neq 0$. The number of rows (or columns) in the remaining matrix tells us the actual number of errors N_e.

4. We find the coefficients $\mathbf{s}_L$ of the error locator polynomial as:

$$\mathbf{s}_L = \mathbf{S}^{-1} \mathbf{s}_s \tag{14.24}$$

There are some iterative algorithms for determination of the matrix inverse, like the Berlekamp–Massey algorithm, but basically any matrix inversion (in finite field arithmetic) will do.

5. We insert all elements of $GF(q)$ into the error polynomial $L(x)$, and find out which are zero. The inverses (again, within finite field arithmetic) indicate the location of errors.

6. Error magnitudes M_i are determined from the equation:

$$\begin{pmatrix} M_1 \\ M_2 \\ \vdots \\ M_{N_e} \end{pmatrix} = \begin{bmatrix} P_1 & P_2 & \cdots & P_{N_e} \\ P_1^2 & P_2^2 & \cdots & P_{N_e}^2 \\ \vdots & \vdots & \cdots & \vdots \\ P_1^{N_e} & P_2^{N_e} & \cdots & P_{N_e}^{N_e} \end{bmatrix}^{-1} \begin{pmatrix} S_1 \\ S_2 \\ \vdots \\ S_{N_e} \end{pmatrix} \tag{14.25}$$

With knowledge of the error location and magnitude, errors in the detected codeword $\hat{X}$ can be corrected.

14.3 Convolutional codes

14.3.1 Principle of convolutional codes

Convolutional codes do not divide (source) datastreams into blocks, but rather add redundancy in a quasi-continuous manner. A convolutional encoder consists of a shift register with L memory cells and N (modulo-2) adders (see Fig. 14.3). Let us assume that at the outset we have a clearly defined state in memory cells – i.e., they all contain 0. When the first data bit enters the encoder, it is put into the first memory cell of the shift register (the other zeros are shifted to the right, and the rightmost zero "falls out" of the register). Then a multiplexer reads out the output of all the adders $n = 1, 2, 3$. We thus get three output bits for one input bit. Then, the next source data bit is put into the register (and the contents of all memory cells are shifted to the right by one cell). The adders then have new outputs, which are again read by the multiplexer. The process is continued until the last source data bit is put into the register. Subsequently, zeros are used as register input, until the last source data bit has been pushed out of the register, and the memory cells are again in a clearly defined (all-zero) state.

A convolutional encoder is thus characterized by the number of shift registers and adders. Adders are characterized by their connections to memory cells. In the example of Fig. 14.3, only element $l = 1$ is connected to the output $n = 1$, so that source data are directly mapped to the coder output. For the second output, the contents of memory cells $l = 1, 2$ are added. For the third output, the contents of elements $l = 1, 2, 3$ are combined.

This coder structure can be represented in different ways. One possibility is via generator sequences: we generate N vectors of length L each. The lth element of the nth vector has value 1 if the lth shift register element has a connection to the nth adder; otherwise it is 0. Generator sequences can immediately be interpreted for building an encoder.

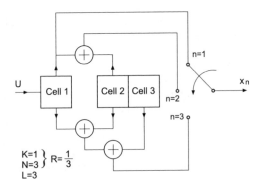

Number of states: $2^{L-1} = 4$

State	Cell 2	Cell 3
A	0	0
B	1	0
C	0	1
D	1	1

Figure 14.3 Example of a convolutional encoder.

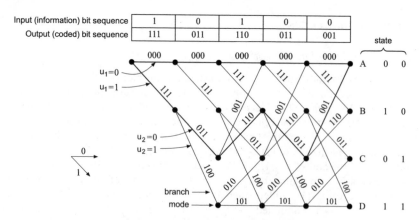

Figure 14.4 Trellis for the convolutional encoder of the previous figure.

Reproduced with permission from Oehrvik [1994] © Ericsson AB.

For the decoder, the trellis diagram is a more useful description method. In this representation, the state of the encoder is characterized by the content of the memory cells. The trellis shows which input bits get the shift register into which state, and which output bits are created consequently. As an example, consider the convolutional encoder of Fig. 14.3. As the first output is directly connected to the input bits, there are effectively only two memory cells, and thus four different states of the memory. Two lines originate from each state: the upper represents source data bit 0, and the lower source data bit 1. It is not possible to get from each state directly into each other state. For example, we can only get from state A to state A or B (but not C or D). This is the redundancy that can be used for the reduction of error probability during decoding. We also find that the trellis is repeated periodically. It is thus unnecessary to plot an infinitely long trellis diagram, even if the input data sequence is infinitely long.

Decoding at the receiver would be very simple were the received data sequence identical to the transmit data sequence. As we would then know which coded sequence was received, we would just have to trace in the trellis the source data from which it was created. When there is additive noise, there are errors in the received bit sequence, and we have to discover from the perturbed receive data what sequence was most likely originally transmitted (*Maximum Likelihood Sequence Estimation*, MLSE). In practice, this estimation is usually done by the Viterbi algorithm that is discussed in the next section. Other implementations are the Fano algorithm, the stack algorithm, and the feedback algorithm (Lin and Costello [2004] and McEliece [2004]).

14.3.2 Viterbi decoder – classical representation

The Viterbi algorithm [Viterbi 1967] is the most popular algorithm for MLSE. The goal of this algorithm is to find the sequence $\widehat{s}$ that was transmitted with the highest likelihood, if the sequence r was received:[6]

$$\widehat{s} = \max_{s} \Pr(r|s) \qquad (14.26)$$

---

[6] This is the definition for Maximum A Posteriori (MAP) detection. However, it is equivalent to MLSE for equiprobable sequences (see Chapter 12).

where maximization is done over all possible transmit sequences **s**. If perturbations of the received symbols by noise are statistically independent,[7] then the probability for the sequence $\Pr(\mathbf{r}|\hat{\mathbf{s}})$ can be decomposed into the product of the probabilities of each symbol:

$$\hat{\mathbf{s}} = \max_{\mathbf{s}} \prod_i \Pr(r_i|s_i) \tag{14.27}$$

Now, instead of maximizing the above product, maximizing its logarithm also finds the optimum sequence – this is true since the logarithm is a strictly monotonous function:

$$\hat{\mathbf{s}} = \max_{\mathbf{s}} \sum_i \log[\Pr(r_i|s_i)] \tag{14.28}$$

The logarithmic transition functions $\log[\Pr(r_i|s_i)]$ are also known as *branch metrics*. It can be shown that in an Additive White Gaussian Noise (AWGN) channel as well as in a flat-fading channel they are proportional to the Euclidean distance $d_i = |r_i - s_i|$ in the signal space diagram. Proportionality constants are of no further importance, as they have no impact on finding the optimum metric:

$$\hat{\mathbf{s}} = \min_{\mathbf{s}} \sum_i |r_i - s_i|^2 \tag{14.29}$$

This equation assumes that attenuation by the channel is the same for all received symbols and that channel attenuation has been compensated by the receiver. If this is not the case, then we have to minimize the expression:

$$\min_{\mathbf{s}} \sum_i |r_i - \alpha_i s_i|^2 \tag{14.30}$$

where α_i is the attenuation of the ith symbol.

The MLSE now determines the total metric $\sum_i |r_i - s_i|^2$ for all possible paths through the trellis – i.e., for all possible input sequences. In the end, the path with the smallest metric is selected. Such an optimum procedure requires large computational effort: the number of possible paths increases exponentially with the number of input bits. The key idea of the Viterbi algorithm is the following: instead of computing metrics for all possible paths (working from "top to bottom") in the trellis, we work our way "from the left to the right" through the trellis. More precisely, we start with a set of possible states of the shift register (A_i, B_i, C_i, D_i, where i denotes the time instant, or considered input bit). Let us now consider all paths that (from the left) lead into state A. We discard any possible path $\mathbf{s}^{(1)}$ if it merges at state A_i with a path $\mathbf{s}^{(2)}$ that has a smaller metric. As the paths run through the same state of the trellis, there is nothing that would distinguish the paths from the point of view of later states. We thus choose the one with the better properties. Similarly, we choose the best paths that run through the states $B_i, C_i,$ and D_i. After having determined the *survivors* for state i, we next proceed to state $i+1$ (or rather, to the tuple of states $A_{i+1}, B_{i+1}, C_{i+1}, D_{i+1}$), and repeat the process. All paths in a trellis ultimately merge in a single, well-defined point, the all-zero state.[8] At this point, there is only a single survivor – the sequence that was transmitted with the highest probability.

---

[7] If the noise is colored – i.e., correlated between the different samples – then the receiver has to use a so-called "whitening filter" (see Chapter 16).

[8] Actually, this only happens if the convolutional encoder appends enough zeros (tail bits) to the source data sequence to force the encoder into the defined state. If this is not done, then the decoder has to consider all possible finishing states, and compare the metrics of the path ending in each and all of them.

Example 14.5 *Example for Viterbi decoding.*

Figure 14.5 shows an example of the algorithm. The basic structure of the trellis is depicted in Fig. 14.5a; the bit sequence to be transmitted is shown in Fig. 14.5b. The metrics shown are the Hamming distances of the received sequence, i.e., *after* the hard decision compared with the theoretically possible bit sequences in the trellis – which is suboptimum. "Soft decisions" will be discussed below.

We assume that at the outset the shift register is in the all-zero state. Figure 14.5c shows the trellis for the first 3 bits. There are two possibilities for getting from state A_0 to state A_4: by transmission of the source data sequence $0, 0, 0$ (which corresponds to the coded bit sequence $000, 000, 000$, and thus via states A_2, A_3), or by transmission of the source data sequence 100 (coded bit sequence $111, 011, 001$ – i.e., via states B_2, C_3). In the former case, the branch metric is 2; in the latter, it is 6. This allows us to immediately discard the second possibility. Similarly, we find that the transition from state A_0 to state B_4 could be created (with greatest likelihood) by the source data sequence $0, 0, 1$, and not by $1, 1, 0$. The following subfigures of Fig. 14.5 show how the process is repeated for ensuing incoming bits.

Storage requirements are decreased by elimination of non-surviving paths, but they are still considerable. It is thus undesirable to wait for the decision as to which sequence was transmitted until the last bits of the source sequence. Rather, the algorithm makes decisions about bits that are "sufficiently" in the past. More precisely, during consideration of states A_i, B_i, C_i, D_i we decide about the symbols of the state tuple $A_{i-L_{Tr}}, B_{i-L_{Tr}}, C_{i-L_{Tr}}, D_{i-L_{Tr}}$, where L_{Tr} is the *truncation depth*. This principle is shown in Fig. 14.6. Data within the window of length L_{Tr} are stored. When moving to the next tuple in the trellis, the leftmost-state tuple moves out of the considered window, and we have to make a final decision about which bits were transmitted there. The decision is made in favor of the state that contains the path that has the smallest metric in the *currently* observed state – i.e., at the right side of the window. While this procedure is suboptimum, performance loss can be kept small by judiciously choosing the length of the window. In practice, a duration:

$$L_{Tr} = 6L \qquad (14.31)$$

has turned out to be a good compromise.

14.3.3 Improvements of the Viterbi algorithm

Soft decoding

The above example used the Hamming distance between received symbols and possible symbols as a metric. This means that received symbols first undergo a hard decision before being used in the decision process. This algorithm thus uses the redundancy of the code (not all combinations of bits are valid codewords), but does not use knowledge about the reliability of the received bits. In some cases, the bits might be close to the decision boundary, in which case they should have less impact on the finally chosen sequence than if they are far away from that boundary. "Soft information" can be taken into account by using (as metric) the Euclidean distance of the demodulated symbol from points in the signal constellation diagram. Performance improves considerably when this is done. A detailed example of a Viterbi algorithm with soft information is given in Chapter 16 (we will see there that MLSE equalization can also be done by means of this algorithm).

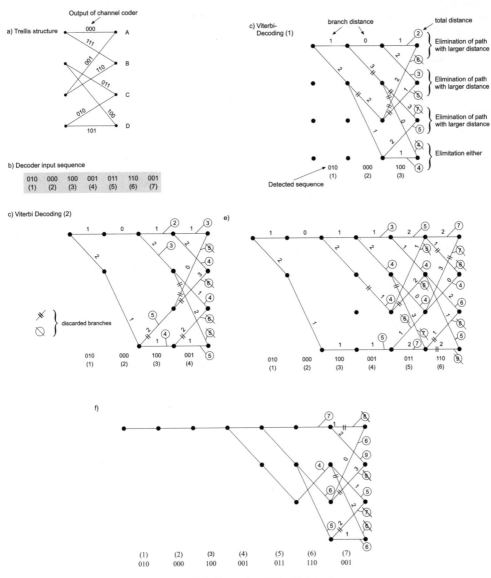

Figure 14.5 Example of Viterbi detection.

Reproduced with permission from Oehrvik [1994] © Ericsson AB.

Tail bits

As mentioned above, it is best if – at the end of the transmitted data sequence – the encoder ends up in a defined state, usually the all-zero state. In that case, it is clear which surviving path has to be chosen. Figure 14.7 shows that this approach significantly decreases the error probability for the latest data bits in the sequence. The drawback is a loss in spectral efficiency, as non-information-bearing symbols (the zeros appended at the end) have to be transmitted. This can become a problem in systems with very short block durations.

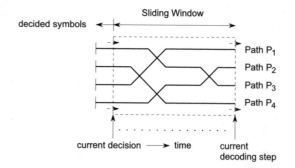

Figure 14.6 Principle of decision using a finite duration sliding window.

Reproduced with permission from Mayr [1996] © B. Mayr.

If the datablocks that are to be transmitted are very long, the principle of tail bits can be used also within a block: at certain, predefined locations, a series of zeros is transmitted, and thus forces the shift register (and the trellis) into a defined state. Such *waists* in the trellis allow a reduction in error probability.

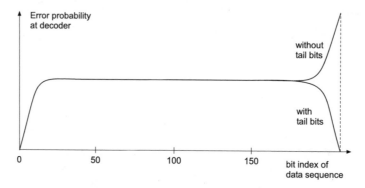

Figure 14.7 Error probability when tail bits are used.

Reproduced with permission from Oehrvik [1994] © Ericsson AB.

Puncturing

Normal convolutional codes can achieve only a certain subset of coding rates, like $R_c = 1/2, 1/3, 1/4$. However, for many applications, we need different code rates that are determined by the source rate and the available bandwidth. For example, the WiMedia ultrawideband standard uses a code rate of 11/32 for its default mode.

An easy way of adapting a code rate is puncturing of a code. It starts out with a code that has a rate that is lower than the desired. Then certain bits of the coded sequence are omitted from the transmission. Since the code contains considerable redundancy, this is not a problem – it is similar to the physical situation when code bits are erased due to fading. It is just required that the punctured bits are appropriately distributed throughout the codeword, so that not all information about a certain bit is eliminated.

14.4 Trellis-coded modulation

14.4.1 Basic principle

A main problem of coding is the reduction in spectral efficiency. Since we have to transmit more bits, the bandwidth requirement becomes larger as we add check bits. This problem can be avoided by the use of higher order modulation alphabets, which allow the transmission of more bits within the same bandwidth. In other words: using a rate $1/3$ code, while at the same time changing the modulation alphabet from Binary Phase Shift Keying (BPSK) to 8-PSK, the number of *symbols* that are transmitted per unit time remains the same.

As we discussed in Chapter 12, increasing the symbol alphabet increases the probability of error; on the other hand, introducing coding decreases the error probability. A simplistic approach to solving the spectral efficiency problem would thus be to add parity check bits to the data bits, and map the resulting coded data to higher order modulation symbols. However, this usually does not give good results. In contrast, *Trellis Coded Modulation* (TCM) adds to the redundancy of the code by increasing the dimension of the signal space, while disallowing some symbol sequences in this enlarged signal space. The important aspect here is that modulation and encoding are designed as a joint process. This allows the design of a modem-plus-codec that shows higher resilience to noise than uncoded systems with the same spectral efficiency.

Example 14.6   *Simple trellis-coded modulation.*

To understand the basic principle of TCM, consider first a simple example in an AWGN channel. We compare an uncoded system with a TCM system, both of which transmit two source data bits per symbol duration. In the uncoded case, Quadrature Phase Shift Keying (QPSK) is used as the modulation format (see Fig. 14.8). Every combination of two bits corresponds to one valid point in the signal constellation diagram. With the energy per bit being E_B, the distance between points in the signal constellation diagram is $2\sqrt{E_B}$. The error probability in an AWGN case can thus be represented (see Chapter 12):

$$BER \approx Q(\sqrt{2\gamma_B}) \qquad (14.32)$$

For TCM, we use a larger symbol alphabet, and thus a higher order modulation format (in our case, 8-PSK). As each symbol can transmit 3 bits, the code rate is $R_c = 2/3$. The allowed symbol sequences are determined by a shift register structure, like in a convolutional code. As with any other convolutional codes, the (coded) transmit symbol depends on the current source bit, as well as the current state of the memory. In our example, the memory is 2 bits long, so that there are four possible states: $00, 01, 10, 11$. Depending on the state, as well as the source bit, different 8-PSK symbols are transmitted. Figure 14.9 shows the possible transitions between states. If, for example, memory is in state 11, and the information symbol 01 is to be transmitted, then the PSK symbol 3 is transmitted, and memory ends up in state 10. We also find from the diagram that so-called *parallel transitions* are possible. These are transitions where we can get from memory state A to memory state B by transmission of different 8-PSK symbols. However, not all combinations of states and 8-PSK symbols are possible: when in state 11, only transmission of symbols $1, 3, 5, 7$ is allowed. Figure 14.10 shows the encoder structure and the trellis diagram for this system.

The trellis diagram also allows determination of the BER of TCM. As a first step, we determine the smallest Euclidean distance between symbol sequences that can lead to errors. Looking at a part of the trellis diagram, we find all allowed paths that lead from state 00 at time 0 to state 00 at time 3. The first possible pair are the parallel transitions using PSK symbols 0 and 4. The Euclidean distance between these two symbols is $d^2 = 8E_B$. Other

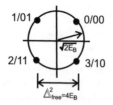

Figure 14.8   Quadrature-phase-shift-keying signal constellation diagram.

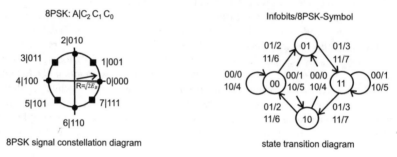

Figure 14.9   8-PSK signal constellation, and state transition diagram for a rate-2/3 8-PSK trellis-coded modulation.

Reproduced with permission from Mayr [1996] © B. Mayr.

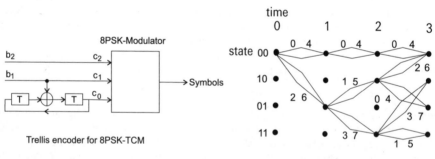

Figure 14.10   Encoder and trellis diagram for a rate-2/3 8-PSK trellis-coded modulation with four trellis states.

Reproduced with permission from Mayr [1996] © B. Mayr.

transitions are not possible by moving just one step to the right in the trellis; rather, a whole sequence is required. For example, a transition from state A_0 (00) to state A_3 (00) is possible by either transmitting the PSK symbols $0 - 0 - 0$ or $2 - 1 - 6$. The Euclidean distance between symbols 0 and 2 is $d^2 = 4E_B$, between 0 and 1 it is $2E_B[2\sin(\pi/8)]^2$, and between 0 and 6 it is again $4E_B$. The sum of the squared magnitudes of the distances between the two paths is thus $(4 + 1.17 + 4) = 9.17$. Other symbol sequences that can lead to errors can be searched for in a similar fashion. It can be shown that the smallest Euclidean distance between sequences that begin and end in the same state is $d^2_{coded} = 8E_B$. This is twice the distance we had for the uncoded case.

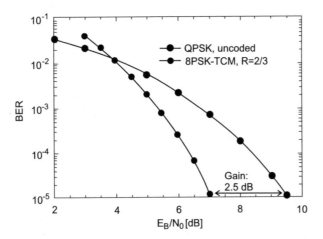

Figure 14.11 Simulated error probability of uncoded quadrature-phase shift keying and rate-2/3 8-PSK trellis-coded modulation in an additive white Gaussian noise channel.

Reproduced with permission from Mayr [1996] © B. Mayr.

Generally, we can define the asymptotic coding gain – i.e., the coding gain at high SNRs – as:

$$G_{\text{coded}} = 10 \log_{10}\left(\frac{d^2_{\text{coded}}}{d^2_{\text{uncoded}}}\right) \tag{14.33}$$

The above example achieves an asymptotic coding gain of 3 dB. In other words, for equal bandwidth requirement, equal source data rate, and equal target BER, the system needs 3 dB less transmit power. For finite target error rates, the coding gain is somewhat smaller; for a BER of 10^{-5}, the coding gain is only 2.5 dB (see Fig. 14.11). It is, however, also noteworthy that at low SNRs, the performance is *worse* than for an uncoded system.

14.4.2 Set partitioning

The previous subsection demonstrated the advantages of TCM, but the underlying code was ad hoc. For 8-PSK, such ad hoc construction is still feasible, and can lead to good results, but it becomes impossible for higher order modulation formats. Ungerboeck [1982] suggested a heuristic method for the construction of good codes, the so-called *set partitioning*. The basic principle of the method is:

1. Double the modulation alphabet.
2. Select the allowable transitions so that the minimal Euclidean distance between sequences is maximized.

For maximization of the distance, the symbol alphabet is partitioned in several steps, and at each step, the minimum distance between symbols of the partitioned sets should increase maximally.

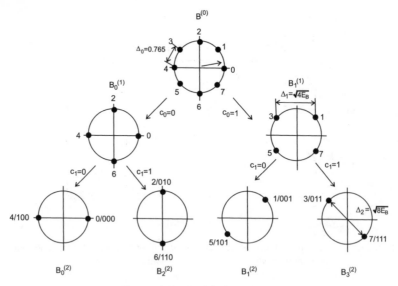

Figure 14.12   Partitioning of 8-PSK.

Example 14.7   *Set partitioning.*

This somewhat abstract principle can best be explained with an example. Consider the 8-PSK constellation of the previous section: the distance between two neighboring points in the signal constellation diagram is $d = \sqrt{E_B}[2\sqrt{2}\sin(\pi/8)] = 1.08\sqrt{E_B}$. As a first step, the existing symbols are partitioned into two sets (see Fig. 14.12). In order to maximize the distance between elements, each set is a QPSK constellation that is rotated $45°$ with respect to the other constellation. This increases the Euclidean distance within a set to $2\sqrt{E_B}$; there is no constellation that would lead to a larger minimum distance. In the next step, the QPSK constellation is partitioned. This results in two BPSK constellations that are rotated $90°$ with respect to each other, and the Euclidean distance within the set is $2\sqrt{2E_B}$. This completes the set partitioning.

Figure 14.13 shows the structure of a TCM encoder with such a set partitioning. We distinguish between N_{symb}, the number of information bits that can be transmitted per

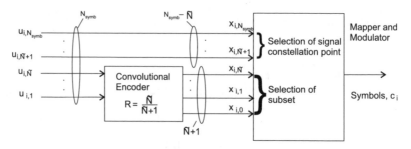

Figure 14.13   Structure of a trellis-coded modulation coder according to Ungerboeck.

Reproduced with permission from Mayr [1996] © B. Mayr.

symbol, and $\tilde{N}$, the number of bits that are mapped into coded bits by a convolutional encoder of rate $R_c = \tilde{N}/(\tilde{N}+1)$. During the encoding process, the source bits $u_1 \cdots u_{\tilde{N}}$ are mapped to coded bits $x_0 \cdots x_{\tilde{N}}$. The remaining source bits $u_{\tilde{N}}+1 \cdots u_{N_{symb}}$ are directly mapped to "uncoded" bits $x_{\tilde{N}}+1 \cdots x_{N_{symb}}$. The uncoded bits are used to select a signal constellation point within a subset; such points have a large Euclidean distance, so that errors are improbable. The other bits select the subsets. The fewer bits are uncoded, the fewer parallel transitions are possible. Decoding is done by means of a Viterbi decoder (see Section 14.3.2).

14.5 Turbocodes

14.5.1 Introduction

Turbocodes are among the most important developments of coding theory since the field was founded. As we have already mentioned in Section 14.1, *very long codes* can approach the Shannon limit (channel capacity). However, the brute force decoding of such long codes is prohibitively complex. Turbocodes were the first practically used codes that came close to the Shannon limit using reasonable effort. Berrou et al. [1993] created *very long codes* by a combination of several parallel, simple codes. The codes are interleaved by a pseudorandom interleaver. The vital trick now lies in the decoder: because the code is a combination of several short codes, the decoder can also be broken up into several simple decoders, and soft information about the decoded bits is exchanged between these parts of the decoder.

The random interleaver in the code approximately realizes the idea of a random code, so that the total codeword has very little structure. This has the following advantages:

- Interleaving increases the effective codelength of the combined code. In other words, it is the interleaver (whose operation can easily be reversed), and not the constituent codes, that determines the length of the codes. This allows the construction of very long codes with simple encoder structures.
- The special structure of the total code – i.e., the composition of separate constituent codes – makes decoding possible with an effort that is essentially determined by the length of the constituent codes.

14.5.2 Encoder

As mentioned above, turbocodes use a combination of several codes. One method of combining codes, serial concatenation, was already discussed in Section 14.2. In this section, we concentrate on parallel codes. The principle is shown in Fig. 14.14. The source datastream is put out directly, and also sent to several encoder branches. Each of the branches initially contains an interleaver that is different from branch to branch. After the interleaver, the datastream (in each parallel branch) is sent through a convolutional encoder, which maps the information vector **u** to the output. The number of these parallel encoders is – in principle – arbitrary. Due to concerns about the data rate, two encoders (resulting in code rate $1/3$) are most common.

It is advantageous to use *Recursive Systematic Convolutional* (RSC) codes for the encoding of each branch. The structure of such a code is outlined in Fig. 14.16. A shift register with feedback is used to compute the encoded bits. As these codes are systematic, there is one output that directly maps the source bits to the output. These bits are not actually transmitted, since the

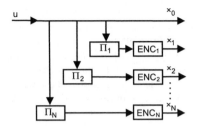

Figure 14.14 Structure of a turbo-encoder. Π denotes interleavers.
Reproduced with permission from Valenti [1995] © M. Valenti.

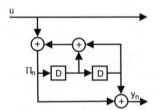

Figure 14.15 Structure of a recursive systematic convolutional encoder. D denotes delay elements.
Reproduced with permission from Valenti [1995] © M. Valenti.

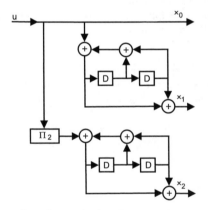

Figure 14.16 Structure of a turboencoder based on recursive systematic convolutional encoders.
Reproduced with permission from Valenti [1999] © M. Valenti.

turbocode transmits the original data sequence anyway (see topmost branch in Fig. 14.14); only encoded bits are actually sent. Therefore, the encoder usually has code rate 1 – i.e., puts out one coded bit per source bit.[9]

Summarizing, we emphasize that for one of the encoders (the direct feedthrough), we use the original source data sequence as input, while in the other case, the sequence is first interleaved. The N sequences (original sequences, plus the redundant bits from the other encoders) are then

---

[9] If we just used this encoder as a normal convolutional encoder, then the code rate would be $1/2$, as the encoder would put out the source bit and the redundant bit for each incoming source bit.

multiplexed. The resulting code is systematic, as it still contains the original data sequence. Since the systematic part is transmitted only once, the code rate of the total system is $1/(N+1)$.

Besides this basic structure, many other turbo-encoders are possible. As research in this area is still very active, we refrain from giving a taxonomy of encoders here.

The code rate of the above encoder is $R_c = 1/3$. However, it can be increased by puncturing. For example, outputs x_1 and x_2 can be used as input to a multiplexer that alternatingly uses a bit from the first encoder x_1 and discards a bit x_2 from encoder 2; or that uses x_2 and discards x_1. This realizes an encoder with code rate $R_c = 1/2$.

14.5.3 Turbodecoder

While encoding is always a comparatively simple process, decoding is the area where problems normally occur. Brute-force implementation of a decoder, where the combined code is viewed as a single, very complicated, code, would require prohibitive computational effort. It is here that turbocodes show their great advantage: it is possible to decode two constituent codes separately, and then combine the information from these two decoders.

Figure 14.17 shows a block diagram of a turbodecoder. It consists of two SISO decoders for the constituent codes, an interleaver, and a de-interleaver. The SISO decoder puts out not only the various bits, but also the confidence it has in a specific decision – i.e., the a posteriori probability. This confidence is described by the *Log Likelihood Ratio* (LLR). A Viterbi-like algorithm, which computes the LLR, was suggested by Bahl et al. [1974] and is well-known as the BCJR algorithm – the initials of the authors of that paper. The LLR is defined as:

$$\log\left[\frac{\Pr(b_i = +1|x)}{\Pr(b_i = -1|x)}\right] \tag{14.34}$$

i.e., the logarithm of the ratios of the probabilities that bit b_i is $+1$ or -1, conditioned on the fact that the value x was received. It can be shown that the LLR is the sum of three components:

- The LLR of the a priori probability for the bits (that LLR is usually 0, as all bits are equally likely).
- The information from the received raw data – i.e., the direct observation. This LLR also depends on the SNR.
- The extrinsic log-likelihood coefficient, which contains the information from decoding. The extrinsic information from the second decoder helps the first decoder, and the first decoder helps the second decoder.

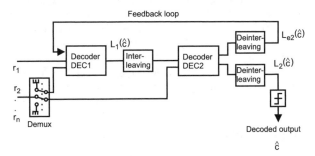

Figure 14.17 Structure of a turbodecoder.

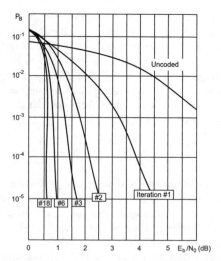

Figure 14.18 Bit error rate of a $R_c = 1/2$ turbocode with interleaver length 64,000 for different numbers of iterations in an additive-white-Gaussian-noise channel.

Reproduced with permission from Sklar [1997] © IEEE.

The first two components also exist for uncoded systems, the last one is the contribution from decoding of the other constituent code. It is important that the extrinsic information does not depend on the input of decoder 1; otherwise, the information in the extrinsic information and in the direct observation would be strongly correlated, and decoder 1 would just confirm its own opinion.

The iteration now starts with the assumption that the extrinsic LLR is zero, and decoder 1 decodes the received signal like a "normal" convolutional decoder (though with soft output). From this soft output, the receiver then computes the extrinsic LLR at decoder 2. The extrinsic information computed from the output of decoder 2 is then used for the next iteration step at decoder 1. Similarly, the extrinsic LLR from decoder 1 is used for the next iteration of decoder 2. This procedure is continued until convergence is achieved.

The BER improves as the number of iterations I increases. In an AWGN channel, $N_{it} = 5$ iterations are usually sufficient to achieve convergence, while a fading channel usually requires $N_{it} = 10$ iterations. The number of operations per information bit can be estimated as:

$$N_{op} \leq N_{it}(62 + 8/R_c)$$ (14.35)

Typically 400–800 operations per information bit are required. This explains why the widespread adoption of turbocodes happened only after the turn of the century. Transmission rates of 400 kbit/s require up to 300 MIPS (Million Instructions Per Second) for decoding – this large amount of signal-processing power has only been available since then.

Turbocodes allow close approximation of channel capacity. As an example, Fig. 14.18 shows the BER of a rate-1/2 turbo-encoder in an AWGN channel. Six iterations are obviously sufficient to closely approach the converged value. After 18 iterations, the BER is 10^{-5} at a 0.7-dB SNR.

14.6 Low-density parity check codes

When turbocodes were announced in 1993, they immediately drew a large amount of richly deserved attention. It seemed that for the first time the Shannon bound could be approached by practical codes. Yet it turns out that the problem had already been solved in the early 1960s! Gallagher [1961] in his thesis had designed linear block codes, called *Low Density Parity Check (LDPC) codes*, which allow the Shannon limit to be closely approached, and also proposed efficient iterative decoding mechanisms. However, this work was largely overlooked, because iterative decoding exceeded the available computer power at that time. However, several papers in the mid-1990s led to the rediscovery of these codes. And by that time the decoding complexity that once had seemed prohibitive looked quite reasonable. Since that time, a large number of papers has been published, and a considerable number of improvements has been proposed.

14.6.1 Definition of low-density parity check codes

LDPC codes are linear block codes (as discussed in Section 14.2). One interesting aspect for them is that they are not defined via the generator matrix $\mathbf{G}$, but rather via the parity check matrix $\mathbf{H}$. This is a key trick, since it is normally decoding that causes the biggest problems, not encoding. It thus makes eminent sense to define a structure that allows for easy decoding! Blocksize, and thus the dimensions of the check matrix, are very large. However, the number of nonzero entries into that matrix is kept low. More precisely, the ratio of the number of nonzero elements to the total number of entries is small; this is the reason why the codes are called "low-density". Following Gallagher, let us define an (N, p, q) LDPC code as a code of length N, whose parity check matrix has p 1's in each column, and q 1's in each row. In order for the code to have good properties, it is necessary that $p \geq 3$. If all rows are linearly independent, then the resulting rate of the code is $(q - p)/q$.

Good parity check matrices can be constructed from a few simple rules. First, subdivide the matrix horizontally into p submatrices of equal size. Then put one "1" into each column of such a submatrix. Let the first submatrix be defined to be, e.g., q concatenated identity matrices, or a structure like:

$$\begin{bmatrix} 1 & 1 & 1 & 1 & 0 & 0 & 0 & 0 & 0 & 0 & 0 & 0 & 0 & 0 & 0 & 0 & 0 & 0 & 0 & 0 \\ 0 & 0 & 0 & 0 & 1 & 1 & 1 & 1 & 0 & 0 & 0 & 0 & 0 & 0 & 0 & 0 & 0 & 0 & 0 & 0 \\ 0 & 0 & 0 & 0 & 0 & 0 & 0 & 0 & 1 & 1 & 1 & 1 & 0 & 0 & 0 & 0 & 0 & 0 & 0 & 0 \\ 0 & 0 & 0 & 0 & 0 & 0 & 0 & 0 & 0 & 0 & 0 & 0 & 1 & 1 & 1 & 1 & 0 & 0 & 0 & 0 \\ 0 & 0 & 0 & 0 & 0 & 0 & 0 & 0 & 0 & 0 & 0 & 0 & 0 & 0 & 0 & 0 & 1 & 1 & 1 & 1 \end{bmatrix} \quad (14.36)$$

Let, then, the other submatrices be random column permutations of this first submatrix. For example, we arrive at the following example of a $(20, 3, 4)$ code [Davey 1999]:

$$\mathbf{H} = \begin{bmatrix} 1 & 1 & 1 & 1 & 0 & 0 & 0 & 0 & 0 & 0 & 0 & 0 & 0 & 0 & 0 & 0 & 0 & 0 & 0 & 0 \\ 0 & 0 & 0 & 0 & 1 & 1 & 1 & 1 & 0 & 0 & 0 & 0 & 0 & 0 & 0 & 0 & 0 & 0 & 0 & 0 \\ 0 & 0 & 0 & 0 & 0 & 0 & 0 & 0 & 1 & 1 & 1 & 1 & 0 & 0 & 0 & 0 & 0 & 0 & 0 & 0 \\ 0 & 0 & 0 & 0 & 0 & 0 & 0 & 0 & 0 & 0 & 0 & 0 & 1 & 1 & 1 & 1 & 0 & 0 & 0 & 0 \\ 0 & 0 & 0 & 0 & 0 & 0 & 0 & 0 & 0 & 0 & 0 & 0 & 0 & 0 & 0 & 0 & 1 & 1 & 1 & 1 \\ 1 & 0 & 0 & 0 & 1 & 0 & 0 & 0 & 1 & 0 & 0 & 0 & 1 & 0 & 0 & 0 & 0 & 0 & 0 & 0 \\ 0 & 1 & 0 & 0 & 0 & 1 & 0 & 0 & 0 & 1 & 0 & 0 & 0 & 0 & 0 & 1 & 0 & 0 & 0 & 0 \\ 0 & 0 & 1 & 0 & 0 & 0 & 1 & 0 & 0 & 0 & 0 & 0 & 1 & 0 & 0 & 0 & 1 & 0 & 0 & 0 \\ 0 & 0 & 0 & 1 & 0 & 0 & 0 & 0 & 0 & 0 & 1 & 0 & 0 & 0 & 1 & 0 & 0 & 0 & 1 & 0 \\ 0 & 0 & 0 & 0 & 0 & 0 & 0 & 1 & 0 & 0 & 0 & 1 & 0 & 0 & 0 & 1 & 0 & 0 & 0 & 1 \\ 1 & 0 & 0 & 0 & 0 & 1 & 0 & 0 & 0 & 0 & 1 & 0 & 0 & 0 & 0 & 0 & 1 & 0 & 0 & 0 \\ 0 & 1 & 0 & 0 & 0 & 0 & 1 & 0 & 0 & 0 & 1 & 0 & 0 & 0 & 0 & 1 & 0 & 0 & 0 & 0 \\ 0 & 0 & 1 & 0 & 0 & 0 & 0 & 1 & 0 & 0 & 0 & 1 & 0 & 0 & 0 & 0 & 0 & 1 & 0 & 0 \\ 0 & 0 & 0 & 1 & 0 & 0 & 0 & 0 & 1 & 0 & 0 & 0 & 0 & 1 & 0 & 0 & 1 & 0 & 0 & 0 \\ 0 & 0 & 0 & 0 & 1 & 0 & 0 & 0 & 0 & 1 & 0 & 0 & 0 & 0 & 1 & 0 & 0 & 0 & 0 & 1 \end{bmatrix} \quad (14.37)$$

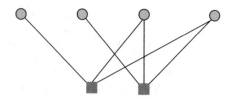

Figure 14.19   Tanner graph for the parity check matrix $H = \begin{bmatrix} 1 & 0 & 1 & 1 \\ 0 & 1 & 1 & 1 \end{bmatrix}$.

This structure obviously follows the rules of having p 1's in each column, and q 1's in each row; it also appears from inspection that the structure is reasonably random.

14.6.2 Encoding of low-density parity check codes

Since LDPC codes are defined via their parity check matrix, the encoding process is more complicated than for "normal" block codes. Remember that in normal block codes, we only have to multiply the codeword vector by the generator matrix. However, for LDPC codes, the generator matrix is not yet known. Fortunately, the computation is not very difficult: Using Gaussian elimination and reordering of columns, we can cast the parity check matrix in the form:

$$\tilde{\mathbf{H}} = \begin{pmatrix} -\mathbf{P}^T & \mathbf{I} \end{pmatrix} \tag{14.38}$$

The corresponding generator matrix is then:

$$\mathbf{G} = \begin{pmatrix} \mathbf{I} & \mathbf{P} \end{pmatrix} \tag{14.39}$$

Note that, due to the process of Gaussian elimination, the generator matrix is typically *not* sparse. This also means that the encoding process requires more operations. Fortunately, all the required operations are simple because they are performed on discrete, known bits. Thus, encoding complexity is usually not an issue.

14.6.3 Decoding of low-density parity check codes

As we have mentioned above, the sparse structure of the parity check matrix is key to decoding that works with reasonable complexity. But it is still far from trivial! Performing an exact maximum likelihood decoding is an N-p hard problem (in other words, we have to check all possible codewords, and compare them with the received signal). It is therefore common to use an iterative algorithm called *belief propagation*. It is this algorithm that we describe in more detail in the following.[10]

Let the received signal vector be **r**. Let us next put the parity check equations into graphical form. In the so-called "Tanner graph" (see Fig. 14.19),[11] we distinguish two kinds of nodes:

1.   *Variable (bit) nodes*: each variable node corresponds to one bit, and we know that it can be either in state 0, or state 1. Variable nodes correspond to the *columns* of the parity check matrix. We denote these nodes by circles.

---

[10] Note that the decoding algorithm can be described based on the syndrome vector (the method we will choose here) as well as the data vector.

[11] There are two types of graphical representations: the "Tanner graph" used here, and the "Forney factor graph" [Loeliger 2004].

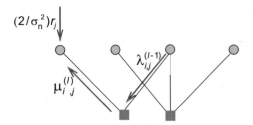

Figure 14.20   Message-passing in a factor graph.

2. *Constraint nodes*: constraint nodes (check nodes) describe the parity check equations; we know that if there are no errors present, the inputs to constraint nodes have to add up to 0. This follows from the definition of the syndrome, which is 0 if no errors are present. Constraint nodes correspond to the *rows* of the parity check matrix. We denote these nodes by squares.

Because there are two different types of nodes, and no connections between nodes of the same type, such a graph is also called a "bipartite" graph. In addition to constraint nodes and variable nodes, there is also external evidence, obtained by observation of the received signal, which has to influence our decisions.

Constraint nodes are connected to variable nodes if the appropriate entries in the parity check matrix are 1 – i.e., constraint node i is connected to variable node j if $H_{ij} = 1$. "Soft" information from the observed signals – i.e., the external evidence – is connected to the variable nodes. We also need to know the probability density function (pdf) of the amplitude of the variables – i.e., the probability that a variable node has a certain state, given the value of the external evidence.

Decoding on such a graph is done by a procedure called *message passing* or *belief propagation*. Each node collects incoming information, makes computations according to a so-called *local rule*, and passes the result of the computation to other nodes. Essentially, the jth variable node tells the constraint nodes it is connected to what it thinks its – i.e., the variable node's – value is, given the external information and the information from the other constraint nodes. This message is denoted λ_{ij}. In turn the ith constraint node tells the jth variable node what *it* thinks the variable node has to be, given the information that the constraint nodes have from all the *other* variable nodes; this message is called μ_{ij}. This is shown in Fig. 14.20: the external evidence is the observed data $\mathbf{r}$.

Let us formulate the decoding strategy mathematically, for an AWGN channel:

1. First, the data bits decide what value they *think* they are, given the external evidence $\mathbf{r}$ only. Knowing the statistics of the noise σ_n^2, the variable nodes can easily compute their probability to be 1 or a 0, respectively, and pass that information to the constraint nodes. Conversely, the constraint nodes cannot pass a meaningful message to the variable nodes yet. Therefore:

$$\mu_{i,j}^{(0)} = 0, \qquad \text{for all } i \tag{14.40}$$

$$\lambda_{i,j}^{(0)} = (2/\sigma_n^2)r_j, \qquad \text{for all } j \tag{14.41}$$

2. Then, the constraint nodes pass a different message to each variable node. Elaborating on the principle mentioned above, let us look specifically at constraint node i: assume that a set of connections ends in node i, which originate from an ensemble $A(i)$ of variable nodes.

Now. each of the checknodes has two important pieces of information: (i) it knows the values (or probabilities) of all data bits connected to this checknode; (ii) furthermore, it knows that all the bits coming into a checknode have to sum up to $0 \bmod 2$ (that is the whole point of a parity check matrix). From these pieces of information, it can compute the probabilities for the value that it thinks data bit j has to have. Since we have an AWGN channel, with a continuous output, and not a binary channel, we have to use LLRs instead of simple probabilities that a bit is reversed, so that the message becomes:

$$\mu_{i,j}^{(l)} = 2\tanh^{-1}\left(\prod_{k \in A(i)-j} \tanh\left(\frac{\lambda_{i,k}^{(l-1)}}{2}\right)\right) \tag{14.42}$$

where $A(i) - j$ denotes "all the members of ensemble $A(i)$ with the exception of j" – i.e., all constraint nodes that connect to the ith variable node, with the exception of the jth node. Superscript $^{(l-1)}$ denotes the $l-1$th iteration – i.e., we use the results from the previous iteration steps.

3.   Next, we update our opinion of what the bit nodes are, based on the information passed by the constraint nodes, as well as the external evidence. This rule is very simple:

$$\lambda_{i,j}^{(l)} = (2/\sigma_n^2)r_j + \sum_{k \in B(j)-i} \mu_{k,j}^{(l)} \tag{14.43}$$

where $B(j) - i$ denotes all variable nodes that connect to the jth constraint node, with the exception of i.

4.   From the above, we can compute the pseudoposterior probabilities that a bit is 1 or 0:

$$L_j = (2/\sigma_n^2)r_j + \sum_i \mu_{i,j}^{(l)} \tag{14.44}$$

based on which tentative decision we make about the codeword. If that codeword is consistent – i.e., its syndrome is 0 – then decoding stops.

Example 14.8   *Decoding of a low-density parity check code.*

Let us now consider a very simple example for this algorithm. Let the parity check matrix be:

$$\mathbf{H} = \begin{bmatrix} 0 & 0 & 1 & 1 & 1 & 1 \\ 1 & 1 & 1 & 1 & 0 & 0 \\ 1 & 1 & 0 & 0 & 1 & 1 \end{bmatrix} \tag{14.45}$$

Let the codeword:

$$\bar{y} = \begin{bmatrix} 0 & 1 & 1 & 0 & 1 & 0 \end{bmatrix} \tag{14.46}$$

be sent through an AWGN channel with $\sigma_n^2 = 0.237$ corresponding to $\gamma = 6.25\,\mathrm{dB}$; the received word be:

$$\bar{r} = \begin{bmatrix} -0.71 & 0.71 & 0.99 & -1.03 & -0.61 & -0.93 \end{bmatrix} \tag{14.47}$$

Then according to step 1 (mentioned above), likelihood values are computed from external evidence as:

$$\overline{\lambda^{(0)}} = \begin{bmatrix} -6.0 & 6.0 & 8.3 & -8.7 & -5.2 & -7.9 \end{bmatrix} \tag{14.48}$$

Hard thresholding of the received likelihood values would result in codeword error with an error at bit position 5:

$$[0 \quad 1 \quad 1 \quad 0 \quad 0 \quad 0] \tag{14.49}$$

Figure 14.21 demonstrates how the message-passing algorithm iterates itself to the correct solution, following the recipe given above.

14.6.4 Performance improvements

It can be shown that the belief propagation algorithm becomes exact if the Tanner graph can be rolled up into a tree structure – i.e., each node is a "parent" or a "child" of another node, but not both at the same time. In other words, there should be no cycles in the Tanner graph. Short cycles, like the one shown in Fig. 14.22, lead to problems with convergence: if nodes start out with a wrong belief, they tend to reinforce it among themselves, instead of being convinced by evidence from other variable nodes that they need to change their self-assessment. The construction of codes without short cycles is one of the most important, and most challenging, tasks in the design of LDPC codes. Note, however, that codes that lead to a *pure* tree structure are usually not good codes, even though they can be decoded exactly by the belief propagation algorithm.

Another important area to improve convergence and performance is the use of irregular codes. This means that column and row weights are not fixed, as we have assumed up to now, but that we rather prescribe only the *mean* weights, and allow some of the columns to have more entries. These nodes associated with these "heavier" columns often converge faster,

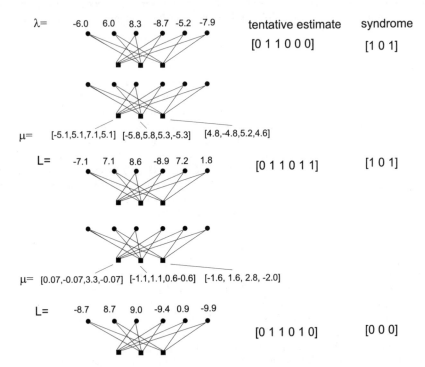

Figure 14.21   Example for the iterations of low-density parity check message-passing.

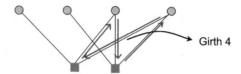

Figure 14.22 Tanner graph from Fig. 14.19, showing a short cycle (with "girth" 4).

and can spread their "secure" knowledge to other nodes, which leads to improved convergence for the other nodes as well.

The performance that can be achieved with LDPC codes is very impressive, and (for large blocksizes) can approach the Shannon capacity within a fraction of a dB. There is also an interesting rule for computational complexity in the decoding of LDPC codes: when the used rate approaches $(1 - \delta)$ of Shannon capacity, then the complexity per bit goes like $(1/\delta) \log_2(1/\delta)$ [Richardson and Urbanke 2003]. For most other codes, the complexity grows with $\exp(1/\delta)$.

14.7 Coding for the fading channel

The structure of errors in a fading channel are different from that in an AWGN channel. The presence of error bursts is noticeable: when the channel shows (instantaneous) high attenuation due to the destructive interference of multipath components, then the error probability is much larger than in constructive interference. Correction of these error bursts can be achieved either by using codes that are especially suited for bursty errors (RS codes, see Section 14.2). Alternatively, interleaving "breaks up" the bursty structure of errors.

14.7.1 Interleaving

Compared with a symbol duration, wireless channels change only slowly. Typically, a mobile station stays in a fading dip (which has spatial extension of about $\lambda/4$) for a duration of 10–100 ms. As a consequence, a large number of bits are strongly attenuated (and thus more susceptible to errors by noise). A normal code usually cannot correct such a large number of errors. To understand the basic principle, consider a simple rate-1/3 repetition code.[12] The probability of a wrong decision in a coded bit is $P_{\text{single}} \sim Q(\sqrt{\gamma})$, where γ is the *instantaneous* SNR. Due to the majority rule, the probability of a wrong final decision is approximately $P_{\text{single}}^2 \sim Q^2(\sqrt{\gamma})$, where we have made use of the fact that the SNR of two subsequent bits is the same. Averaged over different channel states, the BER is then approximately:

$$BER \sim \int_0^\infty pdf_\gamma(\gamma) Q^2(\sqrt{\gamma})\, d\gamma \tag{14.50}$$

When an interleaver is used, then the three bits associated with one source bit are transmitted at large intervals, so that the SNR for each of these transmissions is different (see Fig. 14.23). Therefore, an error occurs only if two of these independent transmissions are in independent fading dips, which is very unlikely (see Fig. 14.24). Mathematically, this can be

---

[12] Each bit is transmitted three times. The receiver first performs a hard decision, and then makes a majority decision. If it decides two or three times that the coded bit was a +1, then it concludes that the source data bit was +1.

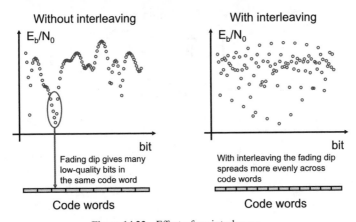

Figure 14.23   Effect of an interleaver.

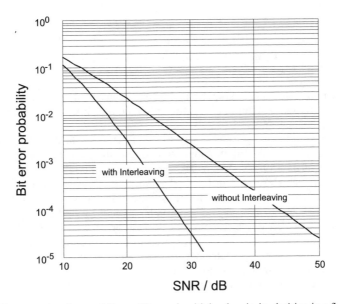

Figure 14.24   Bit error rate of a rate-1/3 repetition code with hard majority decision in a flat Rayleigh-fading channel, with and without interleaving.

written as:

$$BER \sim \int_0^\infty \int_0^\infty pdf_{\gamma_1}(\gamma_1)Q(\sqrt{\gamma_1})pdf_{\gamma_2}(\gamma_2)Q(\sqrt{\gamma_2}) \, d\gamma_1 \, d\gamma_2$$

$$= \left(\int_0^\infty pdf_{\gamma}(\gamma)Q(\sqrt{\gamma}) \, d\gamma \right)^2 \tag{14.51}$$

We stress that an interleaver can reduce the mean BER only in combination with a codec. For an uncoded system, the interleaver still breaks up error bursts (which sometimes can be desirable), but it does not lead to a decrease in the mean BER. Another problem with interleavers lies in the fact that they increase latency of transmission. For speech

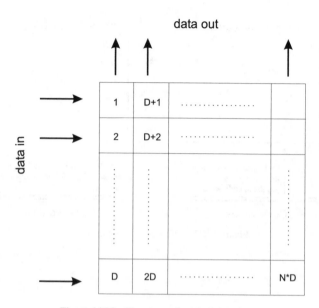

Figure 14.25   Structure of a block interleaver.

communications, it is necessary to keep latency below 50 ms. In that case, it can happen that maximum latency is smaller than the duration of a fading dip, which greatly reduces the effectiveness of the interleaver.

After these basic considerations, we now turn to the structure of the interleaver. Basically, two structures are possible: block interleaving, and convolutional interleaving. For a block interleaver, a block of data of size $N_{\text{interleav}}D_{\text{interleav}}$ is interleaved at once. The structure of the interleaver is a matrix: bits are read in line-by-line, and read out column-by-column. The bits that are originally subsequent are now separated by $D_{\text{interleav}}$. Latency due to this interleaver is $N_{\text{interleav}}D_{\text{interleav}} T_{\text{B}}$ – it has to wait until the matrix has been filled up before it can read out the bits. Latency is thus considerably larger than the separation of bits that can be achieved.

For a convolutional interleaver, the data are interleaved in a continuous stream, similar to a convolutional encoder. This reduces latency compared with the block interleaver (for the same separation of bits). It is common to use block interleavers together with block codes, and convolutional interleavers together with convolutional codes.

14.7.2 Block codes

In fading channels with interleaving, the rules of code design change considerably. The most important contribution is to provide diversity. Due to interleaving, each of the bits (symbols) in a codeword fades independently. The redundancy implicit in the code should thus make it possible to recover the information even when some of the bits are in fading dips and therefore have a very poor SNR. In the absence of interleaving, the effectiveness of the code would be much reduced.

As an example, let us consider a block code with hard decoding: let $K = 12$, $N = 23$, and the minimum distance be $d_{\text{min}} = 7$, and the coding gain 2 dB. Due to the slowness of fading, all 23 codebits see the same channel if there is no interleaving. In other words, each and every codeword has a gain of 2 dB. An uncoded system needs a 26-dB SNR to achieve a BER of 10^{-3}, with coding, this reduces to 24 dB – which is still bad. In other words – even with an encoder, the BER decreases only proportionally to γ^{-1}, just the proportionality constant changes.

The use of interleaving dramatically changes the situation. Errors occur only when there are errors at least at four locations in the codeword (note that $\left\lfloor \dfrac{d_{min} - 1}{2} \right\rfloor = 3$ can be corrected). The BER is approximately proportional to [Wilson 1996]:

$$\sum_{i=4}^{23} K_i \left(\frac{1}{2 + 2\bar{\gamma}_B} \right)^i \left(1 - \frac{1}{2 + 2\bar{\gamma}_B} \right)^{23-i} \approx \frac{1}{\gamma_B^4} \tag{14.52}$$

where the K_i are constants. Quite generally we find that a block code with Hamming distance d_H and "hard" decoding achieves a diversity order $\left\lfloor \dfrac{d_{min} - 1}{2} \right\rfloor + 1$.

When interleaving together with soft decoding is used, the resulting diversity order is almost twice as large – namely, d_{min}. To put this into more mathematical terms, we write the metric that we need to minimize as:

$$\min_{\mathbf{s}} \sum_i |r_i - \alpha_i s_i|^2 \tag{14.53}$$

and the pairwise error probability becomes [Benedetto and Biglieri 1999]:

$$P(\mathbf{s} \rightarrow \mathbf{s}_E) = \prod_{i \in A} \frac{1}{1 + |s_i - s_{E,i}|^2 / (4N_0)} \tag{14.54}$$

where A is the set of indices so that $s_i \neq s_{E,i}$. For a linear code the error probability becomes:

$$P \leq \sum_{w \in W} \left(\frac{1}{1 + R_c \gamma} \right)^w \tag{14.55}$$

where R_c is the code rate, and W the set of nonzero Hamming weights of the code. This shows clearly that the minimum distance determines the diversity order (slope of the BER versus SNR curve).

14.7.3 Convolutional codes

The performance measures and design rules for convolutional codes are quite similar to those of block codes. As a matter of fact, the mathematical description given above for block codes with soft decoding can directly be applied to convolutional codes. Thus, we find that the Euclidean distance of the code is no longer an important metric. Rather, we strive to find goals with good minimum (Hamming) distance properties. If the code sequences differ in many positions, then the probability is small that all of the distinguishing bits are in a fading dip simultaneously. Expressed in other words: we wish that two sequences that can be confused with each other (i.e., start and end in the same state of the trellis) are as long as possible. The minimum length of such sequences is often called the "effective length" of the code.

From these simple considerations, we can derive a few rules about code design for flat-fading channels:

- The minimal Euclidean distance does not occur in the equations for the BER. Thus, the design criteria for AWGN channels and fading channels are quite different. Since Rayleigh-fading and AWGN channels are only limiting cases of the Rician channels that are practically important, it is important to find codes that work in both of these channel types.
- The most important parameter in the fading channel is effective length, which enters exponentially into the BER.

14.7.4 Concatenated codes

An important class of codes for fading channels is the concatenation of convolutional codes with Reed–Solomon (RS) codes. Before 1995, this was actually the most popular code for cases where extremely high coding gains are required. Even today, this method is popular for situations where high data rates or complexity restrictions prevent the use of turbocodes and LDPC codes.

As discussed in Section 14.2.7, RS codes are good at correcting bursts of bit errors – actually much better than at correcting distributed errors. It would seem that this makes them well-suited for fading channels. However, for typical coherence times and bit rates, the length of the fades (in units of bits) in a wireless channel is typically much larger than the correction capability of an RS code. Thus, the way RS codes are applied is the following: first, the data are convolutionally encoded, and the resulting datastream is interleaved, to break up any possible error bursts (long interleavers are much easier to build than codes for long error bursts). While the convolutional code does not see long strings of severely impaired data (as it would without the interleaver), the decoding of a convolutional code results in error bursts, as this is a fundamental property of convolutional codes. These error bursts are then corrected by means of the outer code – namely, the RS code – which is well-equipped to deal with these bursts.

14.7.5 Trellis-coded modulation in fading channels

For TCM, the same design criteria are valid as for convolutional codes. Figure 14.26 shows that the effectiveness of the codes can be very different in different channels. We show a code by Ungerboeck [1982], designed for AWGN channels, and a code from Jamali and Le-Ngoc [1991] designed for fading channels. The superior performance of each code in its "natural" environment is obvious.

TCM is also very helpful in combating the errors due to delay dispersion that were discussed in Section 12.3. Chen and Chuang [1998] showed that these errors are determined essentially by the Euclidean distance between the codewords. Minimization of this metric gives codes that can combat these types of errors (see Fig. 14.27). While the search for codes for flat-fading channels is relatively mature, code designs for frequency-selective channels are rather recent.

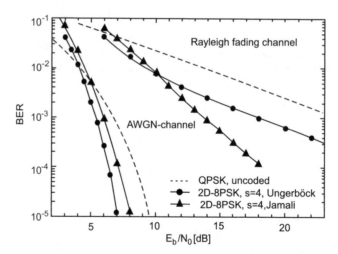

Figure 14.26 Bit error rate in a fading channel: uncoded quadrature-phase shift keying and 8-PSK trellis-coded modulation.

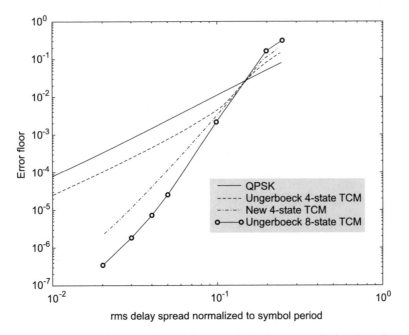

Figure 14.27 Error floor of trellis-coded modulation in frequency-selective channels.
Reproduced with permission from Chen und Chuang [1998] © IEEE.

14.8 Appendix

Please go to *www.wiley.com/go/molisch*

Further reading

More details on channel coding can be found in the many excellent textbooks on this topic; for example, Lin and Costello [2004], MacKay [2002], McEliece [2004], Sweeney [2002], and Wilson [1996]. Steele and Hanzo [1999, ch. 4] give a detailed description of block codes, including the decoding algorithms for RS codes, while Sklar and Harris [2004] give an intuitive introduction; further details can also be found in Lin and Costello [2004]. Convolutional codes are detailed in Johannesson and Zigangirov [1999]. TCM was first invented in the early 1980s by Ungerboeck [1982]; excellent tutorial expositions can be found in Biglieri et al. [1991]. More details about turbocodes can be found in Schlegel and Perez [2003], and also in the excellent tutorial of Sklar [1997]. LDPC codes, originally proposed by Gallagher [1961], were rediscovered by MacKay and Neal [1997], and described in a very understandable way in MacKay [2002]. Richardson and Urbanke [2005] give a detailed description of LDPC codes. Important theoretical foundations were laid by Bahl et al. in the 1970s [Bahl et al. 1974] and Hagenauer in the 1980s in their work of Soft Input Soft Output (SISO) decoding.

15

Speech coding

Gernot Kubin
Signal Processing and Speech Communication Laboratory,
Graz University of Technology, Graz, Austria

15.1 Introduction

15.1.1 Speech telephony as conversational multimedia service

When O. Nußbaumer succeeded in the first wireless transmission of speech and music in the experimental physics lab at Graz University of Technology in 1904, nobody would have predicted the tremendous growth in wireless multimedia communications 100 years after this historical achievement. Many new media types have emerged such as text, image and video, and modern services range from essentially one-way media download, browsing, messaging, application sharing, broadcasting, and real-time streaming to two-way, interactive, real-time conversations by text (chat), speech, and videotelephony. Still, speech telephony is the backbone of all conversational services and continues as an indispensable functionality of any mobile communications terminal. It is for this reason that we will focus our discussion of source-coding methods on speech signals – i.e., on their efficient digital representation for transmission (or storage) applications.

The success story of digital speech coding started with the introduction of digital switching in the Public Switched Telephone Network (PSTN) using Pulse Code Modulated (PCM) speech at 64 kbit/s and continued with a cascade of advanced compression standards from 32 kbit/s in the early 1980s over 16 kbit/s to 8 kbit/s in the late 1990s, all with a focus on long-distance circuit multiplication while maintaining the traditional high-quality level of wireline telephony (*toll quality*). For wireless telephony, the requirements on digital coding were rather stringent with regard to bit rate and complexity from the very beginning in the mid-1980s whereas some compromises in speech quality seemed acceptable because users either had never made the experience of mobile telephony before or, if so, their expectations were biased from the relatively poor quality of analog mobile radio systems. This situation changed quickly, and the first standards introduced at the beginning of the 1990s were completely overturned within 5 years with significant quality enhancements through new coding algorithms, maintaining a bit rate of about 12 kbit/s where the accompanying complexity increase was mitigated by related advances in microelectronics and algorithm implementation.

15.1.2 Source-coding basics

The foundations for source coding were laid by C. Shannon [1959] who developed not only channel-coding theory for imperfect transmission channels but also *rate–distortion theory* for signal compression. The latter theory is based on two components:

Wireless Communications   Andreas F. Molisch
© 2005 John Wiley & Sons, Ltd

- a stochastic source model which allows us to characterize the *redundancy* in source information; and
- a distortion measure which characterizes the *relevance* of source information for a user.

For asymptotically infinite delay and complexity, and certain simple source models and distortion measures, it can be shown that there exists an achievable lower bound on the bit rate necessary to achieve a given distortion level and, vice versa, that their exists an achievable lower bound on the distortion to be tolerated for a given bit rate. While complexity is an ever-dwindling obstacle, delay is a substantial issue in telephony, as it degrades the interactive quality of conversations severely when it exceeds a few 100 ms. Therefore, the main insight from rate–distortion theory is the existence of a three-way tradeoff among the fundamental parameters *rate*, *distortion*, and *delay*. Traditional telephony networks operate in circuit-switched mode where transmission delay is essentially given by the electromagnetic propagation time and becomes only noticeable when dealing with satellite links. However, packet-switched networks are increasingly being used for telephony as well – as in Voice over Internet Protocol (VoIP) systems – where substantial delays can be accumulated in router queues, etc. In such systems, delay becomes the most essential parameter and will determine the achievable rate–distortion tradeoff.

Source coding with a small but tolerable level of distortion is also known as *lossy coding* whereas the limiting case of zero distortion is known as *lossless coding*. In most cases, a finite rate allows lossless coding only for discrete amplitude signals which we we might consider for *transcoding* of PCM speech – i.e., the digital compression of speech signals which have already been digitized with a conventional PCM codec. However, for circuit-switched wireless speech telephony, such lossless coders have two drawbacks: first, they waste the most precious resource – i.e., the allocated radio spectrum – as they invest more bits than necessary to meet the quality expectations of a typical user; second, they often result in a bitstream with a *variable rate* – e.g., when using a Huffman coder – which cannot be matched efficiently to the *fixed rate* offered by circuit-switched transmission.

Variable-rate coding is, however, a highly relevant topic in packet-switched networks and in certain applications of joint source channel coding for circuit-switched networks (see Section 15.4.5). While Shannon's theory shows that, under idealized conditions, source coding and channel coding can be fully separated such that the two coding steps can be designed and optimized independently, this is not true under practical constraints such as finite delay or time-varying conditions where only a joint design of the source and channel coders is optimal. In this case, a fixed rate offered by the network can be advantageously split into a variable source rate and a variable channel code rate.

15.1.3 Speech coder designs

Source-coding theory teaches us how to use models of source redundancy and of user-defined relevance in the design of speech-coding systems. Perceptual relevance aspects will be discussed in later sections; however, the use of the source model gives rise to a generic classification of speech coder designs:

1. *Waveform coders* use source models only *implicitly* to design an adaptive dynamical system which maps the original speech waveform on a processed waveform that can be transmitted with fewer bits over the given digital channel. The decoder essentially inverts encoder processing to restore a faithful approximation of the original waveform. All waveform coders share the property that an increase of the bit rate will asymptotically result in lossless transcoding of the original PCM waveform. For such systems, the definition of a

coding error signal as the difference between the original and the decoded waveform makes sense (although it is no immediate measure of the perceptual relevance of the distortion introduced).

2. *Model-based coders* or *vocoders* rely on an *explicit* source model to represent the speech signal using a small set of parameters which the encoder estimates, quantizes, and transmits over the digital channel. The decoder uses the received parameters to control a real-time implementation of the source model that generates the decoded speech signal. An increase of the bit rate will result in saturation of the speech quality at a nonzero distortion level which is limited by systematic errors in the source model. Only recently, model-based coders have advanced to a level where these errors have little perceptual impact, allowing their use for very-low-rate applications (2.4 kbit/s and below) with slightly reduced quality constraints. Furthermore, due to the signal generation process in the decoder, the decoded waveform is not synchronized with the original waveform and, therefore, the definition of a waveform error is useless to characterize the distortion of model-based coders.

3. *Hybrid coders* aim at the optimal mix of the two previous designs. They start out with a model-based approach to extract speech signal parameters but still compute the modeling error explicitly on the waveform level. This model error or *residual waveform* is transmitted using a waveform coder whereas the model parameters are quantized and transmitted as *side information*. The two information streams are combined in the decoder to reconstruct a faithful approximation of the waveform such that hybrid coders share the asymptotically lossless coding property with waveform coders. Their advantage lies in the explicit parameterization of the speech model which allows us to exploit more advanced models than is the case with pure waveform coders which rely on a single invertible dynamical system for their design.

The model-based view of speech-coding design suggests that decription of a speech-coding system should always start with the decoder that typically contains an implementation of the underlying speech model. The encoder is then obtained as the signal analysis system that extracts the relevant model parameters and residual waveform. Therefore, the encoder has a higher complexity than the decoder and is more difficult to understand and implement. In this sense, a speech-coding standard might specify only the decoder and the format for transmitted datastreams while leaving the design of the best matching encoder to industrial competition.

Further design issues

The reliance on source models for speech coder design naturally results in a dependence of coder performance on the match between this model and the signal to be encoded. Any signal that is not clean speech produced from a single talker near the microphone may suffer from additional distortion which requires additional performance testing and, possibly, design modifications. Examples of these extra issues are the suitability of a coder for *music* (e.g, if put on hold while waiting for a specific party), for severe *acoustic background noise* (e.g., talking from the car, maybe with open windows), for *babble noise* from other speakers (e.g., talking from a cafeteria), for *reverberation* (e.g., talking in hands-free mode), etc.

Further system design aspects include the choice between *narrowband* speech as used in traditional wireline telephony (e.g., an analog bandwidth from 300 Hz to 3.4 kHz with 8-kHz sampling frequency) and *wideband* speech with quality similar to Frequency Modulation (FM) radio (e.g., an analog bandwidth from 50 Hz to 7 kHz with 16-kHz sampling frequency) which substantially enhances user experience with clearly noticeable speech quality improvements over and above the Plain Old Telephone Service (POTS). Second, even with the use of sophisticated channel codes, the *robustness of channel errors* such as individual bit errors, bursts, or entire lost transmission frames or packets is also a source-coding design issue (as discussed below in Section 15.4.5). Finally, within the network path from the talker to the listener, there may be several *codec-tandeming* steps where transcoding from one coding standard to another occurs, each time with a potential further loss of speech quality.

15.2 The sound of speech

While the "sound of music" includes a wide range of signal generation mechanisms as provided by an orchestra of musical instruments, the instrument for generating speech is fairly unique and constitutes the physical basis for speech modeling, even at the acoustic or perception levels.

15.2.1 Speech production

In a nutshell, *speech* communication consists in information exchange using a natural *language* as its code and the human *voice* as its carrier. Voice is generated by an intricate oscillator – the vocal folds – which is excited by sound pressure from the lungs. In the view of wireless engineering, this oscillator generates a nearly periodic, Discrete Multi Tone (DMT) signal with a fundamental frequency f_0 in the range of 100 to 150 Hz for males, 190 to 250 Hz for females, and 350 to 500 Hz for children. Its spectrum slowly falls off towards higher frequencies and spans a frequency range of several 1,000 Hz. From its relative bandwidth, it should be considered an *ultrawideband* signal,[1] which is one of the reasons why the signal is so robust and power-efficient (opera singers use no amplifiers) under many difficult natural environments.

The voice signal travels further down the *vocal tract* – the throat, and the oral and nasal cavities – which can be described as an acoustic waveguide shaping the spectral envelope of the signal by its resonance frequencies known as *formant frequencies*. Finally, the signal is radiated from a relatively small opening (mouth and/or nostrils) in a large sphere (our head) which gives rise to a high-pass radiation characteristic such that the far-field speech signal has a typical spectral rolloff of 20 dB per decade.

Sound generation

Besides the oscillatory voice signal ("phonation"), additional sound sources may contribute to the signal carrier, such as turbulent noise generation at narrow flow constrictions (in "fricative" sounds like [s] in "le**ss**on") or due to impulse-like pressure release after complete flow closures (in "plosive" sounds like [t] in "a**tt**empt"). If the vocal folds do not contribute at all to the sound generation mechanism, the speech signals are called *unvoiced*, otherwise they are *voiced*. Note that in the latter case, nearly periodic and noise-like excitation can still co-exist (in "voiced fricative" sounds like [z] in "pu**zz**le"), such sounds are often referred to as *mixed excitation* sounds.[2] Besides the three above-mentioned sound generation principles (phonation, frication, explosion) there is a fourth one which is often overlooked: *silence*. As an example, compare the words "Mets" and "mess". Their major difference lies in the closure period of the [t] which results in a silence interval between the vowel [ɛ] and the [s] in "Mets" which is absent in "mess". Otherwise, the two pronunciations are essentially the same, so the information about the [t] is carried entirely by the silence interval.

---

[1] This analogy becomes even more striking if we consider that the phase velocities of light and sound are related by a scale factor of approx. 10^6, showing that 1-kHz soundwaves have about the same wavelength as 1-GHz electromagnetic waves.

[2] Note that from our definitions, mixed excitation speech sounds are clearly a subclass of voiced speech.

Articulation

While the sound generation mechanisms provide the speech carrier, the *articulation* mechanism provides its modulation using a language-specific code. This code is both sequential and hierarchical – i.e., spoken language utterances consist of a sequence of phrases built of a sequence of words built of a sequence of syllables built of a sequence of speech sounds (called *phonemes* if discussed in terms of linguistic symbols). However, the articulation process is not organized in a purely sequential way – i.e., the shape of the vocal tract waveguide is not switched from one steady-state pose to another for each speech sound. Rather, the articulatory gestures (lip, tongue, velum, throat movements) are strongly overlapping and interwoven (typically spanning an entire syllable), mostly asynchronous, and continuously evolving patterns, resulting in a continuous modulation process rather than in a discrete shift-keying modulation process. This *co-articulation* phenomenon is the core problem in automatic speech recognition or synthesis where we attempt to map continuously evolving sound patterns on discrete symbol strings (and vice versa).

15.2.2 Speech acoustics

Source filter model

The scientific study of speech acoustics dates back to 1791 when W. von Kempelen, a high-ranked official at Empress Maria Theresa's[3] court, published his description of the first mechanical speaking machine as a physical model of human speech production. The foundations of modern speech acoustics were laid by G. Fant in the 1950s [Fant 1970] and resulted in the *source filter* model for speech signals (see Fig. 15.1).

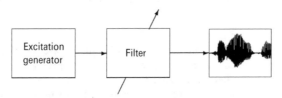

Figure 15.1   Source filter model of speech: The excitation signal generator provides the source which drives a slowly time-varying filter that shapes the spectral envelope according to formant frequencies to produce a speech waveform.

While this model is still inspired from our understanding of natural speech production, it deviates from its physical basis significantly. In particular, all the natural sound sources (which can be located at many positions along the vocal tract and which are often controlled by the local aerodynamic flow) are collapsed into a single source which drives the filter in an independent way. Furthermore, there is only a single output whereas the natural production system may switch between or even combine the oral and nasal branches of the vocal tract. Therefore, the true value of the model does not lie in its accuracy describing human physiology but in its flexibility in modeling speech acoustics. In particular, the typical properties of speech signals as evidenced in its temporal and spectral analyses are well represented by this structure.

Sound spectrograms

An example for the typical time–frequency analysis for speech is shown in Figure 15.2.

---

[3] Empress Maria Theresa was Archduchess of Austria and Queen of Hungary in the 1700s.

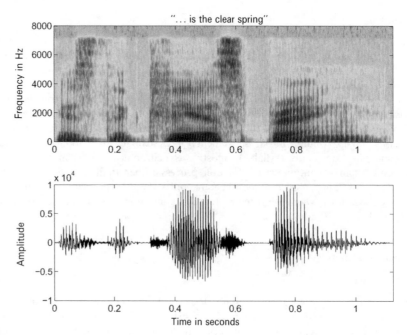

Figure 15.2 Upper graph: Spectrogram of the phrase "*is the clear spring*" spoken by a male speaker, analog bandwidth limited to 7 kHz, sampled at $f_s = 16$ kHz, horizontal axis = time, vertical axis = frequency, dark areas indicate high-energy density. Lower graph: Time domain waveform of the same signal, vertical axis = amplitude.

The lower graph shows the time domain waveform for the phrase "*is the clear spring*", spoken by a male speaker and limited to an analog bandwidth of 7 kHz for 16-kHz sampling. This graph shows the marked alternation between the four excitation source mechanisms where we can note that the "nearly periodic" source of voiced speech can mean an interval of only three fundamental "periods" which show a highly irregular pattern in the case of the second voiced segment corresponding to "*the*". Furthermore, strong fluctuations of the envelope are visible, which often correspond to 20 to 30 dB.

The upper graph shows a *spectrogram* of the same signal which provides the time–frequency distribution of the signal energy where darker areas correspond to higher energy densities. This representation can be obtained either by means of a filterbank or short-time Fourier analysis. It illustrates global signal properties – like anti-aliasing low-pass filtering at 7 kHz – and the observation that a significant amount of speech energy is found above 3.4 kHz (in particular for fricative and plosive sounds) suggesting that conventional telephony is too narrow in its bandwidth and destroys natural speech quality in a significant way.[4] Local properties are found in:

- noise-like signal components visible as broadband irregular energy distributions;
- impulse-like components visible as broadband spikes;
- nearly periodic components visible as narrowly spaced (around 10 ms for a male voice) vertical striations, corresponding to the pulsetrains seen in the time domain;

---

[4] The impact of limited POTS bandwidth on intelligibility is evidenced from the need to use "alpha/bravo/ charlie ..." spelling alphabets when communicating an unknown proper name over the phone.

- formant frequencies visible as slowly drifting energy concentrations due to vocal tract resonances which apparently are very context-dependent as they differ even for the three occurrences of an [i] sound in this utterance.

The first three properties are related to the source or excitation signal of the source–filter model. The fourth property is related to the filter and its resonance frequencies or poles. The slow temporal variation of these poles models the temporal evolution of the formant frequencies.

15.2.3 Speech perception

The ultimate recipient of human speech is the human hearing system, a remarkable receiver with two broadband, directional antennas shaped for spatiotemporal filtering (the outer ear) in terms of the individual, monaural *Head Related Transfer Functions*, HRTFs (functions of both azimuth angle and frequency) which along with interaural delay evaluation give rise to our spatial hearing ability. Second, a highly adaptive, mechanical impedance-matching network (the middle ear) covers a dynamic range of more than 100 dB and a 3,000-channel, phase-locking, threshold-based receiver (hair cells in the inner ear's cochlea) with very low self-noise (just above the level where we could hear our own bloodflow-induced noise) converts the signal to an extremely parallel, low-rate, synchronized representation useful for distributed processing with low-power and imprecise circuits (our nervous system).

Auditory speech modeling

Auditory models in the form of *psychoacoustic* – i.e., behavioral – models of perception are very popular in audiocoding (like those for the ubiquitous MP3 standard) because they allow us to separate relevant from irrelevant parts of the information. For instance, certain signal components may be masked by others to such an extent that they become totally inaudible. Naturally, for high-quality audiocoding, where little prior assumptions can be made about the nature of the signal source and its inherent redundancy, it is desirable to shape the distortion of lossy coders such that it gets masked by the relevant signal components.

In speech coding, the situation is reversed: we have a lot of prior knowledge about the signal source and can base the coder design on a source model and its redundancy, whereas the perceptual quality requirements are somewhat relaxed compared with audiocoding – i.e., in most speech coders some unmasked audible distortion is tolerated (note that we have long learnt to live with the distortions introduced by 3.4-kHz band limitation which would never work when listening to music). Therefore, perceptual models play a lesser role in speech coder design, although a simple *perceptual weighting filter*, originally proposed by Schroeder et al. [1979], has wended its way into most speech-coding standards and allows a perceptually favorable amount of noise shaping. More recent work on invertible auditory models for transparent coding of speech and audio is reviewed in Feldbauer et al. [2005].

Perceptual quality measures

The proof of a speech coder lies in listening. Till today, the best way of evaluating the quality of a speech coder is by controlled listening tests performed with sizable groups of listeners (a couple of dozens or more). The related experimental procedures have been standardized in International Telecommunications Union (ITU-T) Recommendation P.800 and include both *absolute category rating* and *comparative category rating* tests. An important example of the former is the so-called *Mean Opinion Score* (MOS) test which asks listeners to rate the perceived quality on a scale from 1 = poor to 5 = excellent where traditional narrowband speech with logarithmic PCM coding at

64 kbit/s is typically rated with an MOS score in the vicinity of 4.0. With a properly designed experimental setup, high reproducibility and discrimination ability can be achieved. The test can be calibrated by using so-called anchor conditions generated with artifially controlled distortions using the ITU's Modulated Noise Reference Unit (MNRU).

Besides these one-way listening-only tests, two-way *conversational tests* are important whenever speech transmission suffers from delay (including source-coding delay!). Such tests have led to an overall planning tool for speech quality assessment, the so-called E-model standardized by ITU-T Recommendation G.107 which covers various effects from one-way coding and transmission losses to the impact of delay on conversations or the subjective user advantage obtained from mobile service access.

As a means of bypassing tedious listening and/or conversational tests, objective speech quality measures have gained importance. They are often based on perceptual models to evaluate the impact of coding distortions as observed by comparing the original and the decoded speech signal. Such measures are also called *intrusive* because they require transmission of a specific speech signal over the equipment (or network connection) under test in order to make this original signal in its uncoded form available for evaluation at the receiver site. One standardized example is the *Perceptual Evaluation of Speech Quality*, PESQ (compare ITU-T Recommendation P.862).

Non-intrusive or single-ended speech quality measures are still under development. They promise to assess quality without having access to the original signal, just as we do as human listeners when we judge a telephone connection to have bad quality without direct access to the other end of the line. A first attempt (from May 2004) at standardization is the *Single Sided Speech Quality Measure*, 3SQM (ITU-T Recommendation P.563).

15.3 Stochastic models for speech

15.3.1 Short-time stationary modeling

With all the physical insights obtained from speech production and perception, speech coding has only the actual acoustic waveform to work with and, as this is an information-carrying signal, a stochastic modeling framework is called for. At first, we need to decide how to incorporate the time-varying aspects of the physical signal generation mechanism which suggests utilization of a *non-stationary*[5] *stochastic process* model. There are two "external" sources for this non-stationarity best described in terms of the time variations of the acoustic wave propagation channel:

- The movements of the articulators which shape the boundary conditions of the vocal tract at a rate of 10 to 20 speech sounds/second. As the vocal tract impulse response typically has a delay spread of less than 50 ms, a *short-time stationary representation* is adequate for this aspect of non-stationarity.
- The movements of the vocal folds at fundamental frequencies from 100 to 500 Hz give rise to a nearly periodic change in the boundary conditions of the vocal tract at its lower end – i.e., the vocal folds. In this case, comparison with delay spread suggests that the time variation is too fast for a short-time stationary model, Doppler-shift-induced modulation

---

[5] An alternative view would introduce a hypermodel that controls the evolution of time-varying speech production model parameters. A stationary hypermodel could reflect the stationary process of speaking randomly selected utterances such that the overall two-tiered speech signal model would turn out to be stationary.

effects become important and only a *cyclostationary representation* can cope with this effect (compare Subsection 15.3.5).

Most classical speech models neglect cyclostationary aspects; so, let us begin by discussing the class of short-time stationary models. For these models, stochastic properties vary slowly enough to allow the use of a subsampled estimator – i.e., we need to re-estimate the model parameters only once for each speech signal *frame* where in most systems a new frame is obtained every 20 ms (corresponding to $N = 160$ samples at 8 kHz). For some applications – like signal generation which uses the parameterized model in the decoder – the parameters need to be updated more frequently (e.g., for every 5-ms subframe) which can be achieved by interpolation from available frame-based estimates.

Wold's decomposition

Within a frame, the signal is regarded as a sample function of a stationary stochastic process. This view allows us to apply *Wold's decomposition* [Papoulis 1985] which guarantees that *any stationary stochastic process* can be decomposed into the sum of two components: a *regular component* $x_n^{(r)}$ which can best be understood as *filtered noise* and which is not prefectly predictable using linear systems and a *singular component* $x_n^{(s)}$ which is essentially a *sum of sinusoids* that can be perfectly predicted with linear systems:

$$x_n = x_n^{(r)} + x_n^{(s)} \tag{15.1}$$

Note that the sinusoids need not be harmonically related and may have random phases and amplitudes. In the following three subsections, we will show that this result – already established in the theory of stochastic processes by 1938 – serves as the basis for a number of current speech models which only differ in the implementation details of this generic approach. These models are: the Linear Predictive voCoder (LPC), sinusoidal modeling, and Harmonic + Noise Modeling (HNM).

15.3.2 Linear predictive vocoder

The first implementation of Wold's decomposition emphasizes its relationship to the Linear Prediction (LP) theory as first proposed by Itakura and Saito [1968]. The model derives its structure from the representation of the regular component as white noise run through a linear filter, which is causal, stable, and has a causal and stable inverse – i.e., a *minimum-phase filter*. The same filter is used for the generation of the singular component in voiced speech where it mostly consists of a number of harmonically related sinusoids which can be modeled by running a periodic pulsetrain through the linear filter. This allows flexible shaping of the spectral envelope of all harmonics but does not account for their original phase information because the phases are now determined by the phase response of the minimum-phase filter which is strictly coupled to its logarithmic amplitude response (via a Hilbert transform). Furthermore, as there is only one filter for two excitation types (white noise and pulsetrains), the model simplifies their additive superposition to a *hard switch in the time domain*, requiring strict temporal segregation of noise-like and nearly periodic speech (compare the decoder block diagram shown in Fig. 15.3). This simplification results in a systematic model mismatch for mixed excitation signals.

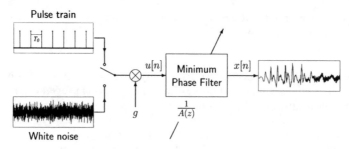

Figure 15.3 Linear predictive vocoder signal generator as used in decoder.

Linear prediction analysis

The LPC encoder has to estimate the model parameters for every given frame, and to quantize and code them for transmission. We will only discuss estimation of the linear prediction filter parameters here; estimation of the fundamental period $T_0 = 1/f_0$ is treated later in this text (Subsection 15.4.3). As a first step, specification of the minimum-phase filter is narrowed down to an *all-pole filter* – i.e., a filter where all transfer function zeros are concentrated in the origin of the z-plane and where only the poles are used to shape its frequency response. This is justified by two considerations:

- The vocal tract is mainly characterized by its resonance frequencies and our perception is more sensitive to spectral peaks than to valleys.
- For a minimum-phase filter, all the poles and zeros lie inside the unit circle. In this case, a zero at position z_0 with $|z_0| < 1$ can be replaced by a geometric series of poles, which converges for $|z| = 1$:

$$(1 - z_0 z^{-1}) = \frac{1}{1 + z_0 z^{-1} + z_0{}^2 z^{-2} + \cdots} \tag{15.2}$$

This allows us to write the signal model in the z-transform domain as:

$$X(z) = \frac{U(z)}{A(z)} \tag{15.3}$$

where the speech signal is represented by $X(z)$, the excitation signal by $U(z)$, and the filter transfer function by $1/A(z)$. In the time domain, this reads as:

$$x_n = -\sum_{i=1}^{m} a_i x_{n-i} + u_n \tag{15.4}$$

where the filter or predictor order is chosen as $m = 10$ for speech sampled at $8\,\mathrm{kHz}$ and the parameters a_i are known as *predictor coefficients*, with normalization[6] $a_0 = 1$.

For a given speech frame, we estimate model parameters $\hat{a}_i$ such that model mismatch is minimized. This mismatch can be observed through the prediction error signal e_n:

$$e_n = x_n - \hat{x}_n = x_n - \left(-\sum_{i=1}^{m} \hat{a}_i x_{n-i} \right) = \sum_{i=1}^{m} (\hat{a}_i - a_i) x_{n-i} + u_n \tag{15.5}$$

For uncorrelated excitation u_n, the prediction error power achieves its minimum iff $\hat{a}_i = a_i$, for $i = 1, \ldots, m$. In this case, the prediction error signal e_n becomes identical to the excitation signal

---

[6] The gain g can be included in the excitation signal amplitude.

u_n. Note that we use the prediction framework only for model fitting, not for forecasting or extrapolation of future signal samples. To apply this estimator to short-time stationary speech data, for every frame with an update rate of N samples, a window of $L \geq N$ samples is chosen, where the greater sign means that the windows have some overlap or lookahead. Typically, a special window function w_n is applied to mitigate artifacts due to the data discontinuity introduced by the windowing mechanism. Asymmetric windows with their peak close to the most recent samples allow a better compromise between estimation accuracy and delay. The window can be applied to the data in two different ways, giving rise to two linear prediction analysis methods:

1. The *autocorrelation method* defines the prediction error power estimate based on a windowed speech signal:

$$\tilde{x}_n = \begin{cases} w_n \cdot x_n, & \text{for } n = 0, 1, \ldots, L - 1 \\ 0, & \text{for } n \text{ outside the window} \end{cases} \tag{15.6}$$

as

$$\sum_{n=-\infty}^{\infty} e_n^2 = \sum_{n=-\infty}^{+\infty} \left(\tilde{x}_n + \sum_{i=1}^{m} \hat{a}_i \tilde{x}_{n-i} \right)^2 \tag{15.7}$$

where data windowing results in an implicit limitation of the infinite sum.

2. The *covariance method* defines the prediction error power estimate via a windowing of the error signal itself without explicit data windowing:

$$\sum_{n=m}^{L-1} (w_n \cdot e_n)^2 = \sum_{n=m}^{L-1} w_n \left(x_n + \sum_{i=1}^{m} \hat{a}_i x_{n-i} \right)^2 \tag{15.8}$$

where the summation bounds are carefully chosen to avoid the use of speech samples outside the window.

In both methods, minimization of the quadratic cost function leads to a system of linear equations for the unknown parameters $\hat{a}_i$. In the autocorrelation method, the system matrix turns out to be a proper correlation matrix with *Toeplitz* structure which allows a computationally efficient solution using the order-recursive *Levinson–Durbin algorithm*. This algorithm reduces the operation count of the estimator from $O(m^3)$ to $O(m^2)$ and guarantees that the roots of the polynomial $\hat{A}(z)$ lie inside the unit circle. In the covariance method, no such structure can be exploited such that higher complexity and additional mechanisms for stabilizing $\hat{A}(z)$ are required. Its advantage is the significantly increased accuracy of estimated coefficients because the absence of explicit data windowing avoids some systematic errors of the autocorrelation method. Finally, to enhance the numerical properties of LP analysis, additional (pre-)processing steps are routinely included such as high-pass prefiltering of input speech to remove unwanted low-frequency components, a bandwidth expansion applied to correlation function estimates, and a correction of the autocorrelation at lag 0 which corresponds to the addition of a weak white noise floor (at $-40\,\text{dB}$ of the data).

15.3.3 Sinusoidal modeling

The second implementation of Wold's decomposition emphasizes the idea of spectral modeling using sinusoids as proposed by MacAuley and Quartieri [1986]. In this model, both signal components are produced by the same *sum of sinusoids*. The noise-like regular component can well be approximated by such a sum (as suggested by spectral representation theory) if the relative phases of the sinusoids are randomized frequently, at least once for each frame. This scheme offers the advantage of retaining the original phase information in the singular component (if the available bit rate allows us to code it, of course) and it is more flexible in combining regular and singular components. Typically, even voiced speech contains a noise-like component in the higher frequency bands which can be modeled by using a fixed-phase model for the lower harmonics and a random-phase model for the higher harmonics. This *hard switch in the frequency domain* assumes the segregation of noise-like and nearly periodic signal components along the frequency axis. While this allows the modeling of mixed excitation speech signals the transitions between excitation signal types are now a priori restricted to the frame boundaries (whereas the LPC allows in principle higher temporal resolution for voicing switch update).

A further development of sinusoidal modeling is the Multi Band Excitation (MBE) coder which allows separate voicing decisions in multiple frequency bands, thereby achieving a more accurate description of mixed excitation phenomena. This also establishes a relationship to the principles of subband coding or transform coding which do not rely on specific source signal models and which are most popular in coding of generic audio.

15.3.4 Harmonic + noise modeling

The third implementation of Wold's decomposition strives at a full realization of the additive superposition of regular and singular components *without enforcing a hard switch in either the time or frequency domains*. Thus, the HNM is maybe the simplest in its decoder structure which just follows Eq. (15.1), whereas the encoder is the most difficult as it has to solve the problem of simultaneous estimation of superimposed continuous spectra (the regular component) and discrete spectra (the singular component, assumed to be harmonically related). So far, use of this model has been confined to applications in speech and audio synthesis, whereas for speech coding it has been recognized that the level of sophistication achieved in the HNM-spectral representation needs to be complemented by a more detailed analysis of time domain variations, too, leading us beyond the conventional short-time stationary model.

15.3.5 Cyclostationary modeling

The short-time stationary model of speech neglects the rapid time variations induced by vocal fold oscillations. For voiced speech, this nearly periodic oscillation not only serves as the main excitation of the vocal tract but also as a cyclic modulation of all signal statistics. Stochastic processes with periodic statistics are known as *cyclostationary processes.* For our purposes, where the fundamental period T_0 and the associated oscillation pattern may slowly evolve over time, we need to adapt this concept to short-time cyclostationary processes. For a given speech frame, the waveform will be decomposed in a cylic mean signal (corresponding to the singular component of Wold's decomposition) and a zero-mean noise-like process with periodically time-varying correlation function. Most importantly, the variance of this noise-like component is a periodic function of time representing the periodic envelope modulation of the noise-like component of voiced speech. This effect is clearly audible for lower pitched male voices and constitutes one of the main improvements of cyclostationary speech models over HNMs. These models were originally introduced as (Prototype) Waveform Interpolation (PWI) coders by Kleijn and Granzow [1991] where the cyclic mean is interpreted as the "slowly evolving waveform" and the periodically time-varying noise-like component is termed the "rapidly evolving waveform". More recent developments work with filterbanks whose channels are adapted to multiples of f_0 by prewarping the signal in the time domain. A key aspect of cyclostationary signal modeling is the extraction of reliable "pitch marks" that allow explicit time domain *synchronization* of subsequent fundamental periods. As all speech properties evolve over time, this synchronization problem is a major challenge and leads to the characterization of speech signals in terms of an underlying self-oscillating nonlinear dynamical system with the added benefit of explaining certain non-additive irregularities of speech oscillations such as jitter or period-doubling phenomena.

15.4 Quantization and coding

Once the signal model and the parameter estimation algorithms have been selected, the encoder has to quantize the parameters and, in hybrid coders, samples of the modeling residual waveform as well.[7] To achieve source-coding efficiency, several techniques beyond the simple uniform quantization scheme found in most analog-to-digital converters have been developed.

15.4.1 Scalar quantization

Scalar quantization refers to the quantization of a single random variable which can be either a waveform sample or a single model parameter. If there is more than one variable to be quantized then they are treated independently of each other. As a general example, the continued (or high-resolution discrete) value x of the random variable X is quantized to the ith quantization interval $[t_i, t_{i+1}]$ if $t_i \leq x < t_{i+1}$, and the index $i = Q(x)$ will be transmitted using a (binary) codeword. The receiver decodes the codeword back to the quantizer index and uses it to address a lookup table that stores a high-resolution reconstruction value $x^{(q)} = Q^{-1}(i)$. The latter operation is sometimes referred to as "inverse quantization" although the static nonlinearity of a quantizer is a many-to-one map and, as such, has no unique inverse function. A typical performance measure for scalar quantizers is the Signal-to-Quantization Noise Ratio (SQNR) based on the mean-square value of the quantization error $q = x^{(q)} - x$:

$$SQNR_{[dB]} = 10 \log_{10} \frac{E(x^2)}{E(q^2)} = 10 \log_{10} \frac{E(x^2)}{E((x^{(q)} - x)^2)} \tag{15.9}$$

where the operator E is the expectation operator performing an average weighted by the probability measure of the variable X. This definition includes both overflow distortions when the input value reaches the boundaries of the quantizer range and granular distortions related to the individual quantizer steps. In most designs, overflow will be avoided by using proper scaling such that granular distortions will be the quality-determining effect. Normalization of error power by input power makes the measure somewhat more relevant to human perception when dealing with waveform quantization.

In uniform quantization, all intervals have the same width $t_{i+1} - t_i = q = $ constant. The most prominent non-uniform scalar quantization law is 8-bit *logarithmic* quantization for PCM speech signals with two variants of the logarithmic characteristic, the μ-law in the US and Japan and the A-law in Europe. Logarithmic quantization has the remarkable property that the SQNR

---

[7] Note that this is the case for *forward-adaptive* speech coders. For extremely low delay, *backward-adaptive* coders avoid the transmission of parameters as side information and rather estimate model parameters making exclusive use of speech samples which have already been decoded. This requires implementation of both encoder and decoder algorithms in the transmitter and works under the assumption of low channel error rates. The parameters estimated from decoded data only implicitly contain some delay relative to those parameters found from the current frame in forward-adaptive schemes. However, due to the short-time stationarity assumption, this delay results in an acceptable loss in modeling performance which is approximately compensated by the bit rate reduction obtained from suppression of the side information channel.

becomes nearly independent[8] of the signal power $E(x^2)$ as long as overflows are avoided and the signal does not fall in the close vicinity of 0 (where the log function is approximated with a linear function, corresponding to the resolution of a 12-bit uniform quantizer). Thus, logarithmic quantizers are designed to cope with the strong temporal variations of the speech signal envelope which are both intrinsic (short-time stationary sequence of highly different speech sounds, soft versus loud voice) and extrinsic (variable distance to the microphone).

Adaptive quantization extends the dynamic range of a quantizer even further than logarithmic quantization such that input power fluctuations of 40 dB and more show no pronounced effect on SQNR performance. It is implemented with an adaptive gain control mechanism, operating on a sample-by-sample backward adaptive basis in one of the most popular variants.

If the random variable X itself has a non-uniform amplitude probability distribution, the uniform quantizer does not achieve the minimal SQNR performance for a given number of quantization intervals I. The set of *optimal quantizer* thresholds $t_i, i = 1, \ldots, I$ and reconstruction values $x^{(q)}(i), i = 1, \ldots, I$ can be found using the iterative Lloyd–Max algorithm which alternates between the two following implicit conditions found from minimizing the mean-square quantization error for fixed input power:

1. The best quantization thresholds lie at equal distance between adjacent reconstruction values:

$$t_i = \frac{x^{(q)}(i) - x^{(q)}(i-1)}{2} \tag{15.10}$$

2. The best reconstruction values lie in the centers of mass of the quantization intervals, computed as a conditional expected value using the input amplitude probability density $f_X(x)$:

$$x^{(q)}(i) = E(x|t_i \le x < t_{i+1}) = \int_{t_i}^{t_{i+1}} f_X(x) \, x \, dx \tag{15.11}$$

In practice, probability distributions are often not known such that the center-of-mass computations are done by evaluating the class-conditional means of a large set of training data samples collected in real-world experiments.

15.4.2 Vector quantization

VQ combines d scalar random variables X_k into a d-dimensional random vector $\vec{X}$. In the d-dimensional space, the scalar quantization intervals turn into multidimensional, convex *Voronoi cells* $V_i, i = 1, \ldots, I$. If an input vector $\vec{x} \in V_i$, then $Q(\vec{x}) = i$ and the value i is mapped to a binary codeword for transmission. Shannon [1959] showed that using VQ with asymptotically increasing dimension d can achieve the rate distortion bound for source coding – i.e., just like in channel coding the best coding performance is obtained when considering large blocks of signal samples simultaneously. VQ is known to have three different advantages over scalar quantization. For this comparison, we will study the number of bits per dimension $\log_2(I)/d$ needed to achieve a certain quantization distortion.

1. The *memory advantage* – i.e., the handling of statistical dependences. In general, a random vector exhibits statistical dependences among its components, which may go beyond linear correlations. Therefore, even if we use linear transforms (such as the Discrete Cosine Transform, DCT, or principal component analysis) to decorrelate the components of the

---

[8] Actually, for logarithmic quantization, the SQNR is essentially independent of the amplitude probability distribution of the input, thereby making it maximally *robust* to unknown or strongly varying signal properties.

random vector, the joint quantization of all components is still more efficient due to the redundancy contained in the remaining (nonlinear) dependences.

2. The *space-filling advantage*. In scalar quantization, there is not much choice in selecting the shape of quantization intervals, whereas, in multiple dimensions, Voronoi cells can be shaped to achieve the best space-filling degree – e.g., according to a dense sphere-packing principle. This allows coverage of a certain volume in space with the same maximal distortion (given by a linear dimension of the Voronoi cell) but fewer cells – i.e., a lower number $\log_2(I)$ of codeword bits. This advantage approaches 0.25 bit/dimension for large vector dimension d. For low-rate speech coding at 4 kbit/s or less (which means less than 0.5 bit/sample) this becomes a very significant contribution.

3. The *shape advantage* is very similar to the gains achieved for optimal non-uniform quantizers of scalar variables where the shape (and size) of the quantizer cells are matched to the probability distribution of the random vector. This advantage is only relevant for resolution-constrained quantizers with a predetermined number of cells I. For entropy-constrained quantizers (where we only care about the average entropy of all codewords but we do neither constrain the number of cells or codewords nor the length of the codewords), there is no shape advantage as this can equally be obtained by lossless entropy coding of the codewords themselves.

The design of optimal vector quantizers proceeds along the same lines as the design of non-uniform quantizers for scalar variables. The generalization of the Lloyd–Max algorithm to multiple dimensions is known as the Linde–Buzo–Gray or LBG algorithm.

Suboptimum vector quantization

The larger the vector dimension d, the better the VQ performance. However, if we design a VQ for a certain *given number of bits per dimension*, this means that the complexity of the VQ grows exponentially with dimension d because I, the number of Voronoi cells, grows exponentially with d. As multidimensional Voronoi cells may have complicated shapes, we rather define them by their reconstruction vectors $\vec{x}_q(i)$ and perform quantization by selecting the index $i = Q(\vec{x})$ which minimizes the quantization distortion $D(\vec{x}_q(i), \vec{x})$ over all $i = 1, \ldots, I$. A typical distortion measure would be the quadratic distortion or Euclidean distance $\|\vec{x}_q(i) - \vec{x}\|^2$. Thus we need to memorize a table or *codebook* of I vectors of dimension d and we need to make a full search through this codebook to minimize distortion, which requires I distortion measure computations where each has a cost proportional to d in case of Euclidean distance. So even per dimension, the memory cost and the number of operations is directly proportional to I and, therefore, suffers from the curse of (= exponential growth with) dimensionality.

To address this issue, simplified schemes for vector quantizer design and implementation have been developed. Surprisingly, it turns out that the suboptimal search for a good vector in a codebook filled with suboptimal entries can still outperform an optimal VQ system of lower dimension. While early attempts at VQ-based waveform quantization worked at 1 bit/dimension with a full search over $I = 2^{10} = 1,024$ entries of dimension $d = 10$, the new suboptimal systems work at 0.875 bit/dimension with a greedy search over $I = 2^{35} = 34.4 \cdot 10^9$ entries of dimension $d = 40$ and still achieve better performance.[9] Examples of such simplified, suboptimal designs are:

- *gain shape quantization* where the codevector is the product of a scalar gain and a unit vector (norm 1);

---

[9] Note that a full search through these 40-dimensional codebooks would result in an increase in complexity *per dimension* by a factor of $2^{23} = 8.4 \cdot 10^6$ when compared with the earlier 10-dimensional codebooks!

- *multistage VQ* where a first codebook provides a coarse approximation and the subsequent $K - 1$ codebooks successively refine the quantization;
- *split VQ* where the high-dimensional vector is split into K pieces of dimension d/K each, which are then quantized independently.

In the two latter cases, complexity is approximately reduced to that of a d/K-dimensional VQ, as non-exponential factors do not matter much for such comparison.

As one of the most successful simplifications – having made it into several international standards – we discuss *algebraic codebooks* where codevector amplitudes are restricted to a ternary alphabet $\{-1, 0, +1\}$ and where, furthermore, only a small percentage of nonzero amplitudes is allowed for (e.g., 10 in a vector of dimension 40). This design is inspired from the earlier techniques of "multipulse excitation" and "Regular Pulse Excitation" (RPE) where the source excitation of a source–filter model was pruned to a few nonzero amplitudes. As Euclidean distance computations consist in the evaluation of inner products, the algebraic structure reduces each evaluation from 40 multiply–add operations to 10 additions only. Furthermore, the selection of nonzero pulses follows a greedy scheme that does not try all possible combinations. Finally, they are encoded efficiently using several interleaved combs of subsampled pulsetrains or polyphases (Interleaved Single Pulse Permutation, ISPP). Note that this codebook is built with a systematic structure in mind which achieves a fairly uniform coverage of the entire vector space, so no training or optimization is done for codevector entries.

Line spectrum pairs

When applying VQ to the parameters of a speech model (and not to waveform samples), special considerations regarding the distortions induced by parameter quantization errors are needed. A common application is the simultaneous quantization of the $m = 10$ linear prediction coefficients a_i which raises the following issues:

1. The quantized parameter vector should still correspond to a minimum-phase filter allowing stable operation of the filter and its inverse.
2. Quantization error is best measured in terms of induced *spectral distortion* which is the mean-square difference between the original spectral envelope and the one described by the quantized parameters, both expressed in dB. From experience, this distortion should be less than 1 dB for most of the parameter vectors to achieve transparent coding quality.
3. Within the short-time stationary model, the transition from one frame to the next should allow simple interpolation mechanisms between two parameter vectors which still maintain stability at all times and avoid unexpected spectral excursions of the interpolated spectral envelopes.

All these issues are best resolved before VQ by applying a nonlinear transform from the predictor coefficients to a vector of line-spectral frequencies which occur in distinct pairs and, therefore, are also referred to as Line Spectrum Pairs (LSPs). They are derived by introducing two polynomials $F_1(z)$ and $F_2(z)$ of order $m + 1$ with even and odd symmetries in their coefficients, respectively:

$$F_1(z) = A(z) + z^{-1} \cdot z^{-m} A(z^{-1}) \tag{15.12}$$

$$F_2(z) = A(z) - z^{-1} \cdot z^{-m} A(z^{-1}) \tag{15.13}$$

where $z^{-m} A(z^{-1}) = \sum_{i=0}^{m} a_{m-i} z^{-i}$ is the *time-reversed predictor polynomial*. We recognize that this first step is invertible as we have $A(z) = (F_1(z) + F_2(z))/2$. However, for a minimum-phase filter $A(z)$, the mirror poynomials $F_1(z)$ and $F_2(z)$ have their roots not inside but exactly on the unit circle such that each of their locations can be fully specified by a single real-valued frequency. Furthermore, the root location sets of the two polynomials follow a joint sorting relationship such that, with increasing frequency, the two sets are strictly interleaved on the unit circle. These properties simplify the search for the polynomial roots significantly and also suggest the combination of pairs of adjacent line spectrum frequencies (one from each mirror polyomial). Furthermore, this sorting relationship helps in interpolation of estimated LSP vectors over time. Finally, the VQ for this vector of ten real numbers is often implemented as a split VQ with three parts sorted by increasing frequencies which contributes to the relative independence of vector parts and minimizes the effects of suboptimal quantization.

15.4.3 Noise shaping in predictive coding

Hybrid speech coders do not rely entirely on the source signal model but observe and transmit the modeling error or residual waveform as well. If we use a predictive signal model, the *residual waveform* is obtained as the prediction error $e_n = x_n - \hat{x}_n$. If we quantize the prediction residual e_n for transmission, it results in $e_n^{(q)} = e_n + q_n$ from which the decoder reconstructs $x_n^{(d)} = \hat{x}_n + e_n^{(q)}$. The *decoder* computes the prediction $\hat{x}_n$ from the delayed reconstructed signal samples $x_{n-i}^{(d)}$:

$$\hat{x}_n = -\sum_{i=1}^{m} a_i x_{n-i}^{(d)} \qquad (15.14)$$

which results in a recursive or *closed-loop* reconstruction filter:

$$x_n^{(d)} = -\sum_{i=1}^{m} a_i x_{n-i}^{(d)} + e_n^{(q)} \qquad (15.15)$$

$$A(z)X^{(d)}(z) = E^{(q)}(z) = E(z) + Q(z) \qquad (15.16)$$

However, when computing the predicted signal sample $\hat{x}_n$ in the *encoder*, we can do this in two different ways:

1. *Open-loop prediction* makes use of the delayed original signal samples as available in the transmitter only:

$$\hat{x}_n = -\sum_{i=1}^{m} a_i x_{n-i} \qquad (15.17)$$

$$E(z) = A(z)X(z) \qquad (15.18)$$

    from which we get the decoder output with Eq. (15.16):

$$X^{(d)}(z) = X(z) + \frac{Q(z)}{A(z)} \qquad (15.19)$$

2. *Closed-loop prediction*, however, uses exactly the same reconstructed samples for prediction in the encoder as in the decoder (Eq. 15.14) to compute:

$$e_n = x_n - \hat{x}_n = x_n + \sum_{i=1}^{m} a_i x_{n-i}^{(d)} \qquad (15.20)$$

This results in the decoder output as:

$$x_n^{(d)} = -\sum_{i=1}^{m} a_i x_{n-i}^{(d)} + e_n^{(q)} \qquad (15.21)$$

$$= -\sum_{i=1}^{m} a_i x_{n-i}^{(d)} + x_n + \sum_{i=1}^{m} a_i x_{n-i}^{(d)} + q_n \qquad (15.22)$$

$$X^{(d)}(z) = X(z) + Q(z) \qquad (15.23)$$

In this setup, the encoder predictor duplicates the *closed-loop* structure of the decoder, hence its name.

Quantization noise $Q(z)$ is modeled with a white spectrum. According to Eq. (15.19), open-loop prediction results in an overall coding distortion $X^{(d)}(z) - X(z)$ with a spectrum shaped like the speech model spectrum $1/A(z)$ which reduces the audibility of this noise due to the masking properties of human hearing. However, filtering of the noise by $1/A(z)$ also results in an overall gain that makes the total noise power identical to that observed in direct quantization of the speech waveform without using prediction. Closed-loop prediction keeps a white noise spectrum without noise amplification according to Eq. (15.23). However, as quantization noise is not shaped like the speech spectrum, it may become audible in the spectral valleys of the latter. A compromise between these two extremal positions is found in the use of a *perceptual noise-shaping filter* [Schroeder et al. 1979]. To this end, quantization noise is observed as the difference between quantizer output and input, $q_n = e_n^{(q)} - e_n$, and a filtered version of this noise is fed back to the input of the quantizer $E(z) \rightarrow E(z) + (W(z) - 1)Q(z)$ (where the zero-delay gain is $w_0 = 1$). Thereby, the open-loop prediction system results in a decoder output equal to $X^{(d)}(z) = X(z) + \dfrac{W(z)}{A(z)} Q(z)$. If we choose the weighting filter $W(z) = A(z/\gamma)$, with $0 < \gamma \leq 1$, we can control the spectral shape of the decoder distortion to be anywhere between a white spectrum ($\gamma = 1$) and the speech model spectrum ($\gamma = 0$). Note that this *bandwidth expansion* for the perceptual weighting filter can be simply obtained by setting $w_i = a_i \gamma^i$.

Long-term prediction

The *Short Term Predictors* (STPs) discussed so far only address the statistical dependences among neighboring signal samples (short-term correlations) which can be related to a non-flat spectral envelope – i.e., the formant structure of speech. The harmonic structure of voiced speech results in an additional spectral fine structure which corresponds to long-term correlations in the speech signal. These correlations are visible in the repetitive nature of the waveform cycles for a nearly periodic signal. The simplest *Long Term Predictor* (LTP) $B(z)$ predicts the current sample x_n from a sample that lies one period T_0 back into the past:

$$\hat{x}_n = b \cdot x_{n-T_0} \qquad (15.24)$$

As both the STP and LTP are stable linear systems, they can be cascaded in any sequence, but it often is more advantageous to first perform STP and then LTP. For LTP, closed-loop prediction is the preferred option. The best parameters b and T_0 can be obtained from analyzing the signal autocorrelation over a range of lags that spans from the shortest fundamental periods (high-pitched female voices) to the longest (low-pitched male voices). Even if the true pitch is not found in some cases (e.g., period doubling), the LTP will still contribute to coder performance by extracting any redundancy related to the autocorrelation maximum.

A specific problem in periodicity modeling with LTP occurs whenever the true fundamental period T_0 is not an integer multiple of the sampling interval T_s ("fractional pitch"). As a workaround, signal interpolation by a factor of 3 to 6 can be used to increase the effective sampling resolution. Another approach is to increase the order of the LTP filter $B(z)$ such that it simultaneously solves the problem of signal interpolation *and* prediction.

15.4.4 Analysis-by-synthesis

The analysis-by-synthesis procedure has been introduced in the context of multipulse-excited linear predictive coding [Atal and Remde 1982] and, later on, it was combined by Atal and Schroeder [1984] and Schroeder and Atal [1985] with vector quantization to result in *Code Excited Linear Prediction* (CELP). Essentially it consists in reformulation of the open-loop/ closed-loop prediction principle with perceptual weighting, which replaces the linear prediction analysis step of the encoder with a linear predictive synthesis step. Thereby, the basic structure of the underlying signal model becomes more directly visible and certain computational advantages can be realized when combined with vector quantization. Note, however, that this analysis-by-synthesis formulation is still exactly equivalent to the noise-shaping predictive coding ideas presented in the previous subsection.

The starting point is to generate or synthesize one (or more) decoded signal sample by selecting a quantized residual signal sample $e_n^{(q)}$ from an inventory of allowed values (codebook). To this end, all entries from the inventory are run through the LP synthesis filter $1/A(z)$ to produce a set of all decoded signal candidates $x_n^{(d)}$. The coding distortion can be computed from the difference $x_n^{(d)} - x_n$. From the above, it is evident that this distortion is proportional to quantization error q_n via the filters $W(z)/A(z)$. To select the best quantized residual signal, the coding distortion is filtered by the inverse weighting filters $A(z)/W(z)$ to obtain the quantization noise $q_n = e_n^{(q)} - e_n$. If we minimize the squared magnitude of this variable we effectively perform a nearest neighbor decision among all possible residual candidates $e_n^{(q)}$ from the codebook to find the entry closest to e_n.

The power of this analysis-by-synthesis interpretation is that it lends itself to any description of the residual waveform inventory as long as it can be enumerated. This is, in particular, the case for vector quantization of the residual waveform where, typically, instead of a single sample a number of $d = 40$ samples are collected into a subframe. The squared magnitude estimate of the perceptually weighted synthesis error is now computed by summing the squares of the d samples. The VQ application suggests the following two computational reorganizations that reduce numerical complexity without compromising the solution:

1. As we process blocks or subframes of d samples at a time, we have to make sure that all filters carry over their filter states from one subframe to the next. This means that the filter output for one subframe will be the sum of a *zero-input response* (the decaying filter states) and a *zero-state response* (the component driven by the current input). Only the latter component provides the information for selecting the best codevector from the VQ codebook. Therefore, it is convenient to first subtract the zero-input response from the speech signal x_n such that codevector selection is done on zero-state responses only.

2. For the zero-state response of each codevector, we have to run the vector through both the LP synthesis filter $1/A(z)$ and the (inverse) weighting filters $A(z)/W(z)$. This filter cascade can be simplified to the overall transfer function $1/W(z)$, resulting in the block diagram shown in Fig. 15.4.

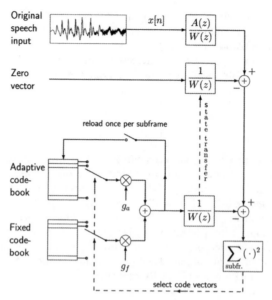

Figure 15.4 Code-excited linear predictive coding of speech; simplified block diagram showing separate treatment of zero-input and zero-state response with adaptive and fixed codebook excitation; weighting filter $W(z) = A(z/\gamma)$.

Adaptive codebook

The analysis-by-synthesis principle can also be carried over to the LTP which tries to explain the current subframe with predictions from (scaled) samples delayed by approximately one fundamental period T_0. As the fundamental period can vary in a certain range, it is advantageous to store subframe-length vectors taken from this range of the residual signal itself in a buffer memory that is called the *adaptive codebook*. One residual subframe is then represented by the weigthed sum of two vectors, one selected from this adaptive codebook and another one selected from the *fixed codebook*. In most systems, the two selections are made sequentially such that first the adaptive codebook entry is selected to model the periodicity of the signal and, based on this choice, the fixed codebook entry is selected to represent the remaining "stochastic" part of the residual (Figure 15.4). This approach is essentially identical to the use of an LTP, with minor differences for high-pitched voices where the fundamental period is shorter than a subframe length. Just as for LTP, good quality requires interpolation of the signal to increase the temporal resolution of the adaptive codebook, at least in the vicinity of the long-term correlation maximum.

15.4.5 Joint source channel coding

Unequal error protection

Once all the source parameters have been computed for an entire speech frame, their codewords need to be assembled for transmission and extra bits will be added for channel-coding purposes. Channel coding usually works with the assumption that the codeword sequence is independent identically distributed (iid) and that all codewords bear the same relevance for the receiver. The last condition is certainly not the case for hybrid speech coders where some codewords represent model parameters such as fundamental period, gains, or spectral envelopes (LSPs) whereas other codewords represent the residual waveform by entries in the fixed VQ codebook. A channel error in one of the more relevant parameters (where T_0 is maybe the most important one) will result in signficant waveform errors in the decoded signal as, e.g., the reconstructed voice may sound much lower or higher than the original. A channel error in one of the less relevant codewords may hardly be noticed as a properly parameterized speech model will generate rather good speech even if the details of the residual waveform are wrong. In conclusion, *"all source bits are equal, but some are more equal than others."* This results in classification of the codewords in different importance or channel sensitivity classes which are handled by different levels of channel error protection, ranging from sophisticated convolutional codes to no error protection at all.

Adaptive multi-rate coding

Both channel and source statistics vary over time. Therefore, the optimum coding strategy has to adapt itself to the time-varying channel state and source state. For circuit-switched communication, fixed overall rates are desired on the channel, so the task of the rate adaptation mechanism is to determine the best tradeoff between the rate budget spent on source coding and on channel error protection such that the total rate remains constant (Adaptive Multi Rate coding, AMR). Because source statistics change much faster (essentially they are different for each frame – i.e., every 20 ms) than channel-state information is updated, current systems will essentially select the best source versus channel rate split based on channel-state information only. Depending on this information, the source coder produces a different number of source bits per frame and the channel coder will use the remaining bits to perform adequate unequal error protection as described in the previous paragraph.

Source-coding optimization

There are two approaches to optimization of source coding in the light of joint source channel coding:

- *Encoder optimization* where, for instance, the *index assignment* of the VQ is optimized such that bit errors in the codewords representing the VQ index will result in the minimum mean perceptual impact, very much like the Gray coding strategy for scalar index assignment. A further encoder technique that strives for equal relevance of all codewords is found in *multiple-description coding*.
- *Decoder optimization* where residual redundancy among the received codewords is used to assist the channel decoder to increase the reliability of the overall decoding process.

15.5 From speech transmission to acoustic telepresence

The ultimate goal of speech transmission or telephony has always been to recreate the acoustic presence of a human talker at a geographically distant location. In that sense, telephony is just a special case of *telepresence* where we attempt to augment our local reality with the virtual presence of one or more remote persons, preferably recreating the remote environment to some extent. Our short discussion starts out from simple add-on functionalities for speech transmission and progresses to the most advanced services of three-dimensional virtual audio.

15.5.1 Voice activity detection

In a symmetric conversation between two persons, each of the participants is silent for about 50% of the time. This fact has long been observed and exploited for multiplexing telephone conversations over the same transmission channel in a technique known as Digital Speech Interpolation (DSI). For wireless communications, multiple access to the same shared radio spectrum requires the reduction of interference created among the users. This can be achieved by transmitting speech frames over the air interface only when the talker is actively speaking, a state which is detected by a Voice Activity Detector. For an inactive speaker, discontinuous transmission (DTX) results in:

1. less power consumption on the mobile terminal (saving of baseband-processing and radio transmitter power), thereby extending battery life;
2. less multiuser interference on the air interface, thereby enhancing mobile access network performance;
3. less network load if packet switching is used, thereby increasing backbone network capacity.

In a typical realization, speech transmission is cut after seven non-active frames, and a "silence descriptor frame" (SID, only 35 bits for 20 ms) is sent as a model for acoustic background noise which is used to regenerate comfort noise (CNG) at the receiver end. This noise model is updated at least every 24 frames and can be seen as a first step to virtual reality rendering of the auditory scene, thereby maintaining coherence between the communication partners by letting the other know implicitly that the call was made while driving a car, etc.

While this DTX method is sometimes referred to as source-controlled rate adaptation, it operates at a very high source description level (i.e., only source on versus source off) and, e.g., is not able to rapidly vary the source rate according to the phonetic contents (e.g., voiced versus unvoiced speech). This is also not the same as AMR coding described above (with VAD, the source rate is either zero or the currently allowed maximum source rate, the latter still being adapted according to the channel state).

15.5.2 Receiver end enhancements

If an entire speech frame is lost due to severe fading on the air interface, the resulting error cannot be corrected with channel codes targeted for individual or bursty bit errors. Rather, *error concealment* methods are required to interpolate such lost frames from received neighboring frames. This usually works very well for substitution of the first lost frame but may lead to unnatural sounding speech if continued over several frames lost in a row. In the latter case, later substituted frames will receive a gradual damping to achieve slow fadeout of the signal amplitude (typically over six frames = 120 ms).

A further receiver end signal enhancement is *adaptive postfiltering* which constitutes filters shaped according to the spectral envelope and LTP filters which will make noise introduced by lossy coding less audible to the listener. *Adaptive playout buffers* or *jitter buffers* help to conceal lost frames on packet-switched networks. If a terminal is equipped with wideband speech Input/Output (I/O) functionality (analog bandwidth from 50 Hz up to 7 kHz), speech received over a narrowband network may be augmented by artificially created frequency bands above 3.4 kHz and below 300 Hz using a synthesis mechanism known as *bandwidth extension*.

15.5.3 Acoustic echo and noise

The acoustic environment of a speaker may not only contain useful background information that adds to the presence of the auditory scene but also noise sources that are considered annoying, or reverberation and (acoustic) echo due to the fact that both speaking parties share a common acoustic space when using hands-free phones. In such situations, echo and noise need not only be controlled for the comfort of the human users but also for the speech-coding systems which, if strongly based on a speech signal model, will fail to handle background noise or echo appropriately. Solutions that perform *joint echo and noise control* will often achieve the best compromise in reducing the two impairments. An uncontrolled echo can have drastic consequences on the stability of the entire communication system; strict performance requirements have been set as standards by ITU-T in recommendations G.167 and G.168.

15.5.4 Service augmentation for telepresence

Speech-enabled services

In a telepresence framework, users will access a much wider range of services than in conventional telephony. Given the shrinking size of mobile terminals, service access benefits a lot from spoken language dialog with the machine offering the service. This can range from simple name dialling by voice (possibly activated by a spoken magic word) to Text To Speech (TTS) synthesis for email reading or *Distributed Speech Recognition* (DSR) where feature vectors for speech recognition are extracted on the mobile terminal, encoded with sufficient accuracy for pattern recognition (but not necessarily for signal regeneration at the central server end), and *transmitted as data* over the wireless link. This allows us to transmit recognition features based on higher quality signals than those used in telephony and it also allows us to use specific source- and channel-coding mechanisms that maximize benefits for the remote speech recognition server (rather than a human listener).

While DSR carries speech information over a data channel, the dual application of *Cellular Text Telephony* (CTT) allows the carriage of textual data over the speech channel so as to provide augmentative communication means for people with hearing or speaking impairments who still can exchange interactive text (not just short messages) in a chat-like style using a modem operating over the digital speech channel.

Personalization

For personalized services, *talker authentication* will be a must which should be distinguished from authentication of the mobile station or the infrastructure itself. The identity of a talker can be established via voice identification or verification techniques and, if needed, an additional personalized watermark might be inserted in the speech signal prior to encoding it.

For *talker privacy* the traditional phonebooth might once be replaced by a virtual talker sphere outside of which active speech cancellation (using wearable loudspeaker arrays) would make the phone conversation hardly audible to bystanders.

Three-dimensional audio

The virtual talker sphere has already introduced the concept of advanced audioprocessing for speech telephony which has seen a tremendous boost from the use of microphone and loudspeaker arrays and virtual/augmented audio techniques that allow the spatial rendering (e.g., ambisonic) of three-dimensional sound fields, potentially converted to binaural headphone listening. The best effects are expected if personalized HRTFs can be used together with real-time tracking of head movements to place virtual sound sources in the context of the real environment, creating the immersive telepresence which blends the presence of virtual participants and local participants for successful teleconferencing.

Ultimately, this will require a significant shift beyond today's speech-coding standards to allow for high-quality, multichannel audio including metadata information. A first move in this direction has been made by standardization of the AMR wideband speech codec which has become the first speech-coding standard ever to be accepted almost simultaneously for wireline communication (ITU), wireless communication (ETSI/3GPP), and Internet telephony (IETF).

Further reading

The classical textbooks on speech coding are Jayant and Noll [1984] and Kleijn and Paliwal [1995], which, at the time of this writing, are unfortunately out of print. The excellent textbook by Vary et al. [1998] is written in German, but an improved edition in English is in preparation. Recently, a second edition of Kondoz [2004] has appeared. Source-coding theory is treated in Berger [1971], Gray [1989], and Kleijn [2005] which is maybe best suited for the needs of speech coding. For speech signal processing in general, we recommend the classic Rabiner and Schafer [1978] and the more recent textbooks of Deller et al. [2000], O'Shaughnessy [2000], and Quatieri [2002]. For extensions to nonlinear speech modeling, see Chollet et al. [2005] and Kubin [1995]. In Gersho and Gray [1992] an extensive discussion of vector quantization is provided. Hanzo et al. [2001] address joint solutions for source and channel coding for wireless speech transmission. The wider context of acoustic signal processing for telecommunications is treated in Gay and Benesty [2000], Hänsler and Schmidt [2004], and Vaseghi [2000]. Bandwidth extension is the main topic of Larsen and Aarts [2004]. An excellent reference for spoken language processing is Huang et al. [2001] and Jurafsky and Martin [2000]. Gibson et al. [1998] treat the compression of general multimedia data.

16

Equalizers

16.1 Introduction

16.1.1 Equalization in the time domain and frequency domain

Wireless channels can exhibit delay dispersion – i.e., Multi Path Components (MPCs) can have different runtimes from the transmitter (TX) to the receiver (RX) (see Chapters 6 and 7). Delay dispersion leads to Inter Symbol Interference (ISI), which can greatly disturb the transmission of digital signals. We already saw in Chapter 12 that even a delay spread that is smaller than the symbol duration can lead to a considerable Bit Error Rate (BER) degradation. If the delay spread becomes comparable with or larger than the symbol duration, as happens often in second- and third-generation cellular systems, then the BER becomes unacceptably large if no countermeasures are taken. Coding and diversity can reduce, but not completely eliminate, errors due to ISI (Chapters 13 and 14). On the other hand, delay dispersion can also be a positive effect. Since fading of the different MPCs is statistically independent, resolvable MPCs can be interpreted as diversity paths. Delay dispersion thus offers the possibility of *delay diversity*, if the RX can separate, and exploit, the resolvable MPCs. Using the fact that the transfer function is the Fourier transform of the impulse response with the Fourier transform pair $\tau \longrightarrow f$, delay diversity can also be interpreted as frequency diversity (see Chapter 13).

Equalizers are RX structures that work both ways: they reduce or eliminate ISI, and at the same time exploit the delay diversity inherent in the channel. The operational principle of an equalizer can be visualized either in the time domain or the frequency domain.

For an interpretation in the frequency domain, remember that delay dispersion corresponds to frequency selectivity. In other words, ISI arises from the fact that the transfer function is not constant over the considered system bandwidth. The goal of an equalizer is thus to reverse distortions by the channel. In other words, the product of the transfer functions of channel and equalizer should be constant.[1] This can be expressed mathematically the following way: let the original signal be $s(t)$; it is sent through a (quasi-static) wireless channel with the impulse response $h(t)$, received, and sent through an equalizer with impulse response $e(t)$. We furthermore assume that the transmit signal uses Pulse Amplitude Modulation, PAM (see Chapter 11), so that:

$$s(t) = \sum_i c_i g(t - iT) \tag{16.1}$$

---

[1] Actually, this is a special form of equalizer, the so-called "zero-forcing" (ZF) linear equalizer, that will be discussed below in more detail.

where the c_i are the complex transmit symbols. We now require that:

$$H(\omega)G(\omega)E(\omega) = \text{const} \tag{16.2}$$

where $E(\omega)$, $H(\omega)$, and $G(\omega)$ are the Fourier transforms of $e(t)$, $h(t)$, and $g(t)$, respectively.

An equivalent formulation in the *time domain* requires that the received signal be free of ISI at the sampling instants. Define $\eta(t)$ as the convolution of the channel impulse response $h(t)$ with the basis pulse $g(t)$:

$$\eta(t) = h(t) * g(t) \tag{16.3}$$

Then we require that:

$$[\eta(t) * e(t)]_{t=iT_\text{s}} = \begin{cases} 1 & i = 0 \\ 0 & \text{otherwise} \end{cases} \tag{16.4}$$

If the channel were known and static, we could build (in hardware) a filter that performs the required equalizations of the transfer function. In wireless communications, however, the channel is (i) unknown and (ii) time-variant. The former problem can be solved by transmission of a *training sequence* – i.e., a sequence of known bits. From the received signal $r(t)$ and knowledge of the shape of the basis pulses $g(t)$, the RX can estimate the channel impulse response $h(t)$. The problem of time variance is solved by repeating the transmission of the training sequence at "sufficiently short" time intervals, so that the equalizer can be adapted to the channel state at regular intervals. The concept is thus known as "adaptive equalization".

Over the years, many different types of equalizers have been developed. The simplest is the linear equalizer, which is usually a tapped-delay-line filter with coefficients that are adapted to the channel state (Section 16.2). Decision feedback filters make use of the fact that the ISI created by past symbols can be computed (and subtracted) from the received signal (Section 16.3). Finally, the optimum way of detection in a delay-dispersive channel is the maximum-likelihood sequence estimation (Section 16.4). Blind equalizers, which do not need a training sequence, have been intensively investigated by researchers, but are not very popular for practical systems.

16.1.2 Modeling of channel and equalizer

The following sections, describing different equalizer structures, will require a discrete time model of the channel and equalizer. We now give such a model, together with the important concept of the *noise-whitening filter* that has great importance for optimum RXs.

The first stage of the RX consists of a filter that limits the amount of received noise power. This filter should make sure that all information is contained in sample values at instances $t_\text{s} + iT_\text{s}$. This is achieved by a filter that is matched to $\eta(t)$ – i.e., the convolution of channel impulse response and basis pulse. The sequence of sample values at the output of the matched filter is then given by:

$$\psi_i = c_i \zeta_0 + \frac{1}{\zeta_0} \sum_{n \neq i} c_n \zeta_{i-n} + \widehat{n}_i \tag{16.5}$$

where ζ_i are the sample values of the Auto Correlation Function (ACF) of $\eta(t)$, and $\widehat{n}_i$ is a sequence of complex Gaussian random variables with ACF $N_0 \zeta_i$ – i.e., the noise filtered by the matched filter.

The z-transform of the sampled ACF ζ_i can be factored as:

$$\Xi(z) = F(z)F^*(z^{-1}) \tag{16.6}$$

This factorization is not unique; but it is advantageous to choose $F^*(z^{-1})$ in such a way that all roots are within the unit circle. In such a case, $1/F^*(z^{-1})$ is a stable, realizable filter.

Now, if the matched filter is concatenated with $1/F^*(z^{-1})$, then the noise at the output of this concatenation is white again, with an ACF given by $N_0\delta_k$. It is therefore also known as the *noise-whitening filter*. The sample values of the impulse response of concatenation of the basis pulse, wireless channel, matched filter, and noise-whitening filter is henceforth denoted as the *discrete time channel*. The impulse response of the discrete time channel is written as f_k and its z-transform as $F(z)$. Note that the impulse response of this channel is causal, so that the output signal can be written as:

$$u_i = \sum_{n=0}^{L_c} f_n c_{i-n} + n_i \tag{16.7}$$

where L_c is the length of the impulse response of the discrete time channel. Therefore, the noise-whitening filter is also often known as the *precursor equalizer*.

We will use the following notation in this chapter:

c_i	ith complex transmit symbol;
$\eta(t)$	convolution of basis pulse and channel impulse response;
e_i	ith equalizer coefficients;
ζ_i	ith sample value of the ACF of $\eta(t)$;
$\widehat{n}_i$	ith sample of a sequence of complex Gaussian random variables with ACF $N_0\zeta_i$;
n_i	ith sample of a sequence of uncorrelated complex Gaussian random variables;
f_i	ith sample of the impulse response of the time-discrete channel;
u_i	ith sample of the output signal of the time-discrete channel;
$\widehat{c}_i$	estimate of transmit symbol c_i;
ε_i	deviation between estimated and true transmit symbol $c_i - \hat{c}_i$.

16.1.3 Channel estimation

A common strategy for data detection with an equalizer is to separate the estimation of **f** and **c**. In a first step, a training sequence (i.e., known **c**) is used to estimate **f**. During the subsequent transmission of the unknown payload data, we assume that the estimated impulse response is the true one, and solve the above equation for **c**.

In this subsection, we discuss estimation of the channel impulse response by means of a training sequence. Channel estimation shows strong similarities to the "channel-sounding" techniques described in Chapter 8. A very simple estimate can be obtained by means of PN-sequences (*Pseudo Noise sequences*) with period N_{per}. The ACF of the PN sequence approximates a Dirac delta function. More precisely, periodic continuation of the sequence $\{b_i\}$, convolved with a time-reversed version of itself,[2] gives a sum of Dirac pulses spaced N_{per} symbols apart:

$$\{b_{-i}\} *_{per} \{b_i\} \approx \sum_{n=-\infty}^{\infty} \delta_{i-nN_{per}} \tag{16.8}$$

If this sequence is sent over a channel with impulse response **f**, the output of the correlator becomes:

$$\{\widehat{f}_i\} = (\{b_{-i}\} *_{per} \{b_i\}) * \{f_i\} \tag{16.9}$$

where $*_{per}$ denotes periodic convolution. Now, if the duration of the channel impulse response is shorter than N_{per}, then $\widehat{\mathbf{f}}$ is simply a periodic repetition of **f** (see Fig. 16.1).

---

[2] Strictly speaking, we need to take the complex conjugate of the time-reversed sequence. However, as the transmit sequence is usually real (+1/−1), bits and complex symbols can be considered to be equivalent.

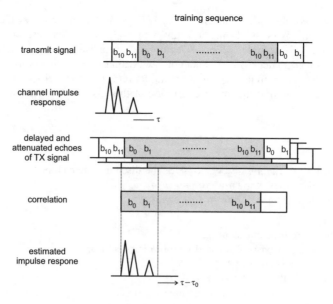

Figure 16.1   The principle of channel estimation by correlation.

In practice, we do not transmit a periodic continuation of the PN sequence, but just a single realization. In order to avoid the possible influence of (unknown) payload bits on correlator output, known "buffer bits" have to be transmitted before and after the sequence.

Channel estimation by means of a training sequence technique has several drawbacks:

1. *A reduction in spectral efficiency*: the training sequence does not convey any payload information. For example, the Global System for Mobile communications (GSM) uses 26 bits in every 148-bit frame for the training sequence (see Chapter 21).

2. *Sensitivity to noise*: in order to keep spectral efficiency reasonable, the training sequence has to be short. However, this implies that the training sequence is sensitive to noise (longer training sequences can average out noise), and also to non-idealities in sounding sequences (remember that the peak-to-offpeak ratio of a PN-sequence increases with the length of such a sequence). If channel estimation is done by means of iterative algorithms, only algorithms with a fast convergence rate can be used; however, such algorithms lead to a high residual error rate.

3. *Outdated estimates*: if the channel changes after transmission of the training sequence, the RX cannot detect this variation. Use of an outdated channel estimate leads to decision errors.

Despite these problems, training-sequence-based channel estimation is used in practically every system. The reason for this is that the alternative (blind techniques, see Section 16.7) requires high computational effort and suffers from significant numerical problems.

16.2 Linear equalizers

Linear equalizers are simple linear filter structures that try to invert the channel in the sense that the product of the transfer functions of channel and equalizer fulfills a certain criterion.

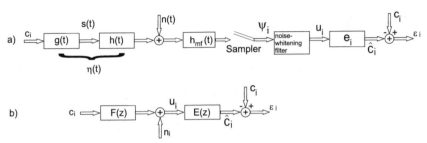

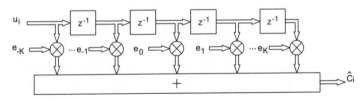

Figure 16.2   Linear equalizer in the time domain (a) and time-discrete equivalent system in the z-transform domain (b).

This criterion can either be achieving a completely flat transfer function of the channel–filter concatenation, or minimizing the mean-squared error at the filter output.

The basic structure of a linear equalizer is sketched in Fig. 16.2. Following the system model of Section 16.1.2, a transmit sequence $\{c_i\}$ is sent over a dispersive, noisy channel, so that the sequence $\{u_i\}$ is available at the equalizer input. We now need to find the coefficients of a Finite Impulse Response (FIR) filter (transversal filter, Fig. 16.3) with $2K+1$ taps. This filter should convert sequence $\{u_i\}$ into sequence $\{\widehat{c}_i\}$:

$$\widehat{c}_i = \sum_{n=-K}^{K} e_n u_{i-n} \tag{16.10}$$

that should be "as close as possible" to the sequence $\{c_i\}$. Defining the deviation ε_i as:

$$\varepsilon_i = c_i - \widehat{c}_i \tag{16.11}$$

we aim to find a filter so that:

$$\varepsilon_i = 0 \quad \text{for } N_0 = 0 \tag{16.12}$$

which gives the *ZF equalizer*, or that:

$$E\{|\varepsilon_i|^2\} \longrightarrow \min \quad \text{for } N_0 \text{ having a finite value} \tag{16.13}$$

which gives the *Minimum Mean Square Error (MMSE) equalizer*.

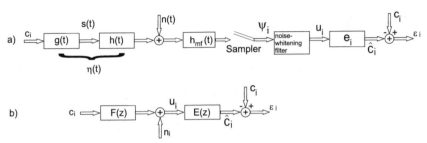

Figure 16.3   Structure of a linear transversal filter. Remember that z^{-1} represents a delay by one sample.

16.2.1 Zero-forcing equalizer

The ZF equalizer can be interpreted in the frequency domain as enforcing a completely flat (constant) transfer function of the combination of channel and equalizer by choosing the equalizer transfer function as $E(z) = 1/F(z)$. In the time domain, this can be interpreted as minimizing the maximum ISI (*peak distortion criterion*). Appendix 16.A (see *www.wiley.com/go/molisch*) shows that these two criteria are identical.

The ZF equalizer is optimum for elimination of ISI. However, channels also add noise, which is amplified by the equalizer. At frequencies where the transfer function of the channel attains small values, the equalizer has a strong amplification, and thus also amplifies the noise. As a

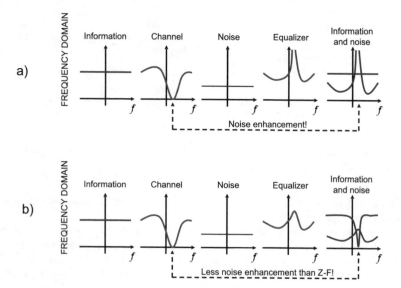

Figure 16.4 Illustration of noise enhancement in zero-forcing equalizer (a), which is remedied in an MMSE linear equalizer (b).

consequence, (i) the noise becomes colored, and (ii) the noise power at the detector input is larger than for the case without an equalizer (see Fig. 16.4).

The Fourier transform $\Xi(e^{j\omega T_s})$ of the sample ACF ζ_i is related to $\widehat{\Xi}(e^{j\omega T})$, the Fourier transform of $\eta(t)$, as:

$$\Xi(e^{j\omega T_s}) = \frac{1}{T_S} \sum_{n=-\infty}^{\infty} \left| \widehat{\Xi}\left(\omega + \frac{2\pi n}{T_S} \right) \right|^2, \qquad |\omega| \leq \frac{\pi}{T_S} \tag{16.14}$$

The noise power at the detector is:

$$\sigma_{\text{n-LE-ZF}}^2 = N_0 \frac{T_S}{2\pi} \int_{-\pi/T_S}^{\pi/T_S} \frac{1}{\Xi(e^{j\omega T_s})} d\omega \tag{16.15}$$

It is finite only if the spectral density Ξ has no (or only integrable) singularities.

16.2.2 The mean-square error criterion

The ultimate goal of an equalizer is minimization, not of the ISI, but of the bit error probability. Noise enhancement makes the ZF equalizer ill-suited for this purpose. A better criterion is minimization of the *Mean Square Error* (MSE) between the transmit signal and the output of the equalizer.

We are thus searching for a filter that minimizes:

$$MSE = E\left\{ |\varepsilon_i|^2 \right\} = E\{ \varepsilon_i \varepsilon_i^* \} \tag{16.16}$$

As shown in Appendix 16.B (see *www.wiley.com/go/molisch*), this can be achieved with a filter whose coefficients $\mathbf{e}_{\text{opt}}$ are given by:

$$\mathbf{e}_{\text{opt}} = \mathbf{R}^{-1}\mathbf{p} \tag{16.17}$$

where $\mathbf{R} = E\{\mathbf{u}_i^* \mathbf{u}_i^T\}$ is the correlation matrix of the received signal, and $\mathbf{p} = E\{\mathbf{u}_i^* c_i\}$ the cross-correlation between the received signal and the transmit signal. Considering the frequency domain, concatenation of the noise-whitening filter with the equalizer $E(z)$ has the transfer function:

$$\tilde{E}(z) = \frac{1}{\Xi(z) + \dfrac{N_0}{\sigma_s^2}} \qquad (16.18)$$

which is the transfer function of the Wiener filter. The MSE is then:

$$\sigma_{\text{n-LE-MSE}}^2 = N_0 \frac{T_S}{2\pi} \int\limits_{-\pi/T_S}^{\pi/T_S} \frac{1}{\Xi(e^{jwT_s}) + \dfrac{N_0}{\sigma_s^2}} dw \qquad (16.19a)$$

Comparison with Eq. (16.15) shows that the noise power of an MMSE equalizer is smaller than that of a ZF equalizer (as illustrated in Fig. 16.4).

Example 16.1 *Equalizer coefficients and noise enhancement for linear equalizers: consider a channel with impulse response $h(\tau) = 0.4\delta(\tau) - 0.7\delta(\tau - T_S) + 0.6\delta(\tau - 2T_S)$, $N_0 = 0.3$, and $g(t) = g_R(t, T_S)$. Compute the noise variance at the output of a ZF equalizer and an MMSE equalizer.*

First we note that $\eta(t) = h(t) * g(t) = 0.4g(t) - 0.7g(t - T_S) + 0.6g(t - 2T_S)$. For the given rectangular pulse, $\sigma_s^2 = 1$. The z-transform of the ACF of $\eta(t)$ is given by:

$$\Xi(z) = 0.24z^2 - 0.7z + 1.01 - 0.7z^{-1} + 0.24z^{-2} = F(z)F^*(z^{-1})$$

Let us then choose $F(z) = 0.4 - 0.7z^{-1} + 0.6z^{-2}$. Thus, $F^*(z^{-1}) = 0.4 - 0.7z^1 + 0.6z^2$ has roots at $0.58 \pm 0.57j$, which are inside the unit circle.

The transfer function of the ZF equalizer is then:

$$E_1(z) = \frac{1}{F(z)} = \frac{1}{0.4 - 0.7z^{-1} + 0.6z^{-2}} \qquad (16.20)$$

The noise variance is given by Eq. (16.15):

$$\sigma_{\text{n-LE-ZF}}^2 = N_0 \frac{T_S}{2\pi} \int\limits_{-\pi/T_S}^{\pi/T_S} \frac{1}{0.24e^{j2\omega T_s} - 0.7e^{j\omega T_s} + 1.01 - 0.7e^{-j\omega T_s} + 0.24e^{-j2\omega T_s}} dw$$

$$= N_0 \frac{1}{2\pi} \int\limits_{-\pi}^{\pi} \frac{1}{0.24e^{j2\omega} + 0.24e^{-j2\omega} - 0.7e^{j\omega} - 0.7e^{-j\omega} + 1.01} dw$$

$$= N_0 \frac{1}{2\pi} \int\limits_{-\pi}^{\pi} \frac{1}{0.48\cos 2\omega - 1.4\cos\omega + 1.01} dw$$

$$\approx 2.94. \qquad (16.21)$$

The effective Signal-to-Noise Ratio (SNR) is thus $1/\sigma_{\text{n-LE-ZF}}^2 = 0.34$.

The MMSE equalizer is given by:

$$E_2(z) = \frac{1}{F(z) + N_0} = \frac{1}{0.7 - 0.7z^{-1} + 0.6z^{-2}} \qquad (16.22)$$

The noise variance is given by Eq. (16.19a). We note that the only difference from the ZF case is the addition of N_o/σ_s^2 in the denominator of the integrand:

$$\sigma_{\text{n-LE-MSE}}^2 = N_0 \frac{1}{2\pi} \int\limits_{-\pi}^{\pi} \frac{1}{0.48\cos 2\omega - 1.4\cos\omega + 1.31} d\omega$$

$$\approx 0.46 \tag{16.23}$$

As expected, the noise variance is lower for the MMSE equalizer than for the ZF equalizer. The effective SNR is [Proakis 1995]:

$$\gamma_\infty = \frac{1 - \sigma_{\text{n-LE-MSE}}^2}{\sigma_{\text{n-LE-MSE}}^2} = 1.17 \tag{16.24}$$

compared with $(1/N_0) = 3.33$ if there were no ISI (and thus, no equalizer). Thus, the necessity to equalize decreases the effective SNR (SNR at the output of the equalizer) by 4.5 dB and 10 dB for the MMSE and ZF equalizer, respectively.

16.2.3 Adaptation algorithms for mean-square-error equalizers

In order to find the optimum equalizer weights, we can directly solve Eq. (16.17). However, this requires on the order of $(2K + 1)^3$ complex operations. To ease the computational burden, iterative algorithms have been developed. The quality of an iterative algorithm is described by the following criteria:

- *Convergence rate*: how many iterations are required to "closely approximate" the final result? It is usually assumed that the channel does not change during the iteration period. However, if an algorithm converges too slowly, it will never reach a stable state – the channel has changed before the algorithm has converged.
- *Misadjustment*: the size of deviation of the converged state of the iterative algorithm from the exact MSE solution.
- *Computational effort per iteration*.

In the following, we discuss two algorithms that are widely used – the Least Mean Square (LMS) and the Recursive Least Square (RLS).

Least-mean-square algorithm

The LMS algorithm; also known as the *stochastic gradient method*, consists of the following steps:

1. Initialize the weights with values $\mathbf{e}_0$.
2. With this value, compute an approximation for the gradient of the MSE. The true gradient cannot be computed, because it is an expected value. Rather, we are using an estimate for $\mathbf{R}$ and $\mathbf{p}$ – namely, their instantaneous realizations:

$$\widehat{\mathbf{R}}_n = \mathbf{u}_n^* \mathbf{u}_n^T \tag{16.25}$$

$$\widehat{\mathbf{p}}_n = \mathbf{u}_n^* c_n \tag{16.26}$$

where subscript n indexes the iterations. The gradient is estimated as:

$$\hat{\mathbf{V}}_n = -2\hat{\mathbf{p}}_n + 2\hat{\mathbf{R}}_n\mathbf{e}_n \tag{16.27}$$

3. We next compute an updated estimate of the weight vector $\mathbf{e}$ by adjusting weights in the direction of the negative gradient:

$$\mathbf{e}_{n+1} = \mathbf{e}_n - \mu\hat{\mathbf{V}}_n \tag{16.38}$$

where μ is a user-defined parameter that determines convergence and residual error.
4. If the stop criterion is fulfilled – e.g., the relative change in weight vector falls below a predefined threshold – the algorithm has converged. Otherwise, we return to step 2.

It can be shown that the LMS algorithm converges if:

$$0 < \mu < \frac{2}{\lambda_{\max}} \tag{16.29}$$

Here $\lambda_{\max}$ is the largest eigenvalue of the correlation matrix $\mathbf{R}$. The problem is that we do not know this eigenvalue (computing it requires larger computational effort than inverting the correlation matrix). We thus have to guess values for μ. If μ is too large, we obtain faster convergence, but the algorithm might sometimes diverge. If we choose μ too small, then convergence is very probable, but slow. Generally, convergence speed depends on the condition number of the correlation matrix (i.e., the ratio of largest to smallest eigenvalue): the larger the condition number, the slower the convergence of the LMS algorithm.

The recursive-least-squares algorithm

In most cases, the LMS algorithm converges very slowly. Furthermore, the use of this algorithm is justified only when the statistical properties of the received signal fulfill certain conditions. The general least-squares criterion, on the other hand, does not require such assumptions. It just analyzes the N subsequent errors ε_i, and chooses weights such that the sum of the squared errors is minimized. This general Least Squares (LS) problem can be solved by a recursive algorithm as well – known as RLS (*Recursive Least Squares*) – the details of which are given in Appendix 16.C (see *www.wiley.com/go/molisch*).

Comparison of algorithms

There are two classes of algorithms for the determination of equalizer coefficients: direct implementation (Wiener filter, LS criterion) and iterative methods (LMS, RLS).

For the Wiener filter, we first have to determine the correlation matrix; this can be the major part of the numerical effort especially if the number of weights is small. The actual inversion of the matrix requires $(2K + 1)^3$ operations. Alternatively, we can use the data matrix directly, and invert it (LS algorithm). Construction of the data matrix requires less numerical effort than the correlation matrix; on the other hand the effort for inversion is much larger and depends on the number of used bits.

When comparing iterative algorithms, we find that the LMS algorithm usually converges too slowly. The RLS algorithm converges faster, but has a larger residual error. Figure 16.5 shows the typical example of the MSE for a Digital Enhanced Cordless Telecommunications (DECT) cordless telephone (see the companion website at *www.wiley.com/go/molisch*). We see that the RLS algorithm has converged after 10 bits, while the LMS algorithm requires almost 300 bits.

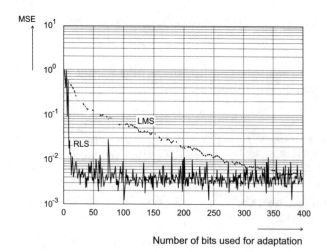

Figure 16.5 Mean-square error as a function of the number of iterations for a decision feedback equalizer (see below). For least mean square: $\mu = 0.03$; for recursive least squares: $\lambda = 0.99, \delta = 10^{-9}$.
From Fuhl [1994].

Due to the temporal variance of the channel, as well as spectral efficiency considerations, fast convergence is more important than an extremely low residual error rate.

On the other hand, the LMS algorithm requires far fewer (complex) operations (see Fig. 16.6; the terms "DFE" and "gradient lattice" will be explained below). One important conclusion from this figure is that for a small number of weights the complexity of the algorithms does not differ significantly. For up to 5–8 weights, the differences are less than 50% (with the exception of the LMS which is disqualified because of its slow convergence). In this regime, convergence, stability, and ease of implementation are the dominant criteria.

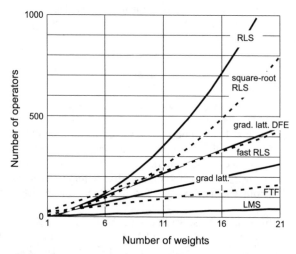

Figure 16.6 Number of operations per iteration step for different algorithms for a decision feedback equalizer (see below).

16.2.4 Further linear structures

Up to now, we have considered transversal FIR filters. However, linear filters can also be realized by means of other structures. One possibility is the use of recursive filters (IIR, *Infinite Impulse Response filters*). They have the advantage that fewer taps are required to achieve equalization. The major drawback is that these filters can show, not only zeros, but also poles in the transfer function, such that they can become unstable. For this reason, IIR filters are rarely used in practice.

A further possible structure is the *lattice filter*. The equations for the equalization algorithms are different from those of transversal filters (details can be found in Proakis [1995]).

16.3 Decision feedback equalizers

A *Decision Feedback Equalizer* (DFE) has a simple underlying premise: once we have detected a bit correctly, we can use this knowledge in conjunction with knowledge of the channel impulse response to compute the ISI caused by this bit. In other words, we determine the effect this bit will have on subsequent samples of the receive signal. The ISI caused by each bit can then be subtracted from these later samples.

The block diagram of a DFE is shown in Fig. 16.7. The DFE consists of a *forward filter* with transfer function $E(z)$, which is a conventional linear equalizer, as well as a *feedback filter* with transfer function $D(z)$. As soon as the RX has decided on a received symbol, its impact on all *future* samples (*postcursor ISI*) can be computed, and (via the feedback) subtracted from the received signal. A key point is the fact that the ISI is computed based on the signal *after* the hard decision; this eliminates additive noise from the feedback signal. Therefore, a DFE results in a smaller error probability than a linear equalizer.

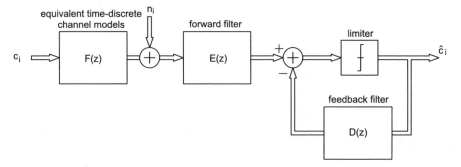

Figure 16.7 Structure of a decision feedback equalizer.

One possible source of problems is *error propagation*. If the RX decides incorrectly for one bit, then the computed postcursor ISI is also erroneous, so that later signal samples arriving at the decision device are even more afflicted by ISI than the unequalized samples. This leads to a vicious cycle of wrong decisions and wrong subtraction of postcursors.

Error propagation does not usually play a role when the BER is small. Note, however, that small error rates are often achieved via coding. It may therefore be necessary to decode the bits, re-encode them (such that the signal becomes a noise-free version of the received signal), and use this new signal in the feedback from the DFE.[3] In the following, we will only consider uncoded systems without error propagation.

16.3.1 MMSE decision feedback equalizer

The goal of the MMSE DFE is again minimization of the MSE, by striking a balance between noise enhancement and residual ISI. As noise enhancement is different in the DFE case from that of linear equalizers, the coefficients for the forward filter are different: as postcursor ISI does not contribute to noise enhancement, we now aim to minimize the sum of noise and (average) *precursor* ISI. Obviously, performance also differs.

The coefficients of the feedforward filter can be computed from the following equation:

$$\sum_{n=-K_{ff}}^{0} e_n \left(\sum_{m=0}^{-l} f_m^* f_{m+l-n} + N_0 \delta_{nl} \right) = -f_{-l}^* \quad \text{for } l,n = -K_{ff}, \ldots, 0 \tag{16.30}$$

where K_{ff} is the number of taps in the feedforward filter. The coefficients of the feedback filter are then:

$$d_n = -\sum_{m=-K_{ff}}^{0} e_m f_{n-m} \quad \text{for } n = 1, \ldots, K_{fb} \tag{16.31}$$

where K_{fb} is the number of taps in the feedback filter.

Assuming some idealizations (the feedback filter must be at least as long as the postcursor ISI; it must have as many taps as required to fulfill Eq. (16.30); there is no error propagation), the MSE at the equalizer output is:

$$\sigma_n^2(DFE - MMSE) = N_0 \exp\left(\frac{T_S}{2\pi} \int_{-\pi/T_S}^{\pi/T_S} \ln\left[\frac{1}{\Xi(e^{j\omega T}) + N_0} \right] d\omega \right) \tag{16.32}$$

16.3.2 Zero-forcing decision feedback equalizer

The ZF DFE is conceptually even simpler. As mentioned in Section 16.1.2, the noise-whitening filter eliminates all precursor ISI, such that the resulting effective channel is purely causal. Postcursor ISI is subtracted by the feedback branch. The effective noise power at the decision device is:

$$\sigma_n^2(DFE - ZF) = N_0 \exp\left(\frac{T_S}{2\pi} \int_{-\pi/T_S}^{\pi/T_S} \ln\left[\frac{1}{\Xi(e^{j\omega T})} \right] d\omega \right) \tag{16.33}$$

This equation demonstrates that noise power is larger than it is in the unequalized case, but smaller than that for the linear zero-forcing equalizer.

---

[3] As the decoder itself shows a delay, this can become a challenging task. Possible solutions to this include the design of joint equalization and decoding, with the exchange of soft information (see also Chapter 14).

Example 16.2 *Using the channel from Example 16.1, compute the noise enhancement for the MMSE DFE and ZF DFE.*

From Example 16.1, remember that $\Xi(e^{j\omega T}) = 0.48\cos 2\omega T_S - 1.4\cos \omega T_S + 1.01$. Inserting this in Eq. (16.32), we obtain:

$$\sigma_n^2(DFE - MMSE) = N_0 \exp\left(\frac{T_S}{2\pi}\int_{-\pi/T_S}^{\pi/T_S} \ln\left[\frac{1}{\Xi(e^{j\omega T}) + N_0}\right] d\omega\right) \tag{16.34}$$

$$= N_0 \exp\left(\frac{1}{2\pi}\int_{-\pi}^{\pi} \ln\left[\frac{1}{0.48\cos 2\omega - 1.4\cos \omega + 1.31}\right] d\omega\right) \tag{16.35}$$

$$\approx 0.33 \tag{16.36}$$

so that the output SNR is 2. Thus, the SNR deteriorates by $2\,dB$ compared with the Additive White Gaussian Noise (AWGN) case. For the ZF DFE, the noise variance at the output is 0.83, so that the SNR deteriorates by $4.4\,dB$.

16.4 Maximum-likelihood sequence estimation – Viterbi detector

The equalizer structures considered up to now influence the decision about which *symbol* has been transmitted. For MLSE, on the other hand, we try to determine the *sequence of symbols* that has most likely been transmitted. This situation shows strong similarities to the decoding of convolutional codes. As a matter of fact, transmission through a delay-dispersive channel can be viewed as convolutional encoding with a code rate $R_c = 1/1$. MLSE estimators give the best performance of all equalizers.

Remember that the output signal of the time-discrete channel can be written as:

$$u_i = \sum_{n=0}^{L_c} f_n c_{i-n} + n_i \tag{16.37}$$

where n is Gaussian white noise with variance σ_n^2. For a sequence of N received values, the joint probability density function of the vector of received signals $\mathbf{u}$ (conditioned on the data vector $\mathbf{c}$ and impulse response vector $\mathbf{f}$) is:[4]

$$pdf(\mathbf{u}|\mathbf{c};\mathbf{f}) = \frac{1}{(2\pi\sigma_n^2)^{N/2}} \exp\left(-\frac{1}{2\sigma_n^2}\sum_{i=1}^{N}\left|u_i - \sum_{n=0}^{L_c} f_n c_{i-n}\right|^2\right) \tag{16.38}$$

The MLSE of $\mathbf{c}$ (for a given $\mathbf{f}$) are the values of the vectors that maximize the joint probability density function (pdf) $pdf(\mathbf{u}|\mathbf{c},\mathbf{f})$. As the variables only occur in the exponent, it is sufficient to minimize:

$$\sum_{i=1}^{N}\left|u_i - \sum_{n=0}^{L_c} f_n c_{i-n}\right|^2 \tag{16.39}$$

---

[4] Here and in the following, we assume that all transmit symbols are equally likely, such that MLSE and maximum-a-posteriori estimation are identical.

As for convolutional decoding, various algorithms exist for determination of the optimum sequence. The RX first generates all possible sequences that can result from convolution of valid transmit sequences with the channel impulse response. We then try to find the sequence that has the smallest distance (best metric) from the received signal. The most straightforward (but also most computationally intensive) method is the *exhaustive search*. In practice, the Viterbi algorithm is used instead.

MLSE as described above is only optimum if the additive noise at MLSE input is white. Therefore, the sample values used at the detector have to be the output of a noise-whitening filter. This filter has to be adapted to the current channel state; and each channel realization requires spectral factorization. Due to these difficulties, the total input filter (matched filter and noise-whitening filter) is often replaced by a simple brickwall filter whose bandwidth is approximately the inverse symbol duration. Note, however, that in this case one sample per symbol no longer provides sufficient statistics.

Example 16.3 *Viterbi equalization.*

This example shows the working of the Viterbi detection of a symbol stream that went through a channel with a discrete time impulse response:

$$\mathbf{f} = \begin{pmatrix} 1 \\ -0.5 \\ 0.3 \end{pmatrix} \tag{16.40}$$

The channel can be viewed as a tapped delay line (shift register) with weights 1, −0.5, and 0.3; see the top part of Fig. 16.8 (the left top part shows the tapped delay line model, the right part shows the "cell" model analogous to the convolutional codes discussed in Chapter 14). For ease of exposition, we chose a channel with a real impulse response, and Binary Phase

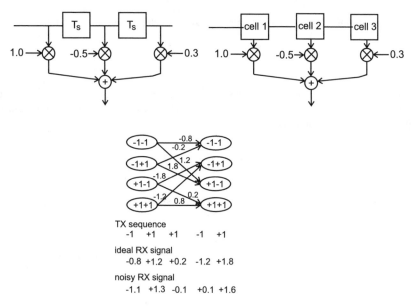

Figure 16.8 Representation of tapped delay line channel (top), transition probabilities (middle), and transmitted and received signals (bottom).

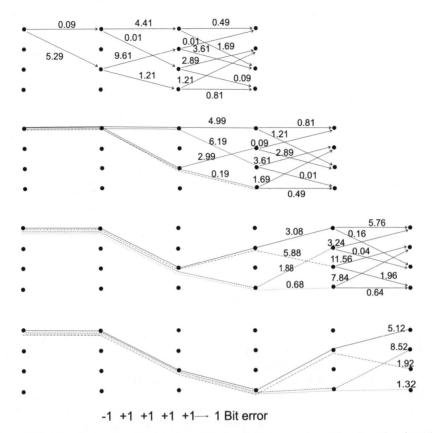

-1 +1 +1 +1 +1— 1 Bit error

Figure 16.9 Viterbi algorithm for detection of the transmit sequence in a delay-dispersive channel.

Shift Keying (BPSK) as the modulation format. The lower part of Fig. 16.8 shows possible transitions in the trellis diagram. We have to consider four states in the trellis, as $L_c = 2$ samples, and the number of possible states in a cell of the equivalent shift register is equal to the size of the modulation alphabet $M = 2$. We assume furthermore that we know the starting state of the trellis – i.e., -1 -1 (e.g., because known bits have been transmitted before the start of our decoding). The bottom part of Fig. 16.9 shows the "unfolding" of the trellis diagram. The numbers next to the transitions are metrics of the considered sequence.

16.5 Comparison of equalizer structures

Figure 16.10 shows a taxonomy of equalizer structures. When selecting an equalizer for a practical system, we have to consider the following criteria:

- *Minimization of the BER*: here MLSE is superior to all other structures. DFEs, though worse than MLSE estimators, are better than linear equalizers. The quantitative difference between the structures depends on the channel impulse response.

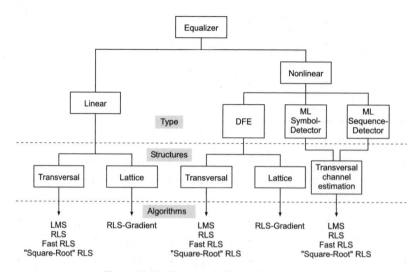

Figure 16.10 Taxonomy of equalizer structures.

Reproduced with permission from Proakis [1991] © IEEE.

- *Can the channel deal with zeros in the channel transfer function?* ZF equalizers have problems, as they invert the transfer function and thus create poles in the equalizer transfer function. Neither MMSE nor MLSE equalizers have this problem.
- *Computational effort*: the effort for linear equalizers and DFEs is not significantly different. Depending on the adaptation algorithm, the number of operations increases linearly, quadratically, or cubically with equalizer length (number of weights). For MLSE, the computation effort increases exponentially with length of the impulse response of the channel. For short-impulse responses (e.g., impulse response is at most four symbol durations long, as in GSM), the computational complexity of MLSE is comparable with that of other equalizer structures.
- *Sensitivity to channel misestimation*: due to the error propagation effect, DFE equalizers are more sensitive to channel estimation errors than linear equalizers. Also, ZF equalizers are more sensitive than MMSE equalizers.
- *Power consumption and cost* can be deduced from the computational effort.

16.6 Fractionally spaced equalizers

In most cases, the RX samples and processes signals at the symbol frequency $(1/T_S)$. This is suboptimum if this sampling rate is lower than the Nyquist rate and the matched filter is matched only to the TX pulse (and not the received, distorted pulse). The spectrum of a signal that was filtered by a raised cosine filter (with rolloff factor α) extends up to a frequency $(1 + \alpha)/2T_S$, such that the Nyquist rate is $(1 + \alpha)/T_S$. A *fractionally-spaced equalizer* is based on sampling at no less than the Nyquist rate. The taps of the equalizers are spaced at $T_S a/b$, where $a < b$ and a and b are integers.

Fractionally spaced equalizers can also be interpreted as performing equalization and matched filtering in one step. They are also less sensitive to errors in the sampling time. The drawback is that the number of required taps is larger than for a symbol-spaced equalizer, and thus the computational effort is higher.

16.7 Blind equalizers

16.7.1 Introduction

"Normal" equalization works in two stages: a training phase, and a detection phase. During the training phase, a known bit sequence is transmitted over the channel, and the distorted version at the RX is compared with a (locally generated) undistorted version; this in turn gives us information about the channel impulse response that is used for equalization (see Section 16.1.3 for a discussion of the pros and cons). In contrast, blind equalization exploits known statistical properties of the transmit signal to estimate both channel and data. Equalizer coefficients are adjusted in such a way that certain statistical properties of the equalizer output match known statistical properties of the transmit signal.

The advantages of blind equalization include:

- The whole timeslot is used for determination of the impulse response, not just a short training sequence.
- Spectral efficiency is improved, because no time is "wasted" on transmission of the training sequence.

The following signal properties can be used for blind equalization:

- *constant envelope*: for many signals (FSK, MSK, GMSK), the envelope (amplitude) is constant;
- *statistical properties*: e.g., cyclostationarity;
- *finite symbol alphabet*: only certain discrete values are valid points in the signal constellation diagram;
- *spectral correlation*: signal spectra are correlated with shifted versions of themselves;
- a combination of these properties.

Well-studied blind algorithms include: (i) *Constant Modulus Algorithm* (CMA), (ii) blind *MLSE*, and (iii) algorithms based on higher order statistics. In the following, we will discuss these classes of algorithms.

16.7.2 Constant modulus algorithm

CMAs are the oldest algorithms for blind equalization. In their simplest form, they use the LMS adaptation. The data-aided LMS algorithm was described in Section 16.2; it is based on minimization of the difference between a desired signal (known at the RX) and the output of an equalizer. In a blind LMS algorithm, the desired signal first has to be generated *from the output of the equalizer*. This is achieved by sending the equalizer output through a nonlinear function. The error signal is then the difference between the output of this nonlinear function and the equalizer output. The difference between a conventional equalizer and the CMA equalizer is shown in Fig. 16.11.

The nonlinear function can either be memoryless, or contain mth order memory. Various types of algorithms (usually named for their inventors) are distinguished by different nonlinearities. The best known are the Sato algorithm (1975) and the Godard algorithm (1980).

The LMS algorithm used for blind adaptation has the same drawbacks as the conventional LMS: the convergence rate is slow (by increasing the stepwidth μ, convergence is improved and the error after convergence is worsened). It is also possible that the algorithm converges to a local minimum. These problems can be solved by the "analytic constant modulus algorithm"

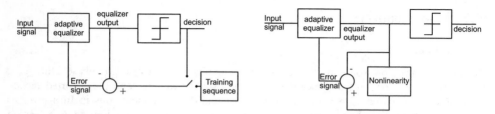

Figure 16.11 Structure of a conventional equalizer (left) and that of a constant-modulus-algorithm-based equalizer (right).

[van der Veen und Paulraj 1996]. This is a non-iterative algorithm that provides exact solutions in the noise-free case, but is also robust with respect to noise.

16.7.3 Blind maximum-likelihood estimation

For conventional MLSE, a training sequence is used for determination of the channel impulse response $\mathbf{f}$. This estimate is then used during the actual transmission of data, so that the Maximum Likelihood (ML) estimate only has to solve for $\mathbf{c}$. For a blind estimate, $\mathbf{c}$ and $\mathbf{f}$ have to be estimated simultaneously. This can be achieved by either of the following methods:

1. The channel impulse response is estimated from $pdf(\mathbf{u}|\mathbf{f}, \mathbf{c})$ by averaging over all possible data sequences. This method has two drawbacks: it requires considerable computation time for the averaging, and it is suboptimum.
2. Alternatively, we can determine the ML estimate for $\mathbf{f}$ for all possible data sequences. Then we select the pair $\mathbf{f}, \mathbf{c}$ that has the best overall metric. This method is even more computationally intensive than the averaging method; however, it is also more accurate. Furthermore, methods for reducing the computational effort have been proposed [Proakis 1995].

16.7.4 Algorithms using second- or higher order statistics

In general, second-order statistics (ACFs) cannot provide information about the phase of the channel impulse response. An exception occurs when the ACF of the received signal is periodic. Cyclostationary properties are thus the basis for blind estimation methods using second-order statistics. Similar to our discussion in Chapter 6, we distinguish between strict-sense and wide-sense cyclostationarity. A process is strict-sense cyclostationary if *all* its statistical properties are invariant to shifts by integer multiples of the sampling period T_{per}. For wide-sense stationarity, only the mean and the ACF have to fulfill this condition:

$$E\{x(t + iT_{per})\} = E\{x(t)\} \tag{16.41}$$

$$E\{x^*(t_1 + iT_{per})x(t_2 + iT_{per})\} = E\{x^*(t_1)x(t_2)\} \tag{16.42}$$

If the signal is oversampled, then the resulting sequence of sample values is guaranteed to be cyclostationary. The actual channel estimate is then based on different correlation matrices of the received signal (for details, see Tong et al [1994]). The use of higher order statistics does not require the applicability of cyclostationarity; however, its accuracy is usually much lower.

16.7.5 Assessment

Historically speaking, blind equalization was first developed in the 1970s for multiterminal computer networks (one central station linked to multiple terminals). During the 1980s and 1990s, a lot of theoretical work was devoted to this topic, as it offers some fascinating mathematical challenges. However, up to now, truly blind equalization has not been able to replace training-sequence-based equalization in practical systems. The main reason seems to be that the difference in computational effort and reliability is significant. In particular, blind equalizers require a long time to converge, and thus do not work well in quickly time-variant wireless channels. Another approach that has been developed in recent years uses a training sequence to obtain an initial estimate of the channel, and then refines these estimates by means of blind (decision-aided) equalization [Loncar et al. 2002].

16.8 Appendices

Please go to *www.wiley.com/go/molisch*

Further reading

The first comprehensive description of adaptive equalizers, which is still worth reading today, was given in Lucky et al. [1968]. The description in this chapter – particularly, that of linear and DFE equalizers – was inspired by the excellent exposition in Proakis [1995], which also gives many more interesting details. Haykin [1991] describes adaptive filters, which is another way of interpreting equalizers. DFEs are also described in Belfiore and Park [1977]. Fractionally spaced equalizers were analyzed, e.g., in Gitlin and Weinstein [1981] and Ungerboeck [1976]. The Viterbi equalizer is an application of the Viterbi algorithm [Viterbi 1967]; the impact of channel estimation errors is discussed in Gorokhov [1998]. Vitetta et al. [2000] give a detailed description of a wide variety of equalizer structures, including the impact of channel-state information (perfect, estimated, or averaged). Spectral factorization techniques are surveyed in Sayed and Kailath [2001]. Blind algorithms for equalization are too numerous to list here; as examples, we just name Giannakis and Halford [1997], Liu et al. [1996], Sato [1975], and van der Veen and Paulraj [1996].

Part IV

MULTIPLE ACCESS AND ADVANCED TRANSCEIVER SCHEMES

In Part III, we described how a single transmitter can communicate with a single receiver. However, for most wireless systems, multiple devices should communicate simultaneously in the same area. It is therefore necessary to provide means for *multiple access*. For most first- and second-generation cellular systems, as well as for cordless phones and wireless local area networks, different devices communicate either on different frequencies, or at different times. The multiaccess schemes based on this principle (called "frequency division multiple access", "time division multiple access", and "packet radio") are discussed in Chapter 17. A different type of multiple access is based on spreading the signal over a large bandwidth and making such a spreading unique for each user. This allows multiple users to be on the air simultaneously, and the receiver can determine from the spreading which part of the "on-air" signal comes from which user. This *spread spectrum* approach, described in Chapter 18, is used, for example, for third-generation cellular systems in the form of *Code Division Multiple Access* (CDMA). The chapter also describes the Rake receiver – a device that enables CDMA systems to deal with a channel's delay dispersion – and shows how multiuser detection can be exploited to significantly increase performance in a multiaccess environment.

As the data rates of wireless systems increase to higher and higher values, both equalizers and Rake receivers become too complex for practical use. Orthogonal Frequency Division Multiplexing (OFDM) is one way of overcoming the problem. In this approach, the datastream is split up into a large number of substreams, each of which is modulated onto a different carrier; for this reason, OFDM is also often called a "multicarrier modulation" method. Chapter 19 describes its principles, and also some variants that combine CDMA with OFDM.

After exploiting the time domain, frequency domain, and code domain for signaling, "space is the final frontier". Multiantenna elements allow exploitation of the spatial domain to increase the data rate, allow more simultaneous users, and/or increase transmission quality. One way of exploiting such multiple antennas – namely, antenna diversity – was treated in Chapter 13; it serves mainly to combat the impact of deep fades on transmission quality. Then, Chapter 20 describes the theory of "smart antennas", which increase the numbers of users that can be served in a transmission system. Furthermore, this chapter also discusses the so-called "multiple-input multiple-output" systems, which have multiple antennas at the transmitter and receiver, and can exploit them to transmit several datastreams in parallel. This in turn allows the data rate to be increased without requiring additional spectrum.

17

Multiple access and the cellular principle

17.1 Introduction

A wireless communications system uses a certain frequency band that is assigned to this specific service. Spectrum is thus a scarce resource, and one that cannot be easily extended. For this reason, a wireless system must make provisions to allow the simultaneous communication of as many users as possible within that band.

The problem of letting multiple users communicate simultaneously can be divided into two parts:

1. If there is only a *single Base Station* (BS), how can it communicate with many Mobile Stations (MSs) simultaneously?
2. If there are *multiple BSs*, how can we assign spectral resources to them in such a way that the total number of possible users is maximized? And how should these BSs be placed in a given geographical area?

As for the first question, there are different methods, called *Multiple Access* (MA) methods, that allow multiple users to talk to a BS simultaneously. In this chapter, we will discuss the following three methods:

- *Frequency Division Multiple Access* (FDMA), where different frequencies are assigned to different users.
- *Time Division Multiple Access* (TDMA), where different timeslots are assigned to different users.
- *Packet Radio* can be viewed as a form of TDMA, where the assignment of timeslots to users is adaptive.

FDMA and TDMA are discussed in Sections 17.2 and 17.3. The fact that these sections are rather brief should not detract from the importance of TDMA and FDMA. However, these MA methods are conceptually easy to understand; furthermore, many of the concepts required for their treatment have been dealt with in Chapters 11, 12, and 16. A variant of TDMA, which is important for wireless data communications, is *packet radio*, which is discussed in Section 17.4. Besides the schemes mentioned above, *Code Division Multiple Access* (CDMA), where each user is assigned a different code, has gained increased popularity in recent years. We will discuss this scheme, as well as other spread spectrum methods, in Chapter 18. Finally, *Space Division*

Wireless Communications   Andreas F. Molisch
© 2005 John Wiley & Sons, Ltd

Multiple Access (SDMA) is an MA format for systems with multiple antennas; it can be combined with all of the other MA methods. It is described in Section 20.1. The so-called "duplexing", which separates transmission and reception at a transceiver, is analyzed in Section 17.5.

The goal of all these methods is to maximize spectral efficiency – i.e., to maximize the number of users per unit bandwidth. As mentioned above, there is also a different (though related) question: how can we design a system so that the number of users per unit bandwidth *and unit area* is maximized? This goal obviously requires multiple BSs, and the assignment of spectral resources to them. All this leads us to the cellular principle, which requires reusing the same spectrum at different BSs; this is discussed in Section 17.6.

17.2 Frequency division multiple access

17.2.1 Multiple access via frequency division multiple access

FDMA is the oldest, and conceptually most simple, multiaccess method. Each user is assigned a frequency (sub)band – i.e., a (usually contiguous) part of the available spectrum (see Fig. 17.1). The assignment of frequency bands is usually done during call setup, and retained during the whole call. FDMA is usually combined with the frequency domain duplex (see Section 17.5), so that two frequency bands (with a fixed duplex distance) are assigned to each user: one for downlink (BS-to-MS) and one for uplink (MS-to-BS) communication.

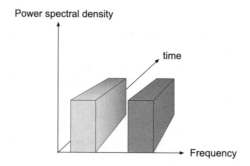

Figure 17.1 Principle of frequency division multiple access.

Pure FDMA is conceptually very simple, and has some advantages for implementation:

- The transmitter (TX) and receiver (RX) require much *less digital signal processing*. However, this is not so important in practice anymore, as the costs for digital processing are continuously decreasing.
- *(Temporal) synchronization is simple.* Once synchronization has been established during the call setup, it is easy to maintain it by means of a simple tracking algorithm, as transmission occurs continuously.

However, pure FDMA also has significant disadvantages, especially when used for speech communications. These problems arise from spectral efficiency considerations, as well as from sensitivity to multipath effects:

- *Frequency synchronization and stability are difficult.* For speech communications, each frequency subband is quite narrow (typically between 5 and 30 kHz). Local oscillators thus must be very accurate and stable; jitters in the carrier frequency result in adjacent channel interference. High spectral efficiency also requires the use of very steep filters to extract the desired signal. Both accurate oscillators and steep filters are expensive, and thus undesirable. If they are not admissible, guard bands can be used to mitigate filter requirements. This, however, reduces the spectral efficiency of the system.

- *Sensitivity to fading.* Since each user is assigned a distinct frequency band, these bands are narrower than for other multiaccess methods (compare TDMA, CDMA) – i.e., 5–30 kHz. For such narrow subbands, fading is flat in practically all environments. This has the advantage that no equalization is required; the drawback is that there is no frequency diversity. Remember that frequency diversity is mainly provided by signal components that are more than one channel coherence bandwidth apart (see Chapters 13 and 16).
- *Sensitivity to random Frequency Modulation (FM).* Due to the narrow bandwidth, the system is sensitive to random FM: the Bit Error Rate (BER) due to random FM is proportional to $(\nu_{max} T_S)^2$ (see Chapter 12). Thus, it is inversely proportional to the square of the bandwidth. On the positive side, appropriate signal-processing schemes can not only mitigate these effects, but even exploit them to obtain time diversity. Note that the situation here is dual to wideband systems, where delay dispersion can be a drawback, but equalizers can turn them into an asset by exploiting frequency diversity.
- *Intermodulation.* The BS needs to transmit multiple speech channels, each of which is active the whole time. Typically, a BS uses 20–100 frequency channels. If these signals are amplified by the same power amplifier, third-order modulation products can be created, which lie at undesirable frequencies – i.e., within the transmit band. We thus need either a separate amplifier for each speech channel, or a highly linear amplifier for the composite signal – each of these solutions makes a BS more expensive.

It is for these reasons that FDMA is mostly used for the following applications:

- *Analogue communications systems.* Here, FDMA is the only practicable MA method.
- *Combination of FDMA with other MA methods.* The spectrum allocated for a service (or a network operator) is divided into larger subbands, each of which is used for serving a *group* of users. Within this group, multiple access is done by means of another MA method – e.g., TDMA or CDMA. Most current wireless systems use an FDMA in that way (see Chapters 21–24).
- *High-data-rate systems.* The disadvantages of FDMA are mostly relevant if each user requires only a small bandwidth – e.g., 20 kHz. The situation can be different for wireless Local Area Networks (LANs), where a single user requires a bandwidth on the order of 20 MHz, and only a few frequency channels are available.

17.2.2 Trunking gain

We will now compute how many subscribers can be covered with an FDMA system by one BS. This seemingly simple question has become a separate branch of communications theory (often called *queuing theory*), with several textbooks dedicated to it (see, e.g., Gross and Harris [1998]). In this subsection, we will describe in a very simplified way how to compute the required number of communications channels so that a given number of users can be served with "sufficient quality". We assume, for present purposes, a pure FDMA system that does not need any channels for signaling. Furthermore, we assume a system that is designed purely for speech communications.

There are two extreme cases in the planning of a cellular network:

1. *Worst case design.* It is assumed that all users want to call simultaneously. If the network operator wants to serve 700 users per cell, it has to provide 700 speech channels. Of course such a network should never be built in practice – this would be like designing a hospital that can treat all inhabitants of a city at the same time.
2. *Best case design.* We know that a typical user uses a phone only 20 min per day; if there are 700 potential users per cell, then 14,000 minutes of call time are actually used. A system with 10 speech channels per cell offers $24 * 60 * 10 = 14,400$ minutes of talk time, and could thus supply the required number of users. However, this computation assumes that all users call sequentially–i.e., new users dial in as soon as old ones have finished their calls – and they do so evenly distributed over a 24-hour period.

Obviously, neither of the two extreme cases is realistic. The art of network design is, to a considerable degree, to predict the call behavior of users, and derive the physical infrastructure (available number of speech channels) that guarantees an acceptable *grade of service*.

Several factors influence this planning process:

1. The number and duration of calls depend on the time of day. We therefore define a *busy hour* (usually around 10h00 and 16h00), which is defined as the hour when most calls are made. The traffic during that busy hour determines the required network capacity.
2. The spatial distribution of users is time-variant. While business districts (city centers) usually see a lot of activity during the daytime, suburbs and entertainment districts experience more traffic during the nighttime.
3. Telephoning habits change over the years. While in the late 1980s, calls from cellular phones were usually limited to a few minutes, now hour-long calls have become quite common.
4. Changing user habits are also related to the offering of new services (e.g., data connections) and new pricing structures (e.g., free calls in the evening hours). The strategy of selling "minute accounts" that have to be used up each month also leads to longer talk times than a pricing strategy of charging per minute.

Based on statistical knowledge of user habits, we can now design a system that *with a certain probability* allows a given number of users per cell to make calls. If, through a statistical fluke, all users want to telephone simultaneously, some of the calls will be blocked.

For the computation of the blocking probability of a simplified system, we make the following assumptions: (i) the times when the calls are placed are statistically independent, (ii) the duration of calls is an exponentially distributed random variable, (iii) if a user is rejected, his next call attempt is made statistically independent of the previous attempt (i.e., behaves like a new user).[1] Such a system is called an *Erlang B* system; the *probability of call blocking* can be shown to be:

$$\mathrm{Pr}_{\mathrm{block}} = \frac{T_{\mathrm{tr}}^{N_C}/N_C!}{\sum_{k=0}^{N_C} T_{\mathrm{tr}}^k/k!} \tag{17.1}$$

---

[1] This obviously does not agree with reality. Typically, a blocked user retries immediately. After being blocked several times within a short time interval, she or he usually gives up and places the next call after a much longer wait time.

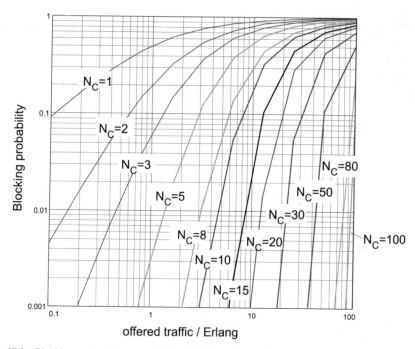

Figure 17.2 Blocking probability in an Erlang-B system. N_C is the number of available speech channels.

where N_C is the number of speech channels (per cell), and T_{tr} is the average offered traffic in units of users (speech channels). Figure 17.2 shows the relationship graphically. We see that the ratio of required channels to offered traffic is very high if N_C is small, especially for low required blocking probabilities. For example, for a required blocking probability of 1%, the ratio of possible offered traffic to available channels is less than 0.1 if $N_C = 2$. If N_C is very large, then the ratio is only slightly less than unity, and becomes almost independent of the required blocking probability. Assuming again a required blocking probability of 1%, the ratio of possible offered traffic to available channels is about 0.9 for $N_C = 50$.

An alternative model assumes that any user that is not immediately assigned a channel is transferred to a waiting loop, and assigned a channel as soon as it becomes available. The *probability that a user is put on hold* is:

$$\mathrm{Pr}_{\mathrm{wait}} = \frac{T_{tr}^{N_C}}{T_{tr}^{N_C} + N_C!\left(1 - \dfrac{T_{tr}}{N_C}\right)\displaystyle\sum_{k=0}^{N_C-1} T_{tr}^k/k!} \tag{17.2}$$

and the average wait-time is:

$$t_{\mathrm{wait}} = \mathrm{Pr}_{\mathrm{wait}} \frac{T_{\mathrm{call}}}{N_C - T_{tr}} \tag{17.3}$$

where T_{call} is the average duration of the call.

Example 17.1 *Consider an Erlang-C system where users are active 50% of the time, and the average call duration is 5 minutes. It is required that no more than 5% of all calls are put into a waiting loop. How many channels are required for $n_{user} = 1, 8, 30$ users? What is the average wait-time in each of these cases?*

Since $T_{tr} = 0.5 \cdot n_{user}$ is the average offered traffic, we need to find N_C that fulfills:

$$0.05 \geq \frac{(0.5 \cdot n_{user})^{N_C}}{(0.5 \cdot n_{user})^{N_C} + N_C! \left(1 - \dfrac{0.5 \cdot n_{user}}{N_C}\right) \displaystyle\sum_{k=0}^{N_C-1} \frac{(0.5 \cdot n_{user})^k}{k!}} \tag{17.4}$$

This equation needs to be solved numerically; the results are given in Table 17.1 below. With $T_{call} = 5$ min, the average wait-time is:

$$t_{wait} = Pr_{wait} \frac{5}{N_C - 0.5 \cdot n_{user}} \tag{17.5}$$

The required number of channels to fulfill inequality and the resulting average wait-time is:

Table 17.1

n_{user}	1	8	30
N_C	3	9	23
Pr_{wait}	0.0152	0.0238	0.0380
t_{wait}	0.0304	0.0238	0.0238

The probability that a call can be placed, and is not blocked, is an important part of service quality: remember that quality of service is defined as 100% minus the percentage of blocked calls, minus ten times the percentage of lost calls (Section 1.3.7). In an FDMA system, a large system load can lead only to blocked calls, but not lost calls, as long as each user stays in the coverage area of his/her BS. However, calls can be dropped when a user with an ongoing call tries to move to a different cell whose BS is already fully occupied. And as we will see in Section 17.6, a fully loaded system also increases interference with neighboring cells, making the links in that cell more sensitive to fluctuations in signal strength, and possibly increasing the number of dropped calls due to insufficient signal-to-noise and interference ratio.

Summarizing, we find that the number of users that can be accommodated with a given quality of service increases faster than linearly with the number of available speech channels. The difference between actual increase and linear increase is called the *trunking gain*. From a purely technical point of view, it is thus preferable to have a large pool of speech channels that serves all users. This situation could be fulfilled, for example, were there only a single operator for cellular systems, owning the complete spectrum assigned to cellular services. The reasons for *not* choosing this approach are political (pricing, monopoly), not technical.[2]

---

[2] An approach envisioned in the U.S.A. in the early days of cellular telephony was to assign the available spectrum to exactly two providers, thus striking a compromise between political requirement (avoiding a monopoly) and technical expediency.

Example 17.2 *In an Erlang-B system, 30 channels are available. A blocking probability of less than 2% is required. What is the traffic that can be served if there is one operator or three operators?*

1. By inserting the required blocking probability $P_{\text{block}} = 0.02$ and the number of channels $N_C = 30$ into Eq. (17.1), we get:

$$0.02 = \frac{T_{\text{tr}}^{30}/30!}{\sum\limits_{k=0}^{30} T_{\text{tr}}^{k}/k!} \tag{17.6}$$

Solving this equation for T_{tr}, we get:

$$T_{\text{tr}} = 21.9 \tag{17.7}$$

2. Similarly, sharing the 30 speech channels among the three operators, each having $N_C = 10$ speech channels, results in an average traffic T_{tr} of each operator of:

$$T_{\text{tr}} = 5.1 \tag{17.8}$$

Hence, the total average traffic that can be handled by all three operators together is:

$$T_{\text{tr,tot}} = 3 \cdot 5.1 = 15.3 \tag{17.9}$$

17.3 Time division multiple access

For TDMA, different users transmit not at different frequencies, but rather at different times (see Fig. 17.3). A time unit is subdivided into N timeslots of fixed duration, and each user is assigned one such timeslot. During the assigned timeslot, the user can transmit with a high data rate (as it can use the whole system bandwidth); subsequently, it remains silent for the next $N - 1$ timeslots, when other users take their turn. This process is then repeated periodically. At first glance, this approach has the same performance as FDMA: a user transmits only during $1/N$ of the available time, but then occupies N times the bandwidth. However, there are some important practical differences:

• Users occupy a larger bandwidth. This allows them to exploit the frequency diversity available within the bandwidth allocated to the system; furthermore, the sensitivity to

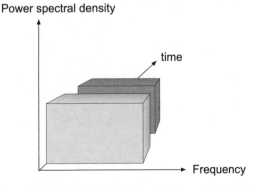

Figure 17.3 Principle behind time division multiple access.

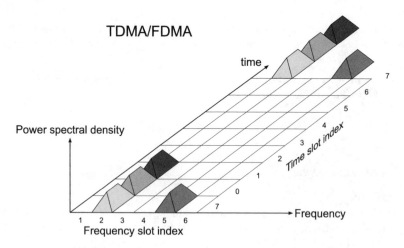

Figure 17.4 How the Global System for Mobile communications (GSM) combines time division multiple access with frequency division multiple access.

random FM is reduced. On the flipside, equalizers are required to combat Inter Symbol Interference (ISI) for most operating environments; this increases the effort needed for digital signal processing.

- There is no need for frequency guard bands, as each user completely fills up the assigned band. However, temporal guard intervals are now required. A TX needs a finite amount of time to ramp up from 0-W output power to "full power" (typically between 100 mW and 100 W). Furthermore, there has to be sufficient guard time to compensate for the runtime of the signal between the MS and BS. It is possible that one MS is far away from the BS, while the one that transmits in the subsequent timeslot is very close to the BS and thus has negligible runtime. As the signals from the two users must not overlap at the BS, the second MS must not transmit during the time it takes the first signal to propagate to the BS.[3]

- Each timeslot might require a new synchronization and channel estimation, as transmission is not continuous. Optimization of timeslot duration is a challenging task. If it is too short, then a large percentage of the time is used for synchronization and channel estimates (in GSM, 17% of a timeslot are used for this purpose). If the timeslot is too long, transmission delays become too long (which users find annoying especially for speech communications), and the channel starts to change during one timeslot. In that case, the equalizer has to track the channel during transmission of a timeslot, which increases implementation effort (this is required, e.g., in IS-136). If the time between two timeslots assigned to one user is larger than coherence time, the channel has changed between these two timeslots, and a new channel estimate is required.

- For interference-limited systems, TDMA has a major advantage: during its period of inactivity, the MS can "listen" to transmissions on other timeslots.[4] This is especially useful for the preparation of handovers from one BS to another, when the MS has to find out whether a neighboring BS would offer better quality, and has communications channels available.

---

[3] The required guard time can be reduced by *timing advance* (as described in Chapter 21).
[4] In a mixed TDMA/FDMA system, the RX can also listen to activity on other frequencies.

TDMA is used in a large number of wireless standards: the worldwide cellular standard GSM (Chapter 21) as well as U.S. standards IS-54 and IS-136, and the cordless standard DECT (Digital Enhanced Cordless Telecommunications, see the companion website at *www.wiley.com/go/molisch*), use it. In contrast, pure FDMA is used mainly in analog cellular and cordless systems.

17.4 Packet radio

Packet radio access schemes break data down into packets, and each of the packets is transmitted over the medium independently. In other words, each packet is like a new user that has to fight for its "own" resources. This allows the transport medium to be exploited much more efficiently when the data traffic from each user is bursty, as is the case for Web browsing, file downloads, and similar data applications.

Packet radio shows two main differences from TDMA and FDMA:

1. Each packet has to fight for its own resources, as described above. The most common methods for resource allocation are ALOHA systems, carrier-sense multiple access, and packet reservation (polling). These methods are described in Sections 17.4.1–17.4.4. In these sections, we consider the case where multiple users try to transmit packets to one BS.
2. Each packet can be routed to the RX in different ways – i.e., via different relay stations. This aspect does not play a major role in cellular systems, where connection can only be to the closest BS,[5] but it does play an important role in wireless ad hoc and sensor networks, where each wireless device can act as a relay for information originating from another wireless device. Appropriate routing is thus a very important aspect of sensor networks; this is described briefly in Section 17.4.5.

17.4.1 ALOHA

The first wireless packet radio system was the ALOHA system of the University of Hawaii; it was used to connect computer terminals in different parts of this archipelago to the central computer in Honululu. We consider in the following the case when multiple MSs try to transmit packets to a given BS; however, the principle also applies to ad hoc and sensor networks.

For an ALOHA system, each user sends packets to the BS whenever the data source makes them available. Now, the situation can arise that several users want to transmit information simultaneously. When two TXs transmit packets at the same time, at least one of these packets suffers so much interference that it becomes unusable, and has to be retransmitted. Such collisions decrease the effective data rate of the system.

If the starting time of packet transmission is chosen completely at random by the TX, then the system is called a *pure* or *unslotted* ALOHA system. A TX does not take into consideration whether other users are already transmitting. Such systems become inefficient when the traffic load is moderate or large, because the probability for collisions becomes large.

Let us determine the possible throughput of an unslotted ALOHA system. For that purpose, we first determine the possible collision time – i.e., the time during which collisions with packets of other users are possible; we assume that all packets have the same length T_p. Figure 17.5 shows that packet A (from TX 2) can suffer collisions with packets that are transmitted by TX 1 either before or after packet A. We assume here that even a short collision leads to such strong interference that the packet has to be retransmitted. In order to completely avoid collisions, a

---

[5] The routing of packets from one BS to another BS, or a fixed line RX, can be either packet-based via a logical connection, or circuit-based.

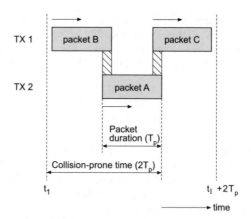

Figure 17.5 Possible collision time in an unslotted ALOHA system.
Reproduced with permission from Rappaport [1996] © IEEE.

packet from TX 1 must start its transmission at least T_p seconds before packet A, or has to start its transmission after packet A has finished – i.e., must start no sooner than T_p seconds after the start of packet A. The total possible collision time is thus $2T_p$.[6]

For mathematical convenience, we assume now that all transmission times are completely random, and different transmitters are independent of each other. The average transmission rate of all TXs is denoted as λ_p packets per second; the offered rate $R = \lambda_p T_p$ is the normalized channel usage, which has to lie between 0 and 1. Under these assumptions, the probability that n packets are transmitted within time duration t is given by a Poisson distribution [Papoulis 1991]:

$$\Pr(n, t) = \frac{(\lambda_p t)^n \exp(-\lambda_p t)}{n!} \tag{17.10}$$

The probability that during time t zero packets are generated is thus:

$$\Pr(0, t) = \exp(-\lambda_p t) \tag{17.11}$$

Effective throughput is the percentage of time during which the channel is used in a meaningful way – i.e., packets are offered, and transmitted successfully. As we have seen above, the possible collision time is twice the packet length, so that the probability of not having a collision is $\exp(-2\lambda_p T_p)$. Effective channel throughput is thus:

$$\lambda_p T_p \exp(-2\lambda_p T_p) \tag{17.12}$$

It can easily be shown that the maximum effective throughput is $1/(2e)$, where e is Euler's number.

In a *slotted* ALOHA system, the BS prescribes a certain slot structure. Each TX has a synchronized clock that makes sure that the start of the transmission time coincides with the beginning of a slot. Thus, partial collisions cannot occur anymore: either two packets collide completely, or they do not collide at all. It is immediately obvious that the possible collision time in such a system is T_p, so that effective throughput is:

$$\lambda_p T_p \exp(-\lambda_p T_p) \tag{17.13}$$

---

[6] In the following, we also assume that T_p contains a guard period that accounts for the different physical runtimes of a packet in a cell.

and the maximum achievable throughput is $1/e$ – i.e., twice as large as in an unslotted ALOHA system.

17.4.2 Carrier-sense multiple access

Basic principle

A TX can determine (*sense*) whether the channel is currently occupied by another user (*carrier*). This knowledge can be used to increase the efficiency of a packet-switched system: if one user is transmitting, no other user is allowed to send a signal. Such a method is called *Carrier Sense Multiple Access* (CSMA). It is more efficient than ALOHA, because it does not disturb other users that are already on the air.

The most important parameters of a CSMA system are detection delay and propagation delay. *Detection delay* is a relative measure for how long it takes a TX to determine whether the channel is currently occupied. It depends essentially on the hardware of the system, but also on the desired false alarm probability and the Signal-to-Noise Ratio (SNR). *Propagation delay* is the measure of how long a data packet takes to get from the MS to the BS. It can happen that at time t_1, TX 1 determines that the channel is free, and thus sends off a packet. At time t_2 another TX senses the channel. If $t_2 - t_1$ is shorter than the time it takes data packet A to get from TX 1 to TX 2, then TX 2 determines that the channel is free, and sends off data packet B. In such a case a collision occurs. This description makes it clear that detection delay and propagation delay should be much smaller than packet duration.

Implementation of carrier-sense multiple access

There are different methods of implementing CSMA. The most popular are [Rappaport 1996]:

- *non-persistent CSMA*: the TX senses the channel. If the channel is busy, the TX waits a random time duration until retransmission.
- *p-persistent CSMA*: this method is applied in slotted channels. When a TX determines that a channel is available, it transmits with probability p in the subsequent frame; otherwise, it transmits one timeslot later.
- *1-persistent CSMA*: the TX constantly senses the channel, until it realizes that the channel is free; then it immediately sends off the packet. This is obviously a special case of p-persistent transmission, with $p = 1$.
- *CSMA with collision detection*: in this method, a node observes whether two TXs start to transmit simultaneously. If that is the case, transmission is immediately terminated. This approach is not commonly used for wireless packet radio.
- *Data Sense Multiple Access* (DSMA): in this approach, the downlink includes a control channel, which transmits at periodic intervals a "busy/available" signal that indicates the state of the channel. If a user finds the channel to be free, it can immediately send off a data packet. Note that for peer-to-peer networks, implementation of the control channel is more difficult than in a scenario with a central node (BS).

17.4.3 Packet reservation multiple access

In *Packet Reservation Multiple Access* (PRMA) each MS can send a request to transmit a data packet. A control mechanism (which can be centralized or non-centralized) answers by telling the MS when it is allowed to send off the packet. This eliminates the risk of collisions of data packets; however, the signals that carry the requests for time can collide. Furthermore, the system sacrifices some transmission capacity for the transmission of reservation requests.

In order to maintain reasonable efficiency, the requests for time must be much shorter than the actual data packets. The method is also known as **SRMA** (*Split-channel Reservation Multiple Access*).

A variant of this method is that a user can keep the transmission medium until it has finished transmission of a datablock. Other methods hand out (time-varying) priorities, in order to avoid a situation where a single user hogs the transmission medium for a long time.

Another important method is *polling*. In this method, a BS asks (polls) one MS after another whether it wants to transmit a data packet. The shortest polling cycle occurs when no MS wants to transmit information; this is also the most inefficient case, as capacity is sacrificed for polling, and no payload is transmitted.

17.4.4 Comparison of the methods

Figure 17.6 shows the efficiency of the various packet-switched methods. The abscissa shows channel usage, the ordinate the average packet delay. We see that ALOHA methods are only suitable if efficient channel use is of no importance. The maximum achievable capacity is only 36% in a slotted ALOHA system. CSMA and polling achieve better results.

When comparing packet radio with TDMA and FDMA, we find that the latter schemes are very useful for the transmission of speech, since this usually requires low latency. Encoded speech data should arrive at their destination no later than about 100 ms after they have been spoken. This can easily be achieved by FDMA and TDMA systems with appropriate slot duration. Also, each TX can be certain of transmitting its data to the RX without significant blocking or delay on the line, since it has a channel (frequency or timeslot) exclusively reserved. On the other hand, TDMA and FDMA waste resources, especially for the transmission of data. Therefore, the fact that a channel (timeslot) is *always* reserved for a single user becomes a major drawback. Even if this user does not have any data to transmit, the multiaccess scheme does not allow anybody else to use that slot in the meantime.

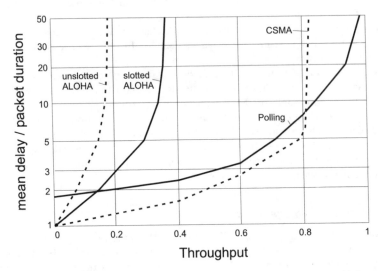

Figure 17.6 Channel usage and mean packet delay for different packet-reservation-multiple-access methods.
Reproduced with permission from Oehrvik [1994] © Ericsson AB.

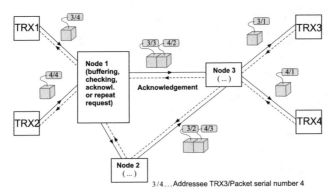

Figure 17.7 Principle behind a packet radio network.

Reproduced with permission from Oehrvik [1994] © Ericsson AB.

17.4.5 Routing for packet radio

We now describe the routing of packets in a packet radio system (see Fig. 17.7); this principle holds both for wireless and wired systems. A packet-switched system builds up a *logical* connection between the TX and RX, but – in contrast to circuit-switched systems – not necessarily a constant *physical* connection. It is only important to get the packets from the TX to the RX in some way; the actual choice of route (the physical connection) can change with time. For this purpose, the message is broken into small pieces (packets), each of which can take a different route to the RX, depending on what transmission paths are currently available. Packets can thus take routes via different network nodes. Each of these nodes determines how it can pass the packet on. If no transmission path is currently available, then the packet is buffered until a path becomes available. This buffering can lead to considerable delays in transmission, and it can also happen that the sequence in which the data arrive is different from the sequence in which they were transmitted. For this reason, packet radio with routing over many nodes cannot easily be used for speech transmissions. However, as the emergence of voice-over-IP telephony shows, it *is* possible to achieve packet-based voice telephony.

A data packet contains payload data as well as *routing information* (see Fig. 17.8). The routing information spells out the origin and the destination of the message; furthermore, additional information about the buffering and the path that the packet has taken can be included.

Generally, routing methods can be classified in the following two categories:

- Source-driven routing: in this case, the header of the packet includes the complete route, and the nodes just follow the instructions for forwarding. The drawback is that the header can become quite long, especially for packets with little payload; this leads to a significant decrease in spectral efficiency.
- Table-driven routing: in this approach, each node stores in a table the nodes to which it should forward packets (depending on the destination address, and the node the packet came from). This method has better spectral efficiency; the drawback is that the tables can become quite large, especially at nodes in the middle of a network.

A related topic is "route discovery" – i.e., determination of which route a packet should take from the transmitter to the receiver. Route discovery is typically done by means of special packets that are broadcast in the network, and record the quality of the links between different nodes. In order to achieve optimum performance, routing has to be changed whenever the channel between nodes changes significantly.

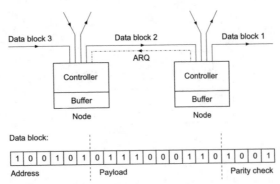

Figure 17.8 Structure of a datablock, and principle behind the buffering mechanism.

Reproduced with permission from Oehrvik [1994] © Ericsson AB.

If a very low packet error rate is required, each node that acts as a relay stores the packets in a buffer and deletes them only after receiving an acknowledgement of successful transmission from the node it forwarded the packet to (see Fig. 17.8).

17.5 Duplexing

Duplexing serves to separate the uplink and downlink. We distinguish between *Time Domain Duplex* (TDD) and *Frequency Domain Duplex* (FDD). In TDD, uplink data are sent at times that are different from downlink transmission times (see Fig. 17.9a). In FDD, uplink and downlink data are sent in different frequency bands (see Fig. 17.9b).

TDD is often used in conjunction with TDMA. In such a case, the available time is divided not into N, but rather $2N$ timeslots, where each of the N users is assigned one timeslot for the uplink, and one for the downlink. We also find that TDD can be used well in conjunction with packet radio schemes. On the other hand, the use of TDD in an FDMA system would counteract many of the advantages inherent in FDMA (continuous transmission and reception, which simplifies synchronization), while retaining the disadvantages.

FDD can be used in combination with any multiaccess method. In most cases, there is a fixed duplexing distance – i.e., a fixed frequency difference between the uplink and the downlink band.

For both TDD and FDD, it is interesting to consider the question as to whether the channel for uplink and downlink are identical. We find that for TDD, the requirement is that the duplexing time is much smaller than the coherence time, while for FDD systems, the

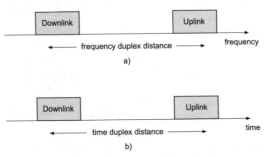

Figure 17.9 Frequency and time duplexing.

requirement is that the frequency duplexing distance is much smaller than the coherence bandwidth (more details are given in Section 20.1.6). When considering practical system parameters, we find that the former condition can be fulfilled quite well, especially for wireless LANs that are operating in a quasi-static environment where neither BS nor MS are moving. The latter condition (for FDD) is practically never fulfilled. However, we find that there are considerable advantages when the channel for uplink and downlink is identical; we will find in Chapter 19 that it simplifies implementation of adaptive modulation, and in Chapter 20 that it is beneficial to implementation of various smart antenna systems.

On the other hand, care has to be taken when designing TDD systems with large cell sizes: there needs to be a *dead time* between transmission and reception. Imagine that a user at distance d (and thus runtime d/c_0) transmits data, and stops the transmission at time T. Then data will arrive at the BS until time $T + d/c_0$. Only now can the BS switch to transmission mode, and send out its own block of data. Since the data have to propagate back to the MS, they start arriving there at time $T + 2d/c_0$. Consequently, the MS has experienced a dead time of $2d/c_0$. This can result in a considerable loss of efficiency for the system.

17.6 Principles of cellular networks

17.6.1 Reuse distance

Let us now turn to the question of how a wireless system can cover a large area, and provide service to as many users as possible within this area. The first mobile radio systems were actually noise-limited systems with few users. Therefore, it was advantageous to put each BS on top of mountains or high towers, so that it could provide coverage for a large area. The next BS was so far away that interference was not an issue. However, this approach severely limited the number of users that could communicate simultaneously. The cellular principle, which we will describe in this section, provides the solution to this problem. For this section, we will use the example of FDMA as the multiaccess scheme within each cell; the same considerations hold for TDMA. CDMA systems will be discussed in Chapter 18.

In a cellular system, the coverage area is divided into many small areas called "cells". In each of these cells, there is one BS that provides coverage for this (and only this) cell area. Now, each frequency channel can be used in *multiple* cells. The question that naturally arises is: can we use each frequency channel in each cell? Typically, the answer is "no". Imagine the situation where user A is at the boundary of its assigned cell, so that distances from the "useful" BS and from a neighboring BS are the same. If the neighboring BS transmits in the same frequency channel (in order to communicate with user B in its own cell), then the Signal-to-Interference Ratio (SIR) seen by user A is $C/I = 0$ dB. This is certainly not enough to sustain reliable communications, especially since 0 dB is the *median* SIR, and fading makes the situation worse 50% of the time.

The solution to this problem is to reuse a frequency channel not in every cell, but only in cells that have a certain minimum distance from each other. The distance between two cells that can use the same frequency channels is called the *reuse distance*, D/R. This reuse distance can be computed from link budgets (as described in Section 3.2). We can also define a *cluster* of cells that all use different frequencies; therefore, there can be no co-channel interference within such a cluster. The number of cells in a cluster is called the *cluster size*. The total coverage area is divided into such clusters.

The cluster size also determines the *capacity* of the cellular system.[7] An operator that has licenses for 35 frequency channels, and uses cluster size 7, can support 5 simultaneous users in

---

[7] Note that we use "capacity" in this chapter for the number of users or communication devices that can be supported simultaneously. We do *not* refer to information-theoretic capacity.

each cell. Maximization of capacity thus requires minimization of cluster size. Cluster size 1 (i.e., using each frequency in each cell) is the ultimate goal; however, we have seen above that this is not possible in an FDMA system due to the required link margins. Analog FDMA systems typically require an SIR of 18 dB, which results in a cluster size of 21 (see below). Digital systems like GSM require less than 10 dB, which decreases the reuse distance to 7 or less. This allows a dramatic increase in capacity, and was one of the most important reasons for the shift from analog to digital cellular systems in the early 1990s.

17.6.2 Cellshape

What shapes do cells normally take on? Let us first consider the idealized situation where pathloss depends only on the distance from the BS, but not the direction. The most natural choice would be a disk (circle), as it provides constant power at the cell boundary. However, disks cannot fill a plane without either gaps or overlaps. Hexagons, on the other hand, have a shape similar to a circle, and they *can* fill up a plane, like in a beehive pattern. Thus, hexagons are usually considered the "basic" cell shape, especially for theoretical considerations.

We stress, however, that hexagonal structures are only possible if:

- The required traffic density is independent of the location. This condition is obviously violated whenever the population density changes.
- The terrain is completely flat, and there are no high edifices, so that pathloss is influenced only by the distance from the BS. This is never fulfilled in Europe, and only rarely in the rest of the world – e.g., the American Midwest, Siberian tundra, deserts. Figure 17.10 shows the terrain, and the power obtained from one BS in a typical hilly terrain. According to the simplified considerations above, lines of equal power should be concentric circles around the

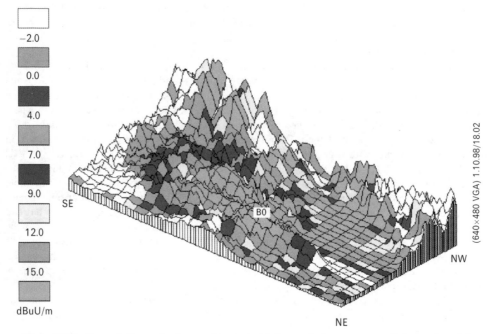

Figure 17.10 Example for received power in a typical hilly terrain (western part of Vienna, Austria).
Reproduced with permission from Buehler [1994] © H. Buehler.

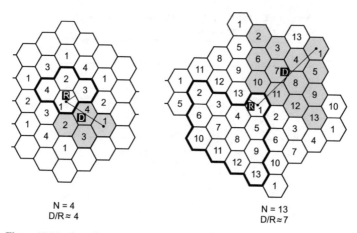

Figure 17.11 Interdependence between reuse distance D/R and cluster size N.
Reproduced with permission from Oehrvik [1994] © Ericsson AB.

BS. We see that the actual result is anything but – the power is not even necessarily a monotonic function of the distance in some directions. Thus, practical cell planning requires computer simulations or measurements of the received power. The modeling techniques described in Chapter 7 are thus a vital basis for realistic cell planning.

17.6.3 Cell planning with hexagonal cells

For the case of hexagonal cells, some interesting conclusions can be drawn about the relationship between link margin and reuse distance. Consider the hexagon whose center is at the origin of the coordinate system. Proceed now i hexagons in the y direction, turn $60°$ counterclockwise, and proceed k hexagons in that new direction (see Fig. 17.12). This gets us to the cell whose center has the following distance from the origin:

$$D = \sqrt{3}\sqrt{(iR + \cos(60°)kR)^2 + (\sin(60°)kR)^2} \tag{17.14}$$

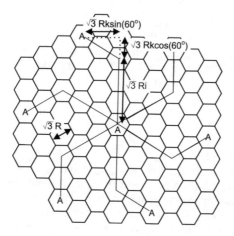

Figure 17.12 Minimum reuse distance.

Table 17.2 Typical cluster sizes and reuse distances, assuming omni-antennas.

N	D/R	
1		CDMA
3	3	
4	3.46	
7	4.58	TDMA system (GSM)
9	5.2	
12		
13	6.24	
16	6.93	
19	7.55	
21	7.94	Analog system (NMT, AMPS)

Note that the distance between the centers of two adjacent hexagons is $\sqrt{3}R$, where R is the distance from the center of a hexagon to its farthest corner. Also note that only integer values of i and k are possible.

The task of frequency planning is to find those values of i and k that make sure that the distance from Eq. (17.14) is larger than the required reuse distance. Of course there is an infinite manifold of such pairs – large values of i and k certainly satisfy the condition. What we want to find, however, is the pair of values that *minimizes* cluster size, and thus maximizes spectral efficiency, while still satisfying the minimum reuse distance.

Due to the hexagonal layout, the relationship between cluster size N and parameters i and k is:

$$N = i^2 + ik + k^2, \qquad i, k = 0, 1, 2, \ldots \tag{17.15}$$

This also establishes that not all integers are possible cluster sizes. Cluster size can only take on the numbers $N = 1, 3, 4, 7, 9, 12, 13, 16, 19, 21 \ldots$ The relationship between reuse distance D/R and cluster size results from Eqs. (17.15) and (17.14) as:

$$D/R = \sqrt{3N} \tag{17.16}$$

see also Fig. 17.11. Table 17.2 shows typical cluster sizes and reuse distances.

Cell planning thus proceeds in the following steps: starting from the specifications for the minimum transmission quality, the link budget (Chapter 3) establishes the minimum distances between the desired BS and interferer. From this relationship, Eq. (17.16) provides the cluster size; note that it has to be the smallest integer number out of the set defined by Eq. (17.15). Using the recipe for obtaining nearest neighbors (move i cells into one direction, turn 60°, move another k cells), the frequencies for each cell can be determined.

Example 17.3 *Cell planning in an interference-limited system.*

The principle, as well as typical values, behind cell planning can be best understood from an example. In order to keep things simple (and work with a real system that uses pure FDMA), we consider the numbers of an analog Advanced Mobile Phone System (AMPS). Each frequency channel is 30 kHz wide, the SIR is 18 dB for satisfactory speech quality. The fading margin (for shadowing plus Rayleigh fading) is set to 15 dB. This implies that at the cell boundary the median values of the signal power must be 33 dB ($2 \cdot 10^3$) stronger than that of the interference power. The distance between the desired BS and the farthest corner of the hexagon is R, and the distance between the interfering BS and this corner is (approximately) $D - R$. Assuming further that power decreases with d^{-4}, we require that:

$$\frac{D - R}{R} = (2 \cdot 10^3)^{1/4} = 6.7 \qquad (17.17)$$

so that the reuse distance must be $D/R = 7.7$. From Eq. (17.16), the reuse distance is $(3N)^{0.5}$, so that the cluster size is 19.8. The smallest $N \geq 19.8$ is $N = 21$. Now consider an operator with a license for 5 MHz of spectrum. Having a total of $5,000/30 = 167$ possible frequency channels, only $167/21 \simeq 8$ can be used in each cell.

17.6.4 Methods for increasing capacity

System capacity is the most important measure for a cellular network. Methods for increasing capacity are thus an essential area of research. In the following, we give a brief overview, often referring to other chapters in this book:

1. *Increasing the amount of spectrum used*: this is the "brute force" method. It turns out to be very expensive, as spectrum is a scarce resource, and usually auctioned off by governments at very high prices. Furthermore, the total amount of spectrum assigned to wireless systems can change only very slowly; changes in spectrum assignments have to be approved by worldwide regulatory conferences, which often takes ten years or more.
2. *More efficient modulation formats and coding*: modulation formats that require less bandwidth (higher order modulation) and/or are more resistant to interference. The former allows an increase in data rate for each user (or an increase in the number of users in a cell while keeping the data rate per user constant). However, the possible benefits of higher order modulation are limited: they are more sensitive to noise and interference (see Chapter 12), so that the reuse distance might have to be increased. The use of interference-resistant modulation allows a reduction in reuse distance. The introduction of near-capacity-achieving codes – turbocodes and low-density parity check codes (see Chapter 14) – is another way of achieving better immunity to interference, and thus increases system capacity.
3. *Better source coding*: depending on required speech quality, current speech coders need data rates between 32 kbit/s and 4 kbit/s. Better models for the properties of speech allow the data rate to be decreased without decreasing quality (see Chapter 15). Compression of data files and music/video compression also allows more users to be served.
4. *Discontinuous transmission*: DTX exploits the fact that during a phone conversation each participant talks only 50% of the time. A TDMA system can thus set up more calls than there are available timeslots. During the call, the users that are actively talking at the moment are multiplexed onto the available timeslots, while quiet users do not get assigned any radio resources.

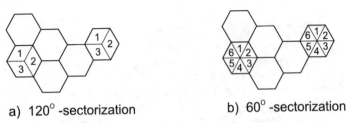

a) 120°-sectorization   b) 60°-sectorization

Figure 17.13 Principle behind sector cells.

5. *Multiuser detection*: this greatly reduces the effect of interference, and thus allows more users per cell for CDMA systems or smaller reuse distances for FDMA systems (for details see Chapter 18).

6. *Adaptive modulation and coding*: employs the knowledge at the TX of the transmission channel, and chooses the modulation format and coding rate that are "just right" for the current link situation. This approach makes better use of available power, and, among other effects, reduces interference (more details can be found in Chapter 19).

7. *Reduction of cell radius*: this is an effective, but very expensive, way of increasing capacity, as a new BS has to be built for each additional cell. For FDMA systems, it also means that the frequency planning for a large area has to be redone.[8] Furthermore, smaller cells also require more handovers for moving users, which is complicated, and reduces spectral efficiency due to the large amount of signaling information that has to be sent during a handover.

8. *Use of sector cells*: a hexagonal (or similarly shaped) cell can be divided into several (typically three) sectors. Each sector is served by one sector antenna. Thus, the number of cells has tripled, as has the number of BS *antennas*. However, the number of BS *locations* has remained the same, because the three antennas are at the same location (see Fig. 17.13).

9. *Use of an overlay structure*: an overlay structure combines cells with different size, and different traffic density. Therefore, some locations may be served by several BSs simultaneously. An *umbrella cell* provides basic coverage for a large area. Within that coverage area, multiple microcells are placed in areas of high traffic density. Within the coverage area of the microcells, most users are served by the microcell BS, but fast-moving users are assigned to the umbrella cell, in order to reduce the number of handovers between cells.

10. *Multiple antennas*: these can be used to enhance capacity via different scenarios:
    (a) diversity (Chapter 13) increases the quality of the received signal, which can be exploited to increase capacity – e.g., by use of higher order modulation formats, or reduction of the reuse distance;
    (b) multiple-input multiple-output systems (Section 20.2) increase the capacity of each link;
    (c) space division multiple access (Section 20.1) allows several users in the same frequency channel in the same cell to be served.

11. *Fractional loading*: this system uses a small reuse distance, but uses only a small percentage of the available timeslots in each cell. This leads to approximately the same average capacity as the "conventional" scheme with large reuse distance and full loading of each cell. However, it has higher flexibility, as throughput can be made higher in some cells when throughput in other cells is low.

---

[8] For the case of hexagonal cells, algorithms are available that allow exact cell splitting without large-scale replanning of frequency assignments [Lee 1986].

17.7 Appendix

Please go to *www.wiley.com/go/molisch*

Further reading

FDMA and TDMA systems are "classical" multiaccess schemes, and nowadays mostly analyzed in the context of specific systems (especially GSM, see Chapter 21), and generically in classical textbooks – e.g., Rappaport [1996]; there is also a nice summary in Falconer et al. [1995]. An interesting performance analysis in interference is given in Xu et al. [2000]; message delays in TDMA and FDMA systems are analyzed in Rubin [1979]. The transition from TDMA/FDMA to CDMA is discussed in Sari et al. [2000]. The issue of queuing is discussed in the excellent textbook of Gross and Harris [1998]. Discussions of traffic distributions can be found, for example, in Ashtiani et al. [2003].

The ALOHA system was first suggested in Abramson [1970], and analyzed for mobile radio applications in Namislo [1984]. A classical description of CSMA is given in Kleinrock and Tobagi [1975] and Tobagi [1980]. PRMA is described in Goodman et al. [1989]. Routing protocols for ad hoc networks are reviewed in Royer and Toh [1999] and Perkins [2001].

The cellular principle, and the basic concepts of cellular radio, are described in detail in Lee [1995]. Though the description is by now somewhat outdated, and tries to cover both analog and digital systems, the principles have remained valid; Alouini and Goldsmith [1999] discuss more advanced aspects, including the effect of fading. Chan [1992] discusses the impact of sectorization. Frullone et al. [1996] give an overview of advanced cell-planning techniques; Woerner et al. [1994] describe issues concerning the simulation of cellular systems and network planning.

18

Spread spectrum systems

Spread spectrum techniques spread information over a very large bandwidth – specifically, a bandwidth that is much larger than the inverse of the data rate. In this chapter, we will discuss various ways of providing multiple access by spreading the spectrum. We start out with the conceptually most simple approach, Frequency Hopping (FH). We then proceed to the most popular form of spread spectrum, Direct Sequence–Code Division Multiple Access (DS-CDMA). Finally, we elaborate on time-hopping impulse radio, a relatively new scheme that has gathered interest in recent years because of its application to ultrawideband systems.

We have stressed in previous chapters how important spectral efficiency is: we want to transmit as much information per available bandwidth as possible. Thus, it might seem like a strange idea to spread information over a large bandwidth in a commercial wireless system. After all, the term "spread spectrum" comes from the military area, where the main interest lies in keeping communications stealthy, safe from intercept, and safe from jamming efforts by hostile transmitters – issues that do not top the list of concerns of cellular operators.[1] It thus seems astonishing that spread spectrum approaches have attained such an important role in wireless communications.

This seeming paradox can be resolved when we recognize that different users can be spread across the spectrum in different ways. This allows multiple users to transmit in the same frequency band simultaneously; the receiver can determine which part of the total contribution comes from a specific user by looking at data with a specific spreading pattern. Thus, capacity (per unit bandwidth) is not necessarily decreased by using spread spectrum techniques, and can even be increased by exploiting its special features.

18.1 Frequency-hopped multiple access

18.1.1 Principle behind frequency hopping

The basic thought underlying *frequency hopping* (FH) is to change the carrier frequency of a narrowband transmission system so that transmission is done in one frequency band only for a short while. The ratio between the bandwidth over which the carrier frequency is hopped, and the narrowband transmission bandwidth is the spreading factor.

---

[1] While security (insensitivity to intercept operations) is important, it can be achieved by cryptographic means, and does not require spectral spreading.

Wireless Communications   Andreas F. Molisch
© 2005 John Wiley & Sons, Ltd

FH originated from military communications; it was invented by actress Hedi Lamar during the Second World War. It was inspired by the problem that emissions from radio transmitters could be used by the enemy to triangulate the position of transmitters, or that transmission could be jammed by the enemy with powerful (narrowband) transmitters. By changing the carrier frequency frequently, the signal is in the vulnerable (observed or jammed) band only for a short while. The FH pattern has to be known to the desired receiver, but unpredictable for the enemy, making them unable to "follow" the FH. In addition to suppressing narrowband interferers, the FH also helps to mitigate the effect of deep fading dips. Sometimes the system transmits on a "good" frequency – i.e., one with low attenuation between transmitter and receiver and low interference – and sometimes on a "bad" frequency – i.e., in a fading dip and/or with high interference. Appropriate coding and interleaving can average out these situations (see Chapter 14).

There are two basic types of FH: "slow" and "fast". *Fast frequency hopping* changes the carrier frequency several times during transmission of one symbol; in other words, transmission of each separate symbol is spread over a large bandwidth. Consequently, the effects of fading or interference can be combated for each symbol separately. It follows from elementary Fourier considerations that transmission of each *part* of a symbol requires more bandwidth than that of a narrowband system. Furthermore, combining of the different contributions belonging to one symbol has to use processing that works faster than at the symbol rate. Fast FH is not in widespread use in commercial wireless systems; it has been mostly edged out by Code Division Multiple Access (CDMA).

Slow frequency hopping transmits one or several symbols on each frequency. This method is often used in conjunction with Time Division Multiple Access (TDMA): each timeslot is transmitted on a given carrier frequency; the next slot then changes to a different frequency. In such a case, the additional effort for synchronization is small, as the receiver has to synchronize for the next slot anyway. In order for the FH to be effective, interleaving and coding has to distribute information belonging to one source bit over several timeslots. Imagine simple repetition coding, where each bit is sent twice, in different timeslots (and thus on different carrier frequencies). If the first timeslot is transmitted in a deep fading dip, chances are that the second is at a frequency where channel attenuation is small; thus the information can be recovered. Slow FH is used, for example, in the Global System for Mobile communications (GSM) (see Chapter 21).

Generally, FH leads to a *whitening* of received signal characteristics. There is implicit averaging over channel attenuation. Furthermore, at each carrier frequency, a different interferer is active, such that FH also leads to an averaging over all interferers. For many types of receivers this is advantageous, as it reduces the probability of "disastrous" scenarios, and thus decreases the required link margins. However, we will see in Sections 18.4 and 20.1 that there are some receiver structures that can actually exploit a specific (known) structure of interference in order to eliminate it. For example, the spatial structure of interferers can be exploited by smart antennas to form antenna patterns that have nulls in the direction the interference is coming from. FH can make such techniques more difficult to apply.

18.1.2 Frequency Hopping for Multiple Access (FHMA)

In the previous section, we looked at FH for the suppression of interference, and increasing frequency diversity. The price that has to be paid is that a larger bandwidth has to be used for transmission, which seems wasteful. In the following, we will show that FH can be used as a multiaccess method that is as spectrally efficient as TDMA and Frequency Division

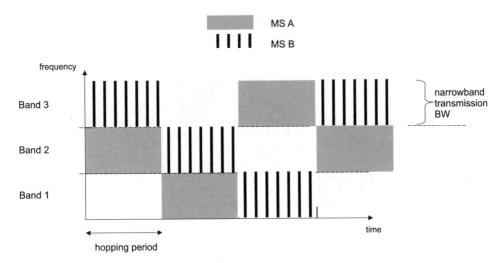

Figure 18.1   Principle behind frequency hopping for multiple access with synchronized users.

Multiple Access (FDMA). For these considerations, we distinguish between synchronized and unsynchronized systems.

Let us consider first the *synchronized* case – e.g., in the downlink of a cellular system – where the Base Station (BS) can always make sure that it emits to all Mobile Stations (MSs) at the same time. Figure 18.1 shows an example with three available bands (carrier frequencies). Clearly, during one time interval, the BS can transmit to three users simultaneously, but for ease of exposition, we assume just two active users (users A and B). During the first time interval (hopping period), the BS transmits to MS A in band 2. At the same time, band 3 is free, so the BS can transmit to MS B in that band. In the next timeframe, the BS now transmits to MS A in band 1, and MS B in band 2. In the third timeslot, MS A is serviced in band 3, and MS B in band 1. Then the whole sequence repeats. The signals for all the MSs use the same hopping sequence, and it can be made sure that there is never a collision between them. Thus, clearly, we have the same capacity as FDMA, with the added benefit of frequency diversity. In order to apply the same concept for the uplink, all MSs have to send their signals in such a way that they arrive at the BS synchronously, and thus recover the situation of Fig. 18.1. This requires information about the runtime from each MS to the BS, which tells each MS exactly when to start transmission (timing advance, see also Chapter 21).

The situation is different when users are not synchronized – such a situation can either occur in simple networks where timing advance is not foreseen, for consideration of intercell interference,[2] or in ad hoc networks. For such a case, it is not a good idea to use the same hopping sequence for all users. Remember that, due to the lack of synchronization, any delay between the signals of different users is possible, including a zero-delay. If all users use the same hopping sequence, then such a zero-delay leads to *catastrophic collisions*, where different users interfere with each other all the time. In order to circumvent this problem, different hopping sequences are used for each user (see Fig. 18.2). These sequences are designed in such a way that during each hopping cycle (i.e., one repetition of the hopping sequence; in our example, three times the hopping period), the duration of exactly one timeslot is disturbed, while the remainder of   the

---

[2] Users that design their uplink signals to arrive synchronously at one BS cannot be synchronous for another BS.

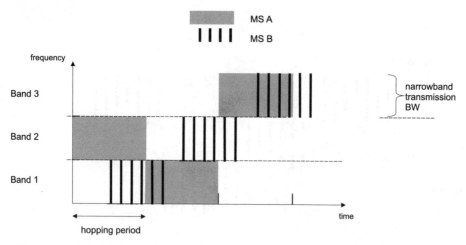

Figure 18.2   Principle behind frequency hopping for multiple access for unsynchronized users.

time is guaranteed to be collision-free.[3] Obviously, the performance of such a system is worse than that of a synchronized system (or an FDMA system). The design of hopping sequences that guarantee the low probability of collisions is also quite tricky, and still an active area of research.

18.2 Code division multiple access

The origins of CDMA can also be traced to military communications research, especially the development of the *Direct Sequence–Spread Spectrum* (DS-SS). In Section 18.2.1, we discuss the basic spreading operation. Sections 18.2.2 and 18.2.3 then describe the principles and the mathematics of how a CDMA system uses DS-SS to enable multiple access. Section 18.2.4 analyzes behavior in frequency-selective channels. Sections 18.2.5 and 18.2.6 discuss synchronization, and the selection of codes for CDMA. More details about the application in cellular networks can be found in Section 18.3.

18.2.1 Basic principle behind the direct-sequence spread spectrum

The DS-SS spreads the signal by multiplying the transmit signal by a second signal that has a very large bandwidth. The bandwidth of this total signal is approximately the same as the bandwidth of the wideband spreading signal. The ratio of the bandwidth of the new signal to that of the original signal is again known as the *spreading factor*. As the bandwidth of the spread signal is large, and the transmit power stays constant, the *power-spectral density* of the transmitted signal is very small – depending on the spreading factor and the BS–MS distance, it can lie below the noise power spectral density. This is important in military applications, because unauthorized listeners cannot determine whether a signal is being transmitted. Authorized listeners, on the other hand, can invert the spreading operation, and thus recover the narrowband signal (whose power-spectral density lies considerably *above* the noise power).

Figure 18.3 shows the block diagram of a DS-SS transmitter. The information sequence (possibly coded) is multiplied by a broadband signal that was created by modulating a

---

[3] By collision-free we mean "free of collisions from user B" – other users might still interfere.

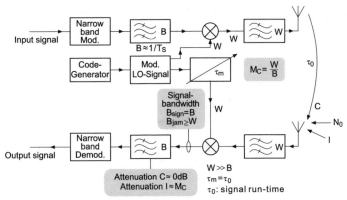

Figure 18.3 Block diagram of a direct-sequence-spread-spectrum transmitter and receiver.
Reproduced with permission from Oehrvik [1994] © Ericsson AB.

sinusoidal carrier signal with a spreading sequence. This can be interpreted alternatively as multiplying each information symbol of duration T_S by a spreading sequence $p(t)$ before modulation.[4] We assume in the following that the spreading sequence is M_C *chips* long, where each chip has the duration $T_C = T_S/M_C$. As the bandwidth is the inverse of the chip duration, the bandwidth of the total signal is now also $W = 1/T_C = M_C/T_S$ – i.e., larger than the bandwidth of a narrowband-modulated signal by a factor M_C. As we assume that the spreading operation does not change the total transmit power, it also implies that the power-spectral density decreases by a factor M_C.

In the receiver, we now have to invert the spreading operation. This can be achieved by correlating the received signal with the spreading sequence – i.e., multiply by the time-inverted, complex-conjugate spreading sequence, and integrate over a suitable interval. This process reverses bandwidth spreading, so that after correlation, the desired signal again has a bandwidth of $1/T_S$. In addition to the desired signal, the received signal also contains noise, other wideband interferers, and possibly narrowband interferers. Note that the effective bandwidth of noise and wideband interferers is not significantly affected by the despreading operation, while narrowband interferers are actually spread over a bandwidth W. As part of despreading the signal passes through a lowpass filter of bandwidth $B = 1/T_S$. This leaves the desired signal essentially unchanged, but reduces the power of noise, wideband interferers, and narrowband interferers by a factor $1/M_C$. At the symbol demodulator, DS-SS thus has the same Signal-to-Noise Ratio (SNR) as a narrowband system: for a narrowband system, the noise power at the demodulator is N_0/T_S. For a DS-SS system, the noise power at the receiver input is $N_0/T_C = N_0M_C/T_S$, which is reduced by narrowband filtering (by a factor of $1/M_C$); thus, at the detector input, it is N_0/T_S. A similar effect occurs for wideband interference.

Let us next discuss the spreading signals for DS-SS systems. In order to perfectly reverse the spreading operation in the receiver by means of a correlation operation, we want the Auto Correlation Function (ACF) of the spreading sequence to be a Dirac delta function. In such a case, the convolution of the original information sequence with the concatenation of spreader and despreader is the original sequence. We thus desire that $ACF(i)$ of $p(t)$ at times iT_C is:

$$ACF(i) = \begin{cases} M_C & \text{for } i = 0 \\ 0 & \text{otherwise} \end{cases} \qquad (18.1)$$

---

[4] This interpretation is valid if the narrowband signal and the wideband signal use the same modulation method.

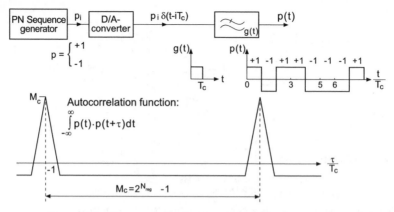

Figure 18.4 Autocorrelation function of a pseudonoise sequence.

Reproduced with permission from Oehrvik [1994] © Ericsson AB.

These ideal properties can only be approximated in practice. One group of suitable code sequences is PN-sequences (*Pseudo Noise sequence = maximum length sequence*). PN-sequences have the following ACF:

$$ACF(i) = \begin{cases} M_C - 1 & \text{for } i = 0 \\ -1 & \text{otherwise} \end{cases} \tag{18.2}$$

see Fig. 18.4.

18.2.2 Multiple access

The Direct Sequence (DS) spreading operation itself – i.e., multiplication by the wideband signal – can be viewed as a modulation method for stealthy communications, and is as such mainly of military interest. CDMA is used on top of it to achieve multiaccess capability. Each user is assigned a different spreading code, which determines the wideband signal that is multiplied by the information symbols. Signals from other users, which were modulated by a different spreading sequence, appear as additional "noise" (wideband interference) at the receiver. Thus, many users can transmit simultaneously in a wide band (see Fig. 18.5).

At the receiver, the desired signal is obtained by correlating the received signal with the spreading signal of the desired user. Other users thus become wideband interferers; after passing through the despreader, the amount of interference power seen by the detector is

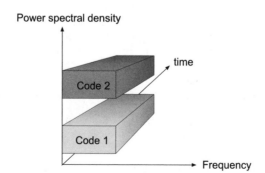

Figure 18.5 Principle behind code division multiple access.

equal to the *Cross Correlation Function* (CCF) between the spreading sequence of the interfering user and the spreading sequence of the desired user. Thus, we ideally wish for:

$$CCF_{j,k}(t) = 0 \quad \text{for } j \neq k \tag{18.3}$$

for all users j and k. In other words, we require code sequences to be orthogonal. Perfect orthogonality can be achieved for at most M_C spreading sequences; this can be immediately seen by the fact that M_C orthogonal sequences span an M_C-dimensional space, and any other sequence of that duration can be represented as a linear combination.

If the spreading sequences are not orthogonal, the receiver achieves finite interference suppression – namely, suppression by a factor ACF/CCF. If the different spreading sequences are shifted PN-sequences, then this suppression factor is M_C.

We can draw some important conclusions from the above description:

- The choice of spreading sequences is an essential factor for the quality of a CDMA system. Sequences have to have good ACFs (similar to a Dirac delta function) and a small cross-correlation. One possible choice is PN-sequences; with a shift register of length N_{reg} it is possible to generate $2^{N_{reg}} - 1$ sequences that have an ACF given by Eq. (18.2) and $ACF/CCF = M_C$. Alternative sequences include Gold and Kasami sequences, which will be discussed in Section 18.2.6.
- CDMA requires accurate power control. If ACF/CCF has a finite value, the receiver cannot suppress interfering users perfectly. When the received interference power becomes much larger than the power from the desired transmitter, it exceeds the interference suppression capability of the despreading receiver. Thus, each MS has to adjust its power in such a way that the powers of all the signals arriving at the BS are approximately the same. Experience has shown that power control has to be accurate within about $\pm 1\,dB$ in order to fully exploit the theoretical capacity of CDMA systems.

18.2.3 *Mathematical representation*

In the following, we put the above qualitative description into a mathematical framework [Viterbi 1995]. We assume for this that Binary Phase Shift Keying (BPSK) modulation is used, and that perfect synchronization between the transmitter and receiver is available (discussed below in more detail). We then obtain four signal components at the receiver:

- *Desired (user) signal.* Let $c_{i,k}$ represent the ith information symbol of the kth user, and $r_{i,k}$ the corresponding receive signal. Assuming that intersymbol interference, interchip interference (to be defined later) and noise are zero-mean processes, then the expected value of the received signal is proportional to the transmit symbol:

$$E\{r_{i,k}|c_{i,k}\} = \sqrt{(E_C)_k}\, c_{i,k} \int_{-\infty}^{\infty} |H_R(f)|^2 df \tag{18.4}$$

where $(E_C)_k$ is the chip energy of the kth user, and $H_R(f)$ the transfer function of the receive filter normalized to $\int_{-\infty}^{\infty} |H_R(f)|^2 df = 1$.
- *Interchip interference*: the receive filter has a finite-duration impulse response, so that the convolution of a chip with this impulse response lasts longer than the chip itself. Thus, the signal after the receive filter exhibits interchip interference (we will see later on that delay dispersion of the channel also leads to interchip interference). If the spreading sequences are zero-mean, then the interchip interference increases the variance of the received signal by:

$$(E_C)_k \sum_{i \neq 0} \left[\int_{-\infty}^{\infty} \cos(2\pi i f T_C)|H_R(f)|^2 df \right]^2 \tag{18.5}$$

- *Noise* increases the variance by $N_0/2$.
- *Co-channel interference*: we assume that the mean of the received signal is not changed by the CCI, as the transmitted chips of interfering users are independent of the data symbols and chips of desired users. The CCI increases the variance by:

$$\sum_{j\neq k} \frac{(E_C)_j}{2T_C} \int_{-\infty}^{\infty} |H_R(f)|^4 df \tag{18.6}$$

If we now assume that interchip interference and CCI are approximately Gaussian, then the problem of computing the error probability reduces to the standard problem of detecting a signal in Gaussian noise:

$$BER = Q\left(\sqrt{\frac{(E_c)_k M_C}{\text{Total variance}}} \right) \tag{18.7}$$

In a fading channel, the energy of the desired signal varies. For the uplink, the power control makes sure that these variations are compensated. For the downlink, power variations are not necessarily compensated, so that the BER has to be averaged over the different powers of the receive signal.

18.2.4 Effects of multipath propagation on code division multiple access

The above, strongly simplified, description of a CDMA system assumed a flat-fading channel. This assumption is violated under all practical circumstances. The basic nature of a CDMA system is to spread the signal over a large bandwidth; thus, it can be anticipated that the transfer function of the channel exhibits variations over this bandwidth.

The effect of frequency selectivity (delay dispersion) on a CDMA system can be understood by looking at the impulse response of the concatenation spreader–channel–despreader. If the channel is slowly time-variant, the effective impulse response can be written as:[5]

$$h_{\text{eff}}(t_i, \tau) = \widetilde{p}(\tau) * h(t_i, \tau) \tag{18.8}$$

where the effective system impulse response $\widetilde{p}(\tau)$ is the convolution of the transmit and receive spreading sequence:

$$\widetilde{p}(\tau) = p_{\text{TX}}(\tau) * p_{\text{RX}}(\tau) = ACF(\tau) \tag{18.9}$$

In the following, we assume an ideal spreading sequence (Eq. 18.1). The despreader output then exhibits multiple peaks: more precisely, one for each Multi Path Component (MPC) that can be resolved by the receiver – i.e., spaced at least T_C apart. Each of the peaks contains information about the transmit signal. Thus, all peaks should be used in the detection process: just using the largest correlation peak would mean that we discard a lot of the arriving signal. A receiver that can use multiple correlation peaks is the so-called *Rake receiver*, which collects ("rakes up") the energy from different MPCs. A Rake receiver consists of a *bank of correlators*. Each correlator is sampled at a different time (with delay τ), and thus collects energy from the MPC with delay τ. The sample values from the correlators are then weighted and combined.

Alternatively, we can interpret the Rake receiver as a tapped delay line, whose outputs are weighted and added up. The tap delays, as well as the tap weights, are adjustable, and matched

---

[5] Note that this expression is identical to that in correlative channel sounders (see Chapter 8). Correlative channel sounders and CDMA systems have the same structure, it is just the goals that are different: the channel sounder tries to find the channel impulse response from the received signal and knowledge of the transmit data, while the CDMA receiver tries to find the transmit data from knowledge of the received signal and the channel impulse response.

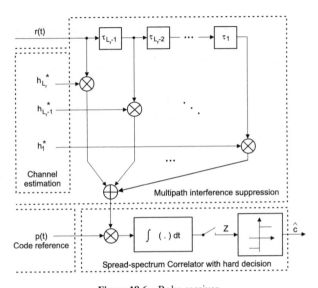

Figure 18.6 Rake receiver.

Reproduced with permission from Molisch [2000] © Prentice Hall.

to the channel. Note that the taps are usually spaced at least one chip duration apart, but there is no requirement for the taps to be spaced at regular intervals. The combination of the receiver filter and the Rake receiver constitutes a filter that is matched to the receive signal. The receive filter is matched to the transmit signal, while the Rake receiver is matched to the channel.

Independent of this interpretation, the receiver adds up the (weighted) signal from the different Rake fingers in a coherent way. As these signals correspond to different MPCs, their fading is (approximately) statistically independent – in other words, they provide delay diversity (frequency diversity). A Rake receiver is thus a diversity receiver, and all mathematical methods for the treatment of diversity remain valid. As for the performance of Rake receiver systems, we can now simply refer to the equations of Sections 13.4–13.5.

Example 18.1 *Performance of a Rake receiver: Compute the Bit Error Rate (BER) of BPSK in (i) a narrowband system and (ii) with a CDMA system that can resolve all multipaths, using a six-finger Rake receiver at a 15-dB SNR, in an International Telecommunications Union (ITU) Pedestrian-A channel.*

The tapped delay line model from an ITU Pedestrian-A channel is:

$$|h(n)|_{dB} = [0 \quad -9.7 \quad -19.2 \quad -22.8] \tag{18.10}$$

$$|h(n)| = [1 \quad 0.3273 \quad 0.1096 \quad 0.0724] \tag{18.11}$$

The average channel gain of the flat-fading channel is:

$$\sum |h(n)|^2 = 1 + 0.33^2 + 0.11^2 + 0.07^2 = 1.1 \tag{18.12}$$

and the transmit SNR has to be:

$$\overline{\gamma}_{TX} = \frac{10^{1.5}}{1.1} = 28.75 \tag{18.13}$$

so that a receive SNR of 15 dB is achieved.

As can be found from Chapter 12, the BER for the flat-fading channel is:

$$\overline{BER} = E[P_{BER}(\gamma_{Flat})] = \int_0^{\pi/2} \frac{1}{\pi} M_{\gamma_{Flat}} \left(-\frac{1}{\sin^2\theta}\right) d\theta \tag{18.14}$$

and

$$\overline{BER} = \int_0^{\pi/2} \frac{1}{\pi} \frac{\sin^2\theta}{\sin^2\theta + \overline{\gamma}_{TX}\sum|h(n)|^2} d\theta = 7.724 \times 10^{-3} \tag{18.15}$$

When combining the signals as done in the Rake receiver we have:

$$\gamma_{Rake} = \gamma_1 + \cdots + \gamma_6 \tag{18.16}$$

Since only four MPCs carry energy, only four Rake fingers are effectively used. If the $\gamma_1, \cdots, \gamma_4$ are independent, the joint pdf of $f_{\gamma_1,\ldots,\gamma_4}(\gamma_1,\ldots,\gamma_4) = f_{\gamma_1}(\gamma_1),\ldots,f_{\gamma_4}(\gamma_4)$ (see also Eq. 13.39):

$$\overline{BER} = \int d\gamma_1 pdf_{\gamma_1}(\gamma_1) \int d\gamma_2 pdf_{\gamma_2}(\gamma_2) \cdots \int d\gamma_4 pdf_{\gamma_4}(\gamma_4) \int_0^{\pi/2} d\theta f_1(\theta) \prod_{i=1}^{N_r} \exp(-\gamma_i f_2(\theta))$$

$$= \int_0^{\pi/2} \frac{1}{\pi} \prod_{k=1}^{4} \int_{\gamma_k} f_{\gamma_k}(\gamma_k) e^{\left(-\frac{\gamma_k}{\sin^2\theta}\right)} d\gamma_k \, d\theta$$

$$= \int_0^{\pi/2} \frac{1}{\pi} \prod_{k=1}^{4} M_{\gamma_k}\left(-\frac{1}{\sin^2\theta}\right) d\theta \tag{18.17}$$

Thus

$$\overline{SER} = \int_0^{\pi/2} \frac{1}{\pi} \prod_{k=1}^{4} \left[\frac{\sin^2(\theta)}{\sin^2(\theta) + \overline{\gamma}_k}\right] d\theta \tag{18.18}$$

For the same transmit SNR $\overline{\gamma}_{TX} = 28.75$, we then get:

$$\overline{BER} = \int_0^{\pi/2} \frac{1}{\pi} \frac{\sin^2\theta}{\sin^2\theta + \overline{\gamma}_{TX}} \frac{\sin^2\theta}{\sin^2\theta + 0.33^2\overline{\gamma}_{TX}} \frac{\sin^2\theta}{\sin^2\theta + 0.1^2\overline{\gamma}_{TX}} \frac{\sin^2\theta}{\sin^2\theta + 0.07^2\overline{\gamma}_{TX}} d\theta$$

$$= 9.9 \times 10^{-4} \tag{18.19}$$

Another consequence of the delay diversity interpretation is the determination of the weights for the combination of Rake finger outputs. The optimum weights are the weights for maximum-ratio combining – i.e., the complex conjugates of the amplitudes of the MPC corresponding to each Rake finger. However, this is only possible if we can assign one Rake finger to each resolvable MPC (the term *all Rake* has been largely used in the literature for such a receiver). $L_r = \tau_{max}/T_C$ taps, where τ_{max} is the maximum excess delay of the channel (see Chapter 6), are required in this case. Especially for outdoor environments, this number can easily exceed 20 taps. However, the number of taps that can be implemented in a practical Rake combiner is limited by power consumption, design complexity, and channel estimation. A Rake receiver that processes only a *subset* of the available L_r resolved MPCs achieves lower complexity, while still providing a performance that is better than that of a single-path receiver. The *Selective Rake (SRake)* receiver selects the L_b best paths (a *subset* of the L_r available resolved MPCs) and then combines the selected subset using maximum-ratio combining. This combining method is "hybrid selection: maximum ratio combining" (as discussed in Chapter 13); however, note that the average power in the different diversity branches is different. It is also noteworthy that the SRake still requires knowledge of the instantaneous values of *all* MPCs so that it can perform appropriate selection. Another possibility is the *Partial Rake (PRake)*, which uses the first L_f MPCs. Although the performance it provides is not as good, it only needs to estimate L_f MPCs.

Another generally important problem for Rake receivers is interpath interference. Paths that have delay τ_i compared with the delay the Rake finger is tuned to are suppressed by a factor $ACF(\tau_i)/ACF(0)$, which is infinite only when the spreading sequence has ideal ACF properties. Rake receivers with non-ideal spreading sequences thus suffer from interpath interference.

Finally, we note that in order for the Rake receiver to be optimal there must be no Inter Symbol Interference (ISI) – i.e., the maximum excess delay of the channel must be much smaller than T_S, though it can be larger than T_C. If there is ISI, then the receiver must have an equalizer (working on the Rake output – i.e., a signal sampled at intervals T_S) in addition to the Rake receiver. An alternative to this combination of Rake receiver and symbol-spaced equalizer is the chip-based equalizer, where an equalizer works directly on the output of the despreader sampled at the chip rate. This method is optimum, but very complex. As we showed in Chapter 16, the computational effort for equalizers increases quickly as the product of sampling frequency and channel delay spread increases.

18.2.5 Synchronization

Synchronization is one of the most important practical problems of a CDMA system. Mathematically speaking, synchronization is an estimation problem in which we determine the optimum sampling time out of an infinitely large ensemble of possible values – i.e., the continuous time. Implementation is facilitated by splitting the problem into two partial problems:

- *Acquisition*: a first step determines in which time interval (of duration T_C or $T_C/2$) the optimum sampling time lies. This is a hypothesis-testing problem: we test a finite number of hypotheses, each of which assumes that the sampling time is in a certain interval. The hypotheses can be tested in parallel or serially.
- *Tracking*: as soon as this interval has been determined, a control loop can be used to fine-tune the sampling time to its exact value.

For the acquisition phase, we use a special synchronization sequence that is shorter than the spreading sequence used during data transmission. This decreases the number of hypotheses that have to be tested, and thus decreases the time that has to be spent on synchronization. Furthermore, the synchronization sequence is designed to have especially good autocorrelation properties. For the tracking part, the normal spreading sequence used for data communications can be employed.

For many system design aspects, it is also important to know whether signals from different users are synchronized with respect to each other:

- *Synchronization within a cell*: the signals transmitted by a BS are always synchronous, as the BS has control over when to transmit them. For the uplink, synchronous arrival of the signals at the BS would require that all MSs arrange their timing advance – i.e., when they start transmitting a code sequence – in such a way that all signals arrive simultaneously. The timing advance would have to be accurate within one chip duration. This is too complicated for most applications, especially since movement of the MS leads to a change in the required timing advance.
- *Synchronization between BSs*: BSs can be synchronized with respect to each other. This is usually achieved by means of GPS (*Global Positioning System*), so that each BS requires a GPS receiver and free line of sight to several GPS satellites. While the former is not a

significant obstacle, the latter is difficult for microcells and picocells. The IS-95 system (see Chapter 22) uses such a synchronization.

18.2.6 Code families

Selection criteria

The selection of spreading codes has a vital influence on performance on a CDMA system. The Pseudo Noise (PN) sequences we used as examples up to now are among, but certainly not the only, possible choices. In general, the quality of spreading codes is determined by the following properties:

- *Autocorrelation*: ideally, $ACF(0)$ should be equal to the number of chips per symbol M_C, and zero at all other instances. For PN-sequences $ACF(0) = M_C - 1 \approx M_C$, and $ACF(n) = -1$ for $n \neq 0$. Good properties of the ACF are also useful for synchronization. Furthermore, as discussed above, autocorrelation properties influence interchip interference in a Rake receiver: the output of the correlator is the sum of the ACFs of the delayed echoes; a spurious peak in the ACF looks like an additional MPC.[6]
- *Cross-correlation*: ideally, all codes should be orthogonal to each other, so that interference from other users can be completely suppressed. In Section 18.2.6.3, we will discuss a family of orthogonal codes. For unsynchronized systems, orthogonality must be fulfilled for arbitrary delays between the different users. Note that bandwidth spreading and separation of users can be done by different codes; in this case we distinguish between spreading codes and scrambling codes.
- *Number of codes*: a CDMA system should allow simultaneous communications of as many users as possible. This implies that a large number of codes has to be available. The number of orthogonal codes is limited by M_C. If more codes are required, worse cross-correlation properties have to be accepted. The situation is complicated further by the fact that codes in adjacent cells have to be different: after all, it is only the codes that distinguish different users. Now, if cell A has M_C users with all orthogonal codes, then the codes in cell B cannot be orthogonal to the codes in cell A; therefore, intercell interference cannot be completely suppressed. The codes in cell B can, however, be chosen in such a way that the interference to users in the original cell becomes noise-like; this approach then requires code planning instead of frequency planning. An alternative approach is the creation of a large number of codes with suboptimum cross-correlation functions (see the next subsection), and assigning them to wherever they are needed. In such a case, no code planning is required; however, system capacity is lower.

Code families

The spreading sequences most frequently used for the uplink of CDMA systems are PN-sequences, Gold sequences, and Kasami sequences. The autocorrelation properties of PN-sequences are excellent; shifted versions of a PN-sequence are again valid codewords with almost ideal cross-correlation properties. Gold sequences are created by the appropriate combination of PN-sequences; the ACFs of Gold sequences can take on three possible values; both the offpeak values of the ACF and the CCF can be upper-bounded.

---

[6] As the ACF is known to the receiver, spurious peaks belonging to the first-detected MPC can be subtracted from the remaining signal, and thus their effect can be eliminated (this is essentially an interference cancellation technique, as discussed in Section 18.4).

An even more general family of sequences is the Kasami sequences. In the following we distinguish between S (*small*), L (*large*), and VL (*very large*) Kasami sequences. The letters describe the number of codes within the family. S-Kasami sequences have the best CCFs, but only a rather small number of such sequences exists. VL-Kasami sequences have the worst CCF, but there is an almost unlimited number of such sequences. Table 18.1 shows the properties of the different code families.

Table 18.1 Properties of code-division-multiple-access codes.

Sequence	Number of codes	Maximum CCF/dB	Comment
PN-Sequence	$2^{N_{reg}} - 1$		Good ACF
Gold	$2^{N_{reg}} + 1$	$\approx -3N_{reg}/2 + 1.5$	
S-Kasami	$2^{N_{reg}/2}$	$\approx -3N_{reg}/2$	Best CCF of all Kasami sequences
L-Kasami	$2^{N_{reg}/2}(2^{N_{reg}} + 1)$	$\approx -3N_{reg}/2 + 3$	
VL-Kasami	$2^{N_{reg}/2}(2^{N_{reg}} + 1)^2$	$\approx -3N_{reg}/2 + 6$	Almost unlimited number

Walsh–Hadamard codes

In the downlink, signals belonging to different users can be made completely orthogonal, as they are all emitted by the same transmitter (the BS). A family of codes that all fulfill these requirements is given by Walsh–Hadamard matrices. Define the $n + 1$-order Hadamard matrix $\mathbf{H}_{had}^{(n+1)}$ in terms of the nth order matrix:

$$\mathbf{H}_{had}^{(n+1)} = \begin{pmatrix} \mathbf{H}_{had}^{(n)} & \mathbf{H}_{had}^{(n)} \\ \mathbf{H}_{had}^{(n)} & \overline{\mathbf{H}}_{had}^{(n)} \end{pmatrix} \qquad (18.20)$$

where $\overline{\mathbf{H}}$ is the modulo-2 complement of $\mathbf{H}$. The recursive equation is initialized as:

$$\mathbf{H}_{had}^{(1)} = \begin{pmatrix} 1 & 1 \\ 1 & -1 \end{pmatrix} \qquad (18.21)$$

The columns of this matrix represent all possible Walsh–Hadamard codewords of length 2; it is immediately obvious that the columns are orthogonal to each other. From the recursion equation we find that $\mathbf{H}_{had}^{(2)}$ is:

$$\mathbf{H}_{had}^{(2)} = \begin{pmatrix} 1 & 1 & 1 & 1 \\ 1 & -1 & 1 & -1 \\ 1 & 1 & -1 & -1 \\ 1 & -1 & -1 & 1 \end{pmatrix} \qquad (18.22)$$

The columns of this matrix are all possible codewords of duration four. Further iterations give additional codewords, each of which is twice as long as that of the preceding matrix.

Orthogonal codes lead to perfect multiuser suppression at the receiver if the signal is transmitted over an Additive White Gaussian Noise (AWGN) channel. Delay dispersion destroys the orthogonality of the codes. The receiver can then either accept the additional interference (described by an *orthogonality factor*), or send the received signal through a chip-spaced equalizer that eliminates delay dispersion before correlation (and thus user separation) is performed.

Example 18.2 *Orthogonality of Orthogonal Variable Spreading Factor (OVSF) codes in frequency-selective channels: the codewords [1 1 1 1] and [1 −1 1 −1] are sent through a channel whose impulse response is (0.8, −0.6). Assuming that [1 1 1 1] was sent, compute the correlation coefficient of the signal arriving at the receiver with the possible transmit signals.*

Assume that the BS communicates with two MSs and transmits spreading codes, $t_1(n) = [1 \quad 1 \quad 1 \quad 1]$ and $t_2(n) = [1 \quad -1 \quad 1 \quad -1]$. $t_1(n)$ and $t_2(n)$ are orthogonal to each other; for example:

$$t_1(n) \cdot t_2(n)^T = 0 \tag{18.23}$$

However, the time dispersive channel (multipath channel) $h(n) = [0.8 \quad -0.6]$ will affect the orthogonality between the two codes. The received signal is the linear convolution of the impulse response and the transmitted signal (assuming that the channel is linear and time-invariant); for example:

$$r(n) = t(n) * h(n) = \sum_{k=-\infty}^{\infty} t(k)h(n-k) \tag{18.24}$$

The received signal when transmitting $t_1(n)$ is then:

$$r_1(0) = 1 \cdot 0.8 = 0.8 \tag{18.25}$$

$$r_1(1) = 1 \cdot (-0.6) + 1 \cdot 0.8 = 0.2 \tag{18.26}$$

$$r_1(2) = 1 \cdot (-0.6) + 1 \cdot 0.8 = 0.2 \tag{18.27}$$

$$r_1(3) = 1 \cdot (-0.6) + 1 \cdot 0.8 = 0.2 \tag{18.28}$$

$$r_1(4) = 1 \cdot (-0.6) = -0.6 \tag{18.29}$$

Thus, by correlating the received signals with $t_1(n)$ and $t_2(n)$ we get for $r_1(n)$:

$$\rho_{r_1 t_1} = \sum_{n=0}^{3} r_1(n)t_1(n) \tag{18.30}$$

$$= [0.8 \quad 0.2 \quad 0.2 \quad 0.2][1 \quad 1 \quad 1 \quad 1]^T = 1.4 \tag{18.31}$$

$$\rho_{r_1 t_2} = \sum_{n=0}^{3} r_1(n)t_2(n) \tag{18.32}$$

$$= [0.8 \quad 0.2 \quad 0.2 \quad 0.2][1 \quad -1 \quad 1 \quad -1]^T = 0.6 \neq 0 \tag{18.33}$$

In a similar manner, the signal that is received when $t_2(n)$ is transmitted is $r_2(n) = [0.8 \quad -1.4 \quad 1.4 \quad -1.4 \quad 0.6]$. The correlation coefficient:

$$\rho_{r_2 t_1} = \sum_{n=0}^{3} r_2(n)t_1(n) = -0.6 \neq 0 \tag{18.34}$$

$$\rho_{r_2 t_2} = \sum_{n=0}^{3} r_2(n)t_2(n) = 5 \tag{18.35}$$

Correlation between the two received signals is:

$$\rho_{r_1 r_2} = \sum_{n=0}^{3} r_1(n)r_2(n) \neq 0 \tag{18.36}$$

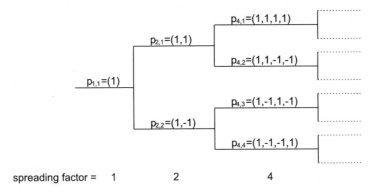

Figure 18.7   Code tree of orthogonal-variable-spreading-factor codes.

An additional challenge arises if different users require different data rates, so that codes of different length need to be used for the spreading. Orthogonal Variable Spreading Factor (OVSF) codes are a class of codes that fulfills these conditions; they are derived from Walsh–Hadamard codes.

Let us first define what we mean by orthogonality for codes of different duration. The chip duration is the same for all codes: it is given by the available system bandwidth, and independent of the data rate to be transmitted. Consider a code A that is two chips long $(1, 1)$, and a code B that is four chips long $(1, -1, -1, 1)$. The output of correlator A has to be zero if code B is at the input of the correlator. Thus, the correlation between code A and the first part of code B has to be zero, which is true: $1 \cdot 1 + 1 \cdot (-1) = 0$. Similarly, correlation of code A with the second part of code B has to be zero $1 \cdot (-1) + 1 \cdot 1 = 0$.

Let us now write all codewords of different Walsh–Hadamard matrices into a "code tree" (see Fig. 18.7). All codes within one level of the tree (same duration of codes) are orthogonal to each other. Codes of different duration A, B are only orthogonal if they are in different branches of the tree. They are *not* orthogonal to each other if one code is a "mother code" of the second code – i.e., code A lies on the path from the "root" of the code tree to code B. Examples of such codes are $p_{2,2}$ and $p_{4,4}$ in Fig. 18.7, whereas codes $p_{2,2}$ and $p_{4,1}$ are orthogonal to each other.

18.3 Cellular code-division-multiple-access systems

18.3.1 Principle behind code division multiple access – revisited

When analyzing the multiaccess capability of a system, we are essentially asking the question: "What prevents us from serving an infinite number of users at the same time?". In a TDMA/FDMA system, the answer is the limited number of available timeslots/frequencies. Users can occupy those slots, and not interfere with each other. But when all possible timeslots have been assigned to users, there are no longer free resources available, and no further users can be accepted into the system.

In a CDMA system, this mechanism is subtly different. We analyze in the following the situation where spreading codes are imperfect, but there is a large number of them. Different users are distinguished by different spreading codes; however, as user separation is not perfect, each user in the cell contributes interference to all other users. Thus, as the number of users increases, the interference for each user increases as well. Consequently, transmission quality decreases gradually (*graceful degradation*), until users find the quality too bad to place (or continue) calls. Consequently, CDMA puts a soft limit on the number of users, not a hard

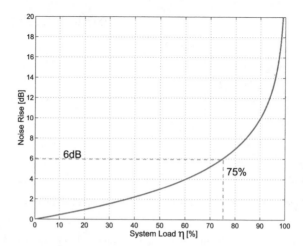

Figure 18.8 Noise rise as a function of system load in a code-division-multiple-access system.
Reproduced with permission from Neubauer et al. [2000] © T. Neubauer.

limit like TDMA. Therefore, the number of users in a system depends critically on the Signal-to-Interference-and-Noise Ratio (SINR) required by the receiver. It also implies that any increase in SINR at the receiver, or reduction in the required SINR, can be immediately translated into higher capacity.

Most interference stems from within the same cell as the desired user, and is thus termed *intracell interference*. Total intracell interference is the sum of many independent contributions, and thus behaves approximately like Gaussian noise. Therefore, it causes effects that are similar to thermal noise. It is often described by *noise rise* – i.e., the increase in "effective" noise power (sum of noise and interference power) compared with the noise alone $(N_0 + I_0)/N_0$. Figure 18.8 shows an example of noise rise as a function of system load; here *system load* is defined as the number of active users, compared with the maximum possible number M_C. We see that noise rise becomes very strong as system load approaches 100%. A cell is thus often judged to be "full" if noise rise is 6 dB. However, as mentioned above, there is no hard limit to the number of active users.

Some interference is from neighboring cells. A key property of a CDMA system is that it uses *universal frequency reuse* (also known as *reuse distance one*). In other words, the same frequency band is used in all cells; users in different cells are distinguished only by different codes. As discussed in Section 18.2.6.1, the amount of interference is mostly determined by the codes that are used in the different cells.

Many of the advantages of CDMA are related to the fact that interference behaves almost like noise, especially in the uplink. This noise-like behavior is due to several reasons:

- The number of users (and therefore, of interferers) in each cell is large.
- Power control makes sure that all signals arriving at the BS have approximately the same strength (see also below).
- Interference from neighboring cells also comes from a large number of users. Spreading codes are designed in such a way that all users from a neighboring cell contribute approximately the same percentage to total intercell interference. Note that this implies that we cannot simply reuse the same codeset in each cell; otherwise, there would be one user in the neighboring cell that would contribute much more interference (the user that uses the same code as the desired user in the desired cell).

Due to this averaging process, interference power shows almost no fluctuations. On the other hand, the power control makes sure that the signal strength from the desired user is always constant. The SINR is thus constant, and no fading margin has to be used in the link budget. However, note that making interference as Gaussian as possible is not always the best strategy; multiuser detection (Section 18.4) actively exploits structure in interference and works best when there are only a few interferers.

All these considerations apply to the uplink, where user separation by means of different spreading codes is not perfect. In the downlink, Walsh–Hadamard codes are popular, so that (at least in theory) different users can be separated completely. However, in this case, the number of users in the cell is limited by the number of Walsh–Hadamard codes. The situation then becomes similar to a TDMA system: if the M_C available Walsh–Hadamard codes are used up, then no further users can be served. Furthermore, there is also interference from the Walsh–Hadamard codes of neighboring cells. The situation can be improved by multiplying the Walsh–Hadamard codes by a *scrambling code*. Walsh–Hadamard codes that are multiplied by the same scrambling code remain orthogonal; codes that are multiplied by different scrambling codes do not interfere catastrophically. Therefore, different cells use different scrambling codes (see also Chapter 23).

In the downlink, intercell interference does *not* come from a large number of independent sources. All the interference comes from the BSs in the vicinity of the considered MS – i.e., a few (at most six) BSs constitute the dominant source of interference. The fact that each of these BSs transmits signals to a large number of users within their cell does not alter this fact: the interfering signal still comes from a single geographical source that has a single propagation channel to the victim MS. Consequently, the downlink might require a fading margin in its link budget.

18.3.2 Power control

As we mentioned above, power control is important to make sure that the desired user has a time-invariant signal strength, and that the interference from other users becomes noise-like. For further considerations, we have to distinguish between power control for the uplink and that for the downlink:

- *Power control in the uplink*: for the uplink, power control is vital for the proper operation of CDMA. Power control is done by a closed control loop: the MS first sends with a certain power, the BS then tells the MS whether the power was too high or too low, and the MS adjusts its power accordingly. The bandwidth of the control loop has to be chosen so that it can compensate for small-scale fading – i.e., has to be on the order of the Doppler frequency. Due to time variance and noise in the channel estimate, there is a remaining variance in the powers arriving at the BS; this variance is typically on the order of 1.5–2.5 dB, while the dynamic range that has to be compensated is 60 dB or more. This variance leads to a reduction in the capacity of a CDMA cellular system of up to 20% compared with the case when there is ideal power control.

  Note that an open control loop (where the MS adjusts its transmit power based on its own channel estimate) cannot be used to compensate for small-scale fading in a Frequency Domain Duplexing (FDD) system: the channel seen by the MS (when it receives signals from the BS) is different from the channel it transmits to (see Section 17.5). However, an open loop can be used in conjunction with a closed loop. The open loop compensates for large-scale variations in the channel (pathloss and shadowing), which are approximately the same at uplink and downlink frequencies. The closed loop is then used to compensate for small-scale variations.

- *Power control in the downlink*: for the downlink, power control is not necessary for CDMA to function: all signals from the BS arrive at one MS with the same power (the channel is the same for all signals). However, it can be advantageous to still use power control in order to keep the total transmit power low. Decreasing the transmit power for all users within a cell by the same amount leaves unchanged the ratio of desired signal power to intracell interference – i.e., interference from signals destined for other users in the cell. However, it does decrease the power of total interference to other cells. On the other hand, we cannot decrease signal power arbitrarily, as the SNR must not fall below a threshold. The goal of downlink power control is thus to minimize the total transmit power while keeping the BER or SINR level above a given threshold. The accuracy of downlink power control need not be as high as for the uplink; for many cases, open loop control is sufficient.

It is worth remembering that the power control of users in adjacent cells does not give constant power of the intercell interference. A user in an adjacent cell is power-controlled by its own BSs – in other words, power is adjusted in such a way that the signal arriving at its desired BS is constant. However, it "sees" a completely different channel to the undesired BS, with temporal fluctuations that the desired BS neither knows nor cares about. Consequently, intercell interference is temporally variant.

Interference power from all users is the same only if all users employ the same data rate. Users with higher data rates contribute more interference power; high-data-rate users can thus be a dominant source of interference. This fact can be understood most easily when we increase the data rate of a user by assigning multiple spreading codes to him. In this case, it is obvious that the interference this user contributes increases linearly with the data rate. While this situation did not occur for second-generation cellular systems, which had only speech users, it is certainly relevant for third-generation cellular systems, which foresee high-data-rate services.

It should also be noted that power control is not an exclusive property of CDMA systems; it can also be used for FDMA or TDMA systems, where it decreases intercell interference and thus improves capacity. The major difference is that power control is *necessary* for CDMA, while it is *optional* for TMDA/FDMA.

Soft handover

As all cells use the same frequencies, an MS can have contact with two BSs at the same time. If an MS is close to a cell boundary, it receives signals from two or more BSs (see Fig. 18.9) and also transmits to all of these BSs. Signals coming from different MSs have different delays, but this can be compensated by the Rake receiver, and signals from different cells can be added coherently.[7] This is in contrast to the hard handover in an FDMA-based system, where an MS can have contact with only one BS at a time, because it can communicate only on one frequency at a time.

Consider now an MS that starts in cell A, but has already established a link to BS B as well. At the outset, the MS gets the strongest signal from BS A. As it starts to move towards cell B, the signal from BS A becomes weaker, and the signal from BS B becomes stronger, until the system decides to drop the link to BS A. Soft handover dramatically improves performance while the MS is near the borderline of the two cells, as it provides diversity (macrodiversity) that can combat large-scale as well as small-scale fading. On the downside, soft handover decreases the available capacity in the downlink: one MS requires resources (Walsh–Hadamard codes) in two cells at the same time, while the user talks – and pays – only once. Furthermore, soft handover increases the amount of signaling that is required between BSs.

---

[7] Note that different cells might use different codes. This is not a major problem; it just means that (for the downlink) different correlators in the fingers of the Rake receiver have to use different spreading sequences.

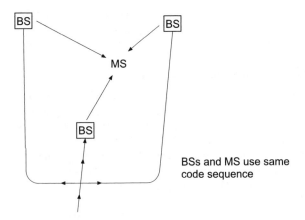

Figure 18.9   Principle behind soft handover.

Reproduced with permission from Oehrvik [1994] © Ericsson AB.

18.3.3 Methods for capacity increases

- *Quiet periods during speech transmission*: for speech transmission, CDMA makes implicit use of the fact that a person does not talk continuously, but rather only about 50% of the time, the remainder of the time (s)he listens to the other participant. In addition, there are pauses between words and even syllables, so that the ratio of "talk time" to "total time of a call" is about 0.4. During quiet periods, no signal, or a signal with a very low data rate, has to be transmitted.[8] In a CDMA system, not transmitting information leads to a decrease in total transmitted power, and thus interference in the system. But we have already seen above that decreasing the interference power allows additional users to place calls. Of course, there can be a worst case scenario where all users in a cell are talking simultaneously, but, statistically speaking, this is highly improbable, especially when the number of users is large. Thus, pauses in the conversation can be used very efficiently by CDMA in order to improve capacity (compare also discontinuous transmission in TDMA systems).
- *Flexible data rate*: in an FDMA (TDMA) system, a user can occupy either one frequency (timeslot), or integer multiples thereof. In a CDMA system, arbitrary data rates can be transmitted by an appropriate choice of spreading sequences. This is not important for speech communications, which operate at a fixed data rate. For data transmission, however, the flexible data rate allows for better exploitation of the available spectrum.
- *Soft capacity*: the capacity of a CDMA system can vary from cell to cell. If a given cell adds more users, it increases interference to other cells. It is thus possible to have some cells with high capacity, and some with lower; furthermore, this can change dynamically, as traffic changes. This concept is known as *breathing cells*.
- *Error correction coding*: the drawback of error correction coding is that the data rate that is to be transmitted is increased, which decreases spectral efficiency. On the other hand, CDMA consciously increases the amount of data to be transmitted. It is thus possible to include error correction coding without decreasing spectral efficiency; in other words, different users are distinguished by different error correction codes (coding by spreading).

---

[8] Actually, most systems transmit *comfort noise* during this time – i.e., some background noise. People speaking into a telephone feel uncomfortable (think the connection has been interrupted) if they cannot hear any sound while they talk.

Note, however, that commercial systems (UMTS, Chapter 23) do not use this approach; they have separate error correction and spreading.

18.3.4 Combination with other multiaccess methods

CDMA has advantages compared with TDMA and FDMA, especially with respect to flexibility, while it also has some drawbacks, like complexity. It is thus obvious to combine CDMA with other multiaccess methods in order to obtain the "best of both worlds". The most popular solution is a combination of CDMA with FDMA: the total available bandwidth is divided into multiple subbands, in each of which CDMA is used as the multiaccess method. Clearly, frequency diversity in such a system is lower than for the case where spreading is done over the whole bandwidth. On the positive side, the processing speed of the transmitters and receivers can be lower, as the chip rate is lower. The approach is used, for example, in IS-95 (which uses 1.25-MHz-wide bands) and Universal Mobile Telecommunications System (UMTS), which uses 5-MHz-wide subbands.

Another combination is CDMA with TDMA. Each user can be assigned one timeslot (as in a TDMA system), while the users in different cells are distinguished by different spreading codes (instead of different frequencies). This approach avoids the near/far problem, since users within the cell now use different timeslots. Another possibility is to combine several timeslots, and build up a narrowband CDMA system within them. This system works best when adding a CDMA component to an existing TDMA system (e.g., the TDD mode of UMTS).

18.4 Multiuser detection

18.4.1 Introduction

Basic idea underlying multiuser detection

Multiuser detection is based on the idea of detecting interference, and exploiting the resulting knowledge to mitigate its effect on the desired signal. Up to 1986, it had been an established belief that interference from other users cannot be mitigated, and that in the best case, interference behaves like AWGN. Correct detection and demodulation in the presence of strong interference was thus considered impossible, just as it is impossible to correctly detect signals in strong noise. The work of Verdu and others – summarized in Verdu [1998] – demonstrated that it is actually possible to exploit the *structure* of multiuser interference to combat its effect. Using such a strategy, interference is *less* detrimental than Gaussian noise. Multiuser detection was intensely researched during the 1990s, and the first practical systems using this approach were introduced in the subsequent decade.

The conceptually simplest version of multiuser detection is serial *interference cancellation*. Consider a system where a (single) interfering signal is much stronger than the desired signal. The receiver first detects and demodulates this strongest signal. This signal has a good SINR, and can thus hopefully be detected without errors. Its effect is then subtracted (cancelled) from the total received signal. The receiver then detects the desired signal within the "cleaned-up" signal. As this cleaned-up signal consists only of the desired signal and the noise, the SINR is good, and detection can be done correctly. This example makes clear that detection at an SIR of less than 0 dB is possible with multiuser detection.

Other multiuser structures include Maximum Likelihood Sequence Estimation (MLSE) detectors, which try to perform optimum detection for the signals of *all* users simultaneously. These receivers show very good performance, but their complexity is usually prohibitive, as the effort increases exponentially with the number of users to be detected. The performance of MLSE can also be approximated by receivers that use the turbo-principle (see Section 14.5).

A general classification of multiuser detection distinguishes between linear and nonlinear detectors. The former class includes the decorrelation receiver and the Minimum Mean Square Error (MMSE) receiver; the latter includes MLSE, interference cancellation, decision feedback receivers, and turboreceivers.

Assumptions

In our basic description here, we make a number of simplifying assumptions:

- The receiver has perfect knowledge of the channel from the interferer to the receiver. This is obviously a best case. Consider the situation in a serial interference cancellation receiver: only if the receiver has perfect channel knowledge of the interfering signal can it perfectly subtract it from the total signal. The stronger the interference, the larger is the impact of any channel estimation error on the subtraction process.
- All users employ CDMA as a multiaccess scheme. We stress, however, that multiuser detection is also possible for other multiaccess methods, like TDMA, and can be used, for example, to decrease the reuse distance in a TDMA network. We will also see later that there is a strong similarity between multiuser detection and detection of spatial-multiplexing systems (Section 20.2).
- All users are synchronized.
- We only treat multiuser detection at the receiver. A related topic would be the design of transmission schemes so that interference at different receivers is mitigated. A surprising result of information theory is that if the transmitter knows the channel and the interference, then the capacity of the interfered channel is the same as the capacity of the non-interfered channel. This capacity can be realized by appropriate coding strategies, called "writing on dirty paper" coding [Peel 2003]. Such coding could be exploited to increase the capacity of the downlink in cellular systems.

18.4.2 Linear multiuser detectors

The block diagram of a linear multiuser system is sketched in Fig. 18.10. It first estimates the signals from different users by despreading using the spreading sequences of the different users.[9] Note that this requires a number of parallel despreaders, each operating with a different spreading sequence. The outputs from these despreaders are then linearly combined. This combination step can be considered as filtering using a matrix filter, and used for elimination of interference. This approach shows a strong similarity to linear equalization for elimination of ISI. Therefore, concepts like zero-forcing, Wiener filtering, etc., which we discussed in Chapter 16, are also encountered in this context.

Decorrelation receiver

The decorrelation receiver is the simplest means of multiuser detection; it is the equivalent of a zero-forcing equalizer. We write the output of the receiver filter (after despreading) as:

$$\mathbf{y} = \mathbf{Rc} + \mathbf{n} \tag{18.37}$$

---

[9] This step might also include the combination of signals from multiple Rake fingers, and the elements of an antenna array.

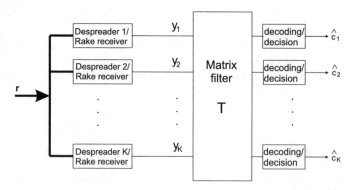

Figure 18.10   Linear multiuser detector.

where the correlation matrix **R** can include possible antenna and/or delay diversity; **n** is the noise vector. Estimation of the symbols is now obtained simply by filtering with a matrix filter $\mathbf{T} = \mathbf{R}^{-1}$:

$$\widehat{\mathbf{c}} = \mathbf{R}^{-1}\mathbf{y} = \mathbf{c} + \mathbf{R}^{-1}\mathbf{n} \tag{18.38}$$

The advantage of this approach is its simplicity, and the fact that it is not necessary to know the received amplitudes. Only the correlation matrix **R** needs to be determined. The drawback lies in noise enhancement (compare, again, the zero-forcing equalizer). The worse the conditioning of the correlation matrix, the more the noise is increased.

MMSE receiver

Just like for the MMSE equalizer, the MMSE multiuser detector strikes a balance between interference suppression and noise enhancement. A measure for total disturbance is the mean quadratic error $E\{|c - \widehat{c}|^2\}$. The matrix filter is thus:

$$\mathbf{T} = \left[\mathbf{R}^{-1} + \sigma_n^2 \mathbf{I}\right]^{-1} \tag{18.39}$$

The MMSE does not lead to complete suppression of interference, but due to its smaller noise enhancement the signal distortions are still smaller than for the decorrelation receiver.

18.4.3 Nonlinear multiuser detectors

Linear multiuser detectors ignore part of the structure of transmit signals: they allow any continuous value for the estimate $\widehat{c}$ of the transmit signal. Thus, they ignore the fact that the transmit signal can only contain members of a finite transmit alphabet. Nonlinear detectors also exploit this information.

Multiuser maximum-likelihood sequence estimation

The structure of multiuser MLSE is the same as that for conventional MLSE (usually decoded by a Viterbi detector). If we have K users, then M^K combinations of transmit symbols can be sent out by the users. The number of possible states in the trellis diagram thus also increases exponentially with the number of users. For this reason, multiuser MLSE is not used in practice. However, it gives important insights into the performance limits of multiuser detection (for more details, see Verdu [1998]).

Successive interference cancellation

Successive Interference Cancellation (SIC) detects users in the sequence of their signal strength. The signal of each user is subtracted from the total signal before the next user is detected (see Fig. 18.11). The SIC is thus a special case of a decision feedback receiver. The receiver works the following way: the sum of all signals is received, and it is despread using the different spreading codes of each user. Then, the strongest signal is detected and decided on, so that we get the original bitstream, unaffected by noise or interference. This bitstream is then respread, and subtracted from the total signal. The "cleaned-up" signal is then sent through the despreaders again, the strongest user within this new signal is detected, respread, and subtracted. The process is repeated until the last user has been detected. Note that error propagation can seriously affect the performance of SIC: if the receiver decides wrongly about one bit, it subtracts the wrong contribution from the total signal, and the residual signal, which is further processed, suffers from more, not less, interference.

There are two possibilities for subtracting interference: "hard" and "soft". For hard subtraction, interference is subtracted completely; for soft subtraction, only a scaled-down version of the signal is subtracted. This does not lead to complete cancellation of the signal, but makes error propagation less of an issue. Another method for reducing error propagation is to make sure that bits are not only demodulated, but also decoded (i.e., undergo error correction decoding), then recoded, remodulated, and respread before subtraction. As the error probability after decoding is much lower than before, this obviously reduces error propagation. On the downside, the decoding process increases the latency of the detection process.

Parallel interference cancellation

Instead of subtracting interference in a serial (user-by-user) fashion, we can also cancel all users simultaneously. To achieve this, the first step makes a (hard or soft) decision for all users based on the total received signal. The signals are then respread, and contributions from all interferers to the total signal are subtracted. Note that a different group of interferers is active for each

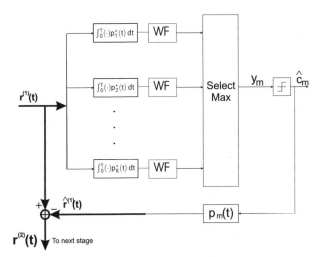

Figure 18.11 One stage of a successive interference cancellation receiver. Received signal **r** from K users is correlated with the spreading sequence of the kth user; the largest signal is selected and its impact—i.e., the respread signal—is subtracted from the received signal. WF denotes the whitening filter. The impact of delay dispersion is not shown.

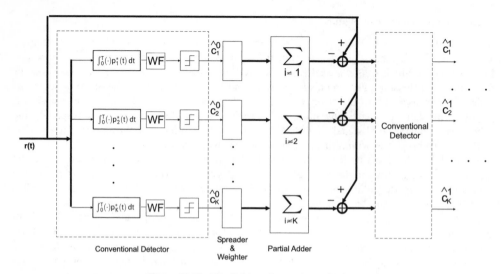

Figure 18.12 Parallel interference cancellation.

user: For user 1, users $2 \cdots K$ act as interferers; for user 2, users 1 and $3 \cdots K$ act as interferers, and so on. The next stage of the canceller then uses the "cleaned-up" signals as a basis for a decision, and again performs remodulation and subtraction. The process is repeated until decisions no longer change from iteration to iteration, or until a certain number of iterations is reached (see Fig. 18.12).

Error propagation can be a significant problem for parallel interference subtraction as well. Note that only the first stage would be necessary if all decisions were correct. However, each signal in the first stage has a bad signal-to-interference ratio, so that there is a high probability of only the strongest signal being decided correctly. Moreover, wrong decisions distort the signal even more, so that later stages perform even worse than the first.

One approach to mitigating these problems is subdivision of the signals into power classes; only signals that are detected reliably are used for interference cancellation. Arriving signals are divided into several groups, depending on their SINR. For a decision in class m, we use feedback from classes $1 \cdots m$ – i.e., the more reliable decisions – but not those of classes $m + 1, m + 2, \ldots$.

It is also possible to use partial cancellation. At each stage, signals are sent through a mapping $x \rightarrow \tanh(\lambda x)$ (instead of a hard-decision device), where the steepness of the mapping curve – i.e., λ – increases from stage to stage. Thus, only the last stages use de facto hard decisions. The *turbo-multiuser detector* further improves on this principle by feeding back the log-likelihood ratios for different bits. This greatly decreases the probability of error propagation: bits that are decided wrongly usually have a considerable uncertainty about them (it is unlikely that noise is so large that we make a wrong decision with great confidence). The turbo-detector does exploit this fact, by assigning a small log-likelihood ratio. It can actually be shown that a turbo-multiuser detector can approach the performance of a multiuser MLSE.

18.5 Time-hopping impulse radio

When the desired spreading bandwidth W is on the order of 500 MHz or higher, it becomes interesting to spread the spectrum by transmitting short pulses whose position or amplitude contains the desired information. This method, often called "impulse radio", allows the use of simple transmitters and receivers. It is mostly used for so-called "ultrawideband" communications systems [diBenedetto et al. 2005].

18.5.1 Simple impulse radio

Let us start out with the most simple possible impulse radio: a transmitter sending out a single pulse to represent one symbol. For the moment, assume that the modulation method is orthogonal pulse position modulation (see Chapter 11). A pulse is either sent at time t, or at time $t + T_d$, where T_d is larger than the duration of a pulse T_C. The detection process in an AWGN channel is then exceedingly simple: we just need an energy detector that determines during which of the two possible time intervals

$$[t + \tau_{\text{run}}, t + \tau_{\text{run}} + T_C] \tag{18.40}$$

or

$$[t + \tau_{\text{run}} + T_d, t + \tau_{\text{run}} + T_C + T_d] \tag{18.41}$$

we get more energy. Here, τ_{run} is the runtime between transmitter and receiver, which we determine via the synchronization process.

Obviously, a pulsed transmission achieves spreading, because the bandwidth of the transmit signal is given by the inverse of pulse duration $1/T_C$. And "despreading" is done in a very simple way: by only recording and using the arriving signal in the intervals given by Eqs. (18.40)–(18.41), we can make our decision about which bit was sent. The SNR occuring at the receiver is E_B/N_0: if the receiver is an energy detector, the observed peak power is $\overline{P}T_S/T_C$, and noise power is N_0/T_C, resulting in an SNR of $\overline{P}T_S/N_0 = E_B/N_0$. This can be interpreted as suppressing the (wideband) noise by a factor T_S/T_C.

However, there are two main drawbacks to such a simple scheme:

1. The peak-to-average ratio (crest factor) of the transmitted signal is very high, giving rise to problems in the design of transmit and receive circuitry.
2. The scheme is not robust to different types of interference. In particular, there are problems with multiaccess interference. In other words, what happens if the desired transmitter sends a 0 such that the signal would arrive at $t + \tau_{\text{run,desired}}$, while another user sends a 0 that arrives at $t + \tau_{\text{run,interfere}}$? It can happen that this interfering pulse arrives precisely at the time the receiver expects a pulse representing a 1 from the desired user. More exactly, this occurs when $t + \tau_{\text{run,interfere}} = t + \tau_{\text{run,desired}} + T_d$. Since it is very difficult to influence the runtimes from different users in a highly accurate way, such a situation occurs with finite probability. We term it "catastrophic collision", since the receiver has no way of making a good decision about the received symbol and error probability will be very high; compare also the catastrophic collisions in frequency-hopping systems (see Section 18.1).

In order to solve these problems, we need to transmit multiple pulses for each symbol. The first idea that springs to mind is to send a regular pulsetrain, where the duration between pulses is T_f. This solves the peak-to-average problem. However, we can easily see that it does not decrease the probability of catastrophic collisions: if the first pulse of the train collides with a pulse from an interfering pulsetrain, then subsequent pulses collide as well. A solution to this problem is provided by the concept of time-hopping impulse radio, described in the following section.

18.5.2 Time-hopping impulse radio

The basic idea of TH-IR is similar to the "pulsetrain" idea described above: we use a whole sequence of pulses to represent a single symbol. More precisely, we divide the available symbol

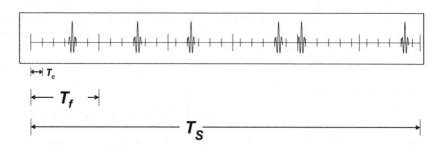

Figure 18.13   Transmit waveform of time-hopping impulse radio for one symbol, $p(t)$, indicating chip, frame, and symbol duration.

time into a number of so-called *frames* of duration T_f, and transmit one pulse within each frame (see Fig. 18.13).[10] The key idea is now to vary the position of the pulses *within* the frames. So for the first frame, we send the pulse in the third chip of the frame; in the second frame, we send the pulse in the eighth chip of the frame, and so on. The positions of the pulses within a frame are determined by a pseudorandom sequence called a "time-hopping sequence". Mathematically speaking, the transmit signal is:

$$s(t) = \frac{\sqrt{E_S}}{\sqrt{N_f}} \sum_i \sum_{j=1}^{N_f} g(t - jT_f - c_j T_C - b_i T_d - iT_S) = \frac{\sqrt{E_S}}{\sqrt{N_f}} \sum_i p(t - b_i T_d - iT_S) \qquad (18.42)$$

where $g(t)$ is the transmitted unit energy pulse of duration T_C, N_f is the number of frames (and therefore also the number of pulses) representing one information symbol of length T_S, and b is the information symbol transmitted. The time-hopping sequence provides an additional timeshift of $c_j T_C$ seconds to the jth pulse of the signal, where c_j are the elements of a pseudorandom sequence, taking on integer values between 0 and $N_c - 1$ (see also Fig. 18.13). In the receiver, we perform matched filtering (matched to the total transmitted waveform $p(t)$), and sample this output at times $t = T_s + \tau_{run}$, and $t = T_s + \tau_{run} + T_d$. A comparison of the sample values at these two times determines which sequence had been sent. As in the case of CDMA, the matched filter operation can also be interpreted as a correlation with the transmit sequence.

In this TH-IR, we have the same suppression of noise as for the "simple" impulse radio system – namely, T_S/T_C; however, now the gain comes from two different sources. One part of the gain stems from the fact that we observe the noise only over a short period of time – i.e., the same type of gain we have in the simple impulse radio. Its value is now T_f/T_C, and is thus smaller than in simple impulse radio. The other type of gain stems from combining the pulses in the different frames: the desired signal components from the different frames add up coherently, while the noise components add up incoherently. Taken together, those two gains provide a total gain of T_S/T_C.

Now, what is the advantage of this approach, compared with a regular pulsetrain? We can ascribe *different* time-hopping sequences to different users. As we will show below, these sequences can be constructed in such a way that pulses collide only in a few ($N_{collide}$) frames, but not the others. And as the receiver correlates the incoming signal with the time-hopping sequence of the desired user, it sees interference only in those frames where there is actually a collision. This leads to a suppression of interference by a factor $N_{collide}/N_f$.

---

[10] Note that the notation "frame", while established in the impulse radio literature, can give rise to confusion. In impulse radio, one data symbol contains several frames. For TDMA or Packet Reservation Multiple Access (PRMA), a block of data is also often called a "frame"; when used in this context, a frame contains several symbols.

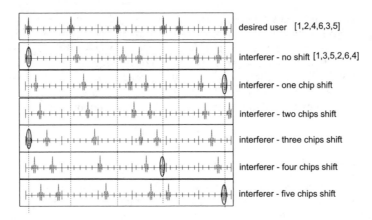

desired user   [1,2,4,6,3,5]

interferer - no shift [1,3,5,2,6,4]

interferer - one chip shift

interferer - two chips shift

interferer - three chips shift

interferer - four chips shift

interferer - five chips shift

Figure 18.14   Interference between two users, for all possible (integer) shifts between the two users. Circles around a pulse indicate collisions.

We have already mentioned that signals between desired user and interferer are not synchronized, and thus can have an arbitrary timeshift against each other. Thus, time-hopping sequences are constructed according to the following criterion: irrespective of the relative shift between the different sequences, the number of collisions between pulses must not exceed a threshold λ (usually, we choose $\lambda = 1$).[11] Note that Multiple Access Interference (MAI) can be suppressed by at least a factor λ/N_f. This way the system designer does not have to worry about runtime effects or synchronization between users; a good suppression of MAI is always guaranteed. Designing such sequences is difficult, especially if we want a large number of sequences with small collisions; exhaustive computer searches are often the best method.

Example 18.3   *The time-hopping sequence of a TH-IR system with $N_f = 6$ is [1, 2, 4, 6, 3, 5]. Find another hopping sequence that has at most one collision for arbitrary shifts.*

By making a systematic search, we find that the sequence [1, 3, 5, 2, 6, 4] fulfills the requirements; this is shown in Fig. 18.14.

Impulse radio can also be given a different, very useful, interpretation: it is a direct-sequence CDMA system (like the one in Section 18.2), where the spreading sequence has a large number of 0's, and a small number of 1's. Compare this with the conventional DS-CDMA systems, where there is an almost equal number of +1's and −1's. The charm of this interpretation is that a lot more research has been performed for conventional DS-CDMA systems than for impulse radio. But, by interpreting TH-IR as a DS-CDMA system, many of these results can be adopted immediately.

Finally, we note that TH-IR can be used not only in conjunction with pulse position modulation, but also with pulse amplitude modulation. Furthermore, the polarity of transmitted pulses within a symbol can be randomized, so that the transmit signal then reads:

$$s(t) = \frac{\sqrt{E_S}}{\sqrt{N_f}} \sum_i b_i \sum_{j=-\infty}^{\infty} d_j g(t - jT_f - c_j T_C - iT_S) \tag{18.43}$$

---

[11] It is interesting that this problem of constructing good time-hopping sequences has strong similarities to constructing good frequency-hopping sequences for FH spread systems (see Section 18.1).

where each pulse is multiplied by a pseudorandom variable d_j that can take on the values $+1$ or -1 with equal probability. Such a polarity randomization has advantages with respect to spectral shaping of the transmit signal. For such a system, the resemblance to DS-CDMA is even more striking: the spreading sequence is now a *ternary* sequence, with an (almost) equal number of $+1$'s and -1's, and a large number of 0's inbetween. The suppression of interference now occurs not just because of a small number of pulse collisions, but also because the interference contributions from different frames (but the same symbol) might cancel out.

18.5.3 Impulse radio in delay-dispersive channels

Up to now, we have discussed impulse radio in AWGN channels. We have found that the transmitter as well as the receiver can be made very simple in these cases. However, TH-IR almost never works in AWGN channels. The purpose of such a system is the use of a very large bandwidth (typically 500 MHz or more), which in turn implies that the channel will certainly show variations over that bandwidth. It is then necessary to build a receiver that can work well in a dispersive channel.

Let us first consider coherent reception of the incoming signal. In this case, we need a Rake receiver, just like the one discussed in Section 18.2.3. Essentially, the Rake receiver consists of multiple correlators or fingers. In each of the fingers, correlation of the incoming signal is done with a delayed version of the time-hopping sequence, where the delay is equal to the delay of the MPC that we want to "rake up". The output of the Rake fingers is then weighted (according to the principles of maximum-ratio combining or optimum combining) and summed up. Because Impulse Radio (IR) systems usually use very large bandwidth, they always use structures whose number of fingers is smaller than the number of available MPCs (SRake, PRake, see Section 18.2.3). In order to get reasonable performance, it might be required to have 20 or more Rake fingers even in indoor environments. Differentially coherent (transmitted reference) or noncoherent schemes thus become an attractive alternative.

In order to understand the principle of *Transmitted Reference* (TR) schemes, remember that the ideal matched filter in a delay-dispersive channel is matched to the convolution of the time-hopping sequence with the impulse response of the channel. For coherent reception, we first have a filter matched to the time-hopping sequence, followed by the Rake receiver, which is matched to the impulse response of the channel. A so-called TR scheme creates this composite matched filter in a different way. A TR transmitter sends out two pulses in each frame: one unmodulated (reference) pulse, and, a fixed time T_{pd} later, the modulated (data) pulse. The sequence of reference pulses is convolved with the channel impulse response when it is transmitted through the wireless channel; this signal thus constitutes a noisy version of the system impulse response (convolution of transmit basis waveform and channel impulse response) – and remember that matched filtering is the same as correlating the received signal with this effective impulse response.

In order to perform a matched filtering on the data-carrying part of the received signal, the receiver just multiplies the received signal by the received reference signal. Let us now have a look at the mathematical expression of the transmit signal for one symbol:

$$p(t) = \sqrt{\frac{1}{2}} \sum_{j=0}^{N_f} d_j [g(t - jT_f - c_j T_C) + b \cdot g(t - jT_f - c_j T_C - T_{pd})] \qquad (18.44)$$

Inspection shows that correlation with the received reference signal can be done by just multiplying the total received signal by a delayed (by time T_{pd}) version of itself and integration. To be more precise: the first step at the receiver involves filtering by a receive filter $h_R(t)$; this filter should be wide enough not to introduce any signal distortions, and just limit the available noise. The filtered received signal $\hat{r}(t)$ is multiplied by a delayed version of itself, and integrated over a finite interval T_{int}, which should be long enough to collect most multipath energy, but short enough not to collect too much noise energy.

TR schemes have the advantage of being exceedingly simple (though implementation of the delay in the receiver may be non-trivial). On the downside, they show poorer performance than coherent schemes for two reasons: (i) reference pulses waste energy, in the sense that they do not carry any information (this results in a 3-dB penalty); (ii) the reference part of the signal is noisy, as is the data-carrying part of the signal. Multiplication of the two signals in the receiver gives rise to noise–noise cross-terms that worsen the SNR. Now remember that impulse radio is a spread spectrum system, so that the SNR of the received signal is negative. Noise–noise cross-terms can thus become large.

Finally, we note that non-coherent detection can be an attractive alternative as well. This is especially true if only a single user is to be served. However, non-coherent receivers have problems with multiaccess interference. Due to multipath propagation, energy is dispersed over several adjacent chips. Thus, energy detection will see much more interference (as exemplified in Fig. 18.15).

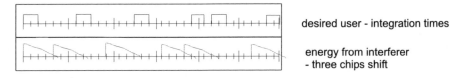

desired user - integration times

energy from interferer
- three chips shift

Figure 18.15 Interference from a delay-dispersed interferer signal (lower row) to the desired signal, whose integration time (two chip durations per frame) is sketched in the upper row.

Further reading

An overview of spread spectrum communications in general can be found in the classic monographs of Dixon [1994] and Simon et al. [1994], which provide good coverage of FH and DS systems. Books that are more tuned to cellular systems, especially CDMA, are Glisic and Vucetic [1997], Goiser [1998], Li and Miller [1998], Viterbi [1995], and Ziemer et al. [1995]; the overview paper [Kohno et al. 1995] gives a shorter description. Scholtz [1982] gives an overview of the history of the spread spectrum (though more from a military perspective, since cellular applications were not yet considered at the time of that article). Milstein [1988] discusses the interference rejection capabilities of various spread spectrum techniques.

Frequency-hopping codes are designed in Maric and Titlebaum [1992]. Estimates for the capacity of CDMA systems were first published in the widely cited paper of Gilhousen et al. [1991]. The effect of multipath on CDMA – namely, the use of Rake receivers – is reviewed in Molisch [2000, pt. V]; particularly, the effect of a finite number of Rake fingers [Win and Chrisikos 2000]. Other important papers on Rake reception include Holtzman and Jalloul [1994] and Bottomley et al. [2000], among many others. For synchronization aspects, the papers by Polydoros and Weber [1984] are still interesting to read. Spreading codes are reviewed in Dinan and Jabbari [1998].

For multiuser detection, the book by Verdu [1998] is a must-read. Also the overview papers ([Duel-Hallen et al. 1995], [Moshavi 1996], and [Poor 2001]) give useful insights. Advanced concepts like blind multiuser detection are described extensively in Wang and Poor [2003]; the interaction between multiuser detection and decoding is explored in Poor [2004].

Time-hopping impulse radio was pioneered by Win and Scholtz in the 1990s. For a description of the basic principles see Win and Scholtz [1998], and more detailed results in Win and Scholtz [2000]. An overview of the different aspects of ultrawideband system design can be found in Roy et al. [2004]; and various aspects of system design, including Rake receivers and polarity randomization, are also discussed in Molisch et al. [2004c]. TR receivers are discussed in Choi and Stark [2002].

19

Orthogonal frequency division multiplexing (OFDM)

19.1 Introduction

Orthogonal Frequency Division Multiplexing (OFDM) is a modulation scheme that is especially suited for high-data-rate transmission in delay-dispersive environments. It converts a high-rate datastream into a number of low-rate streams that are transmitted over parallel, narrowband channels that can be easily equalized.

Let us first analyze why Time Division Multiple Access (TDMA) and Code Division Multiple Access (CDMA) become problematic at very high data rates. If only a single frequency band (single carrier frequency) is used, then the symbol duration T_S has to become very small in order to achieve the required data rate, and the system bandwidth becomes very large. For example, the Global System for Mobile communications (GSM) system (see Chapter 21), which is designed for data rates up to 200 kbit/s, uses 200 kHz bandwidth, while the IEEE 802.11 system (see Chapter 25), with data rates of up to 55 MBit/s uses 20 MHz bandwidth. Now, delay dispersion of a wireless channel is given by nature; its values depend on the environment, but not on the transmission system. Thus, if the symbol duration becomes very small, then the impulse response (and thus the required length of the equalizer) becomes very long *in terms of symbol durations*. The computational effort for such a long equalizer is very large (see Chapter 16), and the probability of instabilities increases. Most multiaccess schemes make the situation even worse: TDMA makes the symbol duration even shorter; while CDMA uses band-spreading, and thus requires a Rake receiver (with a number of fingers that can be up to the spreading factor) in addition to the equalizer (as discussed in Section 18.2.4). OFDM, on the other hand, increases the symbol duration on each of its carriers.

OFDM dates back some 40 years; a patent was applied for in the mid-1960s [Chang 1966]. A few years later, an important improvement – the *cyclic prefix* – was introduced; it helps to eliminate residual delay dispersion. Cimini [1985] was the first to suggest OFDM for wireless communications. But it was only in the early 1990s that advances in hardware for digital signal processing made OFDM a realistic option for wireless systems. Furthermore, the high-data-rate applications for which OFDM is especially suitable emerged only in recent years. Currently, OFDM is used for *Digital Audio Broadcasting* (DAB), *Digital Video Broadcasting* (DVB), and *wireless Local Area Networks* (LANs) (IEEE 802.11a, IEEE 802.11g).

Wireless Communications   Andreas F. Molisch
© 2005 John Wiley & Sons, Ltd

19.2 Principle of orthogonal frequency division multiplexing

OFDM splits the information into N parallel streams, which are then transmitted by modulating N distinct carriers (henceforth called *subcarriers* or *tones*). Symbol duration on each subcarrier thus becomes larger by a factor of N. In order for the receiver to be able to separate signals carried by different subcarriers, they have to be orthogonal. Conventional Frequency Division Multiple Access (FDMA), as described in Section 17.1 and depicted again in Fig. 19.1, can achieve this by having large (frequency) spacing between carriers. This, however, wastes precious spectrum. A much narrower spacing of subcarriers can be achieved. Specifically, let subcarriers be at the frequencies $f_n = nW/N$, where n is an integer, and W the total available bandwidth; in the most simple case, $W = N/T_S$. We furthermore assume for the moment that modulation on each of the subcarriers is Pulse Amplitude Modulation (PAM) with rectangular basis pulses. We can then easily see that subcarriers are mutually orthogonal, since the relationship

$$\int_{iT_S}^{(i+1)T_S} \exp(j2\pi f_k t)\exp(-j2\pi f_n t)\,dt = \delta_{nk} \tag{19.1}$$

holds.

Figure 19.1 shows this principle in the frequency domain. Due to the rectangular shape of pulses in the time domain, the spectrum of each modulated carrier has a $\sin(x)/x$ shape. The spectra of different modulated carriers overlap, but each carrier is in the spectral nulls of all other carriers. Therefore, as long as the receiver does the appropriate demodulation (multiplying by $\exp(-j2\pi f_n t)$ and integrating over symbol duration), the datastreams of any two subcarriers will not interfere.

19.3 Implementation of transceivers

OFDM can be interpreted in two ways: one is an "analog" interpretation, following from the picture of Fig. 19.2a. As discussed in Section 19.2, we first split our original datastream into N parallel datastreams, each of which has a lower data rate. We furthermore have a number of local oscillators available, each of which oscillates at a frequency $f_n = nW/N$, where $n = 0, 1, \ldots, N-1$. Each of the parallel datastreams then modulates one of the carriers. This picture allows an easy understanding of the principle, but is ill-suited for actual implementation – the hardware effort of multiple local oscillators is too high. An alternative implementation

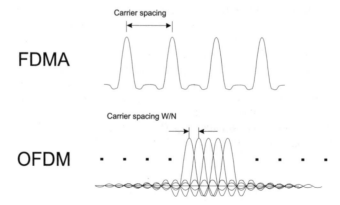

Figure 19.1   Principle behind orthogonal frequency division multiplexing: N carriers within a bandwidth of W.

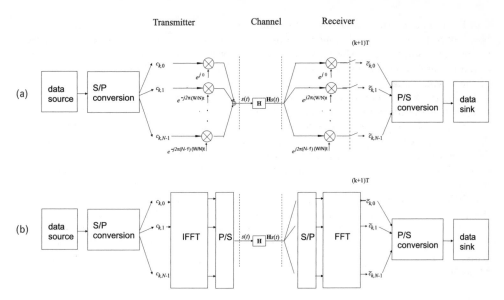

Figure 19.2 Transceiver structures for orthogonal frequency division multiplexing in purely analog technology (a), and using inverse fast Fourier transformation (b).

divides the transmit data into blocks of N symbols. This block of data is subjected to an *Inverse Fast Fourier Transformation* (IFFT), and then transmitted (see Fig. 19.2b). This approach is much easier to implement with digital technology. In the following, we will show that the two approaches are equivalent.

Let us first consider the analog interpretation. Let the complex transmit symbol at time instant i on the nth carrier be $c_{n,i}$. The transmit signal is then:

$$s(t) = \sum_{i=-\infty}^{\infty} s_i(t) = \sum_{i=-\infty}^{\infty} \sum_{n=0}^{N-1} c_{n,i} g_n(t - iT_S) \tag{19.2}$$

where the basis pulse $g_n(t)$ is a normalized, frequency-shifted rectangular pulse:

$$g_n(t) = \begin{cases} \dfrac{1}{\sqrt{T_S}} \exp\left(j2\pi n \dfrac{t}{T_S} \right) & \text{for } 0 < t < T_S \\ 0 & \text{otherwise} \end{cases} \tag{19.3}$$

Let us now – without restriction of generality – consider the signal only for $i = 0$, and sample it at instances $t_k = kT_S/N$:

$$s_k = s(t_k) = \frac{1}{\sqrt{T_S}} \sum_{n=0}^{N-1} c_{n,0} \exp\left(j2\pi n \frac{k}{N} \right) \tag{19.4}$$

Now, this is nothing but the inverse discrete Fourier transform of the transmit symbols. Therefore, the transmitter can be realized by performing an *Inverse Discrete Fourier Transform* (IDFT) on the block of transmit symbols (the blocksize must equal the number of subcarriers). In almost all practical cases, the number of samples N is chosen to be a power of 2, and the IDFT is realized as an IFFT. In the following, we will only speak of IFFTs and FFTs anymore.

Note that the input to this IFFT is made up of N samples (the symbols for the different subcarriers), and therefore the output from the IFFT also consists of N values. These N values now have to be transmitted, one after the other, as temporal samples – this is the reason why we have a P/S (parallel to serial) conversion directly after the IFFT. At the receiver, we can reverse the process: sample the received signal, write a block of N samples into a vector – i.e., an S/P (serial to parallel) conversion – and perform an FFT on this vector. The result is an estimate $\tilde{c}_n$ of the original data c_n.

Analog implementation of OFDM would require multiple local oscillators, each of which has to operate with little phase noise and drift, in order to retain orthogonality between the different subcarriers. This is usually not a practical solution. The success of OFDM is based on fast digital implementation that allows an implementation of the transceivers that is much simpler and cheaper. In particular, highly efficient structures exist for the implementation of an FFT (so-called "butterfly structures"), and the computational effort (per bit) of performing an FFT increases only with $\log(N)$.

OFDM can also be interpreted in the time–frequency plane. Each index i corresponds to a (temporal) pulse; each index n to a carrier frequency. This ensemble of functions spans a grid in the time–frequency plane. Such a grid can be spanned not only by rectangular basis functions, but by any set of basis functions. Optimization of this set of functions according to different criteria (e.g., low sensitivity to time and frequency dispersion) can be found in Kozek and Molisch [1998].

19.4 Frequency-selective channels

In the previous section, we explained how the OFDM transmitter and receiver work in an Additive White Gaussian Noise (AWGN) channel. We could take this scheme without any changes, and just let it operate in a frequency-selective channel. Intuitively, we would anticipate that delay dispersion will have only a small impact on the performance of OFDM: we convert the system into a parallel system of narrowband channels, so that the symbol duration on each carrier is made much larger than the delay spread. But, as we saw in Chapter 12, delay dispersion can lead to appreciable errors even when $S_\tau / T_S < 1$. Furthermore, as we will elaborate below, delay dispersion also leads to a loss of orthogonality between the subcarriers, and thus to *Inter Carrier Interference* (ICI). Fortunately, both these negative effects can be eliminated by a special type of guard interval, called the *Cyclic Prefix* (CP). In this section, we will show how to construct this CP, how it works, and what performance can be achieved in frequency-selective channels.

19.4.1 Cyclic prefix

Let us first define a new base function for transmission:

$$g_n(t) = \exp\left[j2\pi n \frac{W}{N} t\right] \qquad \text{for } -T_{cp} < t < \hat{T}_S \tag{19.5}$$

where again W/N is the carrier spacing, and $\hat{T}_S = N/W$. The symbol duration T_S is now $T_S = \hat{T}_S + T_{cp}$. This definition of the base function means that for duration $0 < t < \hat{T}_S$ the "normal" OFDM symbol is transmitted. During time $-T_{cp} < t < 0$, a copy of the last part of the symbol is transmitted (see Fig. 19.3). This implies that $g_n(t) = g_n(t + N/W)$ – a fact that can easily be verified by substituting into Eq. (19.5). This prepended part of the signal is called the "cyclic prefix".

Now that we know what a CP is, let us investigate why it is beneficial in delay-dispersive channels. When transmitting any datastream over a delay-dispersive channel, the arriving signal is the linear convolution of the transmitted signal with the channel impulse response. The CP

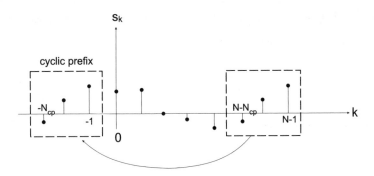

Figure 19.3 Principle of the cyclic prefix. $N_{cp} = NT_{cp}/(N/W)$ is the number of samples in the cyclic prefix.

converts this *linear* convolution into a *cyclical* convolution. During the time $0 < t < \tau_{max}$, where τ_{max} is the maximum excess delay of the channel, the received signal suffers from "real" Inter Symbol Interference (ISI), as echoes of the last part of the preceding symbol interfere with the desired symbol.[1] This "regular" ISI is eliminated by discarding the received signal during this time interval. During the remainder of the symbol, we have *cyclical* ISI; especially, it is the last part of the current (not the preceding) symbol that interferes with the first part of the current symbol. In the following, we show how an extremely simple mathematical operation can eliminate the effect of such a cyclical convolution.

For the following mathematical derivation, we assume that the duration of the impulse response is exactly equal to the duration of the prefix; furthermore, in order to simplify the notation, we assume (without restriction of generality) $i = 0$. In the receiver, there is a bank of filters that are matched to the basis functions *without* the CP:

$$\overline{g}_n(t) = \begin{cases} g_n^*(\widehat{T}_S - t) & \text{for } 0 < t < \widehat{T}_S \\ 0 & \text{otherwise} \end{cases} \tag{19.6}$$

This operation removes the first part of the received signal (of duration T_{cp}) from the detection process; as discussed above, the matched filtering of the remainder can be realized as an FFT operation. The signal at the output of the matched filter is thus convolution of the transmit signal with the channel impulse response and the receive filter:

$$r_{n,0} = \int_0^{\widehat{T}_S} \left[\int_0^{T_{cp}} h(t, \tau) \left(\sum_{k=0}^{N-1} c_{k,0} g_k(t - \tau) \right) d\tau \right] g_n^*(t) \, dt + n_n \tag{19.7}$$

where n_n is the noise at the output of the matched filter. Note that the argument of g_k can attain values between $-T_{cp}$ and $\widehat{T}_S$, which is the region of definition of Eq. (19.5). If the channel can be considered as constant during the time T_S, then $h(t, \tau) = h(\tau)$, and we obtain:

$$r_{n,0} = \sum_{k=0}^{N-1} c_{k,0} \int_0^{\widehat{T}_S} \left[\int_0^{T_{cp}} h(\tau)(g_k(t - \tau)) \, d\tau \right] g_n^*(t) \, dt + n_n \tag{19.8}$$

The inner integral can be written as:

$$\exp\left[j2\pi tk \frac{W}{N} \right] \int_0^{T_{cp}} h(\tau) \exp\left(-j2\pi\tau k \frac{W}{N} \right) d\tau = g_k(t) H\left(k \frac{W}{N} \right) \tag{19.9}$$

---

[1] In the following, we assume $\tau_{max} \leq T_{cp}$.

where $H\left(k\dfrac{W}{N}\right)$ is the channel transfer function at the frequency kW/N. Since, furthermore, the basis functions $g_n(t)$ are orthogonal during the time $0 < t < \widehat{T}_{\mathrm{S}}$:

$$\int_0^{\widehat{T}_{\mathrm{S}}} g_k(t)g_n^*(t)\,dt = \delta_{kn}(t) \tag{19.10}$$

the received signal samples r can be written as:

$$r_{n,0} = H\left(n\dfrac{W}{N}\right)c_{n,0} + n_n \tag{19.11}$$

The OFDM system is thus represented by a number of parallel *non-dispersive*, fading channels, each with its own complex attenuation $H\left(n\dfrac{W}{N}\right)$. Equalization of the system thus becomes exceedingly simple: it just requires division by the transfer function at the subcarrier frequency. In other words, the CP has recovered the orthogonality of the subcarriers.

Two caveats have to be noted: (i) we assumed in the derivation that the channel is static for the duration of the OFDM symbol. If this assumption is not fulfilled, interference between the subcarriers can still occur (see Section 19.7); (ii) discarding part of the received signal decreases the Signal-to-Noise Ratio (SNR), as well as spectral efficiency. For usual operating parameters (CP about 10% of symbol duration), this loss is tolerable.

The block diagram of an OFDM system including the CP is given in Fig. 19.4. The original datastream is S/P converted. Each block of N data symbols is subjected to an IFFT, and then the last $NT_{\mathrm{cp}}/T_{\mathrm{S}}$ samples are prepended. The resulting signal is modulated onto a (single) carrier and transmitted over a channel, which distorts the signal and adds noise. At the receiver, the signal is partitioned into blocks. For each block, the CP is stripped off, and the remainder is subjected to an FFT. The resulting samples (which can be interpreted as the samples in the frequency domain) are "equalized" by means of one-tap equalization – i.e., division by the complex channel attenuation – on each carrier.

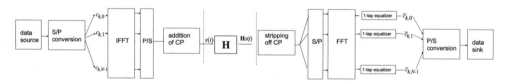

Figure 19.4 Structure of an orthogonal-frequency-division-multiplexing transmission chain with cyclic prefix and one-tap equalization.

19.4.2 Performance in frequency-selective channels

The cyclic prefix converts a frequency-selective channel into a number of parallel flat-fading channels. This is positive in the sense that it gets rid of the ISI that plagues TDMA and CDMA systems. On the downside, an uncoded OFDM system does not show any frequency diversity at all. If a subcarrier is in a fading dip, then error probability on that subcarrier is very high, and dominates the Bit Error Rate (BER) of the total system for high SNRs.

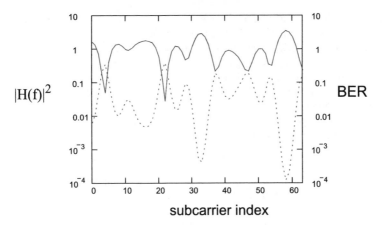

$|H(f)|^2$ BER

subcarrier index

Figure 19.5 Normalized squared magnitude of the transfer function (solid), and bit error rate (dashed), for a channel with taps at $[0, 0.89, 1.35, 2.41, 3.1]$ with amplitudes $[1, -0.4, 0.3, 0.43, 0.2]$. The average signal-to-noise ratio at the receiver is 3 dB; the modulation format is binary-phase shift keying. Subcarriers are at $f_k = 0.05k$, $k = 0.63$.

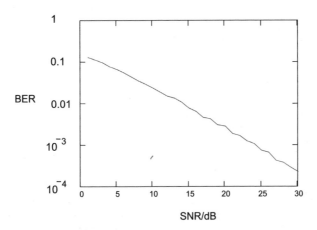

BER

SNR/dB

Figure 19.6 Bit error rate for a channel with taps at $[0, 0.89, 1.35, 2.41, 3.1]$ with mean powers $[1, 0.16, 0.09, 0.185, 0.04]$, each tap independently Rayleigh fading. The modulation format is binary-phase shift keying. Subcarriers are at $f_k = 0.05k$, $k = 0.63$.

Example 19.1 *Bit error rate of uncoded orthogonal frequency division multiplexing.*

Figure 19.5 shows the transfer function and the BER of a Binary Phase Shift Keying (BPSK) OFDM system for specific realization of a frequency-selective channel. Obviously, the BER is highest in fading dips. Note that the results are plotted on a logarithmic scale – while the BER on "good" subcarriers can be as low as 10^{-4}, the BER on subcarriers that are in fading dips are up to 0.5. This also has a significant impact on average error probability: the error probability on bad subcarriers dominates the behavior. Figure 19.6 shows a simulation of the average BER (over many channel realizations) for a frequency-selective channel. We find that the BER decreases only linearly as the SNR increases; closer inspection reveals that the result is the same as in Fig. 12.6.

More generally, we find that uncoded OFDM has the same average BER irrespective of the frequency selectivity of the channel. This can also be interpreted the following way: frequency selectivity gives us different channel realizations on different subcarriers; time variations give us different channel realizations at different times. Doubly selective channels have different realizations on different subcarriers as well as at different times. But, for computation of the average BER, it does not matter how the different realizations are created, as long as the ensemble is large enough.[2]

From these examples, we see that the main problem lies in the fact that carriers with poor SNR dominate the performance of the system. Any of the following approaches circumvents this problem:

- *Coding across the different tones.* Such coding helps to compensate for fading dips on one subcarrier by a good SNR in another subcarrier. This will be described in more detail in Section 19.4.3.
- *Spreading the signal over all tones.* In this approach, each symbol is spread across all carriers, so that it sees an SNR that is the average of all tones over which it is spread. This method will be discussed in more detail in Sections 19.10 and 19.11.
- *Adaptive modulation.* If the transmitter knows the SNR on each of the subcarriers, it can choose its modulation alphabet and coding rate adaptively. Thus, on carriers with low SNR, the transmitter will send symbols using stronger encoding and a smaller modulation alphabet. Also, the power allocated to each subcarrier can be varied. This approach is described in more detail in Section 19.8.

19.4.3 Coded orthogonal frequency division multiplexing

Just as coding can be used to great effect in single-carrier systems to improve performance in fading channels, so can it be gainfully employed in OFDM systems. But now we have data that are transmitted at different frequencies as well as at different times. This gives rise to the question of how coding of the data should be applied.

To get an intuitive feeling for coding across different subcarriers, imagine again the simple case of repetition coding: each of the symbols that are to be transmitted is repeated on K different subcarriers. As long as fading is independent on the different subcarriers, K-fold diversity is achieved. In the most simple case, the receiver first makes a hard decision about symbols on each subcarrier, and then makes a majority decision among the K received symbols about which bit was sent. Of course, practical systems do not use repetition coding, but the principle remains the same.

We could now try and develop a whole theory for coding in OFDM systems. However, it is much easier to just consider the analogy between the time domain and the frequency domain. Remember the main lessons from Section 14.7: enough interleaving should be applied such that fading of coded bits is independent. In other words: we just need independent channel states over which to transmit our coded bits; this will automatically result in a high diversity order. It does not matter whether channel states are created by temporal variations of the channel, or as different transfer functions of subcarriers in a frequency-selective channels. Thus, it is not really necessary to define new codes for OFDM, but it is more a question on how to design appropriate mappers and interleavers that assign the different coded bits in the time–frequency plane. This mapping, in turn, depends on the frequency selectivity as well as the time selectivity of the channel. If the channel is highly frequency-selective, then it might be sufficient to code

---

[2] Note that in a time-invariant, frequency-selective channel, the number of independent channel realizations depends on the ratio of system bandwidth to coherence bandwidth of the channel. If this value is small, there might not be a sufficiently large ensemble to obtain good averaging.

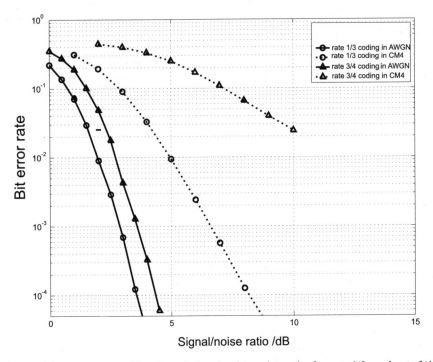

Figure 19.7 Bit error rate as a function of the signal-to-noise ratio for rate-1/3- and rate-3/4-coded orthogonal-frequency-division-multiplexing system. Channel is either additive white Gaussian noise, or channel model 4 of the IEEE 802.15.3a channels. The OFDM system follows the specifications of the WiMedia standard (see Chapter 25).

Reproduced with permission from Ramachandran et al. [2004] © IEEE.

only across available frequencies, without any coding or interleaving along the time axis. This has two advantages: on one hand, this scheme also works in static channels, which occur quite often for wireless LANs and other high-rate data transmission scenarios; on the other, the absence of interleaving in the time domain results in lower latency of the transmission and decoding process.

Figure 19.7 shows a performance example. We see that for AWGN, both rate-1/3- and rate-3/4-coded systems exhibit good performance, with approximately a 1-dB difference. In fading channels, performance is dramatically different. While the rate-1/3 code has good diversity, and therefore the BER decreases fast as a function of the SNR, the rate-3/4 code has very little frequency diversity, and thus bad performance.

19.5 Channel estimation

As for any other coherent wireless system, operation of OFDM systems requires an estimate of the channel transfer function, or, equivalently, the channel impulse response. Since OFDM is operating with a number of parallel narrowband subcarriers, it is intuitive to estimate the channel in the frequency domain. More precisely, we wish to obtain estimates of the N complex-valued channel gains on the subcarriers. Let us denote these channel attenuations as $h_{n,i}$, where n is the subchannel index and i is the time index. Assuming that we know the statistical properties of these channel attenuations, and some structure to the OFDM signal, we can derive good channel estimators.

In the following, we treat three approaches: (i) pilot symbols, which are mainly suitable for an initial estimate of the channel; (ii) scattered pilot tones, which help to track changes in channels over time; and (iii) eigenvalue-decomposition-based methods, which can be used to reduce the complexity of the first two methods.

19.5.1 Pilot-symbol-based methods

The most straightforward channel estimation in OFDM is when we have a dedicated pilot *symbol* containing only known data – in other words, the data on each of the subcarriers is known. This approach is appropriate for initial acquisition of the channel, at the beginning of a transmission burst. Denoting the known data on subcarrier n at time i as $c_{n,i}$, we can find a Least Squares (LS) channel estimate as:

$$h_{n,i}^{LS} = r_{n,i}/c_{n,i}$$

where $r_{n,i}$ is the received value on subchannel n. Applying standard estimation theory, one way of improving this estimate is to use frequency correlation of channel attenuations. Arranging the LS estimates in a vector $\mathbf{h}_i^{LS} = (\,h_{1,i}^{LS} \quad h_{2,i}^{LS} \quad \cdots \quad h_{n,i}^{LS}\,)^T$, the corresponding vector of linear MMSE estimates becomes:

$$\mathbf{h}_i^{LMMSE} = \mathbf{R}_{hh^{LS}}\mathbf{R}_{h^{LS}h^{LS}}^{-1}\mathbf{h}_i^{LS} \tag{19.12}$$

where $\mathbf{R}_{hh^{LS}}$ is the covariance matrix between channel gains and the LS estimate of channel gains, $\mathbf{R}_{h^{LS}h^{LS}}$ is the autocovarance matrix of LS estimates. Given that we have AWGN with variance σ_n^2 on each subcarrier, $\mathbf{R}_{hh^{LS}} = \mathbf{R}_{hh}$ and $\mathbf{R}_{h^{LS}h^{LS}} = (\mathbf{R}_{hh} + \sigma^2\mathbf{I})$. Arranging channel attenuations in a vector $\mathbf{h}_i = (\,h_{1,i} \quad h_{2,i} \quad \cdots \quad h_{n,i}\,)^T$, we can determine:

$$\mathbf{R}_{hh} = E\{\mathbf{h}_i\mathbf{h}_i^\dagger\} = E\{\mathbf{h}_i^*\mathbf{h}_i^T\}^* \tag{19.13}$$

which is independent of time i if the channel is wide-sense-stationary. Note that in order to simplify notation of the subsequent equations, we have defined $\mathbf{R}_{hh}$ as the conjugate of the "usual" correlation matrix.

This estimation approach has its strengths and weaknesses. It produces very good estimates, but computational complexity is high if the number of subcarriers is large: it requires N^2 multiplications – i.e. N multiplications per estimated channel gain (assuming that all correlation matrices and inversions are precalculated). This is quite a large complexity, even if this pilot-symbol-based estimation is usually done only at the beginning of a transmission burst. For this reason, there are several other suboptimal approaches available, where, for example, smoothing Finite Impulse Response (FIR) filters of limited length (much less than N) are applied across LS-estimated attenuations to exploit the correlation between neighboring subchannels.

19.5.2 Methods based on scattered pilots

After obtaining an initial estimate of the channel, we need to track changes in the channel as it evolves with time. In this case we would like to do two things: (i) reduce the number of known bits in an OFDM symbol (this improves spectral efficiency); and (ii) exploit the time correlation of the channel – i.e., the fact that the channel changes only slowly in time. An attractive way of tracking the channel is to use pilot symbols scattered in the OFDM time–frequency grid as illustrated in Figure 19.8, where pilots are spaced by N_f subcarriers and N_t OFDM symbols.[3]

---

[3] We have used a rectangular pilot pattern in the illustration, but other pilot patterns can be used as well.

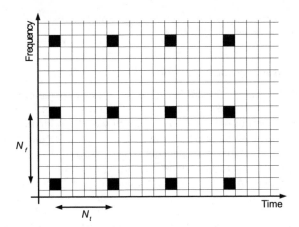

Figure 19.8 Scattered pilots in the orthogonal-frequency-division-multiplexing time–frequency grid. In this case the pattern is rectangular with pilot distances N_f subcarriers in frequency and N_t OFDM symbols in time.

When estimating the channel based on scattered pilots, we can start by performing LS estimation of the channel at pilot positions – i.e. $h_{n,i}^{LS} = r_{n,i}/c_{n,i}$, where $r_{n,i}$ is the received value and $c_{n,i}$ is the known pilot data in pilot position (n, i). From these initial estimates at pilot positions we would like to perform interpolation to obtain estimates of the channel at all other positions. Interpreting the pilots as samples in a two-dimensional space, we can use standard sampling theory to put limits on the required density of our pilot pattern as [Nilsson et al. 1997]:

$$N_f < \frac{N}{N_{cp}}$$

$$N_t < \frac{1}{2(1 + N_{cp}/N)\nu_{max}}$$

Since we need to reduce the effect of noise from the pilots and also help to reduce the complexity of estimation algorithms, it has been argued that a good tradeoff is to place twice as many pilots in each direction as required by the sampling theorem [Nilsson et al. 1997].

In principle, channel interpolation between these pilot positions can be done using the same estimation theory as for the all-pilot symbol case. When estimating a certain channel attenuation $h_{n,i}$ using a set of K pilot positions $(n_j, i_j), j = 1 \ldots K$, we place the LS estimates in a pilot vector $\mathbf{p} = \begin{pmatrix} h_{n_1,i_1}^{LS} & h_{n_2,i_2}^{LS} & \cdots & h_{n_K,i_K}^{LS} \end{pmatrix}^T$ and calculate the LMMSE estimate as:

$$h_{n,i}^{LMMSE} = \mathbf{r}_{hp}\mathbf{R}_{pp}^{-1}\mathbf{p}$$

where $\mathbf{r}_{hp}$ is the correlation (row) vector $E\{h_{n,i}\mathbf{p}^{\dagger}\}$ and $\mathbf{R}_{pp}$ is $E\{\mathbf{p}\mathbf{p}^{\dagger}\}$. The complexity of this estimator grows with the number of pilot tones included in the estimation and requires K multiplications per estimated attenuation, again assuming that all correlation matrices and inversions are precalculated. To obtain good channel estimates, a quite large number of pilots may have to be used.

An alternative approach to the two-dimensional filtering above, where we use pilots in both the frequency direction and time direction at the same time, is to apply separable filters. This implies that we use two one-dimensional filters, one in the time direction and the other in the frequency direction. Many more pilots are thus influencing each estimated channel attenuation, for a given estimator complexity. The resulting increase in performance has been shown to dominate over loss in optimality when going from general two-dimensional filters to separable ones based on two one-dimensional filters.

19.5.3 Methods based in eigendecompositions

The structure of OFDM allows for efficient channel estimator structures. We know that the channel impulse response is short compared with the OFDM symbol length in any well-designed system. This fact can be used to reduce the dimensionality of the estimation problem. In essence, when using the LMMSE estimator in (19.12), we would like to use the statistical properties of the channel to perform the matrix multiplication more efficiently. This can be done using the theory of optimal rank reduction from estimation theory, where an Eigen Value Decomposition (EVD) $\mathbf{R}_{hh} = \mathbf{U}\mathbf{\Lambda}\mathbf{U}^{\dagger}$ results in a new more computationally efficient version of (19.12). The dimension of this space is approximately $N_{\mathrm{cp}} + 1$ – i.e., one more than the number of samples in the cyclic prefix. We can therefore expect that, after the first $N_{\mathrm{cp}} + 1$ diagonal elements in $\mathbf{\Lambda}$, the magnitude should decrease rapidly. Using the SVD to rewrite (19.12) as:

$$\mathbf{h}_i^{\mathrm{LMMSE}} = \mathbf{U}\mathbf{\Delta}\mathbf{U}^{\dagger}\mathbf{h}_i^{\mathrm{LS}}$$

where $\mathbf{\Delta}$ is a diagonal matrix containing the values $\delta_i = \lambda_i/(\lambda_i + 1/\gamma)$ on its diagonal. The diagonal elements δ_i will decrease rapidly after the first $N_{\mathrm{cp}} + 1$ since the λ_i's do. By setting all but the p first λ_i's to zero – i.e. assigning $\delta_i = 0$ for $i > p$ – we get an optimal rank-p estimator for channel gains. The computational complexity of this estimator is $2Np$ multiplications, which is $2p$ per estimated attenuation. This should be compared with the N multiplications per estimated attenuation in the original estimator (19.12). The estimator principle is illustrated in Fig. 19.9.

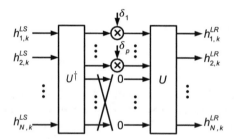

Figure 19.9 The optimal rank-p channel estimator viewed as a transform (U^H) followed by p scalar multiplications and a second transform (U).

In the case when the autocorrelation matrix $\mathbf{R}_{hh}$ is a circulant matrix, the resulting optimal transforms $\mathbf{U}^{\dagger}$ and $\mathbf{U}$ are the IDFT and Discrete Fourier Transform (DFT), respectively, and there are only N_{cp} nonzero singular values. The basic estimator structure stays the same, as shown in Fig. 19.10, while the FFT processor already available in the OFDM receiver can be used to perform channel estimation as well.

In many cases, when the channel correlation matrix is not circulant, the computational efficiency of DFT-based estimators may outweigh the suboptimality of their rank reduction. This general structure of estimators has also been used as one of the two one-dimensional estimators when performing two-dimensional estimation (see above). The gain here is that time direction smoothing can be done between two transforms, leading to a smaller number of filters that have to be applied in parallel. Instead of N filters (one per subcarrier), only p filters are needed in a rank-p estimator (as shown in Fig. 19.11).

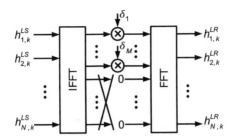

Figure 19.10   Low-rank estimator for channels with circulant autocorrelation, implemented using fast Fourier transforms.

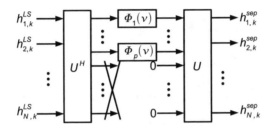

Figure 19.11   Two-dimensional (separable) channel estimation where time domain smoothing is done in the transform domain. This reduces the number of parallel filters needed, from N to p.

19.6 Peak-to-average power ratio

19.6.1 Origin of the peak-to-average-ratio problem

One of the major problems of OFDM is that the peak amplitude of the emitted signal can be considerably higher than the average amplitude. This *Peak-to-Average Ratio* (PAR) issue originates from the fact that an OFDM signal is the superposition of N sinusoidal signals on different subcarriers. On average the emitted power is linearly proportional to N. However, sometimes, the signals on the subcarriers add up constructively, so that the *amplitude* of the signal is proportional to N, and the power thus goes with N^2. We can thus anticipate the (worst case) power PAR to increase linearly with the number of subcarriers.

We can also look at this issue from a slightly different point of view: the contributions to the total signal from the different subcarriers can be viewed as random variables (they have quasi-random phases, depending on the sampling time as well as the value of the symbol with which they are modulated). If the number of subcarriers is large, we can invoke the central limit theorem to show that the distribution of the amplitudes of in-phase components is Gaussian, with a standard deviation $\sigma = 1/\sqrt{2}$ (and similarly for the quadrature components) such that mean power is unity. Since both in-phase and quadrature components are Gaussian, the absolute amplitude is Rayleigh-distributed (see Chapter 5 for details of this derivation). Knowing the amplitude distribution, it is easy to compute the probability that the instantaneous amplitude will lie above a given threshold, and similarly for power. For example, there is a $\exp(-10^{6/10}) = 0.019$ probability that the peak power is 6 dB above the average power.

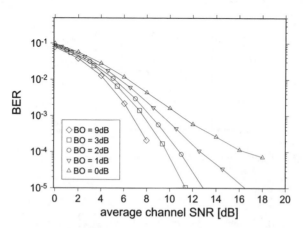

Figure 19.12 Bit error rate as a function of the signal-to-noise ratio, for different backoff levels of the transmit amplifier.

The amplitude distribution is important because it faces us with three choices:

1. Put a power amplifier into the transmitter that can amplify linearly up to the possible *peak* value of the transmit signal. This is usually not practical, as it requires expensive and power-consuming class-A amplifiers. The larger the number of subcarriers N, the more difficult this solution becomes.

2. Use a nonlinear amplifier, and accept the fact that amplifier characteristics will lead to distortions in the output signal. Those nonlinear distortions destroy orthogonality between subcarriers, and also lead to increased out-of-band emissions (*spectral regrowth* – similar to third-order intermodulation products – such that the power emitted outside the nominal band is increased). The first effect increases the BER of the desired signal (see Fig. 19.12), while the latter effect causes interference to other users, and thus decreases the cellular capacity of the OFDM system (see Fig. 19.13). This means that in order to have constant adjacent channel interference we can trade off power amplifier performance against spectral efficiency (note that increased carrier separation decreases spectral efficiency).

3. Use PAR reduction techniques. These will be described in the next subsection.

19.6.2 Peak-to-average-ratio reduction techniques

A wealth of methods for mitigating the PAR problem has been suggested in the literature. Some of the promising approaches are:

1. *Coding for PAR reduction*: under normal circumstances, each OFDM symbol can represent one of 2^N codewords (assuming BPSK modulation). Now, of these codewords only a subset of size 2^K is acceptable in the sense that its PAR is lower than a given threshold. Both the transmitter and the receiver know the mapping between a bit combination of length K, and the codeword of length N that is chosen to represent it, and which has an admissible PAR. The transmission scheme is thus the following: (i) parse the incoming bitstream into blocks of length K; (ii) select the associated codeword of length N; (iii) transmit this codeword via the OFDM modulator. The coding scheme can guarantee a certain value for the PAR. It

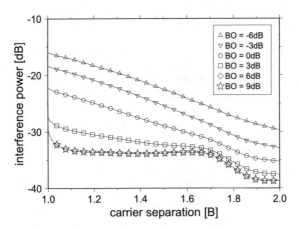

Figure 19.13 Interference power to adjacent bands (OFDM users), as a function of carrier separation, for different values of backoff of the transmit amplifier.

Reproduced with permission from Hanzo et al. [2003] © J. Wiley & Sons, Ltd.

also has some coding gain, though this gain is smaller than for codes that are solely dedicated to error correction.

2. *Phase adjustments*: this scheme first defines an ensemble of phase adjustment vectors $\boldsymbol{\phi}_l$, $l = 1, \ldots, L$, that are known to both the transmitter and receiver; each vector has N entries $\{\phi_n\}_l$. The transmitter than multiplies the OFDM symbol to be transmitted c_n by each of these phase vectors to get:

$$\{\widehat{c}_n\}_l = c_n \exp[j(\phi_n)_l] \tag{19.14}$$

and then selects:

$$\widehat{l} = \arg\min_l (PAR(\{\widehat{c}_n\}_l)) \tag{19.15}$$

which gives the lowest PAR. The vector $\{\widehat{c}_n\}_{\widehat{l}}$ is then transmitted, together with the index $\widehat{l}$. The receiver can then undo phase adjustment, and demodulate the OFDM symbol. This method has the advantage that the overhead is rather small (at least as long as L stays within reasonable bounds); on the downside, it cannot guarantee to keep the PAR below a certain level.

3. *Correction by multiplicative function*: another approach is to multiply the OFDM signal by a time-dependent function whenever the peak value is very high. The simplest example for such an approach is the clipping we mentioned in the previous subsection: if the signal attains a level $s_k > A_0$, it is multiplied by a factor A_0/s_k. In other words, the transmit signal becomes:

$$\widehat{s}(t) = s(t) \left[1 - \sum_k \max\left(0, \frac{|s_k| - A_0}{|s_k|}\right) \right] \tag{19.16}$$

A less radical method is to multiply the signal by a Gaussian function centered at times when the level exceeds the threshold:

$$\widehat{s}(t) = s(t) \left[1 - \sum_n \max\left(0, \frac{|s_k| - A_0}{|s_k|}\right) \exp\left(-\frac{t^2}{2\sigma_t^2}\right) \right] \tag{19.17}$$

Multiplication by a Gaussian function of variance σ_t^2 in the time domain implies convolution with a Gaussian function in the frequency domain with variance $\sigma_f^2 = 1/(2\pi\sigma_t^2)$. Thus, the amount of out-of-band interference can be influenced by the judicious choice of σ_t^2. On the downside, we find that the ICI (and thus BER) caused by this scheme is significant.

4. *Correction by additive function*: in a similar spirit, we can choose an additive, instead of a multiplicative, correction function. The correction function should be smooth enough not to introduce significant out-of-band interference. Furthermore, the correction function acts as additional pseudo-noise, and thus increases the BER of the system.

When comparing the different approaches to PAR reduction, we find that there is no single "best" technique. The coding method can guarantee a maximum PAR value, but requires considerable overhead, and thus reduced throughput. The phase adjustment method has a smaller overhead (depending on the number of phase adjustment vectors), but cannot give a guaranteed performance. Neither of these two methods leads to an increase in either ICI or out-of-band emissions. The correction by multiplicative functions can guarantee performance – up to a point (subtracting the Gaussian functions centered at one point might lead to larger amplitudes at another point). Also, it can lead to considerable Inter Carrier Interference, while out-of-band emissions are fairly well-controlled.

19.7 Intercarrier interference

The CP provides an excellent way of ensuring orthogonality of the carriers in a delay-dispersive (frequency-selective) environment – in other words, there is no ICI due to frequency selectivity of the channel. However, wireless propagation channels are also time-varying, and thus time-selective (frequency-dispersive, due to the Doppler effect, see Chapter 5). Time selectivity has two important consequences for an OFDM system: (i) it leads to random Frequency Modulation (FM, see Chapter 5), which can cause errors especially on subcarriers that are in a fading dip; and (ii) it creates ICI. A Doppler shift of one subcarrier can cause ICI in many adjacent subcarriers (see Fig. 19.14). The impact of time selectivity is mostly determined by the product of maximum Doppler frequency and symbol duration of the OFDM symbol. The spacing between the subcarriers is inversely proportional to symbol duration. Thus, if symbol duration is large, even a small Doppler shift can result in appreciable ICI.

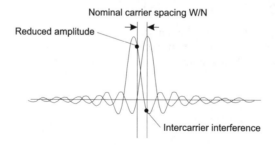

Figure 19.14 Intercarrier interference due to frequency offset.

Delay dispersion can be another source of intercarrier interference if the CP is shorter than the maximum excess delay of the channel. This situation can arise for various reasons. A system might consciously shorten or omit the CP in order to improve spectral efficiency. In other cases, a system may originally be designed to operate in a certain class of environments (and thus a certain range of excess delays), and is later also deployed in other environments that have a larger excess delay. Finally, for many systems, the length of the CP is a compromise between the desire to eliminate ICI, and the need to retain spectral efficiency – in other words, a CP is not chosen to cope with the worst case channel situation.

In the following, we mathematically describe the received signal if ICI occurs either as a result of Doppler shift or insufficient CP. Instead of Eq. (19.11), the relationship between data symbols c_n and receive samples after FFT is now given by:

$$r_k = \sum_{n=0}^{N-1} c_n H_{k,n} + n_k \tag{19.18}$$

where

$$H_{k,n} = \frac{1}{N} \sum_{q=0}^{N} \sum_{l=0}^{L-1} h[q,l] \exp\left[j \frac{2\pi}{N} (qn - nl - qk) \right] \mathcal{H}[q - l + N_{\text{cp}}] \tag{19.19}$$

where $h(n,l)$ is a sampled version of the time-variant channel impulse response $h(t,\tau)$, $\mathcal{H}[\ ]$ denotes the Heaviside function, and L is the maximum excess delay in units of samples $L = \tau_{\text{max}} N / T_{\text{S}}$. Note also that Eq. (19.19) reduces to Eq. (19.11) for the case of a time-invariant channel and a sufficienly long guard interval.

Because ICI can be a limiting factor for OFDM systems, a large range of techniques for fighting it has been developed and can be classified as follows:

- *Optimum choice of carrier spacing and OFDM symbol length.* In this approach, we influence the OFDM symbol length in order to minimize its ICI. It follows from our statements above that short symbol duration is good for reduction of Doppler-induced ICI. On the other hand, spectral efficiency considerations enforce a minimum duration of T_{S}: the CP (which is determined by the maximum excess delay of the channel) should not be shorter than approximately 10% of the symbol duration. The following equation gives a useful guideline on how to choose T_{S}. Let $R(k,l) = P_{\text{h}}(lT_{\text{c}}, kT_{\text{c}})$ be the sampled delay cross power spectral density (see Chapter 6). Define furthermore a function:

$$w(q,r) = \frac{1}{N} \begin{cases} N - |r| & 0 \le q \le N_{\text{cp}} & 0 \le |r| \le N \\ N - q + N_{\text{cp}} - |r| & N_{\text{cp}} \le q \le N + N_{\text{cp}} & 0 \le |r| \le N - q + N_{\text{cp}} \\ N + q - |r| & -N \le q \le 0 & 0 \le |r| \le N + q \\ 0 & \text{elsewhere} \end{cases} \tag{19.20}$$

Then the desired signal power can be approximated as [Steendam and Moeneclaey 1999]:

$$P_{\text{sig}} = \frac{1}{N} \sum_l \sum_k w(k,l) R(k,l) \tag{19.21}$$

and the ICI and ISI powers as:

$$P_{\text{ICI}} = \sum_k w(k,0) R(k,0) - P_{\text{sig}} \tag{19.22}$$

$$P_{\text{ISI}} = \sum_l [1 - w(k,0)] R(k,0) \tag{19.23}$$

and the SINR is:

$$SINR = \frac{\dfrac{E_{\text{S}}}{N_0} P_{\text{sig}} \dfrac{N}{N_{\text{cp}} + N}}{\dfrac{E_{\text{S}}}{N_0} P_{\text{sig}} \dfrac{N}{N_{\text{cp}} + N} \dfrac{P_{\text{ISI}} + P_{\text{ICI}}}{P_{\text{sig}}} + 1} \tag{19.24}$$

The above equations allow an easy tradeoff between the ICI due to the Doppler effect, the ICI due to residual delay dispersion, and SNR loss due to the CP.

- *Optimum choice of OFDM basis signal.* A related approach influences the OFDM basis pulse *shape* in order to minimize ICI. We know that a rectangular temporal signal has a very sharp cutoff in the temporal domain, but has a $\sin(x)/x$ shape in the frequency domain and thus decays slowly. And while in a perfect system each subcarrier is in the spectral nulls of all other subcarriers, the slope of the $\sin(x)/x$ is large near its zeros. Thus, even a small Doppler shift leads to large ICI. By choosing basis pulses whose spectrum decays faster and gentler, we decrease ICI due to the Doppler effect. On the downside, faster decay in the frequency domain is bought by slower decay in the time domain, which increases delay-spread-induced errors. Gaussian-shaped basis functions have been shown to be a useful compromise.

- *Self-interference cancellation techniques*: in this approach, information is modulated not just onto a single subcarrier but onto a group of them. This technique is very effective for mitigation of ICI, but leads to a reduction in spectral efficiency of the system.

- *Frequency domain equalizers*: if the channel and its variations are known, then its impact on the received signal, as described by Eq. (19.18), can be reversed. While this reversal can no longer be done by a single-tap equalizer, there is a variety of suitable techniques. For example, we can simply invert H, or use a minimum mean square error criterion. These inversions can be computationally expensive: as the channel is continuously changing, the inverse matrix has to be recomputed for every OFDM block. However, methods with reduced computational complexity exist. Another approach is to interpret different tones as different users, and then apply multiuser detection techniques (as described in Section 18.4) for detection of the tones. Figure 19.15 shows an example of the effect of different equalization techniques (OPT denotes a linear inversion technique, while PIC and SIC denote multiuser detection).

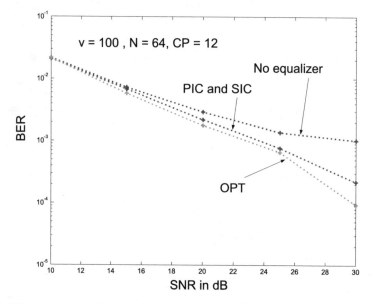

Figure 19.15 Bit error rate as a function of signal-to-noise ratio for an 802.11a-like orthogonal-frequency-division-multiplexing system with 64 carriers and 12 samples CP. Performance is analyzed in channel model F of the 802.11n channel models, with 100-m/s velocity.

In addition to time selectivity and delay dispersion, there is another effect that can destroy orthogonality between carriers: errors in the local oscillator. Such errors can be produced by:

- *Synchronization errors*: as we discussed in Section 19.4, synchronization is critical for retaining orthogonality between carriers. Any errors in the synchronization procedure will be reflected as deviation of the receiver's Local Oscillator (LO) from the optimum frequency, and thus ICI.
- *Phase noise of the transmitter and receiver*. Phase noise, which stems from inaccuracies in the oscillator, leads to deviation of the LO signal from its nominal, strictly sinusoidal shape. The distribution of phase noise is typically Gaussian, and is further characterized by its power-spectral density. Essentially, a narrow spectrum means that phase only changes very slowly, which can be more easily compensated by various receiver algorithms. The effect of phase noise is a spilling of the spectrum of subcarrier signals into adjacent subcarriers, and thus ICI.

Example 19.2 *Consider a system with a 5-MHz bandwidth, 128 tones, and a CP that is 40 samples long. It operates in a channel with an exponential Power Delay Profile (PDP), $\tau_{rms} = 1\,\mu s$, $\nu_{rms} = 500\,Hz$, and an E_S/N_0 of 10 dB. What is the Signal-to-Interference-and-Noise Ratio (SINR) at the receiver? How do results change when the CP is shortened to 12 samples?*

In a first step, we need to find the sampled delay cross power spectral density. For a bandwidth of 5 MHz, the sampling interval is 200 ns. Therefore, the rms delay spread is five samples, and the sampled PDP is described as $\exp(-k/5)$. The Doppler spectrum is assumed to have a Gaussian shape. Assuming furthermore that the Doppler spectrum is independent of delay, we obtain:

$$R(k,l) = \exp(-k/5) \cdot \exp\left(-\frac{l^2}{2 \cdot 10\,000^2}\right) \qquad (19.25)$$

An accurate solution for interference power can then be found by inserting this sampled delay cross power spectral density into Eqs. (19.20)–(19.24). We obtain:

$$\frac{P_{\text{ICI}}}{P_{\text{sig}}} = 2.46 \cdot 10^{-5} \qquad (19.26)$$

$$\frac{P_{\text{ISI}}}{P_{\text{sig}}} = 1.18 \cdot 10^{-5} \qquad (19.27)$$

This shows that ISI and ICI are reasonably balanced, which overall leads to low interference power.

Furthermore, the CP reduces the effective SNR by a factor $128/(128 + 40) = 0.762$. Thus, the total SINR becomes:

$$\frac{7.62}{7.62(2.46 + 1.18) \cdot 10^{-5} + 1} = 7.6 \qquad (19.28)$$

This indicates that the major loss of SINR occurs due to the CP. When we shorten it from 40 to 12 samples, the sum of ISI and ICI increases to:

$$\frac{P_{\text{ICI}} + P_{\text{ISI}}}{P_{\text{sig}}} = 6.2 \cdot 10^{-3} \qquad (19.29)$$

On the other hand, the SNR becomes $10 \cdot 128/(128 + 12) = 9.14$. Thus the effective SNR becomes:

$$\frac{9.14}{9.14 \cdot 6.2 \cdot 10^{-3} + 1} = 8.65 \tag{19.30}$$

This shows that a long CP is not always the best way to improve the SINR. Rather, it is important to correctly balance ISI, ICI, and duration of the CP.

19.8 Adaptive modulation and capacity

Adaptive modulation changes the coding scheme and/or modulation method depending on channel-state information – choosing it in such a way that it always "pushes the limit" of what the channel can transmit. In OFDM, modulation and/or coding can be chosen differently for each subcarrier, and it can also change with time. It can be shown that – at least theoretically – systems using adaptive modulation perform better than systems whose modulation and coding are fixed once and for all.

Let us compare adaptive modulation for OFDM with coded OFDM and multicarrier CDMA. Both of the latter systems try to "smear" the data symbols over many subcarriers, so that each symbol sees approximately the same average SNR. For adaptive modulation, we not only accept the fact that the channel (and thus the SNR) shows strong variations, but even exploit this fact. On subcarriers with good SNR, transmission is done at a higher rate than on subcarriers with low SNR. In other words, adaptive modulation selects a modulation scheme and code rate according to the channel quality of a specific subcarrier.

19.8.1 Channel quality estimation

Adaptive modulation requires that the transmitter knows channel-state information. This requirement sounds trivial, but is quite difficult to realize in practice. It requires the channel to be reciprocal – e.g., the Base Station (BS) learns the channel state while it is in receiving mode; when it then transmits, it relies on the fact that the channel is still in the same state. This can only be fulfilled in systems with Time Domain Duplex (TDD) in slowly time-varying channels. Alternatively, feedback from the receiver to the transmitter can be used (see also Sections 17.5 and 20.1.6). It is also noteworthy that the transmitter has to know the channel-state information *for the time instant when it will transmit* – in other words, it has to look into the future. Channel prediction is thus an important component for many adaptive modulation systems.

19.8.2 Parameter adaptation

Once the channel state is known, the transmitter has to decide how to select the correct transmission parameter for each subcarrier – namely, coding rate and modulation alphabet. Furthermore, we also have to consider how much power should be assigned to each channel. In the following, we will assume that the channel is frequency-selective, but time-invariant. This will make the discussion easier, and the principles can be easily extended to doubly selective channels.

Let us first consider the question of how much power should be assigned to each channel. In order to find the answer, let us first reformulate the question in a more abstract way: "Given a number of parallel subchannels with different attenuations, what is the distribution of transmission power that maximizes capacity?" The answer to this latter question was given by

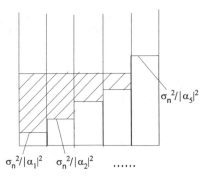

Figure 19.16 Principle behind waterfilling.

Shannon in the 1940s, and is known as "waterfilling". Power allocation P_n of the nth subchannel is:

$$P_n = \max\left(0, \varepsilon - \frac{\sigma_n^2}{|\alpha_n|^2}\right) \tag{19.31}$$

where α_n is the gain (inverse attenuation) of the nth subchannel, σ_n^2 is noise variance, and the threshold ε is determined by the constraint of the total transmitted power P as:

$$P = \sum_{n=1}^{N} P_n \tag{19.32}$$

Waterfilling can be interpreted visually, according to Fig. 19.16. Imagine a number of connected vessels. At the bottom of each vessel is a block of concrete with a height that is proportional to the inverse SNR of the subchannel that we considering. Then take water, and pour it into the vessels; the amount of poured water is proportional to the total transmit power that is available. Because the vessels are connected, the surface level of the water is guaranteed to be the same in all vessels. The amount of power assigned to each subchannel is then the amount of water in the vessel corresponding to this subchannel. Obviously, subchannel 1, which has the highest SNR, has the most water in it. It can also happen that some subchannels that have a poor SNR (like channel 5), do not get any power assigned to them at all. Essentially, waterfilling makes sure that energy is not wasted on subchannels that have poor SNR: in the OFDM context, this means not wasting power on subcarriers that are in a deep fade.

With waterfilling, power is allocated preferably to subchannels that have a good SNR ("give to the rich" principle). This is optimum from the point of view of theoretical capacity; however, it requires that the transmitter can actually make use of the large capacity on good subchannels.

In each subchannel (subcarrier), signaling as close to capacity as possible should be performed.[4] This means that the transmitter has to adapt the data rate according to the SNR that is available (note that waterfilling increases SNR differences between subcarriers). Consequently, the coding rate as well as the constellation size of the modulation alphabet have to be adjusted. For very high SNR, the constellation size, and thus the PAR, has to be very large. A Quadrature Amplitude Modulation (QAM) of alphabet size 64 currently seems to be the largest constellation size that can be used in practical systems, though 256-QAM has been suggested for some special applications. The capacity per subchannel is limited by $\log_2(N_a)$, where N_a is the size of the symbol alphabet. It is thus wasteful to assign more energy to one

---

[4] We assume in the following that near-capacity-achieving codes are used. If this is not the case, we usually try to choose the data rate in such a way that a certain BER can be guaranteed.

stream than can be actually exploited by the alphabet. If the available alphabet is small, a "giving to the poor" principle for power allocation is preferable – i.e., assigning power that cannot be exploited by good subchannels to bad subchannels.

Example 19.3 *Waterfilling: consider an OFDM system with three tones, with $\sigma_n^2 = 1$, $\alpha_n^2 = 1, 0.4, 0.1$, and total power $\sum P_n = 15$. Compute the power assigned to different tones according to (i) waterfilling, (ii) equal power allocation, (iii) predistortion (inverting the channel attenuation), and compute the resulting capacity.*

From Eq. (19.31) we find that $\varepsilon = 9.25$ gives the correct solution: in that case, the powers in the different subchannels are:

$$P_1 = 8.25 \tag{19.33}$$

$$P_2 = 6.75 \tag{19.34}$$

$$P_3 = 0 \tag{19.35}$$

that is, no power is assigned to the channel that suffers from the strongest attenuation. The total capacity can be computed as:

$$C_{\text{waterfill}} = \sum_{n=1}^{N} \log_2(1 + \alpha_n^2 P_n / \sigma_n^2) = 5.1 \text{ bit/s/Hz} \tag{19.36}$$

For equal power allocation:

$$P_1 = P_2 = P_3 = 5 \tag{19.37}$$

so that capacity becomes:

$$C_{\text{equal-power}} = 4.8 \text{ bit/s/Hz} \tag{19.38}$$

For the predistortion case, the powers become:

$$P_1 = 1.1 \tag{19.39}$$

$$P_2 = 2.8 \tag{19.40}$$

$$P_3 = 11.1 \tag{19.41}$$

from which we obtain a capacity of:

$$C_{\text{predistort}} = 3.2 \text{ bit/s/Hz} \tag{19.42}$$

In this example, equal power allocation gives almost as high a capacity as (optimum) waterfilling, while predistortion leads to significant capacity loss.

19.8.3 Signaling of chosen parameters

Most information-theoretic investigations assume a continuum of modulation alphabets that can realize any arbitrary transmission rate. In practice, the transmitter has a finite and discrete set of modulation alphabets available (BPSK, QPSK, 16-QAM, and 64-QAM). There is also a finite set of possible code rates: the different codes are usually obtained from a "mother" code by different amounts of puncturing. Thus, the available data rates form a discrete set.

After the transmitter has decided which transmission mode – i.e., combination of signal constellation and encoder – to use on each tone, it has to communicate that decision to the receiver. There are three possibilities to achieve that task:

- *Explicit transmission*: the transmitter can send, in a predefined and robust format, the index of the transmission mode it intends to use. Transmission of this information itself should always be done in the same mode, and care should be taken that the message is well-protected against errors during transmission.

- *Implicit transmission*: implicit transmission is possible when the transmitter gets its channel-state information from the receiver via feedback. In such a case, the receiver knows exactly what channel-state information is available to the transmitter, and thus the basis on which the decision for a transmission mode is being made. Thus, the receiver just needs to know the decision rule on which the transmitter bases its choice of transmission mode. If the receiver feeds back the mode that the transmitter should use, the situation is even simpler.

  The drawback to this method is that errors in channel-state feedback (from the receiver to the transmitter) not only lead to a wrong choice of transmission mode (which is bad, but usually not fatal), but also to detection and decoding using the wrong code, which leads to very high error rates.

- *Blind detection*: from the received signal, the receiver can try to determine the signal constellation. This can be achieved by considering different statistical properties of the received signal, including the PAR, autocorrelation functions, and higher order statistics of the signal.

19.9 Multiple access

OFDM is a modulation format that allows the transmission of high data rates for a single user; it is usually not seen as a multiaccess format. However, it can be used as such, assigning different subcarriers to different users. Especially, each user could be assigned several subcarriers that are not adjacent to each other, and thus provide considerable frequency diversity. However, such a strategy also raises a lot of problems:

- the administrative effort of such an assignment is large;
- each of the users "sees" different channels, destroying orthogonality between the signals from different users;
- in the uplink, signals from different users have different runtimes, and thus do not arrive at the receiver in a synchronized way.

Instead, it is more common to combine OFDM with TDMA or packet radio as the modulation format. In that way, full-frequency diversity is achieved, while still retaining orthogonality between the subcarriers. This scheme is used, for example, in the IEEE 802.11a standard that will be discussed in Chapter 25. Alternatively, we can also combine OFDM with FDMA, where we assign each user a group of adjacent tones, separated by frequency guard spaces.

19.10 Multicarrier code division multiple access

Multi Carrier CDMA (MC-CDMA) spreads information from each data symbol over all tones of an OFDM symbol. At first glance, it is paradoxical to combine CDMA, which tries to spread a signal over a very large bandwidth, with multicarrier schemes, which try to signal over a very narrowband channel. But we will see in the following that the two methods can actually be combined very efficiently. We have mentioned repeatedly that uncoded OFDM has poor performance, because it is dominated by the high error rate of subcarriers that are in fading dips. Coding improves the situation, but in many cases a low coding rate – i.e., high redundancy – is not desirable. We thus need to find an alternative way of exploiting the frequency diversity of the channel. By spreading a modulation symbol over many tones, MC-CDMA becomes less sensitive to fading on one specific tone.

The basic idea of MC-CDMA is to transmit a data symbol on all available subcarriers simultaneously. In other words, a code symbol c is mapped to a vector $c\mathbf{p}$, where $\mathbf{p}$ is a predetermined vector. This can be interpreted as a repetition code (a symbol is repeated on each tone, but multiplied by a different known constant p_n), or as a spreading action, where each symbol is represented by a code sequence (but the sequence is now in the frequency domain, instead of the time domain). Irrespective of the interpretation, we obtain a bandwidth expansion by a factor of N.

Bandwidth expansion leads to a loss of spectral efficiency. However, we can eliminate this problem by transmitting N different symbols, and thus N different codevectors, simultaneously. The first symbol c_1 is multiplied by codevector $\mathbf{p}_1$, the second symbol c_2 by codevector $\mathbf{p}_2$, and so on. If all the vectors $\mathbf{p}_n$ are chosen to be orthogonal, then the receiver can recover the different transmitted symbols.

Let us now put this into a more compact mathematical form by writing the code vectors into a "spreading matrix" $\mathbf{P}$:

$$\mathbf{P} = [\mathbf{p}_1 \quad \mathbf{p}_2 \quad \cdots \quad \mathbf{p}_N] \tag{19.43}$$

where we will see later that it is advantageous if $\mathbf{P}$ is unitary. The *symbol spreader* performs a matrix multiplication:

$$\tilde{\mathbf{c}} = \mathbf{P}\mathbf{c} \tag{19.44}$$

Now it is this modified signal that is OFDM-modulated – i.e., undergoes an IFFT and has the cyclic prefix prepended – and sent over the wireless channel.

In the receiver, we first perform the same operation as in "normal" OFDM: stripping off the CP, and performing an FFT. Thus, we have again transformed the symbols into the frequency domain. At this point, the received symbols are:

$$\tilde{\mathbf{r}} = \mathbf{H}\tilde{\mathbf{c}} + \mathbf{n} \tag{19.45}$$

where $\mathbf{H}$ is a diagonal matrix with entries $H\left(n\dfrac{W}{N}\right)$ along the diagonal. The next step is "one-tap equalization". Let us assume for the moment that we use zero-forcing equalization (we will see below why this is more relevant in MC-CDMA than in conventional OFDM). Then employing the unitary properties of the spreading matrix, we can perform despreading by just multiplying the received signal by the Hermitian transpose of the spreading matrix to obtain:

$$\mathbf{P}^\dagger \mathbf{H}^{-1} \tilde{\mathbf{r}} = \mathbf{P}^\dagger \mathbf{H}^{-1} \mathbf{H}\mathbf{P}\mathbf{c} + \mathbf{P}^\dagger \mathbf{H}^{-1} \mathbf{n} \tag{19.46}$$

$$= \mathbf{c} + \tilde{\mathbf{n}} \tag{19.47}$$

We have thus recovered the transmit symbols. A summary of the transceiver structure can be found in Fig. 19.17.

Note that noise is no longer white, since the $\tilde{\mathbf{n}}$ includes noise enhancement from zero-forcing equalization. When the receiver uses MMSE equalization instead of zero-forcing, then noise enhancement is not as bad. However, MMSE equalization does not recover the orthogonality of different codewords the way that zero-forcing does. After equalization and despreading, we get:

$$\mathbf{P}^\dagger \frac{\mathbf{H}^*}{|\mathbf{H}|^2 + \sigma_n^2} \tilde{\mathbf{r}} = \mathbf{P}^\dagger \frac{\mathbf{H}^*\mathbf{H}}{|\mathbf{H}|^2 + \sigma_n^2} \mathbf{P}\mathbf{c} + \mathbf{P}^\dagger \frac{\mathbf{H}^*}{|\mathbf{H}|^2 + \sigma_n^2} \mathbf{n} \tag{19.48}$$

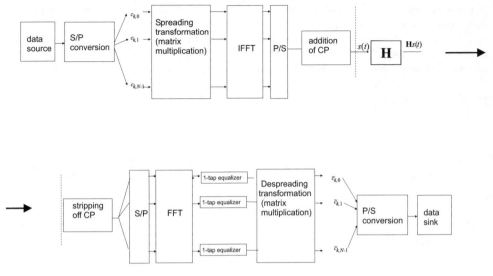

Figure 19.17   Block diagram of a multicarrier code-division-multiple-access transceiver.

The matrix $\mathbf{P}^{\dagger}\mathbf{H}^{*}\mathbf{H}/(|\mathbf{H}|^{2} + \sigma_{n}^{2})\mathbf{P}$ is not diagonal. This means that there is residual crosstalk from one codeword to the other.

What spreading matrices should be used for MC-CDMA systems? One obvious choice is the Walsh–Hadamard matrices that we discussed in Section 18.2. They are unitary, and have an extremely simple structure. All the coefficients are ± 1; furthermore, Walsh–Hadamard matrices of larger size can be computed via recursion from matrices of smaller size. This allows implementation of a Walsh–Hadamard transform with a "butterfly" structure, similar to implementation of an FFT.

How does spreading influence the PAR problem? In most cases, there is no significant impact. The explanation can again be found from the central limit theorem: the output (amplitudes of the I- and Q-components on different subcarriers) of the spreading matrix is approximately Gaussian-distributed, as it is the sum of a large number of variables. The IFFT then just weights and sums those Gaussian variables. But the weighted sum of Gaussian variables is again a Gaussian variable. The amplitude distribution of the transmit signal of MC-CDMA is thus the same as the amplitude distribution of normal OFDM.

19.11 Single-carrier modulation with frequency-domain equalization

A special case of MC-CDMA occurs when the unitary transformation matrix $\mathbf{P}$ is chosen to be the FFT matrix. In such a case, multiplication by the spreading matrix and the IFFT inherent in the OFDM implementation cancel out. In other words, the transmit sequence that is transmitted over the channel is the original data sequence – plus a CP that is just a prepending of a few data symbols at the beginning of each datablock.

This just seems like a rather contrived way of describing the single-carrier system that has already been discussed in Part III. However, the big difference here lies in the existence of the CP, as well as in how the signal is processed at the receiver (see Fig. 19.18). After stripping off the CP, the signal is transformed by an FFT into the frequency domain. Due to the CP, there are no residual effects of ISI or ICI. Then, the receiver performs equalization on each subcarrier (this can be zero-forcing or MMSE equalization), and finally transforms the signal back into the time domain via an IFFT (this is the despreading step of MC-CDMA). The receiver thus performs equalization in the frequency domain. Since FFTs or IFFTs can be implemented efficiently, the computational effort (per bit) for equalization goes only like $\log_2(N)$. This is a considerable advantage compared with the equalization techniques discussed in Chapter 16. On the downside, frequency domain equalization is a linear equalization scheme and therefore does not give optimum performance. Also, the extra overhead of a CP has to be taken into account.

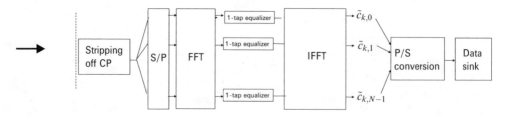

Figure 19.18   Block diagram of a single-carrier frequency-domain-equalization receiver.

Further reading

Several books describe multicarrier schemes, including OFDM and multicarrier CDMA. The most comprehensive account is Hanzo et al. [2003]; another interesting description that also covers applications to various standardized systems is Bahai et al. [2004]. Review papers on that topic include Wang and Giannakis [2000].

OFDM was invented by Chang [1966], and the CP was proposed in Weinstein and Ebert [1971]. A discussion of the CP and its comparison to zero-padding can be found in Muquet et al. [2002]. The performance of a coded OFDM system in a multipath channel is analyzed, for example, in Kim et al. [1999]. ICI is discussed in Cai and Giannakis [2003] and Choi et al. [2001]; the latter also discussed ICI mitigation techniques, as does a number of other papers (see, e.g., Schniter [2004] for one scheme and a review of other papers).

Synchronization and channel estimation are very important topics. Some examples of time and frequency synchronization include Schmidl and Cox [1997], Speth et al. [1999], and van de Beek et al. [1999]. Channel estimation techniques are discussed in Edfors et al. [1998] and Li et al. [1998]. Approximate DFT estimators were introduced in van de Beek et al. [1995] and later analyzed in detail in Edfors et al. [2000]. Filtering after eigentransformation was introduced in Li et al. [1998] and extended to the transmit diversity case in Li et al. [1999]. Methods for PAR techniques are reviewed in May and Rohling [2000]. Multiaccess considerations are described in Rohling and Galda [2005].

For adaptive modulation, the paper by Wong et al. [1999], and especially the overview paper by Keller and Hanzo [2000], give a good account. Yang and Hanzo [2003] review MC-CDMA. Falconer et al. [2002] discuss single-carrier frequency equalization schemes; a combination of this scheme with diversity is detailed in Clark [1998].

20

Multiantenna systems

Since the 1990s, there has been enormous interest in multiantenna systems. As spectrum became a more and more precious resource, researchers investigated ways of improving the capacity of wireless systems without actually increasing the required spectrum. Multiantenna systems offer such a possibility.

When discussing multiantenna systems, we distinguish between *smart antenna systems* (systems with multiantenna elements at one link end only), and *Multiple Input Multiple Output* (MIMO) systems}, which have multiantenna elements at both link ends. Both of these systems will be discussed in this chapter.

20.1 Smart antennas

20.1.1 What are smart antennas?

Let us start out with a general definition: smart antennas are "antennas with multiple elements, where signals from different elements are combined by an adaptive (intelligent) algorithm." This definition is applicable when the smart antenna is at the receiver (RX); for the transmit case, the equivalent statement would use "created" instead of "combined". In most practical situations, smart antennas are located at the BS, and this is also the case we assume in this section, unless otherwise specified.

Intelligence (smartness) is not in the antenna, but rather in signal processing. In the simplest case, combination of antenna signals is a linear combination using a weight vector $\mathbf{w}$. The ways of determining $\mathbf{w}$ essentially differentiates smart antenna systems. We see immediately that there is a strong relationship between multiantenna systems and diversity systems – as a matter of fact, an RX with antenna diversity *is* a smart antenna. We will thus reuse many results from Chapter 13, while concentrating more on system aspects and the impact smart antennas have on multiple access.

The definition of smart antennas as a combiner of different antenna signals stresses the fact that we exploit signals from different *spatial locations*. Alternatively, we can also say that smart antennas exploit the *directional properties of the channel*. Remember from Chapters 8 and 13 that an RX with multiantenna elements can distinguish Multi Path Components (MPCs) with different Directions Of Arrival (DOAs). Thus, one way to interpret smart antennas is as a spatial Rake[1] RX that distinguishes between MPCs with different DOAs, and processes them

---

[1] A device that enables CDMA systems to deal with a channel's delay dispersion (see Section 18.2).

Wireless Communications Andreas F. Molisch
© 2005 John Wiley & Sons, Ltd

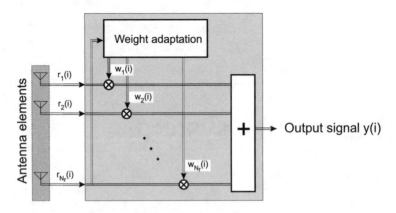

Figure 20.1 Linear combination of antenna signals.

separately. This allows the RX to coherently add up the different MPCs, and thus reduce fading; MPCs from interferers can also be suppressed.

Yet another interpretation is that the smart antenna adaptively forms a beam pattern that enhances (coherently combines) the desired MPCs, while suppressing contributions from interferers.

20.1.2 Purpose

Smart antennas can be used for various purposes:

1. *Increase of coverage*: assume that the smart antenna is at the RX. Now, if the spatial (angular) position of the transmitter (TX) is known, the RX can form an antenna pattern in the direction of the TX (beamforming). This results in higher receive power – e.g., an antenna array with eight elements can increase the Signal-to-Noise Ratio (SNR) by 9 dB compared with a single antenna. In a noise-limited cellular system an improvement in the SNR increases the area that can be covered by one Base Station (BS). Conversely, the coverage range can be kept constant while the transmit power is decreased.

2. *Increase of capacity*: smart antennas can increase the Signal-to-Interference Ratio (SIR), which in turn allows the number of users in the system to be increased. This is the practically most important advantage of smart antennas, and will be discussed in more detail below.

3. *Improvement of link quality*: by increasing signal power and/or decreasing interference power, we can also increase the transmission quality on each single link.

4. *Decrease of delay dispersion*: by suppressing MPCs with large delays, delay dispersion can be reduced. This feature can be especially useful in systems with a very high data rate.

5. *Improvement of user position estimation*: knowledge of the DOAs, especially for the (quasi-) line-of-sight component, improves geolocation. This is of value both for *location-based services* and for the ability to locate users in case of emergency.

It is *not* possible to have all these advantages simultaneously at their fullest extent. For example, we can use the capability of smart antennas to reduce interference in order to *either* improve the quality of a single link *or* have more users in the system or to *slightly* increase both the quality per link and the number of users. For system design, the engineer has to decide which aspect is the most important.

20.1.3 Capacity increase

As mentioned above, increasing the capacity is the most important application of smart antennas. Depending on whether the considered system uses Time Division Multiple Access (TDMA), Frequency Division Multiple Access (FDMA), or Code Division Multiple Access (CDMA), there are different approaches for achieving such capacity gains:

1. *Spatial Filtering for Interference Reduction* (SFIR): this is used in TDMA/FDMA systems to reduce the reuse distance. A conventional TDMA/FDMA system cannot reuse the same frequency in each neighboring cell, since the interference from the adjacent cells would be too strong (Section 17.6). Rather, there is a "cluster" of cells where each cell uses a different frequency, and then a neighboring cluster uses the same set of frequencies. Cluster size is a measure for how spectrally efficient a cellular network is: for Global System for Mobile communications (GSM)-like systems, it is usually between 3 and 7. Smart antennas reduce interference. For this reason we can put cells with the same frequency closer together – in other words, reduce the cluster size. It is immediately obvious that this leads to an improvement in total spectral efficiency: when frequencies can be used in more cells, then the number of users per area is increased proportionately. Simulations have shown that 8 antennas increase capacity increases by a factor of 3 [Kuchar et al. 1997], while field tests with prototypes [Dam et al. 1999] have shown a somewhat smaller gain of approximately 2.

2. *Space Division Multiple Access* (SDMA): this is an alternative way of increasing the capacity of TDMA/FDMA systems. In this method, cluster size (frequency reuse) remains unchanged, while the number of users within a given cell is increased. Multiple users can be served on the *same* time/frequency slot, because the BS distinguishes them by means of their different spatial signatures. To understand this mechanism, imagine a situation where there are multidirectional horn antennas at the BS, each of which can be pointed mechanically at a different user, independently of each other. An actual system realizes these directional antennas as the different beams of an antenna array (remember that an antenna array can form multiple beams simultaneously).

   Not only can antenna arrays form maxima in antenna patterns, but also notches. The latter is important for suppressing co-channel interference. This capability is vital in an SDMA system: if no spatial interference suppression is used, the SIR (in dB) could easily become negative (remember that both the "user" and the "interferer" are users in the cell).

   The capacity gain that can be achieved with SDMA is larger than with SFIR; however, the required modifications within a system, especially at the BS and the BS controller software, are considerably larger.

3. *Capacity increase in CDMA systems*: smart antennas can also be used to increase capacity in CDMA systems. The way this capacity increase is achieved is subtly different from TDMA/FDMA systems. In a CDMA system, all Mobile Stations (MSs) use the same carrier frequency, and are distinguished only by the spreading codes they employ. The suppression of users with other codes is not perfect, and a cell is judged to be "full" when the residual interference from all the other K users becomes comparable with the admissible SIR (refer to Chapter 18 for more details and a discussion of the assumptions):

$$SIR_{\text{threshold}} = \frac{P_{\text{desired}}}{\sum P_{\text{interfere}}} \qquad (20.1)$$

where, for the case of perfect power control, $\sum P_{\text{interfere}} = K \cdot P_{\text{interfere}}$. For typical (voice) CDMA systems, the number of users is on the order of 30 per cell. Thus, there is a large

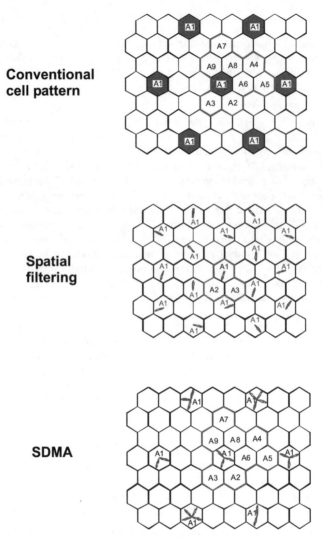

Figure 20.2 Principle behind spatial filtering for interference reduction, and space division multiple access.

number of possible interferers, each of which contributes only a small part of the interference. It is thus *not* possible to suppress interference, since the maximum number of interferers that can be suppressed is smaller than the number of antenna elements. Rather, the goal of the smart antenna is to enhance the received signal power. An N_r-element antenna can increase the received signal power by a factor of N_r, so that approximately:

$$P_{\text{desired}} = M_c \cdot N_r \cdot P_{\text{interfere}} \tag{20.2}$$

where M_c is the spreading gain. In that case, the number of admissible users in the cell becomes:

$$K = \frac{M_c \cdot N_r}{SIR_{\text{threshold}}} \tag{20.3}$$

and thus increases *linearly* with the number of antenna elements. Note that this derivation is based on numerous simplifying assumptions, especially with respect to the channel model. In practice, such an enormous increase in capacity does not occur; typically, capacity can be doubled with an eight-element array. This is comparable with the gains that can be achieved in a TDMA system.

4. *Capacity increase in third-generation CDMA systems*: the above considerations assumed that there is a large number of users in a CDMA system, and that all of them contribute to interference in a similar way. This assumption holds for second-generation CDMA systems – e.g., IS-95[2] – which are mainly tuned to voice users. However, in third-generation networks, high-rate data transmission is considered an important application. Such a high-rate user creates considerably more interference due to their lower spreading factor, such that suppression of such a user by placing a null in its direction becomes desirable. In other words, high-rate users lead to scenarios and results that are more similar to the SFIR approach than the traditional CDMA approach.

Example 20.1 *Consider a CDMA system with spreading factor 128 for voice users, and 4 for data users, and an SIR threshold of 9 dB. There is one data user. How many voice users can be accommodated without smart antennas? How many can be accommodated with a two-element smart antenna?*

The SIR threshold (on a linear scale) is 8. Without smart antennas, the capacity of the system (in units of voice users) can be obtained from:

$$K = \frac{128}{8} = 16 \qquad (20.4)$$

Since the spreading factor for data users is smaller by a factor of 32 compared with voice users, each data user contributes (approximately) 32 times the interference. The data user thus requires a capacity of $K_{\text{data}} = 32$, so that $16 - 32 = 0$ voice users can be accommodated.

Consider now the case with smart antennas. When smart antennas are used solely for enhancement of the desired signal power, the capacity of the cell (in units of voice users) becomes:

$$K = \frac{128 \cdot 2}{8} = 32 \qquad (20.5)$$

so that now $32 - 32 = 0$ voice users can be serviced.

If, however, smart antennas are used to suppress data users, higher capacity can be achieved: as we saw in Chapter 13, we can suppress 1 interferer completely using 2 antennas, while retaining a diversity order of 1 for the desired signal. Thus, in our example, we can suppress the data user while retaining $N_r = 1$ for voice users. Thus, the capacity for voice users becomes:

$$\frac{128}{8} = 16 \qquad (20.6)$$

20.1.4 Receiver structures

In this section, we will describe RX structures that can be used for separation and processing of MPCs. Since our emphasis is on smart antennas at the BS, this implies that we consider the

---

[2] See Chapter 22.

uplink case (see Section 20.1.6 for the downlink). We distinguish between *switched-beam antennas*, *adaptive spatial processing*, *space–time processing*, and *space–time detection*.

Switched-beam antennas

A switched-beam antenna is an antenna array that can form just a small set of patterns – i.e., beams pointing in certain discrete directions. A switch then selects one of the possible beams[3] for downconversion and further processing; it selects the beam that is *best* in the sense that it gives the highest SNR or the highest Signal-to-Interference-and-Noise Ratio (SINR). This approach greatly simplifies RX design.

There are many different ways of realizing switchable beams. For example, the antenna can contain multiple directional elements, oriented in different directions, and the switch selects from them. The most popular variant is a linear array followed by a spatial Fourier transformation. The output of such a spatial Fourier transformation is a number of beams that are all orthogonal to each other, and point in different directions. The spatial Fourier transformation can be realized as a so-called *Butler matrix*. It has a structure that is essentially the butterfly structure that is well-known from software implementations of the Fast Fourier Transform (FFT). The elements of each stage are simple phaseshifters that can be realized in the Radio Frequency (RF) domain.

The main advantage of the switched-beam approach is its simplicity: all processing (spatial FFT and selection) is done in the RF domain, so that only a single signal has to be downconverted to the baseband and processed there. Since downconversion circuits are among the most expensive components in today's wireless systems, this is a significant advantage. The drawback of this scheme is its limited flexibility. The main beam can only be pointed in certain fixed directions, such that gain in the actual direction of the MS might not be the maximum achievable value. Even more importantly, nulls cannot be pointed in arbitrary directions, so that nulling of interferers is basically impossible. For these reasons, switched-beam antennas seem to be more suitable for CDMA applications, where signal enhancement is critical, and not for SDMA or SFIR applications, where interference suppression is essential.

Example 20.2 *For a switched-beam antenna, compute the maximum gain, and the maximum relative loss due to mismatch of DOA and beam direction for a uniform linear array with Butler matrix and number of antenna elements (spaced $\lambda/2$ apart) equal to $N_r = 2, 4, 8$.*

Let us first write down the FFT matrix for N_r elements:

$$W_{\text{FFT},k,n} = w_n(k) = \exp\left(-j\frac{2\pi}{N_r}\right)^{kn} \tag{20.7}$$

The array factor for a uniform linear array was given in Eq. (9.16):

$$M_k(\phi) = \sum_{n=0}^{N_r-1} w_n(k) \exp[-j\cos(\phi)2\pi d_a n/\lambda] \tag{20.8}$$

With $d_a = \lambda/2$, this becomes:

$$M_k(\phi) = \sum_{n=0}^{N_r-1} \exp\left[-j2\pi n\left(\frac{\cos(\phi)}{2} + \frac{k}{N_r}\right)\right] \tag{20.9}$$

---

[3] More precisely, it selects the signal associated with one of the possible beams.

and the magnitude of the array factor is (compare Eq. 9.17):

$$|M_k(\phi)| = \left| \frac{\sin\left[\dfrac{N_r}{2}\left(\pi\cos\phi - 2\pi\dfrac{k}{N_r}\right)\right]}{\sin\left[\dfrac{1}{2}\left(\pi\cos\phi - 2\pi\dfrac{k}{N_r}\right)\right]} \right| \qquad (20.10)$$

The crossover points between patterns at FFT output k and $k+1$ are at:

$$\left| \frac{\sin\left[\dfrac{N_r}{2}\left(\pi\cos\phi - 2\pi\dfrac{k}{N_r}\right)\right]}{\sin\left[\dfrac{1}{2}\left(\pi\cos\phi - 2\pi\dfrac{k}{N_r}\right)\right]} \right| = \left| \frac{\sin\left[\dfrac{N_r}{2}\left(\pi\cos\phi - 2\pi\dfrac{k+1}{N_r}\right)\right]}{\sin\left[\dfrac{1}{2}\left(\pi\cos\phi - 2\pi\dfrac{k+1}{N_r}\right)\right]} \right| \qquad (20.11)$$

Figure 20.3 shows the array factor for different values of N_r. We see that in all cases the value of the array factor of the crossover point is approximately 0.7. The exact values can be found in Table 20.1:

Table 20.1

N_r	Max gain (dB)	Max relative loss due to direction-of-arrival mismatch (dB)
2	3.01	1.51
4	6.02	1.84
8	9.03	1.93

Adaptive spatial processing

The adaptive-spatial-processing approach uses a linear combination of signals (see Fig. 20.1), where antenna weighting and summing is done in the baseband. For this reason, there are no restrictions with respect to weights, and they can also be adapted according to the current channel state. On the downside, this approach requires in general N_r complete downconversion chains for N_r antenna elements, and is thus considerably more expensive, and consumes more power as well.

If weights are adapted based on a training sequence, and change whenever the channel realization changes, then adaptive spatial processing is identical to diversity combining. It is thus possible to suppress $K = N_r - 1$ interferers, or alternatively to achieve high gain in the SNR (up to a factor of N_r) with such an approach (for more details, see Section 13.4.3).

Alternatively, antenna weights can be chosen based on angular spectrum (ADPS, see Chapter 7), but not adapted to the instantaneous amplitude of MPCs arriving from different directions. In this case, interference suppression capabilities are usually smaller. This aspect is discussed in more detail in Section 20.1.5.

Adaptive space–time processing

The adaptive spatial processing described in the previous section first performs spatial processing and then puts out a single signal for further processing. In other words, Rake reception or equalization is done only on the *single* signal that is put out from the spatial processor. This approach works well only if the temporal properties of the signal (delay dispersion) are the same at all antenna elements, or, conversely, if spatial properties are independent of delay. While this situation can occur in some cases (see also Chapter 7 for channel models that correspond to such an assumption), it is certainly not universally valid. If delay and angular properties cannot be

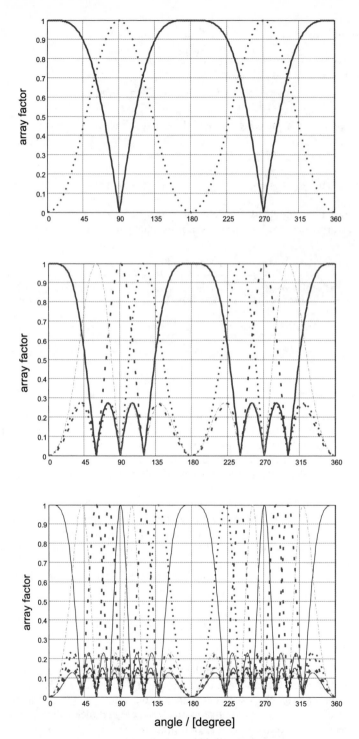

Figure 20.3 Normalized array factor of switched-beam antenna with 2 (top), 4 (middle), and 8 (bottom) elements.

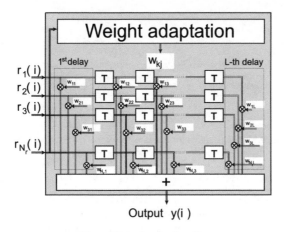

Figure 20.4 Space–time filter.

decomposed multiplicatively, then spatial processing followed by temporal processing is no longer optimum.

In order to fully exploit the possibilities of different MPCs, adaptive *space–time processing* has to be used. The optimum linear RX is a two-dimensional Rake receiver – i.e., a linear combiner that weights all *resolvable* (in the space–time domain) MPCs and adds them up. Figure 20.4 shows such a space–time processor. If the Rake receiver (or equalizer) has L taps, then the total processor has $N_r L$ weights. The output from that processor is sent on to a decoder.

Space–time detection

For optimum reception, space–time processing and decoding/detection have to be done jointly. The special properties of the received signal, including finite-alphabet properties, can be taken into account for detection. The optimum detector is a generalized Maximum Likelihood Sequence Estimation (MLSE) RX. However, these structures are rarely used in practice, because they are too complex, and the performance gain compared with linear processing does not justify the additional effort.

20.1.5 Algorithms for adaptation of antenna weights

Algorithms for adaptation of antenna weights can be broadly classified into spatial reference and temporal reference algorithms.[4] The distinction between these algorithms is subtle, but important. Let us first discuss their common features: both algorithms are based on *linear* weighting and addition of signals from different antenna elements. Both algorithms effectively form a beam pattern – the weights for any linear combining scheme can be associated with a "beam pattern" even when it sometimes looks somewhat strange, with many maxima and minima. The key difference is which information is used for the choice of the linear weights. In spatial reference algorithms weights are chosen according to the spatial structure (DOAs) of the arriving signal combined with information about the array structure. Temporal reference algorithms optimize the SINR after the combiner with the help of a training sequence.

---

[4] In the literature, there is often a distinction between diversity and beamforming algorithms. However, this distinction is problematic as diversity (slope of the effective SNR distribution) and beamforming (change in mean SNR) are effects that, among others, depend on antenna arrangement and channel constellation.

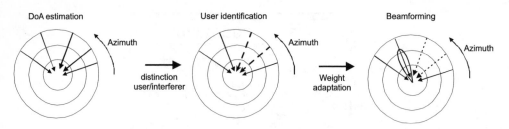

Figure 20.5 Using smart antennas with spatial reference algorithms.

Spatial reference algorithms

In the *Spatial Reference* (SR) case, the antenna tries to form a beam pattern that puts maxima in the direction of the main DOAs of the desired user, and nulls in the direction of MPCs coming from interferers. An SR algorithm thus proceeds in the following way (see Fig. 20.5):

1. In a first step, it determines the main DOAs, ϕ_n, of the MPCs. As we will see below, a main advantage of using these DOAs is that they show only small variations over time and frequency. The directly observable quantities are the signals at the antenna elements; it is thus necessary to extract the DOAs from these signals. This can be done by the same methods as those for channel sounding (see Chapter 8): spatial Fourier transformations (not recommended) or high-resolution algorithms (MVM, ESPRIT, SAGE) can be applied. There are some important practical differences from the evaluation of channel sounder data: the SNR for the reception of data can be considerably worse than that used for channel sounding – this makes DOA determination more difficult. On the other hand, a lot of data are transmitted within a relatively short time. This allows the use of either averaging or tracking algorithms that observe the change in DOAs [Kuchar et al. 2002]. Note that some DOA estimation algorithms (e.g., ESPRIT) do not require a training sequence, since they are based exclusively on the correlation matrix of the received signal, and can thus estimate the DOA from user data that are unknown. Other algorithms, like SAGE, can eschew the training sequence only at the price of greatly complicating the algorithm.[5]
2. Association of DOAs with specific users. In contrast to channel sounding, where we have just a single possible source for the arriving signal, MPCs during communications can stem from different users – the desired user as well as interferers. It is thus necessary to separate desired and interfering users. For this identification problem, it is necessary to have a training sequence, or some other property that is unique for each user.
3. After the DOAs for user and interferer are established, it is possible to form the beam pattern that maximizes the SINR. This beam pattern is used for actual reception of data.

We note that many DOA estimation algorithms require the validity of a certain channel model (e.g., that the received signal can be written as a finite sum of plane waves). If this model is not valid, the performance of the algorithms, and of the ensuing beamforming, deteriorates.

Temporal reference algorithms

Temporal Reference (TR) algorithms are based on the use of a training signal, which serves as a "temporal reference" to which we try to match the output from the smart antenna – hence the name. Antenna weights **w** are adapted directly in such a way that deviation of combiner output

---

[5] However, when SAGE is used in conjunction with a training sequence, it usually outperforms ESPRIT.

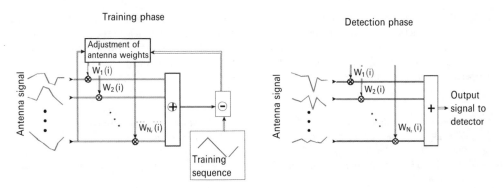

Figure 20.6 Principle behind temporal reference algorithms.

from the (known) training sequence is minimized. As a criterion for deviation we can use the SIR, the minimum mean square error, the Bit Error Rate (BER) during the training sequence, or any other suitable criterion. These criteria and their advantages and disadvantages have already been discussed in Chapter 13 (we stress again that smart antennas with diversity combining at the RX is the "standard" diversity that we discussed in Chapter 13). Linear weights lead to creation of an effective antenna pattern, but this pattern need not have an intuitive interpretation. There are also no assumptions about the existence of discrete DOAs.

A major difficulty with TR algorithms is the fact that the RX must be synchronized to the incoming signal *before* a determination of antenna weights can be done. This is required because the sampling instants for the determination of the training sequence have to be known before the weights can be adapted. However, the SINR for the (spatially unfiltered) receive signal can be very poor, and make synchronization with the desired training sequence difficult.

In summary, TR algorithms proceed according to the following steps (see Fig. 20.6):

1. During a training phase, the smart antenna receives the signal at all antenna elements, and adjusts the weights in such a way that deviation from the known signal is minimized.
2. During transmission of user data, the antenna weights stay fixed, and are used to weight incoming signals before they are combined and decoded/demodulated.

Blind algorithms

TR algorithms rely on the existence of a training sequence, while SR algorithms rely on the existence of a certain spatial structure of the arriving signal. Yet another group of algorithms, dubbed *blind* algorithms, makes neither of those assumptions. Rather, they exploit the statistical properties of the transmit signal.

Let us start out by describing the received signal by the equation:

$$\mathbf{r}_i = \mathbf{h}_d s_i \tag{20.12}$$

where $\mathbf{r}_i$ is the receive signal vector created by the transmit signal s_i related to the ith bit, and $\mathbf{h}_d$ is the desired channel vector – we omit the noise contribution for convenience. A more general representation, which also includes possible intersymbol interference, is obtained by stacking signals corresponding to a large number of bits into matrices:

$$\mathbf{Y} = \mathbf{H}_{\text{stack}}\mathbf{X} \tag{20.13}$$

where the matrix $\mathbf{H}_{\text{stack}}$ is related to channel impulse responses at different antenna elements. The channel description is thus similar to the TR approach in the sense that there is no assumption about spatial structure, and the channel is characterized by impulse response (or transfer function) alone. However, a blind algorithm does *not* learn the values of $\mathbf{H}_{\text{stack}}$ from a training sequence. Rather, it tries to compute a factorization of the matrix $\mathbf{Y}$ according to Eq. (20.13), such that signal matrix $\mathbf{X}$ fulfills certain properties – namely, *known* properties of the transmit signal.

Example 20.3 *Consider a system with one transmit antenna and three receive antennas. Let the impulse response for each of the antenna elements last for two symbol durations, denoted as h_{il}, $l = 0, 1$, $i = 1, 2, 3$. For a five-symbol-long transmit signal, $x(n)$, $n = 1, \ldots, 5$, write the explicit form of $\mathbf{Y} = \mathbf{H}_{\text{stack}}\mathbf{X}$, assuming that impulse responses stay constant over the duration of the transmit signal.*

Writing the nth received symbol at the ith antenna element as $y_i(n)$, we can easily see that:

$$y_i(n) = \sum_{l=0}^{L-1} h_{i,l} x(n-l) \tag{20.14}$$

where $h_{i,l}$ is the channel impulse response to the signal at the ith antenna element in the lth channel delay tap, L is the duration of the channel impulse response, and N is the number of transmitted symbols. Let us then define a matrix $\mathbf{X}$ as [Laurila 2000]:

$$\mathbf{X} = \begin{bmatrix} x(1) & x(2) & \cdots & x(N) \\ x(0) & x(1) & \cdots & x(N-1) \\ \vdots & \vdots & \vdots & \vdots \\ x(-L+2) & x(-L+3) & \cdots & x(N-L+1) \end{bmatrix} \tag{20.15}$$

which in our case becomes:

$$\mathbf{X} = \begin{bmatrix} x(1) & x(2) & x(3) & x(4) & x(5) \\ x(0) & x(1) & x(2) & x(3) & x(4) \end{bmatrix} \tag{20.16}$$

The channel matrix gets the block form:

$$\mathbf{H}_{\text{stack}} = \begin{bmatrix} h_{1,0} & h_{1,1} & \cdots & h_{1,L-1} \\ h_{2,0} & h_{2,1} & \cdots & h_{2,L-1} \\ \vdots & \vdots & \vdots & \vdots \\ h_{N_r,0} & h_{Nl_r,1} & \cdots & h_{N_r,L-1} \end{bmatrix} \tag{20.17}$$

which in our case becomes:

$$\mathbf{H}_{\text{stack}} = \begin{bmatrix} h_{1,0} & h_{1,1} \\ h_{2,0} & h_{2,1} \\ h_{3,0} & h_{3,1} \end{bmatrix} \tag{20.18}$$

Inserting Eqs. (20.16) and (20.18) into Eq. (20.13), we get:

$$\mathbf{Y} = \begin{bmatrix} y_1(1) & y_1(2) & y_1(3) & y_1(4) & y_1(5) \\ y_2(1) & y_2(2) & y_2(3) & y_2(4) & y_2(5) \\ y_3(1) & y_3(2) & y_3(3) & y_3(4) & y_3(5) \end{bmatrix} \tag{20.19}$$

$$= \begin{bmatrix} h_{1,0}x(1) + h_{1,1}x(0) & \cdots & h_{1,0}x(5) + h_{1,1}x(4) \\ h_{2,0}x(1) + h_{2,1}x(0) & \cdots & h_{2,0}x(5) + h_{2,1}x(4) \\ h_{3,0}x(1) + h_{3,1}x(0) & \cdots & h_{3,0}x(5) + h_{3,1}x(4) \end{bmatrix} \tag{20.20}$$

which can easily be seen to be in agreement with Eq. (20.14).

Depending on which signal properties are exploited, we get different algorithms for the determination of **X**. If there is oversampling in the temporal and/or spatial domain, then cyclostationarity of the sampled signal can be used. In other words, signal separation can be based on the second-order statistics of the channel. Another group of algorithms exploits higher order statistics. Also the finite-alphabet property can be exploited: for example, we know that for a Binary Phase Shift Keying (BPSK) signal, the elements of **X** have to be ± 1.

Blind algorithms have a number of advantages:

- No assumptions about the channel are required: the method works even for channels that do not exhibit discrete DOAs or separable MPCs.
- Blind algorithms do not require any calibration of the antenna array, and make no assumptions about the specific structure of the array.
- The detection process can be done in one step, yielding directly the desired user data.
- Blind algorithms eliminate the need for a training sequence, thus increasing the spectral efficiency of the system.

However, these advantages are bought at a price:

- Most of the blind algorithms assume that the channel impulse response does not change over the time it takes to establish the statistics of **Y**. For a good estimate of these statistics, we have to collect samples over a long time, and the assumption of a time-invariant impulse response might be violated. The number of samples that forms the basis of the statistics is thus a compromise between the need to have a large-enough basis for obtaining good statistics and the need to use only samples that correspond to the same channel state. This tradeoff becomes more difficult as the channel varies faster and the data rate decreases.
- Initialization of the estimate for the channel is critical. For this reason, so-called *semi-blind* algorithms have been proposed, which use a very short training sequence at the beginning of transmission. This sequence is chosen to be short enough not to significantly influence the spectral efficiency of the system, but helps to avoid possible convergence problems of the blind algorithms.

These drawbacks are essentially the same as for blind equalization (see Chapter 16), and have prevented widespread use of this method.

20.1.6 Uplink versus downlink

Up to now, we have considered smart antennas at the RX; since we consider cellular systems in which smart antennas are at the BS, this means that we have been looking at the uplink. The next question is how a smart antenna system behaves in the downlink. The answer to this question depends critically on whether we are considering a Time Domain Duplex (TDD) or a Frequency Domain Duplexing (FDD) system.

Time domain duplexing

In a TDD system in a static environment – i.e., no movement of either the TX or the RX – the channel impulse response is the same for uplink and downlink. Let us make this statement more precise. The transfer function from the MS antenna to the mth BS antenna, $(\mathbf{h}_d)_m$, is the same as the transfer function from the mth BS antenna to the MS antenna. Thus, if we have chosen antenna weights in such a way that they constructively add signals from the antenna elements during the uplink, the *same* antenna weights, when used in the downlink, will ensure that the

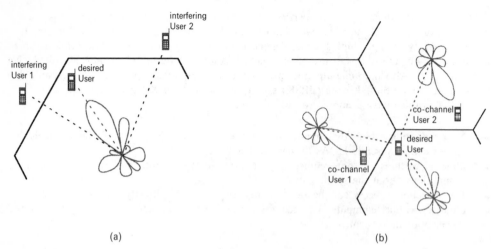

Figure 20.7   Interference situation for the uplink (a) and downlink (b) for a base station with smart antennas.

signals from different BS antenna elements add up constructively at the MS antenna. Furthermore, if BS antenna weights are chosen to supress interference from other MSs during the uplink, the BS will also cause little or no interference to these other MSs in the downlink transmission phase.

However, interference is not completely reciprocal for the uplink and downlink: the interference that the MS sees stems from other *BSs*, not from other MSs. However, antenna weights **w** that are determined during the uplink cannot take this into account – the desired BS does not even "see" these other BSs. Thus, interference suppression for the downlink cannot be as effective as suppression in the uplink.[6]

All we have stated up to now is that the *channel* is reciprocal. However, since antenna weights are determined in the baseband, we also have to require that RF elements and frequency conversion chains – i.e., everything from antenna connectors to the baseband – are reciprocal. This is not self-evident and, in fact, will not be the case for many transceivers. The problem can be eliminated or, at least, migitated by a calibration procedure: during the setup of the call, the BS estimates the channel impulse response both from the reciprocity principle, and by means of feedback from the MS. From the difference of these two measurement methods, the impact of RF circuits can be determined.

Finally, we have to take into account that the channel need not be completely static in the time between uplink and downlink transmission (duplexing time). If the duplexing time is too long (becomes comparable with the coherence time of the channel), then the impulse response of the channel changes due to small-scale fading. The impact of such a decorrelation between the uplink and downlink channel will be discussed in the next subsection.

Frequency domain duplexing

In an FDD system, the uplink and downlink happen at different frequencies; the frequency separation (duplexing distance) for cellular systems is typically on the order of 100 MHz. This is much larger than the coherence bandwidth of the channel (typically on the order of 1 MHz);

---

[6] An exception would be if an MS found itself in a position to notify all the BSs in its surroundings about the details of the interference it sees from them, and the BSs were in a position to cooperate in order to provide as little interference as possible to all users. Such schemes have been proposed, but are rather complex.

as a consequence, small-scale fading of the channel impulse response in the uplink and in the downlink are completely decorrelated. In other words: small-scale fading is created by superposition of MPCs with different phases. The relative phaseshifts depend, among other things, on the carrier frequency. For a large-enough duplexing distance, phaseshifts are sufficiently different in uplink and downlink that superposition of MPCs occurs in different ways.

For this reason, we find that small-scale fading and, thus, the instantaneous impulse response of the channel, are different for uplink and downlink (a similar situation occurs in a TDD system when the duplexing time is much larger than the coherence time of the channel). What remains constant in uplink and downlink is the *average* channel state: in other words, the correlation matrix (averaged over small-scale fading), as well as the DOAs, delays, and mean powers of the MPCs.

It is still possible to obtain performance enhancements in the downlink for FDD. Antenna weights are now chosen based on the average channel state. For example, the beam pattern is chosen so that it points in the direction of the MPC that is strongest *on average*. Now, it is possible that this MPC is not the strongest *instantaneously*, and that it would therefore be better to choose a different beam pattern. However, since we do not know the instantaneous channel, optimizing on average is the best that we can do with the available information.

These considerations also show the usefulness of SR algorithms. Since these algorithms extract average *Channel State Information* (CSI), they can be easily used in FDD systems.

Feedback

Up to now, we have relied on reciprocity (either for the channel impulse response or for DOAs) to provide information about the channel. An alternative approach is to use feedback. Consider an FDD system. Let the BS transmit a training sequence from the first antenna element, then from the second antenna element, and so on. This enables the MS to learn all the separate impulse responses $(\mathbf{h}_d)_m$. It then feeds the information back to the BS on a separate feedback channel. In that case, the BS knows exactly how to adjust weights such that signals from antenna elements add up in the right way at the MS antenna.

This approach is very effective, but has two drawbacks: first, we have to make sure that feedback occurs within a time that is less than the coherence time of the channel. Second, the feedback channel decreases the spectral efficiency of the system. The information carried on that channel is not user data. System designers thus have to consider carefully how often to feed channel-state information back, and with what accuracy. Naturally, the performance of the smart antenna gets better the more accurate the feedback information is, but the performance penalty because of wasted resources can become prohibitive.

20.1.7 Network aspects

In order to really increase capacity by means of smart antennas, it is usually not sufficient to just insert an applique – i.e., replace the single-antenna RF frontend and baseband demodulator with their multiantenna counterparts. Rather, the network has to provide additional functionalities. Consider a TDMA/FDMA system that uses SDMA. It can put two different users on the same time/frequency slot and separate them by their DOAs. However, when the two users get too close, such a separation is no longer possible. In such a case, the network has to swap one of these two users with one that is on a different time/frequency slot and sufficiently far away in the angular domain. In other words, the network has to provide intracell handover capabilities. Thus, the control software of the BS has to be designed specifically for the use of SDMA.

Another major problem is the call setup phase. In this phase, the position and thus DOAs of the various users are not known yet, and the call setup information has to be transmitted in all possible directions simultaneously. Compared with actual data transmission network planning, link budgets, etc., can thus be different for this phase.

20.1.8 Multiuser diversity and random beamforming

Throughout this book, we have assumed that the BS and the MS transmit at predetermined times, irrespective of the channel state. However, it has recently been established that the performance of cellular systems can be improved significantly by scheduling transmission depending on the channel state. It can also be shown that the use of multiple antennas can enhance this idea even further. In the following, we will explain this idea in more detail.

Scheduling and multiuser diversity

Let us first consider the downlink in a system where the BS as well as the MSs have only a single antenna, and multiple MSs communicate with one BS. All links undergo flat Rayleigh fading; we assume for the moment that the mean power is the same for all users. Let us further assume a conventional TDMA system, so that each user is assigned a fixed timeslot, and communicates during that timeslot. The probability for user k to "see" a certain SNR γ is then given by the exponential distribution:

$$\Pr(r_k \leq \gamma) = 1 - \exp\left(-\frac{\gamma}{\bar{\gamma}}\right) \tag{20.21}$$

This probability is independent of user k, and is also independent of the *number* of users K. The scheme, also called *round-robin*, is the best we can do if the TX (BS) does not know anything about the channel. Obviously, it treats users fairly in the sense that each user gets the same amount of time to communicate.

However, we can do better if the BS knows the CSI for the downlink of all users. At any given point in time, chances are that there is a very good link to some users, while there are bad links to other users. The optimum strategy for the BS is now to communicate always with the user who has the best link – i.e., the highest instantaneous path gain – and thus the highest instantaneous SNR γ. As the channel changes, different users k become "instantaneously best". The cumulative distribution function for the active user – i.e., the instantaneously best user – is the same as for selection diversity (see Eq. 13.10). Multiple users thus act like diversity branches – however, we do not need any additional antennas to realize that diversity. This scheme is thus called *multiuser diversity*; it is optimum in the sense that capacity is maximized.

The above description assumed that all users have the same average SNR. In a cellular system, this is typically not the case. Users that are very close to the BS have a much better SNR than users that are far away. Thus, the danger of the above scheme is that the MS closest to the BS gets to use the channel most often, while MSs that are far away are almost never "instantaneously best", and thus are chosen much less frequently. As a consequence, the data rate for users at the cell boundary is much lower than for users near the BS. This problem can be solved by *proportionally fair* scheduling. The BS communicates not with the absolutely best user, but rather with the user whose ratio of instantaneous to average SNR $\gamma/\bar{\gamma}$ is the highest. In such a case, each user is assigned the same amount of time if the *normalized* fading statistics of the different users are identical – e.g., all users are Rayleigh-fading.

A key requirement for scheduling is that the BS has knowledge of the instantaneous downlink channel. As we have discussed in Section 20.1.6, in an FDD system this requires that the MSs feed back their instantaneous SNRs to the BS. The rate of this feedback depends on the coherence time of the channel. Furthermore, it also involves a system design tradeoff.

If feedback is done too rarely, then the BS picks a suboptimum user. If feedback is done too often, then the overhead for this feedback becomes prohibitive. It is also noteworthy that *all* MSs have to feed back their SNRs, even though only one will ultimately be picked for communication.[7]

All the above considerations were for the downlink; however, the scheme works for the uplink as well. In this case, the BS determines the quality of the uplink channel, and then broadcasts which MS is allowed to transmit.

Delay considerations and random beamforming

Another key problem of multiuser diversity is the inherent delay. If the BS always uses proportionally fair scheduling, it retains communication with the chosen user for approximately one coherence time of the channel. If user mobility is low, the coherence time can be quite large. Remember that each user has to wait until its $\gamma/\bar{\gamma}$ becomes instantaneously largest and it can communicate; this waiting time (latency) is roughly KT_{coh}. In a system with many users and large coherence times, this latency can become prohibitively large.

An ingenious solution to this problem, called *random beamforming*, uses multiple antennas at the BS [Viswanath et al. 2002]. Signals transmitted from the different antenna elements of the BS are multiplied by time-varying complex coefficients. This can be interpreted in two (equivalent) ways:

1. Each vector of transmit weight coefficients can be associated with a beam pattern. No matter what beam is formed, it will enhance the channel to *some* MS. Scheduling (as discussed above) then makes sure that the enhanced MS is chosen for communication. Every time the coefficients are changed, the beam pattern changes, and a different MS gets enhanced.
2. The combination of multiple antennas (with weighting coefficients) and physical channel can be viewed as an "equivalent" channel. By varying the coefficients for antenna elements, we are enforcing time variations on the channel, and thus reduce the coherence time T_{coh}. The effective channel exhibits Rayleigh fading with a coherence time that is approximately the smaller of the coherence time of the physical channel and the time over which antenna weights change.

The main difference from transmit diversity or beamforming (as discussed at the beginning of this section) is the following:

*   for *conventional transmit diversity*, the BS needs to know the amplitude and phase of the transfer function from all MSs. It then chooses weights for the signals from different antenna elements in such a way that the different transmit signals add up in an optimum way at the destined MS. Coefficients are changed only if the transfer function between the BS and the chosen MS changes, or if the BS decides to communicate with another MS;
*   for *random beamforming*, the BS chooses the coefficients completely at random and changes them according to system parameters (related to latency), but independently of the channel.

The time over which weighting coefficients are kept constant constitutes a compromise between the resulting latency and system overhead for feeding back channel quality.

---

[7] Some tricks can be used to alleviate this problem: for example, an MS that sees an SNR below a certain threshold does not need to feed its information back since the chance that it will be picked is very low. Again, the choice of this threshold involves an engineering tradeoff: if the threshold is too low, it does not lead to significant savings; if it is too high, then there are times when no MS provides feedback, and the BS thus cannot pick the best user.

Impact on system design

Multiuser diversity is mainly useful for data communications. For voice, stringent delay requirements (total delay less than about 100 ms) preclude the use of "conventional" multiuser diversity, especially when users are stationary or slowly moving (pedestrian speeds), though random beamforming can alleviate this problem somewhat. On the other hand, the scheme seems very well-suited to data communications, where larger delays are acceptable.

From a scientific point of view, it is also noteworthy that multiuser diversity leads to a paradigm change in physical-layer design. Many of the transceiver structures and signal-processing methods we have encountered in Chapters 10–20 aim at combating fading by *reducing* variations in SNR for a specific link – e.g., using diversity antennas. For multiuser diversity, however, we exploit and possibly enhance the variations in SNR.

20.2 Multiple-input multiple-output systems

20.2.1 Introduction

MIMO systems, also known as *Multiple Element Antenna* (MEA) systems, are wireless systems with MEAs at *both* link ends. Originally suggested in Winters [1987], they attracted great attention through theoretical investigations ([Foschini and Gans 1998] and [Telatar 1999]). Since that time, research in these systems has exploded and practical systems based on MIMO have been developed.

The MEAs of a MIMO system can be used for three different purposes: (i) beamforming; (ii) diversity; and (iii) spatial multiplexing (transmission of several datastreams in parallel). The first two concepts are the same as for smart antennas. Having multiple antennas at both link ends leads to some interesting new technical possibilities, but does not change the fundamental effects of this approach. Spatial multiplexing, on the other hand, is a new concept, and has thus drawn the greatest attention. It allows direct improvement of capacity by simultaneous transmission of multiple datastreams. We will show below that the (information-theoretic) capacity *for a single link* increases linearly with the number of antenna elements.

In the early years of MIMO research, the main emphasis was on information-theoretic limits, and this section will also mostly concentrate on these aspects. After 2000, emphasis shifted more to the question of how the theoretical gains of MIMO can be realized in practice. Advances in practical implementation of MIMO systems have also greatly helped in their adoption by international standards organizations. MIMO seems well-poised to be included in the high-speed packet data mode of third-generation cellular systems (see Chapter 23) as well as high-throughput wireless Local Area Networks (LANs) (IEEE 802.11n, see Chapter 24).

20.2.2 How does spatial multiplexing work?

Spatial multiplexing uses MEAs at the TX for transmission of parallel datastreams (see Fig. 20.8). An original high-rate datastream is multiplexed into several parallel streams, each of which is sent from one transmit antenna element. The channel "mixes up" these datastreams, so that each of the receive antenna elements sees a combination of them. If the channel is well-behaved, the received signals represent *linearly independent* combinations. In this case, appropriate signal processing at the RX can separate the datastreams. A basic condition is that the number of receive antenna elements is at least as large as the number of transmit datastreams. It is clear that this approach allows the data rate to be drastically increased – namely, by a factor of $\min(N_t, N_r)$.

Propagation Channel

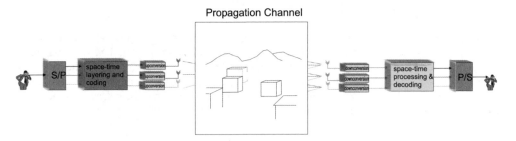

Figure 20.8 Principle behind spatial multiplexing.

For the case when the TX knows the channel, we can also develop another intuition (see Fig. 20.9). With N_t transmit antennas, we can form N_t different beams. We point all these beams at different Interacting Objects (IOs), and transmit different datastreams over them. At the RX, we can use N_r antenna elements to form N_r beams, and also point them at different IOs. If all the beams can be kept orthogonal to each other, there is no interference between the datastreams; in other words, we have established parallel channels. The IOs (in combination with the beams pointing in their direction) play the same role as wires in the transmission of multiple datastreams on multiple wires.

From this description, we can also immediately derive some important principles: the number of possible datastreams is limited by $\min(N_t, N_r, N_s)$, where N_s is the number of (significant) IOs. We have already seen above that the number of datastreams cannot be larger than the number of transmit antenna elements, and that we need a sufficient number of receive antenna elements (at least as many as datastreams) to form the receive beams and, thus, be able to separate the datastreams. But it is also very important to notice that the number of IOs poses an upper limit: if two datastreams are transmitted to the same IO, then the RX has no possibility of sorting them out by forming different beams.

The above intuitive pictures are somewhat simplified. A more exact mathematical treatment follows in subsequent sections.

Propagation Channel

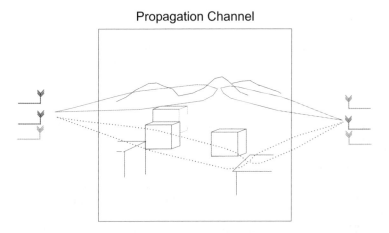

Figure 20.9 Transmission of different datastreams via different interacting objects.

20.2.3 System model

Before going into further details, let us first establish the generic system that will be considered for capacity computations. Figure 20.10 exhibits a block diagram. At the TX, the datastream enters an encoder, whose outputs are forwarded to N_t transmit antennas. From the antennas, the signal is sent through the wireless propagation channel, which is assumed to be quasi-static and frequency-flat if not stated otherwise. By quasi-static we mean that the coherence time of the channel is so long that "a large number" of bits can be transmitted within this time.

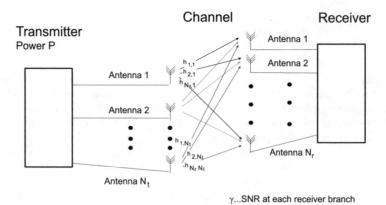

Figure 20.10 Block diagram of a multiple-input multiple-output system.

We denote the $N_r \times N_t$ matrix of the channel as:

$$\mathbf{H} = \begin{pmatrix} h_{11} & h_{12} & \cdots & h_{1N_t} \\ h_{21} & h_{22} & \cdots & h_{2N_t} \\ \vdots & \vdots & \cdots & \vdots \\ h_{N_r 1} & h_{N_r 2} & \cdots & h_{N_r N_t} \end{pmatrix} \tag{20.22}$$

whose entries h_{ij} are attenuations (transfer functions) from the jth transmit to the ith receive antenna.

The received signal vector:

$$\mathbf{r} = \mathbf{H}\mathbf{s} + \mathbf{n} = \mathbf{x} + \mathbf{n} \tag{20.23}$$

is received by N_r antenna elements, where $\mathbf{s}$ is the transmit signal vector and $\mathbf{n}$ is the noise vector.

20.2.4 Channel-state information

Algorithms for MIMO transmission can be catergorized by the amount of CSI that they require. We distinguish the following cases:

1. *Full CSI at the TX and full CSI at the RX*: in this ideal case, both the TX and the RX have full and perfect knowledge of the channel. This case obviously results in the highest possible capacity. However, it is difficult to obtain the full CSI at the TX (as we have discussed in Section 20.1.6).

2. *Average CSI at the TX and full CSI at the RX*: in this case, the RX has full information of the instantaneous channel state, but the TX knows only the average CSI – e.g., the

correlation matrix of the channel impulse response or the angular power spectrum. As we have discussed in Section 20.1.6, this is easier to achieve and does not require reciprocity or fast feedback; however, it does require calibration (to eliminate the non-reciprocity of transmit and receive chains) or slow feedback.

3. *No CSI at the TX and full CSI at the RX*: this is the case that can be achieved most easily, without any feedback or calibration. The TX simply does not use any CSI, while the RX learns the instantaneous channel state from a training sequence or using blind estimation.

4. *Noisy CSI*: when we assume "full CSI" at the RX, this implies that the RX has learned the channel state perfectly. However, any received training sequence will be affected by additive noise as well as quantization noise. It is thus more realistic to assume a "mismatched RX", where the RX processes the signal based on the *observed* channel $\mathbf{H}_{\mathrm{obs}}$, while in reality the signals pass through channel $\mathbf{H}_{\mathrm{true}}$:

$$\mathbf{H}_{\mathrm{true}} = \mathbf{H}_{\mathrm{obs}} + \mathbf{\Delta} \tag{20.24}$$

Some papers have taken this into account by ad hoc modification of noise variance (replacing σ_{n}^2 by $\sigma_{\mathrm{n}}^2 + \sigma_{\mathrm{e}}^2$, where σ_{e}^2 is the variance of the entries of $\mathbf{\Delta}$).

5. *No CSI at the TX and no CSI at the RX*: it is remarkable that channel capacity is also high when neither the TX nor the RX have CSI. We can, for example, use a generalization of differential modulation. For high SNR, capacity no longer increases linearly with $m = \min(N_{\mathrm{t}}, N_{\mathrm{r}})$, but rather increases as $\tilde{m}(1 - \tilde{m}/T_{\mathrm{coh}})$, where $\tilde{m} = \min(N_{\mathrm{t}}, N_{\mathrm{r}}, \lfloor T_{\mathrm{coh}}/2 \rfloor)$, and T_{coh} is the coherence time of the channel in units of symbol duration.

20.2.5 Capacity in non-fading channels

The first key step in understanding MIMO systems is the derivation of the capacity equation for MIMO systems in non-fading channels, often known as "Foschini's equation" [Foschini and Gans 1998]. Let us start with the capacity equation for "normal" (single-antenna) Additive White Gaussian Noise (AWGN) channels. As Shannon showed, the information-theoretic (ergodic) capacity of such a channel is:

$$C_{\mathrm{shannon}} = \log_2\left(1 + \gamma \cdot |H|^2\right) \tag{20.25}$$

where γ is the SNR at the RX, and H is the normalized transfer function from the TX to the RX (as we are for now dealing with the frequency-flat case, the transfer function is just a scalar number). The key statement of this equation is that capacity increases only logarithmically with the SNR, so that boosting the transmit power is a highly ineffective way of increasing capacity.

Consider now the MIMO case, where the channel is represented by matrix Eq. (20.22). Let us then consider a *singular value decomposition*[8] of the channel:

$$\mathbf{H} = \mathbf{W}\mathbf{\Sigma}\mathbf{U}^\dagger \tag{20.26}$$

where $\mathbf{\Sigma}$ is a diagonal matrix containing singular values, and $\mathbf{W}$ and $\mathbf{U}^\dagger$ are unitary matrices composed of the left and right singular vectors, respectively. The received signal is then:

$$\mathbf{r} = \mathbf{H}\mathbf{s} + \mathbf{n} \tag{20.27}$$

$$= \mathbf{W}\mathbf{\Sigma}\mathbf{U}^\dagger\mathbf{s} + \mathbf{n} \tag{20.28}$$

---

[8] Singular value decomposition is similar to eigenvalue decomposition, but exists also for rectangular matrices (more rows than columns, or vice versa). It decomposes any matrix into a product of three matrices: a unitary matrix corresponding to the row space, a diagonal matrix describing the strength of different eigenmodes, and a unitary matrix describing the column space.

Then, multiplication of the transmit data vector by matrix $\mathbf{U}$ and the received signal vector by $\mathbf{W}^\dagger$ diagonalizes the channel:

$$\mathbf{W}^\dagger\mathbf{r}=\mathbf{W}^\dagger\mathbf{W}\mathbf{\Sigma}\mathbf{U}^\dagger\mathbf{U}\tilde{\mathbf{s}}+\mathbf{W}^\dagger\mathbf{n}$$

$$\tilde{\mathbf{r}} = \mathbf{\Sigma}\tilde{\mathbf{s}}+\tilde{\mathbf{n}} \tag{20.29}$$

Note that – because $\mathbf{U}$ and $\mathbf{W}$ are unitary matrices – $\tilde{\mathbf{n}}$ has the same statistical properties as $\mathbf{n}$ – i.e., it is independent identically distributed (iid) white Gaussian noise. The capacity of the system Eq. (20.29) is the same as that of system Eq. (20.23). Now, computation of the capacity of Eq. (20.29) is rather straightforward. The matrix $\mathbf{\Sigma}$ is a diagonal matrix with R_H nonzero entries σ_k, where R_H is the rank of $\mathbf{H}$ (and thus defined as the number of nonzero singular values), and σ_k is the kth singular value of $\mathbf{H}$. We have therefore R_H *parallel* channels, and it is clear that the capacity of parallel channels just adds up.

The capacity of channel $\mathbf{H}$ is thus given by the sum of the capacities of the subchannels (modes of the channel):

$$C = \sum_{k=1}^{R_H}\log_2\left[1 + \frac{P_k}{\sigma_n^2}\sigma_k^2\right] \tag{20.30}$$

where σ_n^2 is noise variance, and P_k is the power allocated to the kth eigenmode; we assume that $\sum P_k = P$ is independent of the number of antennas. This capacity expression can be shown to be equivalent to:

$$C = \log_2\left[\det\left(\mathbf{I}_{N_r} + \frac{\overline{\gamma}}{N_t}\mathbf{H}\mathbf{R}_{ss}\mathbf{H}^\dagger\right)\right] \tag{20.31}$$

where $\mathbf{I}_{N_r}$ is the $N_r \times N_r$ identity matrix, $\overline{\gamma}$ is the mean SNR per RX branch, and $\mathbf{R}_{ss}$ is the correlation matrix of the transmit data (if data at the different antenna elements are uncorrelated, it is a diagonal matrix with entries that describe power distribution among antennas).[9] The distribution of power among the different eigenmodes (or antennas) depends on the amount of CSI at the TX; we also assume for the moment that the RX has perfect CSI. The equations above confirm our intuitive picture that capacity increases linearly with $\min(N_t, N_r, N_s)$, as the number of nonzero singular values R_H is upper-limited by $\min(N_t, N_r, N_s)$.

No channel-state information at the transmitter and full CSI at the receiver

When the RX knows the channel perfectly, but no CSI is available at the TX, it is optimum to assign equal transmit power to all TX antennas, $P_k = P/N_t$, and use uncorrelated datastreams. Capacity thus takes on the form:

$$C = \log_2\left[\det\left(\mathbf{I}_{N_r} + \frac{\overline{\gamma}}{N_t}\mathbf{H}\mathbf{H}^\dagger\right)\right] \tag{20.32}$$

It is worth noting that (for sufficiently large N_s) the capacity of a MIMO system increases linearly with $\min(N_t, N_r)$, irrespective of whether the channel is known at the TX or not.

---

[9] Note that the $\mathbf{H}$ and $\mathbf{R}$ must be normalized to ensure that $\overline{\gamma}$ is the mean SNR.

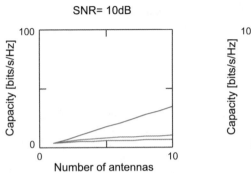
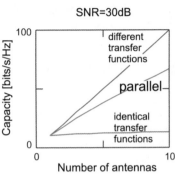

Figure 20.11 Capacity of multiple-input-multiple-output systems in additive-white-Gaussian-noise channels.

Let us now look at a few special cases. To make the discussion easier, we assume that $N_t = N_r = N$:

1. All transfer functions are identical – i.e., $h_{1,1} = h_{1,2} = \cdots = h_{N,N}$. This case occurs when all antenna elements are spaced very closely together, and all waves are coming from similar directions. In such a case, the rank of the channel matrix is unity. Then, capacity is:

$$C_{\text{MIMO}} = \log_2(1 + N\overline{\gamma}) \tag{20.33}$$

We see that in this case the SNR is increased by a factor of N compared with the single antenna case, due to beamforming gain at the RX. However, this only leads to a logarithmic increase in capacity with the number of antennas.

2. All transfer functions are different such that the channel matrix is full-rank, and has N eigenvalues of equal magnitude. This case can occur when the antenna elements are spaced far apart and are arranged in a special way. In this case, capacity is:

$$C_{\text{MIMO}} = N \log_2(1 + \overline{\gamma}) \tag{20.34}$$

and, thus, increases linearly with the number of antenna elements.

3. Parallel transmission channels – e.g., parallel cables. In this case, capacity also increases linearly with the number of antenna elements. However, the SNR per channel decreases with N, so that total capacity is:

$$C_{\text{MIMO}} = N \log_2\left(1 + \frac{\overline{\gamma}}{N}\right) \tag{20.35}$$

Figure 20.11 shows capacity as a function of N for different values of SNR.

Full channel-state information at the transmitter and full CSI at the receiver

Let us next consider the case where both the RX and TX know the channel perfectly. In such a case, it can be more advantageous to distribute power not uniformly between the different transmit antennas (or eigenmodes), but rather assign it based on the channel state. In other words, we are faced with the problem of optimally allocating power to several parallel channels, each of which has a different SNR. This is the same problem that we considered in Section 19.7.2, and therefore the answer is the same: *waterfilling*. This is another nice example of how the same mathematics can be applied to different communications problems: replace the word "subchannel" or "subcarrier in an Orthogonal Frequency Division Multiplexing (OFDM) system" by "eigenmode in a MIMO system", and we can apply the whole discussion of Section 19.7.2 to MIMO systems with CSI at the TX.

20.2.6 Capacity in flat-fading channels

General concepts

In the previous section, we considered capacity for one given channel realization – i.e., for one channel matrix $\mathbf{H}$. In wireless systems, we are, however, faced with channel fading. In this case, entries in channel matrix Eq. (20.22) are random variables. If the channel is Rayleigh-fading, and fading is independent at different antenna elements, the h_{ij} are iid zero-mean, circularly symmetric complex Gaussian random variables with unit variance – i.e., the real and imaginary part each has variance $1/2$. This is the case we will consider for now, unless stated otherwise. Consequently, the power carried by each h_{ij} is chi-square-distributed with 2 degrees of freedom. This is the simplest possible channel model; it requires the existence of "heavy multipath" – i.e., many MPCs of approximately equal strength (see Chapter 5) as well as a sufficient distance between the antenna elements. Since fading is independent, there is a high probability that the channel matrix is full rank and the eigenvalues are fairly similar to each other; consequently, capacity increases linearly with the number of antenna elements. Thus, the existence of heavy multipath, which is usually considered a drawback, becomes a major advantage in MIMO systems.

Because the entries of the channel matrix are random variables, we also have to rethink the concept of information-theoretic capacity. As a matter of fact, two different definitions of capacity exist for MIMO systems:

- *Ergodic (Shannon) capacity.* This is the expected value of the capacity, taken over all realizations of the channel. This quantity assumes an infinitely long code that extends over all the different channel realizations.
- *Outage capacity.* This is the minimum transmission rate that is achieved over a certain fraction of the time – e.g., 90% or 95%. We assume that data are encoded with a near-Shannon-limit-achieving code that extends over a period that is much shorter than the channel coherence time. Low Density Parity Check (LDPC) codes with reasonable blocklengths (e.g., 10,000 bits) get close to the Shannon limit (see Chapter 14). For a data rate of 10 Mbit/s, such a block can be transmitted within 1 ms. This is much shorter than 10 ms, the typical coherence time of wireless channels (see Chapter 5). Thus, each channel realization can be associated with a (Shannon) capacity value. Capacity thus becomes a random variable (rv) with an associated cumulative distribution function (cdf). We thus will look henceforth at this distribution function, or equivalently at the capacity that can be guaranteed for $x\%$ of all channel realizations.

No channel-state information at the transmitter and perfect CSI at the receiver

Now, what is the capacity that we can achieve in a fading channel without CSI? Figure 20.12 shows the result for some interesting systems at a 21-dB SNR. The $(1,1)$ curve describes a Single Input Single Output (SISO) system. We find that the median capacity is on the order of 6 bit/s/ Hz, but the 5% outage capacity is considerably lower (on the order of 3 bit/s/Hz). When using a $(1,8)$ system – i.e., 1 transmit antenna and 8 receive antennas – the mean capacity does not increase that significantly – from 6 to 10 bit/s/Hz. However, the 5% outage capacity increases significantly from 3 to 9 bit/s/Hz. The reason for this is the much higher resistance to fading that such a diversity system has. However, when going to a $(8,8)$ system – i.e., a system with 8

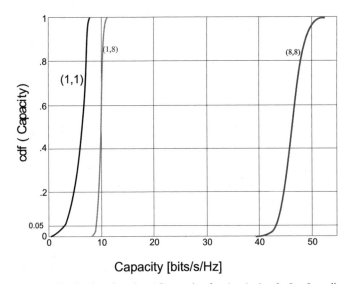

Capacity [bits/s/Hz]

Figure 20.12 Cumulative distribution function of capacity for 1×1, 1×8, 8×8 cycling, and the 8×8 optimum scheme.

Reproduced with permission from Foschini and Gans [1998] © Kluwer.

transmit and 8 receive antennas – both capacities increase dramatically: the mean capacity is on the order of 46 bit/s/Hz, and the 5% outage probability is more than 40 bit/s/Hz.

The exact expression for the *ergodic capacity* was derived in [Telatar 1999] as:

$$E\{C\} = \int_0^\infty \log_2\left[1 + \frac{\bar{\gamma}}{N_t}\lambda\right] \sum_{k=0}^{m-1} \frac{k!}{(k+n-m)!} \left[L_k^{n-m}(\lambda)\right]^2 \lambda^{n-m} \exp(-\lambda)\, d\lambda \qquad (20.36)$$

where $m = \min(N_t, N_r)$ and $n = \max(N_t, N_r)$ and $L_k^{n-m}(\lambda)$ are associated Laguerre polynomials. Exact analytical expressions for the cdfs of capacity are rather complicated; therefore, two approximations are in widespread use:

- Capacity can be well approximated by a Gaussian distribution, such that only the mean – i.e., the ergodic capacity given above – and variance need to be computed.
- From physical considerations, the following upper and lower bounds for capacity distribution have been derived in [Foschini and Gans 1998] for the case $N_t \geq N_r$:

$$\sum_{k=N_t-N_r+1}^{N_t} \log_2\left[1 + \frac{\bar{\gamma}}{N_t}\chi_{2k}^2\right] < C < \sum_{k=1}^{N_t} \log_2\left[1 + \frac{\bar{\gamma}}{N_t}\chi_{2N_r}^2\right] \qquad (20.37)$$

where χ_{2k}^2 is a chi-square-distributed random variable with $2k$ degrees of freedom.[10] These two bounds have very clear physical interpretations. The lower bound corresponds to capacity that can be achieved with a BLAST system; this system and its operating principle will be described below. The upper bound corresponds to an idealized situation where there is a separate array of receive antennas for each transmit antenna; it receives the signal in such a way that there is no interference from other transmit streams.

---

[10] The above equation is a slight abuse of notation, indicating as it does that the cdf of the capacity is bounded by the cdfs of the random variables given by the equations on the left and right side.

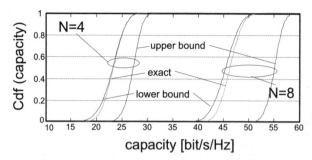

Figure 20.13   Exact capacity, upper bound, and lower bound of a multiple-input-multiple-output system in an independent identically distributed channel at a 21-dB signal-to-noise ratio, with $N_r = N_t = N$ equal to 4 and 8. Reproduced with permission from Molisch and Tufvesson [2005] © Hindawi.

Perfect channel-state information at the transmitter and receiver

The capacity gain by waterfilling (compared with equal-power distribution) is rather small when the number of transmit and receive antennas is identical. This is especially true in the limit of large SNRs: when there is a lot of water available, the height of "concrete blocks" in the vessel has little influence on the total amount that ends up in the vessels. When N_t is larger than N_r, the benefits of waterfilling become more pronounced (see Fig. 20.14). We can interpret this the following way: if the TX has no channel knowledge, then there is little point in having more transmit than receive antennas – the number of datastreams is limited by the number of receive antennas. Of course, we can transmit the same datastream from multiple transmit antennas, but this does not increase the SNR for that stream at the RX; without channel knowledge at the TX, the streams add up incoherently at the RX.

On the other hand, if the TX has full channel knowledge, it can perform beamforming, and direct the energy better toward the receive array. Thus, increasing the number of TX antennas improves the SNR, and (logarithmically) capacity. Thus, having a larger TX array improves capacity. However, this also increases the demand for channel estimation.

20.2.7 Impact of the channel

Up to now, we have discussed the capacity of a channel with flat-fading and iid zero-mean complex Gaussian coefficients. Channels occuring in practice are more complicated, and deviations of the channel from idealized assumptions can have a significant impact on capacity. In the following, we will describe some of the more important effects.

Channel correlation

Correlation of the signals at different antenna elements can significantly reduce the capacity of a MIMO system. This can be shown the following way: the capacity is determined by the distribution of the singular values of the channel matrix. For a given SNR, maximum capacity is achieved when the channel transfer matrix has full rank and the singular values of $\mathbf{H}$ are equally strong. If the coefficients of the channel matrix are iid Rayleigh-fading, then this situation is *approximately* fulfilled (though the ordered eigenvalues still have different values). But if the fading of the channel coefficients is correlated, then the singular value spread – i.e., the difference between the largest and the smallest singular values – becomes much bigger. This, in turn, leads to a reduction in system capacity because some of the "parallel channels" of Eq. (20.30) have extremely low SNRs.

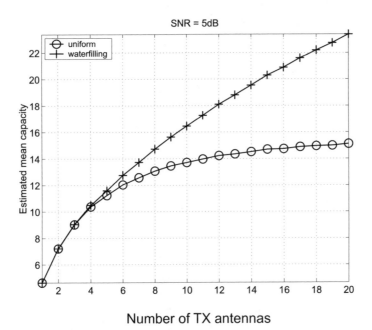

Figure 20.14 Capacity with and without channel-state information at the transmitter with $N_r = 8$ antennas and a signal-to-noise ratio of 5 dB.

Correlation is influenced by the angular spectrum of the channel as well as the arrangement and spacing of antenna elements (see Chapter 13). For antennas that are spaced half a wavelength apart, a uniform angular power spectrum leads approximately to a decorrelation of incident signals. A smaller angular spread of the channel leads to an increase in correlation. Since we are now looking at a MIMO system, we have to consider correlation both at the TX and at the RX. A popular model (the so-called Kronecker model) assumes that correlation at the TX is independent of correlation at the RX (see Section 7.4.7).[11] Realization of the channel matrix can then be obtained as:

$$\mathbf{H}_{\mathrm{kron}} = \frac{1}{\sqrt{\mathrm{tr}(\mathbf{R}_{\mathrm{RX}})}} \mathbf{R}_{\mathrm{RX}}^{1/2} \mathbf{G}_{\mathrm{G}} \mathbf{R}_{\mathrm{TX}}^{1/2} \tag{20.38}$$

where $\mathbf{G}_{\mathrm{G}}$ is a matrix with iid complex Gaussian entries with unit variance.

Analytical computation of capacity is much more complicated in the case of correlated channels. It is easier to obtain simulation results by generating realizations of the channel matrix from Eq. (20.38), and then inserting them into Eq. (20.32) or Eq. (20.31).

Figure 20.15 shows the ergodic capacity of a 4×4 MIMO system with uniform linear arrays at the TX and RX as a function of angular spread at one link end.[12] We see that a small angular spread leads to a drastic reduction in capacity.

---

[11] For a discussion of this assumption, and a more general model, see Weichselberger et al. [2003].

[12] Note that this figure is based on the idealized assumption that there is no mutual coupling between antenna elements; investigations have shown that mutual coupling influences capacity by introducing pattern diversity as well as changing the average power received in different antenna elements.

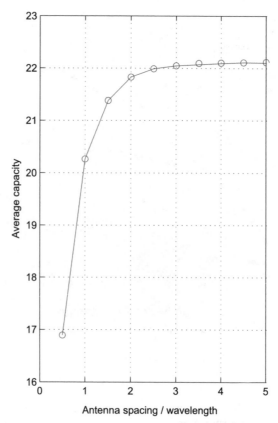

Figure 20.15 Average capacity of a 4 × 4 multiple-input-multiple-output system with a 10° root-mean-square angular spread as seen from the base station as a function of antenna spacing at the BS.

Frequency-selective channels

The previous sections assumed frequency-flat channels. Fortunately, generalization to the frequency-selective case is straightforward. As Shannon has shown, the use of an OFDM-like scheme is optimum for dealing with the frequency-selective channel, converting it into a number of parallel flat-fading channels. Thus, capacity per unit bandwidth is a straightforward generalization of Eq. (20.31):

$$C = \frac{1}{B} \int_B \log_2 \left[\det \left(\mathbf{I}_{N_r} + \frac{\overline{\gamma}}{N_t} \mathbf{H}(f) \mathbf{R}_{ss}(f) \mathbf{H}(f)^\dagger \right) \right] df \qquad (20.39)$$

where B is the bandwidth of the considered system.

This equation also implies that frequency selectivity offers additional diversity that increases the slope of the capacity cdf. If one of the frequency subchannels shows poor capacity, there is a good chance that another subchannel has high quality. Figure 20.16 shows an example of a measured capacity cdf in a microcellular environment. We see that the capacity cdf becomes steeper as the bandwidth increases.

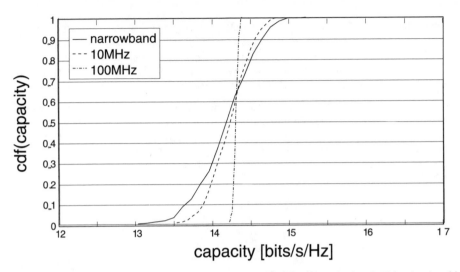

Figure 20.16 Capacity in a measured microcellular channel, for different system bandwidths: 4×4 multiple-input-multiple-output system.

Reproduced with permission from Molisch et al. [2002] © IEEE.

Line-of-sight versus non-line-of-sight

In some situations, there is a Line Of Sight (LOS) connection between the TX and RX, resulting in different fading statistics. As we saw in Chapter 5, the fading statistics of an SISO link becomes Ricean instead of Rayleigh. For a MIMO system, the channel matrix can be written as:

$$\mathbf{H} = \sqrt{\frac{K_{\mathrm{LOS}}}{K_{\mathrm{LOS}}+1}}\hat{\mathbf{H}}_{\mathrm{LOS}} + \sqrt{\frac{1}{K_{\mathrm{LOS}}+1}}\hat{\mathbf{H}}_{\mathrm{res}} \qquad (20.40)$$

where K_{LOS} is the ratio of powers in the LOS to those in residual components, $\hat{\mathbf{H}}_{\mathrm{LOS}}$ is a purely deterministic matrix, and $\hat{\mathbf{H}}_{\mathrm{res}}$ has (uncorrelated or correlated) zero-mean Gaussian entries.[13] If the distance between the TX and RX is large – i.e., larger than the Rayleigh distance (see Chapter 4) – the LOS gives rise to a matrix $\hat{\mathbf{H}}_{\mathrm{LOS}}$ that has rank 1 (a single wave can only be associated with one singular value of the channel matrix!). This in turn implies that the singular value spread of matrix Eq. (20.40) is much larger than for a Non Line Of Sight (NLOS) matrix. Consequently, the capacity of a LOS channel is lower than for a NLOS channel *when assuming equal SNR*. One should however note that the SNR is often better in the LOS case compared with the NLOS case. For a power-limited scenario with realistic channels, the LOS case often gives the highest capacity, despite the imbalance between singular values.

It is also noteworthy that a strong LOS component leads to a larger spread of eigenvalues if the LOS component is a *plane* wave. A *spherical* wave leads to a transfer function matrix that can have full rank if antenna elements are spaced appropriately. The curvature of waves is noticeable up to one Rayleigh distance – i.e., typically a few meters.

---

[13] Both $\hat{\mathbf{H}}_{\mathrm{LOS}}$ and $\hat{\mathbf{H}}_{\mathrm{res}}$ are normalized to $E\{|\hat{\mathbf{H}}|_F^2\} = N_t N_r$, where $|\hat{\mathbf{H}}|_F$ is the Frobenius norm of $\hat{\mathbf{H}}$.

Example 20.4 *Capacity in a channel with LOS. The transmit and receive arrays are uniform linear arrays with element spacing λ between the elements and $N_r = N_t = 8$. The arrays are perpendicular to the LOS connection. The DOA of the NLOS components is uniformly distributed between 0 and 2π. Estimate the mean capacity if the TX does not have CSI for an SNR of 20 dB, and $K_{LOS} = 0$ and 20 dB.*

In a first step, we have to determine the channel matrix. Since the transmit and receive arrays are linear arrays that are oriented perpendicular to the LOS, we find that $\hat{\mathbf{H}}_{LOS}$ is the all-1's matrix if transmit and receive arrays are sufficiently far apart from each other. Furthermore, $\hat{\mathbf{H}}_{res}$ has unit energy, iid complex Gaussian entries because the angular spectrum is uniformly distributed and the antenna elements are more than $\lambda/2$ apart from each other. From Eq. (20.32), capacity is then:

$$C = \log_2\left[\det\left(\mathbf{I}_{N_r} + \frac{\overline{\gamma}}{N_t}\mathbf{H}\mathbf{H}^\dagger\right)\right] \tag{20.41}$$

$$= \log_2\Big[\det\Big(\mathbf{I}_{N_r} + \frac{\overline{\gamma}}{N_t}\Big[\frac{K_{LOS}}{K_{LOS}+1}\hat{\mathbf{H}}_{LOS}\hat{\mathbf{H}}_{LOS}^\dagger$$

$$+ \frac{\sqrt{K_{LOS}}}{K_{LOS}+1}\Big(\hat{\mathbf{H}}_{res}\hat{\mathbf{H}}_{LOS}^\dagger + \hat{\mathbf{H}}_{LOS}\hat{\mathbf{H}}_{res}^\dagger\Big) + \frac{1}{K_{LOS}+1}\hat{\mathbf{H}}_{res}\hat{\mathbf{H}}_{res}^\dagger\Big]\Big)\Big] \tag{20.42}$$

Using Jensen's inequality, the expected value of the capacity can be approximated as:

$$E\{C\} \simeq \log_2\Big[\det\Big(\mathbf{I}_{N_r} + \frac{\overline{\gamma}}{N_t}\Big[\frac{K_{LOS}}{K_{LOS}+1}E\{\hat{\mathbf{H}}_{LOS}\hat{\mathbf{H}}_{LOS}^\dagger\} \tag{20.43}$$

$$+ \frac{\sqrt{K_{LOS}}}{K_{LOS}+1}E\{\hat{\mathbf{H}}_{res}\hat{\mathbf{H}}_{LOS}^\dagger + \hat{\mathbf{H}}_{LOS}\hat{\mathbf{H}}_{res}^\dagger\} + \frac{1}{K_{LOS}+1}E\{\hat{\mathbf{H}}_{res}\hat{\mathbf{H}}_{res}^\dagger\}\Big]\Big)\Big] \tag{20.44}$$

$$= \log_2\Big[\det\Big(\mathbf{I}_{N_r} + \frac{\overline{\gamma}}{N_t}\Big[\frac{K_{LOS}N_t}{K_{LOS}+1}\mathbf{1} + \frac{1}{K_{LOS}+1}E\{\hat{\mathbf{H}}_{res}\hat{\mathbf{H}}_{res}^\dagger\}\Big]\Big)\Big] \tag{20.45}$$

$$= \log_2\Big[\det\Big(\Big[1 + \frac{\overline{\gamma}}{K_{LOS}+1}\Big]\mathbf{I}_{N_r} + \Big[\overline{\gamma}\frac{K_{\%LOS}}{K_{LOS}+1}\mathbf{1}\Big]\Big)\Big] \tag{20.46}$$

where $\mathbf{1}$ is an $N_r \times N_r$ matrix where each entry is 1; such a matrix has only a single nonzero eigenvalue, whose magnitude is N_r. Equation (20.46) can be further approximated as:

$$E\{C\} \simeq N_r\log_2\Big[1 + \frac{\overline{\gamma}}{K_{LOS}+1}\Big] + \log_2\Big[1 + \frac{\overline{\gamma}K_{LOS}N_r}{K_{LOS}+1}\Big] \tag{20.47}$$

Using Eq. (20.47), we find that capacity for $K_{LOS} = 0$ and 20 dB is 54 and 17 bit/s/Hz, respectively. Note that the above is a rather crude approximation, but it gives correct trends. Monte Carlo simulations give 40 and 15 bit/s/Hz, respectively.

Limited number of interacting objects

In the intuitive picture of Section 20.2.1, we have already seen that a certain number of IOs are required to act as relays for datastreams. Relating this to the mathematics of the previous section, we note that a certain number of IOs are needed in order to guarantee that the channel coefficients in matrix $\mathbf{H}$ are independent. While the number of existing IOs is always large, the number of *significant* IOs might be limited in practice. After all, IOs that are too weak to provide appreciable SNR (and thus capacity) are not useful in carrying datastreams.

Figure 20.17 shows some measurement results for capacity as a function of the number of antenna elements. Measurements were taken in a microcellular scenario where the number of

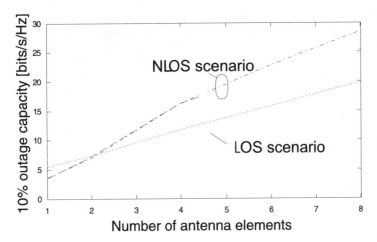

Figure 20.17 10% outage capacity as a function of the number of antenna elements at the transmitter and receiver in measured line-of-sight and non-line-of-sight scenarios.

Reproduced with permission from Molisch et al. [2002] © IEEE.

IOs was rather small. We find that capacity, especially for the LOS scenario, does not increase linearly with the number of antenna elements N when N exceeds 4. This is a clear sign that the number of IOs limits the achievable capacity.

Keyhole channels

There are some special cases where capacity is low even though the signals at antenna elements are uncorrelated. These cases are often referred to as *keyholes* or *pinholes*. An example of a keyhole scenario is a rich scattering environment at both the TX and RX ends; between them there is only one propagation path with just one degree of freedom. Such a scenario can occur when the TX and RX are surrounded by IOs, and the two IO groups are separated by a long stretch of empty space (greenfield). Another scenario is the case when IO areas are connected by a single-mode waveguide or by a diffraction edge. In all these cases, the total transfer function can be written as:

$$\mathbf{H} = \mathbf{R}_{RX}^{1/2}\mathbf{G}_{G,1}\mathbf{R}_{RX-TX}^{1/2}\mathbf{G}_{G,2}\mathbf{R}_{TX}^{1/2} \qquad (20.48)$$

where $\mathbf{G}_{G,1}$ and $\mathbf{G}_{G,2}$ are both iid complex Gaussian matrices, and where $\mathbf{R}_{RX-TX}$ describes the correlation of the channel matrix between the TX and RX environments; in a keyhole channel, this is a low-rank matrix. It can also be seen from this description that the statistics of the entries are no longer Gaussian – this explains why it can be possible to have low correlation and low capacity at the same time. It should, however, be noted that keyhole channels occur very seldom in practice. While Almers et al. [2003] measured one in a controlled environment, it seems to be a rare occurrence in "normal" environments.

Interference channels

In a cellular system, the existence of multiple users – and thus interferers – influences capacity, and decreases the data rate that is possible for a single user. A first investigation into this topic by Catreux et al. [2001] looked at a cellular TDMA system with Minimum Mean Square Error (MMSE) detection. Under these assumptions, it was shown that the *cellular* capacity of a

MIMO system is hardly larger than that of a system with multiple antennas at the BS only (as described in Section 20.1). The reason for this somewhat astonishing result is that in a cellular system with multiple antennas at the BS only, these antennas can be used to suppress adjacent cell interference, and thus decrease reuse distance. For a cellular MIMO system with $N_r = N_t$, the degrees of freedom created by multiple BS antennas are all used for separation of multiple datastreams from a single user, and none for suppression of interfering users. Thus, we can use neither SDMA nor SFIR to increase cellular capacity.

However, it seems possible to combine MIMO with other techniques for multiaccess interference. For example, multiuser interference can be eliminated by BS cooperation. For the uplink, cooperating BSs can be viewed as a giant MIMO system with $N_r N_{BS}$ antenna elements at one link end, where N_{BS} is the number of cooperating BSs. Capacity for such a system can be approximated by inserting the generalized channel matrix into Eq. (20.31). For the downlink, the effect of interfering BSs can be minimized by appropriate preprocessing as long as the BSs cooperate and each BS has the CSI about all MSs.

20.2.8 Layered space–time structure

Up to now, we have only discussed the information-theoretic limits of MIMO systems. The big question is how to realize these capacities in practice. One possibility is joint encoding of the datastreams that are to be transmitted from different antenna elements, combined with maximum-likelihood detection. When this technique is combined with (almost) capacity-achieving codes, it can closely approximate the capacity of a MIMO system. For a small number of antenna elements and for a small modulation alphabet (BPSK or QPSK), such a scheme can actually be gainfully employed. However, for most practical cases, the complexity of a joint MLSE is too high. For this reason, so-called layered space–time structures have been proposed, which allow us to break up the demodulation process into several separate pieces, each of which has lower complexity. These structures are also widely known under the name of *BLAST* (Bell labs LAyered Space Time) *architectures*.

Horizontal BLAST

Horizontal BLAST (H-BLAST) is the simplest possible layered space–time structure.[14] The transmitter first demultiplexes the datastream into N_t parallel streams, each of which is encoded *separately*. Each encoded datastream is then transmitted from a different transmit antenna. The channel mixes up the different datastreams; the RX separates them out by nulling and interference subtraction. In other words, the RX proceeds in the following steps:

- It considers the first datastream as the useful one, and regards the other datastreams as interference. It can then use *optimum combining* for suppression of interfering streams (see Chapter 13). The RX has $N_r \geq N_t$ antenna elements available. If $N_r = N_t$, it can suppress all $N_t - 1$ interfering datastreams, and receive the desired datastream with diversity order 1. If the RX has more antennas, it can receive the first datastream with better quality. But in any case, interference from the other streams can be eliminated.
- The desired stream can now be demodulated and decoded. Outputs from that process are firm decisions on the bits of stream 1. Note that since we have separate encoding of different datastreams, we only need knowledge of the first stream to complete the decoding process.

---

[14] This scheme was originally called *V-BLAST* (for vertical BLAST) but the name was changed later by Foschini et al. [2003].

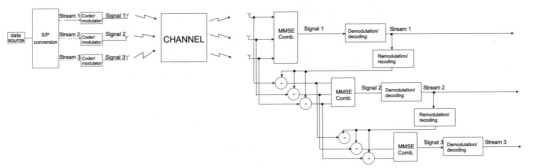

Figure 20.18   Block diagram of a horizontal BLAST transceiver.

- The bits that have thus been decoded are now re-encoded, and remodulated. Multiplying this symbol stream by the transfer function of the channel, we obtain the contribution that stream 1 has made to the total received signal at the different antenna elements.
- We subtract these contributions from the signals at the different antenna elements.
- Now we consider the "cleaned-up" signal and try to detect the second datastream. We again have N_r received signals, but only $N_t - 2$ interferers. Using optimum combining again, we can now receive the desired datastream with diversity order 2.
- The next step is again decoding, recoding, and remodulating the considered datastream (stream 2 now), and subtraction of the associated signal from the total signal at the receive antenna elements obtained in the previous step. This cleans up the received signal even more.
- The process is repeated until the last datastream is decoded.

This scheme is actually very similar to multiuser detection (Section 18.4): if different transmit streams were to come from different users, then H-BLAST would be normal serial interference cancellation. Note also that the encoding scheme does not require "cooperation" between different antenna elements (or users). Similar to serial interference cancellation, H-BLAST also faces the problem of error propagation, especially since the first decoded datastream has the worst quality. In other words: if datastream 1 is decoded incorrectly, then we subtract the "wrong" signal from the remaining signals at the antenna elements. Thus, instead of "cleaning up" the receive signal, we introduce even more interference. This in turn increases the likelihood that the second datastream is decoded incorrectly, and so on. In order to mitigate this problem, stream ordering should be used: the RX should first decode the stream that has the best SINR, then the one with the next best, and so on.

Even with stream ordering, H-BLAST does not achieve channel capacity. However, its simplicity still makes it a very attractive scheme.

Diagonal BLAST

The main problem with H-BLAST is that it does not provide diversity. The first stream, which has diversity order 1, dominates the performance at high SNRs. A better performance can be achieved with the so-called D-BLAST scheme. In this approach, streams are cycled through the different transmit antennas, such that each stream sees all possible antenna elements. In other words, each single transmit stream is subdivided into a number of subblocks. The first subblock of stream 1 is transmitted from antenna 1, the next subblock from antenna 2, and so on.

Decoding can be done stream by stream; again, each decoded block can be subtracted from signals at the other antenna elements and, thus, enhances the quality of the residual signal.

Figure 20.19   Assignment of bitstreams to different antennas for horizontal BLAST and diversity BLAST.

The difference from H-BLAST is that each stream is sometimes in a "good" position in the sense that the other streams have already been subtracted, and thus the SINR is very high, while sometimes it is in a bad position, in the sense that it suffers full interference. Thus, each stream experiences full diversity. As shown by Ariyavisitakul [2000], D-BLAST is also related to the lower capacity bound of Eq. (20.37); let us specialize it to the case when $N_t = N_r$:

$$C_{\text{D-BLAST}} = \sum_{k=1}^{N_t} \log_2\left[1 + \frac{\overline{\gamma}}{N_t}\chi_{2k}^2\right] < C \tag{20.49}$$

Datastreams alternatingly see a channel with diversity order 1 (whose SNR has a pdf that is a chi-square distribution with 2 degrees of freedom χ_2^2), diversity order 2 (chi-square with 4 degrees of freedom, χ_4^2), and so on. Therefore, total capacity can be described by Eq. (20.49).

Structures for channel-state information at the transmitter

When full CSI is available at the TX, transceiver schemes become much simpler, at least conceptually. By multiplying transmitted and received signal vectors by the right and left singular vectors of the channel matrix, respectively, diagonalization of the channel is achieved. Therefore, the different datastreams do not interfere with each other; rather, we just have a number of parallel channels, each of which can be separately encoded and decoded. The difficulties lie instead in the practical aspects of obtaining and using CSI at the TX (as discussed in Section 20.1.6).

20.2.9 Diversity

Multiple antennas can also be used to provide pure diversity. Again, we have to distinguish between systems that have CSI at the TX and systems that do not. The former can achieve beamforming gain in addition to diversity gain, while the latter is restricted to achieving better resistance to fading.

Diversity with channel-state information at the transmitter

Having the CSI at the TX leads to a conceptually simple transceiver structure in the diversity case as well. Since we have a diversity system, the transmit vector is a weighted replica of a single data symbol, $\mathbf{s} = \mathbf{u}c$. Consider again the singular value decomposition of the channel matrix, Eq. (20.28). We then find that the received signal is:

$$\mathbf{r} = \mathbf{W}\Sigma\mathbf{U}^\dagger\mathbf{u}c + \mathbf{n} \tag{20.50}$$

The RX performs a linear combination (summation) of the signals at the different antenna elements, such that the output of the combiner is $\tilde{c} = \mathbf{w}\mathbf{r}$. Choosing $\mathbf{w} = (\mathbf{W}^\dagger)_1$, and $\mathbf{u} = (\mathbf{U})_1$ maximizes the SNR at the RX, where $(\mathbf{W})_1$ is the left singular vector corresponding to the

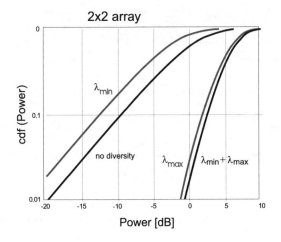

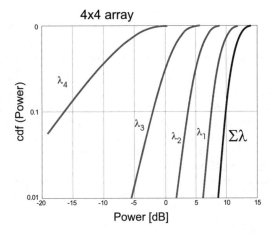

Figure 20.20 Cumulative distribution function of the signal-to-noise ratio in 2×2 (left) and 4×4 (right) multiple-input-multiple-output diversity with independent identically distributed Rayleigh fading at all antenna elements.

largest singular value (and similarly for **U**). In other words, we choose the transmit and receive weights according to the singular value decomposition of the channel. The SNR that can be achieved this way is:

$$\gamma = \frac{P}{\sigma_n^2} \tilde{\sigma}_{max}^2 \tag{20.51}$$

where $\tilde{\sigma}_{max}$ is the largest singular value of matrix **H**. For the case when we have only one transmit antenna, the scheme reduces to maximum-ratio combining; for the case of a single receive antenna, it becomes maximum-ratio transmission. Figure 20.20 shows the cdf of the SNR that can be achieved this way. We can see that the slope of the largest eigenvalue is identical to the slope of the sum of the eigenvalue. In a 4×4 MIMO system, the diversity order is 16, and SNR distribution is almost a step function – in other words, the required fading margin is very small. We also find that there is an enhancement of the average SNR, but it is smaller than $N_r N_t = 16$ (12 dB) (see also Section 20.2.10).

Example 20.5 *Consider a 2×2 MIMO system with a 10-dB average SNR at each of the antenna elements. What is the probability that the SNR is smaller than 7 dB?*

As shown in Andersen [2000], the pdf for the largest eigenvalue of matrix $\mathbf{HH}^\dagger$ in a 2×2 system is:

$$pdf_{\lambda_{\max}}(\lambda) = \exp(-\lambda)[\lambda^2 - 2\lambda + 2] - 2\exp(-2\lambda) \tag{20.52}$$

and the cdf is:

$$cdf_{\lambda_{\max}}(x) = 1 - \exp(-x)(x^2 + 2) + \exp(-2x) \tag{20.53}$$

Here λ describes the normalized SNR. We now want to identify the probability that the largest eigenvalue is 3 dB below the mean value of $\mathbf{HH}^\dagger$; that is:

$$cdf_{\lambda_{\max}}(0.5) = 3 \cdot 10^{-3} \tag{20.54}$$

Of course, we can also ascribe capacity to a diversity system. This capacity follows immediately from SNR statistics:

$$C = \log_2\left[1 + \frac{P}{\sigma_n^2}\max_k(\tilde{\sigma}_k^2)\right] \tag{20.55}$$

Diversity without channel-state information at the transmitter – space–time coding

If the channel is unknown at the TX, then we try to transmit different versions of the datastream from the different transmit antennas. In Chapter 13, we have already encountered some methods that achieve this – especially, delay diversity. Another approach that has gained enormous attention is space–time coding. In this approach, redundancy is introduced by sending from each transmit antenna a differently encoded (and fully redundant) version of the same signal. There are multiple ways in which encoding can be done. In the following, we will first describe the *Alamouti* code, the most popular form of *space–time block codes*. Subsequently, we mention the basic principles of *space–time trellis codes*. The codes we consider here work independently of the number of receive antennas, and can thus be seen as a form of transmit diversity (see also Chapter 13).

Orthogonal space–time block codes The idea behind Space Time Block Codes (STBCs) is to transmit data in a way that guarantees high diversity, while allowing a simple decoding process. The most popular STBC is the Alamouti code [Alamouti 1998]. Its idea is simple yet ingenious. Consider a flat-fading channel, where attenuation from TX antenna 1 is given by h_1 and attenuation from TX antenna 2 to the RX by h_2. Now, transmit the two symbols, c_1 and c_2, from the two transmit antennas at time instant 1:

$$\mathbf{s}_1 = \frac{1}{\sqrt{2}}\begin{pmatrix} c_1 \\ c_2 \end{pmatrix} \tag{20.56}$$

At the second time instant, transmit the vector:

$$\mathbf{s}_2 = \frac{1}{\sqrt{2}}\begin{pmatrix} -c_2^* \\ c_1^* \end{pmatrix} \tag{20.57}$$

The factor $1/\sqrt{2}$ stems from the necessity to keep the energy constant, as we are now using two transmit antennas. Then the received signal can be written as:

$$\mathbf{r} = \begin{pmatrix} r_1 \\ r_2^* \end{pmatrix} = \frac{1}{\sqrt{2}}\begin{pmatrix} h_1 & h_2 \\ h_2^* & -h_1^* \end{pmatrix}\begin{pmatrix} c_1 \\ c_2 \end{pmatrix} + \mathbf{n} = \mathbf{Hc} + \mathbf{n} \tag{20.58}$$

We have thus created a "virtual" MIMO system. It is important to note that the columns of the "virtual" channel matrix $\mathbf{H}$ are orthogonal. Therefore, $\mathbf{H}^\dagger\mathbf{H}$ becomes a scaled identity matrix $\alpha\mathbf{I}$. For decoding, we thus first multiply the received signal vector by $\mathbf{H}^\dagger$. Then we get:

$$\tilde{\mathbf{r}} = \mathbf{H}^\dagger\mathbf{r} = \mathbf{H}^\dagger\mathbf{H}\mathbf{c} + \mathbf{H}^\dagger\mathbf{n} = \alpha\mathbf{c} + \tilde{\mathbf{n}} \tag{20.59}$$

Since the columns of $\mathbf{H}$ are orthogonal, the components of $\tilde{\mathbf{n}}$ are still uncorrelated zero-mean Gaussian, and have variance $\alpha\sigma_n^2$. Therefore, decoding of data c_1 and c_2 becomes decoupled, which greatly reduces the computational effort in the RX.

As far as performance is concerned, we find that α describes the "effective" attenuation by the channel, which is:

$$\alpha = \frac{|h_1|^2 + |h_2|^2}{2} \tag{20.60}$$

Consequently, the diversity order is 2: both h_1 and h_2 would have to be in a fading dip for the effective attenuation to be high. We also see that the scheme can only increase diversity, but not beamforming gain, since it does not have CSI at the transmitter.

There have been many attempts to generalize the Alamouti code to more than two transmit antennas. Unfortunately, it can be shown that orthogonal STBCs for more than two antennas have a rate smaller than 1 – in other words, we cannot even achieve the rate that we could have with a single-antenna system [Tarokh et al. 1999].

Space–time trellis codes STBCs provide full diversity order, but they do not give any beamforming gain, nor do they provide coding gain. Such coding gain can be obtained from Space Time Trellis Codes (STTCs). Given N_t transmit antennas, the STTC maps each symbol from the information source to a *vector* of N_t transmit symbols that are sent from the different antenna elements. Decoding requires a vector Viterbi decoder. The error probability – i.e., the probability of confusing one codeword $\mathbf{C} = (\mathbf{c}_1, \mathbf{c}_2, \ldots, \mathbf{c}_L)$ of length L_c with another codeword $\tilde{\mathbf{C}}$ – is upper-bounded by:

$$P(\mathbf{C} \to \tilde{\mathbf{C}}) \le \left(\prod_{i=1}^{R_e} \lambda_i\right)^{-N_r} \left(\frac{E_S}{4N_0}\right)^{-R_e N_r} \tag{20.61}$$

where R_e and λ_i are the rank and eigenvalues, respectively, of the error matrix:

$$\sum_{i=1}^{L_c} (\mathbf{c}_i - \tilde{\mathbf{c}}_i)(\mathbf{c}_i - \tilde{\mathbf{c}}_i)^\dagger \tag{20.62}$$

The term:

$$\left(\prod_{i=1}^{R_e} \lambda_i\right)^{-N_r} \tag{20.63}$$

represents the coding gain, while the term:

$$\left(\frac{E_S}{4N_0}\right)^{-R_e N_r} \tag{20.64}$$

describes diversity gain. In order to achieve full diversity, the rank of the error matrix should thus be as high as possible (*rank criterion*); in order to achieve high coding gain, the determinant of the error matrix should be maximized (*determinant criterion*). The rank criterion and the determinant criterion provide important guidelines on how to design STTCs.

The main drawback with STTCs is their complexity. The requirement for a vector Viterbi RX has proved to be a major obstacle in their application in practical systems.

20.2.10 Tradeoffs between diversity, beamforming gain, and spatial multiplexing

MIMO systems can be used to achieve spatial multiplexing, diversity, and/or beamforming. However, it is not possible to attain all of those goals simultaneously at their full extent. First, we find that there is a tradeoff between beamforming and diversity gain; this tradeoff also depends on the environment in which we are operating. Consider first an LOS scenario. In this case, it is obvious that the achievable beamforming gain is $N_t N_r$: we form beams at the TX (with gain N_t) and at the RX (with gain N_r), and point them at each other. The gains thus multiply. On the other hand, there is obviously no diversity gain, since there is no fading in an LOS scenario – in other words, the slope of the SNR distribution curve does not change.

In a heavily scattering environment, the slope of SNR distribution changes drastically due to the use of multiantenna elements. If the channel is known at the TX, we can choose our antenna weights in such a way that it is unlikely that the receive signals are in a fading dip; furthermore, even if this happens, it is unlikely that *all* the receive signals are in a fading dip. In other words, it is easy to make sure that *at least one* receive signal has good quality. As discussed in Chapter 13, the diversity order is the slope of the BER versus SNR curve for very high SNRs:

$$d_{\text{div}} = -\lim_{\bar{\gamma} \to \infty} \frac{\log[BER(\bar{\gamma})]}{\log(\bar{\gamma})} \tag{20.65}$$

it can also be related to the slope of the SNR distribution at very low values of the SNR, and is thus a measure for the likelihood that all signals are in a bad fading dip simultaneously. Diversity order in a heavily scattering environment can be shown to be $N_t N_r$. On the other hand, it also turns out that the maximum beamforming gain in such a heavily scattering environment is upper-limited by $(\sqrt{N_t} + \sqrt{N_r})^2$. The reason is that it is not possible to form the transmit beam pattern in such a way that MPCs overlap constructively *at all receive antenna elements simultaneously*. Note the key difference here between achieving full diversity order and beamforming gain.

There is also a fundamental tradeoff between spatial multiplexing and diversity. Zheng and Tse [2003] showed that the optimum tradeoff curve between diversity order and rate r is piecewise linear, connecting the points:

$$d_{\text{div}}(r) = (N_t - r)(N_r - r), \qquad r = 0, \dots, \min(N_t, N_r) \tag{20.66}$$

This implies that maximum diversity order, $N_t N_r$, and maximum rate, $\min(N_t, N_r)$, cannot be achieved simultaneously.

Further reading

An overview of smart antennas is given in the paper by Godara [1997], and the paper collections by Rappaport [1998] and Tsoulos [2001]. The topic of space–time processing, which forms a basis for smart antenna systems, involves many topics we have already discussed in previous chapters (diversity in Chapter 13, Rake receivers in Chapter 18); an overview can be found in Paulraj and Papadias [1997]. Smart antennas for CDMA systems are discussed in Liberti and Rappaport [1999].

For the topic of MIMO systems, the book by Paulraj et al. [2003] gives a good introduction, especially to spatial-multiplexing aspects. Also the review papers by Gesbert et al. [2003] and Diggavi et al. [2004] are "must-reads" for an introduction to the topic. For an understanding of the basic concept of capacity in MIMO systems and derivations of capacity distributions and bounds, the original papers by Foschini and Gans [1998] and Telatar [1999] are still worth reading. Also Andersen [2000] gives important intuitive insights both into capacity and diversity aspects. As for channels that are known at the TX, Raleigh and Cioffi [1998] was the first paper, and still very interesting to read. The possibility of a signaling scheme that allows MIMO communications without CSI at the RX was first pointed out by Marzetta and Hochwald [1999]. Goldsmith et al. [2003] review the information-theoretic capacity of different assumptions about CSI. The impact of channel correlation on capacity is described in Chuah et al. [2002] and Shiu et al. [2000]. MIMO systems with multiple users are discussed in Spencer et al. [2004]. In frequency-selective channels, MIMO is most often combined with OFDM (see Stueber et al. [2004]).

Foschini et al. [2003] is devoted to reviewing different layered space–time structures. MIMO systems with antenna selection are reviewed in Molisch and Win [2004]. For space–time coding, Diggavi et al. [2004] gives an extensive introduction to space–time coding, while Tarokh et al. [1998, 1999] give more mathematical details for STTCs and STBCs, respectively. STBCs are also described in detail in the book by Larsson and Stoica [2003]. Both space–time coding and transmit diversity, as well as other MIMO aspects, are considered in the book by Hottinen et al. [2003].

Part V

STANDARDIZED WIRELESS SYSTEMS

One of the main reasons for the success of wireless systems has been the development of widely accepted standards, especially for cellular communications. These standards ensure that the same type of equipment can be used all over the world, and also for different operators of wireless networks within a country – be they nationwide cellular operators, or a private person who installed an access point for a wireless Local Area Network (LAN) in his/her house, and lets friends access it with their laptops. In this part of the book, we will describe the most important standards for cellular systems, cordless phones, and wireless LANs.

Two things are noteworthy whenever studying wireless standards: the first is that the standard documents themselves are unreadable for anybody but the standards experts. They are not written in a style that any scientist would use – e.g., they do not care about logical derivations, understandability, etc. Rather, they are semilegal documents that only aim for an unambiguous description of *what* should be done in order to ensure compliance with the standard. The reason for choosing a certain modulation format, coding strategy, etc., is only known to those who heard the discussions in a standards meeting, and is more likely to be political than technical anyway. The second point that will occur to even the casual student of a standards document is the huge number of acronyms – another reason why such documents are almost unreadable. The situation is compounded by the fact that each standard uses different acronyms. This is one of the reasons there is a separate list of acronyms for some chapters on the companion website (*www.wiley.com/go/molisch*).

GSM (Global System for Mobile communications), a second-generation cellular standard, is the most successful wireless standard, with more than a billion users worldwide. While little research will be done on this topic in the future, GSM's ubiquity still makes it mandatory for most wireless engineers to have at least a basic understanding of its working. For this reason, Chapter 21 give an introduction to GSM, including both the physical-layer side and some aspects of the networking operations. A rival second-generation system, *IS-95* (often erroneously referred to as CDMA in the newspapers),[1] achieved considerable popularity in the U.S.A. and Korea, and will thus be described in Chapter 22. In the late 1990s, the third-generation cellular system *UMTS* (Universal Mobile Telecommunications System, also known as WCDMA, 3GPP-FDD mode, or simply 3GPP) was standardized, and has been built up since the early 2000s. At the time of this writing, it looks like it will become the dominant cellular

---

[1] Of course, IS-95 is based on Code Division Multiple Access (CDMA), but not every CDMA-based wireless system is an IS-95 system.

system in the world, replacing, in the long run, GSM. Chapter 23 describes both physical-layer aspects and Medium Access Control (MAC)-layer and networking considerations, since they are closely intertwined. Cordless phones are a group of wireless communications devices that have become an important market segment as well. Due to their underlying concept (pure speech communications, linking a mobile station to a permanently assigned base station), they are much simpler. The DECT standard, the dominant standard for such devices, is described on the companion website (*www.wiley.com/go/molisch*). Finally, Chapter 24 describes the *IEEE 802.11* standard (also known as WiFi), which defines devices for wireless communications between computers, and from computers to access points.

This part describes only the most important standards that have been designed for wireless use. Myriads of other standards exist: on one hand, there are other standards for cellular, cordless, and wireless LAN applications; on the other hand, there are standards for many other wireless services that we do not mention here. For example:

- For *second-generation cellular systems, IS-136* Time Division Multiple Access (TDMA) [Coursey 1999], and *PDC* (Pacific Digital Cellular) standards exist, and were used by an appreciable number of users. However, they have never achieved GSM's popularity, as they are used only in specific countries (IS-136 in the U.S.A., PDC in Japan), and started to be phased out in the early 2000s.

- For *third-generation cellular systems*, the 3GPP (Third Generation Partnership Project) standard foresees five different "modes". Actually, each mode is de facto a different standard. In Chapter 23, we only describe the most important of these modes. Furthermore, the *3GPP2* group of standards is a rival standard to 3GPP. 3GPP2 is an evolution of the IS-95 standard. While many features are similar in 3GPP and 3GPP2, the standards are different enough to make them incompatible. 3GPP2 will be briefly discussed in Chapter 22 (a more extensive description is available, e.g., in Garg [2000]).

- For *cordless systems*, the *Personal Handyphone System* (PHS) standard is in widespread use in Japan, while the *Personal Access Communications System* (PACS) standard was used in the U.S.A. ([Noerpel et al. 1996], [Yu et al. 1997]); furthermore, CDMA cordless phones operating in the 2.45-GHz range are now in use in the U.S.A.

- For *wireless LANs*, there had been a battle between IEEE 802.11 and the European *High Performance Local Area Network* (HIPERLAN) standard [Doufexi et al. 2002, Khun-Jush et al. 2002]. However, this battle has been pretty much decided, with 802.11 being the clear winner.

- For *fixed wireless access* – i.e., communications from a base station to fixed user devices (we cannot call them mobile stations now) – the *IEEE 802.16* standard, also known as *WiMax*, promises widespread use ([Anderson 2003], [Eklund et al. 2002], and [Sweeney 2004]). While WiMax makes some provisions for mobility, the *IEEE 802.20* standard (mobile broadband wireless access) puts more emphasis on mobility, and thus it also is a rival to the data services of third-generation cellular systems.

- For *personal area networks*, which allow wireless communications in a range up to about 10 m, the IEEE 802.15 standards have been developed [Callaway et al. 2002]. The *IEEE 802.15.1* standard, also known as *Bluetooth*, is now used for wireless links between headsets and cellphones, and similar applications [Chatschik 2001]. For higher data rates, the *WiMedia-MBOA* standard allows data rates between 100 and 500 Mbit/s, and is used as the physical-layer basis for *wireless USB* and *WiMedia*, two higher layer standards for linking computer components and home entertainment systems, respectively (Chapter 25). An alternative system for that range of data rates is the ultrawideband (UWB) direct-sequence spread spectrum system developed by the *UWB Forum*.

- For *trunking radio* systems, the *TETRA* (TErrestrial Trunked RAdio) standard is widely used in Europe, in addition to a large number of proprietary systems [Dunlop et al. 1999].

21

GSM – Global System for Mobile communications

21.1 Historical overview

The *Global System for Mobile communications* (GSM) is by far the most successful mobile communication system worldwide. Its development started in 1982. The CEPT, predecessor of the *European Telecommunication Standards Institute* (ETSI), founded the *Groupe Speciale Mobile*, with the mandate to develop proposals for a pan-European digital mobile communication system. Two goals were supposed to be achieved:

- First, a better and more efficient technical solution for wireless communications – it had become evident at that time that digital systems would be superior in respect to user capacity, ease of use and number of possible additional services compared with the then-prevalent analog systems.
- Second, a single standard was to be realized all over Europe, enabling roaming across borders. This was not possible before, as *incompatible* analog systems were employed in different countries.

In the following years, several companies developed proposals for such a system. These proposals covered almost all possible technical approaches in different technical areas. For *multiple access*, Time Division Multiple Access (TDMA), Frequency Division Multiple Access (FDMA), and Code Division Multiple Access (CDMA) were suggested. The proposed *modulation techniques* were Gaussian Minimum Shift Keying (GMSK), Frequency Shift Keying (4-FSK), Quadrature Amplitude Modulation (QAM), and Adaptive Differential Pulse Modulation (ADPM). *Transmission rates* varied from 20 kbit/s to 8 Mbit/s. All of the proposed systems were tested both in field tests and with a channel simulator (in Paris in 1986). Apart from technical considerations, marketing and political arguments influenced the decision-making process. FDMA could not be employed, as it would have required antenna diversity at the Mobile Station (MS). Even though the technical feasibility of this diversity had been proven by the Japanese digital system, increased antenna sizes did not make it a desired option. CDMA was ultimately excluded, because the necessary signal processing seemed to be too expensive and unreliable at that time. Therefore, only a TDMA system could survive the selection process. However, the final (TDMA) system was not a proposal from a single company, but rather a compromise system was developed. The reasons for this were of a political and not technical nature: selecting the proposal of one company as the standard would have given this specific company a large competitive advantage. Specific details of the compromise system were

Wireless Communications   Andreas F. Molisch
© 2005 John Wiley & Sons, Ltd

developed by a – now permanent – committee over the following two years and served as the basis for systems implemented in Europe after 1992.

In the early 1990s, it was realized that GSM should have functionalities that had not been included in the original standard. Therefore, the so-called phase-2 specifications, which included these functions, were developed until 1995. Further enhancements, which include packet radio (GPRS, see Appendix 21.6 on the companion website: *www.wiley.com/go/molisch*) and the more efficient modulation of EDGE, have been introduced since then. Because of these extensions GSM is often referred to as the *2.5th generation system*, as its functionalities are beyond those of a second-generation system, but do not enable all third-generation functionalities (UMTS system, compare with Chapter 23).

The success of GSM exceeded all expectations. Though it was originally developed as a European system, it has spread all over the world in the meantime. Australia was the first non-European country that signed the basic agreement (*Memorandum of Understanding*, MoU). Since then, GSM has become *the* worldwide mobile communication standard,[2] with a number of subscribers that exceeded *one billion* in 2004. A few exceptions remain: in Japan and Korea, GSM was never implemented. In the U.S.A., GSM is competing with the CDMA-based IS-95 system. In contrast to most countries where spectral licenses were provided on condition that the network operator would use GSM, the licenses in the U.S.A. were sold without requiring companies to implement a specific system. In 2005, there were two major operators offering GSM-based services, while another two are using rival technologies (see Chapter 22).

There are three versions of GSM, each using different carrier frequencies. The original GSM system uses carrier frequencies around 900 MHz. GSM-1800, which is also called DCS-1800, was added later to support the increasing numbers of subscribers. Its carrier frequencies are around 1,800 MHz, the total available bandwidth is roughly three times larger than the one around 900 MHz, and the maximal transmission power of mobile stations is reduced. Apart from this, GSM-1800 is identical to the original GSM. Thus, signal processing, switching technology, etc. can be reused without changes. The higher carrier frequency, which implies a bigger pathloss, and reduced transmission power reduce the sizes of the cells significantly. This fact, combined with the bigger available bandwidth, leads to a considerable increase in network capacity. A third system, known as GSM1900 or PCS-1900 (*Personal Communication System*) operates on the 1,900-MHz carrier frequency, and is mainly used in the U.S.A.

GSM is an open standard. This means that only the interfaces are specified, not the implementation. As an example we consider the modulation of GSM, which is GMSK. The GSM standard specifies upper bounds for out-of-band emission, phase jitter, intermodulation products, etc. *How* the required linearity is achieved (e.g., by feedforward linearization, by using a class-A amplifier – which is unlikely because of the small efficiency – or by any other method), is up to the equipment manufacturer. Thus, this open standard ensures that all products from different manufacturers are compatible, though they can still differ in quality and price. Compatibility is especially important for service providers. When using proprietary systems, a provider is able to choose the equipment supplier only once – at the beginning of network implementation. For GSM (and other open standards), a provider can first purchase Base Stations (BSs) from one manufacturer but later on buy BSs to extend the capacity of his network from a different manufacturer, which might offer a better price. A provider may also buy some components from one company and other components from another company.

---

[2] Hence the reinterpretation of GSM from "Groupe Speciale Mobile" to "Global System for Mobile communications".

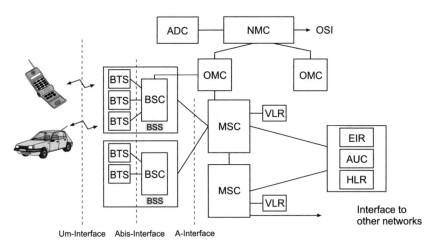

Um-Interface   Abis-Interface   A-Interface

Figure 21.1   Block diagram of a Global System for Mobile communication system.
Adapted with permission from HP [1994] © Hewlett Packard.

21.2 System overview

A GSM system consists essentially of three parts – namely, the *Base Station Subsystem* (BSS), the *Network and Switching Subsystem* (NSS), and the *Operation Support Subsystem* (OSS).

BSS The BSS consists of *Base Transceiver Stations* (BTSs) and the *Base Station Controllers* (BSCs) (see Fig. 21.1). The BTS establishes and maintains the connection to the MSs within its cell. The interface between the MS and the BTS is the air interface, called the *Um-interface* in the GSM context. The BTS hosts, at a minimum, the antennas and the Radio Frequency (RF) hardware of a BS, as well as the software for multiple access. Several – or, rarely, one – BTSs are connected to one BSC; they are either co-located, or connected via landline, directional microwave radio links, or similar connections. The BSC has a control functionality. It is, among other things, responsible for Hand Over (HO) between two BTSs that are connected to the same BSC. The interface between BTS and BSC is called the *Abis-interface*. In contrast to the other interfaces, this interface is not completely specified in the standard.[3] Distribution of the functionalities between BTS and BSC may differ depending on the manufacturer. In most cases, one BSC is connected to several BTSs. Therefore, it is possible to increase the efficiency of implementation by shifting as much functionality as possible to the BSC. However, this implies increased signaling traffic on the link between the BTS and the BSC, which might be undesirable (remember that these links are often rented landline connections). In general the BSS covers a large set of functionalities. It is responsible for channel assignment, maintenance of link quality and HO, power control, coding, and encryption.

NSS The main component of the NSS is the *Mobile Switching Center* (MSC), which controls the traffic between different BSCs (see Fig. 21.1). One function of the MSC is *mobility management*, which comprises all the functions that are necessary to enable true mobility for subscribers. To give but one example, one function of the MSC is the management of HOs that occur when an MS is leaving the area of one BSC and moving into the area covered by another BSC. Other functions are the so-called *paging* and *location update*. All interactions with other

---

[3] Therefore, a set of BTSs always *has* to be combined with a BSC of the same manufacturer.

networks – especially the landline *Public Switched Telephone Network* (PSTN) – are also performed by the MSC.

The NSS includes some databases, too. The *Home Location Register* (HLR) contains all the numbers of the mobile subscribers associated with one MSC and information about the location of each of these subscribers. In the event of an incoming call, the location of the desired subscriber is looked up in the HLR and the call is forwarded to this location.[4] Therefore, we can conclude that from time to time a traveling MS has to send updates of its location to its HLR. The *Visitor Location Register* (VLR) of one MSC contains all the information about mobile subscribers from other networks that are in the area of this MSC and are allowed to roam in the network of this MSC. Furthermore, a temporary number will be assigned to the MS to enable the "host" MSC to establish a connection to the visiting MS.

The *AUthentication Center* (AUC) verifies the identity of each MS requesting a connection. The *Equipment Identity Register* (EIR) contains centralized information about stolen or misused devices.

OSS The OSS is responsible for organization of the network and operational maintenance. More specifically, the OSS mainly covers the following functions:

(i) **Accounting**: How much does a specific call cost for a certain subscriber? There are also plenty of different services and features, from which each subscriber may choose an individual selection included in a specific plan. While this rich choice of services and prices is vital in the marketplace, the administrative support of this individualism is rather complicated. Examples will be discussed in Section 21.10.

(ii) **Maintenance**: The full functionality of each component of the GSM network has to be maintained all the time. Malfunctions may either occur in the hardware or in the software components of the system. Hardware malfunctions are more costly, as they require a technician to drive to the location of the malfunction. In contrast, software is nowadays administrated from a central location. For example, new versions of switching software can be installed in the complete BSS from a central location, and activated all over the network at a specific time. Revision and maintenance software often constitutes a considerable part of the overall complexity of GSM control software.

(iii) **Mobile station management**: Even though all MSs have to pass a type approval, it may happen that "bad apple" devices, which cause systemwide interference, are operating in the network. These devices have to be identified and their further activities have to be blocked.

(iv) **Data collection**: the OSS collects data about the amount of traffic, as well as the quality of the links.

21.3 The air interface

GSM employs a combined FDMA/TDMA approach which further combines with FDD (see Chapter 17). Let us elaborate on these acronyms.

Frequency Division Duplex (FDD) In the first GSM version, frequencies from 890 to 915 MHz and from 935 to 960 MHz were available. The lower band is used for the uplink (connection from the MS to the BS). The upper band is used for the downlink. The frequency spacing between the

---

[4] Actually the call is only forwarded to the BSC in whose area the subscriber is. Routing to and selection of one BTS is the responsibility of the BSC.

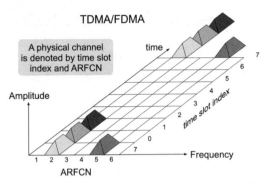

Figure 21.2 Time Division Multiple Access/Frequency Division Multiple Access system.
Adapted with permission from HP [1994] © Hewlett Packard.

uplink and downlink for any given connection is 45 MHz. Therefore, relatively cheap duplex filters are sufficient for achieving very good separation between the uplink and downlink.

For GSM-1800, the frequency ranges are 1,710–1,785 MHz for the uplink, and 1,805–1,880 MHz for the downlink. In North America, 1,850–1,910 MHz are used for the uplink and 1,930–1,990 MHz for the downlink.

Frequency Division Multiple Access (FDMA) Both available frequency bands are partitioned into a 200-kHz grid. The outer 100 kHz of each 25-MHz band are not used,[5] as they are *guard bands* to limit interference in the adjoined spectrum, which is used by other systems. The remaining 124 200-kHz subbands are numbered consecutively by the so-called *Absolute Radio Frequency Channel Numbers* (ARFCNs).

Time Division Multiple Access (TDMA) Due to the very-bandwidth-efficient modulation technique (GMSK, see below), each 200-kHz subband supports a data rate of 271 kbit/s. Each subband is shared by eight users. The time axis is partitioned into timeslots, which are periodically available to each of the possible eight users (Fig. 21.2). Each timeslot is 576.92 μs long, which is equivalent to 156.25 bits. A set of eight timeslots is called a *frame*; it has a duration of 4.615 ms. Within each frame the timeslots are numbered from 0 to 7. Each subscriber periodically accesses one specific timeslot in every frame on one frequency subband. The combination of timeslot number and frequency band is called the *physical channel*. The kind of data that are transmitted over one such physical channel depends on the *logical channel* (see also Section 21.4).

The important features of the air interface are now described in a step-by-step manner.

The assignment of timeslots in the uplink and downlink A subscriber utilizes the timeslots with the same number (index) in the uplink and downlink. However, numbering in the uplink is shifted by three slots relative to the numbering in the downlink. This facilitates the design of the transmitter/receiver, because reception and transmission do not occur at the same time (compare Fig. 21.3).

The modulation technique GSM uses GMSK as a modulation format. GMSK is a variant of Minimum Shift Keying (MSK); the difference is that the data sequence is passed through a filter with a Gaussian impulse response (time bandwidth product $B_G T = 0.3$) (see Chapter 11).

---

[5] This applies to a GSM900 system, and analogously to other frequency bands.

Time slots for uplink and downlink

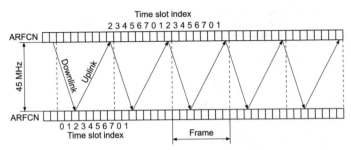

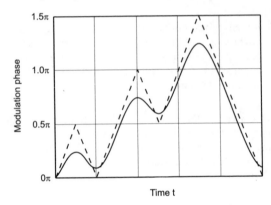

Figure 21.3 The alignment of timeslots in the uplink and downlink.

Adapted with permission from HP [1994] © Hewlett Packard.

Figure 21.4 Phase diagram for the bit sequence 1011011000 for Gaussian minimum shift keying with $B_G T = 0.3$ (solid line) and pure minimum shift keying (dashed line).

This filtering is rather hard. Therefore, the spectrum is rather narrow, but there is a significant amount of Inter Symbol Interference (ISI). On the other hand, the ISI due to delay dispersion of the wireless channel is usually much more severe. Thus, some kind of equalization has to be used anyway. Figure 21.5 illustrates a typical example of a phase trellis of this kind of GMSK and of pure MSK for comparison. The detection method is not specified by the standard. Differential detection, coherent detection, or limiter–discriminator detection might be employed.

Power ramping Were a transmitter to start data transmission right at the beginning of each timeslot, it would have to be able to switch on its signal within a very short time (much shorter than a symbol period). Similarly, at the end of a timeslot, it would have to stop transmitting abruptly, so as not to create interference with the next timeslot. This is difficult to implement in hardware and – even if it could be realized – the sharp transition in the time domain would lead to broadening of the emission spectrum. Therefore, GSM defines a time during which signals are smoothly turned off and on (see Fig. 21.5). Nevertheless, the hardware requirements are still tremendous. In the case when a transmitter is emitting the maximal signal power it has to ramp up from 2×10^{-7} W to 2 W within 28 μs. During the actual transmission of data, on the other hand, signal power may only deviate by 25% (1 dB) from its nominal value.

Principle of "Power rampings"

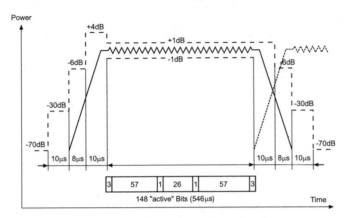

Figure 21.5   Power ramping during a time slot.

Adapted with permission from HP [1994] © Hewlett Packard.

Signal power and power control   GSM provides power control for transmission power. While power control is usually associated with CDMA systems, it also has major benefits in GSM (and other TDMA/FDMA systems):

1. It increases the possible operating time of the batteries. The transmit power amplifier is a major factor in the power consumption of an MS. Therefore, the possible operating time without recharging the batteries depends critically on the emitted signal level. Thus, emitting more power than necessary for maintaining good quality of the received signal at the other link end is a waste of energy.
2. Transmitting at too high a power level increases the interference level in adjacent cells. Because of the cellular concept, every transmitter is a possible interferer for users in other cells that use the same time/frequency slot. However, in contrast to CDMA systems, power control is not *essentially necessary* for operation of the system.

GSM specifies different types of MSs, with different maximum transmission powers though 2-W stations (peak power) are most common. Power control can reduce emitted signal power by about 30 dB; adjustment is done in 2-dB steps. Control is adaptive: the BS periodically informs the MS about the received signal level and the MS uses this information to increase or reduce its transmit power. Maximum power levels of the BSs can vary between 2 and more than 300 W. Also, BSs have a similar power control loop that can decrease output power by some 30 dB.

Out-of-band emission and intermodulation products   The limits for out-of-band emissions are not as severe as, for example, for analog systems. The maximum permitted out-of-band signal power at both BS and MS is roughly $-30\,$dBm, which is a very high value for wireless communications. However, in the band from 890 to 915 MHz (the uplink band), the power emitted by the BS must not exceed $-93\,$dBm. This is necessary because the BS has to receive signals from MSs, with signal levels as low as $-102\,$dBm, in this band. Furthermore, transmit antennas are located close to receive antennas (or even co-located) at the BS, and therefore any out-of-band emission in this band causes severe interference.[6] Similar limits apply for the intermodulation products.[7]

---

[6] Similarly, emissions in the bands used by other systems such as UMTS have strict limits.

[7] Note that intermodulation products can only be found at the BS, as only the BS transmits on several frequencies simultaneously.

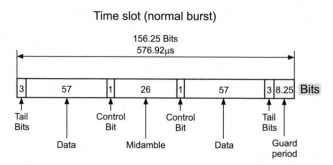

Figure 21.6   Functions of the bits of a normal transmission burst.

Structure of a timeslot   Figure 21.6 illustrates the data contained in a timeslot with a length of 148 bits. However, not all of these bits are payload data. Payload data are transmitted over two blocks of 57 bits. Between these blocks is the so-called *midamble*. This is a known sequence of 26 bits and provides the training for equalization, which will be covered in Section 21.7. Furthermore, the midamble serves as an identifier of the BS. There is an extra control bit between the midamble and each of the two data-containing blocks; the purpose of these control bits will be explained in Section 21.4. Finally, the transmission *burst* starts and ends with three *tail bits*. These bits are known, and enable termination of Maximum Likelihood Sequence Estimation (MLSE) in defined states at the beginning and end of the detection of burst data. This reduces the complexity and increases the performance of decoding (see also Chapter 14). The timeslots end with a *guard period* of 8.25 bits. Apart from "normal" transmission bursts, there are other kinds of bursts. MSs transmit *access bursts* to establish initial contact with the BS. *Frequency correction bursts* enable frequency correction of the MSs. *Synchronization bursts* allow MSs to synchronize to the frame timing of BSs. These bursts will be explained in more detail in Section 21.4.2.

21.4 Logical and physical channels

In addition to the actual payload data, GSM also needs to transmit a large amount of signaling information. These different types of data are transmitted via several *logical channels*. The name stems from the fact that each of the data types is transmitted on specific timeslots that are parts of *physical channels*. The first part of this section discusses the kind of data that is transmitted via logical channels. The second part describes the *mapping* of logical channels to physical channels.

21.4.1 Logical channels

Traffic CHannels (TCHs)

Payload data are transmitted via the traffic channels. The payload might consist of encoded voice data or "pure" data. There is a certain flexibility regarding the data rate: *full-rate traffic channels* (TCH/F) and *half-rate traffic channels* (TCH/H). Two half-rate channels are mapped to the same timeslot, but in alternating frames.

Full-rate traffic channels

- Full-rate voice channels: the output data rate of the voice encoder is 13 kbit/s. Channel coding increases the effective transmission rate to 22.8 kbit/s.
- Full-rate data channels: the payload data with data rates of 9.6, 4.8, or 2.4 kbit/s are encoded with Forward Error Correction (FEC) codes and transmitted with an effective data rate of 22.8 kbit/s.

Half-rate traffic channels

- Half-rate voice channels: voice encoding with a data rate as low as 6.5 kbit/s is feasible. Channel coding increases the transmitted data rate to 11.4 kbit/s.
- Half-rate data channels: payload data with rates of 4.8 or 2.4 kbit/s can be encoded with an FEC code, which leads to an effective transmission rate of 11.4 kbit/s.

Broadcast CHannels (BCHs)

Broadcast channels are only found in the downlink. They serve as *beacon* signals. They provide the MS with the initial information that is necessary to start the establishment of any kind of connection. The MS uses signals from these channels to establish a synchronization in both time and frequency. Furthermore, these channels contain data regarding, for example, cell identity. As the BSs are not synchronized with respect to each other, the MS has to track these channels not only before a connection is established, but all the time, in order to provide information about possible HOs.

Frequency Correction CHannels (FCCHs)   The carrier frequencies of the BSs are usually very precise and do not vary in time, as they are based on rubidium clocks. However, dimension considerations and price considerations make it impossible to implement such good frequency generators in MSs. Therefore, the BS provides the MS with a frequency reference (an unmodulated carrier with a fixed offset from the nominal carrier frequency) via the FCCH. The MS tunes its carrier frequency to this reference; this ensures that both the MS and the BS use the same carrier frequency.

Synchronization CHannel (SCH)   In order to transmit and receive bursts appropriately, an MS not only has to be aware of the carrier frequencies used by the BS but also of its frame timing on the selected carrier. This is achieved with the SCH, which informs the MS about the frame number and the *Base Station Identity Code* (BSIC). Decoding of the BSIC ensures that the MS only joins admissible GSM cells and does not attempt to synchronize to signals emitted by other systems in the same band.

Broadcast Control CHannel (BCCH)   Cell-specific information is transmitted via the BCCH. This includes, for example, *Location Area Identity* (LAI),[8] maximum permitted signal power of the MS, actual available traffic channel, frequencies of the BCCH of neighboring BSs that are permanently observed by the MS to prepare for an HO, etc.

Common Control CHannels (CCCHs)

Before a BS can establish a connection to a certain MS, it has to send some signaling information to all MSs in an area, even though only one MS is the desired receiver. This is

---

[8] A location area is a set of cells, within which the MS can roam without updating any location information in its HLR.

necessary because in the initial setup stage there is no *dedicated* channel established between the BS and a MS. CCCHs are intended for transmission of information to all MSs.

Paging CHannel (PCH)   When a request – e.g., from a landline – arrives at the BS to establish a connection to a specific MS, the BSs within a location area send a signal to all MSs within their range. This signal contains either the permanent *International Mobile Subscriber Identity* (IMSI) or the *Temporary Mobile Subscriber Identity* (TMSI) of the desired MS. The desired MS continues the process of establishing the connection by requesting (via a random access channel) a traffic channel, as discussed below. The PCH may also be used to broadcast local messages like street traffic information or commercials to all subscribers within a cell. Evidently, the PCH is only found in the downlink.

Random Access CHannel (RACH)   A mobile subscriber requests a connection. This might have two reasons. Either the subscriber wants to initiate a connection, or the MS was informed about an incoming connection request via the PCH. The RACH can only be found in the uplink.

Access Grant CHannel (AGCH)   Upon the arrival of a connection request via the RACH, the first thing that is established is a dedicated control channel for this connection. This channel is called the *standalone dedicated control channel*, which is discussed below. This channel is assigned to the MS via the AGCH, which can only be found in the downlink.

Dedicated Control CHannels (DCCHs)

Similar to the traffic channels, the DCCHs are bidirectional – i.e., they can be found in the uplink and downlink. They transmit the signaling information that is necessary during a connection. As the name implies, DCCHs are *dedicated* to one specific connection.

Standalone Dedicated Control CHannel (SDCCH)   After acceptance of a connection request, the SDCCH is responsible for further establishing this connection. The SDCCH ensures that the MS and the BS stay connected during the authentification process. After this process has been finished, a TCH is finally assigned for this connection via the SDCCH.

Slow Associated Control CHannel (SACCH)   Information regarding the properties of the radio link are transmitted via the SACCH. This information need not be transmitted very often, and therefore the channel is called *slow*. The MS informs the BS about the strength and quality of the signal received from serving BSs and neighboring BSs. The BS sends data about the power control and runtime of the signal from the MS to the BS. The latter is necessary for the *timing advance*, which will be explained later.

Fast Associated Control CHannel (FACCH)   The FACCH is used for HOs that are necessary for a short period of time; therefore, the channel has to be able to transmit at a higher rate than the SACCH. Transmitted information is similar to that sent by the SDCCH.

The SACCH is associated with either a TCH or a SDCCH; the FACCH is associated with a TCH.

21.4.2 Mapping between logical and physical channels

The signals of logical channels described above have to be transmitted via physical channels, which are represented by the timeslot number and the ARFCN. In order to better understand

Frames and multiframes

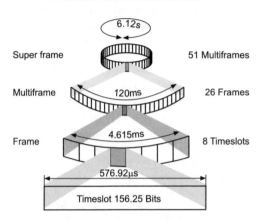

Figure 21.7 Structure of Global System for Mobile communications frames for traffic channels.
Adapted with permission from HP [1994] © Hewlett Packard.

the mapping, we first have to realize that the time dimension is not only partitioned into periodically repeated frames of eight timeslots each, but that these frames and timeslots are the smallest units in the time grid. In fact, multiple frames are combined on different levels to make bigger frames (see Fig. 21.7).

We have already seen above that eight timeslots with a duration of 577 μs each are combined as a frame. The duration of this frame, 4.61 ms, is the basic period of a GSM system. A total of 26 of these frames are combined as a *multiframe*, which has a duration of 120 ms. Furthermore, 51 of these multiframes are contained in one *superframe*, which has a length of 6.12 s. Finally, 2,048 of these superframes are combined into one *hyperframe*, which lasts 3 h and 28 min. The hyperframe is implemented mainly for cryptographic reasons, in order to guarantee privacy over the air interface. Therefore, encryption is applied to the payload data and the period of the encryption algorithm is exactly the length of one hyperframe.

Understanding the multiple frame structure enables us to discuss which timeslot contains which logical channel. Not all timeslots have to be used for the TCH, as the available data rate on the physical channel is $2 \cdot 57$ bits$/4.615$ ms $= 24.7$ kbit/s, while a full rate TCH requires only a 22.8-kbit/s data rate. Therefore, the remaining 1.9 kbit/s may be used for other logical channels.

SACCH As discussed above, 26 frames are combined as a multiframe. Of these 26, only 24 frames are dedicated to the TCH. The 13th (and sometimes the 26th) frame are used by the SACCH. The 26th frame is only employed if two half-rate connections share one physical channel, otherwise the timeslot of the 26th frame is an *idle frame*. The transmission rate of the SACCH is 950 bit/s. The data transmitted via the SACCH is processed differently from the data in the TCH. The bits of four consecutive SACCH bursts are processed together. For this purpose, four multiframes might be combined into a (nameless) higher order frame of length 480 ms. These four SACCH bursts contain 456 bits associated with SACCH data and are used to transmit 184 actual data bits. The data bits are (i) first encoded with a (224, 184) block code, (ii) have four tail bits added, and (iii) then everything is encoded with the regular rate-1/2 convolutional encoder; this leads to the total of $2 \cdot 228$ bit $= 456$ bit.

FACCH An FACCH does not have to be permanently available. It is only necessary in special situations – e.g., when an HO has to be performed. Therefore, no timeslots are *reserved* for the

FACCH. Instead, normal TCH-related bursts of a connection are partly used for FACCH purposes in case this is required. The above-mentioned control bits (stealing bits) between the midamble and the datablocks of a burst indicate whether an FACCH is present in this burst or not – i.e., "steals" bits from the traffic channel. The 184 bits of an FACCH are encoded in the same way as SACCH bits. In order to transmit the resulting 456 bits via the normal TCH timeslots, eight consecutive frames are used: the even payload bits of the first four bursts and the odd bits of the second four bursts are replaced by bits from the FACCH.

Common logical channels   The FACCH and SACCH use the physical channel of the associated connection. This is possible as the physical channel supports a slightly higher data rate than is necessary for one TCH connection. Therefore, it is possible to transmit signaling in timeslots belonging to the same physical channel. However, the other logical signaling channels are not associated with a TCH connection, either because they are required for *establishing* a connection, or because they are used even in the absence of a TCH channel. Therefore, all these channels operate in the first burst of each frame of the so-called "BCCH carrier". This assignment strategy makes sure that one physical channel in each cell is permanently occupied. This leads, of course, to a loss of capacity, especially in cells that use only one carrier. However, there is one option to overcome this: if the cell is full, no new connections can be established. Therefore, no timeslots have to be reserved for signaling related to new connections, and also the first slot of the BCCH carrier can be used for a normal TCH channel.[9] Furthermore, the frames are combined as higher order frames in a different way. A total of 51 frames are combined into a multiframe, which has a duration of 235 ms. Common control channels are unidirectional, with the RACH being the only channel in the uplink, while several common channels exist in the downlink.

RACH   The RACH is necessary only for the uplink. During each multiframe, 8 data bits, encoded into 36 bits, are transmitted via the RACH. These 36 bits are transmitted as an *access burst*. The structure of an access burst has to differ from normally transmitted bursts. At the time the MS requests a connection it is not yet aware of the runtime of the signal from the MS to the BS. This runtime might be in the range from 0 to 100 μs where the maximal value is defined by the maximal cell range of 30 km. Therefore, a larger guard time is necessary to ensure that a random burst does not collide with other bursts in adjacent timeslots. After the connection is established, the BS informs the MS about the runtime and therefore the MS can reduce the size of the guard times by employing *timing advance*, which will be discussed later. A complete random access burst has the following structure. It starts with 8 tail bits, which are followed by 41 synchronization bits. Afterwards, the 36 bits of encoded data and 3 additional tail bits are transmitted. This adds to a total of 88 bits and leaves a guard time of 100 μs at the end, which corresponds to 68.25 bits. As the RACH is the only unassociated control channel in the uplink, the timeslot numbered 0 may be used for a random access burst in every frame.

Common channels in the downlink   The other common channels – such as FCCH, SCH, BCCH, PCH, and AGCH – can only be found in the downlink and have a fixed order in the multiframe. Figure 21.8 illustrates this structure. Remember that only timeslot 0 in each frame carries a CCCH. Of the 51 frames in this multiframe, the last one is always idle. The remaining 50 frames are divided into blocks of 10 frames. Each of these blocks starts with a frame containing the FCCH. Afterwards the SCH is transmitted during the next frame. The first block of frames contains four BCCHs (in frames 3–6) followed by four frames which contain the PCH or AGCH (frames 7–10). The other four blocks of 10 frames also start with the FCCH and SCH frames,

---

[9] Nevertheless, this option is not implemented by most providers.

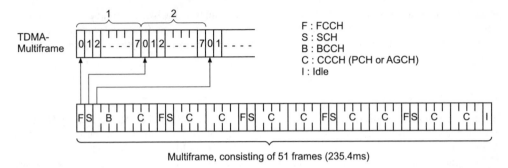

Figure 21.8 Mapping of broadcast channels (FCCH, SCH and BCH) and common control channels to timeslots numbered 0 (compare [CME 20, 1994]).

and then consist either of PCH- or AGCH-carrying frames. The FCCH and the SCH employ bursts that have a special structure (this will be discussed in the next section). As the MSs of neighboring cells continuously evaluate the signal strength of the first timeslot of the frames on the BCCH carrier, the BS always has to transmit some information during these timeslots, even when there is no connection request.

SDCCH The SDCCH may occupy a physical channel by itself, or – in case the common channels do not occupy all the available slots on the BCCH – it may be transmitted during the first timeslots on the BCCH. In the latter case either four or eight SDCCHs share this physical channel.

21.5 Synchronization

Up to now, we have assumed that the BS and the MS are synchronized in time and frequency. However, only the BS is required by the standard to have a high-quality time and frequency reference. For the MS, such a reference would be too expensive. Thus, the MS synchronizes its frequency and time references with those of the BS. This is done in three steps: first, the MS tunes its carrier frequency to that of the BS. Next, the MS synchronizes its timing to the BS by using synchronization sequences. Finally, the timing of the MS is additionally shifted with respect to the timing of the BS to compensate for the runtime of the signal between the BS and MS (*timing advance*).

21.5.1 Frequency synchronization

As mentioned before, the BS uses very precise rubidium clocks, or GPS (Global Positioning System) signals, as frequency references. Due to space and cost limitations, the oscillators at the MS are quartz oscillators, which have much lower precision. Fortunately, this is not a problem, since the BS can transmit its high-precision frequency reference periodically and the MS can adjust its local oscillator based on this received reference. Transmission of the reference frequency is done via the FCCH. As we discussed in the previous section, the FCCH is transmitted during timeslots with index 0 of roughly every tenth frame on the BCCH. An FCCH burst consists of 3 tail bits at the beginning, 142 all-zero bits in the middle, and 3 tail bits at the end. The usual guard period (length equivalent to 8.25 bits) is appended. It should be noted that it is not the carrier frequency that is transmitted as a reference, but rather the carrier modulated with a string of zeros. This equals a sinusoidal signal with a frequency that is the carrier frequency *offset* by the Minimum Shift Keying (MSK) modulation frequency. As this offset is completely deterministic, it does not change the principle underlying the synchronization process.

21.5.2 Time synchronization

Time synchronization information is transmitted from the BS to the MS via the SCH. SCH bursts contain information regarding the current index of the hyperframe, superframe, and multiframe. This is not a lot of information, but has to be transmitted very reliably. This explains the relatively complex coding scheme on the SCH. The MS uses the transmitted reference numbers regarding the multiframes, etc. to set its internal counter. This internal counter is not only a time reference with respect to the timeslot and frame grid, but also serves as a time reference within a timeslot with a quarter-bit precision. This reference is initially adjusted by considering the start of SCH bursts received at the MS. The MS then transmits the RACH burst relative to this internal reference. Based on the reception of the RACH, the BS can estimate the roundtrip time between the BS and the MS and use this information for timing advance (described in the next section).

21.5.3 Timing advance

GSM supports cell ranges of up to 30 km, so that propagation delay between the BS and the MS might be as big as 100 μs. Thus, the following situation might occur: consider user A being at a 30-km distance from the BS, and transmitting bursts in timeslot TS 3 of every frame. User B is located close to the BS and accesses timeslot TS 4. The propagation delay of user A is around 100 μs, whereas the propagation delay of user B is negligible. Therefore, if propagation delay is not compensated, the end of a burst from user A partly overlaps with the beginning of a burst from user B at the BS (this situation is illustrated in Fig. 21.9).

To overcome this problem, the propagation delay from MS to BS is estimated by the BS during the initial phase of establishing a connection. The result is transmitted to the MS, which then sends its bursts *advanced* (with respect to the regular timing structure) to ensure that the bursts *arrive* within the dedicated timeslots at the BS. As the access bursts are transmitted before the MS is aware of the propagation delay, it now becomes clear why they must have a bigger guard period than normal transmission bursts: the guard period must be big enough to accommodate the worst case propagation delay – i.e., an MS at the boundary of a cell with maximum size.

There are some very big cells in rural areas where propagation delays exceed foreseen timing advances. In these cases, it might be necessary to use only every second timeslot as otherwise

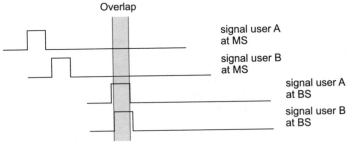

Figure 21.9 Overlapping bursts assuming uncompensated propagation delay.

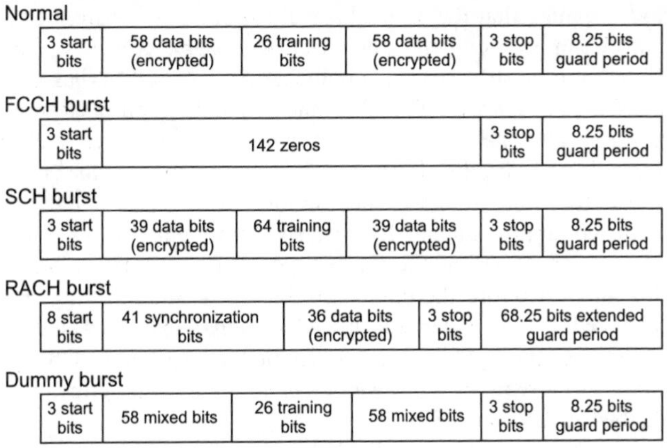

Normal

3 start bits	58 data bits (encrypted)	26 training bits	58 data bits (encrypted)	3 stop bits	8.25 bits guard period

FCCH burst

3 start bits	142 zeros			3 stop bits	8.25 bits guard period

SCH burst

3 start bits	39 data bits (encrypted)	64 training bits	39 data bits (encrypted)	3 stop bits	8.25 bits guard period

RACH burst

8 start bits	41 synchronization bits	36 data bits (encrypted)	3 stop bits	68.25 bits extended guard period

Dummy burst

3 start bits	58 mixed bits	26 training bits	58 mixed bits	3 stop bits	8.25 bits guard period

Figure 21.10 Structure of timeslots in the Global System for Mobile communications.

Reproduced with permission from Rappaport [1996] © IEEE.

timeslots from different users would collide at the BS. This implies a waste of capacity. However, as this might only occur in rural areas with big cell ranges and low subscriber densities, the actual loss for the provider is small.

21.5.4 Summary of burst structures

Finally, Fig. 21.10 provides an overview of the different kinds of bursts and illustrates the functions of their bits.

21.6 Coding

To transmit speech via the physical GSM channel the "speech signals" have to be translated into digital signals. This process should maintain a certain speech quality while keeping the required data rate as low as possible (see also Chapter 15). Different forms of speech coding were considered for GSM, and finally a *Regular Pulse Excited with Long Term Prediction* (RPE-LTP) solution was chosen (see Chapter 15). The digitized speech that is obtained in such a way then has to be protected by FEC in order to remain intelligible when transmitted over typical cellular channels (uncoded BERs of $\sim 10^{-3}$ to 10^{-1}). Both block and convolutional codes are used for this purpose in GSM.

Thus, voice transmission in GSM represents a typical example of the paradox of speech communications. First, redundancy is removed from the source datastream during the speech-coding process, and then redundancy is added in the form of error-correcting coding before transmission. The reason for this approach is that the original redundancy of the speech signal is rather inefficient at ensuring intelligibility of speech when transmitted over wireless channels. In this section, we first describe voice encoding, and subsequently channel coding; these can be seen as important applications of the principles expounded in Chapters 15 and 14, respectively.

21.6.1 Voice encoding

Like most voice encoders (also referred to as *vocoders*), the GSM vocoder is not a classical source-coding processes like, for example, the *Huffman code*. Rather, GSM uses a lossy compression method, meaning that the original signal cannot be reconstructed perfectly, but that the compression and decompression procedures lead to a signal which is similar enough to the original one to allow comfortable voice communications. As GSM has evolved, so has the speech coder. For the first release of GSM, an RPE-LTP approach was used. The idea behind this approach is to consider the human voice as output from a time-varying filter bank which is excited periodically. Both parameters describing the filter bank and the excitation process are transmitted. Since the samples of a voice signal are correlated, any sample can be predicted approximately by linearly combining previous samples. Evidently, correlation reflects the redundancy of the voice signal. However, the correlation properties of the signal vary with time, therefore the filter bank has to be time-varying as well.

Later on, an enhanced speech coder was introduced that improved speech quality without increasing the required data throughput. A more detailed description of GSM speech coding can be found in Appendix 21.A (see *www.wiley.com/go/molisch*), and the principles of speech coding in general are described in Chapter 15.

The data created by the vocoder are divided into different classes, which have different vulnerabilities to bit errors. By this we mean that the bits have different levels of importance for the *perceived* quality of the reconstructed signal. The bits in class 1a are important, as an error in these bits is perceived as a gross distortion of the signal. Therefore, they are protected by a convolutional code and additional block coding. The slightly less important bits of class 1b are protected just by a convolutional code, while the bits associated with class 2 are transmitted without further channel encoding.

Another method to reduce the data rate is *Voice Activity Detection* (VAD). It detects periods when the user is not speaking, and ceases transmission during these periods – this is *discontinuous transmission* (DTX). DTX increases the battery lifetime of the MS and reduces co-channel interference with other users.

21.6.2 Channel encoding

Let us first give an overview of the encoding procedure. Figure 21.11 illustrates the channel coding applied to voice data. For every 20-ms voice signal there are 50 very important bits (class 1a). A block code adds 3 parity bits. This coding is not error *correcting*, but only allows *detection* of bit errors within these 50 bits. The 132 bits of class 1b are attached. After attaching 4 tail bits to determine the final state of the Viterbi decoder a convolutional code with rate-1/2 is applied. This results in 378 bits which are transmitted together with the 78 bits of class 2. Thus, for every 20 ms of the voice signal, 456 bits have to be transmitted. In the following, the details of the different encoder blocks will be discussed.

Block encoding

Block encoding of voice data As discussed above, only class-1a bits of the voice data are encoded using a (53,50) block code. This is a very "weak" block code. It is only supposed to detect bit errors and cannot detect more than three bit errors within the 50 class 1a bits reliably. However, this is sufficient, since a block is completely discarded if an error is detected within the class-1a bits; the receiver then smoothes the resulting signal by "inventing" a block. Figure 21.12 shows the linear shift register representation of the block encoder. As the code is systematic, the 50 data bits pass through the encoder unchanged. However, each of them impacts the state of the shift register. The final state of the shift register determines the 3 parity bits which are attached to the 50 class-1a bits. Class 1a, 1b, and parity check bits are then reordered and interleaved. Finally, four all-zero tail bits are attached, which are needed for the convolutional decoder (see below).

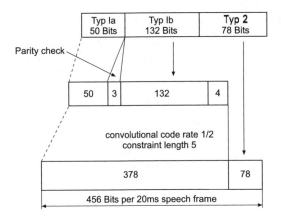

Figure 21.11   Channel coding for voice data in the Global System for Mobile communications.
Reproduced with permission from Rappaport [1996] © IEEE.

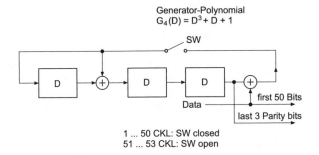

Figure 21.12   Shift register structure for voice block encoding, C1a (53,50) systematic, cyclic block encoder.
Reproduced with permission from Steele and Hanzo [1999] © J. Wiley & Sons, Ltd.

Block encoding of signaling data   As mentioned in Section 21.4.1, the signaling information has to have stronger protection against bit errors than the voice data. While a bit error in voice-related data might lead to an unintelligible audio signal for 20 ms, a bit error in signaling bits can have a more severe impact – e.g., HO to a wrong cell and therefore loss of connection. Thus, higher redundancy is required. For most of the control channels, only 184 signal bits are transmitted within 20 ms (instead of 260 for speech). This allows better error correction. Signaling bits are encoded with a (224,184) Fire code. The Fire code is defined by the generator polynomial:

$$G(D) = D^{40} + D^{26} + D^{23} + D^{17} + D^3 + 1 \tag{21.1}$$

Fire codes are block codes which are particularly capable of correcting burst errors. Burst errors are defined as a series of bit errors, meaning that two or more consecutive bits are wrong; such error bursts occur, for example, when Viterbi decoding fails (see Chapter 14). A total of 4 tail bits are attached to the resulting 224 bits. The result is fed into the convolutional encoder at code-rate-1/2, which is the same as that used for class 1 of the voice signal. For selected logical signaling channels, such as RACH and SCH, different generator polynomials are used. The interested reader is referred to Steele and Hanzo [1999] and the GSM specifications.

Convolutional encoding

Both the class-1 bits of the voice data and all of the signaling information are encoded with a convolutional coder at code-rate-$1/2$ (see Section 14.3). The bits are fed into a 5-bit shift register. For each new input bit, two code bits are calculated according to the generator polynomials:

$$\left.\begin{array}{l} G1(D) = 1 + D + D^3 + D^4 \\ G2(D) = 1 + D^3 + D^4 \end{array}\right\} \tag{21.2}$$

and transmitted. The 4 final tail bits attached to the input sequence ensure that the encoder terminates in the all-zero state at the end of each encoded block.

Interleaving

Due to the nature of fading channels, bit errors may occur in bursts in some transmission blocks – e.g., if those blocks were transmitted during a deep fade. Interleaving orders the bits in such a manner that the burst errors due to the channel are (hopefully) distributed evenly (see Section 14.7.1). Evidently, the more the interleaver distributes corrupted bits, the better. However, latency of the speech signal puts an upper limit on *interleaver depth*: In order to give acceptable speech quality, the delay of the signal should be less than 100 ms.

GSM interleaves the data of two blocks (henceforth called "a" and "b") in the following way: first, each of the blocks is divided into eight subblocks. Specifically, each bit receives an index $i \in \{0, \ldots, 455\}$, and the bits are sorted into subblocks with index $k \in \{0, \ldots, 8\}$ according to $k = i \bmod 8$. Each subblock of block "a" contributes one half of the bits in a transmission burst (114 bits). The other half is associated with subblocks of either a previous or a succeeding block "b". Figure 21.13 illustrates diagonal interleaving.

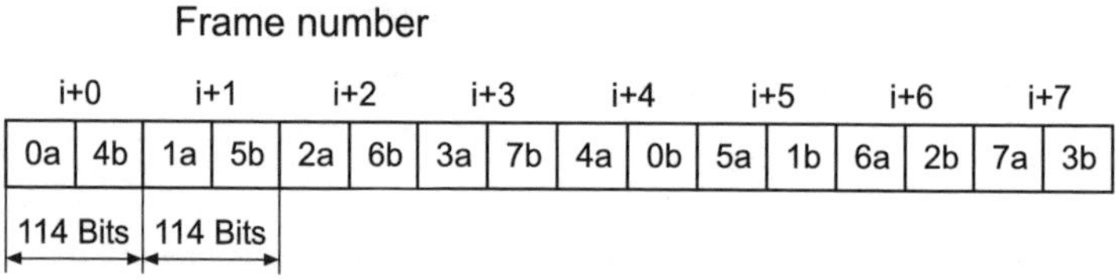

Figure 21.13 Diagonal interleaving for traffic channel/slow associated control channel/fast associated control channel data.

21.6.3 Cryptography

One of the most severe shortcomings of analog mobile communications was the ease with which it could be intercepted. Anybody with a frequency scanner was able to eavesdrop on phone conversations. This posed a threat – e.g., for business people dealing with confidential material. Furthermore, even political scandals have been known to develop as a consequence of eavesdropped conversations.

In a digital system, this problem can be solved by "standard" means: once the audio signal is represented by a bitstream, cryptographic procedures, which had long before been developed for military applications, can be easily applied. For GSM, intercepting a conversation requires a *man-in-the-middle attack*, which involves implementing a BTS, to which the target MS would log on, and forwarding the intercepted signal (to stop the victim noticing the attack) – an exceedingly cumbersome and costly approach. Law enforcement thus typically obtains the collaboration of the network providers, and intercepts conversations not over the air, but *after* the BTS.

Encryption of the transmitted signal is achieved by simply using an XOR operation on the bits on one hand, and a Pseudo Noise (PN)-sequence on the other hand. A PN-sequence is based on feedback linear shift registers and its periodicity is 3.5 hours. Thus, even knowing the sequence does not enable interception, as the listener has to know which part of the sequence is currently in use. The algorithm for encryption of the data, the A5 algorithm, and algorithms involved in authentication, the A3 and A8, were originally only disclosed to members of the MoU. However, they have been reverse-engineered in recent years and successful attacks have been developed. Nevertheless, all these attacks involve a lot of effort and investment. Thus, the GSM air interface still provides the user with a high level of privacy.

21.6.4 Frequency hopping

Slow frequency hopping is an optional feature in GSM, where the carrier frequency is changed for each transmission burst (see also Section 18.1). This helps to mitigate small-scale fading: if the carriers employed are separated by more than one coherence bandwidth of the channel (see Chapter 6), each frame is transmitted over a channel with independent fading realization.[10] As the data belonging to one payload (voice data) packet are interleaved over eight bursts, the probability that all of them are transmitted via bad channels is negligible. This makes it more likely that the packet can be reconstructed at the receiver. A similar effect occurs for (narrowband) interference. In either case (fading and interference), frequency hopping leads to an effective whitening of noise and interference.

The coherence bandwidth of GSM channels can vary from several hundred kHz to a few MHz (see Chapter 7). Given that an operator typically owns only a few MHz of spectrum, and only a subset of frequencies can be used in each cell (see Chapter 17), there can be correlation between the frequency channels used for the hopping. Still, frequency hopping provides some advantages even in this case: co-channel interference from other cells, in particular, is whitened.

In order for the receiver to follow the hopping pattern of the transmitter, both link ends have to be aware of the order in which carrier frequencies are to be used. The control sequence governing this pattern can specify hops over up to 64 different carriers,[11] but it may also specify the "degenerated" case (no hopping) in which one frequency is used over and over.

The BS determines the actual frequency hopping sequence by selecting one of a set of predefined PN-sequences and matching this sequence to the available frequencies for the cell. Furthermore, it informs the MS about the hopping sequence as well as the phase of the sequence – i.e., when to start – during call setup (for details see Steele and Hanzo [1999]). Finally, we note that frequency hopping is *not* applied to the physical channels related to BCHs and CCCHs, as the MS is supposed to "find" them easily.

21.7 Equalizer

Since the symbol duration of GSM is shorter than typical channel delay spreads, ISI occurs, making it necessary to perform equalization (see Chapter 16). However, as GSM is an open standard, neither the structure nor the algorithms of the equalizer are specified. The signal structure just provides the necessary "hooks" (means of implementation), such as a training

---

[10] This is particularly important if the MS is not moving: without frequency hopping, it would "see" the same channel at all times; thus if it is in a fading dip, the Bit Error Rate (BER) would be very high.

[11] As stated above, each cell can only use a subset of the available frequencies. Therefore, the control sequence normally does not specify hops over 64 frequencies but rather, say, 10.

sequence used to estimate the channel impulse response. The most important properties of the training sequence are:

- the training sequence is 26 bits long;
- it is transmitted in the middle of a burst and, hence, is also called the *midamble* – it is preceded by 57 data bits, and followed by another 57 data bits;
- eight different PN-sequences are defined for the midamble – different midambles may be used in different cells, and thus help to distinguish between those cells.

The eight PN-sequences are designed in such a way that their autocorrelation function has a peak of amplitude 26 at zero offset, surrounded by at least five zeros for positive and negative offsets. Therefore, the channel impulse response can be simply estimated by cross-correlating the received midamble with the sequence, as long as the channel impulse response is less than 5 symbol durations long. The cross-correlation thus represents a scaled version of the channel impulse response. This information is used to correct the ISI for all symbols within one burst.[12]

GSM uses a *mid*amble for training as it is supposed to support MS speeds of up to 250 km/h. At this speed the MS covers roughly one-eighth of a wavelength during transmission of a burst ($\sim 500\,\mu s$). The impulse response of the channel shows some variations over this distance. Were the training sequence transmitted at the beginning of the burst (*preamble*), the resulting channel estimate would no longer be sufficiently accurate at the end of the burst. Since training is transmitted in mid-burst, the estimate is still sufficiently accurate at both the start and the end of a burst.

As mentioned above, the GSM standard does not specify any particular equalizer design. Actually, the equalizer is one of the reasons that products from different manufacturers can differ in price and quality. However, most implemented equalizers are Viterbi equalizers. The assumed constraint length of the channel, which relates to the number of states of the trellis, reflects a tradeoff between the complexity and performance of a Viterbi equalizer. Constraint length is identical to the memory of the channel – in other words, the length of the channel impulse response in units of symbol durations. In Chapter 7 we saw that COST 207[13] channel models normally have impulse response lengths of up to 15 μs, which equals 4 symbol durations.[14] We also note that Viterbi equalization can be well combined with convolutional decoding.

We emphasize again that a delay-dispersive fading channel with an appropriate equalizer at the receiver leads to *lower* average bit error probabilities than a flat-fading channel. As the different versions of a symbol arriving at the receiver at different time instants propagate over different paths, their amplitudes undergo independent fading. In other words, delay dispersion leads to delay diversity (see also Chapter 13).

Table 21.1 summarizes the key parameters of GSM.

21.8 Circuit-switched data transmission

When the GSM standard was originally drafted, voice communication was envisioned as the main application. Some data transmission – like the SMS and a point-to-point data transmission

---

[12] Note that the SCH and the RACH use longer training patterns. To simplify implementation, the same algorithm is normally used for equalization of all three different bursts.

[13] COST stands for European COoperation in the field of Scientific and Technical research.

[14] Note that the constraint length of a Viterbi equalizer can be longer due to other effects – e.g., ISI due to the GMSK. Therefore, 4 to 6 is a practical value for constraint length.

Table 21.1 Key parameters of the Global System for Mobile communications.

Parameter	Value

Frequency range
  GSM 900
 	
  GSM 1800
 	
  GSM 1900
	880–915 MHz (uplink)
	925–960 MHz (downlink)
	1710–1785 (uplink)
	1805–1880 (downlink)
	1850-1910 (uplink – U.S.A.)
	1930–1990 (downlink – U.S.A.)

Multiple access	FDMA/TDMA/FDD
Selection of physical channel	Fixed channel allocation/intracell handover/ frequency hopping
Carrier distance	0.2 MHz
Modulation format	GMSK ($B_G T = 0.3$)
Effective frequency usage per duplex speech connection	50-kHz/channel
Gross bit rate on the air interface	271 kbit/s
Symbol duration	3.7 µs
Channels per carrier	8 full slots (13 kbit/s user data)
Frame duration	4.6 ms
Maximal RF transmission power at the MS	2 W
Voice encoding	13 kbit/s RPE-LTP
Diversity	Channel coding with interleaving Channel equalization Antenna diversity (opt.) Frequency hopping (opt.)
Maximal cell range	35 km
Power control	30-dB dynamics

channel with a 9.6-kbit/s data rate – were already included, but were not considered sufficiently important to merit the introduction of much additional complexity. Thus, data transmission was handled in a circuit-switched mode, just like voice transmission. This section provides an overview of the circuit-switched data transmission options in GSM.

In general, the circuit-switched data transmission modes of GSM have severe disadvantages. A main issue is the low data rate of less than 10 kbit/s.[15] Furthermore, the long time needed to set up a connection, as well as the relatively high costs of holding a connection, make it very unattractive, for example, for Internet browsing. There was simply a significant mismatch between the low-data-rate connection-based services offered by GSM, and the new Web

---

[15] The HSCSD mode described below provides higher data rates based on circuit-switched transmission.

applications, which require high data volumes in bursts interrupted by long idle periods.[16] For these reasons, packet-switched (also known as connectionless) transmission (see Section 18.11) was introduced later on. Only SMS text messaging proved to be successful.

21.8.1 Transmission modes

The full-rate TCH provides payload data rates of 9.6, 4.8 and 2.4 kbit/s. Channel coding adds redundancy, so that effectively 22.8 kbit/s are transmitted on a physical channel. In addition, an Automatic Repeat reQuest (ARQ) procedure (see Appendix 14.A at *www.wiley.com/go/molisch*) may be used to further limit the BER. Therefore, we distinguish between two modes:

- In *transparent mode* (T-mode), only FEC is used for error correction. Therefore, data throughput is constant. However, some data packets might suffer from so many transmission errors that the FEC is not strong enough, and a packet error occurs.
- In *non-transparent mode* (NT-mode), an ARQ protocol is added to the FEC. Those packets for which the FEC fails are retransmitted. Therefore, packet errors are negligible; however, data throughput is no longer constant. In a bad channel, the throughput tends to zero, as all time is spent with (unsuccessful) retransmissions.

The ARQ of the NT-mode is a part of the physical layer, making sure that the data source sees an error-free "effective" channel (even if its throughput is low). However, for most data protocols – like TCP/IP – such a high-quality effective channel is not necessary, as these protocols provide ARQ functions themselves, and can thus correct the effect of packet errors. For this reason, the T-mode is more popular, and so, in the following, we will only describe this mode in detail.

21.8.2 Data transmission traffic channels

As mentioned above, the bits of digitized voice have different levels of importance, and are thus protected by FECs of different strength. Bits related to pure data transmission, on the other hand, are assumed to be equally important, as the GSM system is not aware of their meaning. Therefore, they are all encoded in the same way.

TCH/F9.6

For the 9.6-kbit/s mode, a datastream coming from the user is divided into blocks of 48-bit each. A total of 12 bits of auxiliary data (header) are added to each block. The encoder then processes four of these blocks at once. First, 4 all-zero tail bits are appended, resulting in a total of 244 bits. These bits are then encoded using the rate-1/2 convolutional encoder described in Section 21.6.2, resulting in 488 bits. The datastream is *punctured* by deleting every 15th bit starting with the 11th bit. Therefore, 456 bits (equaling four transmission bursts) of encoded data have to be transmitted for each four blocks of the source data. These 456 bits are interleaved over 22 transmission bursts.

TCH/F4.8 and TCH/F2.4

The incoming datastream is again divided into 48-bit blocks, to each of which 12 bits of auxiliary data are attached. A total of 16 all-zero bits are added to the resulting 60-bit block,

---

[16] The mobile communication industry tried to develop an Internet-like application protocol designed to suit the wireless environment better (WAP). However, this was not widely accepted by users.

by inserting 4 all-zero bits after every 15th data bit. Two of the resulting blocks (152 bits total) are processed together by the encoder and the interleaver. The encoder is a code-rate-1/3 convolutional encoder. This results in 456 bits to be transmitted; no puncturing is used. Evidently, the TCH/F4.8 is more robust than the TCH/F9.6 because of the lower code rate and the absence of puncturing. Similarly, the TCH/F2.4 uses a rate-1/6 convolutional code without puncturing.

TCH/H4.8 and TCH/H2.4

The encoding process for the TCH/H4.8 is identical to the one for TCH/F9.6; however, only every second burst is used. This leads to the lower data rate. Similarly, the TCH/H2.4 uses the same encoding as TCH/F4.8, but again uses only every second burst.

TCH/F14.4

The above-described data transmission channels were implemented in phase 2 of GSM. In phase 2+, a circuit-switched high-speed data service was established – i.e., the *High Speed Circuit Switched Data* (HSCSD) service. This service achieves higher data rates by (i) using more timeslots in each frame, and (ii) introducing a new channel with a data rate of 14.4 kbit/s in each frame. The increase in data rate is partly due to a reduction of the overhead of auxiliary data. The incoming datastream is divided into blocks of 36 bits and eight of these blocks are concatenated. *Only* 2 bits of auxiliary data are added to the resulting 288-bit block, and 4 all-zero tail bits are attached at the end. The result is fed into a rate-1/2 convolutional encoder as described in Eq. (21.2). This results in a blocksize of 588 bits. This block is punctured in such a way that 132 bits are deleted, resulting again in a 456-bit block.

21.9 Establishing a connection and handover

In this section we discuss initial establishing of a connection, and the HO procedure, using the logical channels described in Section 20.4. Furthermore, we explore the kind of messages that need to be exchanged during these processes. As a first step, we define various elements of a GSM system that are required for these functionalities.

21.9.1 Identity numbers

An MS or a subscriber can be localized within the network by using identity numbers.[17] An active GSM MS has multiple identity numbers.

Mobile Station ISDN Number (MSISDN) The MSISDN is the unique phone number of the subscriber in the public telephone network. The MSISDN consists of *Country Code* (CC), the *National Destination Code* (NDC), which defines the regular GSM provider of the subscriber, and the subscriber number. The MSISDN should not be longer than 15 digits.

---

[17] Note that we distinguish between a subscriber and the hardware he is using.

International Mobile Subscriber Identity (IMSI) The IMSI is another unique identification for the subscriber. In contrast to the MSISDN, which is used as the phone number of the subscriber within the GSM network *and* the normal public phone network, the IMSI is only used for subscriber identification in the GSM network. It is used by the *Subscriber Identity Module* (SIM), which we explain later, the HLR, and the VLR. It consists again of three parts: the *Mobile Country Code* (MCC, three digits), the *Mobile Network Code* (MNC, two digits), and the *Mobile Subscriber Identification Number* (MSIN, up to ten digits).

Mobile Station Roaming Number (MSRN) The MSRN is a temporary identification that is associated with a mobile if it is not in the area of its HLR. This number is then used for routing of connections. The number consists again of a CC, MNC, and a TMSI, which is given to the subscriber by the GSM network (s)he is roaming into.

International Mobile station Equipment Identity (IMEI) The IMEI is a means of identifying hardware – i.e., the actual mobile device. Let us note here that the three identity numbers described above are all either permanently or temporarily associated with the subscriber. In contrast, the IMEI identifies the actual mobile station used. It consists of 15 digits: six are used for the *Type Approval Code* (TAC), which is specified by a central GSM entity; two are used as the *Final Assembly Code* (FAC), which represents the manufacturer; and six are used as a *Serial Number* (SN), which identifies every MS uniquely for a given TAC and FAC.

21.9.2 Identification of a mobile subscriber

In analog wireless networks, every MS was uniquely identified by a single number that was permanently associated with it. All connections that were established from this MS were billed to its registered owner. GSM is more flexible in this respect. The subscriber is identified by his SIM, which is a plug-in chipcard roughly the size of a postage stamp. A GSM MS can only make and receive calls when such a SIM is plugged in and activated.[18] All calls that are made from the MS are billed to the subscriber whose SIM is plugged in. Furthermore, the MS only receives calls going to the number of the SIM owner. This makes it possible for the subscriber to easily replace the MS, or even rent one for a short time.

As the SIM is of fundamental importance for billing procedures, it has to have several security mechanisms. The following information is saved on it:

- Permanent security information: this is defined when the subscriber signs a contract with the operator. It consists of the IMSI, the authentication key, and the access rights.
- Temporary network information: this includes the TMSI, location area, etc.
- Information related to the user profile: for example, the subscriber can store his/her personal phonebook on the SIM – in this way the phonebook is always available, independent of the MS the subscriber uses.

The SIM can be locked by the user. It is unlocked by entering the *Personal Unblocking Key* (PUK). If a wrong code is entered ten times, the SIM is finally blocked and cannot be reactivated. Removing the SIM, and then plugging it into the same or another MS does not reset the number of wrong trials. This blocking mechanism is an important security feature in case of theft.

The *Personal Identification Number* (PIN) serves a similar function as the PUK. The user may activate the PIN function, so that the SIM requests a four-digit key every time an MS is

---

[18] Emergency calls can be made without a SIM.

switched on. In contrast to the PUK, the PIN may be altered by the user. If a wrong PIN is entered three times, the SIM is locked and may be unlocked only by entering the PUK.

21.9.3 Examples for establishment of a connection

In the following, we give two examples for the steps that are performed when a connection is established. Both the user identification numbers, and the different logical channels (see Section 21.4) play a fundamental role in this procedure.

If a subscriber wants to establish a connection from his MS the following procedure is performed between the MS and the BTS to initialize the connection:

1. The MS requests an SDCCH from the BS by using the RACH.
2. The BS grants the mobile access to an SDCCH via the AGCH.
3. The MS uses the SDCCH to send a request for connection to the MSC. This includes the following activities: the MS tells the MSC which number it wants to call. The authentication algorithm is performed; in this context it is evaluated if the MS is allowed to make a requested call (e.g., an international call). Furthermore, the MSC marks the MS as busy.
4. The MSC orders the BSC to associate a free TCH with the connection. The BTS and the MS are informed of the timeslot and carrier number of the TCH.
5. The MSC establishes a connection to the network to which the call should go – e.g., the PSTN. If the called subscriber is available and answers the call, the connection is established.

A call that is incoming from another network starts the following procedure:

1. A user of the public phone network calls a mobile subscriber, or more precisely, an MSISDN. The network recognizes that the called number belongs to a GSM subscriber of a specific provider, since the National Destination Code in the MSISDN contains information about the network. The PSTN thus establishes a connection to a gateway MSC[19] of the GSM provider.
2. The gateway MSC looks in the HLR for the subscriber's information and the routing information (the current location area of the subscriber).
3. The HLR translates the MSISDN into the IMSI. If call forwarding is activated – e.g., to a voicemail box – the process is altered appropriately.[20]
4. If the MS is roaming, the HLR is aware of the MSC it is connected to, and sends a request for the MSRN to the MSC that is currently hosting the MS.
5. The hosting MSC sends the MSRN to the HLR. The gateway MSC can now access this information at the HLR.
6. As the MSRN contains an identification number of the hosting MSC, the gateway can now forward the call to the hosting MSC. Additional information – e.g., the caller ID – is included.
7. The hosting MSC is aware of the *Location Area* (LA) of the mobile. The LA is the area controlled by one BSC. The MSC contacts this BSC and requests it to page the MS.
8. The BSC sends a paging request to all the BTSs that cover the LA. These transmit the paging information via the broadcast channel.
9. The called MS recognizes the paging information and sends its request for an SDCCH.
10. The BSC grants access to an SDCCH via the AGCH.

---

[19] A gateway MSC is an MSC with a connection to the regular phone network.
[20] This information is found in the HLR as well.

11. Establishment of the connection via the SDCCH follows the same steps as described in bullets 3 and 4 of the "MS-initiated call". If the mobile subscriber answers the incoming call, the connection is established.

21.9.4 Examples of different kinds of handovers

A *handover* is defined as the procedure where an active MS switches the BTS with which it maintains a link; it is a vital part of mobility in cellular communications. Handover is performed when another BTS is capable of providing better link quality than the current one. In order to determine whether another BTS could provide better link quality, the MS monitors the signal strength of the broadcast channel of neighboring BTSs. Since the broadcast channel does not use power control, the MS measures the maximum signal power available from other BTSs. It transmits the results of these measurements to the BSC. Furthermore, the currently active BTS measures the quality of the uplink and sends this information to the BSC. Based on all this information, the BSC decides if and when to initiate an HO. Since the MS contributes to the HO decision, this procedure is called *Mobile Assisted Hand Over* (MAHO).

Let us now consider some more details in this procedure:

- The received signal strength from different BTSs is averaged over a few seconds (the exact value is selected by the network provider); this ensures that small-scale fading does not lead to HO. Otherwise, an MS exposed to a similar signal strength from two BTSs would constantly switch from one BTS to the other by just moving over a small distance.
- Receive power is measured at 1-dB resolution in the range of -103 dBm to -41 dBm. The lower bound reflects the sensitivity of GSM receivers – i.e., the minimum signal power required for communication.
- Furthermore, HO is initialized when the necessary timing advance exceeds the specified limit of 235 µs. If the MS is so far away from the BTS that a bigger timing advance is necessary, HO is made to a closer BTS.
- Even more importantly, an HO is initialized when the signal quality becomes too low due to interference.
- The BS transmits (via the BCCH) several parameters that support the HO procedure.

In the following, we describe three different types of HOs: the most simple involves only BTSs controlled by the same BSC. A more complex case arises if two different BSCs are connected to the same MSC. The most complex case involves different MSCs.

Case 21.1 – HO between BTSs belonging to thee same BSC The steps for this case are illustrated in Fig. 21.14:

1. The BSC orders the new BTS to activate a new physical channel.
2. The BSC uses the FACCH of the link between the MS and the old BTS in order to transmit information about the carrier frequency and timeslot of the physical channel for the new BTS.
3. The MS switches to the new carrier frequency and timeslot and sends Hand Over (HO) access bursts. These bursts are similar to RACH bursts: they are shorter than normal transmission bursts, as the necessary timing advance is unknown and has to be evaluated first by the new BTS.
4. After the new BTS has detected the HO bursts it sends the necessary timing advance and power control information to the MS via the FACCH of the new channel.
5. The MS informs the BSC that the HO was successful.
6. The BSC requests the old BTS to switch off the old channel.

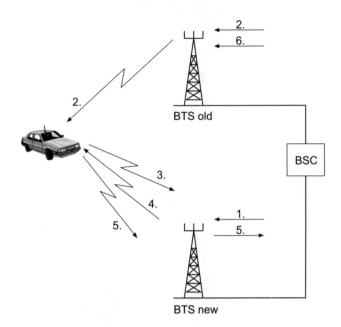

Figure 21.14   Handover between two base transceiver stations of the same base station controller.

Case 21.2 – HO between two BTSs that are controlled by different BSCs but the same MSC   The steps for this case are illustrated in Fig. 21.15:

1. The old BSC informs the MSC that an HO to a specific BTS is necessary.
2. The MSC knows which BSC controls this BTS (the new BTS) and informs this new BSC about the upcoming HO.
3. The new BSC requests the new BTS to activate a physical channel.
4. The new BSC informs the MS about the carrier frequency and timeslot for the new link. This information goes via the MSC, the old BSC, and the old BTS, and is finally transmitted to the MS on the FACCH of the old BTS.
5. The MS switches to this new carrier frequency and timeslot and transmits access bursts (compare item 3 in Case 21.1).
6. After detecting HO bursts the BTS transmits information regarding timing advance and power control to the MS via the FACCH.
7. The MS informs the old BSC about the successful HO (via the new BSC and the MSC).
8. The new BSC instructs the old BSC (via the MSC) to relinquish the connection to the MS.
9. The old BSC instructs the old BTS to deactivate the old physical channel.

Case 21.3 – HO between two BTSs which are associated with two different MSCs   The steps for this case are illustrated in Fig. 21.16:

1. The old BSC informs its own MSC (in the following called "MSC-A") about the necessary HO.
2. MSC-A recognizes that the requested HO involves a BTS associated with another MSC (in the following called "MSC-B") and contacts this MSC-B.

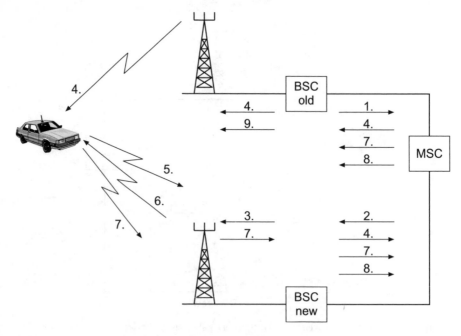

Figure 21.15 Handover between two cells belonging to different base station controllers but the same mobile switching center.

3. MSC-B associates a handover number with the process, so that it is able to reroute the connection. Subsequently, it informs the new BSC about the upcoming HO.
4. The new BSC orders the new BTS to activate a physical channel.
5. MSC-B gets the information about the carrier frequency and timeslot of the new physical channel and forwards this information to MSC-A. Furthermore, it informs MSC-A about the HO number of the connection.
6. A connection between MSC-A and MSC-B is established.
7. MSC-A informs the MS about the carrier frequency and timeslot of the new physical channel. The information goes from MSC-A via the old BSC and the old BTS, whence it is transmitted to the MS via the FACCH.
8. As in Cases 21.1 and 21.2, the MS transmits HO bursts on the new physical channel.
9. After detecting the HO bursts, the new BTS instructs the MS about power control and timing advance.
10. The MS informs MSC-A about the successful HO; this information goes via the new link over the new BTS, the new BSC and MSC-B. From this time on, MSC-A forwards the connection to MSC-B. However, the connection is still maintained by MSC-A. MSC-A acts therefore as a so-called *anchor MSC*.
11. The old physical channel is deactivated.
12. After the connection ends, the new LA of the MS is established. Therefore, the VLR of MSC-B informs the HLR that the MS is now in its area and the HLR updates its entries regarding the location of the MS. Furthermore, the HLR requests the VLR of MSC-A to delete all entries associated with the MS.

We see from these examples that the HO procedure depends on where in the network the switching centers are located.

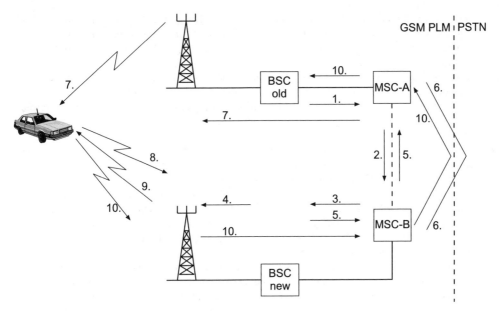

Figure 21.16   Handover between two cells belonging to two different mobile switching centers.

21.10 Services and billing

21.10.1 Available services

In contrast to analog cellular networks, GSM offers a variety of services in addition to regular phone calls. Although providing those services is not a big effort for the network provider, they were a major motivation for customers to switch from analog systems to GSM or another digital mobile phone system. We thus briefly discuss in the following the services offered by GSM, distinguishing between (i) *teleservices*, (ii) *bearer services*, and (iii) *supplementary services*. Teleservices provide a connection between two communication partners, though they may use some additional devices to exchange information via this connection. Bearer services allow the subscriber to access another system. Thus, they provide a connection from the MS to an access point to another network (not a specific terminal in another network). Supplementary services support or manage both teleservices and bearer services.

Teleservices

- *Regular phone calls*: this is still the most common application of GSM.
- *Dual Tone Multi Frequency* (DTMF): this is a signaling method to allow the subscriber to control a device connected to a phoneline using the keypad of the MS. A typical example is the remote checking of an answering machine connected to a regular landline.
- *Emergency calls*: calls to emergency numbers have priority in GSM. If a cell is already full, an emergency call leads to interruption of another connection. Remember also that emergency calls can be made from any MS, even without a valid SIM.
- *FAX*: the protocol that is used to transmit *CCITT group 3* faxes is incompatible with regular GSM connections. The classical fax protocol was developed to transmit a picture that was converted into a specific digital form over an analog phoneline. GSM, on the other hand, provides either voice connections with a voice-encoding function or data transmission

channels for digital data. To send or receive a fax via GSM, adapters are thus required both at the GSM network end and at the MS. The *Terminal Adapter Function* (TAF) provides a general interface between GSM devices and other devices, and is combined with a special fax adapter.

- *SMS*: a short message consists of up to 160 alphanumerical characters. A short message may be transmitted from or to an MS. If it is transmitted to an MS that is switched off, the message is accepted by the GSM system and stored in a dedicated database (*short message service center*). The message is delivered once the MS is switched on and its location is known.
- *Cell broadcast*: a cell broadcast transmits a broadcast message of up to 93 alphanumeric characters to all the MSs within the range of a BTS. This may be used to transmit, for example, news regarding the local (car) traffic situation.
- *Voicemail*: the voicemail service allows the subscriber to forward incoming calls to a service center of the GSM network that acts like an answering machine. After listening to a message from the subscriber the calling party may leave a voice message, which the subscriber can retrieve by connecting to this service center. Typically, a subscriber is notified about new messages via SMS.
- *Fax mail*: this is a service that allows the subscriber to forward incoming faxes to a special service center in the GSM network. The faxes can be retrieved by the subscriber by connecting to this center either with a GSM connection or from the PSTN.

Bearer services

- *Connections to the PSTN*: this allows the user to connect, for example, to modems connected to an analog landline.
- *Connections to the ISDN*: all digital information can be transmitted over the digital network.
- *Connections to the packet-switched networks*: the user may also access packet-switched networks.

Supplementary services

GSM enables the network provider to offer the subscriber a variety of additional, supplementary services:

- *Call forwarding*: the user can select under which conditions calls to his/her mobile number are forwarded to another number: (i) always; (ii) in case the MS cannot be reached; (iii) the user is making or receiving another call; or (iv) the user does not answer after a specified number of rings.
- *Blocking of outgoing calls*: the user or the provider may block outgoing calls under some of the following conditions: (i) all outgoing calls; (ii) all international calls; or (iii) all international calls, with the exception of those to the country of origin in case the subscriber is abroad.
- *Blocking of incoming calls*: this feature is of interest when the subscriber has to pay part of (or all of) the charges for an incoming call. The feature may be activated always or when the user is roaming out of the original network.
- *Advice of charges*: the user may be able to access an estimation of the call charges.
- *Call hold*: the user may put a connection on hold to make or receive another call and then continue with the first connection.

- *Call waiting*: during a call the subscriber may be informed about another incoming call. He/ she may either answer this call by putting the other call on hold or reject it. This feature is available for all circuit-switched connections except emergency calls.
- *Conference calls*: this feature enables connection to multiple subscribers simultaneously. It is only possible for normal voice communications.
- *Caller ID*: the phone number of the incoming call is displayed.
- *Closed groups*: subscribers in GSM, ISDN, and other networks may be defined as a specific user group. Members of this group can, for example, be allowed to make calls only within the group.

21.10.2 Billing

In GSM, billing for to the variety of different subscriber plans is not only an economics issue but also a technical challenge that involves the design of an *Operation Support Subsystem* (OSS). In contrast to the regular public phone system, not all fees have to be paid by the party initiating the calls. Furthermore, accounting for supplementary services has to be done separately. To give an impression of the complexity of the accounting involved we discuss one particular example here.[21]

Example 21.1   *Billing in the Global System for Mobile communications.*

The example involves the following communication parties:

- Subscriber A originates from Austria but is temporarily in Poland.
- Subscriber B is an English subscriber staying in France with a rented MS but his own English SIM.
- Subscriber C is Italian. He is on vacation and the option "If user does not answer forward call to Subscriber B" is activated.
- Subscriber D is a subscriber to a U.S. service, but is currently in Mexico.

Communication now follows the steps below:

1. Subscriber A calls subscriber C in Italy.
2. As subscriber C is not answering, the call is forwarded to subscriber B.
3. Subscriber B is in France and is currently speaking on his MS. Therefore, subscriber A activates the option "automatic call to busy MS". Thus, the MS automatically initiates a call the moment the other MS is no longer busy.
4. After subscriber B finishes his conversation the MS of subscriber A initiates a connection to the MS of subscriber B.
5. This connection is first routed to England, where the HLR of subscriber B is located.
6. From there it gets forwarded to France, where subscriber B is right now.
7. During the conversation subscriber B needs some information from subscriber D. Therefore, he initiates a "conference call" and calls subscriber D.
8. The call to subscriber D is first routed to the U.S.A. and from there to Mexico, where subscriber D is temporarily staying.

Now the question arises as to which subscriber is charged for which fees?

---

[21] Note that we base the discussion of this example partly on European billing procedures. This is fundamentally different from U.S. billing. In the U.S. mobile numbers are similar to landline numbers. Therefore, the calling party just pays the regular fees for a landline call, whereas the mobile subscriber pays the landline-to-mobile fees even for incoming calls.

- Subscriber A has to pay the fees for a call from Poland to Italy. He has to pay both the "international call" charges, and the roaming fees (as he is not in his home country). Furthermore, he has to pay for the service "automatic call to busy MS".
- Subscriber B has to pay for a connection from England to France (for the incoming call), the charges for an international call (from France to the U.S.A.), and the roaming charges (initiating a call while being in a different network). Further he has to pay for the "conference call" feature.
- Subscriber C has to pay for the connection from Italy to England and the "call forwarding" feature involved.
- Subscriber D has to pay for a "received call" in the U.S. (note that in the U.S.A. the called party pays for a received call the same way as for an active call), and the roaming fees from the U.S.A. to Mexico.

We see that for the same conversation different subscribers get charged different fees depending on their roaming. Subscribers do not have to be actively involved in the conversation to be charged (see subscriber C in the above example). This example gives a taste of the complexity of the billing software in the OSS.

21.11 Appendices

Please go to *www.wiley.com/go/molisch*

Further reading

This current chapter is of course only a brief overview of GSM technology. Much more detailed information may be found in the GSM specifications. Note, however, that they encompass 5,000 pages and are also written as a technical specification and not as a textbook, and most engineers only read small sections of them. Another useful source of information is the monograph by Mouly and Pautet [1992]. A detailed GSM chapter can also be found in Steele and Hanzo [1999], Steele et al. [2001], and Schiller [2003]. GPRS is discussed in Cai and Goodman [1997]; GSM network aspects are discussed in Eberspaecher et al. [2001].

22

IS-95 and CDMA 2000

22.1 Historical overview

Direct-sequence spread spectrum communications has a long history, dating back to the middle of the 20th century (see Chapter 18). However, it was deemed to be too complex for commercial applications for a long while. In 1991, the U.S.-based company Qualcomm proposed a system that was adopted by the *Telecommunications Industry Association* (TIA-USA) as *Interim Standard 95* (IS-95). This system became the first commercial Code Division Multiple Access (CDMA) system that achieved wide popularity. In the years after 1992, cellular operators in the U.S.A. started to switch from analog (AMPS) to digital communications. While the market remained fragmented, IS-95 was adopted by a considerable number of operators, and by 2005 was used by two of the four major operators in the U.S.A. It also obtained a dominant market position in South Korea.

The original IS-95 system also did not fully exploit the flexibility inherent in CDMA systems; however, later refinements and modifications did make the system more flexible, and thus ready for data communications. In the late 1990s, the need for further enhancement of data communications capabilities became apparent. The new, third-generation systems needed to be able to sustain high data rates, thus enabling audio- and videostreaming, Web browsing, etc. This would require higher data rates, and flexible systems that could easily support multiple data rates with fine granularity. CDMA seemed well-suited to this approach, and was chosen by all major manufacturers. However, no unique standard evolved. The IS-95 proponents (mostly U.S.-based) developed the so-called CDMA 2000 standard, which is backward-compatible with IS-95, and allows a seamless transition. Operators that were using the Global System for Mobile communications (GSM) standard for their second-generation system are generally opting for the Wideband Code Division Multiple Access (WCDMA) standard described in Chapter 23. CDMA 2000 and WCDMA are quite similar, but have enough differences to make them incompatible.

The main part of this chapter describes the original IS-95 standard. The Appendices (see *www.wiley.com/go/molisch*) then summarize the changes in the CDMA 2000 standard.

22.2 System overview

IS-95 is a CDMA system with an additional Frequency Division Multiple Access (FDMA) component. The available frequency range is divided into frequency bands of 1.25 MHz; duplexing is done in the frequency domain. In the U.S.A., frequencies from 1850–1910 MHz

---

Wireless Communications   Andreas F. Molisch
© 2005 John Wiley & Sons, Ltd

are used for the uplink, and 1930–1990 MHz are used for the downlink band.[1] Within each band, traffic channels, control channels, and pilot channels are separated by the different codes (chip sequences) with which they are spread. IS-95 specifies two possible speech coder rates: 13.3 or 8.6 kbit/s. In both cases, coding increases the data rate to 28.8 kbit/s. The signal is then spread by a factor of 64, resulting in a chip rate of 1.2288 Mchip/s. Theoretically, each cell can sustain 64 speech users. In practice, this number is reduced to 12–18, due to imperfect power control, non-orthogonality of spreading codes, etc.

The downlink signals generated by one Base Station (BS) for different users are spread by different Walsh–Hadamard sequences, and thus orthogonal to each other. This puts an upper limit of 64 channels on each carrier. In the uplink, different users are separated by spreading codes that are not strictly orthogonal. Furthermore, interference from other cells reduces signal quality at the BS and Mobile Station (MS).

BSs use transmit powers between 8 and 50 W, depending on the coverage area required. MSs use peak powers of some 200 mW; accurate power control makes sure that all signals arriving at the BS have the same signal strength. Traffic channels and control channels are separated by different spreading codes. All BSs are synchronized, using signals from GPS (Global Positioning System) to obtain an accurate system time. This synchronization makes it much easier for the MS to detect signals from different BSs and manage the handover from cell to cell.

The requirements for the network and switching system, as well as operating support, servicing, and billing, are quite similar to those in GSM, and will not be repeated here. Similarly, billing considerations are the same as in GSM.

22.3 Air interface

22.3.1 Frequency bands and duplexing

As mentioned in Section 22.2, IS-95 is a CDMA system with an additional FDMA component; it uses Frequency Domain Duplexing (FDD) for separation of uplink and downlink. In the U.S.A., IS-95 operates in the 1850–1990-MHz frequency band. This band, called the Personal Communication System (PCS) band, is divided into units of 50 kHz width, so that the center frequencies are related to the channel numbers n_{ch} as:

$$f_c = (1850 + n_{ch} \cdot 0.05)\text{MHz} \quad \text{for the uplink} \tag{22.1}$$

$$f_c = (1930 + n_{ch} \cdot 0.05)\text{MHz} \quad \text{for the downlink} \tag{22.2}$$

where $n_{ch} = 1, \ldots, 1199$. An IS-95 system requires 1.25 MHz – i.e., 25 of these channels. The preferred center frequencies are $k = 25 n_{ch}$, with $k = 1, \ldots, 47$.[2]

There is also a version operating in the 800-MHz band (historically speaking, this was the first band to be used for IS-95).

22.3.2 Spreading and modulation

IS-95 uses error correction coding, as well as different types of spreading codes, in order to spread the source data rate of 8.6 kbit/s (for rate-set-1) or 13.3 kbit/s (for rate-set-2) to a chip

---

[1] The IS-95 uses the word "reverse link" for the uplink, and "forward link" for the downlink. In order to stay consistent with the notation in our book, we stick here with up- and downlink. However, note that many abbreviations of the IS-95 standard use "F" (for forward) and "R" (for reverse) to indicate whether a channel is used in downlink or uplink.

[2] With the exception of $k = 12, 16, 28, 32, 36$.

rate of 1.2288 Mchip/s. Encoding is usually done with standard convolutional encoders (between rate 1/3 and 3/4), while spreading is done with so-called "M-ary orthogonal keying" and/or multiplication by spreading sequences. More details will be discussed in Section 22.4.

22.3.3 Power control

Power control is one of the key aspects of a CDMA cellular system, and the accuracy of power control determines to a large extent the capacity of the system (see also Chapter 18). For this reason, IS-95 foresees a quite involved power control procedure that is intended to make sure that both large-scale and small-scale fading is compensated by power control. Note that there are two measures for the quality of power control: accuracy and speed. Accuracy indicates how much the received power deviates from the ideal level in a "steady state" where the received power changes slowly, or not at all. The speed of power control determines how quickly power control can adapt to changing channel conditions.

IS-95 foresees both open-loop and closed-loop mechanisms. For the open-loop mechanism, the MS observes the power it receives from downlink signals, reaches a conclusion about the current pathloss, and adjusts its transmit power accordingly. This mechanism can only compensate for shadowing and average pathloss, but not for small-scale fading (see Section 20.1.6). At the start of a call, in particular, open-loop power control is the only available method for controlling MS power.

The closed-loop mechanism uses two different feedbacks: inner-loop and outer-loop power control. In both loops, the BS observes the signal it receives from an MS, and then sends a command to that MS to adjust its power appropriately. In the inner loop, the BS observes the Signal-to-Interference and Noise Ratio (SINR), and as a consequence requests an adjustment of the transmit power of the MS. This is done at 1.25-ms intervals, and the command sent to the MS is to either increase or decrease the power by 1 dB. In the outer loop, the BS appraises the performance of the closed loop using the statistics of the frame quality of uplink transmission. If the frame error rate is too high, closed-loop power control is used to request transmission at a higher power; specifically, the SINR target is adjusted. This is done once at 20-ms intervals (frame).

Power control also exists for the downlink. This rather crude power control allows power to be adjusted only in a rather limited range (about ±6 dB around the nominal value), and at a low speed (once per frame). It is based on the closed-loop scheme, where the MS measures the received signal quality, and requests an adjustment to BS transmit power. The MS sends the ratio of the number of bad frames and number of total frames to the BS in a *power measurement report message*. It is worth noting that power control in the downlink is *not* essential for functioning of the system. The different signals all arrive at the MS after having gone through the same channel, and thus suffer from the same amount of fading.

22.3.4 Pilot signal

Each BS sends out a pilot signal that the MS can use for timing acquisition, channel estimation, and to help with the handover process. The pilot signal always has the same form (a PN-sequence with 32,768 chips – i.e., 26.7 ms long). Pilots from different BSs are offset with respect to each other by 64 chips, which corresponds to a 52.08-μs offset. This is usually long enough that pilots from other BSs are not confused with long-delayed echoes of the desired pilot (though exceptions exist, see Chapter 7). By just correlating the received signal with the pilot Pseudo Noise (PN)-sequence, the MS can determine the signal strength of all the BSs in its surroundings (this will be discussed more in Section 22.7).

22.4 Coding

Before a speech signal can be transmitted over the air interface, it first has to be digitized and encoded. IS-95 foresees the use of different speech coders (vocoders, see Chapter 15). They have different bit rates. As spreading and modulation should not be affected by these different rates, it also implies that error correction coding has to be different: it has to be designed in such a way that output from the channel coder is always 19.2 kbit/s for the downlink and 28.8 kbit/s for the uplink.

22.4.1 Speech coders

IS-95 uses a number of different speech encoders. The original system foresaw a vocoder with 8.6 kbit/s, the IS-96A coder. However, it turned out to have poor speech quality: even in the absence of transmission errors, speech quality was unsatisfactory, and quickly degraded as the frame error rate of transmission increased. For this reason, the CDG-13 vocoder was introduced soon after. This coder (also known as *Qualcomm Code Excited Linear Prediction*, QCELP) is actually a variable-rate speech coder that dynamically selects one of three to four available data rates (13.3, 6.2, 1, and possibly 2.7 kbit/s) for every 20-ms speech frame, depending on voice activity and energy in the speech signal. The encoder then determines the formant, pitch, and codebook parameters, which are required for CELP algorithms (see Chapter 15). The pitch and codebook parameters are determined in an "analysis by synthesis" approach (again, see Chapter 15), using an exhaustive search over all possible parameter values. This search is computationally intensive, and was – especially in the early days of IS-95 – one of the major complexity factors, especially for MSs. For the 13.3-kbit/s mode, each packet (representing 20 ms) consists of 32 bits for Linear Predictive Coding (LPC) information, 4 pitch subframes with 11 bits each, and 4×4 codebook subframes with 12 bits each. The *Enhanced Variable Rate Coder* (EVRC) is based on very similar principles, but uses a smaller number of bits both while the voice user is active (170 bits per 20-ms interval), and during transmission pauses. It furthermore includes adaptive noise suppression, which enhances overall speech quality.

A key aspect of all vocoders is the variable data rate. People are usually silent for approximately 50% of the time (while they are listening to the person at the other end of the line). During that time, the data rate is reduced to about 1 kbit/s. As discussed in Chapter 18, this leads to a significant increase in total capacity.

22.4.2 Error correction coding

Error correction coding for 8.6 kbit/s in the uplink

Forward error correction is different for 8.6 kbit/s and 13.3 kbit/s. However, it is identical for two existing vocoders that are based on 8.6-kbit/s output: the IS-96A vocoder and the EVRC vocoder. These vocoders are associated with rate-set-1, and thus have the following encoding steps:

1. Encoding starts with 172 bits for each 20-ms frame from the vocoder.
2. In a next step, 12 *Frame Quality Indicator* (FQI) bits are added. These bits act as parity check bits, and allow determination of whether the frame has arrived correctly or not.
3. Adding an 8-bit encoder tail brings the number of bits to 192.

4. These bits are then encoded with a rate-1/3 convolutional encoder with constraint length 9. The three generator vectors are:

$$\left. \begin{array}{l} G1(D) = 1 + D^2 + D^3 + D^5 + D^6 + D^7 + D^8 \\ G2(D) = 1 + D + D^3 + D^4 + D^7 + D^8 \\ G3(D) = 1 + D + D^2 + D^3 + D^4 + D^5 + D^8 \end{array} \right\} \qquad (22.3)$$

This brings the bitrate up to 28.8 kbit/s.

Error correction coding for 13.3 kbit/s in the uplink

For the CDG-13 coder, encoding steps are in principle similar, but different numerical values are used:

1. Encoding starts with 267 bits (including some unused bits) for each 20-ms frame.
2. A frame erasure bit is added.
3. A total of 12 FQI bits are added (again, to indicate whether the frame has arrived correctly).
4. Adding an 8-bit tail in order to help the Viterbi decoder. This brings the number of bits per 20-ms frame to 288.
4. These bits are then encoded with a rate-1/2 convolutional encoder with constraint length 9. The two generator vectors are:

$$\left. \begin{array}{l} G1(D) = 1 + D + D^2 + D^3 + D^5 + D^7 + D^8 \\ G2(D) = 1 + D^2 + D^3 + D^4 + D^8 \end{array} \right\} \qquad (22.4)$$

Error correction coding for 8.6 kbit/s in the downlink

Error correction coding is somewhat different in the downlink. It uses the same combination of FQI bits and tail bits as the uplink, but then uses a rate-1/2 convolutional encoder to bring the bit rate to 19.2 kbit/s. The generator vectors are given by Eq. (22.4). Data with rate 19.2 kbit/s are then further processed as described in Section 22.3.4.

Error correction coding for 13.3 kbit/s in the downlink

This mode uses the same encoding steps as the 13.3-kbit/s uplink procedure. However, this leads to a 28.8-kbit/s rate, while only 19.2 kbit/s can be transmitted in one downlink traffic channel. The output from the convolutional encoder is thus punctured, in order to yield the desired bit rate. We can also interpret this as encoding the vocoder output using a rate-3/4 convolutional code. Puncturing eliminates the third and fifth bit of each 6-bit symbol repetition block (see above). This corresponds to eliminating 2 bits created by $G2(D)$, while the bits from $G1(D)$ are completely transmitted.

Interleaving

Output from the convolutional encoder is sent through a block interleaver (see Section 14.7.1) of length 576 for rate set 2. More specifically, the interleaver has a matrix structure similar to the one outlined in Fig. 14.25 with 32 rows and 18 columns. The data are written line by line – i.e., filling first the first column, then the second column, and so on – and read out orthogonally (see, again, Fig. 14.25).

22.5 Spreading and modulation

22.5.1 Long and short spreading codes and Walsh codes

IS-95 uses three types of spreading codes: long spreading codes, short spreading codes, and Walsh codes. These codes play different roles in the uplink and the downlink. In this section, we describe just the codes themselves. In the subsequent two sections, we describe how those codes are used in the uplink and downlink, respectively.

Walsh codes

Walsh codes are strictly orthogonal codes that can be constructed systematically. As we saw in Section 18.2.6, we define the $n + 1$-order Walsh–Hadamard matrix $\mathbf{H}_{had}^{(n+1)}$ in terms of the nth order matrix:

$$\mathbf{H}_{had}^{(n+1)} = \begin{pmatrix} \mathbf{H}_{had}^{(n)} & \mathbf{H}_{had}^{(n)} \\ \mathbf{H}_{had}^{(n)} & \overline{\mathbf{H}}_{had}^{(n)} \end{pmatrix} \tag{22.5}$$

where $\overline{\mathbf{H}}$ is the modulo-2 complement of H. The recursive equation is initialized as:

$$\mathbf{H}_{had}^{(1)} = \begin{pmatrix} 1 & 1 \\ 1 & -1 \end{pmatrix} \tag{22.6}$$

The Walsh codes in IS-95 are the columns of the complement of $\mathbf{H}_{had}^{(6)}$.

Short spreading codes

IS-95 also uses two spreading codes that are PN-sequences, generated with a shift register of length 15, and thus with a periodicity of $2^{15} - 1$. A single zero is inserted in the sequence, in order to increase periodicity to $2^{15} = 32{,}768$ chips, which corresponds to 26.7 ms. The generator polynomials of the sequence are:

$$G_i(x) = x^{15} + x^{13} + x^9 + x^8 + x^7 + x^5 + 1 \tag{22.7}$$

$$G_q(x) = x^{15} + x^{12} + x^{11} + x^{10} + x^6 + x^5 + x^4 + x^3 + 1 \tag{22.8}$$

As we will see later on, each BS uses a time-shifted version of the short spreading code. The *spreading offset index* of the code indicates this timeshift.[3]

Long spreading codes

The third type of codes, called "long spreading codes", is also based on PN-sequences. For the long codes, the shift registers have length 42, so that periodicity is $2^{42} - 1$, corresponding to more than 40 days. The generator polynomial is:

$$G_l = x^{42} + x^{35} + x^{33} + x^{31} + x^{27} + x^{27} + x^{26} + x^{25} + x^{22} + x^{21} + x^{19}$$
$$+ x^{18} + x^{17} + x^{16} + x^{10} + x^7 + x^6 + x^5 + x^3 + x^2 + x^1 + 1 \tag{22.9}$$

Output from the shift register is then modulo-2 added with the *long-code mask*. This long code mask is different for different channels: for access channels (see Section 22.4), it is derived from the paging and access channel numbers and the BS identification. For traffic channels, it can

---

[3] Index zero corresponds to a sequence that had 15 zeros, followed by a 1, at 0h00 on January 5th, 1980.

either be derived from the *Electronic Serial Number* (ESN) (*public mask*) or from an encryption algorithm (*private mask*).

22.5.2 Spreading and modulation in the uplink

Modulation and coding in the uplink are achieved through a combination of steps. The starting point is a bitstream with 28.8 kbit/s, which is obtained by providing error correction coding to the vocoder signal (see Section 22.4):

- The first step involves mapping the data sequence to Walsh code codewords. Remember that each Walsh code is 64 chips long. The transmitter then takes groups of 6 bits $(x_0, \ldots, x_5)$, and maps them to one Walsh code symbol $\tilde{\mathbf{x}} = [\tilde{x}_0, \tilde{x}_1, \ldots, \tilde{x}_{63}]$, according to the rule:

$$j_{\text{Walsh}} = x_0 + 2x_1 + 4x_2 + 8x_3 + 16x_4 + 32x_5 \tag{22.10}$$

$$\tilde{x}_i = 1 + \left(\mathbf{H}_{\text{had}}^{(6)}\right)_{i, j_{\text{Walsh}}} \tag{22.11}$$

  This achieves spreading by a factor 64/6. The chiprate of output from the Walsh encoder is thus 307.2 kchip/s. This can also be seen as *M*-ary orthogonal modulation; in other words, each group of 6 bits is represented by one modulation symbol that is orthogonal to all other admissible modulation symbols. A key advantage of this technique is that it allows non-coherent demodulation.
- The next step involves spreading the Walsh encoder output to 1.2288 Mchip/s. This is achieved by multiplication of the Walsh code output by the long spreading code. Note that (in the uplink) it is the long spreading code that provides channelization, and thus allows different traffic channels and access channels to be distinguished. Output from the long spreading code is sent to the databurst randomizer, whose role will be described later.
- As a last step, the (spread) datastream is separated into in-phase and quadrature-phase components, each of which are separately spread by short training sequences. This multi-plication does not result in a change of chiprate. The chipstream on the I- and Q-branch are then used to modulate the local oscillator with Offset Quadrature Amplitude Modulation (OQAM) (as discussed in Chapter 11).

Figure 22.1 shows a block diagram of uplink transmission. Figure 22.2 shows the data rates involved.

22.5.3 Databurst randomization and gating for the uplink

Up to now we have considered the case when the output of the channel coder actually has a data rate of 28.8 kbit/s – i.e., a source rate of 14.4 or 9.6 kbit/s. However, depending on the source data, a lower rate (14.4 kbit/s, 7.2 kbit/s, or 3.6 kbit/s) can also be the output of a convolutional encoder. In this case, encoded symbols are repeated (several times, if necessary) until a data rate of 28.8 kbit/s is achieved. It is these repeated data that are sent to the block interleaver for further processing.

However, it would waste resources to transmit all of these repeated data at full power. For the uplink, this problem is solved by gating off the transmitter part of the time. If, for example, the coded data rate is 14.4 kbit/s (source data rate 7.2 kbit/s), then the transmitter is turned on only 1/2 of the time. As a consequence, average transmit power is only 1/2 of the full-data-rate case, and the interference seen by other users is only half as large.

Determination of the time for gating off the transmitter has to happen is actually quite complicated. The first issue is that gating has to be coordinated with the interleaver.

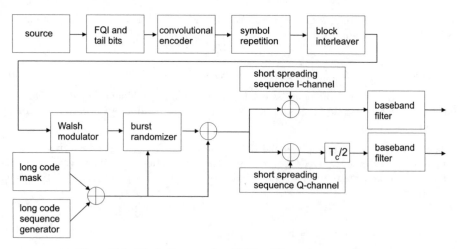

Figure 22.1   Block diagram of an IS-95 mobile station transmitter.

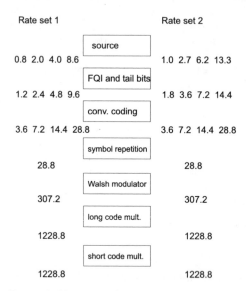

Figure 22.2   Data and chip rates at different interfaces in the uplink transmitter.

For example, for the 7.2-kbit source data rate mode, each 1.25-ms-long group of output symbols is repeated once.[4] Gating thus eliminates one of these two symbol groups.

A decision about which of two symbol groups actually gets transmitted is determined by the long spreading sequence, according to the following algorithm:

• Consider the second-to-last 1.25-ms symbol group in the frame (20-ms period) immediately preceding the currently considered frame.

---

[4] A symbol group of 1.25 ms is also often referred to as a *power control group*, since fast power control can change every 1.25 ms.

- Take the last 14 bits of the long spreading sequence used for spreading this symbol group, label them as $[b_0, b_1, \ldots, b_{13}]$.
- The gating of the symbol groups in the currently considered frame is now determined by these bits. Transmission is done of some of the 15 symbol groups contained in a 20-ms frame:

  —for the 14.4 (9.6) kbit/s source rate mode, always transmit;
  —for the 7.2 (4.8) kbit/s mode, transmit the first of the two identical groups if $b_i = 0$, $i = 0, \ldots, 7$, otherwise transmit the second frame;
  —for the 3.6 (2.4) kbit/s mode, transmit during $b_{2i} + 4i$, $i = 0, \ldots, 3$ if $b_{i+8} = 0$ or transmit during $2 + b_{2i+1} + 4i$ if $b_{i+8} = 1$;
  —for the 1.8 (1.2) kbit/s mode, transmit during $b_{4i} + 8i$, $i = 0, \ldots, 1$ if $b_{i+8} = 0$ and $b_{2i+9} = 0$, or transmit during $2 + b_{4i+1} + 8i$ if $b_{i+8} = 1$ and $b_{2i+9} = 0$, or transmit during $4 + b_{4i+2} + 8i$ if $b_{i+8} = 0$ and $b_{2i+9} = 1$, or transmit during $4 + b_{4i+3} + 8i$ if $b_{i+8} = 1$ and $b_{2i+9} = 1$.

Thus, the gating of sequences is pseudorandom, and different for each user (remember that the long spreading sequence is different for each user). Thus, in a system with low-rate users, interference with other users is "smeared out" – i.e., no user sees interference from all other users at the same time.

22.5.4 Spreading and modulation in the downlink

In the downlink, the different spreading codes are used in a very different way. The starting point is an encoded datastream with a rate of 19.2 kbit/s – i.e., lower by a factor of $1/3$ than for the uplink. This bitstream is then scrambled and spread using the following steps:

- In a first step, the datastream is scrambled using the long spreading sequence. Remember that the long spreading sequence is defined to have a chip rate of 1.2288 Mchip/s. However, for downlink application, we do not want to use it for spreading, only for scrambling. We thus need to reduce the chip rate to 19.2 kchip/s. This is achieved by just using every 64th chip in the sequence. This decimated long code sequence is the modulo-2 added to the data sequence.
- Next, data are spread by using Walsh sequences. In the downlink, Walsh sequences are used for channelization and spreading. Each traffic channel is assigned one Walsh sequence, with a chip rate of 1.2288 Mchip/s. This sequence is periodically repeated (with a period of 64 chips), and multiplied by the scrambled data sequence. We can interpret this alternatively as mapping each data bit either to the (user-specific) Walsh sequence or to its modulo-2 complement.
- Finally, output from the spreader is multiplied separately in the I- and Q-branch by the short spreading sequence. Note that in the downlink, the modulation format is QAM, not O-QAM.

Figure 22.3 shows a block diagram of the BS transmitter (downlink transmission principle), and Fig. 22.4 shows the data rates involved.

In the downlink, there are also provisions for transmitting with data rates lower than 9.6 kbit/s or with a 14.4-kbit/s source rate. Also in this case, symbols are repeated to ensure maximum data rates, but the energy per transmitted symbol is reduced proportionally to the repetition factor, so a constant energy level is transmitted and a desired bit energy level is achieved. This treatment of the lower data rates in the downlink is different from the blanking used in the uplink.

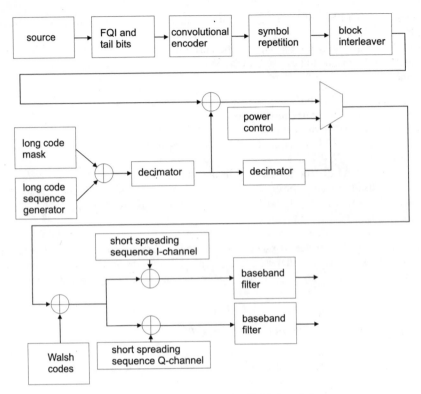

Figure 22.3 Block diagram for IS-95 downlink.

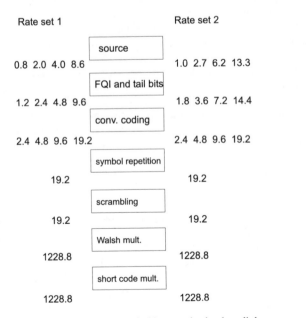

Figure 2.24 Data rates and chip rates in the downlink.

22.5.5 Discussion

When confronted by different spreading schemes for uplink and downlink, the immediate question arises: "Why?" It would seem that a spreading scheme that is good for one direction should also prove to be advantageous for the other direction. However, we have to remember that there is an inherent asymmetry in a cellular system. Signals transmitted from the BS to the various MSs can be transmitted synchronously in a very easy manner – after all, the BS has control over when to transmit those signals. All signals then arrive at a given MS at the same time, having suffered the same attenuation and distortion by the channel.[5] Thus, if the signals are spread by perfectly orthogonal codes at the transmitter, they can be perfectly separated at the receiver (this assumes non-distorting channels). It follows from this reasoning that (perfectly orthogonal) Walsh codes are used for spreading in the downlink. Since 64 users should be able to communicate within one cell, the spreading factor is 64, and the (coded) source data rate must not exceed 19.2 kbit/s. Short spreading sequences can be used for distinction between different BSs.

For the uplink, on the other hand, it is impossible for signals from different users to arrive at the same time. Thus, the use of Walsh codes for user separation is not possible. Walsh codes can have high cross-correlation when they have an arbitrary timeshift relative to each other. Rather, channelization is achieved by PN-sequences, which have low cross-correlation for any timeshift. Walsh codes are used as part of the modulation scheme: since a 64-chip symbol is used to represent 6 bits, inherent redundancy and thus error-correcting capability are present.

22.6 Logical and physical channels

22.6.1 Traffic channels

Traffic channels are the channels on which the voice data for each user are transmitted. We have already discussed them above, and here just reiterate the key data: there are two possible rate sets: rate-set-1, with a $[9.6, 4.8, 2.4, 1.2]$-kbit/s source data rate, and rate-set-2, with a $[14.4, 7.2, 3.6, 1.8]$-kbit/s source data rate. These source data – i.e., output from the vocoder – are possibly repeated, and then convolutionally encoded and interleaved.

The data are subsequently spread and modulated. The spreading and modulation operations are different in the uplink and the downlink, and are described in detail in Section 22.5.

A number of control messages are also transmitted on traffic channels. These include:

- for the uplink: Power Measurement Report, Pilot Strength Measurement, Handoff Completion, Long Code Transition Request, Long Code Transition Response, Data Burst, Request Analog Service;
- for the downlink: Neighbor List, Pilot Measurement Request, Handoff Direction, Long Code Transition Request, Long Code Transition Response, Data Burst, Analog Handoff Direction.

22.6.2 Access channel

The access channel is a channel in the uplink that is used for signaling by MSs without a current call. Access channel messages include security messages (BS challenge, authentication challenge response), page response, origination, and registration.

---

[5] Note that we are discussing here the different signals that arrive at a *given* MS. When considering the received signal at two different MSs, we find that these signals have different delays and different distortions.

The access channel uses a (source) data rate of 4.8 kbit/s. Spreading and modulation is very similar to uplink traffic channels with the same data rate. However, no gating is done, and all repeated symbols are actually transmitted. Since the access channel is spread (channelized) with a long spreading code, a considerable number of access channels can exist. As a matter of fact, up to 32 access channels exist for each paging channel (see below). An MS randomly chooses one of the active access channels before starting an access attempt.

A call initiated by the MS starts with a message on the access channel. The MS sets the initial power (based on the pilot power that it observes), and transmits a probe. If this probe is acknowledged before a timeout, then access was successful. If not, then the MS waits a random time, and then transmits the probe with increased power. This process is repeated until either access is successful, or the probe has reached the maximum admissible power; in the latter case, access is deemed to have failed.

22.6.3 Pilot channels

The pilot channel allows the MS to acquire the timing for a specific BS, obtain the transfer function from BS to MS, and estimate the signal strength for all BSs in the region of interest. The pilot channel is similar to a downlink traffic channel, but shows some important peculiarities:

- It is not power-controlled. The reason for this is that (i) it is used by many MSs (so it would not be clear which MS should determine the power control anyway), and (ii) it is used for estimating the attenuation of various links, which can only be done if transmit power is clearly defined and known to all MSs.
- It uses Walsh code 0 for transmission: this code is the all-zero code.
- It has higher transmit power than traffic channels. Because of its importance, typically 20% of total BS power are assigned to the pilot channel.

The pilot channel is easy to demodulate, because it is just an all-zero sequence spread by the short spreading code. The only difference between pilots transmitted from different BSs is a temporal offset. After an MS has acquired the pilot, it can then more easily demodulate the synchronization channel (see next subsection), as the timing of that synchronization channel is locked to the pilot.

22.6.4 Synchronization channel

The synchronization channel transmits information about system details that are required for the MS to synchronize itself to the network. Examples of such information include network identifier, PN-offset, long-code state, system time (from GPS), local time differential to system time, and the rate at which paging channels operate.

The synchronization channel transmits data at 1.2 kbit/s. After convolutional encoding with rate 1/2 and repetition, the data rate is transmitted. Note that this channel is not scrambled (no application of the long code mask). Each frame of the synchronization channel is aligned at the start of the PN-sequence.

22.6.5 Paging channel

The paging channel transmits system and call information from the BS to the MS. Several paging channels can exist within each cell; each of them is a 9.6-kbit/s channel. The information on the paging channel can include:

- Page message to indicate incoming call.
- System information and instructions:

  —handoff thresholds;
  —maximum number of unsuccessful access attempts;
  —list of surrounding cell PN-offsets;
  —channel assignment messages.

- Acknowledgments to access requests.

22.6.6 Power control subchannel

The power control subchannel provides signaling for compensation of small-scale fading. IS-95 divides signals into *Power Control Groups* (PCGs) of 1.25-ms duration. The BS estimates the Signal-to-Noise Ratio (SNR) for each user for each PCG. It then sends a power control command to the MS within two PCGs, and the MS reacts to it within 500 µs. This command signifies either an increase or a decrease by 1 dB, and thus requires 1 bit. Consequently, the data rate of the power control channel is 800 bit/s.

The power control subchannel is inserted into the traffic channel by simply replacing some of the traffic data symbols. Each PCG contains 24 modulation symbols; however, only the first 16 are candidates for replacement. The exact location is determined by the long-code mask: bits 20–23 of the long-code mask in the previous PCG determine which bit is replaced.

22.6.7 Mapping logical channels to physical channels

In the downlink, mapping is done in a rather straightforward way: different channels are assigned different Walsh codes for spreading. Specifically, the pilot channel uses Walsh-code-0, the paging channels use Walsh-code-1 through Walsh-code-7, the synchronization channel uses Walsh-code-32, and the traffic channels use all other Walsh codes.

In the uplink, only traffic channels and access channels exist. These are assigned different long spreading codes, and are thus also mapped to different coded channels in all cases.

22.7 Handover

One of the most important advantages of CDMA is "soft handover", which helps the performance of MSs especially at the cell edges (see Section 18.3.2). In this section, we will discuss how IS-95 determines the available BSs so as to carry out soft handover, and how handover is actually carried out.

Each BS sends out the same PN-sequence as a pilot signal, just with an offset of 64 chips (see Section 22.3.4). This gives a total of 512 possible pilot signals. By observing these pilots, the MS always knows the signal strengths of all BSs in its environment. However, it is much too complicated and slow to monitor such a huge number of BSs – most of which are below the noise floor anyway. Thus, the MS only observes pilot signals within a given window.

The MS divides the available BSs into groups or *sets*. The *active set* contains pilots that are made available to a specific MS by the Mobile Switching Center (MSC); it can contain up to six pilots. The *candidate set* contains pilots that are strong enough to be demodulated, but are not contained in the active set. The *neighbor set* contains a list of BSs that are "nearby", but do not have sufficient strength to move into the candidate set. A list of the neighbor set is sent by the BS to the MS at regular intervals.

Monitoring of pilots and assignment of BSs to different sets are done the following way: when the MS finds a new pilot (not yet in the candidate set), it compares its strength with a parameter T_ADD. If the pilot is stronger than this threshold, the MS sends a Pilot Strength Measurement Message (PSMM) to the BS; the BS might then order the MS to add the pilot to the candidate set. The MS also monitors the strength of pilots in the candidate set, and if they become stronger than a threshold T_COMP, it moves the pilot from the candidate set to the active set. There are also mechanisms for dropping a pilot from the active or candidate set, if the strength falls below a certain threshold.

When the MS is in soft-handover mode, then it can communicate with members of the active set simultaneously. In the downlink, the MS just combines the signals from different BSs. In the uplink, the BSs in the active set determine which of them gets the best signal quality; this BS is then the one used for demodulation. The former case provides combining gain, while the latter provides selection gain.

22.8 Appendices

Please see companion website *www.wiley.com/go/molisch*

Further reading

A summary of IS-95 can be found in Liberti and Rappaport [1999]. Additional information on CDMA 2000 can be found in the book by Vanghi et al. [2004], and the papers by Tiedemann [2001] and Willenegger [2000]. The 1xEV-DO mode is described in Sindhushayana and Black [2002].

23

WCDMA/UMTS

23.1 Historical overview

This chapter is a short summary of the *Wideband Code Division Multiple Access* (WCDMA) standard for third-generation cellular telephony. This standard is also part of a group of standards that are known as the *Universal Mobile Telecommunication System* (UMTS), *Third Generation Partnership Project* (3GPP), and *International Mobile Telephony* (IMT-2000). In this section, we will review the history of this standard and show its relationship to other third-generation standards.

The success of second-generation cellphones, especially the Global System for Mobile communications (GSM), motivated the development of a successor system. The International Telecommunications Union (ITU) announced goals for a standard for a third-generation system, called IMT-2000:

- better spectral efficiency;
- higher peak data rates – namely, up to 2 Mbit/s – which should result in a choice of channels with a bandwidth of 5 MHz instead of 200 kHz indoor and 384 kbit/s outdoor of GSM;
- supporting multimedia applications, meaning the transmission of voice, arbitrary data, text, pictures, audio and video, which requires increased flexibility in the choice of data rates;
- backward compatibility to second-generation systems.

Europe started research activities towards this goal in the early 1990s, first under the auspices of European Union research programs, and then a more formal process within the European Telecommunications Standards Institute (ETSI). Based on the above list of requirements, different groups developed proposals, which ranged from Orthogonal Frequency Division Multiplexing (OFDM) solutions over broadband Time Division Multiple Access (TDMA) systems to Code Division Multiple Access (CDMA) protocols on the physical layer. During a final poll in January 1998 two drafts were selected: broadband CDMA – also known as *Frequency Division Duplex* (FDD) mode – which is intended as the basic system, and T/CDMA with Joint Detection (JD-TCDMA) – also known as Time Division Duplex (TDD) mode – which is to support high-data-rate applications. The FDD mode and TDD mode are often subsumed under the name WCDMA. FDD mode is the more important part, and will be at the center of our attention in this chapter. WCDMA is mostly used as an abbreviation for radio access technology, while the expression UMTS refers to the complete system, including the core network.

Wireless Communications Andreas F. Molisch
© 2005 John Wiley & Sons, Ltd

These two systems were then included as the European proposal for the IMT-2000 family. A Japanese proposal for a wideband CDMA system was fairly similar to the European proposal, and thus merged. Still, it was not possible to reach agreement for a single third-generation system within the ITU. Besides the Japanese/European proposal, the (mainly U.S.-favored) CDMA 2000 proposal (see Chapter 22) had strong support. Furthermore, Enhanced Data rates for GSM Evolution (EDGE) (see Chapter 21), Digital Enhanced Cordless Telecommunications (DECT) (see the Appendices at *www.wiley.com/go/molisch*), and Universal Wireless Communications (UWC) 136 (a further development of the American second-generation standard IS-136) were advocated. In order to avoid a deadlock, the ITU decided to declare all of these proposals to be valid "modes" of IMT-2000. Of course, having five modes in a single standard is just as bad as having five different standards, but it was a face-saving measure for all participants. After 2000, the Chinese standard TD-SCDMA has also received great attention, and is becoming an important part of third-generation cellular developments.

The further development of the merged Japanese/European standard is being done by the 3GPP industry organization, whose members include ETSI, ARIB (Association of Radio Industries and Businesses, the Japanese standardization organization) and several other members. The original specifications are also known as "Release 99". Subsequent improvements were included in Release 4, Release 5 – especially, the High Speed Downlink Packet Access (HSDPA) data mode – and Release 6 (announced for 2005). While there were advantages to having a broad base of companies contributing to the specifications, 3GPP faced some serious problems: the size of the specifications grew beyond reasonable limits, and became so complicated that it hampered implementation. The resulting high development costs and delays in time to market almost led to the death of the system, and it started to take off only around 2005. The exception is Japan, where implementation proceeded much faster than elsewhere. It is worth noting that Japan did not implement the full UMTS specifications, but rather a simplified system called "FOMA". Development in Japan was also helped by the fact that second-generation Japanese Digital Cellular (JDC) reached its capacity limits earlier and the alternative Personal Communication System (PCS) systems (compare with the DECT material on the companion website) could not meet demand completely.

UMTS development was also hampered by regulatory developments. The first UMTS-related licenses were granted first in 1999 in Finland and the other European countries followed in 2000. Particularly, the bidding process for the licenses in the U.K. and Germany received a lot of public attention, as several billion Euros were finally paid for each license. According to the regulatory authorities the first UMTS networks should have started operating in 2002. However, the real start of operation has been delayed due to technical problems and the fact that the market for mobile high data rate applications had to be developed first. With the end and the start of 2003 and 2004, respectively, the major mobile providers in Europe started to offer subscriptions to UMTS and the appropriate devices.

The remainder of this chapter describes mainly the physical layer of WCDMA/UMTS. Many of the networking functions are similar to GSM, and we simply refer the reader to Chapter 21 for these.

23.2 System overview

23.2.1 Physical-layer overview

We first summarize the physical layer of WCDMA – i.e., communications between the MS and the BS via the air interface. The UMTS standard uses a number of unusual abbreviations. For example, the MS is called *User Equipment* (UE). In order to stay consistent with the notation in the remainder in this book, we will stick to MS. Similarly, the BS is called *Node-B* in the UMTS standard.

The WCDMA air interface uses CDMA for distinguishing between different users, and also between users and some control channels. We distinguish between spreading codes that are responsible for bandwidth expansion of the signals, and scrambling codes that are mainly used for distinguishing signals from different MSs and/or BSs. In addition, WCDMA also shows a timeslot structure: the time axis is divided into units of 10 ms, each of which is subdivided into slots of 0.67 ms. Depending on the location within a timeslot, a symbol might have different meanings.

WCDMA uses a number of logical channels for data and control information, which will be discussed in Section 23.4. They are then mapped to physical channels – i.e., channels that are distinguished by spreading codes, scrambling codes, and positions within the timeslot.

23.2.2 Network structure

To discuss how a mobile network provider may introduce UMTS, we have to distinguish between the *MS*, the *Radio Access Network* (RAN), and the *Core Network* (CN). The MS and the *UMTS Terrestrial Radio Access Network* (UTRAN) communicate with each other via the air interface, as discussed in the previous section. The UTRAN consists of multiple *Radio Network Subsystems* (RNSs), each of which contains several *Radio Network Controllers* (RNCs), each of which controls one or several BSs (Node-Bs).

The CN connects the different RNSs with each other and other networks, like ISDN and data packet networks. The CN can be based on an upgraded GSM CN or might be implemented as a completely new Internet Protocol (IP)-based network. Details about the different functional units of a GSM CN (mobile switching center, home location register, etc.) can be found in Chapter 21. The network functionalities for packet data are similar to the ones of the General Packet Radio Service (GPRS) (see Appendix 21.C at *www.wiley.com/go/molisch*).

Another way of looking at the UMTS architecture is to organize it in two domains:

1. *User equipment domain*, which consists of:

    —*User Service Identity Module* (USIM).
    —*Mobile Equipment* (ME) consisting of:
       ○ *Terminal Equipment* (TE);
       ○ *Terminal Adapter* (TA);
       ○ *Mobile Termination* (MT).

2. *Infrastructure domain*, which consists of:

    —The access network domain consisting of:
       ○ *UMTS Radio Access Network* (UTRAN).
    —The CN domain consisting of:
       ○ *Inter Working Unit* (IWU);
       ○ serving network;
       ○ transit network;
       ○ home network;
       ○ application network.

The physical properties of a link between two network entities and the signals transmitted via this link together with the functions of these signals are collectively referred to as an *interface*. Usually, an interface is standardized. Figure 23.1 shows the relevant interfaces in UMTS.

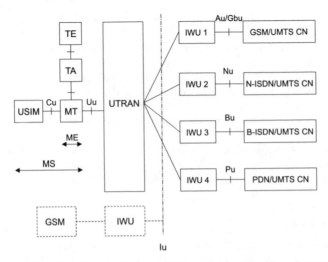

Figure 23.1 Interfaces in Universal Mobile Telecommunications System.

23.2.3 Hierarchical cellular structure

UMTS is intended to achieve global availability and thereby enable roaming worldwide. Therefore, the coverage area in UMTS is divided hierarchically into layers. The higher layers cover a bigger area than the lower layers. The highest layer achieves worldwide coverage by using satellites.[1] The lower layers are the macrolayer, microlayer, and picolayer. They constitute the *UTRAN*. Each layer consists of several cells. The lower the layer the smaller the cells. Thus, the macrolayer is responsible for nationwide coverage with macrocells. Microcells are used for additional coverage in the urban environment and picocells are employed in buildings or "hotspots" such as airports or train stations. This concept is known and has been discussed and partly implemented for a while. However, UMTS is supposed to cover all aspects of this concept worldwide and with the initial rollout. In practice, however, it has been rather different: in the initial phase, only the big cities were covered with a few cells, and large-area coverage is achieved by dual-mode devices that can communicate with either WCDMA or EDGE/GPRS/GSM networks.

23.2.4 Data rates and service classes

The maximal data rate and the highest supported user velocity are different for each hierarchy layer. The macrolayer supports at least 144 kbit/s at speeds up to 500 km/h. In the microlayer data rates of 384 kbit/s at maximal speeds of 120 km/h are achievable. In picocells, maximal user speeds of 10 km/h and maximal data rates of 2 Mbit/s are supported. Figure 23.2 compares data rates and maximal user velocity with other cellular (GSM) or wireless standards.

Maximal Bit Error Rates (BERs) and transmission delays are grouped into sets out of which the user may choose:

- Conversational: this class is mainly intended for speech services, similar to GSM. The delays for this type of service should be on the order of 100 ms or less; larger values are

---

[1] Given that the past difficulties of satellite cellular communications systems (e.g., IRIDIUM), it is questionable whether this layer will actually be implemented.

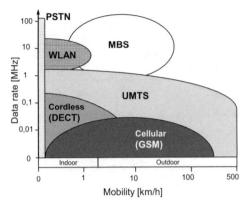

Figure 23.2 Data rates versus mobility for Universal Mobile Telecommunications System, Global System for Mobile communications, Digital Enhanced Cordless Telephone, wireless local area network mobile broadband systems and landline networks.

experienced as unpleasant interruptions by users. BERs should be on the order of 10^{-4} or less.

- Streaming: audio- and videostreaming are viewed as one important application of WCDMA. Larger delays (in excess of 100 ms) can be tolerated, as the receiver typically buffers several seconds of streaming material. BERs are typically smaller, as noise in the audio (music) signal is often considered to be more irksome than in a voice (telephone) conversation.

- Interactive: this category encompasses applications where the user requests data from a remote appliance. The most important category is Web browsing, but also database retrievals and interactive computer games fall into this category. Also for this category, there are upper limits to tolerable delay – the time between choosing a certain website and its actual appearance on the screen should not exceed a few seconds. BERs have to be lower, typically 10^{-6} or less.

- Background class: this category encompasses services where transmission delays are not critical. These services encompass email, Short Message Service (SMS), etc.

23.3 Air interface

23.3.1 Frequency bands and duplexing

UMTS utilizes frequencies in ranges from 1,900 MHz to 2,025 MHz and from 2,110 MHz to 2,200 MHz (see Fig. 23.3). Within these bands there is a dedicated subband for the *Mobile Satellite Service* (MSS).[2] The MSS uplink uses the band from 1,980 MHz to 2,010 MHz and the MSS downlink the band from 2,170 MHz to 2,200 MHz. The remaining parts of bands are split for the two modes of terrestrial operation, UMTS-TDD and UMTS-FDD. UMTS-FDD uses the band from 1,920 MHz to 1,980 MHz for the uplink and the band from 2,110 MHz to 2,170 MHz for the downlink. As the name indicates, TDD mode does not distinguish uplink and downlink by using different carrier frequencies but rather by accessing different timeslots on the same carrier. Therefore, this mode does not require symmetric frequency bands. It can simply use all of the remaining frequencies.

---

[2] Additional bands are currently being investigated, including 800 MHz for Japan, 850 MHz and 1.72/2.1 GHz for the U.S., 1,800 for Australia, etc. and 900 MHz in Europe and 2.6 GHz.

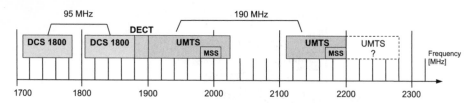

Figure 23.3 Frequency bands allocated to Universal Mobile Telecommunications System.

An exception to these frequency allocations is the U.S.A., where third-generation phones are to be placed in PCS bands – i.e., uplink is in the band from 1,850 to 1,910 MHz, and downlink in the band from 1,930 to 1,990 MHz.

23.3.2 Time-domain-duplexing and frequency-domain-duplexing modes

FDD operating mode is intended for use in macrocells and microcells whereas TDD operating mode is intended for use in picocells. In the TDD mode it is more difficult to handle large propagation delays between the MS and BS, as transmitting and receiving timeslots might overlap. Therefore, it can only be employed in picocells. However, TDD has the advantage that big asymmetries in throughput on the downlink and uplink can be easily supported. This is particularly important for applications like Web browsing for which the MS receives much more information than it transmits. For medium access, TDD mode employs JD-TCDMA and FDD mode WCDMA.

Introduction of the high-speed packet data mode in WCDMA has taken away much of the motivation for TDD mode. Consequently, the deployment of TDD is slow, and (at least up to 2005) only a few local installations exist. We will thus not consider this mode in the remainder of this chapter.

23.3.3 Radio-frequency-related aspects

Power classes and receiver sensitivity

Mobile station MSs are divided into four classes, according to their transmit power. The maximal powers are 33, 27, 24, and 21 dBm, corresponding to power classes 1, 2, 3, and 4. Power is measured before the antenna. Thus, the antenna characteristic does not have an impact on these limits. The receiver of the MS has to be so sensitive that given a received signal power of -117 dBm per 3.84-MHz channel a BER of 10^{-3} is still achievable with a 12.2-kbit/s data rate. Note that this specification includes the effects of forward error correction.

Base station There are no transmission powers specified for the BS; however, typical values for the transmit power are in the 10–40-W range. The receiver in the BS has to be so sensitive that given a received signal power of -121 dBm a BER of 10^{-3} for a 12.2-kbit/s is still feasible. Since noise power in a bandwidth of 12.2 kHz is -133 dBm, this is a quite challenging task.

The specifications also prescribe the resistance of the receiver to blocking – i.e., being able to function even in the presence of strong interferers. Processing gain can be used to decrease the effect of interference. However, if the Radio Frequency (RF) elements of the receiver have a limited dynamic range, then the interferer might drive the receive amplifier into saturation, resulting in a signal from which the desired part can no longer be recovered. Similarly, limits to intermodulation and other RF effects are specified [Richardson 2005].

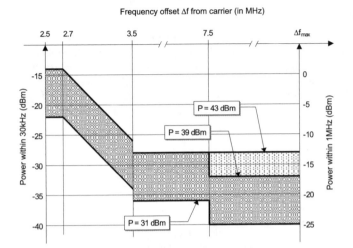

Figure 23.4 Spectrum mask for a base station in Wideband Code Division Multiple Access.

Frequency bands

The regular intercarrier spacing is 5 MHz. This is the nominal distance. In fact the network provider might select the carrier distance to be any multiple of 200 kHz. Therefore, carriers are referred to using the *UTRA Absolute Radio Frequency Channel Number* (UARFCN) which refers to multiples of 200 kHz. The frequency deviation of the local oscillators of the MS is limited to 10^{-7} (0.1 parts per million), which equals roughly 200 Hz.

From a purely theoretical point of view, emissions outside the assigned 5-MHz band should be zero. The basis pulse for modulation is a raised cosine pulse with a rolloff factor of $\alpha = 0.22$. Given a chip rate of 3.84 Mchip/s the signal bandwidth is $(1 + \alpha)/T_C = 4.7$ MHz.[3] However, due to non-ideal filter implementation, emissions outside this band do appear in practice. These emissions include both out-of-band emissions (defined as emissions a distance of 2.5 MHz to 12.5 MHz from the center frequency) and spurious emissions, which refer to emissions farther away from the center frequency.

Out-of-band emissions   There is a set of limitations for out-of-band emissions at a distance of 2.5 MHz to 12.5 MHz:

- **Spectrum emission mask**: Fig. 23.4 shows the emission mask for a BS. For a given distance Δf to the center frequency, there is a limit for the maximum emitted power within a 30-kHz and 1-MHz bandwidth, respectively.
- **Adjacent Channel Leakage Ratio** (ACLR): this is a measure of how much power leaks from the desired band into the adjacent band. This ratio should be better than 45 dB and 50 dB for the first or second adjacent channel (5-MHz or 10-MHz carrier distance), respectively.

---

[3] The mismatch between signal bandwidth and nominal carrier distance can only be explained from a historic perspective. Originally, a chip rate of 4.096 Mchip/s was planned. With $\alpha = 0.22$ this results in a signal bandwidth of 5 MHz.

Spurious emissions

Spurious emissions are emissions far away from the used channel ($\Delta f > 12.5\,\mathrm{MHz}$). They are due to harmonics emissions, intermodulation products, etc. The limits for these emissions are rather soft – e.g. $-30\,\mathrm{dBm}$ for a 1-MHz bandwidth. However, for frequencies which are used by other mobile systems, such as GSM, DECT, Digital Cellular System (DCS-1800), or UTRA-TDD, they are more severe. Especially when a UMTS BS is co-located with a GSM BS, emission limits of $-98\,\mathrm{dBm}/100\,\mathrm{kHz}$ might apply, thus ensuring that UMTS signals do not interfere with GSM signals.

23.4 Physical and logical channels

23.4.1 Logical channels

Similar to GSM, we distinguish between different logical channels in UMTS, which are mapped to physical channels. The logical channels in UMTS are sometimes referred to as *transport channels*. There are two kinds of transport channels: *common transport channels* and *Dedicated transport CHannels* (DCHs).

Common channels

Common channels are relevant to all or at least a group of MSs in a cell. Thus, all of them receive the information transmitted on these channels in the downlink and may access the channels in the uplink. There are different kinds of common channels:

- *Broadcast CHannel* (BCH): the BCH is only found in the downlink. Both cell-specific and network-specific information is transmitted on it. For example, the BS uses this channel to inform all MSs in the cell about free access codes and available access channels. This channel has to be transmitted with relatively high power, as all MSs within the cell have to be able to receive it. Thus, on this channel neither power control nor smart antennas are implemented.
- *Paging CHannel* (PCH): this is also a channel that can be found only in the downlink. It is used to tell an MS about an incoming call. Since attenuation of the channel to the MS, as well as the location of the MS, is not known the PCH is transmitted with high power and without employing smart antennas. Depending on whether the current cell of the MS is known, paging information is either transmitted in only one cell or several cells.
- *Random Access CHannel* (RACH): the RACH is only used in the uplink. The MS uses it to initialize a connection to the BS. It can employ open-loop power control, but no smart antennas, as the BS must be able to receive signals on the RACH from every MS within the cell. As it is a *random* access channel, collisions might occur. Therefore, the structure of bursts in the RACH are different from that of other channels – this will be described below in more detail.
- *Forward Access CHannel* (FACH): the FACH is used to transmit control information to a specific mobile. However, as the FACH is a common channel and therefore received by more than one MS, explicit addressing of the desired MS is required (*in-band identification*, UE-ID at the beginning of the packet). In contrast with this a dedicated control channel has an implicit addressing of the desired MS: the MS is specified by the carrier and the spreading code used. The FACH can also transmit short user information packets. It can employ smart antennas, as information is transmitted to one specific, localized mobile.

- *Common Packet CHannel* (CPCH): the CPCH is an uplink channel, and can be interpreted as the counterpart of the FACH. It can transmit both control and information packets. If the FACH and the CPCH are used together closed-loop power control is possible.
- *Downlink Shared CHannel* (DSCH): the DSCH is a downlink channel similar to the FACH. It sends mainly control data, but also some traffic data, to multiple MSs. Explicit addressing has to be used on this channel, as the same CDMA code is used for all MSs. The reason for using the same code for multiple MSs lies in the limited amount of *short* spreading codes (see next section). Under normal circumstances, one spreading code would be permanently reserved for one MS, even if the traffic is bursty. The cell would thus quickly run out of codes (remember that there are very few codes that can be used for high-data-rate traffic). On the DSCH one short code is used for several mobiles and data are multiplexed in time. The DSCH supports fast power control, use of smart antennas, and rate adaptation from transmission frame to transmission frame. Note that the DSCH uses a different approach in HSDPA – namely, the use of multiple short codes simultaneously.

Dedicated channels

Dedicated channels are present in both the uplink and the downlink. They are used to transmit both higher layer signaling and actual user data. As the position of the MS is known when a dedicated channel is in use, smart antennas, as well as fast power control and adaptation of the data rate on a frame-by-frame basis, can be used:

- *Dedicated transport CHannel* (DCH): this is the only type of dedicated logical channel. MS addressing is inherent as for each MS a unique spreading code is used.

23.4.2 Physical channels

WCDMA transmits control and user data on the same logical channel (in contrast to GSM) – namely, the DCH. However, for physical channels we distinguish between channels for control and user information. Combined transmission of both is called the *Coded Composite Traffic CHannel* (CCTrCH). Note that there are also some physical channels that are not related to a specific logical channel.

Uplink

In the uplink we find dedicated control and user data channels, which are transmitted simultaneously via I- and Q-code multiplexing (see Section 23.6):

- Pilot bits, *Transmit Power Control* (TPC), and *Feed Back Information* (FBI) are transmitted via the *Dedicated Physical Control CHannel* (DPCCH). Furthermore, the *Transport Format Combination Indicator* (TFCI) may be transmitted via the DPCCH (see also Section 23.5.2). The TFCI contains the instantaneous parameter of all data transport channels which are multiplexed on the DPDCH (see below). The spreading factor for the DPCCH is constant – namely, 256. Thus, ten control information bits are transmitted in each slot.
- The actual user data is transmitted via the *Dedicated Physical Data CHannel* (DPDCH). Spreading factors between 4 and 256 can be used. The DPDCH and DPCCH are transmitted at the same time and on the same carrier by using the I- and Q-branch.
- The *Random Access CHannel* (RACH) is not only a logical but also a physical channel (physical RACH or PRACH). A *slotted ALOHA* approach is used for medium access (see Chapter 17). The burst structure of the RACH is completely different from that of the dedicated channel. Data packets may be transmitted via the PRACH, too. Packets can be

transmitted either in pure PRACH mode, meaning that PRACH bursts as described in Section 23.6 are used, or in the *Uplink Common Packet CHannel* (UCPCH). The UCPCH is an extension of the PRACH. It is used in combination with a DPCCH in the downlink. Therefore, fast power control is possible.

- The *Physical Common Packet CHannel* (PCPCH) has a similar burst structure to the PRACH. Information is transmitted after a preamble. Initially several access preambles are transmitted with ascending transmission power until the BS receives the necessary signal strength. After the BS acknowledges reception, another preamble is transmitted to detect eventual collisions with packets from other MSs that try to access the PCPCH at the same time. Before user data are transmitted, a power control preamble of length 0 or 8 timeslots can be transmitted. Only then are the actual data transmitted; the length of this transmission period is a multiple of the frame duration – i.e., 10 ms.

Downlink

Of course, the dedicated data and control channels, DPDCH and DPCCH, can be found in the downlink, too. However, they are multiplexed in a different manner, which is discussed in Section 23.6.

Furthermore, the downlink features the following common control channels:

- *Primary Common Control Physical CHannel* (P-CCPCH): bursts transmitted via the CCPCHs are similar to those of the DDPCH. However, there is no power control associated with the CCPCH and therefore the corresponding bits do not have to be transmitted. The P-CCPCH has an idle period in each frame; this idle period is 256 chips long. The P-CCPCH carries the broadcast channel, and it is thus critical that it can be demodulated by all MSs in the cell. Thus, it uses a constant, high spreading factor – namely, 256.
- *Secondary Common Control Physical CHannel* (S-CCPCH): the main difference between the primary and the secondary CCPCH is that the data rate and spreading factor are fixed in the former, whereas they are variable in the latter. The S-CCPCH carries the FACH and the PCH.
- *Synchronization CHannel* (SCH): the SCH does not relate to any logical channel. Its function is explained in Section 23.7. The SCH is multiplexed with the P-CCPCH onto one timeslot: it sends 256 chips for synchronization. During the above-mentioned idle period of a P-CCPCH burst the SCH is transmitted by sending 256 chips for synchronization.
- *Common Pilot CHannel* (CPICH): this channel has a constant spreading factor of 256. The CPICH consists of a primary and a secondary CPICH. The primary CPICH serves as a reference for phase and amplitude for common channels in the entire cell. The primary CPICH is transmitted all over the cell and the secondary might be transmitted in selected directions. The primary and secondary pilot channel differ in the spreading and scrambling codes used: the primary CPICH always uses the primary scrambling code and a fixed channelization code, so that there is only one such code per cell.

CPICHs are particularly important for establishing a connection as during this period the pilots of the dedicated channels are not available. Furthermore, pilot channels provide an indication of signal strength at the MS and are therefore important for handover procedures. The cell size of a BS can be varied by varying the transmit power of pilot channels. By reducing transmit power the area over which the BS provides the strongest signal is decreased. This reduces the traffic load of a BS.

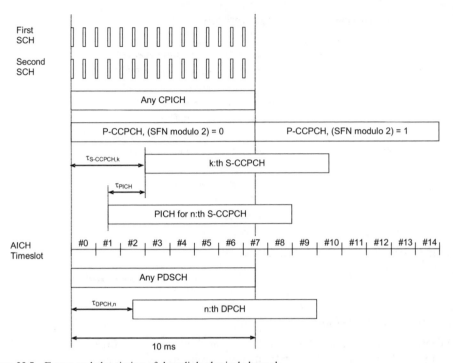

Figure 23.5 Frame and slot timing of downlink physical channels.

- Apart from these channels, the downlink further features the *Physical Downlink Shared CHannel* (PDSCH), which carries the DSCH, the *Acquisition Indication CHannel* (AICH), which provides feedback about whether synchronization was successful or not, and the *Page Indication CHannel* (PICH), which supports paging.

Matching of logical and physical channels

Figure 23.6 illustrates how logical channels, also referred to as transport channels, are matched to physical channels. Details regarding frame and slot timing can be found in the standard.

23.5 Speech coding, multiplexing, and channel coding

23.5.1 Speech coder

The speech coder used in UMTS is an *Adaptive Multirate Coder* (AMR) that also has a strong similarity to the enhanced speech coder used in GSM. AMR codecs are based on *Algebraic Code Excited Linear Prediction* (ACELP) (see Chapter 15). WCDMA-AMR contains eight different encoding modes, with source rates ranging from 4.75 to 12.2 kbit/s, as well as a "background noise" mode.[4]

---

[4] A recently standardized "wideband" AMR that results in even better speech quality, uses nine modes with rates between 6.6 and 23.85 kbit/s.

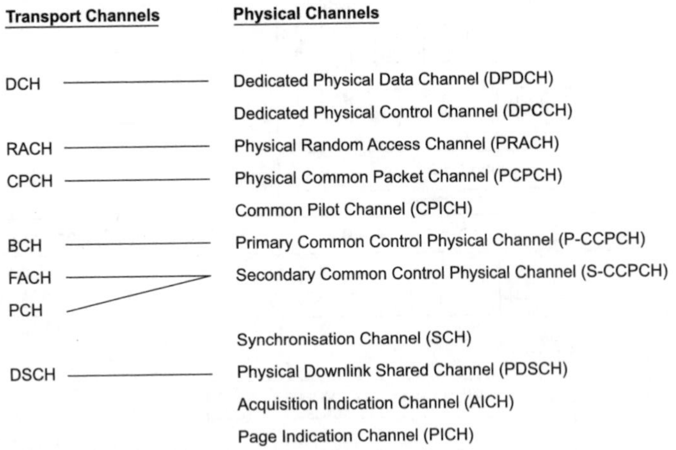

Transport Channels	Physical Channels
DCH ———————	Dedicated Physical Data Channel (DPDCH)
	Dedicated Physical Control Channel (DPCCH)
RACH ———————	Physical Random Access Channel (PRACH)
CPCH ———————	Physical Common Packet Channel (PCPCH)
	Common Pilot Channel (CPICH)
BCH ———————	Primary Common Control Physical Channel (P-CCPCH)
FACH ———————	Secondary Common Control Physical Channel (S-CCPCH)
PCH	
	Synchronisation Channel (SCH)
DSCH ———————	Physical Downlink Shared Channel (PDSCH)
	Acquisition Indication Channel (AICH)
	Page Indication Channel (PICH)

Figure 23.6 Matching of physical and logical channels.

23.5.2 Multiplexing and interleaving

Multiplexing, coding, and interleaving is a very complicated procedure that allows a high degree of flexibility. A datastream coming from upper layers has to be processed before it can be transmitted via the transport channels on the air interface. Transport channels are processed in blocks of 10-, 20-, 40- or 80-ms duration. We first discuss multiplexing and coding in the uplink. The block diagram in Fig. 23.7a illustrates the order of the processes involved in multiplexing and coding:

- When processing a transport block, the first step is to append a *Cyclic Redundancy Check* (CRC) field. This field, which can be 8, 12, 16, or 24 bits long, is used for the purpose of error detection. It is calculated for each block of data for one transmission time interval from the code polynomials:

$$G(D) = D^8 + D^7 + D^4 + D^3 + D + 1 \quad \text{for 8-bit CRC} \tag{23.1}$$

$$G(D) = D^{12} + D^{11} + D^3 + D^2 + D + 1 \quad \text{for 12-bit CRC} \tag{23.2}$$

$$G(D) = D^{16} + D^{12} + D^5 + 1 \quad \text{for 16-bit CRC} \tag{23.3}$$

$$G(D) = D^{24} + D^{23} + D^6 + D^5 + D + 1 \quad \text{for 24-bit CRC} \tag{23.4}$$

and attached at the end of the block.
- Afterwards the datablocks are concatenated or segmented into blocks that have suitable size for channel coding. The blocks should not be too small, because that increases the relative impact of the overhead (tail bits), and makes the performance of turbocodes worse. On the other hand, the blocks should not be too big; otherwise, decoding can become too complicated. For convolutional encoding the blocksize is typically 500 bits, for turbocodes approximately 5,000 bits. Tail bits are appended to help the decoder: 8 tail bits if convolutional encoding is used, and 4 tail bits for turbocoding.
- The blocks are then encoded with convolutional codes or turbocodes, details of which will be discussed in the next subsection.

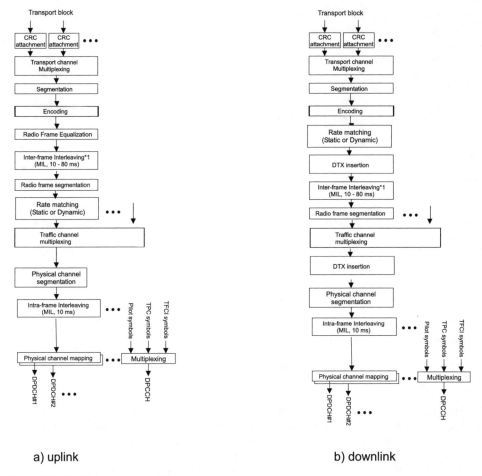

a) uplink

b) downlink

Figure 23.7 Multiplexing and coding.

- The resulting encoded blocks then undergo *radio frame size equalization*. This makes sure that the amount of data is the same for each radio frame.
- In case the block spans more than one frame of length 10 ms, *interframe interleaving* is applied, which interleaves bits over the different frames of this block.
- If necessary, the blocks are then divided into 10-ms transmission blocks, a process called *radio frame segmentation*.
- The encoded block then undergoes *rate matching* – i.e., the rate of the block is then adapted to the desired rate by puncturing or selected bit repetition. Repetition is usually preferred, except for some special high-data-rate cases.
- If multiple data transport channels are transmitted, the resulting transmission blocks or frames are then time-multiplexed. Each block is accompanied by a TFCI. This contains the rate information for the current block, and is therefore very important – if the TFCI is lost, then the whole frame is lost.
- The resulting datastream is then fed into a second interleaver, which interleaves the bits over one radio transmission frame, *intraframe interleaving*. In case multiple physical data

channels are used, the transmission frames are then mapped to these multiple channels. Otherwise, a single dedicated physical channel is used.

The multiplexing operation in the downlink is slightly different, as outlined in Fig. 23.7b, as the order of some of the steps is different. The main difference lies in the insertion of DTX (discontinuous transmission) bits, which indicate when to turn the transmission off. Depending on whether fixed or variable symbol positions are used, DTX indication bits are inserted at different points in the multiplexing/coding chain.

Coding

There are two modes of channel coding:

1. Convolutional codes are used with a coding rate of $1/2$ for common channels and $1/3$ for dedicated channels. Convolutional codes are mainly used for "normal" applications with a data rate of up to $32\,\text{kbit/s}$. The constraint length of the encoders is 9. Figure 23.8 shows the structure of the two encoders. The code polynomials for the rate-$1/2$ encoder are:

$$G1(D) = 1 + D^2 + D^3 + D^4 + D^8 \tag{23.5}$$

$$G2(D) = 1 + D + D^2 + D^3 + D^5 + D^7 + D^8 \tag{23.6}$$

and for the rate-$1/3$ encoder:

$$G1(D) = 1 + D^2 + D^3 + D^5 + D^6 + D^7 + D^8 \tag{23.7}$$

$$G2(D) = 1 + D + D^3 + D^4 + D^7 + D^8 \tag{23.8}$$

$$G3(D) = 1 + D + D^2 + D^5 + D^8 \tag{23.9}$$

2. Turbocodes are mainly used for high-data-rate ($> 32\,\text{kbit/s}$) applications. The code rate is $1/3$. A parallel concatenated code is used (see Chapter 14). Two recursive systematic convolutional encoders are employed (see Fig. 23.9). The datastream is fed into the first one directly, and into the second one after passing an interleaver. Both encoders have a coding rate of $1/2$. Thus, output is the original bit X or X' and the redundancy bits Y or Y', which are output from the recursive shift registers. However, as X equals X' only X, Y, and Y' are transmitted. Thus, the code rate of the turboencoder is $1/3$.

Table 23.1 provides an overview of the different coding modes used in different channels.

23.6 Spreading and modulation

23.6.1 Frame structure, spreading codes, and Walsh–Hadamard codes

WCDMA relies on CDMA for multiple access. However, transmission timing is still based on a hierarchical timeslot structure similar to GSM's: frames of duration $T_f = 10\,\text{ms}$ are divided into 15 timeslots, each of which has a 12-bit-long *System Frame Number* (SFN). Each timeslot has a duration of $0.667\,\text{ms}$ which equals 2,560 chips. The configuration of frames and timeslots is different for uplink and downlink.

WCDMA uses two types of code for spreading and multiple access: *channelization codes* and *scrambling codes* (compare also IS-95, Chapter 22). The former spread the signal by extending the occupied bandwidth in accordance with the basic principle of CDMA. The latter do not lead

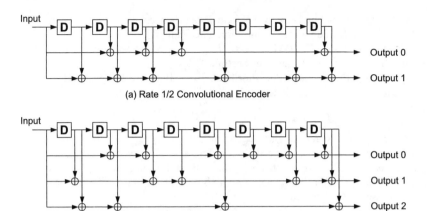

(a) Rate 1/2 Convolutional Encoder

(b) Rate 1/3 Convolutional Encoder

Figure 23.8 Structure of convolutional encoders.

3GPP TSs and TRs are the property of ARIB, ATIS, ETSI, CCSA, TTA, and TTC who jointly own the copyright in them. They are subject to further modifications and are therefore provided "as is" for information purposes only. Further use strictly prohibited.

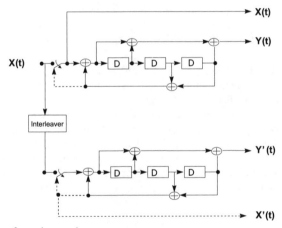

Figure 23.9 Structure of a turboencoder.

3GPP TSs and TRs are the property of ARIB, ATIS, ETSI, CCSA, TTA, and TTC who jointly own the copyright in them. They are subject to further modifications and are therefore provided "as is" for information purposes only. Further use strictly prohibited.

Table 23.1 Forward error correction and different logical channels.

Transport channel type	Coding scheme	Coding rate
BCH		
PCH		1/2
RACH	Convolutional code	
CPCH, DCH, DSCH, FACH		1/3, 1/2
	Turbocode	1/3

to bandwidth expansion but help to distinguish between cells and users. In the following we discuss the codes and modulation for the different channels.

Channelization codes in WCDMA are *Orthogonal Variable Spreading Factor* (OVSF) codes (as discussed in Section 18.2.6, see also Section 22.5.1).

For scrambling codes, a long and a short code exist. Both are complex codes, and are derived from real-valued codes in accordance with the following expression:

$$C_{Scrambler}(k) = c_1(k) \cdot (1 + j \cdot (-1)^k \cdot c_2(2\lfloor k/2 \rfloor)) \tag{23.10}$$

Here, k is the chip index and c_1 and c_2 are real-valued codes.

For short codes c_1 and c_2 are two different members of the very large Kasami sets of length 256 (see Section 18.2.5). It is worth noting that the duration of the short code equals symbol duration only for spreading factor 256. Otherwise, a "short" code in WCDMA is not a short code in the sense of Chapter 18.

The long code is a Gold code, a combination of two Pseudo Noise (PN)-sequences that each have length $2^{25} - 1$. The I- and Q-part, c_1 and c_2 in Eq. (23.10), are versions of the same Gold sequence, shifted relative to each other. The codes are truncated to a length of 10 ms – i.e., one frame.

23.6.2 Uplink

Dedicated channels

Figure 23.10 is a block diagram illustrating spreading and modulation for the uplink:

- Under normal circumstances, user data (DPDCH$_1$) and control data (DPCCH) are transmitted on the in-phase component and the control channel on the quadrature-phase component. First, channelization codes c_d and c_c are applied to the data and the control channel, respectively. As mentioned above these codes actually increase the signal bandwidth. Afterwards, the I- and Q-branch are treated as a complex signal. Processing the control and data channel like this is called I–Q multiplexing.
- If the data rate for the user is very high, then up to five additional data channels may be transmitted in parallel on the I-branch and Q-branch. These channels are then distinguished by applying different spreading codes, $c_{d,k}$ $k \in [1, \dots, 3]$(!). The spreading factor is in this case 4. Therefore, the total data rate is given by:

$$\frac{4 \text{ Mchip/s}}{4 \text{ chips/bit}} \cdot 6 \cdot \frac{1}{3} \tag{23.11}$$

The 4 chips/bit represent the spreading. The factor 6 relates to the maximal six transmitted channels and factor 1/3 relates to channel coding. Thus, the maximal achievable net user data rate is 2 Mbit/s.

- The transmit power of the control channel relative to the data channel(s) is determined by the ratio of spreading factors, β_c/β_d. The received noise power is proportional to the (unspread) bandwidth and thus the data rate. Consequently, the transmit power has to be the higher the lower the spreading factor in order to make sure that the Signal-to-Noise Ratio (SNR) at the receiver is the same for user data and noise data. The spreading factor of control data is always 256.
- Then the complex scrambling code $S_{long,n}$ or $S_{short,n}$ is applied and the signal is fed into the complex modulator. The complex signal usually shows a power imbalance between the I- and Q-factor. In order to deal with this problem, spreading and scrambling codes are designed in such a way that the signal constellation is rotated by 90° between two subsequent chips.

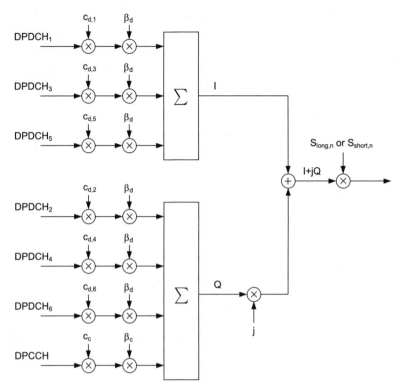

Figure 23.10 Spreading and modulation on the uplink.

- The resulting signal is modulated using Quadrature Amplitude Modulation (QAM) with Nyquist basis pulses. More precisely, the basic pulses are raised cosine pulses with a rolloff factor of $\alpha = 0.22$. Modulation precision is supposed to be better than 17%. Precision is measured by taking the ratio of the power of the difference signal between the real and the ideal signal and the power of the ideal signal – *Error Vector Magnitude* (EVM).

I–Q multiplexing is used in the uplink in order to limit the crest factor. Even when there are no user data to be transmitted – e.g., in no-voice periods – the control channel is continuously active. Therefore, the RF amplifier in the MS does not have to permanently switch on and off. This eases hardware requirements as the amplifier has to cover a smaller dynamic range. Furthermore, amplitude variations from slot to slot or burst to burst could lead to interference with audioprocessing equipment, like the microphone of the MS or the hearing aid of the user, as these activities would be in the audible frequency range.

We next turn to look in more detail at the control channel. It carries the pilot bits, the TPC bits, the FBI, and the TFCI. As the spreading factor is constantly 256, 10 bits of control information are transmitted per timeslot. There are different ways to partition these 10 bits for the pilot, the TPC, the FBI, and the TFCI.

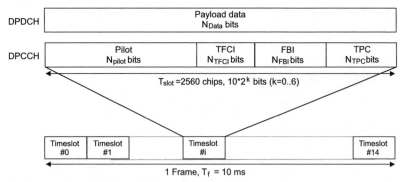

Figure 23.11 Frame structure on the uplink.

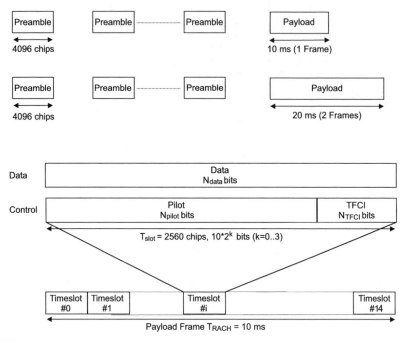

Figure 23.12 Structure of random access transmission.

Random access channel

We next turn our attention to the burst structure of the RACH (see Fig. 23.12). The start time for RACH transmission should be $t_0 + k \cdot 1.33$ ms, where t_0 refers to the start of a regular frame on the BCH. At the beginning of an access burst several preambles of length 4,096 chips are transmitted. The first preamble has power that is determined by "open-loop" power control – i.e., the MS measures the strength of the BCH, and from that concludes the amount of small-scale-averaged pathloss; this in turn determines transmit power (plus a safety margin). If the MS

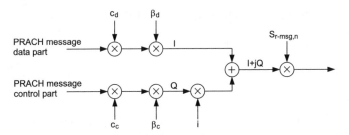

Figure 23.13 Modulation on the physical random access channel.

does not receive an acknowledgement of its access request on the AICH, it transmits the preamble again, with increased power. Afterwards receiving an acknowledgement, the actual access data (a field that is either 10 ms or 20 ms long) are transmitted.

The preamble contains one of a set of 16 predefined signature sequences, which are transmitted with a spreading factor of 256. The "data"-containing message part of length 10 ms is similar to the DPCH divided into 15 timeslots. Within each of those timeslots 8 pilot bits and 2 TFCI bits are transmitted with a spreading factor of 256; this constitutes the control information for layer 1. Furthermore, a "data" message, usually containing control information for layer 2, is transmitted with a spreading factor between 32 and 256. Layer-1 control information and the "data" message are transmitted simultaneously by I–Q multiplexing. Thus, the structure of the message part of the RACH is very similar to a frame of a dedicated physical channel. However, neither a TPC nor an FBI field are transmitted.

Physical common packet channel

The burst structure of the PCPCH is quite similar to that of a transmission on the PRACH. One or several access preambles are transmitted initially with ascending transmission power until the received power at the BS is sufficient and the BS sends an acknowledgment to the MS. Afterwards, one preamble is transmitted whose sole purpose is the detection of collisions with other packets. Furthermore, power control information can be transmitted. Afterwards the actual data are transmitted. This message consists of one or several frames of 10 ms each, each of which is again divided into 15 timeslots.

Base station processing

The BS has to process the received signal periodically in the following way [Holma and Toskala 2000]:

- It receives one frame, despreads it, and stores it with the sampling frequency given by the highest used data rate of this frame. Note that different data rates and spreading factors may be used during one frame.
- For each timeslot: (i) the channel impulse response is estimated using the pilots; (ii) the Signal-to-Interference Ratio (SIR) is estimated; (iii) power control information is transmitted to the MS; and (iv) power control information from the MS is decoded and transmission power is adapted accordingly.
- For every second or fourth timeslot: FBI bits are decoded.[5]

---

[5] Note that two or four FBI blocks make up one complete FBI command, like an update of antenna weights.

- For each 10-ms frame: TFCI information is decoded to get the decoding parameters for the DPDCH.
- for each interleaver period, which is either 10, 20, 40, or 80 ms: user data transmitted via the DPDCH are decoded.

Spreading codes

The uplink uses OVSF codes for spreading. However, they are not used for channelization (distinguishing between users in the uplink). Therefore, different users can use the same spreading codes. As a matter of fact, assignment of spreading codes to the different channels of any MS is predefined. The DPCCH is always spread with the first code, the code with index 0, using spreading factor 256. A DPDCH channel is spread using the code with index $SF/4$, where SF is the spreading factor of the channel. In case multiple data channels are transmitted the spreading factor is 4, when spreading codes with indices 1, 2, or 3 are used; one code is used for both a channel on the I-branch and a channel on the Q-branch. The spreading factor of a traffic channel may vary from frame to frame. There are additional code selection criteria for the PRACH and PCPCH.

Scrambling codes

As mentioned above, signals from different users are distinguished by different scrambling codes. Both, "short" and "long" codes may be used (see Chapter 18). There are no strict rules about when to apply short or long codes. However, it is recommended to use short codes when the BS has multiuser detection capability (see Section 18.4). Short codes limit the computational complexity for multiuser detection, as the size of the cross-correlation matrices involved is smaller. In case no multiuser detection is implemented, long codes should be used as they provide better whitening of interference than short codes.

The BS assigns the specific scrambling code to the MS during connection setup with the *Access Grant Message*.

Codes for random access channels

Random access channels use codes that are different from those in regular channels. These codes are specific to one BS, and two neighboring BSs should not use the same or similar codes. The preamble is transmitted first, which serves for synchronization and identification (see below). The preamble is designed in such a manner that it is very robust against initial uncertainties in frequency synchronization. Furthermore, RACH codes are only transmitted with a binary alphabet, which eases receiver design.

23.6.3 Downlink

Spreading and modulation of the data and control channels for the downlink is different from the uplink. In the downlink, data and control channels are *time-multiplexed* and the resulting single bitstream is then Quadrature Phase Shift Keying (QPSK)-modulated. We may interpret QPSK modulation of a single bitstream as a serial/parallel conversion into two streams on the I- and Q-branch. The resulting I-signal and Q-signal are spread separately using the same channelization code. The complex scrambling code is then applied to the resulting complex signal. Finally, the scrambled complex signal is fed into the complex modulator.

Figure 23.14 illustrates the frame and timeslot structure for the DDPCH. The first datablock containing bits is transmitted. Then the TPC field is sent, followed by the TFCI field. A second

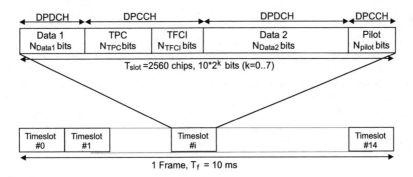

Figure 23.14 Frame and timeslot structure on the downlink.

datafield is transmitted thereafter. The timeslot is concluded with the pilot field. A spreading factor between 4 and 512 can be used. The purpose and properties of the pilot, TFCI, and TPC fields are the same as for the uplink.

The spreading factor for one user does not vary from frame to frame. It is chosen to accommodate the highest occurring data rate of this user; if a lower rate is required instantaneously, DTX is employed – i.e., transmission is blanked out for a while. The MS can be informed about the current data rate by the TFCI field.

The reasons that DTX can be employed in the downlink (in contrast to the uplink) are that (i) the appearance of audible interference resulting from on/off switching of transmit amplifiers is not an issue on the downlink; the common control and synchronization channels, like the BCH or the SCH, are transmitted continuously anyway; and (ii) high values of the crest factor cannot be avoided anyway as transmission of several CDMA signals in parallel always leads to high crest factors.

Mobile station processing

The MS receiver has to perform operations for the downlink similar to those of the BS receiver in the uplink. However, there are some differences [Holma and Toskala 2000]:

- The spreading factor of all channels apart of the DSCH is constant over time.
- There is no need for the FBI field.
- There is one common pilot channel (CPICH) in addition to the pilots of dedicated physical channels, which improves the channel estimate.
- Smart antennas can be used in the downlink.

Channelization codes

The signals for different users are inherently synchronized in the downlink, as they come from the same transmitter – i.e., the BS (see also Chapter 22). Therefore, OVSF codes enable good separation of the signals for different users within one cell. The BS informs the MS about the code used during connection setup. The BS ensures that the same code (or a "mother code" in the OVSF tree) is only used once in the cell. If necessary, the code can be changed during a connection.

Scrambling codes

Long codes for the downlink are the same as in the uplink; short codes are not used for the downlink. There are 512 primary scrambling codes,[6] which are divided into 64 groups of 8 codes each. Primary, secondary, and alternative codes are unambiguously related to each other. Each cell has exactly one primary scrambling code that is used on the CCPCH and CPICH. The other downlink channels might use the same primary or a related secondary code. However, use of more than one scrambling code only makes sense when other measures, like smart antennas, are used to increase cell capacity. Otherwise, one scrambling code in conjunction with spreading codes already achieves maximum user separation.

Overview

Table 23.2 provides an overview of the different codes used.

High-data-rate users

There are basically two possibilities to support high data rates in UMTS:

1. Transmission with a low spreading factor. This is the most straightforward way to increase the data rate. The lowest admissible spreading factor in WCDMA is 4; note that a spreading factor of 1 would lead to a pure Frequency Division Multiple Access (FDMA) system. The drawback with a low spreading factor is that Intersymbol Interference (ISI) becomes significant: with a spreading factor of 4, the symbol duration can be as low as 1 μs. Thus, the receiver has to cope not only with interchip interference, which is effectively dealt with by a Rake receiver structure, but also with severe ISI that needs to be eliminated by an equalizer (Chapter 16). This additional implementation complexity may increase the price of the unit, particularly for an MS.

2. The datastream can be serial/parallel converted and then transmitted using multiple codes, where each code has a sufficiently high spreading factor. This is called "multicode transmission". ISI is not a problem with this approach. However, two other problems appear: (i) the crest factor of a multicode signal is high (this is not a problem on the downlink, as multiple users have to be supported anyway); and (ii) efficiency is lowered as every additionally transmitted code leads to additional overhead.

23.7 Physical-layer procedures

23.7.1 Cell search and synchronization

The search for the signal of the strongest BS and synchronization with this BS has to be done in several steps.

In a first step, the MS *synchronizes with the timeslot* timing of the strongest signal it can observe. This is done by searching for the 256-chip primary synchronization code. Every BS periodically transmits this sequence, and it is identical for the whole system. The MS thus just correlates the incoming signal with this synchronization sequence to identify all available BSs. The output of this correlator shows several peaks at different delays, corresponding to the

---

[6] Additionally, there are 512×25 secondary scrambling codes and 8,192 left and 8,192 right alternative scrambling codes for compressed mode.

Table 23.2 Overview of different channelization and scrambling codes.

	Channelization codes (variable chip rate)		Scrambling codes (constant chip rate)		
	Uplink	Downlink	Uplink		Downlink
Separates	Channels (I/Q, DPDCH) of one user	Channels (DPDCH) and users	Users		Cells
	Assigned to connection	Assigned to connection and user	Assigned to user by BS		Assigned to cell
Reuse	Within the same cell	In all other cells	In other cells		In distant cells (code planning)
Selection	Fixed (given by the SF)	Variable	Variable		Fixed
Length of code	Short	Short	"Short"	Long	Long
Enables	Variable data rates	Different data rates	Multiuser detection		Cell search
Code family	OVSF (real)	OVSF (real)	Complex, based on VL Kasami	Complex, based on segments of Gold codes	Complex, based on segments of Gold codes
Codelength	$4 - 256$	$4 - 512$	256	38,400	38,400
Number of codes	Up to 256		$>10^6$	$\gg$	512 (8,192)

signals of different BSs and the various echoes (MPCs) of these signals. By selecting the highest peak, the MS achieves timeslot synchronization with the strongest signal. In other words, the MS is aware of the commencement of a timeslot in the strongest BS. However, it still does not know about frame timing, as it does not know whether the first or the tenth timeslot is currently received.

Therefore, *frame synchronization* has to take place in the next step. This is done by observing the secondary synchronization channel; in the same step the codegroup of the primary scrambling code used by the BS is determined. As discussed earlier, the CCPCH and CPICH are transmitted in the downlink with one of 512 primary scrambling codes. These 512 primary codes are divided into 64 groups. The secondary synchronization channel transmits a sequence of 15 codes in 15 consecutive timeslots. Each code is a letter of the same alphabet. Thus, we may interpret the code sequence as a word with 15 letters. This codeword is repeated periodically. The MS can thus determine the group used and achieve frame synchronization by observing this codeword. Codewords are chosen in such a manner that a cyclic shift of one codeword never results in another codeword. Therefore, the MS can determine which codeword is transmitted (this indicates the used codegroup) and where the codeword starts (this indicates frame timing).

Let us illustrate this procedure with a simplified example. Assume, that the alphabet size used is five and three codewords of length 4 can be transmitted: (i) *abcd*; (ii) *aedc*; and (iii) *bcde*. Further, assume that *edca* is received. As (i) does not feature the letter *e* and (iii) does not feature the letter *a*, the received word has to be a cyclically shifted version of (ii). Once we know that (ii) is transmitted, we can easily see that frame timing has to be adjusted by one slot.

After the codegroup has been determined, the MS has to determine which code of this group is used on the CCPCH. Therefore, the CCPCH undertakes symbolwise correlation with all possible eight codes of that codegroup. Once this is achieved, the CCPCH can be properly demodulated.

23.7.2 Establishing a connection

A connection initialized by the MS requires the following procedures:

1. The MS synchronizes itself with the strongest BS as described in the previous section.
2. The MS decodes the BCH and thereby acquires information regarding: (i) the spreading code of the preamble and the scrambling code for the message part of the PRACH; (ii) the codes available; (iii) the access timeslots available; (iv) possible spreading factors for messages; (v) the interference level on the uplink; and (vi) the transmit power of the CCPCH.
3. The MS selects a spreading code for the preamble and a scrambling codes for messages.
4. The MS determines the spreading factor for the message.
5. Based on the measured signal strength of the CCPCH and information on the transmit power of the BS the MS estimates the attenuation on the uplink. Based on the attenuation estimate and information regarding the uplink interference level the MS estimates the necessary transmission power for the preamble.
6. The MS randomly selects an access timeslot and a signature from the available set.
7. The MS transmits the preamble. In case of successful acquisition the BS transmits an acknowledgment on the AICH.
8. In case the MS does not receive an acknowledgement from the BS it repeats the preamble with increased transmission power.
9. Once the BS indicates acquisition of the preamble, the MS starts to transmit the access message in the next available timeslot.
10. The MS waits for an access grant message from the network. If it does not receive this message within a predetermined time, it repeats the steps from step 5 on.

23.7.3 Power control

Power control is an essential part of a CDMA system, as it is necessary to control mutual interference (see Chapter 18). Inner-loop power control in WCDMA, in particular, is supposed to adapt to *small-scale* fading for speeds up to 500 km/h. Therefore, the power control procedures in UMTS have to be rather fast. An update of transmit power occurs with every timeslot – i.e., every 0.667 ms. There is also outer-loop power control which continuously adjusts the target SIR for inner-loop power control (see also power control in IS-95, Chapter 22).

Uplink

The uplink uses a closed-loop procedure for power control. The BS estimates the power of the received signal and controls it by transmitting TPC instructions to the MS, which changes its transmit power accordingly. TPC bits are transmitted with the DPCCH, and contain instructions to increase or decrease power. The possible stepsizes are 1 dB or 2 dB with uncertainties of ±0.5 dB or 1 dB. However, transmit power is not decreased below, for example, -44 dBm. Given a maximal transmission power of 33 dBm for class-1 units, the RF amplifier has to cover a dynamic range of 77 dB, which is a rather severe hardware requirement.

As for soft handover, the situation is rather complicated as the MS can receive different TPC bits from the different BSs involved. The UMTS standard specifies an algorithm that determines a means of obtaining a combined power control command from these different commands. The combined instruction is a function of weighted single-power-control commands. The weighting applied is proportional to reliability of the individual signals. Some particular rules apply for special situations. For example, for both compressed mode and the common packet channel a special procedure is implemented, as longer gaps between transmissions and TPC commands occur.

It is not possible to use closed-loop power control on the PRACH, as a dedicated channel has not yet been established between the MS and the BS. Therefore, open-loop power control has to be employed. The MS measures the average received power of the CCPCH over some time to average out small-scale fading effects. This is necessary because, due to the frequency duplex, small-scale fading on the uplink is unrelated to that on the downlink. In other words, it is not possible to determine from the instantaneous received power on the downlink the level of instantaneous received power on the uplink. Average received power allows estimation of just the necessary *average* transmit power for the uplink. However, this is at least a good starting value.

Downlink

All signals on the downlink (to one MS) suffer from the same attenuation. Furthermore, downlink signals are orthogonal, due to the use of OVSF codes for channelization, so that the despreading operation results in a good SIR.[7] Therefore, power control on the downlink aims mainly at maintaining a good SNR.[8] A closed loop is used for power control in the downlink. Each MS measures signal strength and quality and transmits TPC commands to the BS via the DPCCH. As long as at least one MS requests more transmission power the BS increases the power of all channels. Therefore, MSs at cell borders actually control the transmission power of the BS.

One problem is that the MS estimates the SNR *after* the Rake receiver. A cheap device, which employs only a few Rake fingers, captures less received signal energy than a more sophisticated device with more Rake fingers. Therefore, a cheap device tends to requests more transmit power from the BS than an expensive one. Thus, the UMTS network levels out differences in the designs of units by ensuring good receive qualities independent of receiver design quality. Of course, this might limit the motivation of manufacturers to develop and produce good MSs.

Transmit diversity

The downlink may implement the option of transmit diversity with two antennas. It distinguishes between closed-loop diversity, for which the transmitter requires channel-state information, and

---

[7] Note, however, that orthogonality can be destroyed in frequency-selective channels.

[8] This argumentation only considers one cell. Considering intercell interference, the SIR in one cell might be decreased by increasing the transmission power of all channels. However, this in turn decreases the SIR in all neighboring cells.

open-loop diversity, for which the transmitter does not require any information about the channel. Three modes of diversity are specified:

- Orthogonal space–time block coding with two antennas – i.e., the Alamouti code (see Section 20.2).
- The transmit signal might be switched from one antenna to the other. This mode is only used for the SCH.
- Closed-loop transmit diversity: two antennas can transmit the same stream, by applying complex weights for each antenna. The pilot bits on signals from the two antennas are mutually orthogonal sequences. Therefore, the receiver can estimate the channel impulse responses of the two channels separately. Antenna weights are then determined such that signals from the two transmit antennas add up constructively at the MS (see Chapter 13). The computed weights are digitized, and transmitted to the BS via the FBI field of the DPCCH. This mode of diversity is optional for BSs, but MSs have to be able to support it.

23.7.4 Handover

Intrafrequency handover

A connection handover between two BSs on the same carrier frequency is performed as a *soft handover*. The MS has a connection to both BSs during handover (see Chapter 18). Thus, signals from both BSs are used during this time and the Rake receiver processes them similarly to two paths of a multipath signal with two or more fingers. As BSs use different scrambling codes in WCDMA, the Rake receiver in the MS has to be able to apply different codes in each finger.

In preparation for soft handover, the MS has to acquire synchronization with other BSs. This synchronization process is similar to the one described above, apart from the fact that the MS has a priority list for codegroups. This list contains the codegroups used by the neighboring cells for handover, and is continuously updated.

Cell selection for soft handover can be made by different criteria, like signal strength after despreading or wideband power (RSSI). No particular algorithms are specified in the standard. However, it is suggested to divide cells into the active cell and neighboring candidate cells that provide good signal strength.

A so-called *softer handover* is a special case of soft handover, in which the MS switches between two sectors of the same BS. Algorithms and processes are similar to those used for soft handover with the difference that only one BS is involved.

Interfrequency handover

This kind of handover takes place in the following situations:

- Two BSs employ different carriers.
- The MS switches between hierarchy layers in the hierarchical cell structure.
- Handover to other providers or systems – e.g. GSM – is necessary.
- The MS switches from TDD mode to FDD mode or vice versa.

Interfrequency handover is a "hard" handover during which the connection between the MS and the old BS is first interrupted before the MS establishes a new connection with a new BS. There are two ways of measuring signal strength, etc. on other frequencies while the old connection is still active: (i) the MS might feature two receivers, so that one can measure on other frequencies while the first one still receives user data; and (ii) transmission is done in compressed mode, so that data that normally are transmitted within a 10-ms frame are compressed to 5 ms. The time that is freed up can then be used for measurements on other frequencies. Compression can be achieved, for example, by puncturing the datastream or reduction of the spreading factor.

23.7.5 Overload control

The UTRAN has to ensure that it does not accept too many users or provide users with too high data rates, as then the system would be overloaded. Several means can be used for this [Holma and Toskala 2000]:

- No increase in transmit power in the downlink, even when one MS requests it. This of course implies lower signal quality in the downlink. If data packets are transmitted, the Automatic Repeat reQuest (ARQ) rate increases, which in turn lowers throughput. In case of voice transmission, audio quality is decreased.
- The target Signal-to-Interference-and-Noise Ratio (SINR) in the uplink can be decreased. This decreases the transmission power of MSs. It also results in decreased transmission quality for all subscribers.
- Throughput of the data packet channels can be lowered. This decreases the speed of data connections but maintains the quality of voice services.
- The output data rate of UMTS voice encoders can be decreased, as UMTS employs an encoder with a variable rate. This decreases audio quality, but has no impact on data applications.
- Some active links can be transferred to other frequencies.
- Regular phone calls can be handed over to the GSM network.
- Data connections or phone calls can be terminated.

23.8 Appendix

Please go to *www.wiley.com/go/molisch*

Further reading

The most authoritative source for UMTS is, of course, the standard itself, whose most recent version can be found at *www.3gpp.org* However, this material is exceedingly difficult to read. Good summaries can be found in Holma and Toskala [2000] and Richardson [2005].

24

Wireless Local Area Networks

24.1 Introduction

24.1.1 History

In the late 1990s, wired fast Internet connections became widespread both in office buildings and in private residences. For companies, a fast intranet as well as fast connections to the Internet became a necessity. Furthermore, private consumers became frustrated with the long download times of dialup connections for elaborate webpages, music, etc., as connection speeds were limited to 56 kbit/s. They therefore opted for cable connections (several Mbit/s) or Digital Subscriber Lines (DSLs) (up to 1 Mbit/s in the U.S., and more than 20 Mbit/s in Japan) for their computer connections. At the same time, laptop computers started to be widely used in the workplace. This combination of factors spurred demand for wireless data connections – from the laptop to the nearest wired Ethernet port – that could match the speed of wired connections.

In the following years, two rival standards were developed. The ETSI (European Telecommunications Standards Institute) started to develop the HIPERLAN (*HIgh PERformance Local Area Network*) standard, while the IEEE (*Institute of Electrical and Electronics Engineers*) established the 802.11 group – the number 802 refers to all standards of the IEEE dealing with local and metropolitan area networks, and the suffix 11 was assigned for Wireless Local Area Networks (WLANs). In subsequent years, the 802.11 standard gained widespread acceptance, while HIPERLAN became essentially extinct.

Actually, it is not correct to speak of *the* 802.11 standard. 802.11 encompasses a number of different standards, which are not all interoperable. To understand the terminology, we first have to summarize the history of the standard. The "original" 802.11 standard was intended to provide data rates of 1 and 2 Mbit/s; since it operated in the 2.45-GHz ISM (*Industrial, Scientific, and Medical*) band, the frequency regulator in the USA – the Federal Communications Commission (FCC) – required that spectrum-spreading techniques be used. For this reason, the original 802.11 standard defined two modes: (i) frequency hopping, and (ii) direct-sequence spreading; these two modes were incompatible with each other.

It soon became obvious that higher data rates were demanded by users. Two subgroups were formed: 802.11a, which investigated Orthogonal Frequency Division Multiplexing (OFDM)-based schemes, and 802.11b, which attempted to retain the direct-sequence approach. 802.11b became popular first; it defined a standard that allowed an 11-Mbit/s data rate in a 20-MHz channel. Though the scheme was no longer really spread spectrum, the FCC approved its use. The standard was later adopted by an industry group called WiFi (Wireless Fidelity), which was

Wireless Communications   Andreas F. Molisch
© 2005 John Wiley & Sons, Ltd

formed to ensure true interoperability between all WiFi-certified products.[1] WiFi gained widespread market acceptance after 2000. The data rate of 11 Mbit/s still was not sufficient for many applications – especially in light of the fact that actual throughput in practical situations is closer to 3–5 Mbit/s. For this reason, the work of the 802.11a group became of greater interest. 802.11a specified an alternative physical layer that uses OFDM and higher order modulation alphabets, allowing an up to 54-Mbit/s data transfer rate (again, this rate is nominal, and true throughput is lower by about a factor of 2). This mode also works in a different frequency band (above 5 GHz), which is less "crowded" – i.e., has to deal with fewer interferers. The 802.11g group uses the same modulation format in the 2.45-GHz ISM band. Further modifications of the 802.11a standard are provided by the 802.11h and 802.11j standards, which adapt it to European and Japanese regulations, respectively.

In addition, the original MAC (Medium Access Control) has been amended: the 802.11e standard provides modifications to the MAC that allow us to better ensure certain levels for *Quality of Service* (QoS). Additionally, a number of further subgroups of the 802.11 standard-ization group have been formed, all dealing with "amendments" and "additions" to the original standard.[2] Realistically speaking, though, an 802.11a device, using the 802.11e MAC, bears no resemblance to the original 802.11 standard.

Table 24.1   The IEEE 802.11 standards and their main focus.

Standards	Scope
802.11 (original)	Define a wireless LAN standard that includes both MAC and PHY functions
802.11a	Define a high-speed (up to 54 Mbps) physical layer supplement in the 5-GHz band
802.11b	Define a high-speed (up to 11 Mbps) physical layer extension in the 2.4-GHz band
802.11d	Operation in additional regulatory domains
802.11e	Enhance the original 802.11 MAC to support QoS (applies to 802.11a/b/g)
802.11f	Define a recommended practice for interaccess point protocol (applies to 802.11a/b/g)
802.11g	Define a higher rate (up to 54 Mbps) physical layer extension in the 2.4-GHz band
802.11h	Define MAC functions that allow 802.11a products to meet European regulatory requirements
802.11i	Enhance 802.11 MAC to provide improvement in security (applies to 802.11a/b/g)
802.11j	Enhance the current 802.11 MAC and 802.11a PHY to operate in Japanese 4.9-GHz and 5-GHz bands

Due to the multitude of 802.11 standards, we only present the most important. In the following, we give an overview of 802.11a as well as 802.11b physical layers, and the 802.11 MAC layer. More details can be found, as always, in the official standards publications [*www.802wirelessworld.com*] and the multitude of books that have been published on that topic. An excellent summary of the standard can be found in O'Hara and Petrick [2005].

---

[1] Not all 802.11b products are completely interoperable.
[2] By the end of 2004, a subgroup called 802.11T had already been formed, and the question of "how to name groups after 802.11z has been established" turned from a joke to a serious concern.

24.1.2 Applications

The main markets for WLANs are:

- Wireless networks in office buildings and private homes, to allow unhindered Internet access from anywhere within a building. Access points for WLANs need to follow the specifications, but also allow a certain amount of vendor leeway. For example, multiple antennas can be used at the access point, without leading to incompatibilities. At the "mobile" end, WLAN cards have turned into a mass market with very little distinction between products from different vendors, and are often built into laptops. As a consequence, research tends to focus on methods for production cost reduction (implementation with smaller chip area, low-cost semiconductor technology), while research on access points includes a broader field of topics.
- "Hotspots" – i.e., wireless access points that allow the public to connect to the Internet. These hotspots are often set up in coffee shops, hotels, airports, etc. Several providers also have a "nationwide" or even "continentwide" network of hotspots, so that subscribers can log in at many different locations.[3] However, it must be stressed that coverage from these networks is *much* lower than for cellular networks. For this reason, research is ongoing on how to seamlessly integrate WLANs with cellular networks or large-area networks.

24.1.3 Relationship between the medium-access-control layer and the physical layer

Before going into details of the MAC and the PHYsical (PHY) layer, we first have to establish some notation used by the 802.11 community. The data payload received from upper layers is attached to headers and trailers at both the MAC and PHY layer before it gets transmitted on the air. For example, each *MAC Service Data Unit* (MSDU) received from the *Logic Link Control* (LLC) layer is appended with a 30-byte-long MAC header and a 4-byte-long *Frame Check Sequence* (FCS) trailer to form the *MAC Protocol Data Unit* (MPDU). The same MPDU, once handed over to the physical layer, is then called the *Physical layer Service Data Unit* (PSDU). And, then a *Physical Layer Convergence Procedure* (PLCP) preamble and header, and proper tail bits and pad bits are attached to the PSDU to finally generate the *Physical layer Protocol Data Unit* (PPDU) for transmission. The relationships among MSDU, MPDU, PSDU, and PPDU are illustrated in Fig. 24.1.

From just this brief paragraph, the reader will have seen that – as with most standards – the alphabet soup of the numerous acronyms is a major hurdle to understanding this standard. For this reason, there is a list of acronyms and their meaning in the frontmatter of this book (see p. xxv) and a separate list of abbreviations for some chapters on the companion website (see *www.wiley.com/go/molisch*).

24.2 802.11a – Orthogonal-frequency-division-multiplexing-based local area networks

In an attempt to attain higher data rates and avoid overcrowding of the 2.45-GHz band, the 802.11 Working Group (WG) published their 802.11a standard defining a physical layer for high-speed data communications in the 5-GHz band in 1999. The standard is based on OFDM.

---

[3] In most cases, users are charged a per-minute fee or a flat rate for 24 hours of usage. Nationwide networks often have an option for monthly or annual subscriptions.

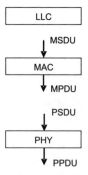

Figure 24.1 Relations among the MAC service data unit, MAC protocol data unit, physical-layer service data unit, and physical-layer protocol data unit.

Reproduced with permission from IEEE 802.11 © IEEE.

Its main properties are the following:

- use of the 5.15–5.825-GHz band (in the U.S.A.);
- 20-MHz channel spacing;
- data rates include 6, 9, 12, 18, 24, 36, 48, and 54 Mbps, where support of 6, 12, and 24 Mbps is mandatory;
- OFDM with 64 subcarriers, out of which 52 are user-modulated with Binary or Quadrature Phase Shift Keying (BPSK/QPSK), 16-Quadrature Amplitude Modulation (QAM), or 64-QAM.
- forward error correction, using convolutional coding with coding rates of 1/2, 2/3, or 3/4 as Forward Error Correction (FEC) coding.

24.2.1 Frequency bands

In the U.S.A., the frequency bands 5.15–5.25, 5.25–5.35, and 5.725–5.825 GHz, called the *Unlicensed National Information Structure* (U-NII) bands, are used for 802.11a. These channels are numbered, starting every 5 MHz, according to the formula:

$$\text{Channel center frequency} = 5{,}000 + 5 \times n_{\text{ch}}(MHz) \tag{24.1}$$

where $n_{\text{ch}} = 0, 1, \ldots, 200$. The transmit powers in the 5.15–5.25, 5.25–5.35, and 5.725–5.825-GHz bands are limited to 40, 200, and 800 mW, respectively.

Obviously, each 20-MHz channel used by 802.11a occupies four channels in the U-NII band. Recommended channel usage is given in Table 24.3

The overall channel plan for 802.11a in the U.S.A. is shown in Figure 24.3. In Japan, the assigned carrier frequencies are slightly lower.

24.2.2 Modulation and coding

802.11a uses OFDM as its modulation format, enabling high data rates. The principles of OFDM were described in Chapter 19, so here we simply analyze details specific to 802.11a.

In 802.11a, OFDM with 64 tones is specified. Powers of 2 are habitually used as numbers for OFDM carriers, as they allow the most efficient implementation via Fast Fourier Transforms (FFTs). However, only 52 of the 64 tones are actually used (modulated and transmitted), while the other 12 tones are null-carriers that do not carry any useful information; useful tones are indexed from −26 to 26, without a DC component. Of these 52 tones, 4 are used as pilot tones –

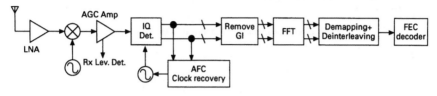

Figure 24.2   Block diagram of a 802.11a transceiver.
Reproduced with permission from IEEE 802.11 © IEEE.

Table 24.2   Important parameters of the 802.11a PHY layer.

Information data rate	6, 9, 12, 18, 24, 36, 48, 54 Mbit/s
Modulation	BPSK, QPSK, 16-QAM, 64-QAM
FEC	$K = 7$ convolutional code
Coding rate	1/2, 2/3, 3/4
Number of subcarriers	52
OFDM symbol duration	4 µs
Guard interval	0.8 µs
Occupied bandwidth	16.6 MHz

Table 24.3   Frequency assignment for 802.11a in the U.S.A.

Bands (GHz)	Allowed power	Channel numbers (n_{ch})	Channel center frequency (MHz)
U-NII lower band (5.15–5.25)	40 mW (2.5 mW/MHz)	36	5,180
		40	5,200
		44	5,220
		48	5,240
U-NII middle band (5.25–5.35)	200 mW (12.5 mW/MHz)	52	5,260
		56	5,280
		60	5,300
		64	5,320
U-NII upper band (5.725–5.825)	800 mW (50 mW/MHz)	149	5,745
		153	5,765
		157	5,785
		161	5,805

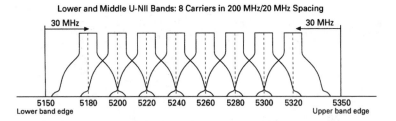

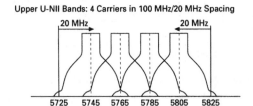

Figure 24.3 802.11a channel plan.

namely, tones number -21, -7, 7, 21. The pilot should be BPSK-modulated by a pseudorandom sequence to prevent generation of spectral lines.

The other 48 tones carry the PSDU data. BPSK, QPSK, 16-QAM, or 64-QAM are all admissible modulation alphabets, depending on the channel state. Note, however, that the standard does *not* foresee truly adaptive modulation in the sense that the modulation alphabet can differ from tone to tone. Rather, the system uses an average "transmission quality" criterion to adapt the data rate to the current channel state. Rate adaptation is achieved by modifying either the modulation alphabet or the rate of the error correction code (see below), or both.

The duration of an OFDM symbol is $4\,\mu s$, including a cyclic prefix of $0.8\,\mu s$. This is sufficient to accommodate the maximum excess delay of most indoor propagation channels, including factory halls and other challenging environments.

For FEC, 802.11a uses a convolutional encoder with coding rates $1/2$, $2/3$, or $3/4$, depending on the desired data rate. The generator vectors are $G1 = 133$ and $G2 = 171$ (in octal notation), for the rate-$1/2$ coder shown in Fig. 24.4. Higher rates are derived from this "mother code" by puncturing.

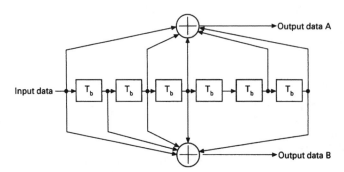

Figure 24.4 Convolutional encoder $(K = 7)$.

All encoded data bits are interleaved by a block interleaver with a block size equal to the number of bits in a single OFDM symbol. The interleaver works in two steps (permutations). The first permutation ensures that adjacent coded bits are mapped onto non-adjacent subcarriers. The second ensures that adjacent coded bits are mapped alternately onto both less and more significant bits of the constellation and, thereby, long runs of low-reliability bits are avoided.

Table 24.4 summarizes the rates that can be achieved with different combinations of alphabets and coding rates, as well as the OFDM modulation parameters.

Table 24.4 Data rates in 802.11a.

Data rate (Mbit/s)	Modulation	Coding rate	Coded bits per subcarrier	Coded bits per OFDM symbol	Data bits per per OFDM symbol
6	BPSK	1/2	1	48	24
9	BPSK	3/4	1	48	36
12	QPSK	1/2	2	96	48
18	QPSK	3/4	2	96	72
24	16-QAM	1/2	4	192	96
36	16-QAM	3/4	4	192	144
48	64-QAM	2/3	6	288	192
54	64-QAM	3/4	6	288	216

24.2.3 Headers

For transmission, a preamble and a PLCP header are prepended to the encoded PSDU data[4] that are received from the MAC layer, creating a PPDU. At the receiver, the PLCP preamble and header are used to aid in demodulation and data delivery. A PPDU frame format is shown in Fig. 24.5.

The PLCP header is transmitted in the SIGNAL field of the PPDU. It incorporates the RATE field, a LENGTH field, a TAIL field, and so on:

- RATE (4 bits): indicates transmission data rate.
- LENGTH (12 bits): indicates the number of octets in the PSDU.
- Parity (1 bit): parity check.
- Reserved (1 bit): future use.
- TAIL (6 bits): convolutional-coding tail.
- SERVICE (16 bits): initialization of the scrambler.

24.2.4 Synchronization and channel estimation

Synchronization is achieved by means of the PLCP preamble field. It consists of 10 short symbols and 2 long symbols (Fig. 24.6).

---

[4] Plus pilot tones.

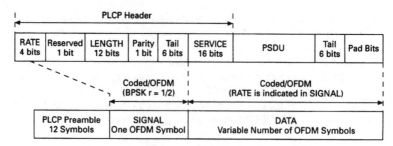

Figure 24.5   PHY-protocol-data-unit frame format.

Reproduced with permission from IEEE 802.11 © IEEE.

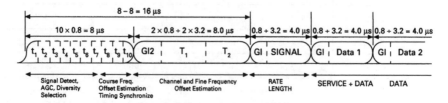

Figure 24.6   Orthogonal-frequency-division-multiplexing training structure.

Reproduced with permission from IEEE 802.11 © IEEE.

The training sequence starts out with 10 short symbols of duration 0.8 μs that allow the receiver to detect the signal, adjust the Automatic Gain Control (AGC), and perform a coarse-frequency offset estimation. These short symbols consist of just 12 tones, which are modulated by elements of the following sequence:

$$S_{-26,26} = \sqrt{(13/6)}\{0,0,1+j,0,0,0,-1-j,0,0,0,1+j,0,0,0,-1-j,0,0,0,-1-j,0,0,0,1+j,$$
$$0,0,0,0,0,0,0,-1-j,0,0,0,-1-j,0,0,0,1+j,0,0,0,1+j,0,0,0,1$$
$$+j,0,0,0,1+j,0,0\} \tag{24.2}$$

Multiplication by a factor of $\sqrt{(13/6)}$ is done so as to normalize the average power of the resulting OFDM symbol, which utilizes 12 of the 52 subcarriers.

These symbols are followed by 2 long training symbols that serve for both channel estimation and finer frequency offset estimation, preceded by a guard interval. A long OFDM training symbol consists of 53 subcarriers (including a 0 value at DC), which are modulated by elements of sequence L, given by:

$$L_{-26,26} = \{1,1,-1,-1,1,1,-1,1,-1,1,1,1,1,1,1,-1,-1,1,1,-1,1,-1,1,1,1,1,0,$$
$$1,-1,-1,1,1,-1,1,-1,1,-1,-1,-1,-1,-1,1,1,-1,-1,1,-1,1,-1,1,1,1,1\} \tag{24.3}$$

The PLCP preamble is followed by the SIGNAL and DATA fields. The total training length is 16 μs. The dashed boundaries in the figure denote repetitions due to periodicity of the inverse Fourier transform.

Table 24.5 summarizes the most important parameters for 802.11a mode.

Table 24.5 Parameters of 802.11a.

Parameter	Value
Number of data subcarriers	48
Number of pilot subcarriers	4
Subcarrier spacing	0.3125 MHz
IFFT/FFT period	3.2 μs
Preamble duration	16 μs
Duration of OFDM symbol	4.0 μs
Guard interval for signal symbol	0.8 μs
Guard interval for training symbol	1.6 μs
Short training sequence duration	8 μs
Long training sequence duration	8 μs

24.3 802.11b – Wireless Fidelity

The initial version of the 802.11 standard defined frequency hopping and direct sequence spreading, but each of these methods enabled data rates of up to 2 Mbps only. In 1999, the 802.11 WG added the 802.11b extension of the specifications, also known as High Rate Direct Sequence PHY (HR/DS or HR/DSSS). As a pure PHY-layer extension, this mode uses the same MAC as the original 802.11 – it just allows faster transfer of the data. The supported rates are 1, 2, 5.5, and 11 Mbps. To provide higher rates, 8-chip *Complementary Code Keying* (CCK) is employed as the modulation scheme. The chipping rate is 11 MHz, which is the same as the DSSS system described in 802.11, thus providing the same occupied channel bandwidth.

24.3.1 Frequency band

The 802.11b standard operates in the 2.45-GHz ISM band. More precisely, this band extends from 2.4 GHz to 2.4835 GHz in Europe and the U.S.A., and from 2.471 GHz to 2.497 GHz in Japan. In the U.S.A., the frequency band in 2.4 GHz has been divided into 5-MHz-wide channels by the FCC with center frequencies as shown in Table 24.6.

In Europe, channels with center frequencies of 2.467 and 2.472 GHz may be used as well.[5]

802.11b uses a spread-spectrum-like modulation with a chip rate of 11 MHz and thus occupies a channel bandwidth of 22 MHz. With channel spacing of 5 MHz, 802.11b networks must be separated by five channels to prevent interference. The maximum allowed transmit power is 1,000 mW in the U.S., 100 mW in Europe. In Japan, power-spectral density (and not power) is prescribed, and must not exceed 10 mW/MHz.

---

[5] Spain and France have different regulations.

Table 24.6 Frequencies for 802.11.

Channel ID	Center frequency (MHz)
1	2,412
2	2,417
3	2,422
4	2,427
5	2,432
6	2,437
7	2,442
8	2,447
9	2,452
10	2,457
11	2,462

24.3.2 Modulation and coding

The 802.11b standard uses two different types of modulation: for low data rates, and for transmission in the header and preamble, direct-sequence spreading with Barker sequences is used; and for transmission of PSDU data with higher data rates, CCK is used.

Direct spreading with Barker sequences

For the header and preamble, direct-sequence spreading (as described in Chapter 18) is used. Information symbols – i.e., symbols occuring at a rate of 1 or 2 Mbit/s – are used to modulate an 11-chip Barker sequence:

$$S_{\text{Barker}} = \{+1, -1, +1, +1, -1 + 1 + 1 + 1, -1, -1, -1\} \tag{24.4}$$

Modulation here uses differential phase shift keying (see Chapter 11), where for 1-Mbit/s mode +1 corresponds to counterclockwise rotation by $180°$.[6]

Complementary code keying

For higher data rates, CCK is used. Its basic principle is that there are a number of N-bit-long codewords, and, depending on the data symbol to be transmitted, one of these possible codevectors is transmitted. Since codewords can be complex, we have in principle 4^N codewords available. Should all of these 4^N codewords be valid, then the transmission scheme becomes just the standard QPSK, using $1:1$ mapping between 16-bit groups of data symbols and codewords. The point of CCK is to define a *subset* of codewords with "good" properties (to be discussed below) as valid codewords, while other codewords are not valid.

---

[6] For 2-Mbit/s mode, DQPSK is used, with shifts of $90°$ for the different bit pairs.

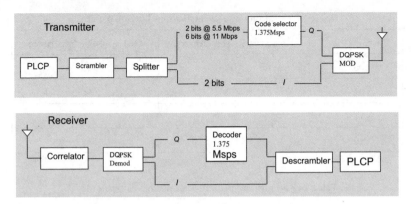

Figure 24.7   Structure of complementary-code-keying modulation and demodulation in 802.11b.
Reproduced with permission from IEEE 802.11 © IEEE.

In the 802.11b standard, the length of the codewords is $N = 8$, so that more than 65,000 codewords are possible. Of these, only 64 codewords are actually admissible, representing 6 bits.[7] These 64 codewords are further Differential Quadrature Phase Shift Keying (DQPSK)-modulated – i.e., a phase rotation of 0, 90, 180, or 270° is applied, depending on the 2 additional bits. Thus, a total of 8 bits is transmitted in one codeword. With the chipping rate remaining at 11 Mchip/s, the symbol rate now increases to 1.375 Msymbols/s. The resulting data rate is 11 Mbit/s.

Let us now return to the 64 admissible codewords, and how they are selected. The sequences are complementary code sequences. A complementary series is defined as a pair of equally long sequences composed of two types of elements which have the property that the number of pairs of like elements with any given separation in one series is equal to the number of pairs of unlike elements with the same separation in the other series. A *set* of complementary sequences is defined as fulfilling the following equation:

$$\sum_{k=1}^{K} R^{(k)}(j) = \begin{cases} 0 & \text{for } j \neq 0 \\ KN & \text{for } j = 0 \end{cases} \tag{24.5}$$

where

$$R^{(k)}(j) = \sum_{l=1}^{N-j} s_l^{(k)} s_{l+j}^{(k)} \tag{24.6}$$

where s_j is the entry of the lth codeword, and superscript (k) indexes the kth sequence. Equation (24.5) also shows us the reason for using complementary sequences: they have excellent auto-correlation and cross-correlation properties, and thus provide good performance in multipath environments. Of course, for optimal reception and exploitation of all available energy, a Rake receiver (or equivalent structure) is required, as described in Chapter 18.

Let us now provide exact details about generation of the sequences: first, define a sequence of phases:

$$c = \{\varphi_1 + \varphi_2 + \varphi_3 + \varphi_4, \varphi_1 + \varphi_3 + \varphi_4, \varphi_1 + \varphi_2 + \varphi_4, \varphi_1 + \varphi_4 + \pi,$$

$$\varphi_1 + \varphi_2 + \varphi_3, \varphi_1 + \varphi_3, \varphi_1 + \varphi_2 + \pi, \varphi_1\} \tag{24.7}$$

A total of 8 bits (b_0 to b_7) are transmitted per symbol, where b_0 is transmitted first. The first bit pair (b_0, b_1) encodes the phase φ_1 using DQPSK. The DQPSK encoder is specified in

---

[7] We are speaking here only of the 11 Mbit/s standard. For the 5.5-Mbit/s mode, only 4 codewords are valid.

Table 24.7   Encoding table for the first bit pair pattern.

Bit pair b_0, b_1	Even symbols phase change	Odd symbols phase change
00	0	π
01	$\pi/2$	$3\pi/2$
11	π	0
10	$3\pi/2$	$\pi/2$

Table 24.8   Encoding patterns for later bit pair patterns.

Bit pair	Phase change
00	0
01	$\pi/2$
10	π
11	$3\pi/2$

Table 24.7. The phase change for φ_1 is relative to phase φ_1 of the preceding symbol. All odd-numbered symbols of the PSDU are given an extra 180° (π) rotation. Symbol numbering starts with "0" for the first symbol of the PSDU. Data bit pairs (d_2, d_3), (d_4, d_5), and (d_6, d_7) encode φ_2, φ_3, and φ_4, respectively, based on QPSK as specified in Table 5.4. Note that this table is binary (not Grey) coded.

Coding and scrambling   As an option, the 802.11b standard also foresees convolutional encoding with a 64-state, rate-1/2 binary convolutional encoder.[8]

Furthermore, all data bits are scrambled, in order to avoid long strings of 0's or 1's. Scrambling is done using a 7-bit polynomial $G(z) = z^{-7} + z^{-4} + 1$. We will see below that such scrambling is also used for the SYNCH field in the header of a datablock.

24.3.3 Headers

Two different preambles and headers are defined: the mandatorily supported Long Preamble and header, which interoperates with legacy 1-Mbps and 2-Mbps 802.11 DSSS PHYs, and an optional Short Preamble and header. At the receiver, the PLCP preamble and header are processed to aid demodulation and delivery of the PSDU. The optional Short Preamble and header is intended for applications where maximum throughput is desired and interoperability with legacy and non-short-preamble-capable equipment is not a consideration – i.e., it is expected to be used only in networks of like equipment, all of which can handle the optional mode:

- Long PLCP PPDU: Fig. 24.8 shows the format for the interoperable (long) PPDU. The preamble and header are transmitted with a data rate of 1 Mbit/s, which is independent of the data rate used for actual PSDU data. In other words, the preamble and header are

---

[8] A cover sequence is used to map encoder outputs to QPSK symbols.

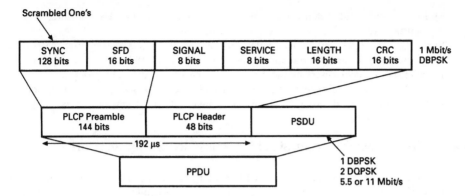

Figure 24.8 Long physical-layer-convergence-procedure physical-layer-protocol-data-unit format.

Reproduced with permission from IEEE 802.11 © IEEE.

interoperable with the preamble of the original 802.11 standard; it is only the PSDU data that are transmitted at the higher data rate. Note that this long preamble/header is almost 200 μs long, and can thus lead to considerable overhead: for transmission of 1 kByte of data at 11 Mbit/s, we only need about 700 μs, so that the header leads to overhead of more than 25%. The situation is even worse for smaller packet sizes.

- Short PLCP PPDU: due to the inefficiency of the long preamble, a short PLCP preamble and header are defined (as options). The Short Preamble and header may be used to minimize overhead and, thus, maximize network data throughput. They have a shorter SYNCH field, and use 2 Mbit/s for the header, thus cutting the duration of the total preamble/header in half. The format of the PPDU is depicted in Fig. 24.9.
- In both cases, header sequences and training sequences consist of the following fields:

    —SYNCH: this field is used by the receiver to detect the signal, and acquire carrier tracking and timing. The transmitted sequence is a scrambled sequence of logical "1", where scrambling is again done using the polynomial $G(z) = z^{-7} + z^{-4} + 1$.
    —SFD: this field marks the start of a frame, and always contains the same bit sequence.

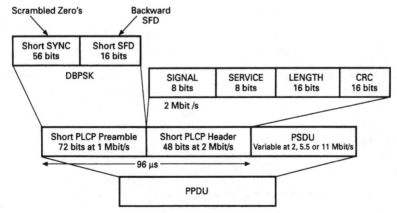

Figure 24.9 Short physical-layer-convergence-procedure physical-layer-protocol-data-unit format.

Reproduced with permission from IEEE 802.11 © IEEE.

—SIGNAL: this field defines the type of modulation in the PSDU data. The bits transmitted are a representation of the used bit rate, divided by 100 kbit/s.

—SERVICE: this field actually uses only 3 of the 8 available bits, describing aspects related to hardware implementation, coding, and duration of the PSDU.

—LENGTH: this field describes the duration of the PSDU in microseconds.

—CRC: the header is protected by an error detection code to see whether the header data have arrived in good order. If an error is detected, the frame might be terminated.

24.4 Packet transmission in 802.11 wireless local area networks

There are nine MAC services specified by IEEE 802.11. These include distribution, integration, association, reassociation, disassociation, authentication, deauthentication, privacy, and MSDU delivery. Six of the services are used to support MSDU delivery between STAs (STAtions) (used as a generic expression for 802.11 devices, both BSs and MSs). Three of the services are used to control 802.11 WLAN access and confidentiality. Each of the services is supported by one or more MAC frame types. The IEEE 802.11 MAC uses three types of messages: *data*, *management*, and *control*. Some of the services are supported by MAC management messages and some by MAC data messages. All messages gain access to the WM (Wireless Medium) via the IEEE 802.11 MAC medium access method which includes both contention-based and contention-free channel access methods: *Distributed Coordination Function* (DCF) and *Point Coordination Function* (PCF). In the following, the 802.11 MAC functions and services will be described.

24.4.1 General medium-access-control structure

The 802.11 MAC uses a temporal superframe structure with *Contention Period* (CP)[9] and *Contention Free Period* (CFP) alternately as shown in Fig. 24.10. Superframes are separated by periodic management frames, the so-called "beacon frames". During the CP, DCF is used for channel access, while PCF is used for channel access during the CFP.

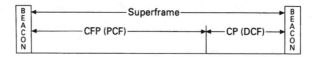

Figure 24.10   Superframe structure of 802.11.
Reproduced with permission from IEEE 802.11 © IEEE.

802.11 MAC uses different interframe gaps, denoted as *Inter Frame Spaces* IFS), in order to control medium access – i.e., to give STAs in specific cases a higher or lower priority. These IFSs are (in the order shortest to longest):

- *Short Inter Frame Space* (SIFS);
- *Priority Inter Frame Space* (PIFS);
- *Distributed Inter Frame Space* (DIFS);
- *Extended Inter Frame Space* (EIFS).

Their actual values depend on PHY parameters.

---

[9] Note that it is only in this section that we use the abbreviation CP for contention period (and not for cyclic prefix). Since we are talking about the MAC layer only, no confusion can arise.

24.4.2 Frame formats

The MAC frame format comprises a set of fields that occur in a fixed order in all frames. Figure 24.11 depicts the general MAC frame format. The fields Address 2, Address 3, Sequence Control, Address 4, and Frame Body are only present in certain frame types.

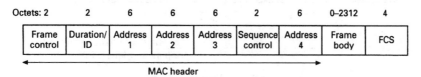

Figure 24.11   Medium-access-control frame format (a typical MPDU).
Reproduced with permission from O'Hara and Petrik [2005] © IEEE.

When the MSDUs handed down to the MAC become too large, it becomes difficult to transmit them in one block: obviously, the probability of a block error – i.e., that one of the bits in the block is in error – increases with duration of the block.[10] As each block error might lead to the necessity of retransmission, this is highly undesirable. Thus, MSDUs have to be fragmented in order to increase transmission reliability. This fragmentation is done when MSDU size exceeds the fragmentation threshold. In this case, the MSDU will be broken into multiple fragments with an MPDU size equal to a fragmentation threshold, and a special field (the "more_fragments" field) is set to 1 in all but the last fragment. The receiving STA acknowledges each fragment individually. The channel is not released until the complete MSDU has been transmitted successfully or until an acknowledgement has been received for a fragment. In the latter case, the source STA will recontend for the channel following the normal rules and retransmit the non-acknowledged fragment, as well as all the subsequent ones.

24.4.3 Packet radio multiple access

Carrier-sense multiple access

The DCF employs *Carrier Sense Multiple Access with Collision Avoidance* (CSMA/CA), as described in Chapter 17, plus a random backoff mechanism. Support for DCF is mandatory for all STAs. In DCF mode, each STA checks whether the channel is idle before attempting to transmit. If the channel has been sensed idle for a DIFS period, transmission can begin immediately. If the channel is determined to be busy, the STA will defer until the end of the current transmission. After the end of the current transmission, the STA will select a random number called a "backoff timer", in the range between 0 and a *Contention Window* (CW). This is the time the WM has to be free before the STA might try to transmit again. The size of the CW increases (up to a limit) every time a transmission has to be deferred. If transmission is not successful, the STA thinks that a collision has occurred. Also in this case, the CW is doubled, and a new backoff procedure starts again. The process will continue until transmission is successful (or discarded). The basic access method and backoff procedure are shown in Figs. 24.12 and 24.13, respectively.

Physical and virtual carrier-sense functions are used to determine the state of the channel. When either function indicates a busy channel, the channel should be considered busy; otherwise, it should be considered idle. A physical carrier-sense mechanism is provided by the PHY. A virtual carrier-sense mechanism is provided by the MAC. This mechanism is referred to

---

[10] Note that, in general, this effect may be offset by the fact that larger blocks allow the use of better codes, like highly efficient LDPC codes (see Chapter 14). However, for codes used in 802.11 this is not relevant.

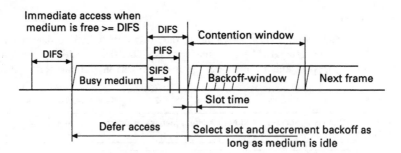

Figure 24.12   Basic access method.

Reproduced with permission from IEEE 802.11 © IEEE.

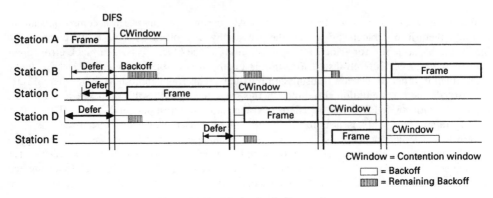

Figure 24.13   Timing backoff procedure.

Reproduced with permission from IEEE 802.11 © IEEE.

as the *Network Allocation Vector* (NAV). The NAV maintains a prediction of future traffic on the medium based on duration information that is announced in the DURATION field in the transmitted frames.

Polling

PCF is an optional medium access mode for 802.11. It provides contention-free frame transfer, based on polling (see Chapter 17). The *Point Coordinator* (PC) resides in the BS (access point). All STAs inherently obey the medium access rules of the PCF and set their NAV at the beginning of each CFP. The PCF relies on the PC to perform polling, and enables polled STAs to transmit without contending for the channel. When polled by the PC, an STA transmits only one MPDU, which can be to any destination. If the transmitted dataframe is not in turn acknowledged, the STA does not retransmit the frame unless it is polled again by the PC, or it decides to retransmit during the CP. An example of PCF frame transfer is given in Fig. 24.14. At the beginning of each CFP, the PC senses and makes sure the channel is idle for one PIFS before sending the beacon frame. All STAs adjust their NAVs according to the broadcast CFP duration value in the beacon. After one SIFS time of the beacon, the PC may send out a *Contention Free Poll* (CF-Poll), data, or data plus a CF-Poll. Each polled STA can get a chance to transmit to another STA or respond to the PC after one SIFS with an acknowledgment (plus possibly data).

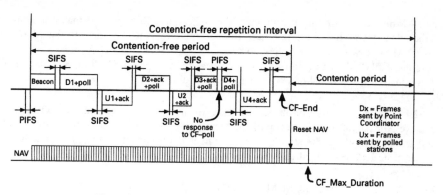

Figure 24.14   Point-coordination-function frame transfer.
Reproduced with permission from IEEE 802.11 © IEEE.

As discussed above, the DCF and PCF coexist in a manner that permits both to operate concurrently.[11] The two access methods alternate, with a CFP followed by a CP. Since the PCF is built on top of the DCF, there are no conflicts between DCF and PCF when they coexist in the system. All STAs will inherently obey the medium access rules of the PCF. STAs will stay silent during CFP unless they are polled.

24.5 Alternative wireless local area networks and future developments

24.5.1 Other 802.11 and high-performance-local-area-network modes

As we mentioned in the introduction, there is a multitude of "802.11-named" standards, most of which have not gained widespread popularity. Among these standards, the original 802.11 standard with its 1-Mbit/s direct-sequence spreading mode is a typical example. Furthermore, the frequency-hopping mode of this standard never gained popularity. Finally, the standard also defined a mode for infrared communications between computers; this application never gained significant popularity as well.

In all the discussions above, we have concentrated on WLANs that have one BS, plus a number of MSs that connect to this BS. The 802.11 group of standards also establishes modes for peer-to-peer communications. This approach has not gathered widespread popularity either.

We have also mentioned the HIPERLAN standards developed by ETSI. The HIPERLAN II standard, in particular, bears considerable similarity to the 802.11a PHY, though MAC is based on TDMA instead of CSMA. While a number of research papers have been published on this standard, it has not gained practical relevance, and even its previous proponents have switched to using 802.11a.

24.5.2 802.11n and multiple-input–multiple-output techniques

For many applications, the data rate offered by 802.11a is not sufficient for several reasons:

- New applications, like the transmission of high-definition TV, have increased the required data rates.

---

[11] Within the same Basic Service Set (BSS).

- The 54-Mbit/s data rate that is nominally the highest describes only the maximum data rate of the PHY layer, but does not take into account the inefficiencies of the MAC layer (typical MAC efficiencies are below 50%).
- The highest PHY data rate can only be achieved for very good SNIRs (remember that the rate requires the use of 64-QAM with rate-3/4 coding).

The 802.11n subgroup was established in order to define a new high-throughput layer that could achieve at least 100 Mbit/s on top of the MAC – i.e., considering both the efficiency of the PHY and the MAC. While the standard is still under development at the time of writing this book, some basic features can already be discerned:

- The PHY layer is based on MIMO-OFDM – i.e., a combination of OFDM and spatial multiplexing based on multiple antenna systems (MIMO, see Section 20.2):
  —the OFDM component is similar to the 802.11a standard;
  —the standard will use multiple antennas at both link ends (at least two antennas at each link end), consequently spatial multiplexing of at least two streams is enabled;
  —a high-rate datastream is encoded *before* being multiplexed onto different transmit antennas. While the receiver structure is not prescribed, the transmission scheme is cued to MMSE receivers, not the higher performing (but more complex) BLAST (Bell labs LAyered Space Time) receivers.
- The MAC layer is based on the 802.11e MAC, and will provide some modifications to further enhance efficiency.

24.5.3 Personal area networks

WLANs provide links over a range of up to 30–100 m. Personal Area Networks (PANs) are intended for even shorter connection lengths – namely, up to about 10 m. Restricting coverage can result in one of two major benefits: increased data rate, or reduced complexity. The former advantage is achieved – e.g., by the WiMedia–MBOA (*Multiband OFDM Alliance*) specifications for high-data-rate communications, which have the following main properties:

- It uses OFDM, where each OFDM symbol has a bandwidth of more than 500 MHz, with 128 tones, and a duration of some 300 ns, with a guard interval of 60 ns.
- QPSK is used as the modulation format.
- The center frequency of each OFDM symbol is different, following a time–frequency code, with three center frequencies (3,432 MHz, 3,960 MHz, and 4,488 MHz) available.
- The signal follows the regulations of the FCC (the frequency regulator of the U.S.A.) for ultrawideband signals, and thus can operate without license.[12]
- Payload data are encoded with a convolutional code with a rate between 1/3 and 3/4, and interleaved over different OFDM symbols. Depending on the data rate, repetition coding might also be applied.
- The allowed data rates are 110 Mbit/s, 200 Mbit/s, and 480 Mbit/s.
- The 480-Mbit/s mode employs a form of multicarrier CDMA (see Chapter 19), with spreading over two OFDM subcarriers.

An alternative method for high-data-rate transmission, developed by the UWB Forum, uses direct-sequence CDMA with a chip rate of about 1.5 Gchip/s.

---

[12] At the time of the writing, 2005, the FCC had still not actually provided a type approval.

At the other end of the PAN spectrum, the Bluetooth standard is used for data rates of 1 Mbit/s. It is intended to provide low-cost solutions for, for example, communication between a cellphone and a headset. Further developments can be expected in this area in the coming years. Research will concentrate on lowering the complexity of devices on one hand, and increase in data rates on the other hand. It can also be anticipated that high-throughput WLANs will start to compete with high-rate PANs.

24.6 Appendix

Please go to *www.wiley.com/go/molisch*.

Further reading

The official standards documents for the 802.11 standard can be found online at *www.802wirelessworld.com* An excellent summary is given in O'Hara and Petrick [2005]. CCK is described in Pearson [2000] and Andren et al. [2000]. The principles of the WiMedia–MBOA specifications are described in several submissions to the IEEE 802.15.3a Standards Group; however, the final version is not publicly available.

25

Exercises

Peter Almers, Ove Edfors, Fredrik Floren, Anders Johanson, Johan Karedal, Buon Kiong Lau, Andreas F. Molisch, Andre Stranne, Fredrik Tufvesson, and Shurjeel Wyne

25.1 Chapter 1: Applications and requirements of wireless services

1. What was the timespan between:
   (a) invention of the cellular principle, and deployment of the first widespread cellular networks;
   (b) start of the specification process of the GSM system and its widespread deployment;
   (c) specification of the IEEE 802.11b (WiFi) system and its widespread deployment?

2. Which of the following systems cannot transmit in both directions (duplex or semi-duplex): (i) cellphone, (ii) cordless phone, (iii) pager, (iv) trunking radio, (v) TV broadcast system?

3. Assuming that speech can be digitized with 10 kbit/s, compare the difference in the number of bits for a 10-s voice message with a 128-letter pager message.

4. Are there conceptual differences between (i) wireless PABX and cellular systems, (ii) paging systems and wireless LANs, (iii) cellular systems with closed user groups and trunking radio?

5. What are the main problems in sending very high data rates from an MS to a BS that is far away?

6. Name the factors that influence the market penetration of wireless devices.

25.2 Chapter 2: Technical challenges of wireless communications

1. Does the carrier frequency of a system have an impact on (i) small-scale fading, (ii) shadowing? When moving over a distance x, will variations in the received signal power be greater for low frequencies or high frequencies? Why?

2. Consider a scenario where there is a direct path from BS to MS, while other multipath components are reflected from a nearby mountain range. The distance between the BS and MS is 10 km, and the distance between the BS and mountain range, as well as the MS and mountain range, is 14 km. The direct path and reflected components should arrive at the

receiver within 0.1 times the symbol duration, to avoid heavy intersymbol interference. What is the required symbol rate?

3. Why are low carrier frequencies problematic for satellite TV? What are the problems at very high frequencies?

4. In what frequency ranges can cellphones be found? What are the advantages and drawbacks?

5. Name two advantages of power control.

25.3 Chapter 3: Noise- and interference-limited systems

1. Consider a receiver that consists (in this sequence) of the following components: (i) an antenna connector and feedline with an attenuation of 1.5 dB; a low-noise amplifier with a noise figure of 4 dB and a gain of 10 dB, and a unit gain mixer with a noise figure of 1 dB. What is the noise figure of the receiver?

2. Consider a system with 0.1-mW transmit power, unit gain for the transmit and receive antennas, operating at 50-MHz carrier frequency with 100-kHz bandwidth. The system operates in a suburban environment. What is the receive SNR at a 100-m distance, assuming free space propagation? How does the SNR change when changing the carrier frequency to 500 MHz and 5 GHz? Why does the 5-GHz system show a significantly lower SNR (assume the receiver noise figure is 5 dB independent of frequency)?

3. Consider a GSM uplink. The MS has 100-mW transmit power, and the sensitivity of the BS receiver is -105 dBm. The distance between the BS and MS is 500 m. The propagation law follows the free space law up to a distance of $d_{break} = 50$ m, and for larger distances the receive power is similar to $(d/d_{break})^{-4.2}$. Transmit antenna gain is -7 dB; the receive antenna gain is 9 dB. Compute the available fading margin.

4. Consider a wireless LAN system with the following system specifications:

    $f_c = 5$ GHz
    $B = 20$ MHz
    $G_{TX} = 2$ dB
    $G_{RX} = 2$ dB
    Fading margin = 16 dB
    Pathloss = 90 dB
    $P_{TX} = 20$ dBm
    TX losses: 3 dB
    Required SNR: 5 dB

    What is the maximum admissible RF noise figure?

5. Consider an environment with propagation exponent $n = 4$. The fading margin between the median and 10% decile, as well as between the median and 90% decile is 10 dB each. Consider a system that needs an 8-dB SIR for proper operation. How far do serving and interfering BSs have to be apart so that the MS has sufficient SIR 90% of the time at the cell boundary? Make a worst case estimate.

25.4 Chapter 4: Propagation mechanisms

1. Antenna gain is usually given in relation to an isotropic antenna (radiating/receiving equally in all directions). It can be shown that the effective area of such an antenna is $A_{iso} = \lambda^2/4\pi$. Compute the antenna gain G_{par} of a circular parabolic antenna as a function of its radius r, where the effective area is $A_e = 0.55A$ and A is the physical area of the opening.

2. When communicating with a geostationary satellite from Earth, the distance between transmitter and receiver is approximately 35,000 km. Assume that Friis' law for free space loss is applicable (ignore any effects from the atmosphere) and that stations have parabolic antennas with gains 60 dB (Earth) and 20 dB (satellite), respectively, at the 11-GHz carrier frequency used.
    (a) Draw the link budget between transmitted power P_{TX} and received power P_{RX}.
    (b) If the satellite receiver requires a minimum received power of -120 dBm, what transmit power is required at the Earth station antenna?

3. A system operating at 1 GHz with two 15-m-diameter parabolic antennas at a 90-m distance are to be designed.
    (a) Can Friis' law be used to calculate the received power?
    (b) Calculate the link budget from transmitting antenna input to receiving antenna output assuming that Friis' law *is* valid. Compare P_{TX} and P_{RX} and comment on the result.
    (c) Determine the Rayleigh distance as a function of antenna gain G_{par} for a circular parabolic antenna as in Problem 1.

4. A transmitter is located 20 m from a 58-m-high brick wall ($\varepsilon_r = 4$), while a receiver is located on the other side, 60 m away from the wall. The wall is 10 cm thick and can be regarded as lossless. Let both antennas be of height 1.4 m and using a center frequency of 900 MHz.
    (a) Considering TE waves, determine the fieldstrength at the receiver caused by transmission through the wall, $E_{through}$.
    (b) The wall can be regarded as a semi-infinite thin screen. Determine the fieldstrength at the receiver caused by diffraction over the wall, E_{diff}.
    (c) Determine the ratio of the magnitudes of the two fieldstrengths.

5. Show that for a wave propagating from medium 1 to medium 2, the reflection coefficients for TE and TM can be written as:

$$\left.\begin{array}{l} \rho_{TM} = \dfrac{\varepsilon_r \cos \Theta_e - \sqrt{\varepsilon_r - \sin^2 \Theta_e}}{\varepsilon_r \cos \Theta_e + \sqrt{\varepsilon_r - \sin^2 \Theta_e}} \\[4mm] \rho_{TE} = \dfrac{\cos \Theta_e - \sqrt{\varepsilon_r - \sin^2 \Theta_e}}{\cos \Theta_e + \sqrt{\varepsilon_r - \sin^2 \Theta_e}} \end{array}\right\} \qquad (25.1)$$

if medium 1 is air and medium 2 is lossless with a dielectric constant ε_r.

6. Waves are propagating from the air towards a lossless material with a dielectric constant ε_r.
    (a) Determine an expression for the angle that results in totally transmitted waves – i.e., $|\rho_{TM}| = 0$. Is there a corresponding angle for TE waves?
    (b) Assuming that $\varepsilon_r = 4.44$, plot the magnitude for TE and TM reflection coefficents. Determine the angle for which $|\rho_{TM}| = 0$.

7. For propagation over a perfectly conducting ground plane the magnitude of total received field E_{tot} is given by:

$$|E_{tot}(d)| = E(1m) \frac{1}{d} 2 \frac{h_{TX}h_{RX}}{d} \frac{2\pi}{\lambda} \qquad (25.2)$$

Show that if transmit power is P_{TX} and transmit and receive antenna gains are G_{TX} and G_{RX}, respectively, the received power P_{RX} is given by:

$$P_{RX}(d) = P_{TX}G_{TX}G_{RX}\left(\frac{h_{TX}h_{RX}}{d^2}\right)^2 \qquad (25.3)$$

8. Assume that we have a base station with a 6-dB antenna gain and a mobile station with antenna gain of 2 dB, at heights 10 m and 1.5 m, respectively, operating in an environment where the ground plane can be treated as perfectly conducting. The lenghts of the two antennas are 0.5 m and 15 cm, respectively. The base station transmits with a maximum power of 40 W and the mobile with a power of 0.1 W. The center frequency of the links (duplex) are both at 900 MHz, even if in practice they are separated by a small duplex distance (frequency difference).
   (a) Assuming that Eq. (4.24) holds, calculate how much received power is available at the output of the receive antenna (base station antenna and mobile station antenna, respectively), as a function of distance d.
   (b) Plot the received powers for all valid distances d – i.e., where Eq. (4.24) holds *and* the farfield condition of the antennas is fulfilled.

9. The following system is used to give an estimate of the exposure to electromagnetic waves from a Base Station (BS) and a Mobile Station (MS): consider communication between the BS and MS separated by a distance d in the 900-MHz band. The ground plane is treated as perfectly conducting, with antenna heights $h_{BS} = 10$ m and $h_{MS} = 1.5$ m. The antenna gains are $G_{BS} = 6$ dB and $G_{MS} = 2$ dB, respectively. On a straight line between the BS and MS, at a distance of 3 m from the MS and a height $h_{ref} = 1.5$ m, there is a Reference Antenna (RA), picking up signals from both, which can be used to measure exposure. The RA has a gain G_{ref}. Assume that Eq. (4.24) can be used to describe transmission between the BS and MS as well as between the BS and RA, although transmission between the MS and RA is better described by Friis' law due to the short distance.
   (a) Assume that the base station has a 10-dB lower requirement on the received power available at the antenna output and determine expressions for (as functions of distance d and receiver sensitivity level $P_{RX,MS}^{min}$ of the mobile station):
      i. Required transmit power by the base station, $P_{TX,BS}$.
      ii. Required transmit power by the mobile station, $P_{RX,BS}$.
      iii. Received power in the reference antenna from the base station, $P_{RX,ref}^{BS}$.
      iv. Received power in the reference antenna from the mobile station, $P_{RX,ref}^{MS}$.
   (b) Use the expressions from (a) to determine the difference in dB between $P_{RX,ref}^{MS}$ and $P_{RX,ref}^{BS}$ as a function of d. Plot the result for $d = 50 - 5,000$ m.

10. Communication is to take place from one side of a building to the other as depicted in Fig. 25.1, using 2-m-tall antennas. Convert the building into a series of semi-infinite screens and determine the fieldstrength at the receive antenna caused by diffraction using Bullington's method for (a) $f = 900$ MHz, (b) $f = 1,800$ MHz, and (c) $f = 2.4$ GHz.

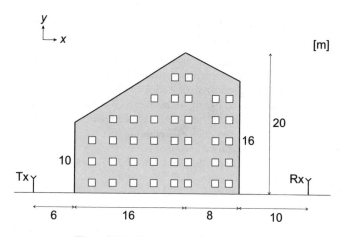

Figure 25.1   The geometry for Problem 8.

25.5 Chapter 5: Statistical description of the wireless channel

1. A mobile receiver travelling at $25\,\text{m/s}$ receives the following two multipath components:

$$\left.\begin{array}{l} E_1(t) = 0.1\cos\left[2\pi \cdot 2 \cdot 10^9 t - 2\pi \upsilon_{\text{max}}\cos(\gamma_1)t + 0\right]\text{V}\cdot\text{m}^{-1} \\[2mm] E_2(t) = 0.2\cos\left[2\pi \cdot 2 \cdot 10^9 t - 2\pi \upsilon_{\text{max}}\cos(\gamma_2)t + 0\right]\text{V}\cdot\text{m}^{-1} \end{array}\right\} \tag{25.4}$$

Assume that the receiver moves directly opposite to the direction of propagation of the first component and exactly in the direction of propagation of the second component. Compute the power per unit area at the receive antenna at time instants $t = 0$, 0.1, and $0.2\,\text{s}$. Compute the average power per unit area received over this interval.

Hint: The power per unit area is given by the magnitude of the average Poynting vector, $S_{\text{avg}} = \dfrac{1}{2}\cdot\dfrac{E^2}{Z}$, where averaging is performed over one period of the electric field. E is the amplitude of the total electric field and $Z_0 = 377\,\Omega$ for air.

2. Show that the Rice distribution with a large Rice factor, $K_r = \dfrac{A^2}{2\sigma^2}$, can be approximated by a Gaussian distribution with mean value A.

3. A mobile communication system is to be designed to the following specifications: the instantaneous received amplitude r must not drop below 10% of a specified level r_{min} at the cell boundary (maximum distance from base station) for 90% of the time. The signal experiences both small-scale Rayleigh fading and large-scale log-normal fading, with $\sigma_F = 6\,\text{dB}$. Find the required fading margin for the system to work.

4. A radio system is usually specified in such a way that a receiver should be able to handle a certain amount of Doppler spread in the received signal, without losing too much in performance. Assuming that only the mobile receiver is moving and that the maximal Doppler spread is measured as twice the maximal Doppler shift. Further, assume that you are designing a mobile communication system that should be able to operate at both $900\,\text{MHz}$ and $1,800\,\text{MHz}$.
    (a) If you aim at making the system capable of communicating when the teminal is moving at $200\,\text{km/h}$, which maximal Doppler spread should it be able to handle?
    (b) If you design the system to be able to operate at $200\,\text{km/h}$ when using the 900-MHz

band, at what maximal speed can you communicate if the 1,800-MHz band is used (assuming the same Doppler spread is the limitation)?

5. Assume that, at a certain distance, we have a deterministic propagation loss of 127 dB and large-scale fading, which is log-normally distributed with $\sigma_F = 7$ dB.
    (a) How large is the outage probability (due to large-scale fading) at that particular distance, if our sytem is designed to handle a maximal propagation loss of 135 dB?
    (b) Which of the following alternatives can be used to lower the outage probability of our system, and why are they/are they not possible to use?
        i. Increase the transmit power.
        ii. Decrease the deterministic pathloss.
        iii. Change the antennas.
        iv. Lower the σ_F.
        v. Build a better receiver.

6. Assume a Rayleigh-fading environment in which the sensitivity level of a receiver is given by a signal amplitude r_{min}, then the probability of outage is $P_{out} = \Pr\{r \leq r_{min}\} = cdf(r_{min})$.
    (a) Considering that the squared signal amplitude r^2 is proportional to the instantaneous received power C through the relation $C = K \cdot r^2$, where K is a positive constant, determine the expression for the probability of outage expressed in terms of receiver sensistivity level C_{min} and mean received power $\overline{C}$.
    (b) The fading margin (level of protection against fading) is expressed as:

    $$M = \frac{\overline{C}}{C_{min}} \tag{25.5}$$

    or (in dB):

    $$M_{dB} = \overline{C}_{dB} - C_{min\,dB} \tag{25.6}$$

    Determine a closed-form expression for the required fading margin, in dB, as a function of outage probability P_{out}.

7. For an outage probability P_{out} of up to 5% in a Rayleigh-fading environment, we assume that we can use the approximation:

    $$P_{out} = \Pr\{r \leq r_{min}\} \approx \frac{r_{min}^2}{2\sigma^2} \tag{25.7}$$

    where r_{min} is the sensitivity level of the receiver.
    (a) Determine a closed-form expression for the required fading margin, in dB, as a function of outage probability P_{out}, when the above approximation is used. Denote the resulting approximation of the fading margin as $\widetilde{M}_{dB}$.
    (b) Determine the largest error between the approximate $\widetilde{M}_{dB}$ and the exact M_{dB} values for outage probabilities $P_{out} \leq 5\%$.

8. In some cases, the "mean" of the received signal amplitude r is expressed in terms other than $\overline{r^2}$ – i.e., terms that are proportional to the mean received power. A common value to be found in propagation measurements is the median value r_{50} – i.e., the value showing the signal is below 50% of the time, and above 50% of the time.
    Now, assume that we have small-scale fading described by the Rayleigh distribution.
    (a) Derive the expression for $cdf(r)$ when given the median value r_{50} of the received signal.
    (b) Derive the expressions for the required fading margins in dB (for a certain P_{out}), expressed both in relation to $\overline{r^2}$ (the mean power) and in relation to r_{50}^2. Call these fading margins $M_{mean|dB}(P_{out})$ and $M_{median|dB}(P_{out})$.

(c) Compare the expressions for $M_{\text{mean}|\text{dB}}(P_{\text{out}})$ and $M_{\text{median}|\text{dB}}(P_{\text{out}})$ and try to find a simple relation between them.

9. For Rayleigh fading, derive expressions for the level crossing rate $N_r(r_{\text{min}})$ and the average duration of fades $\text{ADF}(r_{\text{min}})$ in terms of the fading margin $M = r^2/r_{\text{min}}^2$ rather than parameters Ω_0 and Ω_2 (compare Eqs. 5.49 and 5.50).

10. Assume that we are going to design a wireless system with maximal distance $d_{\text{max}} = 5\,\text{km}$ between the Base Station (BS) and Mobile Station (MS). The BS antenna is at a 20-m elevation and the MS antenna is at 1.5-m elevation. The carrier frequency of choice is 450 MHz and the environment consists of rather flat and open terrain. For this particular situation we find a propagation model called Egli's model, which says that the propagation loss is:

$$\Delta L_{|\text{dB}} = 10 \log\left(\frac{f_{|\text{MHz}}^2}{1{,}600}\right) \tag{25.8}$$

in addition to what is predicted by the theoretical model for propagation over a ground plane. The model is only valid if the distance is smaller than the radio horizon:

$$d_{\text{h}} \approx 4100 \left(\sqrt{h_{\text{BS}|\text{m}}} + \sqrt{h_{\text{MS}|\text{m}}}\right)_{|\text{m}} \tag{25.9}$$

The calculated value of propagation loss is a *median* value. The narrowband link is subjected to both large-scale log-normal fading, with $\sigma_F = 5$ dB, and small-scale Rayleigh fading. In addition, there is a receiver sensitivity level of $C_{\text{min}|\text{dBm}} = -122$ for the MS (at the antenna output). Both the BS and MS are equipped with gain 2.15-dB $\lambda/2$ dipoles.

(a) Draw a link budget for the downlink – i.e., for the link from the BS to MS. Start with the input power $P_{\text{TX}|\text{dB}}$ to the antenna and follow the budget through to the received median power $C_{\text{median}|\text{dB}}$, at the MS antenna output. Then, between $C_{\text{median}|\text{dB}}$ and $C_{\text{min}|\text{dB}}$ there will be a fading margin $M_{|\text{dB}}$.

(b) The system is considered operational (has coverage) if the instantaneous received power is not below $C_{\text{min}|\text{dB}}$ more than 5% of the time. Calculate the fading margin needed as a consequence of small-scale fading.

(c) We wish for the system to have a 95% boundary coverage – i.e., to be operational at 95% of the locations at the maximal distance d_{max}. Calculate the fading margin needed as a consequence of large-scale fading to fulfill this requirement.

(d) Calculate the required transmit power $P_{\text{TX}|\text{dB}}$, by adding the fading margins obtained in (b) and (c) to a total fading margin $M_{|\text{dB}}$, which is to be inserted in the link budget. Remember that the propagation model gives a median value and make the necessary compensation, if necessary.

Note: Adding the two fading margins from (b) and (c) does not give the lowest possible fading margin. In fact the system is slightly overdimensioned, but it is much simpler than combining the statistics of the two fading characteristics (giving the Suzuki distribution).

11. Let us consider a simple interference-limited system, where there are two transmitters Tx A and Tx B, both with antenna heights 30 m, at a distance of 40 km. They both transmit with the same power, use the same omnidirectional $\lambda/2$ dipole antennas, and use the same carrier frequency, 900 MHz. Tx A is transmitting to Rx A located at a distance d in the direction of Tx B. Transmission from Tx B is interfering with the reception of Rx A, which requires an average (small-scale averaged) carrier-to-interference ratio of $(C/I)_{\text{min}} = 7\,\text{dB}$. Incoming signals to Rx A (wanted and interfering) are both subject to independent 9-dB log-normal

large-scale fading. The propagation exponent in the environment we are studying is $\eta = 3.6$ – i.e., the received power decreases as $d^{-\eta}$.

(a) Determine the fading margin required to give a 99% probability that (C/I) is not below $(C/I)_{min}$.

(b) Using the fading margin from (a), determine the maximal distance d_{max} between Tx A and Rx A.

(c) Can you, by studying the equations, give a quick answer to what happens to the maximal distance d_{max} (as defined above) if Tx A and Tx B were located at 20 km from each other?

25.6 Chapter 6: Wideband and directional channel characterization

1. Explain the difference between *spreading function* and *scattering function*.

2. Give examples of situations where the information contained in the time frequency correlation function $R_H(\Delta t, \Delta f)$ can be useful.

3. Assume that the measured impulse response of Fig. 6.6 can be approximated as stationary. Furthermore, approximate the power delay profile as consisting of two clusters, each having exponential decay on a linear scale; that is:

$$P(\tau) = \begin{cases} a_1 e^{-b_1 \tau} & 0 \leq \tau \leq 20\ \mu s \\ a_2 e^{55 \cdot 10^{-6} - b_2 \tau} & 55\ \mu s \leq \tau \leq 65\ \mu s \\ 0 & \text{elsewhere} \end{cases} \quad (25.10)$$

First, find the coefficients a_i and b_i and then calculate the time-integrated power, the average mean delay, and the average rms delay spread, all given in Eqs. (6.37)–(6.39). If the coherence bandwidth is approximated by:

$$B_{coh} \approx \frac{1}{2\pi S_\tau} \quad (25.11)$$

where S_τ is the rms delay spread, would you characterize the channel as flat- or frequency-selective for a system bandwidth of 100 kHz? Use Fig. 6.4 to justify your answer.

4. As described in Section 6.4.5, the GSM system has an equalizer of 4 symbol durations, corresponding to 16 μs – i.e., multipath components arriving within that window can be processed by the receiver. Calculate the interference quotient for a window of duration 16 μs and starting at $t_0 = 0$ for the data measured in Fig. 6.6 using the approximation in Problem 3. Perform this calculation first for the whole PDP as specified in Problem 3 and then again while ignoring all components arriving after 20 μs. Compare the two results.

5. The coherence time T_c gives a measure of how long a channel can be considered to be constant, and can be approximated as the inverse of the Doppler spread. Obtain an estimate of the coherence time in Fig. 6.7.

6. In wireless systems one has to segment the transmitted stream of symbols into blocks, also called *frames*. In every frame one usually also inserts symbols that are known by the receiver, so-called *pilot symbols*. In this manner the receiver can estimate the current value of the channel, and thus coherent detection can be performed. Therefore, let the

receiver in every frame be informed of the channel gain for the first symbol in the frame, and assume that the receiver then believes that this value is valid for for the whole frame. Using the definition of coherence time in Problem 5 and assuming Jakes' Doppler spectrum, estimate the maximum speed of the receiver for which the assumption that the channel is constant during the frame is still valid. Let the framelength be 4.6 ms.

7. In a CDMA system the signal is spread over a large bandwidth by multiplying the transmitted symbol by a sequence of short pulses, also called *chips*. The system bandwidth is thus determined by the duration of a chip. If the chip duration is 0.26 µs and the maximum excess delay is 1.3 µs, into how many delay bins do the multipath components fall? If the maximum excess delay is 100 ns, is the CDMA system wideband or narrowband?

25.7 Chapter 7: Channel models

1. Give a physical interpretation of the log-normal distribution (is it realistic?).

2. Assume that we are calculating propagation loss in a medium-size city, where a base station antenna is located at a height of $h_b = 40$ m, the mobile station at $h_m = 2$ m. The carrier frequency of the transmission taking place is $f = 900$ MHz and the distance between the two is $d = 2$ km.
    (a) Calculate the predicted propagation loss L_{Oku} between isotropic antennas by using the formula for free space attenuation, in combination with Okumura's measurements (Figs. 7.12–7.13).
    (b) Calculate the predicted propagation loss L_{O-H} between isotropic antennas by using parameterization of Okumura's measurements provided by Hata – i.e., use the Okumura–Hata model.
    (c) Compare the results. If there are differences, where do you think they come from and are they significant?

3. Assume that we are calculating propagation loss at $f_0 = 1,800$ MHz in a medium-size city environment with equally spaced buildings of height $h_{Roof} = 20$ m, at a building-to-building distance of $b = 30$ m and $w = 10$-m-wide streets. The propagation loss we are interested in is between a base station at height h_b and a mobile station at height $h_m = 1.8$ m, with a distance of $d = 800$ m between the two. The mobile station is located on a street which is oriented at an angle $\varphi = 90°$ relative to the direction of incidence (direction of incoming wave). For this purpose, the COST 231–Walfish–Ikegami model is a suitable tool.
    (a) Check that the given parameters are within the validity range of the model.
    (b) Calculate the propagation loss when the base station antenna is located 3 m above the rooftops – i.e., when $h_b = 23$ m.
    (c) Calculate the propagation loss when the base station antenna is located 1 m below the rooftops – i.e., when $h_b = 19$ m, and comment on the difference from (b).

4. When evaluating GSM systems, a wideband model is required since the system is designed in such a way that different delays in the channel are experienced by the receiver. The COST 207 model is created for GSM evaluations:
    (a) Draw power delay profiles for the tap delay line implementations Rural Area (RA), Typical Urban (TU), Bad Urban (BU), and Hilly Terrain (HT) using Tables 7.3–7.6 in the text. Use the dB scale for power and the same scale for the delay axis of all four plots so that you can compare the four PDPs.

(b) Convert the delays into pathlengths, using $d_i = c\tau_i$, where c is the speed of light, and make notations of these pathlengths in the graphs drawn in (b). Using these pathlengths, try to interpret the distribution of power in the four scenarios (from where does the power come?).

5. For the COST 207 (created for GSM evaluations):
   (a) Find the rms delay spread of the COST 207 environments RA, TU.
   (b) What is the coherence bandwidth of the RA and TU channels?
   (c) Could two different function PDPs have the same rms delay spread?

6. The correlation between the elements of an antenna array are dependent on angular power distribution, element response, and element spacing. An antenna array consists of two omnidirectional elements separated by distance d and placed in a channel with azimuthal power spectrum $f_\phi(\phi)$.
   (a) Derive an expression for correlation.
   (b) [MATLAB] Plot the correlation cofficient between two elements for different antenna spacings for 100 MPCs with a uniform angular distribution $(0, 2\pi)$.

26.8 Chapter 8: Channel sounding

1. Assume a simple direct RF pulse channel sounder that generates a signal which is a sequence of narrow probing pulses. Each pulse has duration $t_{on} = 50\,\text{ns}$, and the pulse repetition period is $20\,\mu\text{s}$. Determine the following:
   (a) Minimum time delay that can be measured by the system.
   (b) Maximum time delay that can be measured unambiguously.

2. As the wireless propagation channel may be treated as a linear system, why are m-sequences (PN-sequences) used for channel sounding? *Hint*: elaborate on the auto-and cross-correlation properties of white noise, and discuss its similiarities to PN-sequences.

3. A maximal length sequence (m-sequence) is generated from a Linear Feedback Shift Register (LFSR) with certain allowed connections between the memory elements and the modulo-2 adder. In Fig. 25.2 an LFSR is shown with $m = 3$ memory elements, connected so as to generate an m-sequence. Determine the following:
   (a) The period M_c of the m-sequence.
   (b) The complete m-sequence $\{C_m\}$, given that the memory is initialized with $a_{k-1} = 0, a_{k-2} = 0, a_{k-3} = 1$.

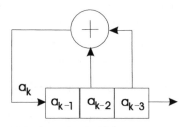

Figure 25.2   A maximal LFSR sequence generator.

4. Autocorrelation refers to the degree of correspondence between a sequence and a shifted copy of itself. If the ± 1 sequence $\left\{ \acute{C}_m \right\}$ is defined as $\acute{C}_m = 1 - 2C_m$, where $\{C_m\} \epsilon (0, 1)$ is an m-sequence, then the autocorrelation function $R_{\acute{C}_m}(\tau)$ defined as:

$$R_{\acute{C}_m}(\tau) = \frac{1}{M_c} \sum_{m=1}^{M} \acute{C}_m \acute{C}_{m+\tau} = \begin{cases} 1, & \tau = 0, \pm M_c, \pm 2M_c, \ldots \\ -\dfrac{1}{M_c}, & \text{otherwise} \end{cases} \tag{25.13}$$

is also periodic. If the code waveform $p(t)$ is the square-wave equivalent of sequence $\{\acute{C}_m\}$ with pulse duration T_c then determine the ACF for all values of τ.
(a) Plot the ACF for $T_c = 1$, $M_c = 7$, and $-8 \le \tau \le 8$.
(b) Neglecting the effects of system noise, what is the dynamic range over which the system can detect received signals.

5. In an STDCC system the maximum measurable Doppler shift is given by:

$$v_{\max} = \frac{1}{2K_{\text{scal}} M_c T_c} \tag{25.14}$$

Given that $K_{\text{scal}} = 5{,}000$, $M_c = 31$, $T_c = 0.1\,\mu s$, and the carrier frequency is 900 MHz, determine the following:
(a) The maximum permissible velocity of the MS.
(b) The longest time delay that can be measured.
(c) If we increase the m-sequence length to $M_c = 63$, what happens to the above quantities?

6. Plane waves from two separate sources impinge on a uniform linear array as shown in Fig. 25.3. The array consists of four isotropic antenna elements with interelement spacing of $\dfrac{\lambda}{2}$ where λ is the carrier wavelength. The covariance matrix of array outputs is provided for two different cases. Determine the angles of arrival of the two waves in each case, using conventional (Fourier–Bartlett) beamforming.

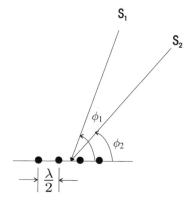

Figure 25.3 Plane waves S_1 and S_2 impinging on a four-element ULA. The direction of propagation is indicated by arrows perpendicular to the wavefronts.

(a)

$$\mathbf{R}_{rr} = \begin{bmatrix} 1.2703 & 0.3559+0.6675i & 0.4215-0.1392i & 0.9818+0.7086i \\ 0.3559-0.6675i & 1.3252 & 0.3489+0.6493i & 0.4805-0.1864i \\ 0.4215+0.1392i & 0.3489-0.6493i & 1.2475 & 0.3314+0.5787i \\ 0.9818-0.7086i & 0.4805+0.1864i & 0.3314-0.5787i & 1.2344 \end{bmatrix} \quad (25.15)$$

(b)

$$\mathbf{R}_{rr} = \begin{bmatrix} 0.9198 & 0.2125+0.7644i & -0.5295+0.1202i & 0.1355-0.5488i \\ 0.2125-0.7644i & 0.8957 & 0.1661+0.7244i & -0.5104+0.0526i \\ -0.5295-0.1202i & 0.1661-0.7244i & 0.8323 & 0.1444+0.6853i \\ 0.1355+0.5488i & -0.5104-0.0526i & 0.1444-0.6853i & 0.8283 \end{bmatrix}$$

$$(25.16)$$

7. Repeat the estimation problem discussed in Problem 6 using the ESPRIT algorithm as outlined in Appendix 8.A (see *www.wiley.com/go/molisch*).

8. Let M samples of the impulse response of a frequency flat channel be measured during a time interval t_{meas} – the channel is assumed to be time-invariant during this interval. Furthermore, it is assumed that the measurement noise in these samples is iid with zero mean and variance σ^2. Prove that averaging over M samples of the impulse response enhances the SNR by a factor M.

25.9 Chapter 9: Antennas

1. A base station for GSM 900 MHz has output power of 10 W; the cell is in a rural environment, with a decay exponent of 3.2 for distances larger than 300 m. The cell coverage radius is 37 km when measured by the MS in the absence of a user. Calculate how much smaller the coverage radius becomes for the median user in Fig. 9.1 when she/he places the cellphone in close proximity to the head.

2. Sketch possible antenna placements on a PDA (Personal Digital Assistant) for a 2.4-GHz antenna for WiFi. Assume that the antenna is a patch antenna with effective permittivity $\varepsilon_r = 2.5$. Take care to minimize the influence of the user's hand.

3. Consider a helical antenna with $d = 5$ mm operating at 1.9 GHz. Calculate the diameter D such that polarization becomes circular.

4. Consider an antenna with a pattern:

$$G(\phi, \theta) = \sin^n(\theta/\theta_0) \cos(\theta/\theta_0) \quad (25.17)$$

where $\theta_0 = \pi/1.5$.
    (a) What is the 3-dB beamwidth?
    (b) What is the 10-dB beamwidth?
    (c) Wat is the directivity?
    (d) Compute the numerical values of (i), (ii), (iii) for $\theta_0 = \pi/1.5$, $n = 5$.

5. Compute the input impedance of a half-wavelength slot antenna; assume uniform current distribution. *Hint*: the input impedance of antenna structure A and its complement B (a structure that has metal whenever structure A has air, and has air whenever structure A has metal) are related by:

$$Z_A Z_B = \frac{Z_0^2}{4} \quad (25.18)$$

where Z_0 is the free-space impedance $Z_0 = 377 \ \Omega$.

6. Let the transmit antenna be a vertical $\lambda/2$ dipole, and the receive antenna a vertical $\lambda/20$ dipole.
    (a) What are the radiation resistances at transmitter and receiver?
    (b) Assuming ohmic losses due to $R_{ohmic} = 10\,\Omega$, what is the radiation efficiency?

7. Consider an antenna array with N elements where all elements have the same amplitude, and a phaseshift:

$$\Delta = -\frac{2\pi}{\lambda} d_a \qquad (25.19)$$

    (a) What is the gain in the endfire direction?
    (b) What is the gain in the endfire direction if:

$$\Delta = -\left[\frac{2\pi}{\lambda} d_a + \frac{\pi}{N}\right] \qquad (25.20)$$

25.10 Chapter 10: Structure of a wireless communication link

1. A directional coupler has a directivity of 20 dB, an insertion attenuation of 1 dB, and a coupling attenuation of 20 dB. The reflection coefficients of all ports (with respect to the nominal impedance of $Z_0 = 50\,\Omega$) are 0.12, and the phaseshift in the main path is 0, in the coupled path $\pi/2$, and 0 between the decoupled paths.
    (a) Compute the S-matrix of this directional coupler.
    (b) Is the directional coupler passive, reciprocial, symmetrical?

2. The input impedance of an antenna $Z = 70 - 85j\,\Omega$ is transformed by putting it in series with an inductance L. The antenna operates at 250 MHz.
    (a) What is the transformed impedance if $L = 85\,\text{nH}$?
    (b) Which value does L have to take on to make the transformed impedance real?

3. Consider the concatenation of two amplifiers with gains G_1, G_2 and noise figures F_1, F_2. Show that in order to minimize total noise figure, the amplifier with the smaller value of:

$$M = \frac{F - 1}{1 - 1/G} \qquad (25.21)$$

    has to be placed first.

4. For the measurement of received power for AGCs, it is common to use diodes, assuming that the output (diode current) is proportional to the square of the input (voltage). However, true characteristics are better described by an exponential function:

$$i_D \exp(U/U_t) - 1 \qquad (25.22)$$

    where the voltage U_t is a constant. This function can be approximated by a quadratic function only as long as the voltage is low.
    (a) Show that the diode current contains a DC component.
    (b) What spectral components are created by the diode? How can they be eliminated?
    (c) How large is the maximum diode current if the error (due to the difference between ideal quadratic and exponential behavior) in measured power is to remain below 5%?

5. An amplifier has cubic characteristics:

$$U_{out} = a_0 + a_1 U_{in} + a_2 U_{in}^2 + a_3 U_{in}^3 \qquad (25.23)$$

Let there be two sinusoidal input signals (with frequencies f_1, f_2) at the input, both with a power of -6 dBm. At output, we measure desired signals (at f_1, f_2) with 20-dBm power, and third-order intermodulation products (at $2f_2 - f_1$ and $2f_1 - f_2$) with a power of -10 dBm. What is the intercept point? In other words, at what level of the input signal do intermodulation products become as large as desired signals?

6. A sawtooth received signal is to be quantized by a linear ADC (linear spacing of quantization steps).
    (a) What is the minimum number of quantization steps if quantization noise (variance between ideal and quantized received signal) has to stay below 10, 20, 30 dB?
    (b) How does the result change when – due to a badly working AGC – the peak amplitude of the received signal is only half as large as the maximum amplitude that the ADC could quantize?

25.11 Chapter 11: Modulation formats

1. Both GSM and DECT use GMSK, but with different Gaussian filters ($B_G T = 0.3$ in GSM, $B_G T = 0.5$ in DECT). What are the advantages of having a larger bandwidth time product? Why is the lower one used in GSM?

2. Derive the smallest frequency separation for orthogonal binary frequency shift keying.

3. In the EDGE standard (high-speed data in GSM networks), $\frac{3\pi}{8}$-shifted 8-PSK is used. Sketch transitions in the signal space diagram for this modulation format. What is the ratio between average value and minimum value of the envelope? Why is it that $\frac{\pi}{4}$-shifted 8-PSK cannot be used?

4. MSK can be interpreted as offset QAM with specific pulseshapes. Show that MSK has a constant envelope using this interpretation.

5. Consider two functions:

$$f(t) = \begin{cases} 1 & 0 < t < T/2 \\ -2 & T/2 < t < T \end{cases} \qquad (25.24)$$

and

$$g(t) = 1 - (t/T) \qquad (25.25)$$

for $0 < t < T$.
    (a) Find a set of expansion functions, using Gram–Schmidt orthogonalization.
    (b) Find points in the signal space diagram for $f(t)$, $g(t)$, and the function that is unity for $0 < t < T$.

6. For the bit sequence of Fig. 11.6, plot $p_D(t)$ for *differentially encoded* BPSK.

7. Relate the mean signal energy of 64-QAM to the distance between points in the signal space diagram.

8. A system should transmit as high a data rate as possible within a 1-MHz bandwidth, where out-of-band emissions of -50 dBm are admissible. The transmit power used is 20 W. Is it better to use MSK or BPSK with root-raised cosine filters with $\alpha = 0.35$? *Note*: this

question only concentrates on spectral efficiency, and avoids other considerations like the peak-to-average ratio of the signal.

25.12 Chapter 12: Demodulation

1. Consider a point-to-point radio link between two highly directional antennas in a stationary environment. The antennas have antenna gains of 30 dB, distance attenuation is 150 dB, and the receiver has a noise figure of 7 dB. The symbol rate is 20 Msymb/s and Nyquist signaling is used. It can be assumed that the radio link can be treated as an AWGN channel without fading. How much transmit power is required (disregarding power losses at transmitter and receiver ends) for a maximum BER of 10^{-5}:
    (a) When using coherently detected BPSK, FSK, differentially detected BPSK, or non-coherently detected FSK?
    (b) Derive the exact bit and symbol error probability expressions for coherently detected Gray-coded QPSK. Start by showing that the QPSK signal can be viewed as two antipodal signals in quadrature.
    (c) What is the required transmit power if Gray-coded QPSK is used?
    (d) What is the penalty in increased BER for using differential detection of Gray-coded QPSK in (c)?

2. Use the full union bound method for upper-bounding the BER for higher order modulation methods.
    (a) Upper-bound the BER for Gray-coded QPSK by using the full union bound.
    (b) In which range of E_b/N_0 is the difference between the upper bound in (a) and the exact expression less than 10^{-5}?

3. Consider the 8AMPM modulation format shown in Fig. 25.4. What is the average symbol energy expressed in d_{min} (all signal alternatives assumed to be equally probable)? What is the nearest neighbor union bound on the BER?

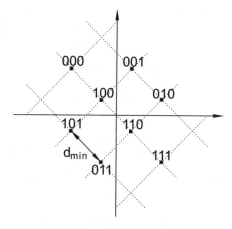

Figure 25.4   8-AMPM constellation.

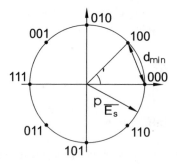

Figure 25.5

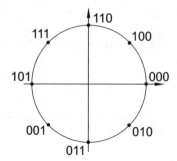

Figure 25.6

4. Use the nearest-neighbor-union-bound method to calculate approximate BERs for higher order modulation methods.
    (a) Use the nearest neighbor union bound to calculate the approximate upper bound on BER for 8-PSK as shown in Fig. 25.5.
    (b) Use the nearest neighbor union bound to calculate the approximate upper bound on BER for Gray-coded 8-PSK as shown in Fig. 25.6. How large is the approximate gain from Gray coding at a BER of 10^{-5}?

5. Obtain simulated BER results for Gray-coded QPSK and 8-PSK by creating a MATLAB™ program. Run simulations for E_b/N_0 in the range from 0 to 15 dB. In which region does the nearest neighbor union bound on BER (see Problem 3) appear to be tight?

6. Calculate the nearest neighbor union bound for Gray-coded 16-QAM. Assuming that the BER must not exceed 10^{-5}, what are the useful ranges of E_b/N_0 for adaptive switching between the two modulation schemes QPSK and 16-QAM for maximum achievable data rate?

7. Assume that M-ary orthogonal signaling is used. Based on the union bound, how small does E_b/N_0 need to be to allow for completely error-free transmissions as $M \to \infty$?

8. Consider the point-to-point radio link introduced in Problem 1. By how much must transmit power be increased to maintain the maximum BER of 10^{-5} if the channel is flat-Rayleigh-fading:
    (a) When coherently detected BPSK and FSK, and DPSK and non-coherent FSK are used?

(b) What is the required increase in transmit power if the channel is Ricean with $K_r = 10$ and DPSK is used? What are the expected results for $K_r \to 0$?

9. Consider a mobile radio link with a carrier frequency of $f_c = 1,200$ MHz, a bit rate of 3 kbit/s, and a required maximum BER of 10^{-4}. The modulation format is MSK with differential detection. A maximum transmit power of 10 dBm (EIRP) is used with 5-dB gain antennas and receivers with noise figures of 9 dB.
(a) Assuming that the channel is flat-Rayleigh-fading and that base station and mobile station heights are 40 and 3 m, respectively, what is the achievable cell radius in a suburban environment according to the Okumura–Hata pathloss model?
(b) Assume that the channel is frequency-dispersive Rayleigh-fading and characterized by a classical Jakes' Doppler spectrum. What is the maximum mobile terminal speed for an irreducable BER due to frequency dispersion of 10^{-5}?

10. Consider a mobile radio system using MSK with a bit rate of 100 kbit/s. The system is used for transmitting IP packets of up to 1,000 bytes. The packet error rate must not exceed 10^{-3} (without the use of an ARQ scheme).
(a) What is the maximum allowed average delay spread of the mobile radio channel?
(b) What are typical values of average delay spread in indoor, urban, and rural environments for mobile communication systems?

25.13 Chapter 13: Diversity

1. To illustrate the impact of diversity, we look at the average BER for BPSK and MRC, given approximately by Eq. (13.35). Assume that the average SNR is 20 dB.
(a) Calculate the average BER for $N_r = 1$ receive antennas.
(b) Calculate the average BER for $N_r = 3$ receive antennas.
(c) Calculate the SNR that would be required in a one-antenna system in order to achieve the same BER as a three-antenna system at 20 dB.

2. Assume a situation where we have the possibility of adding one or two extra antennas, with uncorrelated fading, to a receiver operating in a Rayleigh-fading environment. The two diversity schemes under consideration are RSSI-driven selection diversity or Maximum Ratio Combining (MRC) diversity.
(a) Assume that the important performance requirement is that the instantaneous $\bar{E}_b/N_0$ cannot be below some fixed value more than 1% of the time (1% outage). Determine the required fading margin for one, two, and three antennas with independent Rayleigh fading and the corresponding "diversity gains" for the two- and three-antenna cases when using RSSI-driven selection diversity and MRC diversity.
(b) Assume that the important performance requirement is that the average BER is 10^{-3} (when using BPSK). Determine the required average $\bar{E}_b/N_0$ for one, two, and three antennas with independent Rayleigh fading and the corresponding "diversity gains" for the two- and three-antenna cases when using RSSI-driven selection diversity and MRC diversity.
(c) Perform (b) and (c) again, but with 10% outage in (b) and an average BER of 10^{-2} in (c). Compare the "diversity gains" with the ones obtained earlier and comment on the differences!

3. Let a receiver be connected to two antennas, for which the SNRs are independent and exponentially distributed using the same average SNR. RSSI-driven selection diversity is employed and the outage probability is P_{out}. We are interested in the fading margin.

(a) Derive an expression in terms of P_{out} for the fading margin when only one antenna is used

(b) Derive an expression in terms of P_{out} for the fading margin when both antennas are used

(c) Use the two results above to calculate the diversity gain for an outage probability of 1%.

4. In a wideband CDMA system the Rake receiver can take advantage of the multipath diversity arising from the delay dispersion of the channel. Under certain assumptions the Rake receiver acts as a maximal ratio combiner where branches correspond to the delay bins of the CDMA system. Assume a rectangular power delay profile and that the number of Rake fingers are equal to the number of resolvable multipaths. Furthermore, assume that each MPC is Rayleigh-fading. We require that the instantaneous BER exceeds 10^{-3} only 1% of the time when using BPSK on a channel with an average SNR of 15 dB. How many resolvable multipaths must the channel consist of in order for the BER requirement to be fulfilled?

5. In order to reduce the complexity a hybrid selection MRC scheme can be used instead of full MRC. If we have five antennas but only use the three strongest, what is the loss in terms of mean SNR compared with full MRC?

6. Consider a scenario where the transmitter is equipped with two antennas and the receiver only one. The transmitter sends two symbols in the following fashion (see Fig. 25.7). In the first symbol interval, symbol s_1 is the transmitter from the first antenna and symbol s_2 from the second antenna. In the second symbol interval s_2^* is transmitted from the first antenna and $-s_1^*$ from the second antenna. The (complex-valued) attenuation between the first transmit antenna and the receive antenna is h_1, and the corresponding attenuation for the second transmit antenna is h_2. The attenuations are assumed to be constant over both signaling intervals. At the receiver AWGN is added; n_1 in the first symbol interval and n_2 in the second (see also Section 20.2).

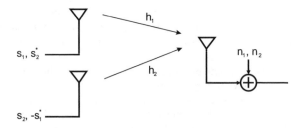

Figure 25.7 Principle of Alamouti codes.

(a) Derive the output from the receive antenna for both symbol intervals. Let the output for the first interval be r_1, and r_2 for the second.

(b) Let the following operation be performed at the receiving end:

$$\left. \begin{aligned} \hat{s}_1 &= h_1^* r_1 - h_2 r_2^* \\ \hat{s}_2 &= h_2^* r_1 + h_1 r_2^* \end{aligned} \right\} \tag{25.26}$$

What is achieved by this operation?

7.   Consider an N_r-branch antenna diversity system using the MRC rule. All branches are subject to Rayleigh fading and the fading at one branch is independent of fading at all other branches. Derive the pdf of the SNR at the output of the combiner for the following cases:

(a) The average SNR per branch is $\bar{\gamma}$, for all branches.

(b) The average SNR for the ith branch is $\bar{\gamma}_i$. Let the $\bar{\gamma}_i$'s be distinct.

8.   In CDMA systems each symbol is spread over a large bandwidth by multiplying the symbol using a spreading sequence. If s is the transmitted symbol and $\xi(t)$ is the spreading sequence, the received baseband signal can be written:

$$r(t) = \sum_{m=1}^{M} \alpha_m \xi(t - \tau_m)s + n(t) \tag{25.27}$$

where M is the number of resolvable multipaths, α_m is the complex gain of the mth multipath, τ_m the delay of the mth multipath, and $n(t)$ is an AWGN process. Assume that gains $\{\alpha_m\}$ and delays $\{\tau_m\}$ can be estimated perfectly, and that the spreading sequence has perfect autocorrelation properties. Let there be one matched filter per multipath and a device that combines the outputs of the matched filters. Derive an expression for the SNR at the output of the combiner.

9.   As described in Section 13.5.2 in the text, the receiver in a switched diversity system with two antennas takes as its input the signal from one antenna as long as the SNR is above some threshold. When the SNR falls below the threshold, the receiver switches to the second antenna, regardless of the SNR at the second antenna. If the antennas fade independently and according to the same distribution, it can be shown that the cdf of the SNR at the receiver is:

$$cdf_\gamma(\gamma) = \begin{cases} \Pr(\gamma_1 \leq \gamma_t \text{ and } \gamma_2 \leq \gamma), & \text{for } \gamma < \gamma_t \\ \Pr(\gamma_t \leq \gamma_1 \leq \gamma \text{ or } [\gamma_1 \leq \gamma_t \text{ and } \gamma_2 \leq \gamma]), & \text{for } \gamma \geq \gamma_t \end{cases} \tag{25.28}$$

where γ is the SNR after the switching device (i.e., the SNR at the receiver), γ_1 is the SNR of the first antenna, γ_2 is the SNR of the second antenna, and γ_t is the switching threshold.

(a) For Rayleigh fading where both antennas have the same mean SNR, give the cdf and pdf of γ.

(b) If we use mean SNR as our performance measure, what is the optimum switching threshold and resulting mean SNR. What is the gain in dB of using switched diversity compared with that of a single antenna? Compare this gain with that of maximal ratio combining and selection diversity.

(c) If instead our performance measure is average BER, what is the optimum switching threshold for binary non-coherent FSK? What is the average BER for an SNR of 15 dB? Compare this with the case of a single antenna. Remember that the BER for binary non-coherent FSK is:

$$\text{BER} = \frac{1}{2}\exp\left(-\frac{\gamma}{2}\right) \tag{25.29}$$

10. In maximal ratio combining each branch is weighted with the complex conjugate of that branch's complex fading gain. However, in practice the receiver must somehow estimate fading gains in order to multiply received signals by them. Assume that this estimation is based on pilot symbol insertion, in which case the weights become subject to a complex Gaussian error. If an N_r-branch diversity system with Rayleigh fading and a mean SNR Γ on each branch is used, the PDF of the output SNR becomes [Tomiuk et al. 1999]:

$$pdf_\gamma(\gamma) = \frac{(1-\rho^2)^{N_r-1} e^{-\gamma/\bar{\gamma}}}{\bar{\gamma}} \sum_{n=0}^{N_r-1} \binom{N_r-1}{n} \left[\frac{\rho^2\gamma}{(1-\rho^2)\bar{\gamma}}\right]^n \frac{1}{n!} \qquad (25.30)$$

where ρ^2 is the normalized correlation coefficient between fading gain on a branch and its estimate – i.e., the weight.

(a) Show that the PDF above can be written as a weighted sum of N_r ideal-maximal-ratio SNR PDFs.

(b) What happens when fading gains and weights become fully uncorrelated?

(c) What happens when fading gains and weights become fully correlated?

(d) The average error rate for an ideal N_r-branch MRC channel, for a certain modulation scheme, is denoted $P_e(\bar{\gamma}, N_r)$. Using the result from (a), give a general expression for the average error rate for N_r-branch MRC with imperfect weights.

(e) For many schemes, the average error rate for ideal MRC for large mean SNRs can be approximated as:

$$\tilde{P}_e(\bar{\gamma}, N_r) = \frac{C(N_r)}{\bar{\gamma}^{N_r}} \qquad (25.31)$$

where $C(s)$ is a constant specific to the modulation scheme. Using the result from (d), find out what happens to the average error rate for N_r-branch MRC with imperfect weights when the mean SNR grows very large and $\rho < 1$.

11. Consider an N_r-branch diversity system. Let the signal on the kth branch be $\tilde{s}_k = s_k e^{-j\phi_k}$, noise power on each branch be N_0, and noise be independent between the branches. Each branch is phase-adjusted to zero phase, weighted with α_k, and then the branches are combined. Give expressions for the branch SNR, the combined SNR, and then derive the weights α_k that maximize the combined SNR. With optimal weights, what is the combined SNR expressed in terms of the branch SNRs?

25.14 Chapter 14: Channel coding

1. Let us consider a linear cyclic $(7,3)$ block code with generator polynomial $G(x) = x^4 + x^3 + x^2 + 1$.

(a) Encode the message $U(x) = x^2 + 1$ systematically, using $G(x)$.

(b) Calculate the syndrome $S(x)$ when we have received (probably corrupted) $R(x) = x^6 + x^5 + x^4 + x + 1$.

(c) A close inspection reveals that $G(x)$ can be factored as $G(x) = (x+1)T(x)$, where $T(x) = x^3 + x + 1$ is a primitive polynomial. This implies, we claim, that the code can correct all single errors and all double errors, where the errors are located next to each other. Describe how you would verify this claim.

2. We have a binary systematic linear cyclic $(7,4)$ code. The codeword $X(x) = x^4 + x^2 + x$ corresponds to the message $U(x) = x$. Can we, with just this information, calculate the codewords for all messages? If so, describe in detail how it is done and which code properties you use.

3. Show that for a cyclic (N, K) code with generator polynomial $G(x)$ there is only one codeword with degree $N - K$ and that codeword is the generator polynomial itself.

4. The polynomial $x^{15} + 1$ can be factored into irreducible polynomials as:

$$x^{15} + 1 = (x^4 + x^3 + 1)(x^4 + x^3 + x^2 + x + 1)$$
$$\cdot (x^4 + x + 1)(x^2 + x + 1)(x + 1)$$

Using this information, list all generator polynomials generating binary cyclic $(15, 8)$ codes.

5. Assume that we have a linear $(7, 4)$ code, where the codewords corresponding to messages $\mathbf{u} = [1000]$, $[0100]$, $[0010]$, and $[0001]$ are given as:

	Message				Codeword						
1	0	0	0	→	1	1	0	1	0	0	0
0	1	0	0	→	0	1	1	0	1	0	0
0	0	1	0	→	0	0	1	1	0	1	0
0	0	0	1	→	0	0	0	1	1	0	1

(a) Determine all codewords in this code.
(b) Determine the minimum distance, d_{min}, and how many errors, t, the code can correct.
(c) The code above is not in systematic form. Calculate the generator matrix $\mathbf{G}$ for the corresponding systematic code.
(d) Determine the parity check matrix $\mathbf{H}$, such that $\mathbf{HG}^T = 0$.
(e) Is this code cyclic? If so, determine its generator polynomial.

6. We have a linear systematic $(8, 4)$ block code, with the following generator matrix:

$$\mathbf{G} = \begin{bmatrix} 1 & 0 & 0 & 0 & 1 & 1 & 0 & 1 \\ 0 & 1 & 0 & 0 & 0 & 1 & 1 & 1 \\ 0 & 0 & 1 & 0 & 1 & 1 & 1 & 0 \\ 0 & 0 & 0 & 1 & 1 & 0 & 1 & 1 \end{bmatrix}$$

(a) Determine the codeword corresponding to the message $\mathbf{u} = [1011]$, the parity matrix $\mathbf{H}$, and calculate the syndrome when the word $\mathbf{y} = [010111111]$ is received.
(b) By removing the fifth column of $\mathbf{G}$ we get a new generator matrix $\mathbf{G}^*$, which generates a $(7, 4)$ code. The new code is, in addition to its linear property, also cyclic. Determine the generator polynomial by inspection of $\mathbf{G}^*$. It can be done by observing a property of cyclic codes – which?

7. Show that the following inequality (the Singleton bound) is always fulfilled for a linear (N, K) block code:

$$d_{min} \leq N - K + 1$$

8. Show that the Hamming bound

$$2^{N-K} \geq \sum_{i=0}^{t} \binom{N}{i}$$

needs to be fulfilled if we want syndrome decoding to be able to correct t errors using a linear binary (N, K) code. *Note*: A code that meets the Hamming bound with equality is called a *perfect code*.

9. Show that Hamming codes are perfect $t = 1$ error-correcting codes.

10. Consider the convolutional encoder in Fig. 14.3.
    (a) If binary antipodal transmission is used, we can represent coding and modulation together in a trellis where 1's and 0's are replaced by +1's and −1's instead. Draw a new version of the trellis stage in Fig. 14.5a, using this new representation.
    (b) A signal is transmitted over an AWGN channel and the following (soft) values are received:

$$-1.1; 0.9; -0.1 \quad -0.2; -0.7; -0.6 \quad 1.1; -0.1; -1.4 \quad -0.9; -1.6; 0.2$$
$$-1.2; 1.0; 0.3 \quad 1.4; 0.6; -0.1 \quad -1.3; -0.3; 0.7$$

If these values were detected as binary digits, before decoding, we would get the binary sequence in Fig. 14.5b. This time, however, we are going to use soft Viterbi decoding with squared Euclidean metric. Perform soft decoding such that it corresponds to the hard decoding performed in Fig. 14.5d–f. After the final step, are there the same survivors as for the hard-decoding case?

11. Block codes on fading channels. The text indicates (the expression is stated for a special case in Section 14.7.2) that the BER of a t-error correcting (N, K) block code over a properly interleaved Rayleigh-fading channel with hard decoding is proportional to:

$$\sum_{i=t+1}^{N} K_i \left(\frac{1}{2 + 2\bar{\gamma}_B} \right)^i \left(1 - \frac{1}{2 + 2\bar{\gamma}_B} \right)^{N-i}$$

where the K_i's are constants and $\bar{\gamma}_B$ is the average SNR. The text also states that in general a code with minimum distance $d_{\min}$ achieves a diversity order of $\left\lfloor \frac{d_{\min} - 1}{2} \right\rfloor + 1$. Prove this statement using the expression above for proportionality of the BER.

25.15 Chapter 15: Speech coding

1. Explain the main drawbacks to lossless speech coders.

2. Describe the three basic types of speech coders.

3. What are the most relevant spectral characteristics of speech?

25.16 Chapter 16: Equalizers

1. To mitigate the effects of multipath propagation, we can use an equalizer at the receiver. A simple example of an equalizer is the linear zero-forcing equalizer. Noise enhancement is, however, one of the drawbacks to this type of equalizer. Explain the mechanism behind noise enhancement and name an equalizer type where this is less pronounced.

2. State the main advantage and disadvantage of blind equalization and name three approaches to designing blind equalizers.

3. The "Wiener–Hopf" equation was given by $\mathbf{Re}_{opt} = \mathbf{p}$. For real-valued white noise n_m with zero mean and variance σ_n^2, calculate the correlation matrix $\mathbf{R} = E\{\mathbf{u}^*\mathbf{u}^T\}$ for the following received signals:
   (a) $u_m = a\sin(\omega m) + n_m$;
   (b) $u_m = bu_{m-1} + n_m$; $a, b \in R, b \neq \pm 1$.

4. Consider a particular channel with the following parameters:

$$\mathbf{R} = \begin{bmatrix} 1 & 0.576 & 0.213 \\ 0.576 & 1 & 0.322 \\ 0.213 & 0.322 & 1 \end{bmatrix}, \qquad \mathbf{p} = \begin{bmatrix} 0 & 0.278 & 0.345 \end{bmatrix}, \qquad \text{and} \qquad \sigma_S^2 = 0.7$$

   Write the MSE equation for this channel in terms of real-valued equalizer coefficients.

5. An infinite-length ZF equalizer can completely eliminate ISI, as long as the channel transfer function is finite in the transform domain – i.e., Eq. (16.40). Here, we investigate the effect of using a finite-length equalizer to mitigate ISI.
   (a) Design a five-tap ZF equalizer for the channel transfer function described by Table 25.1 – i.e., an equalizer that forces the impulse response to be 0 for $i = -2, -1, 1, 2$ and 1 for $i = 0$.

Table 25.1 Channel transfer function.

n	f_n
−4	0
−3	0.1
−2	−0.02
−1	0.2
0	1
1	−0.1
2	0.05
3	0.01
4	0

   Hint: A 5×5 matrix inversion is involved.
   (b) Find the output of the above equalizer and comment on the results.

6. When transmitting 2ASK (with alternatives −1 and +1 representing "0" and "1", respectively) over an AWGN channel, ISI is experienced as a discrete time equivalent channel $F(z) = 1 + 0.5z^{-1}$ (see Fig. 25.8). When transmitting over this channel, we assume that the initial channel state is −1, and the following noisy sequence is received when transmitting 5 consecutive bits. Transmission continues after this, but we only have the following information at this stage:

$$0.66 \quad 1.59 \quad -0.59 \quad 0.86 \quad -0.79$$

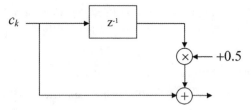

Figure 25.8 Block diagram of the channel $F(z) = 1 + 0.5z^{-1}$.

(a) What would the equalizer filter be if we apply a ZF linear equalizer?
(b) What is the memory of this channel
(c) Draw one trellis stage with states, input symbols, and output symbols shown.
(d) Draw a full trellis for this case and apply the Viterbi algorithm to find the maximum-likelihood sequence estimate of the transmitted 5-bit sequence.

7. In general, the MSE equation is a quadratic function of equalizer weights. It is always positive, convex, and forms a hyperparabolic surface. For a two-tap equalizer, the MSE equation takes the form:

$$Ae_1^2 + Be_1e_2 + Ce_2^2 + De_1 + Ee_2 + F$$

with $A, B, C, D, E \in \mathbb{R}$.

For the following data, make a contour plot of the hyperbolic surface formed by the MSE equation:

$$R = \begin{bmatrix} 1 & 0.651 \\ 0.651 & 1 \end{bmatrix}$$

$$\mathbf{p} = \begin{bmatrix} 0.288 & 0.113 \end{bmatrix}^T, \quad \text{and} \quad \sigma_S^2 = 0.3$$

8. As stated in the text, the choice of the μ parameter strongly influences the performance of the LMS algorithm. It is possible to study the convergence behavior in isolation by assuming perfect knowledge of R and $\mathbf{p}$.
(a) For the data provided in Exercise 16.7, plot the convergence of the LMS algorithm for $\mu = 0.1/\lambda_{\max}, 0.5/\lambda_{\max}, 2/\lambda_{\max}$ with initial value $\mathbf{e} = \begin{bmatrix} 1 & 1 \end{bmatrix}^T$ and make comparisons between the results:

$$R = \begin{bmatrix} 1 & 0.651 \\ 0.651 & 1 \end{bmatrix}$$

$$\mathbf{p} = \begin{bmatrix} 0.288 & 0.113 \end{bmatrix}^T, \quad \text{and} \quad \sigma_S^2 = 0.3$$

Hint: For the real-valued problem, the gradient of the MSE equation is given by

$$\frac{\partial}{\partial e_n} \text{MSE} = \nabla_n = -2\mathbf{p} + 2R\mathbf{e}_n.$$

(b) Plot the convergence paths of the equalizer coefficients for all three cases.

25.17 Chapter 17: Multiple access and the cellular principle

1. An analog cellular system has 250 duplex channels available (250 channels in each direction). To obtain acceptable transmission quality the relation between reuse distance (D) and cell radius (R) has to be at least $D/R = 7$. The cell structure is designed with a cell

radius of $R = 2$ km. During a busy hour the traffic per subscriber is on average one call of 2-min duration. The network setup is modeled as an Erlang-B loss system with the blocking probability limited to 3%.

(a) Calculate:
  i. The maximal number of subscribers per cell.
  ii. The capacity of the network in Erlangs/km^2. (Assume that the cell area is $A_{cell} = \pi R^2$.)

(b) The analog system above is modernized for digital transmission. As a consequence, the channel separation has to be doubled – i.e. only 125 duplex are now available. However, digital transmission is less sensitive to interference and acceptable quality is obtained for $D/R = 4$. How is the capacity of the network affected by this modernization (in terms of Erlangs/km^2).

(c) To increase the capacity of the network in B the cells are made smaller, with a radius of only $R = 1$ km. How much is capacity increased (in terms of Erlangs/km^2) and how many more base stations are required to cover the same area?

2. A system specifies a blocking level to be less than 5% for 120 users each with an activity level of 10%. When a user is blocked it is assumed to be cleared immediately – i.e., the system is an Erlang-B system. Assume two scenarios: (i) one operator and (ii) three operators. How many channels are needed for the two scenarios?

3. A system specifies a blocking level to be less than 5% for 120 users, each with an activity level of 10%. When a user is blocked it is assumed to be placed in an infinite queue – i.e., the system is an Erlang-C system. Assume two scenarios: (i) one operator and (ii) three operators.
(a) How many channels are needed for the two scenarios (compare with the previous problem)?
(b) What is the average waiting time if the average call duration is 5 min?

4. TDMA requires a temporal guard interval.
(a) The cell radius of a mobile system is specified as 3,000 m and the longest impulse response in the cell is measured as 10 μs. What is the minimum temporal guard interval needed to avoid overlapping transmissions?
(b) How is the temporal guard interval reduced in GSM?

5. Consider a cellular system of hexagonal structure and a reuse distance D (distance to the closest co-channel base station), a cell radius of R, and a propagation exponent η. It is assumed that all six co-channel BSs transmit independent signals with the same power as the base station in the studied cell.
(a) Show that the downlink-carrier-to-interference ratio is bounded by:

$$\left(\frac{C}{I}\right) > \frac{1}{6}\left(\frac{R}{D-R}\right)^{-\eta} \tag{25.32}$$

(b) Figure 25.9 shows an illustration of the worst case uplink scenario, where communication from MS 0 to BS 0 is affected by interference from other co-channel mobiles in first-tier co-channel cells. The worst interference scenario is when co-channel mobiles (MS 1 to MS 6) communicating with their respective base stations (BS 1 to BS 6) are at their respective cell boundaries, in the direction of BS 0. Calculate the C/I at BS 0, given that all mobiles transmit with the same power, and compare your expression with the one above.

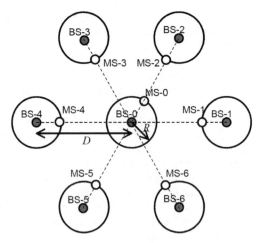

Figure 25.9 Cell structure.

6. Consider a system with a requirement for successful demodulation such that $SNIR = 7\,\text{dB}$ operating in an environment with 10-dB SNR. The received power at the MS is equal for both the signal of interest and the interferer. However, the interferer is suppressed by $10\,\text{dB}$ from, for example, spreading gain. Compute the maximum effective throughput in a slotted ALOHA system.

25.18 Chapter 18: Spread spectrum systems

1. Two MSs are communicating with the same BS. The information to be sent by the two MSs is

$$s_1 = \begin{bmatrix} 1 & -1 & 1 & 1 \end{bmatrix} \tag{25.33}$$

$$s_2 = \begin{bmatrix} -1 & 1 & -1 & 1 \end{bmatrix} \tag{25.34}$$

spread by spreading sequences c_1 and c_2. The radio channels between the MSs and BS are described by impulse responses h_1 and h_2.
    (a) Plot the received signal before and after despreading for a $h_1 = h_2 = [1]$ using PN spreading sequences c_1 and c_2 of length 4, 16, and 128. Find the information sent in the despreaded sequence.
    (b) Plot the received signal before and after despreading for $h_1 = h_2 = \begin{bmatrix} 1 & 0.5 & 0.1 \end{bmatrix}$ using PN-sequences c_1 and c_2 of length 4, 16, and 128. Find the information sent in the despreaded sequence.
    (c) Plot the received signal before and after despreading for $h_1 = \begin{bmatrix} 1 & 0.5 & 0.1 \end{bmatrix}$, $h_2 = \begin{bmatrix} 1 & 0 & 0.5 \end{bmatrix}$ using PN-sequences c_1 and c_2 of length 4, 16, and 128. Find the information sent in the despreaded sequence.
    (d) Repeat (a), (b), and (c) for Hadamard sequences of length 16.

2. Consider a frequency-hopping system with four possible carrier frequencies, and hopping sequence 1, 2, 3, 4. What are hopping sequences of length four that have only one collision with this sequence (for arbitrary integer shifts between sequences)?

3. Consider a system with frequency hopping plus simple code $(7, 4$ Hamming code) with generator matrix:

$$
\mathbf{G} = \begin{bmatrix} 1 & 0 & 0 & 0 & 1 & 1 & 0 \\ 0 & 1 & 0 & 0 & 1 & 0 & 1 \\ 0 & 0 & 1 & 0 & 0 & 1 & 1 \\ 0 & 0 & 0 & 1 & 1 & 1 & 1 \end{bmatrix} \tag{25.35}
$$

Assume BPSK modulation, and compute the BER with and without frequency hopping. An interleaver maps every symbol to alternating frequencies (seven are available). Every frequency is independently Rayleigh-fading. Let the receiver use hard decoding. Plot the BER as a function of average BER (use MATLAB).

4. Consider a Matched Front End Processor (MFEP) with Equivalent Low Pass (ELP) impulse response:

$$
f_{\mathrm{M}}(t) = \begin{cases} s^*(T_{\mathrm{s}} - t), & 0 < t < T_{\mathrm{s}} \\ 0, & \text{otherwise} \end{cases} \tag{25.36}
$$

where $s(t)$ denotes the ELP-transmitted signal having energy $2E_{\mathrm{s}}$, and T_{s} is the symbol duration. Show that for a slowly varying WSSUS channel, the correlation function of the MFEP output is:

$$
R_{\mathrm{y}}(t_1, t_2) = \begin{cases} \displaystyle\int_{-\infty}^{+\infty} P_{\mathrm{h}}(0, \tau) \widetilde{R}_{\mathrm{s}}^*(t_1 - T_{\mathrm{s}} - \tau) \widetilde{R}_{\mathrm{s}}(t_2 - T_{\mathrm{s}} - \tau) d\tau & \text{for } |t_2 - t_1| < \dfrac{2}{B_{\mathrm{s}}} \\ 0 & \text{otherwise} \end{cases} \tag{25.37}
$$

where

$$
\widetilde{R}_{\mathrm{s}}(t - T_{\mathrm{s}} - \tau) \triangleq \int_{0}^{+\infty} s(t - \alpha - \tau) f_{\mathrm{M}}(\alpha)\, d\alpha \tag{25.38}
$$

What is the autocorrelation of the noise?

5. Consider a channel with three taps that are Nakagami-m-fading, and have mean powers 0.6, 0.3, 0.1, and m-factors of 5, 2, and 1.
   (a) What is the diversity order – i.e., the slope of the BER versus SNR curves at high SNRs – when maximum ratio combining is applied?
   (b) Give a closed-form equation for the average BER of BPSK in such a channel.
   (c) Plot the BER as a function of average SNR, and compare it with pure Rayleigh fading (other parameters identical).

6. A CDMA handset is at the boundary of the cell, and thus in soft handover. It operates in a rich multipath environment (and thus sees a large number of resolvable MPCs). The shadowing standard deviation is $\sigma_{\mathrm{F}} = 5\,\mathrm{dB}$ and mean received SNR is $8\,\mathrm{dB}$. What is the probability of outage if the handset requires a 4-dB SNR to operate? *Hint*: for computing the distribution of the sum of log-normally distributed variables, convert to natural logarithms, and match the first and second moments of the desired approximation and of the given sum of powers.

7. Consider a CDMA system, operating at 1,800 MHz, with a cellsize of 1 km, circularly shaped cells, and users distributed uniformly in the cell area. The pathloss model is the following: free space up to a distance of 100 m, and $n = 4$ beyond that. In addition, Rayleigh fading is superimposed; neglect shadow fading. Let power control ensure that the signal received at the desired BS is constant at -90 dBm. Simulate the average power received from these handsets at the *neighboring* base station (use MATLAB).

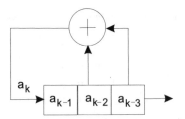

Figure 25.10   A maximal LFSR sequence generator.

8.   Consider a CDMA system with three users, each of which is spread using a PN-sequence. The spreading sequences are generated from the shift register of Fig. 25.10, where the initializations are [100], [110], and [101], respectively. Let the three transmitters send out sequences [1 −1 1 1], [−1 −1 −1 1], [−1 1 −1 1]. The gains from the transmitters to the receiver h_{1j} are 1, 0.6, and 11.3 and let there be no power control. Consider the noise sequence [generate 21 noise samples with variance 0.3]. Compute the received signal both before and after despreading. What bit sequences are detected for the three users? How do the results change when perfect power control is implemented? Use MATLAB for the simulation.

9.   MATLAB: Consider again the CDMA signal from Problem 8. Write a MATLAB program that performs zero-forcing multiuser detection, and one that performs serial interference cancellation. What are the detected signals in the two cases?

10. Derive the power-spectral density of TH-IR with 2-PPM and short spreading sequences – i.e., the duration of the spreading sequence is equal to the symbol duration.

25.19   Chapter 19: Orthogonal frequency division multiplexing (OFDM)

1.   Consider an eight-tone OFDM system transmitting a time domain baseband signal $s[n] = \{1\ \ 4\ \ 3\ \ 2\ \ 1\ \ 3\ \ 1\ \ 2\}$ over a channel with an impulse response of $h[n] = \{2\ \ 0\ \ 2\ \ -1\}$.
     (a) Draw the block diagram of a baseband OFDM system.
     (b) What is the minimum required length of the cyclic prefix?
     (c) Plot the signal vector for:
          i. $s[n]$;
          ii. $s_C[n] = s[n]$ with cyclic prefix;
          iii. $y_C[n] =$ after passing through the channel $h[n]$;
          iv. $y[n] =$ after cyclic prefix deletion;
          v. $Y[k] =$ after DFT;
          vi. $\hat{S}[k] = Y[k]/H[k]$;
          vii. $\hat{s}[n] =$ IDFT of $\hat{S}[k]$.

2.   Consider an OFDM system with mean output power normalized to unity. Let the system have a power amplifier that amplifies with a cutoff characteristic – i.e., amplifies linearly only between amplitude levels $-A_0$ and A_0, and otherwise emits levels $-A_0$ and A_0, respectively. How much larger than unity must A_0 be so that the probability of cutoff is less than (a) 10%, (b) 1%, (c) 0.1%?

3. Show that in the presence of insufficient cyclic prefix and intersymbol interference, the resulting signal can be described as:

$$\mathbf{Y}^{(i)} = \mathbf{Y}^{(i,i)} + \mathbf{Y}^{(i,i-1)} = \mathbf{H}^{(i,i)} \cdot \mathbf{X}^{(i)} + \mathbf{H}^{(i,i-1)} \cdot \mathbf{X}^{(i-1)} \tag{25.39}$$

where $\mathbf{Y}^{(i,i-1)}$ is the ISI term and $\mathbf{Y}^{(i,i)}$ contains the desired data disturbed by ICI. Derive $\mathbf{H}^{(i,i)}$ as described in Eq. (19.19), and obtain equations for the ISI matrix $\mathbf{H}^{(i,i-1)}$.

4. Consider an eight-point OFDM system with Walsh–Hadamard spreading, operating in a channel with impulse response $h = [0.5 + 0.2j, \ -0.6 + 0.1j, \ 0.2 - 0.25j]$ and $\sigma_n^2 = 0.1$. Assume that the system prepends a four-point cyclic prefix:
   (a) Assuming data vector [10110001] and BPSK modulation, sketch:
      i. the data vector;
      ii. the WH-transformed signal;
      iii. the transmit signal (after Fourier transform and prepending of the cyclic prefix);
      iv. the received signal (disregarding noise);
      v. the received signal after Fourier transformation and zero-forcing equalization;
      vi. the received signal after Walsh–Hadamard transformation.
   (b) What is the noise enhancement for a zero-forcing receiver?
   (c) Simulate the BER of the system using zero-forcing equalization? Use MATLAB for the simulations.

5. Consider an OFDM system with a 128-point FFT where each OFDM symbol is 128-μs-long. It operates in a slowly time-variant (i.e., negligible Doppler), frequency-selective channel. The PDP of the channel is $P_h = \exp(-\tau/16 \, \mu s)$, and the average SNR is 8 dB. Compute the duration of the cyclic prefix that maximizes the SINR at the receiver.

6. Consider a channel with $\sigma_n^2 = 1$, $\alpha_n^2 = 1, 0.1, 0.01$, and total power $\sum P_n = 100$. Compute the capacity when using waterfilling. What (approximate) capacity can be achieved if the transmitter can only use BPSK?

7. Consider an OFDM system with coding across the tones, where the code is a block code with Hamming distance $d_H = 7$.
   (a) If all tones that carry bits of this code fade independently, what is the diversity order that can be achieved?
   (b) In a channel with a 5-μs rms delay spread and exponential power delay profile, what spacing between these tones is necessary so that fading is independent? Approximate "independent" as "correlation coefficient less than 0.3".

8. Consider an FDMA system, using raised-cosine pulses ($\alpha = 0.35$) on each carrier, and using BPSK modulation.
   (a) What is the spectral efficiency if signals on carriers are to be completely orthogonal?
   (b) What is the spectral efficiency of an OFDM system with BPSK if the number of carriers is very large (i.e., no guardbands required)?

25.20 Chapter 20: Multiantenna systems

1. State three advantages of using a smart antenna system, as opposed to the conventional single-antenna system.

2. Consider the uplink of a CDMA system with spreading factor $M_C = 128$, and $SIR_{threshold} = 6 \, dB$. How many users could be served in the system if the BS had only a single antenna? Now, let the BS have $N_r = 2, 4$, or 8 antenna elements. How many users can be served according to the simplified equations of Section 20.1.1? By how much is that

number reduced if the angular power spectrum of the desired user is Laplacian with $APS(\phi) = (6/\pi)\exp(-|\phi - \phi_0|/(\pi/12))$, for $\phi_0 = \pi/2$?

3. A MIMO system can be used for three different purposes, one of which is ground-breaking and has contributed most to its popularity. List all three and explain its most popular use in greater detail.

4. For the following realization of a channel for a 3×3 MIMO system, calculate the channel capacity with and without channel knowledge at the TX for a mean SNR per receive branch ranging from 0 dB to 30 dB, in steps of 5 dB. Comment on the results:

$$\mathbf{H} = \begin{bmatrix} -0.0688 - j1.1472 & -0.9618 - j0.2878 & -0.4980 + j0.5124 \\ -0.5991 - j1.0372 & 0.5142 + j0.4967 & 0.6176 + j0.9287 \\ 0.2119 + j0.4111 & 1.1687 + j0.5871 & 0.9027 + j0.4813 \end{bmatrix}$$

5. Section 20.2.1 states that the number of possible datastreams for spatial multiplexing is limited by the number of transmit/receive antennas (N_t, N_r) and the number of significant scatterers N_S. Using an appropriate channel model, demonstrate the limitation due to N_S for the case of a 4×4 MIMO system. The four-element arrays are spaced apart by 2λ.

6. The Laplacian function:

$$f(\theta) = \frac{1}{\sqrt{2}\sigma_S} e^{-\sqrt{2}|\theta|/\sigma_S}, \quad \theta \in (-\pi, \pi] \tag{25.40}$$

has been proposed as a suitable power angular spectrum at the BS. On the other hand, due to the richness of scatterers in its surroundings, a uniform angular distribution is often assumed for the MS; that is:

$$f(\theta) = \frac{1}{2\pi}, \quad \theta \in (-\pi, \pi] \tag{25.41}$$

Using the Kronecker model, study the effect of angular spread (at the BS) on the 10% outage capacity of the channel for different array spacings (at the BS) for the case of a 4×4 MIMO system. The SNR is assumed to be 20 dB and uniform linear arrays are used. Arrays at the MS are spaced apart by $\lambda/2$.

7. Consider a downlink scenario where a 3×3 MIMO system is used to boost the throughput of an indoor wireless LAN. Uniform linear arrays (spaced apart by $\lambda/2$) are used at both the user's mobile station (a PDA) and a wall-mounted base station. Calculate the additional transmit power needed (in percent) to compensate for the loss of expected (or mean) channel capacity when the user moves from a Line Of Sight (LOS) region into a Non LOS (NLOS) region (see Fig. 25.11).

In the LOS region, the user receives 6 mW of power in the LOS path and 3 mW in all the other paths combined. Noise power at the receiver is constant at 1 mW. Here we assume

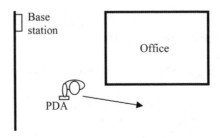

Figure 25.11   User moving from LOS to NLOS region.

that LOS consists of a single path and the statistics of NLOS components do not vary when the receiver moves from the LOS to the NLOS scenario. Heavy scattering in the environment ensures that NLOS paths follow a Rayleigh distribution. For simplicity, pathloss is neglected and equal power transmission is assumed. Repeat the problem for 5% outage capacity to be maintained. Comment on the difference from the expected capacity case.

8. Generate channel capacity cdf's for an iid MIMO channel with $N_R = 4$ and $N_T = [1, \ldots, 8]$. The channel matrix is normalized as $\|\mathbf{H}\|_F^2 = N_r N_t$.
    (a) What is the 10%, 50%, and 90% outage capacity gain compared with a SISO system, assuming that no CSI is available at the TX?
    (b) Comment on the gain from using more transmit antennas than receive antennas?

9. Use the Kronecker model and plot the 10% outage capacity as a function of $r = 0, 0.1, \ldots, 1$ for marginal correlation matrices of:

$$\mathbf{R}_R = \mathbf{R}_T = \begin{bmatrix} 1 & r & r^2 \\ r & 1 & r \\ r^2 & r & 1 \end{bmatrix} \tag{25.42}$$

10. Show that:

$$\sum_{k=1}^{M} \log\left(1 + \frac{\bar{\gamma}}{N_t} \sigma_k^2\right) = \log\det\left(\mathbf{I}_{N_r} + \frac{\bar{\gamma}}{N_t}\mathbf{H}\mathbf{H}^\dagger\right) \tag{25.43}$$

where σ_k, $\bar{\gamma}$, N_t, and M are the kth singular value of $\mathbf{H}$, the average received SNR, the number of transmit elements, and the number of nonzero singular values of $\mathbf{H}$, respectively.

25.21 Chapter 21: GSM – Global System for Mobile communications

1. Consider a GSM operator that has licenses for both 900- and 1,800-MHz bands. How should they be used in the buildup of the network?

2. Which is true: the following combinations of devices *must* be bought from the same vendor: (i) BTS–MS, (ii) BTS–BSC, (iii) BSC–MSC?

2. Assuming that directions of arrival are uniformly distributed at the MS, how large is the correlation coefficient (for a GSM 1,800 system) between the channel in the middle and the end of the burst when the MS moves at 250 km/h? How large is the correlation coefficient between the channels at the beginning and end of the burst?

3. Consider a GSM system at 1,900 MHz operating in a Typical Urban (TU) environment. What is the correlation coefficient between two channels that are separated by (i) one carrier frequency, (ii) one auctioned frequency block (5 MHz), or (iii) one duplex frequency?

4. Explain the difference between the fast and slow associated control channel. When are they used?

5. Explain the difference between block FEC for voice and control data. Why are different schemes used?

6. What is the common property of all midambles? Why are different midambles defined?

7. A user has repeatedly entered the wrong PIN code into his phone, so that it is now blocked. Is there a possibility to unblock the phone?

8. Consider the following billing problem: subscriber A is Swedish but temporarily based in Denmark. Subscriber B is in Finland. Subscriber C is in France, but forwards all calls to subscriber D in England. A calls B, and wants to conference in subscriber C. As C has call forwarding on, the call goes to D, and that subscriber is then active in the phone conference. Who pays which fees?

9. Compare the required SNR to achieve 10^{-2} BER of uncoded GMSK and 8-PSK in an AWGN channel. Note that GMSK is used for GSM data transmission, while 8-PSK is used in EDGE.

25.22 Chapter 22: IS-95 and CDMA 2000

1. What are the spreading factors for the uplink and the downlink for rate set 1 and rate set 2 in IS-95?

2. What is the maximum speed of the MS that the power control of IS-95 can follow in the (i) 800-MHz band and (ii) the 1,900-MHz band?

3. What is the typical percentage of total power transmitted from a BS that a pilot in IS-95 is using? If that percentage is halved, by what percentage could the cellsize be increased? How would that affect capacity?

4. What code rates are used for convolutional codes in IS-95? Where are they applied?

5. How are physical channels separated in IS-95? Compare this approach with GSM.

6. How is the transmission of low data rates handled in IS-95 in the uplink and downlink? Comment on the difference.

7. How is frequency diversity obtained in the multicarrier mode of CDMA 2000 in the uplink and downlink? What are the advantages and drawbacks of the two approaches?

25.23 Chapter 23: WCDMA/UMTS

1. What service classes are defined in UMTS? For what purposes are they used, and how does that affect admissible BERs and delays?

2. Two operaters are building up systems in two adjacent 5-MHz bands. Let an MS in operator A's system be at the cell boundary, operating at 3 dB above sensitivity level. What is the minimum required pathloss from the BS of operator B in order to still function? Assume that all antennas have omniradiation patterns.

3. Compare the way pilot signals are transmitted in WCDMA/UMTS to the method of IS-95.

4. What is the maximum data rate in an UMTS uplink? How is that achieved?

5. How often is feedback information transmitted? Assuming radiation incident only in the horizontal plane, and a uniformly distributed azimuthal power spectrum, what is the maximum velocity admissible so that envelope correlation between channel-state observation and its use is higher than 0.9? How does this result change when radiation is incident isotropically (both in azimuth and elevation)?

25.24 Chapter 24: Wireless Local Area Networks

1. In 802.11a, what is the loss of spectral efficiency due to (i) not all subcarriers carrying data, (ii) cyclic prefix, (iii) training sequence and signaling field (assume that 16 OFDM symbols are transmitted).

2. What are the maximum data rates of 802.11, 802.11b, and 802.11a.

3. What are the mechanisms that 802.11a uses to achieve high maximum throughput?

4. Explain the principle of complementary code keying. What is the difference from block encoding and QPSK modulation?

5. In 802.11b, how long does a PSDU have to be in order to make spectral efficiency of a PPDU at least 80% when using (i) 1-Mbit/s, (ii) 11-Mbit/s data transmission?

References

Abdi et al. 2000 A. Abdi, K. Wills, H. A. Barger, M. S. Alouini, and M. Kaveh, "Comparison of the level crossing rate and average fade duration of Rayleigh, Rice and Nakagami fading models with mobile channel data," *Proc. VTC*, Fall 2000, 1850–1857 (2000).

Abramowitz and Stegun 1965 M. Abramowitz and I. A. Stegun, *Handbook of Mathematical Functions*, National Bureau of Standards, Washington, DC (1965).

Abramson 1970 N. Abramson, "The ALOHA system: Another alternative for computer communications," *Proc. Fall 1970 AFIPS Computer Conference* (1970).

Adachi and Parsons 1989 F. Adachi and J. D. Parsons, "Error rate performance of digital FM mobile radio with postdetection diversity," *IEEE Transactions on Communications*, **37**, 200–210 (1989).

Adachi and Ohno 1991 F. Adachi and K. Ohno, "BER performance of QDPSK with postdetection diversity reception in mobile radio channels," *IEEE Transactions on Vehicular Technology*, **40**, 237–249 (1991).

Alamouti 1998 S. M. Alamouti, "A simple transmit diversity technique for wireless communications," *IEEE Journal of Selected Areas Communications*, **16**, 1451–1458 (1998).

Almers et al. 2003 P. Almers, F. Tufvesson, and A. F. Molisch, "Measurement of keyhole effect in wireless multiple-input–multiple-output (MIMO) channels," *IEEE Communications Lett.*, **7**, 373–375 (2003).

Alouini and Goldsmith 1999 M. S. Alouini and A. J. Goldsmith, "Area spectral efficiency of cellular mobile radio systems," *IEEE Transactions on Vehicular Technology*, **48**, 1047–1066 (1999).

Andersen 1991 J. B. Andersen, "Propagation parameters and bit errors for a fading channel," *Proc. Commsphere '91*, Paper 8.1 (1991).

Andersen 1997 J. B. Andersen, "UTD multiple-edge transition zone diffraction," *IEEE Transactions on Antennas Propagation*, **45**, 1093–1097 (1997).

Andersen 2000 J. B. Andersen, "Antenna arrays in mobile communications: Gain, diversity, and channel capacity," *IEEE Antennas Propagat. Mag.*, **42**(April), 12–16 (2000).

Andersen 2002 J. B. Andersen, "Power distributions revisited," *COST 273*, TD(02)004 (2002).

Andersen and Hansen 1977 J. B. Andersen and F. Hansen, "Antennas for VHF/UHF personal radio: A theoretical and experimental study of characteristics and performance," *IEEE Transactions on Vehicular Technology*, **VT-26**, 349–357 (1977).

Andersen et al. 1990 J. B. Andersen, S. L. Lauritzen, and C. Thommesen, "Distribution of phase derivatives in mobile communications," *Proceedings of the IEE*, Part H, **137**, 197–204 (1990).

Andersen et al. 1995 J. B. Andersen, T. S. Rappaport, and S. Yoshida, Propagation measurements and models for wireless communications channels, *IEEE Communications Magazine*, **33**(1), 42–49 (1995).

Anderson 2003 H. R. Anderson, *Fixed Broadband Wireless System Design*, John Wiley & Sons, Ltd. (2003).

Anderson 2005 J. B. Anderson, *Digital Transmission Engineering* (2nd edn.), Prentice Hall (2005).

Anderson et al. 1986 J. B. Anderson, T. Aulin, and C. E. Sundberg, *Digital Phase Modulation*, Plenum Press (1996).

Andren et al. 2000 C. Andren, K. Halford, and M. Webster, "CCK: The new IEEE 802.11 standard for 2.4 GHz wireless LANs," *Proc. International IC, Taipei*, pp. 25–30 (2000).

Andrews et al. 2001   M. R. Andrews, P. P. Mitra, and R. de Carvalho, "Tripling the capacity of wireless communications using electromagnetic polarization," *Nature*, **409**, 316–318 (2001).

Annamalai et al. 2000   A. Annamalai, C. Tellambura, and V. K. Bhargava, "A general method for calculating error probabilities over fading channels," *Proc. Int. Conf. Communications 2000*, pp. 36–40 (2000).

Ariyavisitakul 2000   S. L. Ariyavisitakul, "Turbo space-time processing to improve wireless channel capacity," *IEEE Transactions Commun.*, **48**, 1347–1359 (2000).

Ashtiani et al. 2003   F. Ashtiani, J. A. Salehi, and M. R. Aref, "Mobility modeling and analytical solution for spatial traffic distribution in wireless multimedia networks," *IEEE Journal on Selected Areas in Communications*, **21**, 1699–1709 (2003).

Asplund et al. 2004   H. Asplund, A. A. Glazunov, A. F. Molisch, K. I. Pedersen, and M. Steinbauer, "The COST 259 directional channel model – II. Macrocells," *IEEE Transactions Wireless Communications*, submitted.

Atal and Hanauer 1971   B. S. Atal and S. L. Hanauer, "Speech analysis and synthesis by linear prediction of the speech wave," *Journal of the Acoustical Society of America*, **50**(2), 637–655 (1971).

Atal and Remde 1982   B. Atal and J. Remde, "A new model of LPC excitation for producing natural-sounding speech at low bit rates," *Proc. IEEE International Conference on Acoustics, Speech, and Signal Processing, ICASSP '82, Paris*, pp. 614–617 (1982).

Atal and Schroeder 1984   B. S. Atal and M R. Schroeder, "Stochastic coding of speech signals at very low bit rates," *Proc. IEEE Int. Conf. Communications ICC '84, Amsterdam*, pp. 1610–1613 (1984).

Badsberg et al. 1995   M. Badsberg, J. Bach Andersen, and P. Mogensen, "Exploitation of the terrain profile in the Hata model," *COST 231*, TD(95)9 (1995).

Bahai et al. 2004   A. R. S. Bahai, B. R. Saltzberg, and M. Ergen, *Multi-carrier Digital Communications: Theory and Applications of OFDM* (2nd edn.), Springer-Verlag (2004).

Bahl et al. 1974   L. R. Bahl, J. Cock, F. Jelink, and J. Raviv, "Optimal decoding of linear codes for minimum symbol error rate," *IEEE Transactions Information Theory*, **20**, 248–287 (1974).

Balanis 1997   C. A. Balanis, *Antenna Theory: Analysis and Design* (2nd edn.), John Wiley & Sons, Inc. (1997).

Barry et al. 2003   J. R. Barry, D. G. Messerschmidt, and E. A. Lee, *Digital Communications* (3rd edn.), Kluwer (2003).

Bass and Fuks 1979   F. G. Bass and I. M. Fuks, *Wave Scattering from Statistically Rough Surfaces*, Pergamon (1979).

Belfiore and Park 1977   C. A. Belfiore and J. H. Park, "Decision feedback equalization," *Proceedings of the IEEE*, **67**, 1143–1156 (1977).

Bello 1963   P. A. Bello, "Characterization of randomly time-variant linear channels," *IEEE Transactions Communications*, **11**, 360–393 (1963).

Bello and Nelin 1963   P. Bello and B. D. Nelin, "The effect of frequency selective fading on the binary error probabilities of incoherent and differentially coherent matched filter receivers," *IEEE Transactions Communications*, **11**, 170–186 (1963).

Benedetto and Biglieri 1999   S. Benedetto and E. Biglieri, *Principles of Digital Transmission: With Wireless Applications*, Kluwer (1999).

Berger 1971   T. Berger, *Rate Distortion Theory: A Mathematical Basis for Data Compression*, Prentice Hall (1971).

Bergljung 1994   C. Bergljung, "Diffraction of electromagnetic waves by dieletric wedges," Ph.D. thesis, Lund Institute of Technology, Lund, Sweden (1994).

Berrou et al. 1993   C. Berrou, A. Glavieux, and P. Thitimajshima, "Near Shannon limit error-correcting coding and decoding: Turbocodes," *Proc. IEEE International Conference on Communications, ICC '93* (1993).

Bertoni 2000   H. L. Bertoni, *Radio Propagation for Modern Wireless Systems*, Prentice Hall (2000).

Bettstetter et al. 1999   C. Bettstetter, H. J. Voegel, and J. Eberspecher, "GSM Phase 2+ General Packet Radio Service GPRS: Architecture, protocols, and air interface," *IEEE Communications Surveys*, Third Quarter 1999, **2**(3) (1999).

Bi et al. 2001   Q. Bi, G. L. Zysman, and H. Menkes, "Wireless mobile communications at the start of the 21st century," *IEEE Communications Magazine*, **39**(1), 110–116 (2001).

Biglieri et al. 1991   E. Biglieri, *Introduction to Trellis-coded Modulation with Applications*, Macmillan (1991).

Blanz and Jung 1998   J. J. Blanz and P. Jung, "A flexibly configurable spatial model for mobile radio channels," *IEEE Transactions Communications*, **46**, 367–371 (1998).

Blaunstein 1999 N. Blaunstein, *Radio Propagation in Cellular Networks*, Artech House (1999).

Bottomley et al. 2000 G. E. Bottomley, T. Ottosson, and Y. P. E. Wang, "A generalized RAKE receiver for interference suppression," *IEEE Journal on Selected Areas in Communications*, **18**, 1536–1545 (2000).

Bowman et al. 1987 J. J. Bowman (ed.), *Electromagnetic and Acoustic Scattering by Simple Shapes*, Hemisphere (1987).

Braun and Dersch 1991 W. R. Braun and U. Dersch, "A physical mobile radio channel model," *IEEE Transactions on Vehicular Technology.*, **40**, 472–482 (1991).

Brennan and Cullen 1998 C. Brennan and P. Cullen, P., "Tabulated interaction method for UHF terrain propagation problems," *IEEE Transactions on Antennas Propagation*, **46**, 738–739 (1998).

Buehler et al. 1993 H. Buehler, E. Bonek, and B. Nemsic, "Estimation of heavy time dispersion for mobile radio channels using a path tracing concept," *Proc. 43rd IEEE Vehicular Technology Conference, Secaucus, NJ*, pp. 257–260 (1993).

Burr 2001 A. Burr, *Modulation and Coding for Wireless Communications*, Prentice Hall (2001).

Cai and Giannakis 2003 X. Cai and G. B. Giannakis, "Bounding performance and suppressing intercarrier interference in wireless mobile OFDM," *IEEE Transactions Communications*, **51**, 2047–2056 (2003).

Cai and Goodman 1997 J. Cai and D. J. Goodman, "General packet radio service in GSM," *IEEE Communications Magazine*, **35**(10), 122–131 (1997).

Calcev et al. 2004 G. Calcev et al., "A Wideband Spatial Channel Model for System-Wide Simulations," *IEEE Transactions on Vehicular Technology*, submitted.

Callaway et al. 2002 E. Callaway, P. Gorday, L. Hester, J. A. Gutierrez, M. Naeve, B. Heile, and V. Bahl, "Home networking with IEEE 802.15.4: A developing standard for low-rate wireless personal area networks," *IEEE Communications Magazine*, **40**(8), 70–77 (2002).

Cardieri and Rappaport 2001 P. Cardieri and T. Rappaport, "Statistical analysis of co-channel interference in wireless communications systems," *Wireless Commun. Mobile Computing*, **1**, 111–121 (2001).

Catreux et al. 2001 S. Catreux, P. F. Driessen, and L. J. Greenstein, "Attainable throughput of an interference-limited multiple-input multiple-output cellular system," *IEEE Transactions Commun.*, **48**, 1307–1311 (2001).

Cattermole 1986 K. W. Cattermole, *Mathematical Foundations for Communication Engineering: Statistical Analysis and Finite Structures*, Halsted Press (1986).

Chan 1992 G. K. Chan, "Effects of sectorization on the spectrum efficiency of cellular radio systems," *IEEE Transactions Veh. Technol.*, **41**, 217–225 (1992).

Chang 1966 R. W. Chang, "Synthesis of band-limited orthogonal signals for multichannel data transmission," *Bell Systems Technical Journal*, **46**, 1775–1796 (1966).

Chatschik 2001 B. Chatschik, "An overview of the Bluetooth wireless technology," *IEEE Communications Magazine*, **39**(12), 86–94 (2001).

Chen and Chuang 1998 Y. Chen and J. C. I. Chuang, "The effects of time-delay spread on unequalized TCM in a portable radio environment," *IEEE Transactions on Vehicular Technology*, **46**, 375–380 (1998).

Chennakeshu and Saulnier 1993 S. Chennakeshu and G. J. Saulnier, "Differential detection of $\pi/4$-shifted-DQPSK for digital cellular radio," *IEEE Transactions on Vehicular Technology*, **42**, 46–57 (1993).

Choi and Stark 2002 J. D. Choi and W. E. Stark, "Performance of ultra-wideband communications with suboptimal receivers in multipath channels," *IEEE Journal on Selected Areas in Communications*, **20**, 1754–1766 (2002).

Choi et al. 2001 Y.-S. Choi, P. J. Voltz, and F. Cassara, "On channel estimation and detection for multicarrier signals in fast and frequency selective Rayleigh fading channel," *IEEE Transactions Communications*, **49**, 1375–1387 (2001).

Chollet et al. 2005 G. Chollet, A. Esposito, M. Faundez-Zanuy, and M. Marinaro, *Nonlinear Speech Modeling and Applications*, Springer–Verlag [*Lecture Notes in Computer Science*, No. 3445] (2005).

Chuah et al. 2002 C. N. Chuah, D. Tse, J. M. Kahn, and R. Valenzuela, "Capacity scaling in MIMO wireless systems under correlated fading," *IEEE Transactions Inform. Theory*, **48**, 637–650 (2002).

Chuang 1987 J. Chuang, "The effects of time delay spread on portable radio communications channels with digital modulation," *IEEE Journal on Selected Areas in Communications*, **5**, 879–888 (1987).

Cimini 1985 L. J. Cimini, "Analysis and simulation of a digital mobile channel using orthogonal frequency division multiplexing," *IEEE Transactions on Communications*, **33**, 665–675 (1985).

Clark 1998 M. V. Clark, "Adaptive frequency-domain equalization and diversity combining for broadband wireless communications," *IEEE Journal on Selected Areas in Communications*, **16**, 1385–1395 (1998).

Clarke 1968   R. Clarke, "A statistical theory of mobile radio reception," *Bell System Technical Journal.*, **47**, 957–1000 (1968).

CME 20 1994   Ericsson, *Course Handouts for Course CME 20, 1994*, Ericsson, Vienna (1994).

Collin 1985   R. E. Colling, *Antennas and Radiowave Propagation*, McGraw-Hill (1985).

Collin 1991   R. E. Collin, *Field Theory of Guided Waves*, IEEE Press (1991).

COST 231   E. Damosso und L. Correira, *Digital Mobile Radio: The View of COST 231*, European Union (1999).

Coursey 1999   C. C. Coursey, *Understanding Digital PCS: The TDMA Standard*, Artech House (1999).

Cox 1972   D. C. Cox, "Delay-Doppler characteristics of multipath propagation at 910 MHz in a suburban mobile radio environment," *IEEE Transactions on Antennas Propagation*, **20**, 625–635 (1972).

Cramer et al. 2002   R. J. Cramer, R. A. Scholtz, and M. Z. Win, "Evaluation of an Ultra-Wide-Band propagation channel," *IEEE Transactions on Antennas Propagation*, **50**, 541–550 (2002).

Crohn et al. 1993   I. Crohn, G. Schultes, R. Gahleitner, and E. Bonek, "Irreducible Error Performance of a Digital Portable Communication System in a Controlled Time-Dispersion Indoor Channel," *IEEE Journal on Selected Areas in Communications*, **11**, 1024–1033 (1993).

Cullen et al. 1993   P. J. Cullen, P. C. Fannin, and A. Molina, "Wide-band measurement and analysis techniques for the mobile radio channel," *IEEE Transactions on Vehicular Technology*, **42**, 589–603 (1993).

Dam et al. 1999   H. Dam, M. Berg, R. Bormann, M. Frerich, F. Ahrens, T. Henß, S. Andersson, and D. Almquist, "Functional test of adaptive antenna base stations for GSM," *3rd EPMCC Eur. Personal Mobile Communications Conf.*, Paris (1999).

Damosso and Correia 1999   E. Damosso and L. Correia (eds.), *Digital Mobile Communications: The View of COST 231*, European Union (1999).

Davey 1999   M. C. Davey, "Error correction using low-density parity check codes," Ph.D thesis, Cambridge University (1999).

David and Benkner 1997   K. David and T. Benkner, *Digital Mobile Radio Systems*, Teubner (1996) [in German].

David and Nagaraja 2003   H. A. David and H. N. Nagaraja, *Order Statistics*, John Wiley & Sons (2003).

de Santo and Brown 1986   J. A. de Santo and G. S. Brown, *Progress in Optics* (Vol 23, edited by E. Wolf), North-Holland (1986).

de Weck 1992   J. P. de Weck, "Real-time characterization of wideband mobile radio channels," Dissertation, Technical University Vienna (1992).

Deller et al. 2000   J. R. Deller, Jr., J. H. Hansen, and J. G. Proakis, *Discrete-time Processing of Speech Signals*, IEEE Press (2000).

Deygout 1966   J. Deygout, "Multiple knife edge diffraction of microwaves,"*IEEE Transactions on Antennas Propagation*, **14**, 480–489 (1966).

diBenedetto et al. 2005   M. G. diBenedetto, T. Kaiser, A. F. Molisch, I. Oppermann, C. Politano, and D. Porcino, *UWB Communication Systems: A Comprehensive Overview*, EURASIP Publishing (2005).

Dietrich et al. 2001   C. B. Dietrich, K. Dietze, J. R. Nealy, and W. L. Stutzman, "Spatial, polarization, and pattern diversity for wireless handheld terminals," *IEEE Transactions on Antennas Propagation*, **49**, 1271–1281 (2001).

Diggavi et al. 2004   S. N. Diggavi, N. Al-Dhahir, A. Stamoulis, and A. R. Calderbank, "Great expectations: The value of spatial diversity in wireless networks," *Proceedings of the IEEE*, **92**, 219–270 (2004).

Dinan and Jabbari 1998   E. H. Dinan and B. Jabbari, "Spreading codes for direct sequence CDMA and wideband CDMA cellular networks," *IEEE Communications Magazine*, **36**(9), 48–54 (1998).

Divsalar and Simon 1990   D. Divsalar and M. K. Simon, "Multiple-symbol differential detection of MPSK," *IEEE Transactions Communications*, **38**, 300–308 (1990).

Dixon 1994   R. C. Dixon, *Spread Spectrum Systems with Commercial Applications*, John Wiley & Sons, Inc. (1994).

Doufexi et al. 2002   A. Doufexi, S. Armour, M. Butler, A. Nix, D. Bull, J. McGeehan, and P. Karlsson, "A comparison of the HIPERLAN/2 and IEEE 802.11a wireless LAN standards," *IEEE Communications Magazine*, **40**(5), 172–180 (2002).

Duel-Hallen et al. 1995   A. Duel-Hallen, J. Holtzman, and Z. Zvonar, "Multiuser detection for CDMA systems," *IEEE Personal Communications Magazine*, **2**(2), 46–58 (1995).

Dunlop et al. 1999   J. Dunlop, D. Girma, and J. Irvine, *Digital Mobile Communications and the TETRA System*, John Wiley & Sons, Ltd. (1999).

Durgin 2003   G. Durgin, *Space-time Wireless Channels*, Cambridge University Press, 2003.

Eberspaecher et al. 2001 J. Eberspaecher, H. J. Voegel, and C. Bettstetter, *GSM Switching, Services, and Protocols*, John Wiley & Sons, Ltd. (2001).

Edfors et al. 1998 O. Edfors, M. Sandell, J. J. van de Beek, S. K. Wilson, and P. O. Borjesson, "OFDM channel estimation by singular value decomposition," *IEEE Transactions Communications*, **46**, 931–939 (1998).

Edfors et al. 2000 O. Edfors, M. Sandell, J. J. van de Beek, S. K. Wilson, and P. O. Börjesson, "Analysis of DFT-based channel estimators for OFDM," *Personal Wireless Communications*, **12**, 55–70 (January 2000).

Eggers 1998 P. C. F. Eggers, "Generation of base station DOA distributions by Jacobi transformation of scattering areas," *El. Lett.*, **34**, 24–26 (1998).

Eklund et al. 2002 C. Eklund, R. B. Marks, K. L. Stanwood, and S. Wang, "IEEE standard 802.16: A technical overview of the WirelessMANTM air interface for broadband wireless access," *IEEE Communications Magazine*, **40**(6), 98–107 (2002).

Epstein and Petersen 1953 J. Epstein and D. W. Peterson, "An experimental study of wave propagation at 850 MC," *Proceedings of the IEEE*, **41**, 595–611 (1953).

Erätuuli and Bonek 1997 P. Erätuuli and E. Bonek, "Diversity arrangements for internal handset antennas," *8th IEEE International Symposium on Personal, Indoor and Mobile Radio Communications (PIMRC '97), Helsinki, September 1–4*, pp. 589–593 (1997).

Erceg et al. 2004 V. Erceg et al., *TGn Channel Models* (IEEE Document 802.11-03/940r4), available online at *www.802wirelessworld.com* (2004).

Ertel et al. 1998 R. B. Ertel, P. Cardieri, K. W. Sowerby, T. S. Rappaport, and J. H. Reed, "Overview of spatial channel models for antenna array communication systems," *IEEE Personal Communications*, **5**(1), 10–22 (1998).

ETSI 1992 ETS 300 175-1: "Radio Equipment and Systems (RES); Digital European Cordless Telecommunications (DECT) Common interface Part 1: Overview," Part 2: Physical Layer, ETSI (October 1992).

ETSI 1998 ETSI, "Terminal and Smart Cards Concepts, UMTS 22.07," and "General UMTS Architecture, UMTS 23.0" (1998).

Failli 1989 E. Failli, (ed.), *Digital Land Mobile Radio Communications: COST 207*, European Union (1989).

Falconer et al. 1995 D. D. Falconer, F. Adachi, and B. Gudmundson, "Time division multiple access methods for wireless personal communications," *IEEE Communications Magazine*, **33**(1), 50–57 (1995).

Falconer et al. 2002 D. Falconer, S. L. Ariyavisitakul, A. Benyamin-Seeyar, and B. Eidson, "Frequency domain equalization for single-carrier broadband wireless systems," *IEEE Communications Magazine*, **40**(4), 58–66 (2002).

Fant 1970 G. Fant, *Acoustic Theory of Speech Production* (2nd edn.), Mouton (2000).

Featherstone and Molkdar 2002 W. Featherstone and D. Molkdar, "Capacity benefits of GPRS coding schemes CS-3 and CS-4," *3G Mobile Communications Technologies*, 287–291 (2002).

Feldbauer et al. 2005 C. Feldbauer, G. Kubin, and W. B. Kleijn, "Anthropomorphic coding of speech and audio: A model inversion approach," *EURASIP Journal on Applied Signal Processing*, **9**, 1334–1349 (2005).

Fehlhauer et al. 1993 T. Fehlhauer, P. W. Baier, W. König, and W. Mohr, "Optimized wideband system for unbiased mobile radio channel sounding with periodic spread spectrum signals," *IEICE Transactions on Communications*, **E76-B**, 1016–1029 (1993).

Felsen and Marcuvitz 1973 L. B. Felsen and V. Marcuvitz, *Radiation and Scattering of Waves*, Prentice Hall (1973).

Fleury 1990 B. Fleury, "Characterisierung von Mobil- und Richtfunkkanälen mit schwach stationären Fluktuationen und unkorrelierter Streuung (WSSUS)," Dissertation, ETH Zuerich, Switzerland (1990) [in German].

Fleury 1996 B. H. Fleury, "An uncertainty relation for WSS processes and its application to WSSUS systems," *IEEE Transactions Communications*, **44**, 1632–1634 (1996).

Fleury 2000 B. H. Fleury, "First- and second-order characterization of direction dispersion and space selectivity in the radio channel," *IEEE Transactions Information Theory*, **46**, 2027–2044 (2000).

Fleury et al. 1999 B. H. Fleury, M. Tschudin, R. Heddergott, D. Dahlhaus, and K. I. Pedersen, "Channel parameter estimation in mobile radio environments using the SAGE algorithm," *IEEE Journal on Selected Areas in Communications*, **17**, 434–450 (1999).

Foschini 1996 G. J. Foschini, "Layered space-time architecture for wireless communication in a fading environment when using multi-element antennas," *Bell Lab Techn. J.*, Autumn, 41–59 (1996).

Foschini and Gans 1998 G. J. Foschini and M. J. Gans, "On limits of wireless communications in a fading environment when using multiple antennas," *Wireless Personal Communications*, **6**, 311–335 (1998).

Foschini et al. 2003 G. J. Foschini, D. Chizhik, M. J. Gans, C. Papadias, and R. A. Valenzuela, "Analysis and performance of some basic space-time architectures," *IEEE Journal on Selected Areas in Communications*, **21**, 303–320, (2003).

Frullone et al. 1996 M. Frullone, G. Riva, P. Grazioso, and G. Falciasecca, "Advanced planning criteria for cellular systems" *IEEE Personal Communications*, **3**(6), 10–15 (1996).

Fuhl and Molisch 1996 J. Fuhl and A. F. Molisch, "Capacity enhancement and BER in a combined SDMA/TDMA system," *Proc. VTC '96, Atlanta, GA, April*, pp. 1481–1485 (1996).

Fuhl et al. 1998 J. Fuhl, A. F. Molisch, and E. Bonek, "Unified channel model for mobile radio systems with smart antennas," *IEE Proc. Radar, Sonar and Navigation*, **145**, 32–41 (1998).

Fujimoto et al. 1987 K. Fujimoto, "A review of research on small antennas," *Journal of the Institute of Electronics, Information and Communication Engineers*, **70**, 830–838 (1987).

Fujimoto and James 2001 K. Fujimoto and J. R. James, *Mobile Antenna Systems Handbook* (2nd edn.), Artech House (2001).

Gahleitner 1993 R. Gahleitner, "Radio wave propagation in and into urban buildings," Ph.D. thesis, Technical University Vienna (1993).

Gallagher 1961 R. Gallagher, "Low density parity check codes," Ph.D. thesis, Massachusetts Institute of Technology, 1961

Garg 2000 V. K. Garg, *IS-95 CDMA and cdma2000: Cellular/PCS Systems Implementation*, Prentice Hall (2000).

Gay and Benesty 2000 S. L. Gay and J. Benesty, *Acoustic Signal Processing for Telecommunication*, Kluwer Academic (2000).

Gersho and Gray 1992 A. Gersho and R. M. Gray, *Vector Quantization and Signal Compression*, Kluwer Academic (1992).

Gesbert et al. 2002 D. Gesbert, H. Boelcskei, and A. Paulraj, "Outdoor MIMO wireless channels: Models and performance prediction," *IEEE Transactions Communications*, **50**(12), 1926–1935 (2002).

Gesbert et al. 2003 D. Gesbert, M. Shafi, D. S. Shiu, P. J. Smith, and A. Naguib, "From theory to practice: An overview of MIMO space-time coded wireless systems," *IEEE Journal on Selected Areas in Communications*, **21**, 281–302 (2003).

Ghazi-Moghadam and Kaveh 1998 V. Ghazi-Moghadam and M. Kaveh, "A CDMA interference cancelling receiver with an adaptive blind array," *IEEE Journal on Selected Areas in Communications*, **16**, 1542–1554 (1998).

Giannakis and Halford 1997 G. B. Giannakis and S. D. Halford, "Blind fractionally spaced equalization of noisy FIR channels: Direct and adaptive solutions," *IEEE Transactions Signal Processing*, **45**, 2277–2292 (1997).

Gibson et al. 1998 J. D. Gibson, T. Berger, T. Lookabaugh, R. Baker, and D. Lindbergh, *Digital Compression for Multimedia: Principles and Standards*, Morgan Kaufman (1998).

Gilhousen et al. 1991 K. S. Gilhousen, I. M. Jacobs, R. Padovani, A. J. Viterbi, L. A. Weaver, and C. E. Wheatley, "On the capacity of a cellular CDMA system," *IEEE Transactions on Vehicular Technology*, **40**, 303–312 (1991).

Gitlin and Weinstein 1981 R. D. Gitlin and S. B. Weinstein, "Fractionally-spaced equalization: An improved digital transversal equalizer," *Bell System Technical Journal*, **60**, 275–296 (1981).

Glassner 1989 A. S. Glassner, *An Introduction to Ray Tracing*, Morgan Kaufmann (1989).

Glisic and Vucetic 1997 S. Glisic and B. Vucetic, *Spread Spectrum CDMA Systems for Wireless Communications*, Artech House (1997).

Godara 1997 L. C. Godara, "Applications of antenna arrays to mobile communications, I: Performance improvement, feasibility, and system considerations," *Proceedings of the IEEE*, **85**, 1031-1060; and "Application of antenna arrays to mobile communications, II: Beam-forming and direction-of-arrival considerations," *Proceedings of the IEEE*, **85**, 1195-1245 (1997).

Godara 2001 L. C. Godara, *Handbook of Antennas in Wireless Communications*, CRC Press (2001).

Godard 1980 D. N. Godard, "Self-recovering equalization and carrier tracking in two-dimensional data communication systems," *IEEE Transactions on Communications*, **28**, 1867–1875, (1980).

Goiser 1998 A. Goiser, *Handbuch der Spread-Spectrum Technik*, Springer-Verlag (1998) [in German].

Goiser et al 2000 A. Goiser, M. Z. Win, G. Chrisikos, and S. Glisic, "Code Division Multiple Access," in A. F. Molisch (ed.), *Wireless Wideband Digital Communications*, Prentice Hall (2000).

Goldsmith et al. 2003 A. Goldsmith, S. A. Jafar, N. Jindal, and S. Vishwanath, "Capacity limits of MIMO channels," *IEEE Journal on Selected Areas in Communications*, **21**, 684–702 (2003).

Gonzales 1984 G. Gonzalez, *Microwave Transistor Amplifiers: Analysis and Design*, Prentice Hall (1984).

Goodman et al. 1989 D. J. Goodman, R. A. Valenzuela, K. T. Gayliard, and B. Ramamurthi, "Packet reservation multiple access for local wireless communications," *IEEE Transactions on Communications*, **37**, 885–890 (1989).

Goransson et al. 2000 B. Goransson, B. Hagerman, S. Petersson, and J. Sorelius, "Advanced antenna systems for WCDMA: Link and system level results," *Proc. PIMRC '2000*, pp. 62–66 (2000).

Gorokhov 1998 A. Gorokhov, "On the performance of the Viterbi equalizer in the presence of channel estimation errors," *IEEE Signal Processing Letters*, **5**, 321–324 (1998).

Gray 1989 R. M. Gray, *Source Coding Theory*, Kluwer Academic (1989).

Greenstein et al. 1997 L. J. Greenstein, V. Erceg, Y. S. Yeh, and M. V. Clark, "A new path-gain/delay-spread propagation model for digital cellular channels," *IEEE Transactions on Vehicular Technology*, **46**, 477–485 (1997).

Gross and Harris 1998 D. Gross and C. M. Harris, *Fundamentals of Queuing Theory* (3rd edn.), John Wiley & Sons (1998)

Haardt and Nossek 1995 M. Haardt and J. A. Nossek, "Unitary ESPRIT: How to obtain increased estimation accuracy with a reduced computational burden," *IEEE Transactions on Signal Processing*, **43**, 1232–1242 (1995).

Haensler and Schmidt 2004 E. Haensler and G. Schmidt, G. (eds.), *Acoustic Echo and Noise Control*, John Wiley & Sons (2004).

Hagenauer and Hoeher 1989 J. Hagenauer and P. Hoeher, "A Viterbi algorithm sith soft-decision outputs and its applications," *IEEE Globecom*, 1680–1686 (1989).

Hammerschmidt 2000 J. Hammerschmidt, "Dissertation," TU München, München (2000) [in German].

Hansen 1998 R. C. Hansen, *Phased Array Antennas*, John Wiley & Sons (1998).

Hanzo et al. 2000 L. Hanzo, W. Webb, and T. Keller, *Single- and Multi-carrier Quadrature Amplitude Modulation: Principles and Applications for Personal Communications, WLANs and Broadcasting*, John Wiley & Sons, Ltd. (2000).

Hanzo et al. 2001 L. Hanzo, F. C. A. Somerville, and J. P. Woodard, *Voice Compression and Communications*, IEEE Press/Wiley Interscience (2001).

Hanzo et al. 2003 L. Hanzo, M. Muenster, B. J. Choi, and T. Keller, *OFDM and MC-CDMA for Broadband Multi-User Communications, WLANs and Broadcasting*, John Wiley & Sons, Ltd. (2003).

Harrington 1993 R. F. Harrington, *Field Computation by Moment Methods*, John Wiley & Sons/IEEE Press (1993).

Hashemi 1979 H. Hashemi, "Simulation of the urban radio propagation channel," *IEEE Transactions on Vehicular Technology*, 28213–28225 (1979).

Hashemi 1993 H. Hashemi, "Impulse response modeling of indoor radio propagation channels," *IEEE Journal on Selected Areas in Communications*, **11**, 943, 1993.

Hata 1980 M. Hata, "Empirical formula for propagation loss in land mobile radio services," *IEEE Transactions on Vehicular Technology*, **29**, 317–325 (1980).

Haykin 1991 S. Haykin, *Adaptive Filter Theory*, Prentice Hall (1991).

Heavens 1965 O. S. Heavens, *Optical Properties of Thin Film Solids*, Dover (1965).

Hirade et al. 1979 K. Hirade, M. Ishizuka, F. Adachi, and K. Ohtani, "Error-rate performance of digital FM with differential detection in land mobile radio channels," *IEEE Transactions on Vehicular Technology*, **28**, 204–212 (1979).

Hirasawa and Haneishi 1991 K. Hirasawa and M. Haneishi (eds.), *Analysis, Design, and Measurements of Small and Low-profile Antennas*, Artech House (1992).

Hirata and Shiozawa 2003 A. Hirata and T. Shiozawa, "Correlation of maximum temperature increase and peak SAR in the human head due to handset antennas," *IEEE Transactions on Microwave Theory Technology*, **51**, 1834–1841 (2003).

Hoeher 1992 P. Hoeher, "A statistical discrete-time model for the WSSUS multipath channel," *IEEE Transactions on Vehicular Technology*, **41**, 461–468 (1992).

Holma and Toskala 2000 H. Holma und A. Toskala (eds.), *WCDMA for UMTS: Radio Access for Third Generation Mobile Communications*, John Wiley & Sons (2000).

Holtzman and Jalloul 1994 J. M. Holtzman and L. M. Jalloul, "Rayleigh fading effect reduction with wideband DS/CDMA signals," *IEEE Transactions on Communications*, **42**, 1012–1016 (1994).

Hoppe et al. 2003   R. Hoppe, P. Wertz, F. M. Landstorfer, and G. Wölfle, "Advanced ray optical wave propagation modelling for urban and indoor scenarios including wideband properties," *European Transactions on Telecommunications*, **14**, 61–69 (2003).

Hottinen et al. 2003   A. Hottinen, O. Tirkkonen, and R. Wichman, *Multi-antenna Transceiver Techniques for 3G and Beyond*, John Wiley & Sons, Ltd. (2003).

HP 1994   Hewlett-Packard, *Schulungsunterlagen GSM*, Hewlett-Packard (1994) [in German].

Huang et al. 2001   X. Huang, A. Acero, and H. W. Hon, *Spoken Language Processing*, Prentice Hall (2001).

IEEE 802.11   Institute of Electrical and Electronics Engineers, "Standard 802.11," particularly the following documents: IEEE std 802.11-1999, Part 11: "Wireless LAN Medium Access Control (MAC) and Physical Layer (PHY) specifications" (1999); IEEE 802.11e draft/D9.0, Part 11: Wireless Medium Access Control (MAC) and physical layer (PHY) specifications: "Medium Access Control (MAC) Quality of Service (QoS) Enhancements," (2004); IEEE std 802.11a-1999, Part 11: Wireless LAN Medium Access Control (MAC) and Physical Layer (PHY) specifications: "High-speed Physical Layer in the 5 GHz Band" (1999); IEEE std 802.11b-1999, Part 11: Wireless LAN Medium Access Control (MAC) and Physical Layer (PHY) specifications: "Higher-Speed Physical Layer Extension in the 2.4 GHz Band" (1999).

IEGMP 2000   Independent Expert Group on Mobile Phones, *Mobile Phones and Health*, National Radiological Protection Board (UK), available online at *http://www.iegmp.org.uk/* (2000).

Ikegami et al. 1984   F. Ikegami, S. Yoshida, T. Takeuchi, and M. Umehira, "Propagation factors controlling mean field strength on urban streets," *IEEE Transactions on Antennas Propagation*, 822–829 (1984).

Isidoro et al. 1995   F. A. Isidoro, P. B. Cardoso, T. T. Guerra, and L. M. Correira, "Adaptation of the Okumura–Hata model to the city of Lisbon for short distances," *COST 231*, TD(95)3 (1995).

Itakura and Saito 1968   F. Itakura and S. Saito "Analysis synthesis telephony based on the maximum likelihood principle," *Proc. 6th Int. Congress on Acoustics, Tokyo*, pp. C-17–C-20 (1968).

ITU 1997   International Telecommunications Union, *Guidelines for Evaluation of Radio Transmission Technologies for IMT-2000* (Recommendation ITU-R M.1225), International Telecommunications Union (1997).

Jakes 1974   W. C. Jakes, *Microwave Mobile Communications* (reprint), IEEE Press (1974).

Jamali and Le-Ngoc 1991   S. Jamali and T. Le-Ngoc, "A new 4-state 8PSK TCM scheme for fast fading, shadowed mobile radio channels," *IEEE Transactions on Vehicular Technology*, **40**, 216–222 (1991).

Jayant and Noll 1984   N. S. Jayant and P. Noll, *Digital Coding of Waveforms: Principles and Applications to Speech and Video*, Prentice Hall (1984).

Jelinek et al. 2004   M. Jelinek, R. Salami, S. Ahmadi, B. Bessetle, P. Gournay, and C. Laflamme, "On the architecture of the cdma2000 variable-rate multimode wideband (VMR-WB) speech coding standard," *Proc. IEEE Int. Conf. Acoustics, Speech, and Signal Processing*, 281–284, (2004)

Johannesson and Zigangirov 1999   R. Johannesson and K. S. Zigangirov, *Fundamentals of Convolutional Coding*, John Wiley & Sons/IEEE Press (1999).

Jurafsky and Martin 2000   D. Jurafsky and J. H. Martin, *Speech and Language Processing*, Prentice Hall (2000).

Kattenbach 1997   R. Kattenbach, "Characterisierung zeitvarianter Indoor Mobilfunkkanäle mittels ihrer System- und Korrelationsfunktionen [Characterization of time-variant indoor radio channels by means of their system and correlation functions]," Dissertation at der Universität GhK Kassel, published by Shaker-Verlag, Aachen (1997) [in German].

Kattenbach 2002   R. Kattenbach, "Statistical modeling of small-scale fading in directional radio channels," *IEEE Journal on Selected Areas in Communications*, **20**, 584–592 (2002).

Keller 1962   J. B. Keller, "Geometrical theory of diffraction," *J. Opt. Soc. Amer.*, **2**, 116–130 (1962).

Keller and Hanzo 2000   T. Keller and L. Hanzo, "Adaptive multicarrier modulation: A convenient framework for time-frequency processing in wireless communications," *Proceedings of the IEEE*, **88**, 611–640 (2000).

Kermoal et al. 2002   J. P. Kermoal et al., "A stochastic MIMO radio channel model with experimental validation," *IEEE Journal on Selected Areas in Communications*, **20**, 1211 (2002).

Khun-Jush et al. 2002   J. Khun-Jush, P. Schramm, G. Malmgren, and J. Torsner, "HiperLAN2: Broadband wireless communications at 5 GHz," *IEEE Communications Magazine*, **40**(6), 130–136 (2002).

Kim et al. 1999   Y. H. Kim, I. Song, H. G. Kim, T. Chang, and H. M. Kim; "Performance analysis of a coded OFDM system in time-varying multipath Rayleigh fading channels," *IEEE Transactions on Vehicular Technology*, **48**, 1610–1615 (1999).

Kivekaes et al. 2004 O. Kivekäs, J. Ollikainen, T. Lehtiniemi, and P. Vainikainen, "Bandwidth, SAR, and efficiency of internal mobile phone antennas," *IEEE Transactions Electromagnetic Comp.*, **46**, 71–76 (2004).

Kleijn 2005 W. B. Kleijn, *'Information Theory and Source Coding* (unpublished course notes), Royal Institute of Technology (KTH) (2005).

Kleijn and Granzow 1991 W. B. Kleijn and W. Granzow, "Methods for waveform interpolation in speech coding,," *Digital Signal Processing*, **1**, 215–230 (1991).

Kleijn and Paliwal 1995 W. B. Kleijn and K. K. Paliweh, *Speech Coding and Synthesis*, Elsevier (1995).

Kleinrock and Tobagi 1975 L. Kleinrock and F. Tobagi, "Packet switching in radio channels: Part I – Carrier sense multiple access modes and their throughput-delay characteristics," *IEEE Transactions on Communications*, **23**, 1400–1416 (1975).

Klemenschits and Bonek 1994 T. Klemenschits and E. Bonek, "Radio coverage of road tunnels at 900 and 1800 MHz by discrete antennas," *Proc. PIMRC '94*, pp. 411–415 (1994).

Kohno et al. 1995 R. Kohno, R. Meidan, and L. B. Milstein, "Spread spectrum access methods for wireless communications," *IEEE Communications Magazine*, **33**(1), 58–67 (1995).

Kondoz 2004 A. M. Kondoz, *Digital Speech: Coding for Low Bit Rate Communication Systems* (2nd edn.), John Wiley & Sons, Ltd. (2004).

Kouyoumjian and Pathak 1974 R. G. Kouyoumjian and P. H. Pathak, "A uniform geometrical theory of diffraction for an edge in a perfectly conducting surface," *Proceedings of the IEEE*, **62**, 1448–1461 (1974).

Korn 1992 I. Korn, "GMSK with frequency-selective Rayleigh fading and cochannel interference," *IEEE Journal on Selected Areas in Communications*, **10**, 506–515 (1992).

Kozek 1997 W. Kozek, "Matched Weyl–Heisenberg expansions of nonstationary environments," Dissertation, Technical University Vienna (1997).

Kozek and Molisch 1998 W. Kozek and A. F. Molisch, "Nonorthogonal pulseshapes for multicarrier communications in doubly dispersive channels," *IEEE Journal on Selected Areas in Communications*, **16**, 1579–1589 (1998).

Kraus and Marhefka 2002 J. D. Kraus and R. J. Marhefka, *Antennas: For All Applications* (3rd edn.), McGraw-Hill (2002).

Kreuzgruber et al. 1993 P. Kreuzgruber, P. Unterberger, and R. Gahleitner, "A ray splitting model for indoor propagation associated with complex geometries," *Proc. 43rd IEEE Vehicular Technology Conference, Secaucus, NJ*, pp. 227–230 (1993).

Krim and Viberg 1996 H. Krim und M. Viberg, "Two decades of array signal processing: The parametric approach," *IEEE Signal Processing Magazine*, **13**, 67–94 (1996).

Kubin 1995 G. Kubin, "Nonlinear processing of speech," in W. B. Kleijn and K. K. Paliwal (eds.), *Speech Coding and Synthesis* (pp. 557–610), Elsevier (1995).

Kuchar et al. 1997 A. Kuchar, J. Fuhl, and E. Bonek, "Spectral efficiency enhancement and power control of smart antenna system," *EPMCC '97, Bonn, September 30–October 2* (1997).

Kuchar et al. 2000 A. Kuchar, J. P. Rossi, and E. Bonek, "Directional macro-cell channel characterization from urban measurements," *IEEE Transactions on Antennas PropagationIEEE Transactions on Antennas Propagation*, **48**, 137–146 (2002).

Kuchar et al. 2002 A. Kuchar, M. Taferner, and M. Tangemann, "A real-time DOA-based smart antenna processor," *IEEE Transactions on Vehicular Technology*, **51**, 1279–1293 (2002).

Kunz and Luebbers 1993 K. S. Kunz and R. J. Luebbers, *The Finite Difference Time Domain Method for Electromagnetics*, CRC Press (1993).

Larsen and Aarts 2004 E. R. Larsen and R. M. Aarts, *Audio Bandwidth Extension: Application of Psychoacoustics, Signal Processing and Loudspeaker Design*, John Wiley & Sons (2004).

Larsson and Stoica 2003 E. G. Larsson and P. Stoica, *Space-time Block Coding for Wireless Communications*, Cambridge University Press (2003).

Laurent 1986 P. A. Laurent, "Exact and approximate construction of digital phase modulations by superposition of amplitude modulated pulses," *IEEE Transactions on Communications*, **34**, 150–160 (1986).

Laurila et al. 1998 J. Laurila, A. F. Molisch, and E. Bonek, "Influence of scatter distribution on power delay profiles and azimuthal power spectra of mobile radio channels," *Proc. ISSSTA '98*, pp. 267–271 (1998).

Laurila 2000 J. Laurila, "Semi-blind detection of co-channel signals in mobile communications," Dissertation TU Wien, Vienna (2000).

Lawton and McGeehan 1994 M. C. Lawton and J. P. McGeehan, "The application of a deterministic ray launching algorithm for the prediction of radio channel characteristics in small-cell environments," *IEEE Transactions on Vehicular Technology*, **43**, 955–969 (1994).

Lee 1973 W. C. Y. Lee, "Effects on correlations between two mobile base-station antennas," *IEEE Transactions on Communications*, **21**, 1214–1224 (1973).

Lee 1982 W. C. Y. Lee, *Mobile Communications Engineering*, McGraw-Hill (1982).

Lee 1986 W. C. Y. Lee, *Mobile Communications Design Fundamentals*, Sams Publishing (1986).

Lee 1995 W. C. Y. Lee, *Mobile Cellular Telecommunications: Analog and Digital Systems*, McGraw-Hill (1995).

Li et al. 1996 J. Li, J. F. Wagen, and E. Lachat, "ITU model for multi-knife-edge diffraction," *IEEE Proceedings on Microwaves, Antennas and Propagation*, **143**, 539–541 (1996).

Li et al. 1998 Y. G. Li, L. C. Cimini, and N. R. Sollenberger, "Robust channel estimation for OFDM systems with rapid dispersive fading channels," *IEEE Transactions on Communications*, **46**, 902–915 (1998).

Li et al. 1999 Y. G. Li, N. Seshadri, and S. Ariyavisitakul, "Channel estimation for OFDM systems with transmitter diversity in mobile wireless channels," *IEEE Journal on Selected Areas in Communications*, **17**(3), 461–471 (1999).

Li and Miller 1998 J. S. Lee and L. E. Miller, *CDMA Systems Engineering Handbook*, Artech House, (1998).

Liberti and Rappaport 1996 "A geometrically based model for line-of-sight multipath radio channels," *Proc. IEEE Vehicular Technology Conf.*, pp. 844–848 (1996).

Liberti and Rappaport 1999 J. C. Liberti and T. S. Rappaport, *Smart Antennas for Wireless Communications: IS-95 and Third Generation CDMA Applications*, Prentice Hall (1999).

Liebenow and Kuhlman 1993 U. Liebenow and P. Kuhlmann, "Determination of scattering surfaces in hilly terrain," *COST 231*, TD(93), 119 (1993).

Lin 2003 J. C. Lin, "Safety standards for human exposure to radio frequency radiation and their biological rationale," *IEEE Microwave Magazine*, **4**(4), 22–26 (2003).

Lin and Costello 2004 S. Lin and D. J. Costello, *Error Control Coding* (2nd edn.), Prentice Hall (2004)

Liu et al. 1996 H. Liu, G. Xu, L. Tong, and T. Kailath, "Recent developments in blind channel equalization: From cyclostationarity to subspaces," *Signal Processing*, **50**, 83–99 (1996).

Lo 1999 T. K. Y. Lo, "Maximum ratio transmission," *IEEE Transactions on Communications*, **47**, 1458–1461 (1999).

Loeliger 2004 H. A. Loeliger, "An introduction to factor graphs," *IEEE Signal Processing Magazine*, 28–41, January (2004).

Loncar et al. 2002 M. Loncar, R. Müller, T. Abe, J. Wehinger, and C. Mecklenbräuker, "Iterative equalizer using soft-decoder feedback for MIMO systems in frequency-selective fading," *Proc. of URSI General Assembly* (2002).

Lucky et al. 1968 R. W. Lucky, J. Salz, and E. J. Weldon Jr., *Principles of Data Communication*, McGraw-Hill (1968).

MacAulay and Quatieri 1986 R. J. McAulay and T. F. Quatieri, "Speech analysis/synthesis based on a sinusoidal representation," *IEEE Transactions on Acoustics, Speech, and Signal Processing*, **34**, 744–754 (1986).

Mailloux 1994 R. J. Mailloux, *Phased Array Antenna Handbook*, Artech House (1994).

Manholm et al. 2003 L. Manholm, M. Johansson, and S. Petersson, "Antennas with electrical beamtilt for WCDMA: Simulations and implementation," *Swedish National Conference on Antennas* (2003).

Marcuse 1991 D. Marcuse, *Theory of Dielectric Optical Waveguides* (2nd edn.), Academic Press (1992).

Mardia et al. 1979 K. V. Mardia, J. T. Kent, and J. M. Bibby, *Multivariate Analysis*, Academic Press (1979).

Maric and Titlebaum 1992 S. V. Maric and E. L. Titlebaum, "A class of frequency hop codes with nearly ideal characteristics for use in multiple-access spread-spectrum communications and radar and sonar systems," *IEEE Transactions on Communications*, **40**, 1442–1447 (1992).

Martin 1998 U. Martin, "Spatio-temporal radio channel characteristics in urban macrocells," *IEE Proc. Radar, Sonar and Navigation*, **145**, 42–49 (1998).

Marzetta and Hochwald 1999 T. L. Marzetta and B. M. Hochwald, "Capacity of a mobile multiple-antenna communication link in Rayleigh at fading," *IEEE Transactions Inform. Theory*, **45**, 139–157 (1999).

Matz and Hlawatsch 1998 G. Matz and F. Hlawatsch, "Time-frequency transfer function calculus (symbolic calculus) of linear time-varying systems (linear operators) based on a generalized underspread theory," *Journal of Mathematical Physics* (Special Issue on Wavelet and Time-Frequency Analysis), **39**, 4041–4070 (1998).

Matz et al. 2002 G. Matz, A. F. Molisch, F. Hlawatsch, M. Steinbauer, and I. Gaspard, "On the systematic measurement errors of correlative mobile radio channel sounders," *IEEE Transactions on Communications*, **50**, 808–821 (2002).

Matz 2003 G. Matz, "Characterization of non-WSSUS fading dispersive channels," *Proc. ICC '03*, pp. 2480-2484 (2003).

Markel and Gray 1976 J. P. Markel and A. M. Gray Jr., *Linear Prediction of Speech*, Springer-Verlag (1976).

May and Rohling 2000 T. May and H. Rohling, "Orthogonal Frequency Division Mutlple Access," in A. F. Molisch (ed.), *Wideband Wireless Digital Communications*, Prentice Hall (2000).

Mayr 1996 B. Mayr, *Modulationsangepasste Codierung*, lecture notes, TU Vienna (1996).

McEliece 2004 R. McEliece, *The theory of Information and Coding* (student edn.), Cambridge University Press (2004).

MacKay 2002 D. J. C. MacKay, *Information Theory, Inference and Learning Algorithms*, Cambridge University Press (2002)

MacKay and Neal 1997 D. J. C. MacKay and R. M. Neal, "Near Shannon limit performance of low density parity check codes," *Electronics Letters*, **33**, 457–458 (1997).

McNamara 1990 D. A. McNamara, C. W. I. Pistorius, and J. A. G. Malherbe, *Introduction to the Uniform Geometrical Theory of Diffraction*, Artech House (1990).

Mengali and D'Andrea 1997 U. Mengali and A. N. D'Andrea, *Synchronization Techniques for Digital Receivers*, Plenum Press (1997).

METAMORP 1999 METAMORP Consortium, *Final Report and Deliverables*, C/1-1 and C/2-2, METAMORP Consortium (1999).

Meurling and Jeans 1994 J. Meurling and R. Jeans, *The Mobile Phone Book*, Communications Week International (1994).

Meyr and Ascheid 1990 H. Meyr and G. Ascheid, *Digital Communication Receivers, Phase-, Frequency-Locked Loops, and Amplitude Control*, John Wiley & Sons (1990).

Meyr et al. 1997 H. Meyr, M. Moeneclaeye, and S. A. Fechtel, *Digital Communication Receivers, Vol. 2: Synchronization, Channel Estimation, and Signal Processing*, John Wiley & Sons (1997).

Milstein 1988 L. B. Milstein, "Interference rejection techniques in spread spectrum communications," *Proceedings of the IEEE*, **76**, 657–671 (1988).

Mogensen 1996 P. E. Mogensen and J. Wigart, "On antenna- and frequency diversity in GSM related systems GSM-900, DCS-1800, and PCS1900," *Proc. PIMRC '96*, pp. 1272–1276 (1996).

Mogensen et al. 1997 P. E. Mogensen, K. I. Pedersen, P. L. Espensen, B. Fleury, F. Frederiksen, K. Oelsen, and S. L. Larsen. "Preliminary measurement results from an adaptive array testbed for GSM/UMTS," *Proc. IEEE VTC '97*, pp. 1592–1596 (1997).

Molisch 1999 A. F. Molisch, "A new method for the computation of the error probability of differentially detected FSK and PSK in mobile radio channels: The case of minimum shift keying," *Wireless Personal Communications*, **9**, 165–178 (1999).

Molisch 2000 A. F. Molisch (ed.), *Wideband Wireless Digital Communications*, Prentice Hall (2000).

Molisch 2002 A. F. Molisch, "Modeling of directional mobile radio channels," *Radio Science Bulletin*, **302**, September, 16–26 (2002).

Molisch 2004 A. F. Molisch, "A generic model for the MIMO wireless propagation channel," *IEEE Proc. Signal Proc.*, **52**, 61–71 (2004).

Molisch 2005 A. F. Molisch, "Ultrawideband propagation channels," in M. G. DiBenedetto, T. Kaiser, A. F. Molisch, I. Oppermann, C. Politano, and D. Porcino (eds.), *UWB Communications Systems: A Comprehensive Overview*, EURASIP Publishing (2005).

Molisch and Steinbauer 1999 A. F. Molisch and M. Steinbauer, "Condensed parameters for characterizing wideband mobile radio channels," *International Journal of Wireless Information Networks*, **6**, 133–154 (1999).

Molisch and Tufvesson 2004 A. F. Molisch and F. Tufvesson, "Multipath propagation models for broadband wireless systems," in M. Ibnkahla (ed.), *Digital Signal Processing for Wireless Communications Handbook* (pp. 2.1–2.43), CRC Press (2004).

Molisch and Tufvesson 2005 A. F. Molisch and F. Tufvesson, "MIMO channel capacity and measurements," in T. Kaiser (ed.), *Smart Antennas: State of the Art*, Hindawi (2005).

Molisch and Win 2004 A. F. Molisch and M. Z. Win, "MIMO systems with antenna selection," *IEEE Microwave Magazine*, March, 46–56 (2004).

Molisch et al. 1995 A. F. Molisch, J. Fuhl, and E. Bonek, "Pattern distortion of mobile radio base station antennas by antenna masts and roofs," *Proc. 25th Europ. Microwave Conf., Bologna*, pp. 71–76 (1995).

Molisch et al. 1996 A. F. Molisch, J. Fuhl, and P. Proksch, "Error floor of MSK modulation in a mobile-radio channel with two independently-fading paths," *IEEE Transactions on Vehicular Technology*, **45**, 303–309 (1996).

Molisch et al. 1998 A. F. Molisch, H. Novak, J. Fuhl, and E. Bonek, "Reduction of the error floor of MSK by selection diversity," *IEEE Transactions on Vehicular Technology*, **47**, 1281–1291 (1998).

Molisch et al. 2002 A. F. Molisch, M. Steinbauer, M. Toeltsch, E. Bonek, and R. Thoma, "Capacity of MIMO systems based on measured wireless channels," *IEEE Journal on Selected Areas in Communications*, **20**, 561–569 (2002).

Molisch et al. 2003a A. F. Molisch, J. R. Foerster, and M. Pendergrass, "Channel models for ultrawide-band personal area networks," *IEEE Personal Communications Magazine*, **10**, 14–21 (2003).

Molisch et al. 2003b A. F. Molisch, A. Kuchar, J. Laurila, K. Hugl, and R. Schmalenberger, "Geometry-based directional model for mobile radio channels: Principles and implementation," *European Transactions Telecomm.*, **14**, 351–359 (2003).

Molisch et al. 2004a A. F. Molisch et al., *802.15.4a Channel Modeling Group Final Report* (802.15.4/535), IEEE Press (2004).

Molisch et al. 2004b A. F. Molisch, H. Asplund, M. Steinbauer, R. Heddergoot, and T. Zwick, "The COST 259 directional channel model, I: Philosophy and general aspects," *IEEE Transactions on Wireless Communications*, accepted pending revisions.

Molisch et al. 2004c A. F. Molisch et al., "A low-cost time-hopping impulse radio system for high data rate transmission," *EURASIP Journal of Applied Signal Processing*, special issue on UWB (2004).

Molnar et al. 1996 B. G. Molnar, I. Frigyes, Z. Bodnar, and Z. Herczku, "The WSSUS channel model: Comments and a generalisation," *Proc. Globecom '96*, pp. 158–162 (1996).

Moroney and Cullen 1995 D. Moroney and P. Cullen, "A fast integral equation approach to UHF coverage estimation," in E. delRe (ed.), *Mobile and Personal Communications*, Elsevier (1995).

Moshavi 1996 S. Moshavi, "Multi-user detection for DS-CDMA communications," *IEEE Communications Magazine*, October, 124–136 (1996).

Motley and Keenan 1988 A. J. Motley and J. P. Keenan, "Personal communication radio coverage in buildings at 900 MHz and 1700 MHz," *Electronics Letters*, **24**, 763–764 (1988)

Mouly and Pautet 1992 M. Mouly and M. B. Pautet, *The GSM System for Mobile Communications*, Telecom Publishing (1992).

Muirhead 1982 R. J. Muirhead, *Aspects of Multivariate Statistical Theory*, John Wiley & Sons, Inc. (1982).

Muquet et al. 2002 B. Muquet, Z. Wang, G. B. Giannakis, M. de Courville, and P. Duhamel, "Cyclic prefixing or zero padding for wireless multicarrier transmissions?" *IEEE Transactions on Communications*, **50**, 2136–2148 (2002).

Murota and Hirade 1981 K. Murota and K. Hirade, "GMSK modulation for digital mobile radio telephony," *IEEE Transactions on Communications*, **29**, 1044–1050 (1981).

Nakagami 1960 M. Nakagami, "The M-distribution: A general formula of rapid fading," *Statistical Methods of Radio Wave Propagation*, Pergamon Press (1960).

Namislo 1984 N. Namislo, "Analysis of mobile radio slotted ALOHA networks," *IEEE Journal on Selected Areas in Communications*, **2**, 583–588 (1984).

Narayanan et al. 2004 R. M. Narayanan, K. Atanassov, V. Stoiljkovic, and G. R. Kadambi, "Polarization diversity measurements and analysis for antenna configurations at 1800 MHz," *IEEE Transactions on Antennas Propagation*, **52**, 1795–1810 (2004).

Neubauer et al. 2001 Th. Neubauer, H. Jaeger, J. Fuhl, and E. Bonek, "Measurement of the background noise floor in the UMTS FDD uplink band," *Europ. Personal and Mobile Communications Conf.* (2001).

Nilsson et al. 1997 R. Nilsson, O. Edfors, M. Sandell, and P. O. Börjesson, "An analysis of two-dimensional pilot-symbol assisted modulation for OFDM," *Proc. IEEE International Conference on Personal Wireless Communications, Bombay, December*, pp. 71–74 (1997).

Noerpel et al. 1996 A. R. Noerpel, Y. B. Lin, and H. Sherry, "PACS: Personal communications system: A tutorial," *IEEE Personal Communications*, June, 32–43 (1996).

Norklit and Andersen 1998 O. Norklit and J. B. Andersen, "Diffuse channel model and experimental results for array antennas in mobile environments," *IEEE Transactions on Antennas Propagation*, **46**, 834–840 (1998).

Novak 1999 H. Novak, "Switched beam adaptive antenna demonstrator for UMTS data rates," *Virginia Tech Symposium on Wireless Communications, Blacksburg, VA*, pp. 85–93 (1999).

O'Hara and Petrick 2005 B. O'Hara and A. Petrick, *The IEEE 802.11 Handbook: A Designer's Companion* (2nd edn.), IEEE Standards Publications (2005).

O'Shaughnessy 2000 D. O'Shaughnessy, *Speech Communication: Human and Machine* (2nd edn.), IEEE Press (2000).

Oehrvik 1994 S. O. Oehrvik, *Radio School*, Ericsson AB (1994).

Ogawa et al. 2001 K. Ogawa, T. Matsuyoshi, and K. Monma, "An analysis of the performance of a handset diversity antenna influenced by head, hand, and shoulder effects at 900 MHz, II: Correlation characteristics," *IEEE Transactions on Vehicular Technology*, **50**, 845–853 (2001).

Ogawa and Matsuyoshi 2001 K. Ogawa and T. Matsuyoshi, "An analysis of the performance of a handset diversity antenna influenced by head, hand, and shoulder effects at 900 MHz, I: Effective gain characteristics," *IEEE Transactions on Vehicular Technology*, **50**, 830–844 (2001).

Ogilvy 1991 J. A. Ogilvie, *Theory of Wave Scattering from Random Rough Surfaces*,' IOP Publishing (1991).

Okumura et al. 1968 Y. Okumura, E. Ohmori, T. Kawano, and K. Fukuda, "Field strength and its variability in VHF and UHF land mobile services," *Review of the Electrical Communications Laboratory*, **16**, 825–873 (1968).

Oppenheim and Schaefer 1985 A. V. Oppenheim and R. W. Schaefer, *Discrete-time Signal Processing*, Prentice Hall (1989).

Paetzold 2002 M. Paetzold, *Mobile Fading Channels: Modelling, Analysis, and Simulation*, John Wiley & Sons (2002).

Pajusco 1998 P. Pajusco, "Experimental characterization of D.O.A at the base station in rural and urban area," *Proc. IEEE VTC '98*, pp. 993–998 (1998).

Papoulis 1985 A. Papoulis, "Predictable processes and Wold's decomposition: A review," *IEEE Transactions Acoustics, Speech, and Signal Processing*, **33**, 933–938 (1985).

Papoulis 1991 A. Papoulis, *Probability, Random Variables, and Stochastic Processes* (3rd edn.), McGraw-Hill (1991).

Parry 2002 R. Parry, "CDMA 2000, 1xEV," *IEEE Potentials*, October/November, 10–13 (2002).

Parsons 1992 J. D. Parsons, *The Mobile Radio Channel*, John Wiley & Sons, Ltd. (1992, 2nd edn 2001).

Parsons et al. 1991 J. D. Parsons, D. A. Demery, and A. M. D. Turkmani, "Sounding techniques for wideband mobile radio channels: A review," *Proceedings of the Institute of Elect. Eng. —I*, **138**, 437–446, October (1991).

Paulraj and Papadias 1997 A. J. Paulraj and C. B. Papadias, "Space-time processing for wireless communications," *IEEE Personal Communications*, **14**(5), 49–83 (1997).

Paulraj et al. 2003 A. Paulraj, D. Gore, and R. Nabar, *Multiple Antenna Systems*, Cambridge University Press (2003).

Pawula et al. 1982 R. F. Pawula, S. O. Rice, and J. H. Roberts, "Distribution of the phase angle between two vectors perturbed by Gaussian noise," *IEEE Transactions on Communications*, **30**, 1828–1841 (1982).

Pearson 2000 B. Pearson, *Complementary Code Keying Made Simple* (Application Note AN 9850.1), Intersil (2000).

Pedersen et al. 1997 K. Pedersen, P. E. Mogensen, and B. Fleury, "Power azimuth spectrum in outdoor environments," *IEEE Electronics Letters*, **33**, 1583–1584 (1997).

Pedersen et al. 1998 G. F. Pedersen, J. O. Nielsen, K. Olensen, and I. Z. Kovács, "Measured variation in performance of handheld antennas for a large number of test persons," *COST 259*, TD(98)025 (1998).

Peel 2003 C. B. Peel, "On dirty-paper coding," *IEEE Signal Processing Magazine*, **20**(3), 112-113 (2003).

Perkins 2001 C. E. Perkins, *Ad Hoc Networking*, Addison-Wesley (2001).

Petrus et al. 2002 P. Petrus, J. H. Reed, and T. S. Rappaport, "Geometrical-based statistical macrocell channel model for mobile environments," *IIEEE Transactions on Communications*, **50**, 495 (2002).

Plenge 1997 C. Plenge, "Leistungsbewertung öffentlicher DECT-Systeme," Dissertation, RWTH Aachen (1997) [in German].

Polydoros and Weber 1984 A. Polydoros and C. L. Weber, "A unified approach to serial search spread-spectrum code acquisition, Part 1: General theory," *IEEE Transactions on Communications*, **32**, 542–549; "Part II: Matched filter receiver," *IEEE Transactions on Communications*, **32**, 550–560 (1984).

Poor 2001 H. V. Poor, "Turbo multiuser detection: A primer," *Journal of Communications Networks*, **3**, 196–201 (2001).

Poor 2004 H. V. Poor, "Iterative multiuser detection," *IEEE Signal Processing Magazine*, **21**, 81–88 (2004).

Pozar 2000 D. M. Pozar, *Microwave and RF Design of Wireless Systems*, John Wiley & Sons, Inc. (2000)

Proakis 1968 J. G. Proakis, "On the probability of error for multichannel reception of binary signals," *IEEE Transactions on Communications*, **16**, 68–71 (1968).

Proakis 1995 J. G. Proakis, *Digital Communications* (3rd edn.), McGraw-Hill (1995, 4th edn 2000).

Quatieri 2002 T. F. Quatieri, *Discrete-time Speech Signal Processing*, Prentice Hall (2002).

Qiu 2002 R. C. Qiu, "A study of the ultra-wideband wireless propagation channel and optimum uwb receiver design," *IEEE Journal on Selected Areas in Communications*, **20**, 1628–1637 (2002).

Qiu 2004 R. C. Qiu, "A generalized time domain multipath channel and its application in ultra-wideband (UWB) wireless optimal receiver design, Part II: Physics-based system analysis," *IEEE Transactions on Wireless Communications*, **3**, 2312–2324 (2004).

Rabiner 1994 L. Rabiner, "Applications of voice processing to telecommunications," *Proceedings of the IEEE*, **82**, 199–228 (1994).

Rabiner and Schafer 1978 L. R. Rabiner and R. W. Schafer, *Digital Processing of Speech Signals*, Prentice Hall (1978).

Raleigh and Cioffi 1998 G. Raleigh and J. M. Cioffi, "Spatial-temporal coding for wireless communications," *IEEE Transactions on Communications*, **46**, 357–366 (1998).

Ramachandran et al. 2004 I. Ramachandran, Y. P. Nakache, P. Orlik, J. Zhang, and A. F. Molisch, "Symbol spreading for ultrawideband systems based on multiband OFDM," *Proc. Personal, Indoor, Mobile Radio Symp.*, pp. 1204–1209 (2004).

Ramo et al. 1967 S. Ramo, J. R. Whinnery, and T. van Duzer, *Fields and Waves in Communication Electronics*, John Wiley & Sons (1967).

Rappaport 1995 T. S. Rappaport, *Cellular Radio and Personal Communications* (self-study course), IEEE/ EAB (1995).

Rappaport 1996 T. S. Rappaport, *Wireless Communications: Principles and Practice*, IEEE Press (1996; 2nd edn. 2001).

Rappaport 1998 T. S. Rappaport (ed.), *Smart Antennas: Adaptive Arrays, Algorithms, and Wireless Position Location*, IEEE Press (1998).

Rasinger et al. 1990 J. Rasinger, A. L. Scholtz, and E. Bonek, "A new enhanced-bandwidth internal antenna for portable communication systems," *Proc. 40th IEEE Vehicular Technology Conference, Orlando*, pp. 7–12 (1990).

Razavi 1997 B. Razavi, *RF Microelectronics*, Prentice Hall (1997).

Rice 1947 S. O. Rice, "Statistical properties of a sine wave plus random noise," *Bell System Technical Journal*, **27**, 109–157 (1947).

Richardson 2005 A. Richardson, *WCDMA Design Handbook*, Cambridge University Press (2005).

Richardson and Urbanke 2005 T. Richardson and R. Urbanke, *Modern Coding Theory*, to appear (2005); see also Ecole Polytechnique Fédérale de Lausanne lecture notes, available online at *http://lthcwww.epfl.ch/ index.php*

Richardson et al. 2001 T. J. Richardson, M. A. Shokrollahi, and R. L. Urbanke, "Design of capacity-approaching irregular low-density parity-check codes," *IEEE Transactions on Information Theory*, **47**, 619–637 (2001).

Rohling and Galda 2005 H. Rohling and D. Galda, "OFDM transmission technique: A strong candidate for the next generation mobile communications," *Radio Science Bulletin* (in press).

Rokhlin 1990 V. Rokhlin, Rapid solution of integral equations of scattering theory in two dimensions, *Journal of Comp. Phys.*, **96**, 414–439 (1990).

Roy and Fortier 2004 S. Roy and P. Fortier, "A closed-form analysis of fading envelope correlation across a wideband basestation array," *IEEE Transactions on Wireless Communications*, **3**, 1502–1507 (2004).

Roy et al. 1986 R. Roy, A. Paulraj, and T. Kailath, "ESPRIT: A subspace rotation approach to estimation of parameters of cisoids in noise," *IEEE Transactions Acoustics, Speech, and Signal Processing*, **34**, 1340–1342 (1986).

Roy et al. 2004 S. Roy, J. R. Foerster, V. S. Somayazulu, and D. G. Leeper, "Ultrawideband radio design: the promise of high-speed, short-range wireless connectivity," *Proceedings of the IEEE*, **92**, 295–311 (2004).

Royer and Toh 1999 E. M. Royer and C. K. Toh, "A review of current routing protocols for ad hoc mobile wireless networks," *IEEE Personal Communications*, **6**(2), 46–55 (1999).

Rubin 1979 I. Rubin, "Message delays in FDMA and TDMA communication channels," *IEEE Transactions on Communications*, **27**, 769-777 (1979).

Saleh and Valenzuela 1987 A. Saleh and R. A. Valenzuela, "A statistical model for indoor multipath propagation," *IEEE Journal on Selected Areas in Communications*, **5**, 128-137 (1987).

Sari et al. 2000 H. Sari, F. Vanhaverbeke, and M. Moeneclaey, "Extending the capacity of multiple access channels," *IEEE Communications Magazine*, **38**(1), 74-82 (2000).

Sato 1975 Y. Sato, "A method of self recovering equalization for multilevel amplitude modulation," *IEEE Transactions on Communications*, **23**, 679-682 (1975).

Sayed and Kailath 2002 A. H. Sayed and T. Kailath, "A survey of spectral factorization methods," *Journal of Numerical Linear Algebra with Applications*, **8**, 467-496 (2001).

Sayeed 2002 A. M. Sayeed, "Deconstructing multiantenna fading channels," *IEEE Transactions on Signal Processing*, **50**, 2563 (2002).

Sayre 2001 C. W. Sayre, *Complete Wireless Design*, McGraw-Hill (2001).

Schiller 2003 J. Schiller, *Mobile Communications* (2nd edn.), Addison-Wesley (2003).

Schlegel and Perez 2003 C. B. Schlegel and L. C. Perez, *Trellis and Turbo Coding*, John Wiley & Sons, Inc. (2003).

Schmidl and Cox 1997 T. M. Schmidl and D. C. Cox, "Robust frequency and timing synchronization for OFDM," *IEEE Transactions on Communications*, **45**, 1613-1621 (1997).

Schmidt 1986 R. Schmidt, "Multiple emitter location and signal parameters estimation," *IEEE Transactions on Antennas Propagation*, **34**, 276-280 (1986).

Schniter 2004 P. Schniter, "Low-complexity equalization of OFDM in doubly selective channels," *IEEE Transactions on Signal Processing*, **52**, 1002-1011 (2004).

Schroeder and Atal 1985 M. Schroeder and B. Atal, "Code-excited linear prediction (CELP): High-quality speech at very low bit rates," *Proc. IEEE Int. Conf. Acoustics, Speech, and Signal Processing, ICASSP '85*, pp. 937-940 (1985).

Schroeder et al. 1979 M. R. Schroeder, B. S. Atal, and J. L. Hall, "Optimizing digital speech coders by exploiting masking properties of the human ear," *Journal of the Acoustical Society of America*, **66**, 1647-1652 (1979).

Scholtz 1982 R. A. Scholtz, "The origins of spread spectrum communications," *IEEE Transactions on Communications*, **30**, 822-854 (1982).

Shafi et al. 2005 M. Shafi, M. Zhang, P. J. Smith, A. L. Moustakas, A. F. Molisch, F. Tufvesson, and S. H. Simon, "The use of cross polarised antennas for MIMO systems," *IEEE Journal on Selected Areas in Communications* (in press).

Shannon 1948 C. E. Shannon, "A mathematical theory of communication," *Bell System Technical Journal*, **27**, 379-423 and 623-656 (July and October 1948)

Shannon 1949 C. E. Shannon, "Communication in the presence of noise," *Proc. IRE*, **37**, 10-21 (1949).

Shannon 1959 C. E. Shannon, "Coding theorems for a discrete source with a fidelity criterion," *IRE National Convention Record*, **4**, 142-163 (1959).

Shiu et al. 2000 D. Shiu, G. J. Foschini, M. J. Gans, and J. M. Kahn, "Fading correlation and its effect on the capacity of multielement antenna systems," *IEEE Transactions on Communications*, **48**, 502-513 (2000).

Simon et al. 1994 M. K. Simon, J. K. Omura, R. A. Scholtz, and B. K. Levitt, *Spread Spectrum Communications Handbook* (rev. edn.), McGraw-Hill (1994).

Simon and Alouini 2000 M. K. Simon and M. S. Alouini, *Digital Communications over Fading Channels*, John Wiley & Sons, Inc. (2000, 2nd edn 2005).

Sindhushayana and Black 2002 N. T. Sindhushayana and P. J. Black, " Forward link coding and modulation for CDMA2000 1XEV-DO (IS-856)," *Proc. IEEE PIMRC 2002*, pp. 1839-1846 (2002).

Sklar 1997 B. Sklar, "A primer on turbo code concepts," *IEEE Communications Magazine*, **35**, December, 94-102 (1997).

Sklar 2001 B. Sklar, *Digital Communications: Fundamentals and Applications* (2nd edn.), Prentice Hall (2001).

Sklar and Harris 2004 B. Sklar and F. J. Harris, "The ABCs of linear block codes," *IEEE Signal Processing Magazine* **21**(4), 14-35 (2004).

Spencer et al. 2004 Q. H. Spencer, C. B. Peel, A. L. Swindlehurst, and M. Haardt, "An introduction to the multi-user MIMO downlink," *IEEE Communications Magazine*, **42**(10), 60-67 (2004).

Speth et al. 1999   M. Speth, S. A. Fechtel, G. Fock, and H. Meyr, "Optimum receiver design for wireless broad-band systems using OFDM. I," *IEEE Transactions on Communications*, **47**, 1668–1677 (1999); "Part II: A case study," *IEEE Transactions on Communications*, **49**, 571–578 (2001).

Steele and Hanzo 1999   R. Steele and L. Hanzo, *Mobile Communications* (2nd edn.), John Wiley & Sons, Ltd. (1999).

Steele et al. 2001   R. Steele, C. C. Lee, and P. Gould, *GSM, cdmaOne and 3G Systems*, John Wiley & Sons (2001).

Steendam and Moeneclaey 1999   H. Steendam and M. Moeneclaey, "Analysis and optimization of the performance of OFDM on frequenty-selective time-selective fading channels," *IEEE Transactions on Communications*, **47**, 1811–1819 (1999).

Stein 1964   S. Stein, "Unified analysis of certain coherent and noncoherent binary commpuications systems," *IEEE Transactions on Information Theory*, **11**, 239–246 (1964).

Steinbauer and Molisch 2001   M. Steinbauer and A. F. Molisch, (eds.), "Spatial channel models," in L. Correia (ed.), *Wireless Flexible Personalized Communications*, John Wiley & Sons, Ltd. (2001).

Steinbauer et al. 2000   M. Steinbauer, A. F. Molisch, A. Burr, and R. Thomae, "Capacity of MIMO channels based on measurements," *Europ. Conf. Wireless Techn., Paris, October*, pp. 52–55 (2000).

Strang 1988   G. Strang, *Linear Algebra and Its Application* (3rd. edn.), Harcourt Brace Jovanovich (1988).

Stueber 1996   G. Stueber, *Principles of Mobile Communication*, Kluwer (1996, 2nd edn 2001).

Stueber et al. 2004   G. L. Stueber, J. R. Barry, S. W. McLaughlin, Y. G. Li, M. A. Ingram, and T. G. Pratt, "Broadband MIMO-OFDM wireless communications," *Proceedings of the IEEE*, **92**, 271–294 (2004).

Stutzmann and Thiele 1997   W. L. Stutzman and G. A. Thiele, *Antenna Theory and Design* (2nd edn.), John Wiley & Sons, Inc. (1997).

Suzuki 1977   H. Suzuki, "A statistical model for urban radio propagation," *IEEE Transactions on Communications*, **25**, 673–680 (1977).

Suzuki 1982   H. Suzuki, "Canonic receiver analysis for M-ary angle modulations in Rayleigh fading environment," *IEEE Transactions on Vehicular Technology*, **31**, 7–14 (1982).

Swarts et al. 1998   F. Swarts, P. van Rooyan, I. Oppermann, and M. P. Lötter, *CDMA Techniques for Third Generation Mobile Systems*, Kluwer (1998).

Sweeney 2002   P. Sweeney, *Error Control Coding: From Theory to Practice*, John Wiley & Sons (2002).

Sweeney 2004   D. Sweeney, *WiMax Operator's Manual: Building 802.16 Wireless Networks*, Apress, (2004).

Swoboda 1973   J. Swoboda, *Codierung zur Fehlerkorrektur und Fehlererkennung*, Oldenbeurg (1973) [in German].

Symes 2003   P. Symes, *Digital Video Compression*, McGraw-Hill (2003).

Taga 1990   T. Taga, "Analysis for mean effective gain of mobile antennas in land mobile radio environments," *IEEE Transactions on Vehicular Technology*, **39**, 117–131 (1990).

Taga 1993   T. Taga, "Characteristics of space-diversity branch using parallel dipole antennas in mobile radio communications," *Electron. Commun. Jpn.*, Pt. 1, **76**, 55–65 (1993).

Tarokh et al. 1998   V. Tarokh, N. Seshadri, and A. R. Calderbank, "Space-time coding for high data rate wireless communication: Performance criterion and code construction," *IEEE Transactions on Information Theory*, **44**, 744–765 (1998).

Tarokh et al. 1999   V. Tarokh, H. Jafarkhani, and A. R. Calderbank, "Space-time block codes from orthogonal designs," *IEEE Transactions on Information Theory*, **45**, 1456–1467 (1999).

Telatar 1999   I. E. Telatar, "Capacity of multi-antenna Gaussian channels," *European Transactions Telecomm.*, **10**, 585–595 (1999).

3GPP   Third Generation Partnership Project, *Specification Documents TS 21.211, 25.104, 25.212, 25.212*, 3GPP.

Thjung and Chai 1999   T. T. Tjhung and C. C. Chai, "Fade statistics in Nakagami-lognormal channels," *IEEE Transactions on Communications*, **47**, 1769–1772 (1999).

Thomae et al. 2000   R. S. Thomae, D. Hampicke, A. Richter, G. Sommerkorn, A. Schneider, A., U. Trautwein, and W. Wirnitzer, "Identification of time-variant directional mobile radio channels," *IEEE Transactions Instrum. Meas.*, **49**, 357–364 (2000).

Thomae et al. 2005   R. S. Thomae, M. Landmann, A. Richter, and U. Trautwein, "Multidimensional high-resolution channel sounding," in T. Kaiser (ed.), *Smart Antennas in Europe: State-of-the-Art* (EURASIP Book Series, 27 pp.), Hindawi Publishing (to appear 2005).

Tiedemann 2001   E. Tiedemann, "Cdma2000 1X: New capabilities for CDMA networks," *IEEE Vehicular Technology Society News*, **48**(4), 4–12, November (2001).

Tobagi 1980 F. Tobagi, "Multiaccess protocols in packet communication systems," *IEEE Transactions on Communications*, **28**, 468–488 (1980).

Tomiuk et al. 1999 B. R. Tomiuk, N. C. Beaulieu, and A. A. Abu-Dayya, "General forms for maximal ratio diversity with weighting errors," *IEEE Transactions on Communications*, **47**(4), April, 488–492 (1999).

Tong et al. 1994 L. Tong, G. Xu, and T. Kailath, "Blind identification and equalization based on second-order statistics: A time domain approach," *IEEE Transactions on Information Theory*, **40**, 340–349 (1994).

Tong et al. 1995 L. Tong, G. Xu, B. Hassibi, and T. Kailath, "Blind identification and equalization based on second-order statistics: A frequency domain approach," *IEEE Transactions on Information Theory*, **41**, 329–334 (1995).

Tsoulos 2001 G. V. Tsoulos, *Adaptive Antennas for Wireless Communications*, IEEE Press (2001).

Turin et al. 1972 G. L. Turin, F. D. Clapp, T. L. Johnston, S. B. Fine, and D. Lavry, "A statistical model of urban multipath propagation," *IEEE Transactions on Vehicular Technology*, **21**, 1–9 (1972).

Tuttlebee 1997 W. H. W. Tuttlebee (ed.), *Cordless Telecommunications Worldwide*, Springer-Verlag (1997).

Ungerboeck 1976 G. Ungerboeck, "Fractional tap-spacing equalizer and consequences for clock recovery in data modems," *IEEE Transactions on Communications*, **24**, 856–864 (1976).

Ungerboeck 1982 G. Ungerboeck, "Channel coding with multilevel/phase signals," *IEEE Transactions on Information Theory*, **28**, 55–67 (1982).

Valenti 1999 M. C. Valenti, "Iterative detection and decoding for wireless communications," Ph.D. thesis, Virginia Tech University (1999).

Valenzuela 1993 R. A. Valenzuela, "A ray tracing approach to predicting indoor wireless transmission," *Proc. VTC '93*, pp. 214–218 (1993).

van de Beek et al. 1995 J.-J. van de Beek, O. Edfors, M. Sandell, S. K. Wilson, and P. O. Börjesson, "On channel estimation in OFDM systems," *Proc. IEEE Vehic. Technol. Conf., Chicago, July* (Vol. 2, pp. 815-819) (1995).

van de Beek et al 1999 J.-J. van de Beek, P. O. Borjesson, M.-L. Boucheret, D. Landstrom, J. M. Arenas, P. Odling, C. Ostberg, M. Wahlqvist, and S. K. Wilson, "A time and frequency synchronization scheme for multiuser OFDM," *IEEE Journal on Selected Areas in Communications*, **17**, 1900–1914 (1999).

van der Plassche 2003 R. van der Plassche, *Cmos Integrated Analog-to-digital and Digital-to-analog Converters* (2nd edn.), Kluwer (2003).

van der Veen and Paulraj 1996 A.J. van der Veen and A. Paulraj, "An Analytical Constant Modulus Algorithm," *IEEE Transactions on Signal Processing*, **44**, 1136–1155 (1996).

van der Veen et al. 1997 M. C. van der Veen, C. B. Papadias, and A. Paulraj, "Joint angle and delay estimation (JADE) for multipath signals arriving at an antenna array," *IEEE Communications Letters*, **1**, 12–14 (1997).

Vanghi et al. 2004 V. Vanghi, A. Damnjanovic, and B. Vojcic, *The CDMA2000 System for Mobile Communications*, Prentice Hall (2004).

Varshney and Kumar 1991 P. Varshney and S. Kumar, "Performance of GMSK in a land mobile radio channel," *IEEE Transactions on Vehicular Technology*, **40**, 607–614 (1991).

Vary et al. 1998 P. Vary, U. Heute, and W. Hess, *Digitale Sprachsignalverarbeitung [Digital Speech Signal Processing]*, B.G. Teubner (1998) [in German].

Vaseghi 2000 S. V. Vaseghi, *Advanced Digital Signal Processing and Noise Reduction* (2nd edn.), John Wiley & Sons, Ltd. (2000).

Vaughan and Andersen 2003 R. Vaughan and J. B. Andersen, *Channels, Propagation and Antennas for Mobile Communications*, IEEE Press (2003).

Verdu 1998 S. Verdu, *Multiuser Detection*, Cambridge Universit Press (1998).

Viswanath et al. 2002 P. Viswanath, D. N. C. Tse, and R. Laroia, "Opportunistic beamforming using dumb antennas," *IEEE Transactions on Information Theory*, **48**, 1277–1294 (2002).

Viterbi 1967 A. Viterbi, "Error bounds for convolutional codes and an asymptotically optimum decoding algorithm," *IEEE Transactions on Information Theory*, **13**, 260–269 (1967).

Viterbi 1995 A. J. Viterbi, *CDMA: Principles of Spread Spectrum Communication*, Addison-Wesley Wireless Communications Series (1995).

Vitetta et al. 2000 G. Vitetta, B. Hart, A. Mammela, and D. Taylor, "Equalization techniques for single-carrier, unspread digital modulation," in A. F. Molisch (ed.), *Wireless Digital Communications*, Prentice Hall (2000).

Waldschmidt et al. 2004 C. Waldschmidt, S. Schulteis, and W. Wiesbeck, "Complete RF system model for analysis of compact MIMO arrays," *IEEE Transactions on Vehicular Technology*, **53**, 579–586 (2004).

Walfish and Bertoni 1988 J. Walfish and H. L. Bertoni, "A theoretical model of UHF propagation in urban environments," *IEEE Transactions on Antennas Propagation*, 822–829 (1988).

Wallace and Jensen 2004 J. W. Wallace and M. A. Jensen, "Mutual coupling in MIMO wireless systems: A rigorous network theory analysis," *IEEE Transactions on Wireless Communications*, **3**, 1317–1325 (2004).

Wang and Giannakis 2000 Z. Wang and G. B. Giannakis, "Wireless multicarrier communications," *IEEE Signal Processing Magazine*, **17**(3), 29–48 (2000).

Wang and Poor 2003 X. Wang and H. V. Poor, *Wireless Communication Systems: Advanced Techniques for Signal Reception*, Prentice Hall (2003).

Watkinson 2001 J. Watkinson, *MPEG Handbook*, Focal Press (2001).

Weichselberger et al. 2003 W. Weichselberger, M. Herdin, H. Özcelik, and E. Bonek, "A stochastic MIMO channel model with joint correlation of both link ends," *IEEE Transactions on Wireless Communications*, to appear.

Weinstein and Ebert 1971 S. Weinstein and P. Ebert, "Data transmission by frequency-division multiplexing using the discrete fourier transform," *IEEE Transactions on Communications*, **19**, 628–634 (1971).

Weisstein 2004 Eric W. Weisstein, *Lindeberg-Feller Central Limit Theorem*, MathWorld (available online at *http://mathworld.wolfram.com/Lindeberg-FellerCentralLimitTheorem.html*).

Willenegger 2000 S. Willenegger, "CDMA2000 physical layer: An overview," *IEEE Journal of Communications and Networks*, **2**(1), 5–17, March (2000).

Wilson 1996 S. G. Wilson, *Digital Modulation and Coding*, Prentice Hall (1996).

Win 2000 M. Z. Win, "Impact of spreading bandwidth and selection diversity order on Rake reception," in A. F. Molisch (ed.), *Wideband Wireless Digital Communications*, Prentice Hall (2000).

Win and Chrisikos 2000 M. Z. Win and G. Chrisikos, "Impact of spreading bandwidth and selection diversity order on selective Rake reception," in A. F. Molisch (ed.), *Wideband Digital Communications* (pp. 424–454), Prentice Hall (2001).

Win and Scholtz 1998 M. Z. Win and R. A. Scholtz, "Impulse radio: How it works," *IEEE Communications Letters*, **2**, 36–38 (1998).

Win and Scholtz 2000 M. Z. Win and R. A. Scholtz, "Ultra-wide bandwidth time-hopping spread-spectrum impulse radio for wireless multiple-access communications," *IEEE Transactions on Communications*, **48**, 679–691 (2000).

Win and Winters 1999 M. Z. Win and J. H. Winters, "Analysis of hybrid selection/maximal-ratio combining of diversity branches with unequal SNR in Rayleigh fading," *Proc. VTC '99, Spring*, pp. 215–220 (1999).

Winters 1984 J. H. Winters, "Optimum combining in digital mobile radio with cochannel interference," *IEEE Journal on Selected Areas in Communications*, **2**, 528–539 (1984).

Winters 1987 J. H. Winters, "On the capacity of radio communications systems with diversity in Rayleigh fading environments," *IEEE Journal on Selected Areas in Communications* (1987).

Winters 1994 J. H. Winters, "The diversity gain of transmit diversity in wireless systems with Rayleigh fading," *Proc. IEEE International Conf. Communications*, 1121–1125 (1994).

Wittneben 1993 A. Wittneben, "A new bandwidth efficient transmit antenna modulation diversity scheme for linear digital modulation," *Proc. IEEE International Conf. Communications*, 1630–1634 (1993).

Woerner et al. 1994 B. D. Woerner, J. H. Reed, and T. S. Rappaport, "Simulation issues for future wireless modems," *IEEE Communications Magazine*, **32**(7), 42–53 (1994).

Wolf 1978 J. K. Wolf, "Efficient maximum-likelihood decoding of linear block codes using a trellis," *IEEE Transactions on Information Theory*, **24**, 76–80 (1978).

Wong et al. 1999 C. Y. Wong, R. S. Cheng, K. B. Lataief, and R. D. Murch, "Multiuser OFDM with adaptive subcarrier, bit, and power allocation," *IEEE Journal on Selected Areas in Communications*, **17**, 1747–1758 (1999).

Wozencraft and Jacobs 1965 J. M. Wozencraft and I. M. Jacobs, *Principles of Communication Engineering*, John Wiley & Sons (1965).

Xiong 2000 F. Xiong, *Digital Modulation Techniques*, Artech House (2000).

Xu et al. 2000 Z. Xu, A. N. Akansu, and S. Tekinay, "Cochannel interference computation and asymptotic performance analysis in TDMA/FDMA systems with interference adaptive dynamic channel allocation," *IEEE Transactions on Vehicular Technology*, **49**, 711–723.

Xu et al. 2002 H. Xu, D. Chizhik, H. Huang, and R. A. Valenzuela, "A wave-based wideband MIMO channel modeling technique," *Proc. 13th IEEE Int. Symp. Personal, Indoor Mobile Radio Communications*, 1626–1630 (2002).

Yang and Giannakis 2002 L. Yang and G. B. Giannakis, "Multistage block-spreading for impulse radio multiple access through ISI channels," *IEEE Journal on Selected Areas in Communications*, **20**, 1767–1777 (2002).

Yang and Hanzo 2003 L. L. Yang and L. Hanzo, "Multicarrier DS-CDMA: A multiple access scheme for ubiquitous broadband wireless communications," *IEEE Communications Magazine*, 41(10), 116–124 (2003).

Ying and Anderson 2003 Z. Ying and J. Anderson, "Multi band, multi antenna system for advanced mobile phone," *Swedish National Conference on Antennas* (2003).

Yongacoglu et al. 1988 A. Yongacoglu, D. Makrakis, and K. Feher, "Differential detection of GMSK using decision feedback," *IEEE Transactions on Communications*, **36**, 641–649 (1988).

Yu et al. 1997 C. C. Yu, D. Morton, C. Stumpf, R. G. White, J. E. Wilkes, and M. Ulema, "Low-tier wireless local loop radio systems: Parts 1 and 2," *IEEE Communications Magazine*, March, 84–98 (1997).

Yu and Ottersten 2002 K. Yu and B. Ottersten, "Models for MIMO propagation channels: A review," *Wireless Communications and Mobile Computing*, **2**, 653 (2002).

Zheng and Tse 2003 L. Zheng and D. L. C. Tse, "Diversity and multiplexing: A fundamental tradeoff in multiple-antenna channels," *IEEE Transactions on Information Theory*, **49**, 1073–1096 (2003).

Ziemer et al. 1995 R. E. Ziemer, R. L. Peterson, and D. E. Borth, *Introduction to Spread Spectrum Communications*, Prentice Hall (1995).

Zienkiewicz and Taylor 2000 O. C. Zienkiewicz and R. L. Taylor, *Finite Element Method: Vol. 1: The Basis*, Butterworth-Heinemann (2000).

Index